METHODS IN MOLECULAR BIOLOGY

Series Editor
John M. Walker
School of Life and Medical Sciences
University of Hertfordshire
Hatfield, Hertfordshire, AL10 9AB, UK

For further volumes:
http://www.springer.com/series/7651

NF-kappa B

Methods and Protocols

Edited by

Michael J. May

Department of Animal Biology, University of Pennsylvania School of Veterinary Medicine, Philadelphia, PA, USA

 Humana Press

Editor
Michael J. May
Department of Animal Biology
University of Pennsylvania School of Veterinary Medicine
Philadelphia, PA, USA

Additional material to this book can be downloaded from http://extras.springer.com.

ISSN 1064-3745   ISSN 1940-6029   (electronic)
Methods in Molecular Biology
ISBN 978-1-4939-2421-9   ISBN 978-1-4939-2422-6   (eBook)
DOI 10.1007/978-1-4939-2422-6

Library of Congress Control Number: 2015931925

Springer New York Heidelberg Dordrecht London

Cover illustration: The cover images show the formation in activated T cells of cytosolic Bcl10 clusters named POLKADOTS that provide a platform for the assembly of the terminal signaling complex that ultimately mediates NF-κB activation. These images are from Chapter 12 by Paul and Schaefer titled "*Visualizing TCR induced POLKADOTS formation and NF-κB activation in the D10 T cell clone and mouse primary effector T cells*".

Printed on acid-free paper

Humana Press is a brand of Springer
Springer Science+Business Media LLC New York is part of Springer Science+Business Media (www.springer.com)

Preface

Activation of the transcription factor nuclear factor-kappa B (NF-κB) is essential for normal immune and inflammatory responses, cell survival and proliferation, and the maintenance of cellular and tissue homeostasis. In addition to these crucial physiological functions, aberrant NF-κB activation occurs in a wide range of chronic inflammatory and autoimmune diseases, solid tumors, leukemia, and lymphomas. Due to this central pathophysiological role in a diverse array of diseases, understanding the precise mechanisms underlying normal and deregulated NF-κB signaling has become a major field of biomedical research. An overarching goal of this extraordinary research effort is the identification of targets in the many emerging NF-κB signaling pathways that are amenable to pharmacological blockade or manipulation. The reward for these endeavors will be the emergence of novel classes of highly specific anti-inflammatory, immunomodulatory, or anticancer drugs.

Almost 30 years of basic research has resulted in the NF-κB activation pathway becoming the paradigm for inducible receptor to nuclear signal transduction. Indeed, efforts to unravel NF-κB signaling mechanisms have been the breeding ground for seminal discoveries with far-reaching impact in the signal transduction field. These contributions include defining the role of ubiquitination-mediated protein degradation in signal transduction and the more recent revelations of the many nondegradative ubiquitination mechanisms as key posttranslational modifications controlling signaling cascades. As the number of original papers focusing on the mechanisms and functions of normal and pathophysiological NF-κB signaling increases, it is clear that the field of NF-κB research remains highly active, continues to expand, and still exists at the forefront of driving the techniques employed by the wider signal transduction community.

In this volume of Methods in Molecular Biology, prominent researchers in the NF-κB field have contributed essential insight into the methods and techniques required to dissect the complex mechanisms of NF-κB activation, regulation, and function. This will provide a timely and invaluable resource for researchers seeking to perform experiments aimed at understanding the role of NF-κB signaling in health and disease.

Part I (Standard Approaches to Detect NF-κB Pathway Activation) contains three chapters describing the now "classic" methods to assay NF-κB pathway activation in cultured cells or tissues. Fittingly the book begins with a chapter by Ramiswami and Hayden describing the electrophoretic mobility shift assay (EMSA) that was the technique initially employed to identify regulators of kappa light chain expression in B cells that led to the first description of NF-κB. Following this is a chapter by Starokadomskyy and Burstein that describes the methods used to detect IκB degradation; then Collins and colleagues provide the approaches required to measure NF-κB transcriptional activity using luciferase reporter assays. These three powerful methodologies have been and remain the mainstay techniques used by researchers studying the activation status of NF-κB signaling in a multitude of experimental systems.

The chapters in Part II (Detection and Analysis of NF-κB Signaling) expand the range of methods used to study NF-κB activation and introduce the techniques used to specifically analyze classical and alternative NF-κB signaling pathways. Following the transcriptional regulation theme of Chapter 3, this section begins with a description by Colleran et al. of chromatin immunoprecipitation (ChIP) as it is used to detect DNA binding of classical NF-κB dimers. Next are two chapters highlighting the methods employed to study classical NF-κB activation in the context of specific cell types and activation mechanisms. In the first of these, Mihalas and Meffert describe techniques to assess the levels of IKK activity in neurons following excitatory neurotransmission-induced signaling. Jiang and Lin then provide insight into the multiple approaches that can be employed to study epidermal growth factor (EGF)-induced NF-κB activation. The next four chapters by Remouchamps and Dejardin, Qu and Xiao, McCorkell and May, and Gray and May introduce the noncanonical NF-κB pathway and provide detailed descriptions of the methods and protocols used to examine this alternative mechanism in various cell types induced by separate stimuli. Part II concludes with a chapter by Jackson and Miyamoto outlining genetic complementation of NEMO-deficient cells and examining how this approach can be used to study nuclear to cytoplasmic NF-κB signaling in response to genotoxic stress.

Part III (Methods to Study the Control of NF-κB Signaling) details a series of methods to explore the mechanisms that control separate NF-κB signaling pathways. The first two chapters dissect the mechanisms of T cell receptor (TCR)-induced NF-κB activation beginning with Paul and Schaefer (Chapter 12) who describe elegant imaging techniques used to study the formation of POLKADOTS that are signaling platforms necessary for TCR-induced IKK activation. Nagel and Krappmann then demonstrate an innovative fluorogenic cleavage assay used to measure the TCR-induced paracaspase activity of MALT1, thus allowing assessment of this key signaling intermediate as a potential therapeutic target for the treatment of inflammatory and autoimmune diseases and lymphomagenesis. Analysis of signal transduction proteins involved in NF-κB activation continues in Chapters 14 and 15 in which Reichardt and colleagues and Varfolomeev et al. discuss the methodologies used to study the roles of the TRAF and c-IAP proteins, respectively, in NF-κB activation. Again, these techniques provide the methodological foundation for studies addressing the potential therapeutic effects of targeting these crucial signaling intermediates.

In the remainder of Part III, procedures to assess the mechanisms and function of post-translational modifications of critical NF-κB pathway signaling proteins and the NF-κB proteins themselves are presented. This section begins with chapters by Shembade and Harhaj, and Sasaki and colleagues, each describing separate approaches to study nondegradative ubiquitination in NF-κB signaling including analysis of key protein-protein interactions and linear ubiquitination of NEMO. The next two chapters describe the techniques of DELFIA and microscale thermophoresis (Vincendeau et al.) and a sophisticated fluorescence spectroscopy approach (Dubosclard et al.), each of which is employed to specifically study NEMO-ubiquitin binding interactions. Methods to examine ubiquitin-dependent degradation of IκBα are then discussed by Chong et al. who describe an in vitro biochemical approach to reconstitute the ubiquitination system. The topic of ubiquitination controlling NF-κB signaling continues in Chapters 21 and 22 in which Collins et al. and Li and colleagues describe protocols to examine the direct ubiquitination of NF-κB proteins. Both chapters describe separate approaches to assess the ubiquitin-dependent degradation of active NF-κB that is a crucial negative regulatory mechanism preventing deregulated sustained NF-κB activity. In the remaining chapters in Part III, techniques are described to dissect the mechanisms of methylation and acetylation of NF-κB proteins. In Chapter 23, Lu and Stark demonstrate how to perform immunoprecipitations followed by mass spectrometry to identify methylation sites on p65, and then Chen and Chen conclude this section by describing experimental methods to monitor the in vitro and in vivo functions of acetylated or methylated forms of NF-κB.

Approaches to study the role of aberrant NF-κB signaling in diseases including cancer and inflammation and the effects of blocking this activity are explored in depth in Part IV (Analyzing and Targeting NF-κB Activity in Disease). The section begins with a chapter by Wessel and Hanson describing how to quantitate active NF-κB in dermal fibroblast from human skin biopsies obtained from immunodeficiency patients. The next four chapters address the role of dysregulated NF-κB signaling in cancer starting with a comprehensive description by Gilmore and Gélinas of methods to qualitatively evaluate the in vitro transforming activity of Rel proteins. This is followed by a discussion by Bassères and Baldwin of how to apply an RNA interference strategy to target NF-κB signaling in lung cancer cells; then Allen and Van Waes provide a detailed description of methods to immunohistochemically analyze NF-κB in human tumor tissue samples. In Chapter 29, Gaurnier-Hausser and Mason describe approaches to obtain, process, and analyze NF-κB activity in canine malignant lymphoid tissue including techniques to determine the effects of inhibiting classical NF-κB signaling in canine lymphoma using the NEMO-binding domain (NBD) peptide. Importantly, this chapter emphasizes the exciting potential of using outbred animal patient populations as highly representative, clinically relevant models of human diseases.

Methodologies to determine the effects of exogenously inhibiting classical NF-κB in vitro and in vivo in disease models are described in detail in Chapters 30–34. In each of these chapters, the authors provide approaches targeting the NBD; however, the techniques discussed can be considered a blueprint of methods that can be modified to assess other inhibitors of NF-κB signaling in these disease models. This section begins with a protocol from McCorkell and May to analyze the effects of the NBD peptide on vascular inflammation in vivo. Application of the NBD is explored further by Swarnker and Abu-Amer who describe in detail how to synthesize and purify the peptide and apply this approach to study osteoclast differentiation. In the next chapter, Zhao et al. describe methods employed to study NF-κB signaling in mouse models of aging, and they demonstrate how to assess the effects of blocking NF-κB activation in these mice using a novel NBD-protein transduction domain (PTD) fusion peptide. In Chapter 33, Sehnert and colleagues outline the development of an exciting new "sneaking ligand" approach that allows specific in vivo delivery of the NBD to only activated vascular endothelial cells at sites of inflammation. This strategy is expanded in the next chapter in which Sehnert et al. describe how to test the effects of this novel endothelial cell-targeting approach on the development of arthritis in mice. Part IV concludes with a chapter by Shaked and colleagues who describe detailed in vitro and in vivo approaches to analyze NF-κB activation in mouse intestinal epithelial cells using an innovative genetic model of intestinal inflammation.

Part V (Bioinformatics and Modeling Approaches to Study NF-κB) contains three chapters that focus on applying powerful bioinformatics and mathematical methodologies to NF-κB biology. In the first of these chapters, Siggers and colleagues describe the use and computational bioinformatics analysis of protein binding microarrays that provide a high-throughput method to measure proteins binding to distinct DNA sequences. In Chapter 37, Finnerty and Gilmore describe a combination of bioinformatics and phylogenetic approaches to study NF-κB evolution based on menu-driven, open-source computer programs that are readily accessible and can be used without training in phylogenetic methods. The volume then concludes with a chapter by Mitchell et al. describing the state-of-the-art in silico and mathematical modeling approaches that are leading to the development of a highly sophisticated, expansive and ultimately predictive "wiring diagram" of NF-κB activation.

In light of the continuously expanding fields of basic, applied, and translational biomedical research in which NF-κB signaling plays a crucial role, it was not possible to generate a methods volume encompassing all of the techniques, approaches, and protocols used in the wide arena of NF-κB biology. However, the 38 highly detailed protocols ranging from basic but powerful biochemical approaches, to complex computational and mathematical modeling, provide the insight necessary for researchers to expertly analyze NF-κB signaling in their own fields of biomedical or biological investigation. The promise of identifying powerful new drugs targeting aberrant NF-κB signaling pathways in many debilitating and destructive diseases remains the overarching goal driving much of the research in this field. It is certain that in the pursuit of this goal, the techniques described here will be central to the discovery effort. It is therefore my hope that this volume provides the essential "go-to" resource for current and future workers pursuing this goal, and that the book finds its home on the many laboratory benches at which the boundaries of understanding of NF-κB biology are being relentlessly pushed forward.

In closing, I would like to extend my most sincere thanks and appreciation to all of the authors who contributed to this volume for their extraordinary willingness to share their expertise and for their patience as this book came to life.

Philadelphia, PA, USA   *Michael J. May, Ph.D.*

Contents

PART IV   ANALYZING AND TARGETING NF-κB ACTIVITY IN DISEASE

Contributors

YOUSEF ABU-AMER • *Department of Orthopedic Surgery-Research, Washington University School of Medicine, St. Louis, MO, USA; Department of Cell Biology and Physiology, Washington University School of Medicine, St. Louis, MO, USA*

FABRICE AGOU • *Unité de Signalisation Moléculaire et Activation Cellulaire, Département de Biologie Cellulaire et Infection, Institut Pasteur, Paris, France*

CLINT T. ALLEN • *Head and Neck Surgery Branch, National Institute on Deafness and Other Communication Disorders (NICDC), Bethesda, MD, USA*

ALBERT S. BALDWIN • *Lineberger Comprehensive Cancer Center, University of North Carolina, Chapel Hill, NC, USA*

BRIAN BARRON • *Department of Biology, Boston University, Boston, MA, USA*

DANIELA S. BASSÈRES • *Department of Biochemistry, Chemistry Institute, University of São Paulo, São Paulo, Brazil*

HARALD BURKHARDT • *Division of Rheumatology, Department of Internal Medicine II and Fraunhofer IME-Project-Group Translational Medicine and Pharmacology, Johann Wolfgang Goethe University Frankfurt am Main, Frankfurt am Main, Germany*

EZRA BURSTEIN • *Departments of Internal Medicine and Molecular Biology, UT Southwestern Medical Center, Dallas, TX, USA*

RUAIDHRÍ J. CARMODY • *Institute of Infection, Immunology and Inflammation, College of Medical, Veterinary and Life Sciences, University of Glasgow, Glasgow, UK*

JINJING CHEN • *Department of Biochemistry, College of Medicine, University of Illinois at Urbana-Champaign, Urbana, IL, USA*

LIN-FENG CHEN • *Department of Biochemistry, College of Medicine, University of Illinois at Urbana-Champaign, Urbana, IL, USA*

GENHONG CHENG • *David Geffen School of Medicine at the University of California Los Angeles, Los Angeles, CA, USA*

ROBERT A. CHONG • *Department of Oncological Sciences, The Icahn School of Medicine at Mount Sinai, New York, NY, USA*

AMY COLLERAN • *Institute of Infection, Immunology and Inflammation, College of Medical, Veterinary and Life Sciences, University of Glasgow, Glasgow, UK*

PATRICIA E. COLLINS • *Institute of Infection, Immunology and Inflammation, College of Medical, Veterinary and Life Sciences, University of Glasgow, Glasgow, UK*

EMMANUEL DEJARDIN • *Laboratory of Molecular Immunology and Signal Transduction, GIGA-Research, University of Liège, Liège, Belgium*

STEFAN DÜBEL • *Institute of Biochemistry, Biotechnology and Bioinformatics, Technische Universität Braunschweig, Braunschweig, Germany*

VIRGINIE DUBOSCLARD • *Unité de Signalisation Moléculaire et Activation Cellulaire, Département de Biologie Cellulaire et Infection, Institut Pasteur, Paris, France*

JOHN R. FINNERTY • *Department of Biology and Program in Bioinformatics, Boston University, Boston, MA, USA*

ELISABETH FONTAN • *Unité de Signalisation Moléculaire et Activation Cellulaire, Département de Biologie Cellulaire et Infection, Institut Pasteur, Paris, France*

HIROAKI FUJITA • *Department of Molecular and Cellular Physiology, Graduate School of Medicine, Kyoto University, Kyoto, Japan*

ANITA GAURNIER-HAUSSER • *Department of Professional Studies in the Health Sciences, Drexel University, Philadelphia, PA, USA*

CÉLINE GÉLINAS • *Center for Advanced Biotechnology and Medicine, RBHS-Robert Wood Johnson Medical School, Rutgers, The State University of New Jersey, Piscataway, NJ, USA; Department of Biochemistry and Molecular Biology, RBHS-Robert Wood Johnson Medical School, Rutgers, The State University of New Jersey, Piscataway, NJ, USA; Rutgers Cancer Institute of New Jersey, New Brunswick, NJ, USA*

THOMAS D. GILMORE • *Center for Advanced Biotechnology and Medicine, RBHS-Robert Wood Johnson Medical School, Rutgers, The State University of New Jersey, Piscataway, NJ, USA; Department of Biochemistry & Molecular Biology, RBHS-Robert Wood Johnson Medical School, Rutgers, The State University of New Jersey, Piscataway, NJ, USA; Rutgers Cancer Institute of New Jersey, New Brunswick, NJ, USA*

TATIANA GONCHAROV • *Department of Early Discovery Biochemistry, Genentech, Inc., South San Francisco, CA, USA*

CAROLYN M. GRAY • *Department of Animal Biology, The University of Pennsylvania School of Veterinary Medicine, Philadelphia, PA, USA*

MONICA GUMA • *Department of Pharmacology and Medicine, University of California San Diego School of Medicine, La Jolla, CA, USA*

KAMYAR HADIAN • *Assay Development and Screening Platform, Helmholtz Zentrum München, Institute of Molecular Toxicology and Pharmacology, Neuherberg, Germany*

ERIC P. HANSON • *Immunodeficiency and Inflammation Unit, Arthritis and Rheumatism Branch, National Institutes of Health, Bethesda, MD, USA*

EDWARD W. HARHAJ • *Department of Oncology, Sidney Kimmel Comprehensive Cancer Center, Johns Hopkins School of Medicine, Baltimore, MD, USA*

MATTHEW S. HAYDEN • *Department of Dermatology, College of Physicians and Surgeons, Columbia University, New York, NY, USA*

ALEXANDER HOFFMANN • *Institute for Quantitative and Computational Biosciences, University of California Los Angeles, Los Angeles, CA, USA; Department of Microbiology, Immunology, and Molecular Genetics, University of California Los Angeles, Los Angeles, CA, USA; Signaling Systems Laboratory, University of California San Diego, La Jolla, CA, USA; San Diego Center for Systems Biology, University of California San Diego, La Jolla, CA, USA*

KAZUHIRO IWAI • *Department of Molecular and Cellular Physiology, Graduate School of Medicine, Kyoto University, Kyoto, Japan*

SHAWN S. JACKSON • *Department of Oncology, University of Wisconsin-Madison, Madison, WI, USA*

CHANGYING JIANG • *Division of Basic Science Research, Department of Molecular and Cellular Oncology, The University of Texas MD Anderson Cancer Center, Houston, TX, USA*

MICHAEL KARIN • *Department of Pharmacology and Medicine, University of California San Diego School of Medicine, La Jolla, CA, USA*

JORDAN KOVACEV • *Department of Oncological Sciences, The Icahn School of Medicine at Mount Sinai, New York, NY, USA*

DANIEL KRAPPMANN • *Research Unit Cellular Signal Integration, Helmholtz Zentrum München, Institute of Molecular Toxicology and Pharmacology, Neuherberg, Germany*

HAIYING LI • *Department of Internal Medicine, UT Southwestern Medical Center, Dallas, TX, USA*

XUESEN LI • *Department of Metabolism and Aging, The Scripps Research Institute, Jupiter, FL, USA*

XIN LIN • *Division of Basic Science Research, Department of Molecular and Cellular Oncology, The University of Texas MD Anderson Cancer Center, Houston, TX, USA*

TAO LU • *Department of Pharmacology and Toxicology, Indiana University School of Medicine, Indianapolis, IN, USA; Department of Biochemistry and Molecular Biology, Indiana University School of Medicine, Indianapolis, IN, USA*

NICOLA J. MASON • *Department of Pathobiology, University of Pennsylvania School of Veterinary Medicine, Philadelphia, PA, USA*

MICHAEL J. MAY • *Department of Animal Biology, The University of Pennsylvania School of Veterinary Medicine, Philadelphia, PA, USA; The Mari Lowe Center for Comparative Oncology, The University of Pennsylvania School of Veterinary Medicine, Philadelphia, PA, USA*

KELLY A. MCCORKELL • *Department of Animal Biology, The University of Pennsylvania School of Veterinary Medicine, Philadelphia, PA, USA*

SARA MCGOWAN • *Department of Metabolism and Aging, The Scripps Research Institute, Jupiter, FL, USA*

MOLLIE K. MEFFERT • *Department of Biological Chemistry, The Johns Hopkins University School of Medicine, Baltimore, MD, USA; The Solomon H. Snyder Department of Neuroscience, The Johns Hopkins University School of Medicine, Baltimore, MD, USA*

ANCA B. MIHALAS • *Department of Biological Chemistry, The Johns Hopkins University School of Medicine, Baltimore, MD, USA*

SIMON MITCHELL • *Signaling Systems Laboratory, San Diego Center for Systems Biology, University of California San Diego, La Jolla, CA, USA*

SHIGEKI MIYAMOTO • *Department of Oncology, University of Wisconsin-Madison, Madison, WI, USA*

DANIEL NAGEL • *Research Unit Cellular Signal Integration, Helmholtz Zentrum München, Institute of Molecular Toxicology and Pharmacology, Neuherberg, Germany*

MISA NAKAI • *Department of Molecular and Cellular Physiology, Graduate School of Medicine, Kyoto University, Kyoto, Japan*

LAURA J. NIEDERNHOFER • *Department of Metabolism and Aging, The Scripps Research Institute, Jupiter, FL, USA*

CHRISTINE O'CARROLL • *Institute of Infection, Immunology and Inflammation, College of Medical, Veterinary and Life Sciences, University of Glasgow, Glasgow, UK*

ZHEN-QIANG PAN • *Department of Oncological Sciences, The Icahn School of Medicine at Mount Sinai, New York, NY, USA*

SUMAN PAUL • *Department of Microbiology and Immunology, Uniformed Services University, Bethesda, MD, USA; Center for Neuroscience and Regenerative Medicine, Uniformed Services University, Bethesda, MD, USA*

ASHLEY PENVOSE • *Department of Biology, Boston University, Boston, MA, USA*

ZHAOXIA QU • *Department of Microbiology and Molecular Genetics, University of Pittsburgh, Pittsburgh, PA, USA*

SITHARAM RAMASWAMI • *Department of Dermatology, College of Physicians and Surgeons, Columbia University, New York, NY, USA*

ANNA D. REICHARDT • *David Geffen School of Medicine at the University of California Los Angeles, Los Angeles, CA, USA*

CAROLINE REMOUCHAMPS • *Laboratory of Molecular Immunology and Signal Transduction, GIGA-Research, University of Liège, Liège, Belgium*

PAUL D. ROBBINS • *Department of Metabolism and Aging, The Scripps Research Institute, Jupiter, FL, USA*

Jose Pindado • *David Geffen School of Medicine at the University of California Los Angeles, Los Angeles, CA, USA*

Yoshiteru Sasaki • *Department of Molecular and Cellular Physiology, Graduate School of Medicine, Kyoto University, Kyoto, Japan*

Brian C. Schaefer • *Department of Microbiology and Immunology, Uniformed Services University, Bethesda, MD, USA; Center for Neuroscience and Regenerative Medicine, Uniformed Services University, Bethesda, MD, USA*

Bettina Sehnert • *Department of Rheumatology and Clinical Immunology and Centre of Chronic Immunodeficiency, University Medical Centre and University of Freiburg, Freiburg, Germany*

Helena Shaked • *Department of Pharmacology, University of California San Diego School of Medicine, La Jolla, CA, USA*

Noula Shembade • *Department of Microbiology and Immunology, Sylvester Comprehensive Cancer Center, Miller School of Medicine, The University of Miami, Miami, FL, USA*

Trevor Siggers • *Department of Biology, Boston University, Boston, MA, USA*

George R. Stark • *Department of Molecular Genetics, Cleveland Clinic, Cleveland, OH, USA*

Petro Starokadomskyy • *Department of Internal Medicine, UT Southwestern Medical Center, Dallas, TX, USA*

Gaurav Swarnkar • *Department of Orthopedic Surgery-Research, Washington University School of Medicine, St. Louis, MO, USA; Department of Cell Biology and Physiology, Washington University School of Medicine, St. Louis, MO, USA*

Rachel Tsui • *Institute for Quantitative and Computational Biosciences, University of California Los Angeles, Los Angeles, CA, USA; Department of Microbiology, Immunology, and Molecular Genetics, University of California Los Angeles, Los Angeles, CA, USA*

Eugene Varfolomeev • *Department of Early Discovery Biochemistry, Genentech, Inc., South San Francisco, CA, USA*

Michelle Vincendeau • *Research Unit Cellular Signal Integration, Helmholtz Zentrum München, Institute of Molecular Toxicology and Pharmacology, Neuherberg, Germany*

Reinhard E. Voll • *Department of Rheumatology and Clinical Immunology and Centre of Chronic Immunodeficiency, University Medical Centre and University of Freiburg, Freiburg, Germany*

Domagoj Vucic • *Department of Early Discovery Biochemistry, Genentech, Inc., South San Francisco, CA, USA*

Carter Van Waes • *Head and Neck Surgery Branch, National Institute on Deafness and Other Communication Disorders (NICDC), Bethesda, MD, USA*

Alex W. Wessel • *Immunodeficiency and Inflammation Unit, Arthritis and Rheumatism Branch, National Institutes of Health, Bethesda, MD, USA*

Kenneth Wu • *Department of Oncological Sciences, The Icahn School of Medicine at Mount Sinai, New York, NY, USA*

Gutian Xiao • *Department of Microbiology and Molecular Genetics, University of Pittsburgh, Pittsburgh, PA, USA*

Shivam A. Zaver • *David Geffen School of Medicine at the University of California Los Angeles, Los Angeles, CA, USA*

Jing Zhao • *Department of Metabolism and Aging, The Scripps Research Institute, Jupiter, FL, USA*

Jochen Zwerina • *Department of Internal Medicine, University of Erlangen-Nürnberg, Erlangen, Germany*

Part I

Standard Approaches to Detect NF-κB Pathway Activation

Chapter 1

Electrophoretic Mobility Shift Assay Analysis of NF-κB DNA Binding

Sitharam Ramaswami and Matthew S. Hayden

Abstract

The discovery and characterization of the nuclear factor-kappa B (NF-κB) family of transcription factors was predicated on the technical ability to detect protein binding to defined sequences of DNA. Proteins capable of binding to specific sequences of nucleic acid are detected through the use of the electrophoretic mobility shift assay (EMSA), also called a gel shift assay. While newer techniques, including chromatin immunoprecipitation (ChIP), are widely used to assess NF-κB binding to the promoters and enhancers of specific genes, the EMSA remains a powerful experimental tool to quickly test for the presence of NF-κB that is capable of binding DNA. In this way, the EMSA is a useful general readout of the activation state of the NF-κB pathway and an essential tool for the investigation of this important transcription factor family.

Key words Transcription factor, NF-kappa B, TNF, Electrophoretic mobility shift assay, In-gel fluorescence

1 Introduction

Electrophoretic mobility shift assay (EMSA) or gel shift assay is a widely utilized technique for the investigation of the study of transcription factor function and, more generally, the interaction of proteins with nucleic acids. The use of this technique for quantitative studies of DNA-protein interactions was described as early as 1981 to study the *E.coli* lac operon regulatory system [1, 2]. Since then, EMSA has been widely used to study the function of DNA-binding proteins, especially in studies involving the specificity of transcription factors for specific DNA sequence motifs as well as in characterizing the transcription factor subunits binding to consensus sequences [3–9]. The EMSA technique is based on the principle that in a non-denaturing gel, molecules of different molecular weight and charge have distinct mobility. DNA, being negatively charged, will migrate toward the anode and will be retarded by the sieving properties of the gel in a manner dependent on the size

Michael J. May (ed.), *NF-kappa B: Methods and Protocols*, Methods in Molecular Biology, vol. 1280, DOI 10.1007/978-1-4939-2422-6_1, © Springer Science+Business Media New York 2015

and conformation of the DNA. When a DNA-protein complex is formed, it will migrate in a distinct manner, usually at a rate that is slower due to the increase in size and decrease in negative charge to size ratio of the DNA-protein complex relative to the naked DNA probe. The resulting upward shift in the mobility of the DNA caused by its binding to the protein can be detected through a variety of means. Its ease of use and high sensitivity, 10^{-18} mol of DNA [2], make it a powerful tool to not only study transcription factor activity but also in the studies of DNA replication, repair, and chromosomal recombination.

Nuclear factor-kappa B (NF-κB) comprises a family transcription factors that play an important role in transcribing genes that govern immune responses, cell survival, proliferation, and differentiation. Since its discovery in 1986, initially describing it as a B lymphocyte-specific immunoglobulin κ-light chain enhancer-binding protein [4], NF-κB has been shown to be present and has an important biological function in almost every nucleated cell type examined so far. The pathway is evolutionarily conserved [10] and widely implicated in the pathogenesis of numerous disease states [11–13]. The early studies of DNA-binding activities of NF-κB to its corresponding consensus sequence, induced by lipo-polysaccharide- or phorbol ester-treated cells, were performed using EMSA [14].

In mammals, five DNA-binding members of NF-κB family have been identified. These include p65 (also called RelA), c-Rel, RelB, p50 (NF-κB1), and p52 (NF-κB2) [15]. All NF-κB proteins have a Rel homology domain (RHD), which imparts DNA-binding ability and dimer formation. In most resting cells, NF-κB is present in the cytoplasm, bound to its inhibitor, inhibitor of kappa B (IκB) [16]. NF-κB exists in cells as homodimers or heterodimers. The most common NF-κB dimer is the p65/p50 heterodimer, which is bound to its inhibitor IκBα. Different stimuli can lead to the activation of NF-κB by activating its upstream kinase complex, IκB kinase (IKK). Upon activation, IKK phosphorylates IκBα, which leads to its ubiquitination and degradation by the cellular proteasomal complex. The NF-κB dimer can now translocate to the nucleus, bind to its consensus sequence on promoters and enhancers of target genes, and activate transcription of pro-inflammatory and pro-survival genes [17]. One of the first genes activated by NF-κB is its own inhibitor, IκBα [18]. The newly formed IκBα protein translocates to the nucleus, binds to the promoter-bound NF-κB, and returns it to the cytoplasm in an inactivate state, thus terminating transcription. This process is called post-induction repression [19–23].

Here, we describe a protocol that uses Alexa Fluor 488-labeled double-stranded DNA oligonucleotide to study tumor necrosis factor alpha (TNF)-induced NF-κB DNA binding in HeLa cells. The binding of TNF to the receptor TNFR1 leads to receptor multimerization and formation of a large signaling complex containing the adapter protein TRADD (TNF receptor-associated death domain protein), RIP1 (receptor-interacting protein 1), and TRAF2 (TNF receptor-associated factor 2) [24]. TRAF and RIP1 recruit the IKK kinase TAK1 and the IKK complex, resulting in activation of IKK kinase activity, phosphorylation of IκBα, and NF-κB activation as described above. The NF-κB EMSA tests for two properties that are indicative of NF-κB transcriptional activity: nuclear localization and the capacity to bind double-stranded DNA with a κB sequence. However, it should be stressed that NF-κB transcriptional activity is also further regulated by posttranslation modifications, such as phosphorylation, and binding to various co-regulatory proteins [25]. Thus, although the EMSA is an excellent means of assessing one of the crucial regulatory steps in NF-κB activation, it should not be used in isolation.

The traditional NF-κB EMSA is performed with a radiolabeled κB probe. This method remains the method of choice in most labs, as the sensitivity of radiolabeled probes is superb. However, because many biomedical research labs have reduced or eliminated the use of radioactivity and fluorescent gel imagers have improved in sensitivity and become more readily available, we describe here a protocol for performing an NF-κB EMSA using fluorescently labeled probes. The use of fluorophore-labeled κB sequence double-stranded DNA oligonucleotide is an excellent alternative to the use of radioactive isotopic method of EMSA analysis and visualization. The assay can also be completed in a few hours without performing gel transfer and prolonged film exposure. The assays can be multiplexed to allow control and κB probes to be run simultaneously, and the results are readily quantified with manufacturer or publicly available software such as NIH ImageJ [26].

2 Materials

2.1 Cell Culture

1. HeLa cells (human cervical epithelial cells; CCL-2) obtained from the American Type Culture Collection (Manassas, VA, USA).

2. DMEM, high glucose medium supplemented with 10 % fetal bovine serum (FBS) (*see* **Note 1**).

3. 1× phosphate buffered saline (PBS) without calcium and magnesium.

4. TrypLE Express 1× dissociation reagent (Gibco, Bethesda, MD, USA).

5. Human TNF. Make a working stock of concentration 10 μg/ml using sterile 1× PBS. Aliquot in conveniently sized single use volumes and store at –80 °C.

6. 6 cm^2 culture plates and 6-well tissue culture plates.

2.2 Preparation of Cytoplasmic and Nuclear Extracts

1. Relaxation buffer [RB]: 10 mM HEPES, pH 7.5, 10 mM KCl, 3 mM NaCl, 3 mM $MgCl_2$, 1 mM EDTA, 1 mM EGTA.

2. 1 % Igepal CA-630 (NP-40; Sigma, St. Louis, MO, USA): In a sterile 15-ml conical centrifuge tube, dilute 0.1 ml of 100 % Igepal CA-630 in 9 ml of sterile molecular biology grade water. Mix well by inverting the tube several times. Avoid vortexing, which will cause foaming and make the solution difficult to mix. Store at room temperature. If turbidity or sediment develops during storage, heat briefly to 40 °C, and mix by inverting.

3. Nuclear extract buffer [NAR-C]: 20 mM HEPES, pH7.9, 0.4 M NaCl, 0.1 mM EDTA. Store at 4 °C.

4. Protease and phosphatase inhibitors for mammalian cell extracts: pepstatin A, phenylmethylsulfonyl fluoride (PMSF), bestatin, leupeptin, aprotinin, beta-glycerophosphate (disodium salt), and sodium fluoride (*see* **Note 2**).

5. 1× PBS-prechilled on ice.

6. Cell scrapers (*see* **Note 3**).

7. 1.5-ml microcentrifuge tubes.

2.3 Preparing Gel for EMSA

1. 10× Tris-borate EDTA [TBE] buffer: 1.3 M Tris, 450 mM boric acid, and 25 mM EDTA, pH-8.4. Store at room temperature. This concentrate can easily be diluted to 0.5× with molecular biology grade water before use.

2. 40 % acrylamide/bis-acrylamide solution (37:1).

3. Ammonium persulfate (APS; 10 % w/v). Dissolve 100 mg of APS in 1 ml of water. Prepare fresh before use.

4. *N,N,N',N'*,-Tetramethylethane-1,2-diamine (TEMED). Store at room temperature.

5. 50-ml conical centrifuge tubes.

2.4 For Performing EMSA

1. Alexa Fluor 488-labeled NF-κB double-stranded DNA oligonucleotide 5'-AGAG<u>GGGACTTTCC</u>GAGG-3' containing one NF-κB binding site (underlined). Both the sense and antisense oligonucleotides should be ordered with 5' Alexa Fluor 488 modification (50 nmol, Life Technologies, Carlsbad, CA, USA) (*see* **Note 4**).

2. Alexa Fluor 680-labeled CREB double-stranded DNA oligonucleotide 5'-AGAGATTGCCTGACGTCAGAGAGCTAG-3'. For the CREB probe, we used a 5' Alexa Fluor 680-labeled sense strand and an unlabeled reverse complement oligonucleotide (*see* **Note 4**).

3. 10× binding buffer [10× BB]: 100 mM Tris, pH 7.5, 500 mM NaCl, 10 mM DTT, 5 mM EDTA, 10 mM $MgCl_2$. Aliquot and store at −20 °C.

4. 100 mM DTT (Pierce/Thermo Scientific): Dissolve 15.4 mg of DTT in 1 ml water. Aliquot and store at –20 °C. Avoid multiple freeze/thaw cycles (*see* **Note 5**).

5. Poly-dI.dC (deoxyinosinic-deoxycytidylic acid): 1 µg/µl in nuclease free water. Aliquot and store at –20 °C.

6. 100 mM $MgCl_2$: Dissolve 20.3 mg of $MgCl_2$ in 1 ml of water. Store at 4 °C gel loading dye, 6×: 15 % Ficoll 400 (Fisher) with 0.15 % Orange G (Sigma). Dissolve 1.5 g of Ficoll and 15 mg Orange G in 10 ml of water.

3 Methods

In this section, we describe the protocol for analysis of NF-κB DNA binding in HeLa cells stimulated with TNF. The stimulation of HeLa cells with TNF results in rapid nuclear localization of NF-κB and rapid induction of NF-κB activation as assessed by the EMSA protocol described below (Fig. 1).

3.1 Cell Culture

1. HeLa cells are grown in 6 cm^2 dishes in 4 ml of DMEM containing 10 % FBS at 37 °C and 5 % CO_2 until nearly confluent. Aspirate the medium, and wash cells twice with 4 ml of 1× PBS, followed by incubating in 2 ml of TrypLE for 2–4 min until the cells detach (verify microscopically). Harvest the cells by gently washing the plate with 10 ml of prewarmed, 37 °C, DMEM with 10 % FBS. Generate a single cell suspension by gently pipetting up and down, and transfer the cells to a 6-well tissue culture plate; 2 ml of cell suspension per well. The cells should be used for stimulation 12–24 h after splitting and before they reach full confluence.

2. The cells are stimulated at specific times by adding 2 µL of TNF stock to get a final concentration of 10 ng/ml. Gently rock the plate to mix the TNF, and return the plate to 37 °C incubator for the appropriate period of time.

3.2 Preparation of Cytoplasmic and Nuclear Extracts

All steps must be performed at 4 °C or on ice, unless stated otherwise. All equipment and tubes should be prechilled prior to harvesting the cells.

1. After the required time of stimulation, aspirate and discard the medium from the cell culture plates. Rinse the cells twice with 2 ml of ice-cold 1× PBS.

2. Add 500 µl of 1× PBS into each well. Scrape to gently lift off cells from the plate, and transfer to prechilled 1.5-ml microcentrifuge tubes (*see* **Note 3**).

3. Add 300 µl of 1× PBS into each plate; collect any remaining cells, and transfer them into the corresponding microcentrifuge tubes.

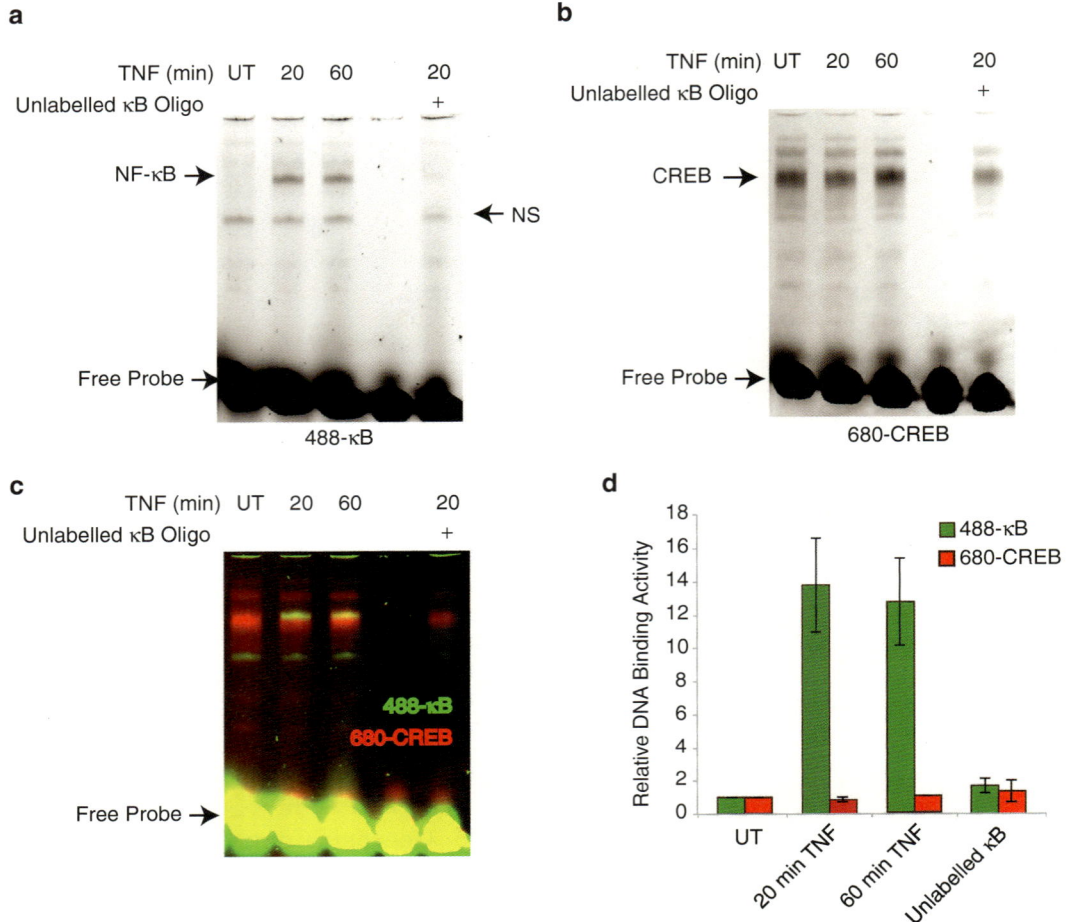

Fig. 1 Nuclear extracts (5 μg) from HeLa cells treated for the indicated times with 10 ng/ml TNF were subjected to EMSA according to the protocol described in the text. After electrophoresis, the same gel was imaged for both Alexa Fluor 488 (**a**) and Alexa Fluor 680 (**b**) as described. Superimposed multichannel image (**c**) of the above with pseudocolored NF-κB (*green*) and CREB (*red*). The specific NF-κB (and CREB) signal was quantitated from three independent binding reactions using Image Lab. Mean relative abundance, with standard error of the mean, is shown (**d**). *NS* nonspecific bands

4. Centrifuge the tubes containing the cell suspension in a refrigerated microcentrifuge at $500 \times g$ for 5 min.

5. Carefully aspirate and discard the supernatants. Gently flick the tubes to dislodge the cell pellet.

6. Add 125 μl of ice-cold relaxation buffer [RB] containing protease inhibitors. Mix well but gently by slowly pipetting up and down.

7. Incubate the cell suspensions on ice for 20 min. The relaxation buffer is hypotonic; the cells will swell in this buffer.

8. Lyse the cells by adding 0.05 volumes of 10 % Igepal CA-630. Vortex each tube individually vigorously for 15 s to lyse the cells, and place the tube back on ice. Vortex each sample again for an additional 5 s, and quickly move to **step 9**. Do not leave the cells in RB containing Igepal CA-630 for longer than 20 min.

9. Centrifuge for 5 min at $1,000 \times g$.

10. Carefully remove the supernatant with a 200-μl micropipette and transfer to a new, prechilled 1.5-ml microcentrifuge tube. Label as Cytoplasmic Extract (CE). Keep on ice. Measure the protein concentration (*see* **Note 6**). Add 40 μl of 6× SB to cytoplasmic extracts, and boil immediately for 12 min on a boiling water bath. Store at –80 °C.

11. Wash the nuclear pellets by adding 200 μl of RB containing the protease inhibitors.

12. Centrifuge as above (**step 9**), remove, and discard the supernatant.

13. To obtain the nuclear extract, resuspend the nuclear pellets in 30 μl of ice-cold nuclear extract buffer (NAR-C) containing protease inhibitors. Mix well by pipetting. Incubate on ice for 20 min; mix by flicking the tubes every 4–5 min to prevent contents from settling down.

14. After 20 min, vortex the tubes vigorously for 15 s. Centrifuge for 10 min at $16,000 \times g$ at 4 °C. Transfer the supernatants, this is the nuclear extract, to prechilled tubes using a 200-μl micropipette, remove a small amount of the nuclear extract (2 μl) in order to determine the protein concentration (*see* **Note 6**). It is preferable to continue directly to the binding reactions for EMSA (Subheading 3.5). Alternately add glycerol to a final concentration of 5 %, aliquot to avoid freeze-thawing samples, flash-freeze in a dry ice isopropanol (or ethanol) bath, and store the extracts at –80 °C.

3.3 Annealing Oligonucleotides for EMSA

1. Dissolve the oligonucleotides in nuclease free water to a concentration of 100 μM.

2. Add 5 μl each of the forward and reverse oligos to an amber-colored 1.5-ml microcentrifuge tube, and add 90 μl of nuclease free water to get 5 μM; mix well.

3. Heat in a heat block at 90 °C for 10 min. Leave the oligonucleotides in the heat block, and turn off power or remove block to bench top, allowing the block to slowly cool to room temperature. This annealed oligonucleotide stock is the double-stranded DNA probe. Keep fluorophore-labeled oligos in amber-colored tubes, and minimize exposure to light at all steps.

4. Dilute the stock to get a final concentration of 500 nM for use in binding reactions.

3.4 Preparing Non-denaturing Gel for EMSA

Below are the instructions for casting and running 5 % non-denaturing polyacrylamide gels with glass plates (1.5-mm spacer plates and short plates) using Bio-Rad Mini-PROTEAN Electrophoresis apparatus. Alternately purchase precast non-denaturing TBE polyacrylamide gels.

1. Clean the back and front glass plates using a mild detergent, and then wash in distilled water. Wipe the plates using a 70 % ethanol solution, and leave it to air dry.

2. Assemble the glass plates according to the manufacturer's instructions.

3. To a 50-ml centrifuge tube, add 16.5 ml of water, 1 ml of 10× TBE, 2.5 ml of 40 % acrylamide/bis-acrylamide solution, 50 μl of 10%APS, and 5 μl of TEMED. Mix immediately by gently pipetting using a serological pipette (*see* **Note 7**).

4. Pour the solution into the gel chamber, and insert the 10-well comb. The gel should polymerize in 30–45 min. Polymerization can be assessed by observing the remaining acrylamide solution in the 50-ml conical tube.

5. Prepare 500 ml of 0.5× TBE using the 10× stock.

6. After the gel polymerizes, assemble the cassette in the tank, and pour the running buffer into inner and outer chambers.

7. Remove the comb; gently rinse the wells with running buffer using a syringe and needle or micropipette to flush out non-polymerized acrylamide.

8. Pre-run the gel at 100 V for 30 min.

3.5 Preparing Binding Reactions and Performing the EMSA

1. Prepare binding reactions by mixing the following components in the order specified (*see* **Notes 5** and **8**) to a final volume of 20 μl:

Ultra pure water	xμl
10× binding buffer	2 μl
100 mM DTT	0.5 μl
Poly-dI.dC	1 μl
Nuclear extract	xμl
(5–10 μg) (*see* **Note 9**)	

2. Gently mix by tapping the side of the tubes. Briefly spin the sample to the bottom.

3. Incubate the tubes at 37 °C for 5–7 min.

4. Add 1 μl each of 500 nM NF-κB and CREB probe into each tube. Mix and spin briefly (*see* **Note 10**).

5. Incubate the tubes at room temperature for 30 min in the dark.

6. Add 4 µl of 6× orange dye to each 20-µl reaction tube. Pipette gently to mix.

7. Load samples into the wells of the pre-run polyacrylamide gel.

8. Run the gel 100 V at constant voltage until the orange dye migrates to the bottom of the gel. During electrophoresis, cover the running apparatus with an inverted box to protect from light. Electrophoresis should take approximately 40–45 min when using 10 cm gels.

9. After electrophoresis, transfer the gel carefully to the imaging system, and visualize using appropriate excitation/emission settings. We use a Bio-Rad ChemiDoc MP with blue epi-illumination (470 nm; 30-nm bandpass) and a 530/28-nm emission filter to image the Alexa Fluor 488 κB probe and red epi-illumination (625 nm; 30-nm bandpass) and 695/55-nm emission filter to image the Alexa Fluor 680 CREB probe (*see* **Note 11**).

10. If desired, the intensity of the bands can be quantified using appropriate software such as Bio-Rad Image Lab or NIH ImageJ using appropriate background subtraction. Results can be plotted as fold change relative to unstimulated.

4   Notes

1. We routinely test lots of FBS from various manufacturers for effects on constitutive and inducible NF-κB activation, selecting those with the least effect on constitutive activation and with robust activation in response to multiple relevant stimuli. DMEM substituted with 10 % FBS should be stored at 4 °C. Warm the medium in a 37 °C water bath for at least 30 min prior to use. Maintain sterile working conditions through the use of a biological safety cabinet.

2. It is advisable to make small aliquots of protease inhibitors to prevent repeated freeze-thawing. Protease inhibitors must be added immediately before use as some, notably PMSF, rapidly lose efficacy once diluted in aqueous solutions.

3. Use scrapers with a thin, soft, pliable, edge to minimize cell damage. We prefer Sarstedt TC cell scrapers. The blade of the scraper can be cut with a razor blade for use in smaller wells. While using scrapers to lift cells off the well plate, a separate one should be used for each sample. Alternately rinse in 1× PBS between uses.

4. Oligonucleotides should be HPLC purified. Most manufacturers will HPLC purify labeled oligonucleotides as a required step in the synthesis process. Residual salt in insufficiently desalted oligonucleotides can affect annealing.

5. 100-mM DTT is an optional component used specifically in this EMSA involving the NF-κB and CREB oligonucleotides described. Refer to **Note 8** for more details.

6. We use Micro BCA Protein Assay Kit (Thermo Scientific) to measure protein concentration. Measure concentration of cytoplasmic extract along with the nuclear extract. Keep on ice till then. An aliquot of these extracts could be used to perform immunoblotting to test the purity of the extract preparation.

7. Extreme care should be taken when handling acrylamide; unpolymerized acrylamide is a potential neurotoxin.

8. The binding reaction described here is optimized for the NF-κB and CREB probes described. The 37 °C incubation with DTT was performed as it was found to greatly facilitate CREB-DNA binding. In our hands, this step also yields a slight improvement in NF-κB κB DNA binding. Therefore, this step may not be necessary for all DNA-protein binding. Also, to optimize DNA-protein binding of other transcription factors, several optional components such as Salmon Sperm DNA, glycerol, Igepal CA 630, KCl, $MgCl_2$, and EDTA should be tested and optimized.

9. The optimal amount of protein required varies between experiments and cell types. If protein concentrations vary widely between samples, normalize using NAR-C.

10. When using an unlabeled probe to test the specificity of DNA-protein binding, add a 50- to 200-fold excess of unlabeled probe after the addition of nuclear extract, and incubate at 37 °C as described. Alternately, incubate for 10–15 min at room temperature, and follow subsequent steps as written.

11. Exposure times should be adjusted manually. The free, unbound, probe will be in vast excess, and the imaging software will, therefore, capture images that are too short to image the shifted probe signal. We typically capture 60 and 240 s exposures using a Bio-Rad ChemiDoc MP imaging system. The use of illumination sources and the order of exposures should be adjusted to minimize photobleaching.

References

1. Garner MM, Revzin A (1981) A gel electrophoresis method for quantifying the binding of proteins to specific DNA regions: application to components of the Escherichia coli lactose operon regulatory system. Nucleic Acids Res 9(13):3047–3060

2. Fried M, Crothers DM (1981) Equilibria and kinetics of lac repressor-operator interactions

by polyacrylamide gel electrophoresis. Nucleic Acids Res 9(23):6505–6525

3. Roche PJ, Hoare SA, Parker MG (1992) A consensus DNA-binding site for the androgen receptor. Mol Endocrinol 6(12):2229–2235

4. Sen R, Baltimore D (1986) Multiple nuclear factors interact with the immunoglobulin enhancer sequences. Cell 46(5):705–716

5. Tanaka N, Kawakami T, Taniguchi T (1993) Recognition DNA sequences of interferon regulatory factor 1 (IRF-1) and IRF-2, regulators of cell growth and the interferon system. Mol Cell Biol 13(8):4531–4538

6. Tanaka H, Dong Y, Li Q, Okret S, Gustafsson JA (1991) Identification and characterization of a cis-acting element that interferes with glucocorticoid-inducible activation of the mouse mammary tumor virus promoter. Proc Natl Acad Sci U S A 88(12):5393–5397

7. Ko LJ, Yamamoto M, Leonard MW, George KM, Ting P, Engel JD (1991) Murine and human T-lymphocyte GATA-3 factors mediate transcription through a cis-regulatory element within the human T-cell receptor delta gene enhancer. Mol Cell Biol 11(5):2778–2784

8. Singh H, Sen R, Baltimore D, Sharp PA (1986) A nuclear factor that binds to a conserved sequence motif in transcriptional control elements of immunoglobulin genes. Nature 319(6049):154–158

9. Staudt LM, Singh H, Sen R, Wirth T, Sharp PA, Baltimore D (1986) A lymphoid-specific protein binding to the octamer motif of immunoglobulin genes. Nature 323(6089):640–643. doi:10.1038/323640a0

10. Gilmore TD, Wolenski FS (2012) NF-kappaB: where did it come from and why? Immunol Rev 246(1):14–35. doi:10.1111/j.1600-065X.2012.01096.x

11. Baker RG, Hayden MS, Ghosh S (2011) NF-kappaB, inflammation, and metabolic disease. Cell Metabol 13(1):11–22. doi:10.1016/j.cmet.2010.12.008

12. DiDonato JA, Mercurio F, Karin M (2012) NF-kappaB and the link between inflammation and cancer. Immunol Rev 246(1):379–400. doi:10.1111/j.1600-065X.2012.01099.x

13. Sun SC, Chang JH, Jin J (2013) Regulation of nuclear factor-kappaB in autoimmunity. Trends Immunol 34(6):282–289. doi:10.1016/j.it.2013.01.004

14. Sen R, Baltimore D (1986) Inducibility of kappa immunoglobulin enhancer-binding protein Nf-kappa B by a posttranslational mechanism. Cell 47(6):921–928

15. Hayden MS, Ghosh S (2012) NF-kappaB, the first quarter-century: remarkable progress and outstanding questions. Gene Dev 26(3):203–234. doi:10.1101/gad.183434.111

16. Baeuerle PA, Baltimore D (1996) NF-kappa B: ten years after. Cell 87(1):13–20

17. Ghosh S, Hayden MS (2008) New regulators of NF-kappaB in inflammation. Nat Rev Immunol 8(11):837–848

18. Sun SC, Ganchi PA, Ballard DW, Greene WC (1993) NF-kappa B controls expression of inhibitor I kappa B alpha: evidence for an inducible autoregulatory pathway. Science 259(5103):1912–1915

19. Arenzana-Seisdedos F, Thompson J, Rodriguez MS, Bachelerie F, Thomas D, Hay RT (1995) Inducible nuclear expression of newly synthesized I kappa B alpha negatively regulates DNA-binding and transcriptional activities of NF-kappa B. Mol Cell Biol 15(5):2689–2696

20. Arenzana-Seisdedos F, Turpin P, Rodriguez M, Thomas D, Hay RT, Virelizier JL, Dargemont C (1997) Nuclear localization of I kappa B alpha promotes active transport of NF-kappa B from the nucleus to the cytoplasm. J Cell Sci 110(Pt 3):369–378

21. Turpin P, Hay RT, Dargemont C (1999) Characterization of IkappaBalpha nuclear import pathway. J Biol Chem 274(10):6804–6812

22. Rodriguez MS, Thompson J, Hay RT, Dargemont C (1999) Nuclear retention of IkappaBalpha protects it from signal-induced degradation and inhibits nuclear factor kappaB transcriptional activation. J Biol Chem 274(13):9108–9115

23. Ghosh CC, Ramaswami S, Juvekar A, Vu H-Y, Galdieri L, Davidson D, Vancurova I (2010) Gene-specific repression of proinflammatory cytokines in stimulated human macrophages by nuclear IκBα. J Immunol 185(6):3685–3693

24. Hayden MS, Ghosh S (2014) Regulation of NF-kappaB by TNF family cytokines. Semin Immunol 26(3):253–266. doi:10.1016/j.smim.2014.05.004

25. Huang B, Yang XD, Lamb A, Chen LF (2010) Posttranslational modifications of NF-kappaB: another layer of regulation for NF-kappaB signaling pathway. Cell Signal 22(9):1282–1290. doi:10.1016/j.cellsig.2010.03.017

26. Schneider CA, Rasband WS, Eliceiri KW (2012) NIH Image to ImageJ: 25 years of image analysis. Nat Methods 9(7):671–675

Chapter 2

Detection of IκB Degradation Dynamics and IκB-α Ubiquitination

Petro Starokadomskyy and Ezra Burstein

Abstract

The NF-κB signaling pathway is a primary regulator of inflammation that has been implicated in the pathogenesis of immune disorders and cancer. This signaling network is strictly regulated; in a nonactivated state, NF-κB transcription factors are sequestered in the cytoplasm by the IκB family of proteins. Various pro-inflammatory stimuli result in the phosphorylation and subsequent ubiquitination of IκBs. These events lead to rapid degradation of IκB and allow translocation of the transcription factors to the nucleus. Therefore, ubiquitination and degradation of IκBs are critical steps in NF-κB pathway activation and can serve as a quantitative parameter to assess pathway activation. In this article, we present a detailed protocol for the quantification of in vivo ubiquitination and turnover of IκB-α in response to a variety of cellular stimuli.

Key words NF-κB, RelA, IκB, Immunoprecipitation, Ubiquitin, Ubiquitination, Ubiquitin-like proteins (UBL)

1 Introduction

The NF-κB transcription factors regulate expression of a large number of genes, which in turn drive a variety of physiological processes including innate and adaptive immunity, apoptosis, and cellular differentiation [1]. Under basal conditions, NF-κB transcription factors are sequestered in the cytoplasm by the inhibitory binding of the so-called classical IκB proteins (IκB-α, IκB-β, and IκB-ε) [2]. In response to a variety of stimuli such as inflammatory cytokines or microbial products, these IκB proteins undergo rapid degradation, enabling NF-κB dimers to translocate to the nucleus and activate gene transcription [3].

The degradation results from two modification events: phosphorylation and subsequent ubiquitination of the IκB proteins [1]. These modifications are executed by corresponding enzyme complexes and lead to proteasomal degradation of IκB. The IκB kinase complex (IKK complex), consisting of the adaptor protein NEMO

Michael J. May (ed.), *NF-kappa B: Methods and Protocols*, Methods in Molecular Biology, vol. 1280, DOI 10.1007/978-1-4939-2422-6_2, © Springer Science+Business Media New York 2015

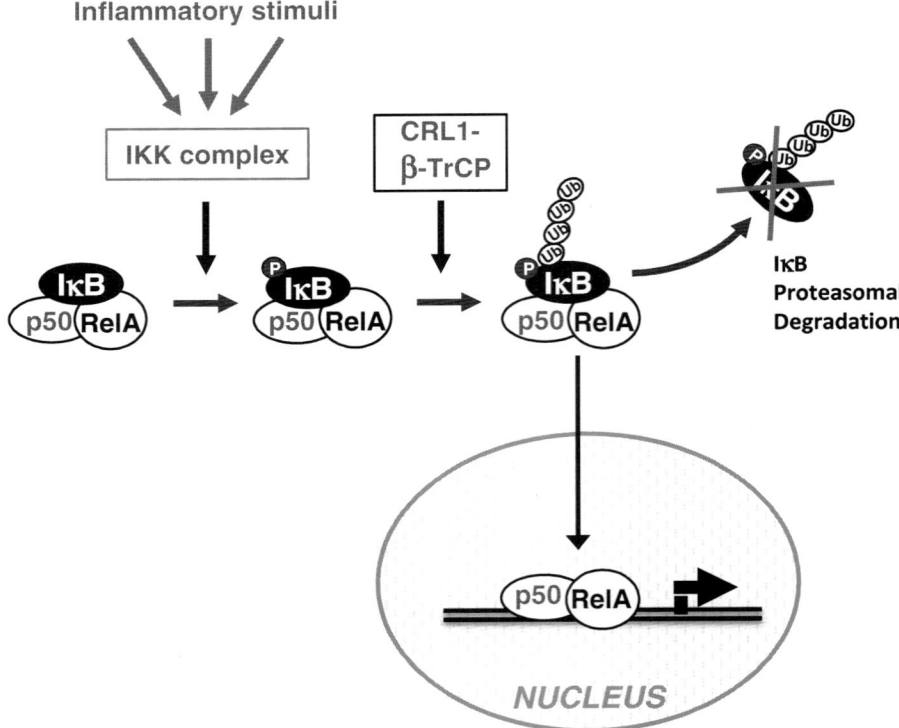

Fig. 1 Schematic representation of IκB-α regulation by phosphorylation and ubiquitination

and the catalytic subunits IKK1 and IKK2, is responsible for the IκB phosphorylation in the classical pathway. Two serine residues in the amino-termini of IκB-α, IκB-β, and IκB-ε are targeted, and the resulting phosphopeptide can be recognized by β-TrCP. This protein serves as the substrate recognition subunit of a multimeric ubiquitin ligase responsible for IκB ubiquitination (Fig. 1), which is known as CRL1-β-TrCP (also containing Skp1, Cul1, and Rbx1) [1].

The interruption of IκB degradation results in significant defects of the NF-κB signaling pathways, and inherited mutations that affect these steps result in developmental and immune defects. Acquired alterations in these pathways are seen in immune disorders and a variety of malignancies. Therefore, enzyme complexes which are involved in phosphorylation and ubiquitination of IκB proteins are key regulators of this pathway and possible targets for drug development. For all these reasons, methods for detecting IκB modifications and degradation are essential tools for understanding the NF-κB pathway. In this regard, IκB phosphorylation and degradation can be detected using a number of commercially available antibodies. However, monitoring IκB ubiquitination in vivo is more challenging due to rapid degradation of the target.

In this article, we review the methods to monitor IκB degradation and a protocol to monitor IκB-α ubiquitination in cells, first developed by Ben-Neriah and colleagues [4] and subsequently used in our laboratory, to study degradation and ubiquitination kinetics within the context of disease-causing mutations affecting the NF-κB pathway.

1.1 Short Method Review

We will describe the method to detect the degradation and endogenous ubiquitination of IκB-α that we typically employ in the laboratory [5], including the antibodies that we find most useful in this regard. Given the rapid degradation of ubiquitinated IκB-α, the detection of this modified form by Western blot of whole cell lysates is not possible in our hands. In addition, direct immunoprecipitation of endogenous IκB-α with a variety of commercially available antibodies proved to be problematic due to low efficiency. Therefore, we utilized a method first described by Ben-Neriah and colleagues [4] for the detection of ubiquitinated IκB-α, which exploits the interaction of phospho-IκB-α (pIκB-α) with the NF-κB subunit p65/RelA. Briefly, cells are treated with the corresponding stimulus (e.g., tumor necrosis factor, TNF, or others) and are then lysed under native conditions. Immunoprecipitation of RelA results in the coprecipitation of NF-κB-associated proteins, including IκB-α. This purified material is then analyzed by Western blotting, which allows for the detection of ubiquitinated IκB-α species using an antibody directed against pIκB-α. Using this approach, monitoring of phosphorylation, ubiquitination, and degradation of IκB-α is feasible and simple (Fig. 2).

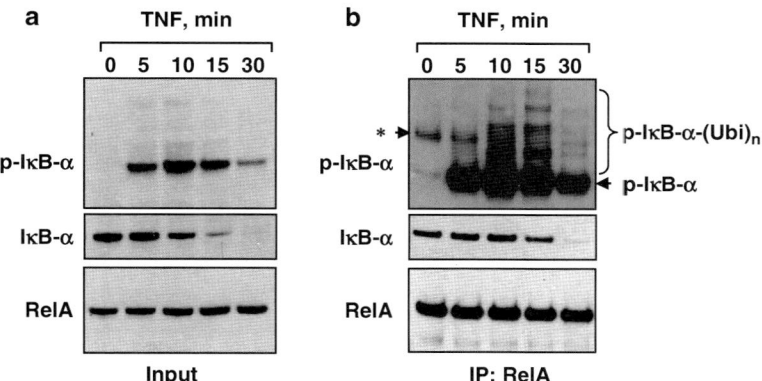

Fig. 2 Purification of ubiquitinated IκB-α from different cell lines as examples of the applicability of the described protocol. Cell lysates from 293 HEK, treated with TNF for indicated times. (**a**) The dynamic of phosphorylation and degradation of IκB-α in total lysates. (**b**) The signal dependent ubiquitination of IκB-α associate with RelA. Maximal ubiquitination rate is observed between 10 and 15 min of stimulation and after dramatically decreased due to proteasomal degradation of IκB-α (30 min sample). RelA serves as loading controls. Asterisk, nonspecific bands

2 Materials

2.1 Reagents

1. Anti-RelA antibody (Santa Cruz Biotechnology, Santa Cruz, CA, USA).
2. Anti-phospho-IκB-α antibody (phospho-Ser32/36; Cell Signaling, Danvers, MA, USA).
3. Anti-IκB-α antibody (Millipore, Billerica, MA, USA).
4. Anti-β-Actin antibody (Sigma, St. Louis, MO, USA).
5. Control rabbit IgG (Santa Cruz Biotechnology).
6. Bradford protein assay.
7. Complete Mini, EDTA free protease/phosphatase inhibitor tablets (Roche, Madison, WI, USA).
8. Dithiotheitol (DTT).
9. Western Lightning Plus-ECL reagent (PerkinElmer, Waltham, MA, USA).
10. Ethylenediaminetetraacetic Acid (EDTA).
11. Glycerol.
12. HEPES.
13. Isopropanol.
14. MegaCD40L soluble recombinant CD40L (Enzo Life Sciences, Farmingdale, NY, USA) (*optional, see* Subheading 3.2).
15. MG-132 (Boston Biochem, Boston, MA, USA).
16. NaCl.
17. Nonfat dry milk.
18. NuPAGE LDS Sample Buffer 4× (Life Technologies, Carlsbad, CA, USA).
19. PBS (phosphate buffered saline), 1× without calcium and magnesium.
20. Phenylmethylsulfonyl fluoride (PMSF).
21. Phorbol 12-myristate 13-acetate (PMA) (*optional, see* Subheading 3.2).
22. Protein A Agarose (Life Technologies).
23. Pre-stained molecular weight marker mix for SDS-PAGE.
24. HRP-linked secondary antibodies (anti-Rabbit IgG and anti-Mouse IgG).
25. Sodium orthovanadate.
26. TNF.
27. Triton X-100.
28. Tween 20.

2.2 Equipment	1. Benchtop microcentrifuge with cooling system for 1.5-mL microcentrifuge tubes.

1. Benchtop microcentrifuge with cooling system for 1.5-mL microcentrifuge tubes.

2. Biophotometer.

3. Cell lifter, 2-cm blade.

4. Filter paper.

5. Inverted microscope.

6. Hypercassette for exposing film.

7. Laminar flow hood and CO_2 incubators for cell culture.

8. Microcentrifuge tubes, 1.5 mL.

9. Rocking platform.

10. Small plastic container for western membrane developing.

11. Serological pipettes and pipette aid.

12. Tissue culture plates, 10-cm (for attached cells) or tissue culture flask, 75 mm (for suspension cells).

13. Tube rotator.

14. X-ray film.

2.3   Cell Lines

The cell line choice will depend on the specific conditions of the experiment. We have utilized this protocol successfully with a variety of cell lines including human embryonic kidney (HEK) 293 cells, HeLa cells, murine, and human immortalized fibroblasts or lymphoblastoid cell lines (LCL). We have not utilized these methods with primary tissues, but they could be adapted to such use.

2.4   Buffers and Stock Solutions

Prepare all solutions using ultrapure water (prepared by purifying deionized water to attain a sensitivity of 18 MΩ cm at 25 °C) and analytical grade reagents. Prepare and store all reagents at room temperature (RT), unless indicated otherwise. Diligently follow all waste disposal regulations when disposing waste materials. The following solutions can be made ahead of time:

1. 1 M DTT (store at –20 °C).

2. 0.5 M EDTA.

3. 1 M HEPES, pH 7.2 (store at 4 °C).

4. 5 M NaCl.

5. 100 mM PMSF in isopropanol (store at –20 °C).

6. 100 mM sodium orthovanadate (store at –20 °C, *see* **Note 1**).

7. 3× NuPAGE LDS Sample Buffer with 0.3 M DTT: mix 800 μL of 4× NuPAGE LDS loading buffer and 400 μL of 1 M DTT (store at –20 °C).

8. PBS-T buffer: PBS with 0.02 % Tween.

9. 5 % solution of nonfat dry milk in 1× PBS-T (store at 4 °C).

10. Lysis buffer: 25-mM HEPES, 100-mM NaCl, 1-mM EDTA, 10 % (v/v) glycerol, 1 % (v/v) Triton X-100. This buffer can be made ahead of time and stored at RT. Just prior to use, add the following to make "complete" lysis buffer: 1-mM PMSF, 10-mM DTT, 1-mM sodium orthovanadate, protease/phosphatase inhibitor Complete Mini tablet (1 tablet per 10 mL of lysis buffer). Complete lysis buffer can be stored at −20 °C for 4 weeks.

3   Methods

3.1   Cell Culture

Culture conditions and cell numbers will depend on the specific cell line being utilized. For HEK 293, HeLa, and primary human fibroblasts, 75 % confluent 10-cm dishes are sufficient to carry out this protocol. We have also utilized suspension cells (patient-derived lymphoblastoid cell lines, LCL) and found that ~5–7.5 × 10^6 cells per sample are similarly sufficient. In general, we recommend seeding the cells in fresh growth medium 24–48 h prior to cell stimulation.

3.2   Cell Stimulation

Treat cells with appropriate stimuli. For HEK 293, HeLa, U2OS, or NIH 3T3, we usually use recombinant commercial TNF at a concentration of 1,000 U/mL. LCL cells are treated with CD-40L (*see* **Note 2**) at concentration 25–50 ng/mL or with PMA (250 nM). Cells are treated for 5–30 min, depending on specific differences between cell lines (*see* **Note 3**). In order to promote the accumulation of the ubiquitinated form of the IκB-α, cells might be simultaneously treated with a proteasome inhibitor (*see* **Note 4**).

3.3   Preparation of Cell Lysates

1. Keep the plates on ice during this procedure. For cells grown in monolayer, aspirate medium from tissue culture plates, and gently wash at room temperature with 5 mL of 1× PBS. For suspension cells (e.g., LCLs), pellet them for 5 min at 300 g at 4 °C in a conical tube. Remove all supernatant by aspiration, and wash with 5 mL of cold 1× PBS. Repeat the centrifugation, and aspirate the supernatant.

2. Lyse cells with lysis buffer (typically 1.0 mL per 10-cm plate). Scrape the plates with a cell scraper, and collect the lysates and debris in microcentrifuge tubes (*see* **Note 5**).

3. Incubate tubes on ice for 10 min.

4. Remove cell debris by centrifugation of the lysate at $15,000 \times g$ for 10 min at 4 °C. Transfer supernatant to a fresh microcentrifuge tube, and determine protein concentration by Bradford assay. Optimal protein concentration is about 1 mg/mL, and if

yield is much higher, lysate should be diluted with cold lysis buffer. Similarly, concentration of protein in different samples should be equalized (*see* **Note 6**). Set aside an aliquot (about 50 µL or 5 % of the supernatant) for IκB-α phosphorylation and degradation analysis.

3.4 Lysates Preclearing

This is an optional step to reduce nonspecific binding of proteins to Protein A agarose (*see* **Note 7**).

1. Equilibrate Protein A agarose: add required amount of the resin (10 µL of 50 % bead slurry per sample) to 200–500 mL of lysis buffer (a minimum of 10× volume of beads). Incubate for 5 min with rotation at 4 °C; afterwards, precipitate beads by centrifugation ($300 \times g$ for 2 min at 4 °C). Repeat washing two more times.

2. Add equilibrated Protein A agarose to the lysates. Incubate at 4 °C for 30–60 min on a tube rotator. Spin for 10 min ($300 \times g$) at 4 °C. Discard pellet (beads), and use cleared supernatant for immunoprecipitation.

3.5 RelA Immunoprecipitation

1. Divide the cell lysate samples into two equal aliquots (approximate 500 µL each), and add 250 ng of primary antibody against RelA or rabbit IgG to the control sample (*see* **Note 8**). Incubate on tube rotator for 3–12 h at 4 °C.

2. Add Protein A agarose beads to each sample (20 µL of 50 % bead slurry, equilibrated with lysis buffer). Incubate samples under rotary agitation for 2–12 h at 4 °C.

3. Microcentrifuge at $300 \times g$ for 2 min at 4 °C. Aspirate the supernatant and discard it.

4. Wash pellet three times with 500 µL of 1× lysis buffer on tube rotator for 5 min, precipitate the beads every time by centrifugation ($300 \times g$ for 2 min at 4 °C). Keep on ice during the washes.

5. Finally, carefully remove as much wash buffer as possible from the beads using a fine tip (e.g., gel loading tips). The pellet contains the RelA-IκB-α complex, which is now ready for analysis by SDS-PAGE followed by Western immunoblotting.

3.6 SDS-PAGE Separation and Western Blotting

1. Resuspend beads in 3× gel loading buffer (1 µL of loading buffer per 1 µL of bead volume). Heat sample for 10 min at 80–95 °C for complex dissociation. Now samples can be stored at −80 °C for several weeks before proceeding with SDS-PAGE separation.

2. Before SDS-PAGE separation, centrifuge tubes for 1 min at maximum speed ($>10,000 \times g$) at room temperature. Load maximal amount of immunoprecipitated sample onto a SDS-PAGE gel (*see* **Note 9**). In addition, load input lysates (40–60 µg per sample).

3. The transfer of proteins to nitrocellulose membrane should be performed using a wet or semidry transfer system. After the transfer, proceed to immunodetection steps to detect ubiquitinated forms of IκB-α.

3.7 Immunodetection

After transfer, membrane blocking is performed for 0.5–1 h at room temperature with 5 % solution of fat-free milk in PBS-T buffer with constant gentle agitation.

3.7.1 Immunodetection of pIκB-α

1. Incubate the membrane with anti-phospho (Ser32/36)-IκB-α (dilution 1:1,000 in PBS-T with 2.5 % of nonfat milk) for 1 h at room temperature with gentle agitation (*see* **Note 10**).

2. Wash the membrane with PBS-T buffer three times for 5 min.

3. Apply a secondary HRP-conjugated antibody (dilution 1:4,000 in PBS-T with 2.5 % of nonfat milk) or 1 h at room temperature with gentle agitation.

4. Wash three times for 5 min with PBS-T buffer.

5. Perform ECL-detection using ECL-detection kit according to manufacturer's instruction. A ladder of phosphorylated and ubiquitinated forms of IκB-α will be detectable in late exposures of IP samples (usually, 30–60 min with X-ray film). This step allows detecting phosphorylation and ubiquitination dynamics of IκB-α in samples (Fig. 2a and b, upper blots).

3.7.2 Immunodetection of IκB-α Degradation

1. Wash the membrane for 10 min in PBS-T buffer, and repeat for a total of three times in order to remove chemicals from previous step (*see* **Note 11**).

2. Repeat blocking in 5 % nonfat milk solution in PBS-T for 15 min at room temperature.

3. Re-probe the membrane with anti-IκB-α antibody (dilution 1:1,000 in PBS-T with 2.5 % of nonfat milk).

4. Repeat the procedures as described above. In order to detect the degradation rate of IκB-α, analyze relative amount of the protein in the input samples (Fig. 2a and b, middle blots).

3.7.3 Immunodetection of RelA

1. Wash the membrane for 10 min in PBS-T buffer and repeat for a total of three times in order to remove chemicals from previous steps.

2. Re-probe the same membrane with anti-RelA antibody (dilution 1:1,000 in PBS-T with 2.5 % of nonfat milk), and repeat the procedures as described above. Because RelA is heavier than IκB-α (65 kDa vs. 36 kDa), we do not perform membrane stripping procedure. This step is required to confirm that the amount of immunoprecipitated RelA is equal between loaded samples (Fig. 2a and b, lower blots).

4   Notes

1. Prepare a 100-mM sodium orthovanadate solution in double distilled water. Set pH to 9.0 with HCl, and boil until colorless. Cool down to room temperature, and again set pH to 9.0. Repeat this cycle until the solution remains at pH 9.0 after boiling and cooling. Bring up to the initial volume with water. Store in aliquots at –20 °C.

2. Among the different variants of recombinant CD40L that are commercially available, we recommend to use the recombinant chimeric form known as MegaCD40L (Enzo Life Sciences). This protein consists of two trimeric extracellular domains of mouse CD40L (CD154), which are artificially linked via the collagen domain of ACRP30/adiponectin. In our hands, this form of the CD40L was very effective in stimulating the NF-κB pathway in human lymphocytes.

3. The dynamic of IκB-α degradation varies in different cell lines. We found that the optimal time of stimulation for most cell lines, including HEK 293 and HeLa cells, is 15–20 min. In general, we recommend using more than two time points (e.g., 0, 5, 10, 15, and 30 min of stimulation, Fig. 2). On the other hand, LCLs show very quick IκB-α degradation, and the optimal time to detect ubiquitinated forms of IκB-α is around 5–10 min of stimulation [5].

4. In the case of proteasomal blockade, we routinely use MG-132, at a concentration of 40 μM for 3–5 h. In the case of ligand stimulation, we use standard time points for the ligand. Alternatively, co-expression of the known E3 ligase for the target protein might be useful.

5. The volumes in this protocol are given for a single 10-cm plate ($2–4 \times 10^6$ adherent cells) or $5–7 \times 10^6$ LCLs (due to the smaller volume of the latter). For larger-scale experiments, all the volumes should be increased proportionally.

6. The supernatant is the cell lysate. If necessary, lysate can be stored at –80 °C for several months.

7. This step removes proteins, which may potentially bind nonspecifically to the agarose beads during immunoprecipitation. It is worthwhile to evaluate the amount of nonspecific proteins under your exact experimental conditions. This step may result in a lower level of background and improve signal-to-noise ratio.

8. The amount of applied antibody may vary, so we advise to begin with an average amount as a guideline. Usually, we use 1 μg of antibody per total lysate from a 10-cm plate (HeLa, HEK 293) or 5×10^6 LCL cells and 2 μg per total lysate from 15-cm plate (HeLa, HEK 293) or 10×10^6 LCL cells. There is no need to further increase the amount of cells for the detection of IκBα ubiquitination.

9. We routinely use premade NuPAGE Bis–Tris Mini Gels from Life Technologies. Briefly, 12-wells 4–12 % gradient gel (1.0 mm) are loaded with 20–25 μL of samples (~50 μg of total protein per lane); run conditions are set up according to the manufacturer's recommendations.

10. The membrane can be incubated with the antibodies overnight (either with a primary or a secondary). In many cases, this step will be reached at the end of the day, and thus, we routinely performed the binding overnight at 4 °C without any apparent detriment to the procedure.

11. In this case, there is no need for a stripping procedure, since the used antibodies are generated from different animal species (anti-phospho (Ser32/36)-IκB-α antibody is raised from a mouse, while anti-IκB-α is raised from a rabbit). Hence, the corresponding bands are detected by different secondary antibodies and are not overlapped.

References

1. Hayden MS, Ghosh S (2012) Genes Dev 26: 203–234
2. Basak S, Kim H, Kearns JD, Tergaonkar V, O'Dea E, Werner SL, Benedict CA, Ware CF, Ghosh G, Verma IM, Hoffmann A (2007) Cell 128:369–381
3. Hoffmann A, Natoli G, Ghosh G (2006) Oncogene 25:6706–6716
4. Lee J, Mo JH, Katakura K, Alkalay I, Rucker AN, Liu YT, Lee HK, Shen C, Cojocaru G, Shenouda S, Kagnoff M, Eckmann L, Ben-Neriah Y, Raz E (2006) Nat Cell Biol 8:1327–1336
5. Starokadomskyy P, Gluck N, Li H, Chen B, Wallis M, Maine GN, Mao X, Zaidi IW, Hein MY, McDonald FJ, Lenzner S, Zecha A, Ropers HH, Kuss AW, McGaughran J, Gecz J, Burstein E (2013) J Clin Invest 123: 2244–2256

Chapter 3

Measurement of NF-κB Transcriptional Activity and Identification of NF-κB *cis*-Regulatory Elements Using Luciferase Assays

Patricia E. Collins, Christine O'Carroll, and Ruaidhrí J. Carmody

Abstract

The NF-κB family of transcription factors is activated in response to numerous environmental stimuli and coordinates the transcriptional response to immunoreceptors such as the Toll-like receptors, cytokine receptors, and antigen receptors, growth factors, survival factors, and stress signals such as ultraviolet irradiation and oxidative stress. The transcriptional targets of these various pathways include approximately 500 experimentally indentified genes, and it is highly likely that many others remain to be discovered. A genome-wide analysis of NF-κB–chromatin interactions has revealed a surprisingly large number of NF-κB binding sites across the entire genome, many of which are found in intergenic regions and many more do not appear to be associated with changes in transcription of nearby genes. Assessing the consequences of NF-κB binding at genomic sites is therefore essential to determine the functional role of NF-κB in regulating the expression of specific genes. Luciferase-based reporter assays provide a robust and flexible method to test the contribution of specific NF-κB sites to the regulation of gene transcription. The methods described in this chapter may be applied to any promoter sequence and used in a variety of cell lines and conditions to provide critical information on the regulation of gene expression by NF-κB.

Key words NF-κB, Promoter, Genomic DNA, Luciferase, Transfection

1 Introduction

The NF-κB family of transcription factors is activated by a wide range of stimuli including cytokines, antigen receptors, survival and growth factors, genotoxic and oxidative stress, and mitogens [1]. The pleiotropic nature of NF-κB is reflected in its transcriptional targets that presently number approximately 500 (see www.nf-kb. org) and include mediators of inflammation and the immune response, regulators of cell death and proliferation, and stress response genes. NF-κB is composed of five subunits (p65/RelA, c-Rel, p50, p52, and RelB) capable of forming hetero- and homo-dimers to generate potential 15 possible dimers. Genetic studies employing knockout mice have demonstrated that individual

Michael J. May (ed.), *NF-kappa B: Methods and Protocols*, Methods in Molecular Biology, vol. 1280,
DOI 10.1007/978-1-4939-2422-6_3, © Springer Science+Business Media New York 2015

NF-κB subunits have unique biological functions [2], likely based on expression levels in specific cell types [2], activation by specific pathways, and dimer-specific roles in gene regulation [3, 4].

The DNA sequences that specifically bind NF-κB dimers are often referred to as κB sites. Typically, κB sites are 10 bp in length and possess the following consensus sequence: 5'-GGGRNW YYCC-3' (where R is an A or G, N is any base, W is an A or T, and Y is a C or T) [5]. However, such a consensus sequence may be too constrained in capturing the full extent of NF-κB DNA-binding activity, and a number of studies have identified NF-κB binding on noncanonical κB sequences contained in the promoters of target gene such as *Il12b* [6], *Il2* [7], and *Csf2* [7]. More recently, protein-binding microarrays coupled with surface plasmon resonance analysis have provided an unprecedented large and detailed database of κB sequences and their relative affinities for 8 different NF-κB dimers [8]. This study identified distinct classes of κB sites based on the dimer-binding specificity and also identified a large number of noncanonical κB sequences. The κB site preferences for NF-κB dimers may explain, at least in part, the subunit-specific regulation of gene expression as found in genetic models. Moreover, the database generated provides a valuable tool for the analysis of genomic NF-κB regulatory elements.

Perhaps, one of the most important recent technical advances in the study of NF-κB-regulated gene expression has been the chromatin immunoprecipitation (ChIP) assay [9]. This assay permits the measurement of NF-κB binding to DNA in situ and when performed in conjunction with next-generation sequencing (ChIP-seq) provides a genome-wide map of NF-κB and chromatin interactions [10]. Studies employing ChIP-seq techniques have yielded the remarkable finding that NF-κB binds widely across the genome, including intergenic regions [11–13], and that NF-κB binding at many promoters does not correlate with a change in gene expression [12]. Thus, the binding of NF-κB to a κB site is not a sufficient evidence to conclusively prove a role for NF-κB in the regulation of the transcription of an associated gene, and an analysis of the κB site and its contribution to gene expression is required.

The analysis of κB sites in the context of a promoter or gene enhancer is a critical step in assessing the contribution of NF-κB and the κB sequence to gene expression. In this chapter, we will describe the use of plasmid-based promoter reporter assays to measure NF-κB-regulated gene expression and identify the contribution of specific κB sites to gene expression. These assays, in addition to providing critical information on the regulatory sequences controlling target gene expression, can also be used as convenient readouts of NF-κB activity in a format that is readily scalable for compound, cDNA, or shRNA library screening.

2   Materials

1. Prepare all reagents using deionized water unless otherwise stated. All reagents are stored at room temperature unless specified (*see* **Note 1**).

2.1   Special Equipment

1. NanoDrop UV spectrometer.
2. Centrifuge capable of handling large volumes (>100 ml).
3. Water bath.
4. Plate rocker/shaker.
5. Thermal cycler.
6. Luminometer.

2.2   Cell Culture

1. RAW267.4 murine macrophage cell line.
2. Cell culture medium of Dulbecco's Modified Eagle's Medium (DMEM) supplemented with 100 μg/ml streptomycin, 100 units/ml penicillin, 2 mM L-glutamine, and 10 % (volume/volume) fetal bovine serum (FBS). Store at 4 °C until required. Pre-warm medium to 37 °C prior to culturing of cells.
3. 24-well tissue culture dishes.
4. Cell scraper.

2.3   Promoter Cloning

1. Sterile DNase-free pipette tips.
2. Sterile DNase-free 1.5-ml microfuge tubes.
3. Sterile DNase-free 0.2-ml thin-walled microfuge tubes.
4. Primer pair to amplify genomic sequence of interest (10-μM working stock) stored at –20 °C.
5. Ultrapure molecular biology grade water.
6. Genomic DNA isolated from cells using a DNA tissue extraction kit and stored at –20 °C.
7. Expand High Fidelity enzyme system including Deoxynucleoside Triphosphate Set (10 mM) (Roche, Madison, WI, USA).
8. pSTBlue-1 AccepTor Vector Kit containing NovaBlue Singles Competent Cells and SOC medium (Novagen, Madison, WI, USA).
9. pGL3-Basic reporter plasmid stored at –20 °C (Promega, Madison, WI, USA).
10. Wizard SV Gel and PCR Clean-Up System stored at –20 °C (Promega).
11. Isopropyl thiogalactoside (IPTG) solution: 100 μM in H_2O, sterilized by filtration through a 0.22-μM filter, and aliquots can be stored at –20 °C.

12. 5-Bromo-4-chloro-3-indolyl-beta-d-galactopyranoside (Xgal) solution: 20 mg/ml in DMSO stored at –20 °C and protected from light.

13. Restriction enzymes with reaction buffer stored at –20 °C (NEB, Ipswich, MA, USA).

14. T4 ligase with reaction buffer, stored at –20 °C (NEB).

15. QuikChange II Site-Directed Mutagenesis Kit (Stratagene, La Jolla, CA, USA) containing PfuUltra HF, *Dpn* I, and XL1-Blue competent cells and NZY+ broth stored at –70 °C or below.

16. DH5α chemically competent cells stored at –70 °C or below.

17. LB broth (pH 7.0) made by dissolving 10 g of tryptone, 5 g yeast extract, and 5 g NaCl in 1 l of deionized water. Autoclave the solution and store at 4 °C until required.

18. Ampicillin solution (50 mg/ml) prepared by dissolving 0.5 g ampicillin sodium salt in 10 ml of deionized water. Using aseptic technique, pass this solution through a 0.22-μM filter to sterilize, aliquot, and store at –20 °C.

19. LB agar plates (pH 7.0) containing 50 μg/ml ampicillin made by dissolving 5 g of tryptone, 2.5 g of yeast extract, and 2.5 g of NaCl and 7.5 g of agar in 500 ml of water. Autoclave to sterilize the mixture and allow to cool to about 55 °C (*see* **Note 2**). Using aseptic technique, pipette 25 μl of 50 mg/ml stock of ampicillin into the agar solution. Mix thoroughly and pour into petri dishes, allow to solidify at room temperature, and store at 4 °C until required.

20. 500-ml conical flasks sterilized by autoclaving.

21. Sterile 15-ml BD Falcon round-bottom tubes and 50 ml conical BD Falcon tubes.

2.4 Agarose Gel Electrophoresis

1. Agarose.

2. 0.5× TBE buffer: 45 mM Tris–borate, 1 mM EDTA, prepared from a 5× stock containing 54 g Tris base, 27.5 g boric acid, and 20 ml of 0.5 M EDTA per liter (*see* **Note 3**).

3. 6× Orange G loading buffer: 30 % glycerol, 0.2 % w/v Orange G in water.

4. DNA standard ladder.

5. SYBR Safe DNA gel stain (Life Technologies, Carlsbad, CA, USA).

6. Gel documentation system including UV light source.

2.5 Transfection

1. Sterile DNase-free 1.5-ml microfuge tubes.

2. Sterile DNase-free pipette tips.

3. pGL3-Basic firefly luciferase reporter plasmid and *Renilla* luciferase reporter plasmid (pRL-TK) (Promega) stored at –20 °C.

4. Lipopolysaccharides (LPS) from *E. coli* (O55:B5) stored in 100 ng/ml aliquots in a cell culture medium and stored at −20 °C.

5. TurboFect Transfection Reagent (Thermo Scientific, Waltham, MA, USA) stored at 4 °C.

6. Serum-free DMEM stored at 4 °C.

2.6 Cell Lysis and Reporter Assay Components

1. 5× Passive Lysis Buffer (Promega) stored at −20 °C.

2. 1.5-ml microfuge tubes.

3. Phosphate buffered saline (PBS) pH 7.4: NaCl 137 mM, KCl 2.7 mM, Na_2HPO_4 10 mM, KH_2PO_4 1.8 mM.

4. Firefly luciferase assay buffer: 25-mM glycylglycine, 15-mM K_2PO_4 pH 8, 4-mM ethylene glycol tetraacetic acid (EGTA), 15-mM $MgSO_4$ 1, 0.1-mM coenzyme A, 1-mM dithiothreitol (DTT), 2-mM adenosine triphosphate (ATP), 75-μM luciferin (*see* **Note 4**).

5. *Renilla* luciferase assay buffers : 1.1-M NaCl, 2.2-mM ethylenediaminetetraacetic acid (EDTA), 220-mM K_2PO_4 pH 5.1, 0.44 mg/ml bovine serum albumin (BSA), 1.3-mM NaN_3, 1.43-μM coelenterazine (*see* **Note 5**).

3   Methods

Identification of the genes regulated by NF-κB is critical for understanding the physiological and pathological roles of NF-κB. Mouse genetic models of NF-κB have established a fundamental role for NF-κB in mediating transcriptional responses in a wide range of pathways of importance to human health and disease [2]. Transcriptomic techniques such as microarrays or RNA sequencing can be employed to determine the global impact of defective or aberrant NF-κB activity on transcriptional responses. The data generated from such profiling approaches can in turn be bioinformatically interrogated to identify overrepresented transcription factor binding sites in the regulatory regions of groups of genes associated with a particular transcriptional response [14]. Such analyses can identify NF-κB as a potential regulator of specific transcriptional programs or genes; however, additional experiments are required to define the contribution of NF-κB to the expression of specific genes. Techniques such as chromatin immunoprecipitation (ChIP) can confirm the binding of NF-κB subunits to specific sites on chromatin, but as recent genome-wide ChIP analyses have shown, the binding of NF-κB to promoter regions is not always associated with changes in the transcription [12]. Thus, an analysis of the promoter region, specifically the NF-κB binding site, is required to conclusively determine an NF-κB target gene.

The protocols detailed here allow the analysis of specific NF-κB binding sites in promoter sequences to be assessed in a cell type and culture condition relevant to the regulation of the gene of interest. We describe a procedure for cloning the region of a promoter containing a putative or known NF-κB binding site from genomic DNA to generate a recombinant reporter plasmid incorporating the luciferase gene (Fig. 1). We also describe a method to mutate a promoter sequence in order to determine the contribution of a specific NF-κB binding site to gene expression. Finally, we describe a method to measure the luciferase reporter activity following transient transfection of a cell line. The techniques described may be applied to any promoter sequence and can also be used to generate the tools to perform a high-throughput analysis of reporter activity.

3.1 Promoter Cloning (Fig. 1)

3.1.1 Designing Primers to Clone Promoter Sequence Containing NF-κB Binding Site

1. Identify the promoter sequence of the gene of interest using the UCSC genome browser (http://genome.ucsc.edu/).

2. Identify NF-κB binding sites in the promoter sequence using the NF-κB PBM dataset (http://thebrain.bwh.harvard.edu/nfkb/) or other predictive software.

3. Use software tools such as the NEBcutter V 2.0 (http://tools.neb.com/NEBcutter2/) to identify restriction enzyme sites present in the multiple cloning site of the pGL3-Basic plasmid but not in the promoter region of interest.

4. Design primer sequences using software such as Primer3 (http://bioinfo.ut.ee/primer3/) or Vector NTI (Invitrogen, Carlsbad, CA, USA) to amplify the promoter region of interest and incorporating the restriction sites identified in the previous step.

3.1.2 Genomic DNA Isolation

1. Isolate genomic DNA using DNeasy Tissue extraction kit according to the manufacturer's guidelines (QIAGEN).

2. Quantify the DNA using a NanoDrop UV spectrometer, and calculate the A_{260}/A_{280} ratio to ensure DNA is of sufficient quality for downstream experiments (see **Note 6**). The DNA can be stored at $-20°$ C before proceeding to the next step.

3.1.3 PCR Amplification and Purification

1. Thaw all Expand High Fidelity reagents on ice, vortex, and centrifuge briefly to collect the mixture at the bottom of the tube before use.

2. In separate sterile microfuge tubes, prepare two reaction mixes on ice as outlined in Table 1.

3. In a 0.2-ml thin-walled PCR tube, combine both reaction mixtures on ice and vortex to mix thoroughly. Centrifuge briefly to collect the mixture at the bottom of the tube using a microfuge.

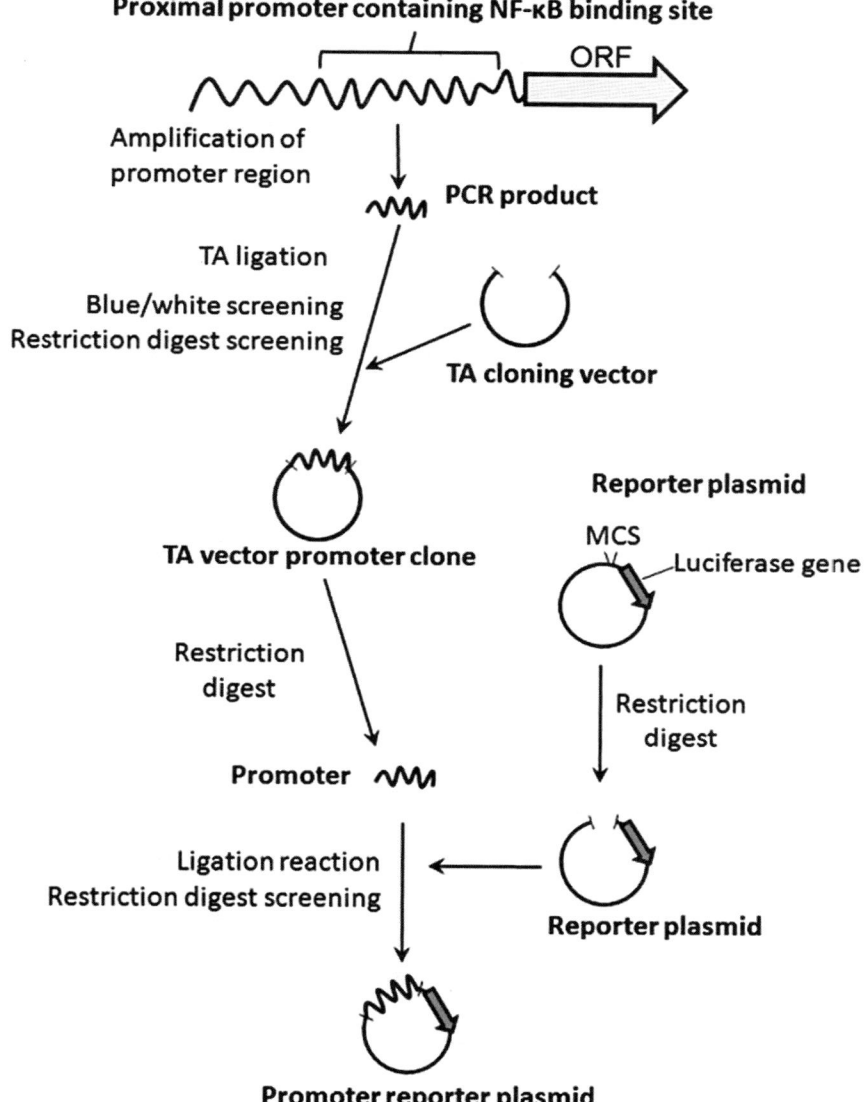

Fig. 1 Schematic overview of promoter cloning workflow. The promoter sequence of interest upstream of an open reading frame (ORF) is amplified by PCR using primers incorporating restriction sites compatible with the multiple cloning site (MCS) of the pGL3-Basic reporter plasmid. The PCR product is ligated into the pSTBlue-1 TA cloning vector and recombinant plasmids identified by blue/white colony and restriction digest screening. The promoter fragment is cut from a positive TA vector clone by restriction digest and ligated with the reporter plasmid digested with the same restriction enzymes. Recombinant promoter reporter plasmid is identified by restriction digest initially, followed by plasmid sequencing

4. Include a negative control by repeating **steps 1–3** without the addition of genomic template DNA.

5. Place sample in a thermal cycling block programmed as per Table 2 and initiate cycling (*see* **Note 7**).

Table 1
Reaction components for PCR

Component	Volume (µl)
Reaction 1	
dNTP mix (10 mM each)	1
Primer (forward) (10 µM)	1.5
Primer (reverse) (10 µM)	1.5
Template DNA (250 ng)	X
Molecular biology grade water	Up to 25
Total volume	25
Reaction 2	
Molecular biology grade water	19.25
Expand High Fidelity buffer 10×	5
Expand High Fidelity enzyme mix	0.75
Total volume	25

Table 2
Thermal cycling profile

Stage	Step	Temperature	Time	Cycles
1	Initial denaturation	94 °C	2 min	1
	Denaturation	94 °C	15 s	
2	Annealing	Primer specific	30 s	10
	Elongation	72 °C	See **Note 7**	
	Denaturation	94 °C	15 s	
	Annealing	Primer specific	30 s	
3	Elongation	72 °C	+5 s per each successive cycle (*see* **Note 7**)	20
4	Final elongation	72 °C	7 min	1
5	Cooling	4 °C	Hold	

3.1.4 Agarose Gel Electrophoresis

1. Prepare a 1 % agarose gel by weighing out 0.5 g of agarose and dissolve in 50 ml of 0.5× TBE in a glass bottle (*see* **Note 8**).

2. Heat in a microwave until the agarose dissolves and the solution is no longer turbid. Using a protective glove, carefully swirl the agarose solution to ensure a uniform mixture. Once partially cooled (not solid), pipette 5 µl of SYBR Safe DNA gel stain (10,000×) and mix thoroughly by swirling.

3. Pour warm agarose solution into a clean casting tray and insert a comb to form sample wells. Confirm that no air bubbles are present, and allow gel to set at room temperature for approximately 30 min.

4. Once the gel is completely set, remove comb and mount the gel in the electrophoresis tank; add enough 0.5× TBE buffer to completely cover the gel ensuring wells are submerged.

5. Prepare samples for electrophoresis by mixing 20 μl of amplified PCR product with 4 μl of 6× loading dye (*see* **Note 9**).

6. Carefully pipette 5 μl of DNA molecular weight ladder into the first lane, and continue by pipetting 20 μl of sample mixture into separate wells while taking care not to pierce the bottom of the well.

7. Apply the lid of the electrophoresis tank, and attach electrical leads to the power pack, and run the gel at 100 V.

8. Once the loading dye has migrated three-quarters of the length of the gel, disconnect the electrodes, remove the gel, and visualize it using a UV imager. Successful amplification will produce a readily detectable DNA fragment at the predicted size.

9. Using a UV light, excise the amplified PCR product from the gel using a sterile blade or scalpel, and transfer to a sterile 1.5-ml microfuge tube (*see* **Note 10**).

10. Extract DNA from the agarose slice using Wizard SV Gel and PCR Clean-Up System. DNA can be stored at –20 °C before proceeding to the next step.

3.1.5 TA Cloning

1. Spread 40 μl of IPTG and 40 μl of Xgal on top of LB agar plates containing ampicillin with a hockey stick spreader. Allow the plates to dry for 30 min to an hour before use.

2. Prepare a ligation reaction using the pSTBlue-1 AccepTor Vector Kit. Include a positive and negative control ligation reaction using the controls provided with the kit.

3. Incubate the ligation reactions at 16 °C for 2 h.

4. For each transformation, remove a tube of NovaBlue Singles Competent Cells from –70 °C and place immediately on ice. Thaw cells on ice for 2–5 min.

5. Gently resuspend the cells by flicking. Do not vortex the tube.

6. Pipette 1 μl of the ligation reaction directly to the cells. Mix gently using the pipette tip and incubate on ice for 5 min.

7. Following incubation, quickly place the tubes in a 42 °C water bath for exactly 30 s.

8. Immediately place the tube back on ice for 2 min before adding 250 μl of room temperature SOC medium (*see* **Note 11**).

9. Pipette 20–100 μl of the transformation mixture onto an LB agar plate containing ampicillin, IPTG, and Xgal. Spread evenly onto the agar plate using a sterile bent glass rod or disposable hockey stick spreader, and allow the excess liquid to be absorbed.

10. Place inverted in a 37 °C incubator overnight (15–18 h).

11. Following overnight incubation, remove the plate from the incubator; inspect for visible blue and white colonies (*see* **Note 12**).

3.1.6 DNA Isolation and Screening by Restriction Enzyme Analysis

1. Using aseptic technique, isolate a single white colony using a sterile pipette tip or inoculation loop and transfer into a 50-ml centrifuge tube containing 5 ml of LB broth supplemented with 50 μg/ml ampicillin (*see* **Note 13**).

2. Incubate at 37 °C shaking overnight (15–18 h).

3. Centrifuge culture at $6,000 \times g$ for 10 min to pellet the bacterial cells.

4. Purify plasmid DNA using Wizard Plus SV Minipreps DNA Purification System. Quantify the DNA and calculate the A_{260}/A_{280} ratio (*see* **Note 6**). DNA can be stored at –20 °C before proceeding to the next step.

5. To determine if the PCR insert was successfully cloned into the pSTBlue-1 vector, screen the isolated DNA by a restriction enzyme digest. Using restriction enzymes based on the restriction sites incorporated in the primers as designed in Subheading 3.1.1, prepare a reaction on ice as outlined in Table 3 (*see* **Note 14**).

6. Incubate at 37 °C or other recommended temperatures for 1 h (*see* **Note 17**).

7. Separate the digested DNA by agarose gel electrophoresis as described under Subheading 3.1.4 (*see* **Note 18**).

Table 3
Restriction enzyme digestion reaction setup

Component	Volume (μl)
10× Restriction enzyme buffer (*see* **Note 15**)	5
100× Molecular biology grade BSA (*see* **Note 16**)	0.5
DNA (1 μg)	X
Restriction enzyme 1 (10 units)	X
Restriction enzyme 2 (10 units)	X
Nuclease-free water	To 50

8. Visualize the gel using a UV light source. A successful recombination is determined by the presence of two DNA fragments corresponding to the expected size of the amplified PCR product (promoter) and the linear pSTBlue-1 vector.

9. Once a positive clone is identified, the promoter can be subcloned from the pSTBlue-1 plasmid into the pGL3-Basic vector.

3.1.7 Subcloning of Promoter into Luciferase Reporter

1. To prepare the promoter insert for subcloning into the pGL3-Basic luciferase reporter vector, the promoter fragment must be purified from a restriction digest of a positive pSTBlue-1 vector clone as detailed under Subheading 3.1.6. A restriction digest of 1–5 μg of plasmid will generally be sufficient to complete the subcloning steps.

2. Separate the digested DNA using agarose gel electrophoresis, and excise the DNA fragment corresponding to the promoter as outlined under Subheading 3.1.4.

3. Purify the DNA from the agarose gel slice using Wizard SV Gel and PCR Clean-Up System (Promega).

4. 1–5 μg of pGL3-Basic luciferase reporter vector must be digested with the same restriction enzymes. Repeat **steps 1–3** above using the pGL3-Basic vector.

5. Quantify the promoter fragment and digested pGL3-Basic using a NanoDrop spectrometer, and, with the following formula, calculate the required quantities of insert (promoter) fragment and linearized vector (digested pGL3-Basic) required for the ligation reaction (*see* **Note 19**).

$$\frac{Size(kb)insert}{Size(kb)vector} \times ng\ of\ vector = ng\ of\ insert\ required\ for\ a\ 1:1\ molar\ ratio$$

6. Using the reaction conditions detailed in Table 4, set up on ice a ligation reaction between the linearized pGL3-Basic vector and the promoter insert (*see* **Note 20**).

7. Incubate the ligation reaction at 16 °C overnight or room temperature for 10 min (*see* **Note 21**).

8. Transform one tube of NovaBlue Singles Competent Cells with 1 μl of the ligation reaction as described under Subheading 3.1.5.

9. Isolate plasmid DNA and screen by restriction enzyme analysis as described under Subheading 3.1.6.

10. Positive clones identified by the restriction enzyme analysis should always be confirmed by sequencing. Once confirmed, a large-scale plasmid preparation should be performed to isolate high-quality plasmid DNA for use in transfections.

Table 4
Ligation reaction setup

Component	Volume (µl)
10× T4 DNA ligase buffer	2
Vector DNA (100 ng)	X
Insert DNA	X (*see* **Note 19**)
T4 DNA ligase	1
Nuclease-free water	To 20

3.2 Quick Transformation of Competent E. coli for Routine Plasmid Propagation

1. Use aseptic technique throughout.

2. Remove an aliquot of DH5α chemically competent cells from the freezer, immediately place on ice, and thaw cells on ice for 2–5 min.

3. Gently resuspend the cells by flicking. Do not vortex the tube.

4. Pipette 1–2 µl of plasmid DNA (10–20 ng/µl) directly to the cells. Mix gently and incubate on ice for 1 min.

5. Following incubation, quickly place the tubes in a 37 °C water bath for 60 s.

6. Immediately place the tubes back on ice for 1 min before adding 100 µl of room temperature SOC medium to each tube.

7. Pipette 20–100 µl of the transformation mixture onto an LB agar plate containing ampicillin.

8. Spread evenly onto the agar plate using a sterile bent glass rod or disposable hockey stick, and allow the excess liquid to be absorbed.

9. Place inverted in a 37 °C incubator overnight (15–18 h).

10. Following overnight incubation, remove the plate from the incubator and inspect for visible colonies (*see* **Note 12**).

11. Using a sterile inoculating loop or pipette tip, pick a single colony, and inoculate a 2 ml starter culture of LB broth containing 50 µg/ml ampicillin. Incubate the starter culture at 37 °C shaking for 8 h.

12. Prepare a culture in a 500-ml sterile flask by inoculating 100 ml of LB broth containing 50 µg/ml ampicillin with 100 µl of starter culture. Incubate overnight at 37 °C shaking.

13. Pellet bacteria by centrifugation at $6,000 \times g$ for 10 min.

14. Purify plasmid DNA using PureYield Plasmid Midiprep System (Promega).

Table 5
Mutagenesis reaction setup

Component	Volume (μl)
10× reaction buffer	5
dsDNA template (5–50 ng)	X
Oligonucleotide primer #1 (100 ng/μl)	1.25
Oligonucleotide primer #2 (100 ng/μl)	1.25
dNTP mix	1
Nuclease-free water	To 50

Table 6
Mutagenesis thermal cycling conditions

Stage	Temperature (°C)	Time	Cycle
1	95	30 s	1
2	95	30 s	12–18
	55	1 min	(*See* **Note 23**)
	68	1 min/kb of plasmid	

3.3 Site-Directed Mutagenesis

1. Design complimentary mutagenic oligonucleotide primers using the QuikChange Primer Design program: (http://www.genomics.agilent.com/primerDesignProgram.jsp) (*see* **Note 22**).

2. In a thin-walled 0.2-ml PCR tube, prepare the sample reaction on ice as per Table 5.

3. Add 1 μl of PfuUltra HF DNA polymerase (2.5 units/μl). Place sample in a thermal cycling block programmed as per Table 6 and initiate cycling.

4. Following amplification, add 1 μl of *Dpn* I restriction enzyme to each reaction and mix well by pipetting. Briefly spin reactions in a microfuge and incubate at 37 °C for 1 h to digest the template DNA (*see* **Note 24**).

5. Thaw XL1-Blue cells on ice and gently flick (1–2 times) to resuspend cells, and on ice quickly aliquot 50 μl of cells to a prechilled 15-ml BD Falcon round-bottom tube.

6. Pipette 1 μl of *Dpn* I-digested DNA directly to the cells, swirl to gently mix, and incubate on ice for 30 min.

7. Place tubes in a 42 °C water bath for exactly 45 s and return to ice for 2 min.

8. Add 500 μl of pre-warmed (42 °C) NZY+ broth, and incubate transformation reactions for 1 h, shaking at 37 °C.

9. Pipette 250 µl of the transformation mixture onto each of two LB agar plates containing ampicillin. Spread evenly onto the agar plate using a sterile bent glass rod or disposable hockey stick, and allow the excess liquid to be absorbed.

10. Place inverted in a 37 °C incubator overnight (15–18 h).

11. Following overnight incubation, remove the plate from the incubator and inspect for visible colonies (*see* **Note 12**).

12. Select a minimum of 5 colonies, and isolate plasmid DNA using Wizard Plus SV Minipreps DNA Purification System as outlined in **steps 1–4** under Subheading 3.1.6.

13. A successful mutagenesis is determined by sequencing of the selected clones using the appropriate plasmid sequencing primers.

3.4 Transfection and Treatment Conditions

1. All steps are performed in sterile culture.

2. RAW 264.7 cells are grown in DMEM containing FCS 10 % (v/v), glutamine, penicillin, and streptomycin. Maintain cells at 37 °C in a humidified environment with 5 % CO_2.

3. Cells are passaged by first washing in a pre-warmed (37 °C) medium and detached using a cell scraper.

4. Detachment of cells from tissue culture flask should be visible by eye, but complete detachment should be confirmed using a microscope.

5. On the day prior to transfection, seed 2×10^5 RAW267.4 cells per well of a 24-well tissue culture plate in a total volume of 500 µl DMEM supplemented with serum, glutamine and antibiotics (*see* **Note 25**). When seeding cells, take into consideration that each transfection is performed in triplicate to measure experimental variation. Incubate overnight at 37 °C in a humidified incubator with 5 % CO_2.

6. The firefly luciferase reporter containing the promoter of interest (pGL3-Basic promoter) and the *Renilla* luciferase (pRL-TK) plasmids should be transfected together at a ratio of 10:1 (*see* **Note 26**).

7. Immediately prior to transfection, prepare the transfection mix in a 1.5-ml microfuge tube by diluting the pGL3-Basic promoter and pRL-TK plasmids in a serum-free DMEM (*see* **Note 27**).

8. Add TurboFect Transfection Reagent at a ratio of 2 µl to 1 µg of DNA, mix well by pipetting or vortexing, and incubate for 20 min at room temperature to allow the plasmid DNA to complex with the transfection reagent. Typically, 100 ng of pGL3-Basic promoter and 10 ng of pRL-TK in 50 µl of transfection mix are transfected per single well in a 24-well plate (*see* **Note 26**).

9. Following incubation, pipette the transfection reagent/DNA mixture dropwise per well (*see* **Note 27**). Ensure the dropwise motion is spread across the well, and gently rock the plate to ensure an even distribution.

10. Following 24 h transfection, cells are either left unstimulated or stimulated with 100 ng/ml of LPS for 8 h (*see* **Note 28**).

3.5 Luciferase Reporter Assay

1. These directions are for performing a dual luciferase assay using a GloMax 20/20 Luminometer (Promega) (*see* **Note 29**).

2. Thaw firefly luciferase assay and *Renilla* luciferase assay buffers, and add luciferin and coelenterazine (*see* **Notes 4** and **5**).

3. Prepare a 1× solution of passive lysis buffer from the 5× stock by diluting 1 volume of 5× passive lysis buffer with 4 volumes of water and mixing well just prior to use.

4. At the experimental end point, gently aspirate the culture medium, and wash the cells with cold PBS without detaching. For LPS treatment of RAW 264.7 cell, 8 h of stimulation is sufficient.

5. Gently aspirate the PBS and ensure no residual liquid remains in the wells.

6. Dispense 100 μl of 1× passive lysis buffer to each well ensuring the entire monolayer of cells is covered, and incubate at room temperature for 15 min on a rocker.

7. While the cells are lysing, dispense 100 μl of firefly luciferase assay buffer into the appropriate number of clear 1.5-ml microfuge tubes dependent on the number of samples to be analyzed. Protect from light.

8. Pipette 10 μl of the cell lysate directly into the pre-dispensed 100 μl of firefly luciferase assay buffer, and mix by pipetting two to three times.

9. Place the tube into the luminometer, close the lid, and initiate reading; if the luminometer is not connected to a computer, record the reading (*see* **Note 30**).

10. Remove the sample tube from the luminometer, and add 100 μl of *Renilla* luciferase assay buffer and mix well by pipetting. Repeat **step 9**.

11. Repeat **steps 8–10** for each sample to be measured.

12. Normalize the firefly luciferase activity with the *Renilla* luciferase measurement. Data can be presented as a fold change in reporter activity using the ratio of firefly luciferase to *Renilla* luciferase of an unstimulated control sample as a comparator (*see* Fig. 2).

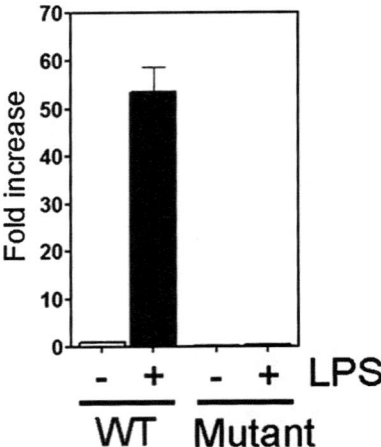

Fig. 2 LPS-induced luciferase reporter activity in RAW 264.7 cells transfected with a pGL3-Basic plasmid containing the mouse IL-23p19 promoter sequence (WT) or the IL23p19 promoter sequence in which the NF-κB binding site has been mutated (mutant) [15]. Cells were transfected with WT or mutant IL-23p19 reporter plasmid along with pRL-TK *Renilla* luciferase plasmid at a ratio of 10:1. 24 h following transfection, the cells were stimulated with 100 ng/ml LPS for 8 h. The cells were then lysed and luciferase activity measured. Promoter reporter activity was normalized to the *Renilla* luciferase activity and expressed as fold increase relative to unstimulated cells transfected with the WT reporter plasmid. Assays were performed in triplicate and data presented as mean±SEM

4   Notes

1. Specific kits are outlined in this protocol; however, a number of alternative kits are commercially available from a range of suppliers.

2. Amplicillin is degraded at high temperatures; ensure agar has cooled to 55 °C before the addition of the antibiotic.

3. Prepare a 5× solution of TBE in 1 l of deionized water: Weigh out 27.5 g of boric acid and 54 g of Tris base and dissolve in deionized water. Add 20 ml of 0.5 M EDTA pH 8.0 and mix the solution thoroughly. The pH of the stock buffer should be ~8.3. Prepare a 0.5× working TBE solution by performing a 1 in 10 dilution of 5× TBE with deionized water prior to use.

4. Firefly luciferase assay buffer is prepared as per [16] and stored in aliquots at −20 °C without the addition of luciferin; add luciferin just prior to use and protect from light. This buffer is also commercially available.

5. *Renilla* luciferase assay buffer is prepared as per [16] and stored in aliquots at −20 °C without the addition of coelenterazine; add coelenterazine just prior to use and protect from light. This buffer is also commercially available.

6. The purified DNA should have an A_{260}/A_{280} ratio of 1.8.

7. Annealing and elongation conditions are specific to the fragment of interest to be amplified and melting temperature of the primers designed under Subheading 3.1.1. For PCR products up to 3 kb, elongation temperature should be 72 °C; for products longer than 3 kb, elongation temperature should be 68 °C. Elongation time depends on fragment length: 45 s for 0.75 kb, 1 min for 1.5 kb, 2 min for 3 kb, 4 min for 6 kb, and 8 min for 10 kb.

8. This volume is dependent on the size of the casting tray. Prepare a volume of agarose gel solution suitable for the desired casting tray. Once poured, the gel should be 3–5 mm in thickness.

9. The volume of DNA that can be loaded is dependent on the dimensions of the well and should be determined before loading.

10. Follow safety procedures when working with UV, ensuring no exposure of the skin or eyes to UV source.

11. Cells may be incubated at 37 °C for up to 2 h at this point before plating onto agar plates to increase the number of colonies obtained, but it is not essential.

12. If required, plates can be stored at 4 °C short term before proceeding to the next step.

13. Screen 5–10 colonies to increase chances of identifying a positive clone. Select white colonies only and avoid blue colonies.

14. The reaction volume can be scaled up or down depending on the quantity of DNA to be digested. Prepare a master mix if a larger number of plasmids are to be digested.

15. If performing a double digest, ensure both enzymes are active in the reaction buffer by consulting the supplier datasheets. Otherwise, a sequential digest may be required if there is no buffer in which both enzymes exhibit >50 % activity.

16. The addition of BSA may be optional depending on the restriction enzymes used. Consult supplier product sheet.

17. Incubation temperature is dependent on the restriction enzymes used.

18. Run undigested plasmid DNA on the same gel as a control for restriction enzyme activity.

19. Ligation reactions are generally carried out with an excess of insert DNA. A 3:1 molar ratio of insert to vector is commonly used, but this ratio can be varied from 10:1 to 1:1.

20. Include a negative control ligation in which the insert is replaced by water. This will determine the amount of background colonies formed as a result of incomplete vector digestion or vector religation.

21. Ligation can also be carried out at 4 °C overnight. One strategy can be to split a ligation reaction over three temperatures and determine efficiency following transformation step.

22. Individual primers must be designed to anneal to the same sequence on opposite strands of the plasmid DNA with both primers containing the mutation. Ideally, primers should be between 25 and 45 bases in length with the desired mutation in the middle of the primer. The purity of the primers is critical for high-efficiency mutagenesis. Primers should be purified either by high-performance liquid chromatography (HPLC) or by polyacrylamide gel electrophoresis (PAGE). An A/T to G/C or G/C to A/T substitution of 3–4 bases of the NF-κB binding site should be sufficient to abolish NF-κB binding. To confirm that the mutation abolishes NF-κB binding, an electrophoretic mobility shift assay (EMSA) can be performed using wild-type and mutated oligo probes.

23. The number of cycles is dependent on the desired mutation. Point mutations require 12 cycles, a single amino acid change requires 16, and multiple amino acid changes, deletion, or insertions require 18.

24. *Dpn* I will digest only bacterial template DNA and not the amplified DNA. *Dpn* I-digested DNA can be stored at –20 °C until required for transformation.

25. Cell-seeding density is dependent on the cell type used. Optimization of cell density may be required for other cell types. Cells should be 70–90 % confluent on the day of transfection.

26. The total amount of DNA to be transfected can be increased to incorporate additional plasmids (e.g., expression plasmids for regulators of the NF-κB pathway). If co-transfection with additional plasmids is performed, ensure the total DNA concentration is maintained constant with the addition of an empty vector to control transfected cells.

27. Prepare enough transfection mix to perform transfections in triplicate.

28. LPS is a potent activator of NF-κB in RAW 264.7 cells. In addition to LPS, RAW cells can be stimulated with other TLR ligands (e.g., CpG, PAM3CSK) and cytokines (e.g., TNFα, IL-1β) to induce NF-κB activation.

29. The GloMax 20/20 is a single-sample luminometer; for a high-throughput analysis, multi-sample and plate-reading luminometers are available.

30. Preprogram the luminometer to perform a 2-s premeasurement delay, followed by a 10-s measurement period for each reporter assay.

References

1. Ghosh S, Hayden MS (2012) Celebrating 25 years of NF-kappaB research. Immunol Rev 246:5–13

2. Gerondakis S, Grossmann M, Nakamura Y, Pohl T, Grumont R (1999) Genetic approaches in mice to understand Rel/NF-kappaB and IkappaB function: transgenics and knockouts. Oncogene 18:6888–6895

3. Natoli G (2006) Tuning up inflammation: how DNA sequence and chromatin organization control the induction of inflammatory genes by NF-kappaB. FEBS Lett 580:2843–2849

4. Hoffmann A, Leung TH, Baltimore D (2003) Genetic analysis of NF-kappaB/Rel transcription factors defines functional specificities. EMBO J 22:5530–5539

5. Hoffmann A, Natoli G, Ghosh G (2006) Transcriptional regulation via the NF-kappaB signaling module. Oncogene 25:6706–6716

6. Sanjabi S, Williams KJ, Saccani S, Zhou L, Hoffmann A, Ghosh G, Gerondakis S, Natoli G, Smale ST (2005) A c-Rel subdomain responsible for enhanced DNA-binding affinity and selective gene activation. Genes Dev 19:2138–2151

7. Natoli G, Saccani S, Bosisio D, Marazzi I (2005) Interactions of NF-kappaB with chromatin: the art of being at the right place at the right time. Nat Immunol 6:439–445

8. Siggers T, Chang AB, Teixeira A, Wong D, Williams KJ, Ahmed B, Ragoussis J, Udalova IA, Smale ST, Bulyk ML (2011) Principles of dimer-specific gene regulation revealed by a comprehensive characterization of NF-kappaB family DNA binding. Nat Immunol 13:95–102

9. Ren B, Dynlacht BD (2004) Use of chromatin immunoprecipitation assays in genome-wide location analysis of mammalian transcription factors. Methods Enzymol 376:304–315

10. Furey TS (2012) ChIP-seq and beyond: new and improved methodologies to detect and characterize protein-DNA interactions. Nat Rev Genet 13:840–852

11. Martone R, Euskirchen G, Bertone P, Hartman S, Royce TE, Luscombe NM, Rinn JL, Nelson FK, Miller P, Gerstein M, Weissman S, Snyder M (2003) Distribution of NF-kappaB-binding sites across human chromosome 22. Proc Natl Acad Sci U S A 100:12247–12252

12. Antonaki A, Demetriades C, Polyzos A, Banos A, Vatsellas G, Lavigne MD, Apostolou E, Mantouvalou E, Papadopoulou D, Mosialos G, Thanos D (2011) Genomic analysis reveals a novel nuclear factor-kappaB (NF-kappaB)-binding site in Alu-repetitive elements. J Biol Chem 286:38768–38782

13. Xing Y, Yang Y, Zhou F, Wang J (2013) Characterization of genome-wide binding of NF-kappaB in TNFalpha-stimulated HeLa cells. Gene 526:142–149

14. Yan Q, Carmody RJ, Qu Z, Ruan Q, Jager J, Mullican SE, Lazar MA, Chen YH (2012) Nuclear factor-kappaB binding motifs specify Toll-like receptor-induced gene repression through an inducible repressosome. Proc Natl Acad Sci U S A 109:14140–14145

15. Carmody RJ, Ruan Q, Liou HC, Chen YH (2007) Essential roles of c-Rel in TLR-induced IL-23 p19 gene expression in dendritic cells. J Immunol 178:186–191

16. Dyer BW, Ferrer FA, Klinedinst DK, Rodriguez R (2000) A noncommercial dual luciferase enzyme assay system for reporter gene analysis. Anal Biochem 282:158–161

Part II

Detection and Analysis of NF-κB Signaling

Chapter 4

Assessing Sites of NF-κB DNA Binding Using Chromatin Immunoprecipitation

Amy Colleran, Patricia E. Collins, and Ruaidhrí J. Carmody

Abstract

The NF-κB transcription factor is in fact a family of related proteins which dimerize to form at least 12 distinct complexes which regulate the expression of hundred of genes of importance to a range of physiological and pathological processes. The binding of NF-κB to the regulatory regions and promoters of target genes is influenced by a number of factors including the sequence of DNA-binding sites, the post-translational modification of NF-κB, and the interaction of cofactors and co-regulators of transcription. In addition, the binding of NF-κB to promoters is highly dynamic and the recruitment of specific subunits to specific binding sites may occur with distinct kinetics. Moreover, genome-wide analysis of NF-κB chromatin binding indicates that the majority of DNA-binding events are not associated with changes in transcriptional activity. Thus, the analysis of NF-κB recruitment and activity at specific binding sites is of critical importance in understanding the regulation of transcription. In this chapter we describe a chromatin immunoprecipitation assay to investigate the in situ binding of NF-κB to specific sites in the genome.

Key words NF-κB, Chromatin immunoprecipitation, PCR, LPS, Macrophage

1 Introduction

NF-κB is a family of transcription factors, comprising of five subunits: p65/RelA, p50, p52, c-Rel, and RelB. Each NF-κB subunit is capable of dimerizing with itself or another subunit and, although in theory 15 combinations are possible, only 12 have been identified in vivo [1]. The physiological relevance of all dimers has not been fully explored and some dimers may be limited to certain cell types [2]. Although the abundance of a particular NF-κB dimer is dependent on cell type and in some cases is stimulus specific, in general the p65:p50 heterodimer is the most predominant form of NF-κB expressed in most cell types [1]. Of the 12 experimentally identified NF-κB dimers, only one, the p65:RelB dimer, appears unable to bind DNA [3], leaving 11 possible NF-κB complexes which may bind to the promoter and/or regulator regions of target genes. The p50 and p52 subunits are generated from the limited

Michael J. May (ed.), *NF-kappa B: Methods and Protocols*, Methods in Molecular Biology, vol. 1280,
DOI 10.1007/978-1-4939-2422-6_4, © Springer Science+Business Media New York 2015

proteasomal processing of the precursor proteins p105 and p100 respectively. The p50 and p52 subunits are also distinguished from the other NF-κB subunits by the absence of a transactivation domain (TAD) in their C-terminal region. Since homodimers of these subunits may bind DNA but lack a TAD, they are often considered as repressors of NF-κB-mediated transcription through competition with TAD-containing NF-κB dimers for binding to κB sites [4]. However, several studies have shown that p50 and p52 homodimers may also enhance the expression of certain genes, possibly through interaction with co-activators or other transcription factors [5, 6]. Recent high-throughput analysis of NF-κB DNA-binding sites using protein-binding arrays has identified dimer specificity for binding to certain sequences which indicates dimer-selective roles in the regulation of specific NF-κB target genes [7].

Early studies have demonstrated a gradual exchange of NF-κB dimers on the promoters of genes induced during dendritic cell maturation which appears to fine-tune the transcriptional response over time [8]. More recent studies demonstrate a highly dynamic interchange of a number of transcription factors, including NF-κB dimers, at inducible gene promoters [9]. Although this interchange appears important for the expression kinetics of NF-κB target genes, how such interchange is controlled is not fully clear at present. It is however becoming apparent that NF-κB bound at promoter regions is subject to a distinct set of regulatory events than non-DNA-bound NF-κB. Recent studies have demonstrated that promoter-bound NF-κB undergoes polyubiquitination and proteasomal degradation, a process which serves to terminate the transcriptional response [10]. The stability of DNA-bound NF-κB is controlled by a balance of ubiquitination and deubiquitination, and key components of this ubiquitin regulatory apparatus are also found associated with NF-κB at target promoters [11]. How NF-κB bound at target promoters is specifically recognized by these regulatory factors is not currently known, but a conformation change in NF-κB upon DNA binding is likely important [12]. Genome-wide analyses of NF-κB binding have revealed that once in the nucleus NF-κB binds extensively across the entire genome, primarily at off-target or transcriptionally inactive sites [13]. Estimates from these experimental data indicate that only ~10 % of NF-κB DNA binding occurs at gene promoters of which only ~10 % are transcribed following NF-κB binding. Thus the binding of NF-κB to specific DNA sequences is not always associated with a transcriptional event.

The measurement of the binding of NF-κB dimers to target promoters is critical for the understanding of the regulation of gene transcription. The development of chromatin immunoprecipitation assays for the analysis of NF-κB binding to transcriptional regulatory elements offers the potential to answer many of the outstanding questions: How is DNA-bound NF-κB regulated?

What controls the dynamic exchange of NF-κB dimers at gene promoters? And what is the impact of the selective recruitment of specific NF-κB dimers to gene promoters? In this chapter we describe a chromatin immunoprecipitation (ChIP) assay protocol which will allow the investigation of such questions.

2 Materials

2.1 Cell Culture and Treatment with LPS

1. Dulbecco's modified Eagle's medium (DMEM) supplemented with 10 % (v/v) fetal calf serum, glutamine (2 mM), penicillin (10 U/ml), and streptomycin (10 μg/ml).

2. Lipopolysaccharides (LPS) from *E. coli* (O55:B5) stored in 100 ng/ml aliquots in cell culture medium and stored at −20 °C.

3. RAW 264.7 macrophage cell line.

4. 10 cm tissue culture dishes.

5. Cell scraper.

2.2 Cell Fixation, Lysis and Sonication

1. 100 mM Phenylmethanesulfonyl fluoride (PMSF) (*see* **Note 1**).

2. 1× PBS prepared from a 10× stock (4 °C).

3. 1× PBS prepared from a 10× stock (4 °C) and supplemented with 50 μM PMSF immediately prior to use (*see* **Note 2**).

4. 1× Glycine stop solution prepared from a 10× stock (−20 °C).

5. Lysis buffer (−20 °C): 1× protease inhibitor cocktail and 0.5 mM PMSF.

6. Protease inhibitor cocktail 200× (−20 °C).

7. Dounce homogenizer (2 ml capacity) with tight-fitting pestle.

8. Fixation solution prepared freshly by the addition of 270 μl of a 37 % formaldehyde (formalin) solution to cell culture medium and maintained at room temperature.

9. Rocking platform (*see* **Note 3**).

10. Sonicator with microtip (*see* **Note 4**).

11. 15 ml conical tubes.

2.3 Chromatin Immunoprecipitation

We perform all chromatin immunoprecipitations using the ChIP IT Express kit (Active Motif, Carlsbad, CA, USA) (*see* **Note 5**). Unless specified, the components of the kit are stored at −20 °C.

1. Shearing Buffer containing 0.5 mM PMSF and 1× protease inhibitor cocktail.

2. Elution Buffer AM2.

3. ChIP Buffer 1.

4. ChIP Buffer 2.

5. Reverse Cross-linking Buffer.

6. RNase A (10 μg/μl).

7. 5 M NaCl.

8. 100 mM PMSF.

9. Proteinase K (0.5 μg/ml).

10. Proteinase K Stop solution.

11. Protease inhibitor cocktail 200×.

12. Protein G magnetic beads (4 °C).

13. Magnetic stand.

14. Rotator for 1.5 ml tubes.

15. Anti-p65 antibody, polyclonal rabbit IgG (Santa Cruz Biotechnology, Santa Cruz, CA, USA).

16. Normal Rabbit IgG.

2.4 Column Purification of Chromatin

We use the QIAquick PCR Purification Kit (Qiagen, Valencia, CA, USA) to purify chromatin.

1. Buffer PB.

2. Buffer PE.

3. Buffer EB.

4. QIAquick columns.

5. Vacuum manifold.

6. 1.5 ml Eppendorf tubes.

2.5 Agarose Gel Electrophoresis

1. Agarose.

2. 0.5× TBE buffer: 45 mM Tris-borate, 1 mM EDTA prepared from a 5× stock containing 54 g of Tris base, 27.5 g of boric acid, and 20 ml of 0.5 M EDTA per liter.

3. 6× Orange G loading buffer: 30 % Glycerol, 0.2 % v/v Orange G in water (*see* **Note 6**).

4. Ethidium bromide (10 mg/ml) (*see* **Note 7**).

5. DNA standard ladder.

6. Gel documentation system including UV light source.

2.6 Polymerase Chain Reaction

1. dNTP mix (10 mM).

2. 10× PCR reaction buffer.

3. Taq Enzyme (5 U/μl).

4. Primers for the mouse IL-6 promoter Forward: 5′-CGTTTA TGATTCTTTCGATGCTAAACG-3′, Reverse 5′-GTGGGCT CCAGAGCAGAATGAG-3′ (*see* **Note 8**).

2.7 Real-Time Quantitative Polymerase Chain Reaction (QPCR)

1. PerfeCTa SYBR Green Supermix (Quanta BioSciences, Gaithersburg, MD, USA).

2. 96 well PCR plate, semi skirted.

3. Clear adhesive cover for PCR plate.

4. Applied Biosystems 7500 real-time PCR system (Grand Island, NY, USA).

5. Primers for the mouse IL-6 promoter Forward: 5'-CGTTT ATGATTCTTTCGATGCTAAACG-3', Reverse 5'-GTGGGC TCCAGAGCAGAATGAG-3' (*see* **Note 9**).

3 Methods

Chromatin immunoprecipitation (ChIP) is an indispensible tool for the analysis of NF-κB binding at specific sites across the genome. ChIP allows the measurement of NF-κB DNA binding in situ and incorporates the properties and features of chromatin and promoter regions in a native conformation. In this regard, it offers advantages to other techniques such as electrophoretic mobility shift assay (EMSA), which measures total NF-κB-binding activity using short double-stranded oligonucleotides, and plasmid-based reporter assays, which do not represent native chromatin conformation. Although a number of variations of the basic protocol have been developed, the key steps of most ChIP assays involve the fixation of protein-DNA interactions using chemical cross-linking, the shearing of genomic DNA into short fragments, the immunoprecipitation of protein-DNA complexes using antibodies specific for the factor of interest, and the detection of the specific DNA sequences co-purified with the immunoprecipitated protein. The detection of the immunoprecipitated chromatin is typically performed using primers designed to sites flanking a region of interest that is generally known or at least predicted to bind the factor of interest. However, the true power of ChIP to measure in situ NF-κB DNA binding is realized when it is coupled to next-generation sequencing (ChIP-seq). By sequencing all immunoprecipitated chromatin, ChIP-seq offers an unbiased analysis of all chromatin interactions of the protein of interest across the entire genome. However such analysis comes with additional experimental costs as well as a significant burden in data analysis that requires bioinformatics skills.

ChIP assays have been developed over the past 30 years and are currently available in kit form from a number of commercial sources. In this chapter we describe the ChIP of NF-κB p65 using a commercially available kit. We describe the optimization of chromatin shearing using sonication (Fig. 1) and the analysis of immunoprecipitated chromatin using sequence-specific primers by conventional (Fig. 2) and quantitative real-time PCR (qPCR) (Fig. 3).

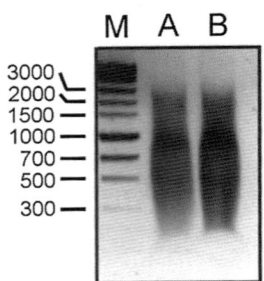

Fig. 1 Agarose gel electrophoresis of chromatin sheared by sonication. Following sonication and clean up, 5 μl (*lane A*) and 10 μl (*lane B*) of sheared chromatin were loaded on a 1 % agarose gel containing ethidium bromide. Five microliters of DNA standard ladder was also loaded (*lane M*) and the ladder band sizes are shown in base pairs. The gel was visualized using a UV source and Gel Doc instrument equipped with a digital camera

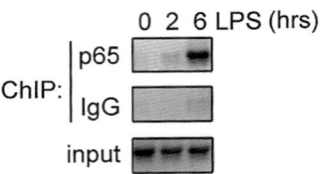

Fig. 2 NF-κB p65 subunit binding to the IL-6 promoter by ChIP assay. RAW 264.7 cells were stimulated with LPS (100 ng/ml) for the indicated time prior to analysis by ChIP assay using control IgG or anti p65 antibodies. Immunoprecipitated chromatin was used in a PCR reaction employing primers flanking the NF-κB site of the IL-6 promoter. PCR products were analyzed by agarose gel electrophoresis (2 % gel) and visualized by ethidium bromide and a Gel Doc instrument containing a UV light source and a digital camera

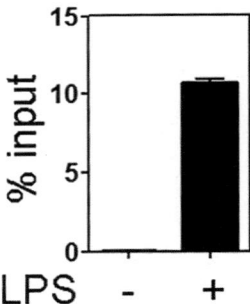

Fig. 3 NF-κB p65 subunit binding to the IL-6 promoter by ChIP assay. RAW 264.7 cells were stimulated with LPS (100 ng/ml) for 6 h prior to analysis by ChIP assay using anti-p65 antibodies. Immunoprecipitated chromatin was used in a real-time PCR reaction employing primers flanking the NF-κB site of the IL-6 promoter. Amplification of the IL-6 promoter is expressed as a ratio of the chromatin used in the ChIP assay (input)

3.1 Stimulation of RAW 264.7 Cells

All steps must be carried out in sterile culture.

1. RAW 264.7 cells are grown in DMEM containing FCS 10 % (v/v), glutamine, penicillin, and streptomycin. Maintain cells at 37 °C in a humidified environment with 5 % CO_2.

2. Cells are passaged by first washing in pre-warmed (37 °C) serum-free medium; then cells are detached using a cell scraper. Detachment of cells from tissue culture flask should be visible by eye but complete detachment should be confirmed microscopically (*see* **Note 10**).

3. Seed 10×10^6 cells in 10-cm tissue culture dishes in a total of 10 ml of DMEM supplemented with serum, glutamine, and antibiotics and incubate overnight at 37 °C in a humidified environment with 5 % CO_2.

4. The next day add 10 μl of a 100 ng/μl solution of LPS to the cells and mix by gently swirling the dish. Incubate at 37 °C for 60 min to activate NF-κB (*see* **Note 11**).

3.2 Fixing Cells

1. Decant medium from the cells and add 10 ml of fixation solution to each dish and incubate on a rocking platform for 10 min at room temperature.

2. Decant fixing solution and wash cells with 5 ml of ice-cold 1× PBS.

3. Add 5 ml of 1× glycine stop solution and incubate on a rocking platform for 5 min at room temperature.

4. Decant the glycine stop solution and wash cells with 5 ml of ice-cold 1× PBS.

5. Add 5 ml of ice-cold 1× PBS containing 50 μM PMSF to the cells and scrape cells using a cell scraper. Transfer the cells to a clean 15 ml conical tube.

6. Centrifuge the cells at $720 \times g$ for 10 min and decant the supernatant (*see* **Note 12**).

3.3 Shearing by Sonication

1. Add 500 μl of ice-cold lysis buffer to the cell pellet. Pipette up and down gently 3–4 times and incubate on ice for 30 min.

2. Transfer the cells to an ice-cold Dounce homogenizer and using the tight-fitting pestle gently dounce for ten strokes to release nuclei (*see* **Note 13**).

3. Transfer the cells to a 1.5 ml microcentrifuge tube and centrifuge at $2,500 \times g$ for 10 min at 4 °C.

4. Remove the supernatant and resuspend the nuclei pellet in 350 μl of shearing buffer supplemented with protease inhibitors and place on ice.

5. Shear the DNA by sonication using six repeats of a 30 s burst at 30 % power and 50 % duty cycle. Sonication should be

performed on ice with a 1 min rest between each pulse to avoid sample overheating (*see* **Note 14**).

6. Centrifuge the sheared chromatin at $16,000 \times g$ for 10 min at 4 °C and transfer the supernatant to a fresh 1.5 ml microcentrifuge tube on ice (*see* **Note 15**).

7. Assess the efficiency of sonication by diluting 50 μl of chromatin 1:4 with water and adding 10 μl of 5 M NaCl. The chromatin is then incubated at 65 °C for 4 h to reverse cross-links.

8. 1 μl of RNase A is added and chromatin incubated at 37 °C for 15 min.

9. 10 μl proteinase K is added and sample is incubated at 42 °C for an additional 1.5 h.

10. To 16 μl of chromatin add 4 μl of 6× loading dye and load 10 μl in a single lane of a 1 % agarose gel as described in Subheading 3.9. Sheared chromatin should yield a smear of DNA fragments in the region of 200–2,000 bp (Fig. 1).

3.4 Immunoprecipitation

1. Reserve 10 μl of the sheared chromatin in a separate microcentrifuge tube to be used as input and processing as with immunoprecipitated samples in Subheadings 3.8 and 3.9.

2. Set up the immunoprecipitation reactions as shown in Table 1.

3. Mix components well by pipetting and incubate on an end-to-end rotator overnight at 4 °C.

3.5 Washing and Elution of Immunoprecipitated Chromatin (See Note 18)

1. Spin the tubes using a short burst on a benchtop microfuge to collect any liquid from inside the cap and place on the magnetic stand.

2. Ensure that the magnetic beads have moved to the side of the tube nearest the magnet before carefully aspirating and discarding the supernatant (*see* **Note 18**).

Table 1
Immunoprecipitation reaction volumes for ChIP

Reagent	Volume per immunoprecipitation
Protein G magnetic beads	25 μl
ChIP Buffer 1	20 μl
Sheared chromatin generated in Subheading 3.3	100 μl
Protease inhibitor cocktail	1 μl
H$_2$O	Adjust to final volume
Antibody (*see* **Notes 16** and **17**)	2 μg
Total volume	200 μl

3. Remove the tube from the magnetic stand and add 800 μl of ChIP buffer 1.

4. Pipette gently up and down to resuspend the magnetic beads and replace on the magnetic stand.

5. Repeat **steps 2–4** twice more using 800 μl of ChIP buffer 2.

6. After the third wash, aspirate as much of the supernatant as possible without dislodging the beads.

3.6 Chromatin Elution and Preparation for Analysis

1. Resuspend the magnetic beads in 50 μl of Elution Buffer and incubate on a rotator for 15 min at room temperature.

2. Spin the tube briefly using a short burst on a benchtop microfuge to collect the liquid from inside the cap, add 50 μl of Reverse Cross-linking buffer, and pipette gently up and down to mix.

3. Place the tube on the magnetic stand to collect the magnetic beads and transfer the eluted chromatin to a clean tube.

4. To a clean tube, add 10 μl of the input DNA sample that was reserved in Subheading 3.4 along with 88 μl of the ChIP 2 buffer and 2 μl of 5 M NaCl.

5. Incubate both the ChIP and input DNA samples at 65 °C for 4 h.

6. Cool the tubes to room temperature, spin briefly to collect the liquid at the bottom, and add 2 μl of proteinase K.

7. Mix well and incubate at 37 °C for 60 min.

8. Cool the tubes to room temperature then add 2 μl of proteinase K stop solution and mix well (*see* **Note 19**).

3.7 Column Purification of Chromatin

1. Add 500 μl of Buffer PB to the chromatin samples; mix and apply to a QIAquick purification column attached to a vacuum manifold.

2. Apply vacuum until the solution has passed through the column then promptly turn off vacuum.

3. Add 750 μl of Buffer PE to each column then apply vacuum until the solution has passed through the column then promptly turn off vacuum.

4. Transfer the column to a 2 ml tube and centrifuge for 1 min at $14,000 \times g$.

5. Discard the collection tube and place the column in a clean 1.5 ml tube.

6. Elute the chromatin by adding 30–50 μl Buffer EB to the center of the column followed by centrifugation at $14,000 \times g$ for 1 min (*see* **Note 20**).

Table 2
PCR reaction volumes

Reagent	Volume per reaction (µl)
Chromatin	5.0
Forward primer (10 µM)	0.75
Reverse primer (10 µM)	0.75
dNTPs (10 mM)	1.0
H_2O	14.75
10× PCR buffer	2.5
Taq Enzyme (5 U/µl)	0.25
Total	25.0

3.8 End Point PCR Analysis

1. Assemble the PCR reactions as outlined in Table 2.
2. Dilute the input chromatin samples 1:10 prior to addition to PCR reaction.
3. Program a thermocycler as follows: 94 °C for 3 min, then 36 cycles of 94 °C for 20 s, 59 °C for 30 s, and 72 °C for 30 s (*see* **Note 21**).

3.9 Agarose Gel Electrophoresis

1. Assemble a gel tray and comb ready for pouring a gel.
2. Weigh out 1–2 g of agarose (1–2 % gel) and add to 100 ml of TBE in a microwave-resistant flask (*see* **Note 22**).
3. Microwave on medium setting for 1–5 min until agarose is completely dissolved.
4. Allow the agarose to cool at room temperature for 5 min.
5. Add 0.5 µg/ml ethidium bromide and swirl gently to mix.
6. Pour the gel into the gel tray and allow gel to cool to room temperature until it has solidified.
7. Place solidified gel into an electrophoresis tank and fill with 0.5× TBE buffer containing 0.5 µg/ml ethidium bromide until gel is completely covered.
8. In the first lane, carefully pipette 5 µl of DNA standard ladder.
9. Add 5 µl of 6× gel loading buffer to each PCR sample, mix well, and carefully pipette 20 µl into a well of the gel.
10. Run the gel at 100 V until the dye front has migrated three quarters of the length of the gel.
11. Switch off the power pack, disconnect the electrodes, remove the gel, and visualize DNA fragments on a UV source connected to a camera system (Fig. 2) (*see* **Note 23**).

Table 3
Real-time PCR reaction volumes

Reagent	Volume per reaction (μl)
Chromatin	5.0
Forward primer	0.75
Reverse primer	0.75
2× SYBR green PCR mix	12.5
H_2O	6.0
Total	25.0

3.10 Real-Time PCR Analysis

1. Dilute the input chromatin samples 1:10 prior to addition to PCR reaction.

2. Assemble PCR reactions in triplicate for each sample in a 96-well PCR plate as shown in Table 3.

3. Affix the plate seal and centrifuge the plate briefly to collect the reaction at the bottom of the wells using a bench centrifuge equipped with plate holder attachments.

4. Program the real-time PCR instrument as follows: 95 °C for 20 s, then 40 cycles of 95 °C for 3 s and 60 °C for 30 s.

5. Data obtained for chromatin immunoprecipitation should be normalized to the input and may be expressed as a percentage of input signal (Fig. 3) (*see* **Note 24**).

4 Notes

1. PMSF is a harmful compound and should be handled with care.

2. PMSF has a very short half-life in aqueous solution and should be only added immediately prior to use.

3. Add the formalin to the cell culture medium immediately prior to use.

4. A variety of sonicators are commercially available. We use a Bandelin Sonopuls Ultrasonic Homogenizer (Bandelin Electronics, Berlin, Germany). When selecting a microtip, ensure that it will fit into a 1.5 ml tube without touching the sides of the tube.

5. A number of commercially available kits are available from a range of suppliers.

6. Orange G dye does obscure any DNA bands when visualizing gels by UV light.

7. Ethidium bromide is a carcinogen and should be handled accordingly.

8. Primer analysis programs (e.g., http://genome.cse.ucsc.edu/cgi-bin/hgPcr or http://www.ncbi.nlm.nih.gov/tools/primer-blast/) will assist in the design of primers to the promoter of interest. Primers should produce a single amplicon approximately 150–400 bp in size.

9. Primers that work in standard PCR may not necessarily work in qPCR conditions and so should be tested. Primers that dimerize will cause high background when SYBR green is used and should be avoided. Amplicons for qPCR should be approximately 50–150 bp in size.

10. RAW 264.7 cells are a mouse macrophage cell line. These cells are resistant to trypsin removal from tissue culture plates.

11. LPS is a potent activator of NF-κB in RAW 264.7 cells. In addition to LPS, RAW cells can be stimulated with other TLR ligands (e.g., CpG, PAM3CSK) and cytokines (e.g., TNFα, IL-1β) to induce NF-κB activation.

12. The cell pellet can be stored at –80 °C before proceeding to the next step.

13. The release of nuclei can be verified by light microscopy, but with certain cell types, including RAW 264.7, this may not always be obvious.

14. These sonication settings have been optimized for RAW 264.7 cells using the sonicator described in Subheading 2.2. Sonication conditions should be optimized for specific cell types and sonicators used. Generally, altering the number of times the sample is sonicated while maintaining power output and duration of pulses is the best approach.

15. The chromatin can be stored at –80 °C before proceeding to the next step.

16. The selection of antibody is one of the most critical factors contributing to the success of the ChIP assay. Many suppliers offer ChIP-validated antibodies that have been tested for use in ChIP assays. The amount of antibody used will vary from antibody to antibody but should generally be in the range of 1–10 μg.

17. Control immunoprecipitations should be used to account for background chromatin binding. Use an equivalent amount of nonimmune (normal) species- and isotype-matched antibodies.

18. Don't allow the magnetic beads to dry out as this can affect the yield of chromatin following the elution step.

19. The samples can be stored at –20 °C before proceeding to the next step.

20. Phenol chloroform purification of the eluted DNA may be used instead of columns.

21. The cycle conditions will vary according to the melting temperature of the primers used and the abundance of the target sequence in the immunoprecipitated chromatin.

22. A 2 % gel should be used when analyzing PCR fragments of 150–400 bp in size. A 1 % gel should be used when assessing the shearing efficiency as described in Subheading 3.3.

23. A primer dimer band may be visible below the amplified fragment.

24. The 1 in 10 dilution of the input should be taken into account when calculating the relative immunoprecipitated signal.

References

1. Hoffmann A, Natoli G, Ghosh G (2006) Transcriptional regulation via the NF-kappaB signaling module. Oncogene 25:6706–6716

2. Oeckinghaus A, Ghosh S (2009) The NF-kappaB family of transcription factors and its regulation. Cold Spring Harb Perspect Biol 1:a000034

3. Marienfeld R, May MJ, Berberich I, Serfling E, Ghosh S, Neumann M (2003) RelB forms transcriptionally inactive complexes with RelA/p65. J Biol Chem 278:19852–19860

4. Carmody RJ, Ruan Q, Palmer S, Hilliard B, Chen YH (2007) Negative regulation of toll-like receptor signaling by NF-kappaB p50 ubiquitination blockade. Science 317:675–678

5. Viatour P, Bentires-Alj M, Chariot A, Deregowski V, de Leval L, Merville MP, Bours V (2003) NF- kappa B2/p100 induces Bcl-2 expression. Leukemia 17:1349–1356

6. Barre B, Perkins ND (2010) The Skp2 promoter integrates signaling through the NF-kappaB, p53, and Akt/GSK3beta pathways to regulate autophagy and apoptosis. Mol Cell 38:524–538

7. Siggers T, Chang AB, Teixeira A, Wong D, Williams KJ, Ahmed B, Ragoussis J, Udalova IA, Smale ST, Bulyk ML (2012) Principles of dimer-specific gene regulation revealed by a comprehensive characterization of NF-kappaB family DNA binding. Nat Immunol 13:95–102

8. Saccani S, Pantano S, Natoli G (2001) Two waves of nuclear factor kappaB recruitment to target promoters. J Exp Med 193:1351–1359

9. Garber M, Yosef N, Goren A, Raychowdhury R, Thielke A, Guttman M, Robinson J, Minie B, Chevrier N, Itzhaki Z, Blecher-Gonen R, Bornstein C, Amann-Zalcenstein D, Weiner A, Friedrich D, Meldrim J, Ram O, Cheng C, Gnirke A, Fisher S, Friedman N, Wong B, Bernstein BE, Nusbaum C, Hacohen N, Regev A, Amit I (2012) A high-throughput chromatin immunoprecipitation approach reveals principles of dynamic gene regulation in mammals. Mol Cell 47:810–822

10. Saccani S, Marazzi I, Beg AA, Natoli G (2004) Degradation of promoter-bound p65/RelA is essential for the prompt termination of the nuclear factor kappaB response. J Exp Med 200:107–113

11. Colleran A, Collins PE, O'Carroll C, Ahmed A, Mao X, McManus B, Kiely PA, Burstein E, Carmody RJ (2013) Deubiquitination of NF-kappaB by Ubiquitin-Specific Protease-7 promotes transcription. Proc Natl Acad Sci U S A 110:618–623

12. Matthews JR, Nicholson J, Jaffray E, Kelly SM, Price NC, Hay RT (1995) Conformational changes induced by DNA binding of NF-kappa B. Nucleic Acids Res 23:3393–3402

13. Antonaki A, Demetriades C, Polyzos A, Banos A, Vatsellas G, Lavigne MD, Apostolou E, Mantouvalou E, Papadopoulou D, Mosialos G, Thanos D (2011) Genomic analysis reveals a novel nuclear factor-kappaB (NF-kappaB)-binding site in Alu-repetitive elements. J Biol Chem 286:38768–38782

Chapter 5

IKK Kinase Assay for Assessment of Canonical NF-κB Activation in Neurons

Anca B. Mihalas and Mollie K. Meffert

Abstract

Nuclear factor kappa B (NF-κB) is a potent transcription factor highly expressed in the central nervous system (CNS) where it has been shown to be required for multiple behavioral paradigms of learning and memory in both mammalian and invertebrate systems. NF-κB dimers are found in neuronal cell bodies, are also present at synapses, and can participate in the activity-dependent regulation of gene expression in response to excitatory neurotransmission. Multiple serine-directed phosphorylation events are critical in the canonical NF-κB activation pathway, including activation of the IκB kinase complex (IKK) and phosphorylation and degradation of the inhibitor of NF-κB (IκB). In this chapter, we describe methods for immunoprecipitation (IP) of the IKK complex from dissociated cultured murine hippocampal neurons, followed by in vitro kinase assay to evaluate excitatory neurotransmission-induced IKK activation by monitoring phosphorylation of a GST-IκBα substrate. These methods can also be successfully implemented in subcellular-reduced brain preparations, such as biochemically isolated synapses.

Key words IKK in vitro kinase assay, GST-IκBα, Murine neurons, Excitatory neurotransmission, Synaptic activity

1 Introduction

The nuclear factor kappa B (NF-κB) family of transcription factors was initially characterized as key regulators of genes involved in both innate and adaptive immune responses [1–5]. Dysregulated NF-κB signaling has been linked to cancer and inflammatory and autoimmune disorders [6–9]. Over the past decade, research from multiple laboratories, including our own, has revealed a conserved role for NF-κB in the regulation of synaptic plasticity, learning, and memory [10–21]. While NF-κB is ubiquitously expressed in all tested cell types of the central nervous system (CNS), previous work has demonstrated that NF-κB is present at neuronal synapses and can mediate synaptic activity-dependent regulation of gene expression in response to excitatory neurotransmission [17, 22, 23]. Synaptic input can alter the complement of expressed genes and

Michael J. May (ed.), *NF-kappa B: Methods and Protocols*, Methods in Molecular Biology, vol. 1280,
DOI 10.1007/978-1-4939-2422-6_5, © Springer Science+Business Media New York 2015

represents a critical pathway employed by the nervous system to affect the enduring adaptation required in development, differentiation, and plasticity. Neuronal NF-κB is classified as an activity-dependent transcription factor that can be induced by multiple stimuli, including excitatory neurotransmitters [17, 24, 25], cytokines (e.g., tumor necrosis factor alpha (TNFα)) [5, 26], and growth factors (BDNF, NGF) [27, 28]. In addition to these physiological activators, NF-κB has also been shown to become activated during brain injury, ischemia, and oxidative stress.

The mammalian NF-κB family (also known as the Rel family) consists of five members which can hetero- and homo-dimerize: p50 (product of the *NF-κB1* gene), p52 (product of the *NF-κB2* gene), p65 (also known as *Rel A*), *c-Rel*, and *RelB* [29]. The expression of all NF-κB family members has been reported in brain tissue, although the predominant neuronal dimers under basal conditions appear to be p65:p50 and p50:p50 [17, 24, 30]. All Rel family members contain a conserved Rel homology domain that includes an immunoglobulin-fold DNA-binding domain, a dimerization domain, and a nuclear localization sequence (NLS). RelA, RelB, and c-Rel have a transactivation domain (TAD) in their carboxyl-termini.

Dimers of NF-κB are held latent in the cytoplasm by non-covalent interactions with a class of inhibitors called inhibitor of NF-κB (IκB) proteins. Canonical pathway activation of NF-κB is mediated by the IκB kinase complex (IKK), which is composed of two catalytic subunits, IKKα and IKKβ, and a regulatory subunit, IKKγ. IKK phosphorylates the inhibitor of kappa B protein (IκB) and leads to its subsequent ubiquitination and degradation by the 26S proteasome [31]. IκB degradation exposes the DNA-binding domain and NLS of NF-κB and permits stable translocation to the nucleus (Fig. 1) followed by binding to consensus κB-binding sites of target genes and the regulation of gene expression. While IKK activation is well characterized as a pathway mediating NF-κB activation, it should be noted that at least one IKK subunit (IKKα) has also been demonstrated to regulate gene expression through a distinct mechanism of chromatin remodeling in multiple tissues, including brain [32]. In some settings, it has been determined that the kinase activity of IKKα is also requisite for this non-NF-κB-related function [33].

In this chapter, we describe a protocol for the assay of neuronal NF-κB activation produced by excitatory synaptic stimulation and include instructions for the production of dissociated murine neuronal cultures. Synaptic excitation is produced by transient competitive inhibition of GABA_A receptors (using bicuculline), which results in enhanced excitatory synaptic responses and the effective production of NF-κB activation in mature neuronal cultures with fully developed inhibitory GABA networks. Activation of the NF-κB pathway can be monitored using

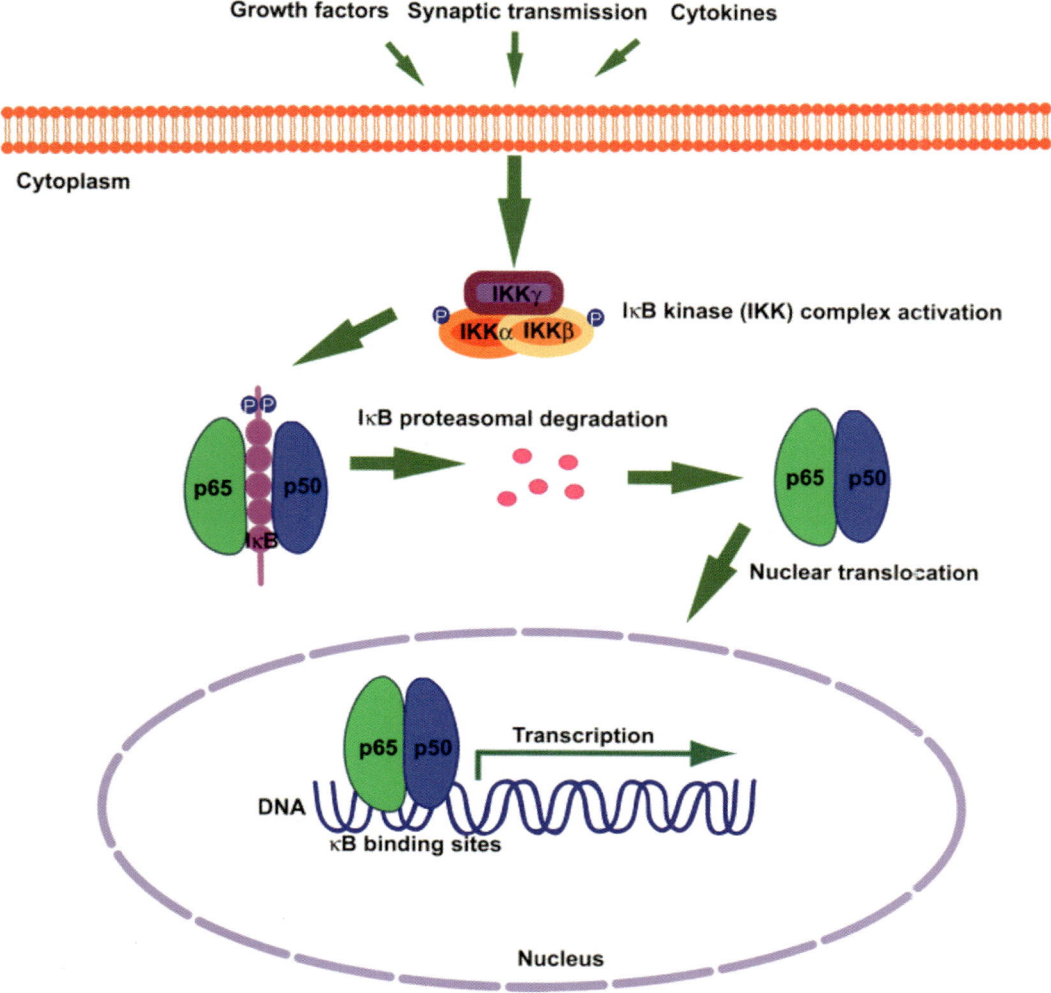

Fig. 1 Canonical pathway of NF-κB activation. Dimers of NF-κB are held latent in the cytoplasm by IκB. Extracellular stimuli (e.g., synaptic transmission, cytokines, and growth factors) induce IKK complex activation leading to site-specific phosphorylation of IκBα on conserved serine residues. Phosphorylated IκBα is targeted for ubiquitination and directed to the 26S proteasome for degradation. NF-κB is then able to stably translocate to the nucleus and regulate transcription

multiple approaches. In this protocol, we describe an approach for the measurement of IKK activation in neurons by in vitro kinase assay of IKK. This protocol derives from previous protocols [34–36], but is optimized for use in mouse neuronal tissue. IKK complexes are immunoprecipitated from control or stimulated neurons and presented with an IκBα-based substrate. We also include a protocol for the production and purification of this recombinant in vitro IKK substrate.

2 Materials

2.1 Neuronal Culture Stocks

1. 1 mM AraC (cytosine β-D-arabinofuranoside; Sigma-Aldrich, St. Louis, MO, USA): in sterile nanopure H_2O. Sterile filter, make 1 ml aliquots in Eppendorf tubes, and store at –20 °C.

2. 50× B27 supplement for optimal growth of neurons (Life Technologies, Carlsbad, CA, USA): make 1 ml aliquots in Eppendorf tubes and store at –20 °C.

3. 200 mg/ml BSA: BSA fraction V in sterile nanopure H_2O. Sterile filter, make 1 ml aliquots in Eppendorf tubes, and store at –20 °C.

4. 25 mg/ml cysteine: L-cysteine in sterile nanopure H_2O. Sterile filter, make 200 μl aliquots, and store at –20 °C.

5. 5 mg/ml DNase I (Worthington #LS002004, Code D, 5 mg): in sterile nanopure H_2O. Sterile filter, make 400 μl aliquots, and store at –20 °C.

6. 5 ng/ml basic FGF (bFGF, Life Technologies): make 50 μg/ml aliquots in Neurobasal A (Life Technologies #10888-022), and store at –20 °C.

7. 0.2 mg/ml glucose: in nanopure H_2O. Sterile filter and store at room temperature.

8. 200× glutamine: 200 mM L-glutamine.

9. 10× HBSS without $Ca^{2+}/Mg^{2+}/NaHCO_3$: 100 ml bottle (Life Technologies).

10. 10× HBSS with Ca^{2+}/Mg^{2+}, EDTA, $NaHCO_3$: 100 ml bottle as above, add 0.1 ml of 0.5 M EDTA (0.5 mM final), and add 9.6 ml of 524 mM $NaHCO_3$ (50 mM final).

11. 1 M HEPES, pH 7.5.

12. 0.2 units/μl papain (Worthington Biochemical Corporation, Lakewood, NJ, USA): resuspend 100 mg vial in sterile nanopure H_2O at 4 °C overnight, sterile filter into a 15 ml conical tubes, and store at 4 °C.

13. 100× pen/strep/glutamine: 1:1 mix of penicillin (10,000 units), streptomycin (10 mg), and 200 mM L-glutamine. Sterile filter, make 1 ml aliquots in Eppendorf tubes, and store at –20 °C.

14. 40× pen/strep/pyruvate/glucose: 200 μl pen/strep, 200 μl 200 mg/ml glucose, 400 μl 100 mM sodium pyruvate, 200 μl sterile nanopure H_2O. Sterile filter, make 1 ml aliquots in Eppendorf tubes, and store at –20 °C.

15. 100 mM sodium pyruvate: store at 4 °C.

16. 50 ml of plating medium: 48.5 ml of Neurobasal A, 1 ml of 50× B-27 supplement, 0.5 ml of 100× pen/strep/L-glutamine, 5 μl of 5 ng/ml bFGF.

17. 50 ml of growth medium: 48.75 ml of Neurobasal A, 1 ml of 50× B-27 supplement, 0.25 ml of L-glutamine, 5 µl of 5 ng/ml bFGF (*see* **Note 1**).

18. Trituration medium: 20 ml of DMEM, 200 µl of 200 mg/ml BSA (2 mg/ml final), 0.5 ml of 40× pen/strep/glucose/pyruvate, 200 µl of 5 mg/ml DNase (0.05 mg/ml final).

19. Centrifugation medium: 5 ml of trituration medium, 400 µl of 200 mg/ml BSA (16 mg/ml final); add three drops of sterile 0.1 N NaOH or until medium is slightly pink.

20. Papain solution: 1.5 ml of 6.67× HBSS/NaHCO₃/EDTA, 128 µl of 25 mg/ml cysteine (2.13 mg/ml final), 0.25 ml of 40× pen/strep/glucose/pyruvate, 8 ml of H_2O. Store on Ice. Immediately prior to use, warm to room temperature, add 1/2 to flask, and gently aerate with 95 % O_2/5 % CO_2. Add 0.2 units/µl papain later.

21. Dissection solution: 2 ml of 10× HBSS without Ca^{2+}/Mg^{2+}/NaHCO₃, 200 µl of 1 M HEPES (1 mM final), 0.5 ml of 40× pen/strep/glucose/pyruvate, 17.3 ml of H_2O.

22. 0.1 M sodium borate buffer: add 38.14 g of sodium borate (Na_3BrO_3) to 1 L of H_2O. Adjust pH to 8.4 with HCl. Sterile filter or autoclave.

23. Poly-L-lysine (PLL: 0.5 mg/ml) solution: add 10 mg of PLL (Sigma) to 100 ml of 0.1 M sodium borate buffer (pH 8.4). Sterile filter and make 5 ml aliquots in 50 ml conical tubes (concentration 333 nM and final concentration 33 nM after adding 45 ml of sodium borate buffer). Freeze at –20 °C. Add 45 ml of sodium borate buffer to the 5 ml of PLL aliquots. Add the PLL solution to the wells (500 µl/well of 24-well plate). Incubate for 4 h at 37 °C; overnight is better. Wash 1× in a large volume with 1× PBS or sterile water. Vacuum dry and plate immediately or store in the tissue culture incubator until plating.

24. 100× stop solution for bicuculline stimulation for neurons: 500 mM sodium HEPES, use 0.3 M NaOH to pH to 7.4, 100 mM kynurenic acid, anhydrous (Sigma K-3375), 1 M $MgCl_2$, sterilized. The kynurenate solution must be filtered with a prewashed 0.2 µm nylon filter to remove any undissolved material and bacteria (*see* **Note 2**), followed by aliquoting (1 ml per 1.5 ml tube) and storage at –20 °C.

25. 30 mM bicuculline (Sigma): dissolve in water and store in 50 µl aliquots at –20 °C.

2.2 GST-IκBα and In Vitro Kinase Assay Stocks

1. LB: 20 g of LB in 1 L of nanopure water, autoclaved.

2. 1× phosphate buffered solution (PBS), pH 7.3: 137 mM NaCl, 2.7 mM KCl, 4.3 mM Na_2HPO_4·7H_2O, 1.4 mM KH_2PO_4, autoclaved and stored at room temperature. For washing neurons at harvesting, add 0.9 mM $MgCl_2$ to PBS.

3. Glutathione buffer: 50 mM Tris, pH 7.5, 10 mM glutathione (Sigma).

4. 1 mM IPTG (Isopropyl β-D-1-thiogalactopyranoside).

5. Mini EDTA-free protease inhibitor cocktail (Roche, Madison, WI, USA): add fresh, prior to use, 1 tablet/10 ml lysis buffer.

6. Lysozyme.

7. Glutathione Sepharose beads (GE Healthcare, Mickleton, NJ, USA).

8. Poly-Prep chromatography elution columns (Bio-Rad, Hercules, CA, USA).

9. Bradford protein assay kit (Bio-Rad).

10. Kinase reaction buffer: 10 mM ATP ("cold"), 5 mCi of [γ-P^{32}]-ATP (10 mCi/ml), 1 mg of GST-IκBα$_{(1-62)}$. Make fresh prior to use in radioactive designated area.

11. 1× kinase buffer: 10 mM HEPES, pH 7.9, 5 mM $MgCl_2$, 1 mM $MnCl_2$. Store at room temperature. Add fresh, prior to use, 50 mM DTT and phosphatase inhibitors: 12.5 mM beta-glycerophosphate, 2 mM NaF, 50 mM Na_3VO_4. You can make 50× mixed stock of phosphatase inhibitors stored in small aliquots at –20 °C.

2.3 Mini SDS-PAGE Gel, Immunoblotting, and Immunoprecipitation (IP)

1. Stacking gel: 1.67 ml of 0.5 M Tris, pH 6.8, 0.83 ml of 30 % acrylamide/bis solution (Bio-Rad), 0.067 ml of 10 % SDS, 0.333 ml of 1.5 % ammonium persulfate, 3.77 ml of nanopure water, 0.005 ml of TEMED (N,N,N',N'-tetramethyl-ethylenediamine).

2. 12 % resolving gel: 1.25 ml of 3 M Tris, pH 8.8, 4 ml acrylamide–bisacrylamide (30 %:0.8 %), 0.1 ml of 10 % SDS, 0.5 ml of 1.5 % ammonium persulfate, 4.13 ml of nanopure water, 0.005 ml of TEMED.

3. 1× TBST: 50 mM Tris–HCl, pH 7.4, 150 mM NaCl, 0.1 % Tween 20. Make a 10× stock and store at room temperature.

4. 6× SDS sample loading buffer: 1 M Tris, pH 6.8, 10 % SDS, 30 % glycerol, 0.012 % bromophenol blue. Store in 10 ml aliquots in 15 ml conical tubes at room temperature. Prior to use, heat gently to bring into solution, and add 20 µl of 3 M DTT to 100 µl of 6× sample buffer.

5. 1× running buffer: 25 mM Tris–HCl, 200 mM glycine, 0.1 % SDS. Make 5× stock and store at room temperature.

6. 1× transfer buffer: 25 mM Tris, 192 mM glycine, 10 % methanol. Make 10× stock, without methanol, and store at room temperature. Add the methanol prior to use, when you make the 1× working solution.

7. IP lysis buffer: 20 mM Tris, pH 7.5, 150 mM NaCl, 1.0 % Triton X-100, 1.0 mM EDTA. Store at room temperature.

Add fresh phosphatase inhibitors: 20 mM beta-glycerophosphate, 10 mM NaF, 100 μM Na_3VO_4. Add fresh mini EDTA-free protease inhibitor cocktail prior to use.

8. Primary antibody: mouse anti-IKKα monoclonal (Novus Biologicals, Littleton, CO, USA).

9. Secondary antibody anti-mouse HRP (Santa Cruz Biotechnology, Santa Cruz, CA, USA): 1:5,000 dilution.

10. Enhanced chemiluminescence reagents such as the Pierce ECL Western Blotting Substrate (Thermo Scientific, Waltham, MA, USA).

2.4 Labware and Instruments

1. Pipetters and pipette tips.

2. Eppendorf 1.5 and 15 ml conical tubes.

3. 24-well culture plates and culture flasks (75 cm^2).

4. Eppendorf 1.5 and 15 ml tube centrifuges with temperature control.

5. Tube rotator at 4 °C.

6. Gel casting apparatus and accessories, gel running and transfer apparatus accessories, and power source for gel running and transfer. We use the Mini-Protean gel system (Bio-Rad).

7. Hot block for Eppendorf tubes.

8. Radiation badge, Geiger counter, radioactivity designated area, and instruments.

9. CO_2 and 95 % O_2/5 % CO_2 gas tanks.

10. Tissue culture room, tissue culture hood, and tissue culture incubator set at 37 °C and 5 % CO_2.

11. Ethanol-based flame lamp.

12. Dissection scope, instruments, and dissection designated area (*see* **Note 3**).

13. UV sterilizer.

14. PhosphorImager instrument.

3 Methods

3.1 Murine Hippocampal Neuron Dissociation and Culture

3.1.1 Dissection and Digestion of Murine Hippocampi

For additional details and visual aid, *see* refs. 37 and 38:

1. Prepare fresh trituration, centrifugation, papain, and dissection solutions.

2. Make a hole (using soldering iron) in the top of a sterile 75 cm^2 culture flask, and connect the hole by tubing to oxygen/CO_2 (95/5 %).

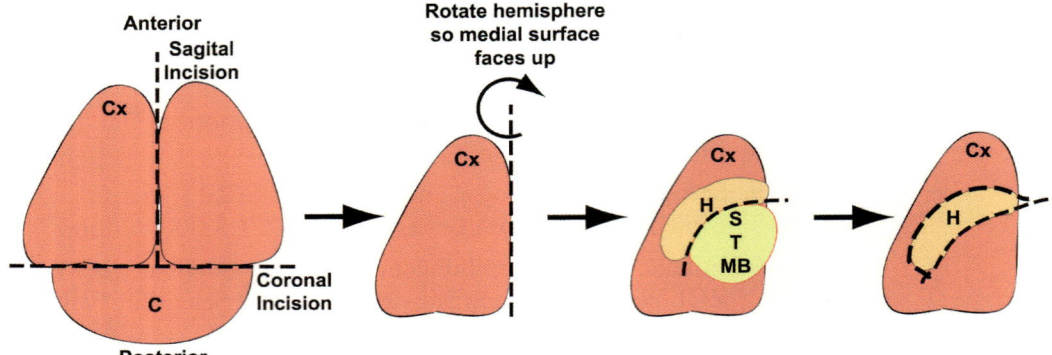

Fig. 2 Diagram depicting hippocampal dissection from neonatal mouse pups. Place the whole brain with the caudal surface toward the bottom of the dissection dish. Make an initial coronal incision to remove the cerebellum (C). Cut the remaining of the brain sagitally on the midline to separate the two brain hemispheres. Rotate the hemisphere placing the medial surface facing up. Remove the striatal (S), thalamic (T), and midbrain (MB) structures to expose the hippocampus (H). Dissect out the hippocampus

3. Place 3×35 mm culture dishes containing dissection medium on ice: one for the whole brains, another for dissected hippocampi, and a third for the final hippocampi with meningeal membranes removed.

4. Dissect the hippocampi from postnatal P0 mouse pups using a dissecting scope (Fig. 2) (*see* **Note 4**).

5. While dissecting, keep the hippocampi on ice in dissection solution.

6. Remove the meninges using fine tweezers under the dissecting microscope.

7. When the dissection is finished, transfer the tissue to the flask connected by tubing to oxygen/CO_2 (95/5 %), taking care to transfer as little dissection solution as possible, and add 100 U papain.

8. Digest for ~10–15 min at room temperature, depending upon papain age and temperature.

9. Add 80 U of papain and 100 µl of 5 mg/ml DNase to the remaining papain solution and add to the digestion mix in the flask.

10. Digest for another 10–15 min, and monitor by eye watching for feathering of the tissue edges.

3.1.2 Trituration and Plating

1. Label three 15 ml conical tubes, #1–#3.

2. Once the digestion is over, move the flask to a tissue culture (TC) hood. Allow the tissue to settle in one corner, and then remove the tissue to #1 conical tube using a 5 ml plastic pipette.

3. Allow the tissue to settle and remove as much of the digestion solution as possible from over the tissue. Add 4 ml of trituration solution with mixing, allow the tissue to settle, and again remove as much solution as possible from over the tissue.

4. Triturate the tissue gently up and down four times with a sterile Pasteur pipette, and then add 3 ml of trituration solution and allow chunks to settle.

5. Remove solution from the top, which is free of chunks, and transfer to #2 conical tube.

6. Fire-polish a sterile Pasteur pipette to slightly decreased diameter and repeat the trituration. Again add 3 ml of trituration solution, allow chunks to settle, and remove the top solution to #2 conical tube.

7. Repeat the above trituration procedure fire-polishing pipettes to progressively smaller diameters.

8. Transfer the solution from conical tube #2 to conical tube #3 using a small-bore Pasteur pipette. Retriturate bits at the bottom of #2 if necessary.

9. Layer 5 ml of centrifugation solution underneath the triturated cells in conical tube #3.

10. Centrifuge to pellet cells for 10 min ($150 \times g$, at 20 °C).

11. Remove the supernatant and resuspend the cell pellet in plating medium (with pen/strep), and count cells with hemocytometer to plate at ~235,000 cells/cm^2.

12. Put plated cells in an incubator.

13. Approximately 5–7 h later suck off about half of the medium and replace (*see* **Note 5**).

14. The next morning (no more than 16 h post-plating), gently swirl the dishes, remove by suction 1/2–1/3 of the medium, and replace with fresh growth medium (without pen/strep) (*see* **Notes 6** and **7**).

3.2 IKK Substrate (GST-IκBα) Preparation

1. Transform 100 ng plasmid expressing recombinant-tagged GST-IκBα$_{(1–62)}$ [39] into 100 μl bacteria (BL21) (*see* **Note 8**).

2. Starter culture: inoculate 10 ml of LB plus ampicillin (final concentration 0.1 mg/ml) with transformed GST-IκBα$_{(1–62)}$ in BL21 bacteria by shaking at 30 °C overnight.

3. Add the starter culture into 250 ml of LB plus ampicillin and shake at 37 °C.

4. When the bacterial culture reaches $OD_{600} = 0.6–0.8$ (optical density at 600 nm using UV/visible light spectrophotometer), induce GST-IκBα$_{(1–62)}$ expression in the cultured bacteria with a final concentration of 1 mM IPTG by shaking at room temperature for 2 h.

5. Spin down the bacteria for 20 min at $4,000 \times g$ at 4 °C.

6. Resuspend the bacterial pellet with 12.5 ml of cold PBS with protease inhibitors.

7. Add lysozyme to a final concentration of 0.5 mg/ml and incubate at room temperature for 15 min.

8. Sonicate to reduce viscosity.

9. Add Triton X-100 to a final concentration of 1 %.

10. Spin at $10,000 \times g$ for 10 min at 4 °C.

11. To the supernatant, add 400 μl of a 50 % slurry of Glutathione Sepharose beads that have been pre-equilibrated in PBS; incubate with rotation at 4 °C for 3 h.

12. Spin down the tube from **step 11** at $1,000 \times g$ at room temperature for 1 min to collect the beads from the lid. Collect all beads and supernatant and load onto an elution column.

13. Wash the column with 8 ml of PBS.

14. Elute in 200 μl fractions with glutathione buffer. After adding each 200 μl aliquot of elution buffer, plug the bottom of the column, and let it sit for 10 min at room temperature. Collect 6–10 elution fractions.

15. Determine the protein concentration of each fraction by Bradford protein assay.

16. Aliquot each fraction separately and store at –80 °C; make small aliquots to avoid freeze-thaw degradation.

17. Run a sample of each fraction on a 12 % polyacrylamide mini SDS-PAGE gel, followed by Coomassie blue staining to confirm the size of the purified substrate that should migrate at 37 kDa.

3.3 IKK Complex Immunoprecipitation

1. Use days in vitro (DIV) 21 hippocampal dissociated neuronal culture plated in 24-well plates.

2. Stimulate the neurons with a $GABA_A R$ blocker, bicuculline, at a final concentration of 50 μM for 30 s (*see* **Notes 9** and **10**).

3. Stop the stimulation with 1× stop solution (using 100× kynurenate magnesium stock solution) by incubation at 37 °C for 10 min.

4. The stop solution is washed by removal of 50 % of the medium and replacement with fresh growth medium, and then incubate the neurons at 37 °C 3 h before harvesting.

5. Put the plates on ice and wash once with ice-cold PBS with added 0.9 mM $MgCl_2$ (*see* **Note 11**).

6. Lyse the cells with 100 μl of cold lysis buffer per well from a 24-well plate.

7. Incubate the cells on ice for 10 min, scrape the cells into a 1.5 ml tube, and leave for 10 more minutes on ice.

8. Spin the lysate at $15,600 \times g$ for 10 min at 4 °C and transfer the supernatant to a new 1.5 ml tube.

9. Add 2 μg of anti-IKKα antibody to at least 1,000 μg of lysate and rotate for 3 h at 4 °C (*see* **Note 12**).

10. Add 20 μl of prewashed protein A/G Sepharose (50 % slurry, protein A/G Sepharose should be prewashed three times in PBS with added protease inhibitors), and bind for 2 h rotating at 4 °C (*see* **Note 13**).

11. Wash three times with 1 ml of cold lysis buffer (rotate each wash at 4 °C for 5 min, then briefly spin down at $15,600 \times g$ in a 4 °C microfuge, and remove the supernatant before the next wash) and one time with 1 ml of cold kinase buffer. Use a 23 G bent needle attached to a vacuum line to remove the supernatant, being careful not to remove the beads.

12. Remove the kinase buffer fully with a needle (gauge 27) attached to a vacuum line after the last wash.

3.4 In Vitro Kinase Assay Followed by Immunoblot and PhosphorImaging

Use all standard precautions in conformity with radiation safety regulations. Use a lab coat designated for radiation use only. Always wear a radiation badge when handling radiolabeled phosphate. Monitor frequently for radioactive spillage:

1. To each tube from Subheading 3.3, add 30 ml of kinase reaction buffer.

2. Incubate for 30 min at 30 °C with gentle agitation.

3. Terminate the reaction with 10 ml of 5× SDS-PAGE sample buffer, and boil the samples for 2–3 min.

4. Run the samples and pre-stained molecular weight marker on 12 % polyacrylamide mini SDS-PAGE gel, and transfer to nitrocellulose or PVDF (run the gel at a constant 90 V and transfer overnight at 35 constant voltage, 4 °C).

5. Wrap the blot in Saran wrap and expose in a PhosphorImager cassette overnight or as required. Quantitate using a PhosphorImager.

6. Unwrap the blot (rehydrate if necessary), and block for 1 h in TBST with 5 % nonfat milk at room temperature.

7. Incubate the blot for 1 h in TBST with 5 % milk and 1:1,000 dilution of anti-IKKα antibody at room temperature.

8. Wash five times for 5 min each with excess of TBST.

9. Incubate the blot with the secondary anti-mouse HRP antibody in 10 ml of TBST with 5 % milk for 1 h at room temperature.

10. Wash five times for 5 minutes each with excess of TBST.

11. Place the blot on Saran wrap, develop by ECL, and expose the developed blot to film.

4   Notes

1. Basic fibroblast growth factor (bFGF) is optimally in the growth medium only at the time of plating and no longer than the first 3 days of culture, to avoid excessive glia proliferation.

2. Kynurenate may be difficult to dissolve. The solution is typically left in the refrigerator for several days, shaking it once or twice a day, or just warm at room temperature starting in the morning. Very small residual particles may be tolerated.

3. Dissection can take place on the lab bench provided the area is cleansed with 70 % ethanol, sterile bench techniques are practiced, and dissecting instruments are sterilized with UV.

4. The use of pups on the day of birth is optimal to ensure the survival of the greatest numbers of neurons. It is better to dissect the minimum number of hippocampi that yield the size of the culture required (generally from dissecting up to 15 pups), since a shorter dissection duration allows better survival of the dissociated neurons. If required, for the inhibition of glial proliferation, AraC (10 μM) may be included in the culture medium at DIV 3 for 48 h.

5. During changing culture medium, sufficient volume to cover the surface of the cells must always be maintained.

6. Plating in medium containing pen/strep may be used to reduce the possibility of bacterial infection, but cultures must be switched to *WITHOUT* pen/strep as soon as possible and no later than 18 h after plating to reduce the toxicity to neurons. To avoid contamination, always use sterile techniques when handling neuronal cultures, including keeping a separate set of pipette tips that is opened only in the TC hood.

7. The use of glial conditioned medium can be useful to improve neuronal viability, especially when the B27 lot is less than optimal or when it is desired to maintain a low-density culture (e.g., for immunohistochemistry).

8. Recombinant GST-IκBα$_{(1-62)}$ can be cloned into a bacterial expression vector: pGEX (GE Healthcare) by in-frame insertion of an amino-terminal fusion with GST to the first 62 amino acid residues of human IκBα).

9. High basal synaptic activity may dampen the apparent activation of NF-κB, particularly if the cultures are of high density. If necessary, basal synaptic activity may be inhibited prior to stimulation using several possible approaches: (a) reduction of B27 (to 50 %) in growth medium, up to 8 h prior to stimulation, with stimulation in normal growth medium, or (b) blocking NMDA receptors using 150 μM APV and/or voltage-gated L-type sodium channels using 10 μM nimodipine, up to 8 h

prior to stimulation, and then wash the neurons and stimulate in normal growth medium.

10. Inhibitory GABA networks may not be fully formed in cultures at $DIV \leq 18$ and may not allow sufficient induction of endogenous excitation using a $GABA_AR$ blocker. In this case, other stimuli are more suitable, including glutamate or KCl.

11. $MgCl_2$ is used to reduce neuronal activation during harvesting.

12. Anti-IKKγ (NEMO) can be used instead.

13. Use a pipette tip with the end cut to pipette the 50 % slurry. Mark the level of the 50 % slurry on the tube exterior with a marker, and wash beads in a large volume of cold PBS by inverting the tubes repeatedly. Spin down the tubes at $850 \times g$ at 4 °C, and then remove the supernatant to the marked level.

References

1. Bonizzi G, Karin M (2004) The two NF-kappaB activation pathways and their role in innate and adaptive immunity. Trends Immunol 25:280–288

2. Li Q, Verma IM (2002) NF-kappaB regulation in the immune system. Nat Rev Immunol 2:725–734

3. Hayden MS, Ghosh S (2011) NF-kappaB in immunobiology. Cell Res 21:223–244

4. Sen R, Baltimore D (1986) Multiple nuclear factors interact with the immunoglobulin enhancer sequences. Cell 46:705–716

5. Verhelst K, Carpentier I, Beyaert R (2011) Regulation of TNF-induced NF-kappaB activation by different cytoplasmic ubiquitination events. Cytokine Growth Factor Rev 22: 277–286

6. Courtois G, Gilmore TD (2006) Mutations in the NF-kappaB signaling pathway: implications for human disease. Oncogene 25:6831–6843

7. Karin M (2006) Nuclear factor-kappaB in cancer development and progression. Nature 441: 431–436

8. Lawrence T (2009) The nuclear factor NF-kappaB pathway in inflammation. Cold Spring Harb Perspect Biol 1:a001651

9. DiDonato JA, Mercurio F, Karin M (2012) NF-kappaB and the link between inflammation and cancer. Immunol Rev 246:379–400

10. Albensi BC, Mattson MP (2000) Evidence for the involvement of TNF and NF-kappaB in hippocampal synaptic plasticity. Synapse 35: 151–159

11. Burrone J, O'Byrne M, Murthy VN (2002) Multiple forms of synaptic plasticity triggered by selective suppression of activity in individual neurons. Nature 420:414–418

12. Christoffel DJ et al (2011) IkappaB kinase regulates social defeat stress-induced synaptic and behavioral plasticity. J Neurosci 31:314–321

13. Freudenthal R, Romano A (2000) Participation of Rel/NF-kappaB transcription factors in long-term memory in the crab Chasmagnathus. Brain Res 855:274–281

14. Freudenthal R, Romano A, Routtenberg A (2004) Transcription factor NF-kappaB activation after in vivo perforant path LTP in mouse hippocampus. Hippocampus 14:677–683

15. Kaltschmidt B et al (2006) NF-kappaB regulates spatial memory formation and synaptic plasticity through protein kinase A/CREB signaling. Mol Cell Biol 26:2936–2946

16. Meffert MK, Baltimore D (2005) Physiological functions for brain NF-kappaB. Trends Neurosci 28:37–43

17. Meffert MK et al (2003) NF-kappa B functions in synaptic signaling and behavior. Nat Neurosci 6:1072–1078

18. O'Mahony A et al (2006) NF-kappaB/Rel regulates inhibitory and excitatory neuronal function and synaptic plasticity. Mol Cell Biol 26:7283–7298

19. O'Riordan KJ et al (2006) Regulation of nuclear factor kappaB in the hippocampus by group I metabotropic glutamate receptors. J Neurosci 26:4870–4879

20. Russo SJ et al (2009) Nuclear factor kappa B signaling regulates neuronal morphology and cocaine reward. J Neurosci 29:3529–3537

21. Schmeisser MJ et al (2012) IkappaB kinase/ nuclear factor kappaB-dependent insulin-like growth factor 2 (Igf2) expression regulates synapse formation and spine maturation via Igf2 receptor signaling. J Neurosci 32:5688–5703

22. Boersma MC et al (2011) A requirement for nuclear factor-kappaB in developmental and plasticity-associated synaptogenesis. J Neurosci 31:5414–5425

23. Kaltschmidt C, Kaltschmidt B, Baeuerle PA (1993) Brain synapses contain inducible forms of the transcription factor NF-kappa B. Mech Dev 43:135–147

24. Guerrini L, Blasi F, Denis-Donini S (1995) Synaptic activation of NF-kappa B by glutamate in cerebellar granule neurons in vitro. Proc Natl Acad Sci U S A 92:9077–9081

25. Kaltschmidt C, Kaltschmidt B, Baeuerle PA (1995) Stimulation of ionotropic glutamate receptors activates transcription factor NF-kappa B in primary neurons. Proc Natl Acad Sci U S A 92:9618–9622

26. Wajant H, Scheurich P (2011) TNFR1-induced activation of the classical NF-kappaB pathway. FEBS J 278:862–876

27. Carter BD et al (1996) Selective activation of NF-kappa B by nerve growth factor through the neurotrophin receptor p75. Science 272:542–545

28. Yeiser EC et al (2004) Neurotrophin signaling through the p75 receptor is deficient in traf6-/- mice. J Neurosci 24:10521–10529

29. Vallabhapurapu S, Karin M (2009) Regulation and function of NF-kappaB transcription factors in the immune system. Annu Rev Immunol 27:693–733

30. Kaltschmidt C et al (1994) Constitutive NF-kappa B activity in neurons. Mol Cell Biol 14:3981–3992

31. Kanarek N, Ben-Neriah Y (2012) Regulation of NF-kappaB by ubiquitination and degradation of the IkappaBs. Immunol Rev 246:77–94

32. Lubin FD, Sweatt JD (2007) The IkappaB kinase regulates chromatin structure during reconsolidation of conditioned fear memories. Neuron 55:942–957

33. Yamamoto Y et al (2003) Histone H3 phosphorylation by IKK-alpha is critical for cytokine-induced gene expression. Nature 423:655–659

34. DiDonato JA et al (1997) A cytokine-responsive IkappaB kinase that activates the transcription factor NF-kappaB. Nature 388:548–554

35. Pomerantz JL, Denny EM, Baltimore D (2002) CARD11 mediates factor-specific activation of NF-kappaB by the T cell receptor complex. EMBO J 21:5184–5194

36. Zandi E et al (1997) The IkappaB kinase complex (IKK) contains two kinase subunits, IKKalpha and IKKbeta, necessary for IkappaB phosphorylation and NF-kappaB activation. Cell 91:243–252

37. Beaudoin GM 3rd et al (2012) Culturing pyramidal neurons from the early postnatal mouse hippocampus and cortex. Nat Protoc 7:1741–1754

38. Nunez J (2008) Primary culture of hippocampal neurons from p0 newborn rats. J Vis Exp 19:e895

39. Geleziunas R et al (1998) Human T-cell leukemia virus type 1 Tax induction of NF-kappaB involves activation of the IkappaB kinase alpha (IKKalpha) and IKKbeta cellular kinases. Mol Cell Biol 18:5157–5165

Chapter 6

Analysis of Epidermal Growth Factor-Induced NF-κB Signaling

Changying Jiang and Xin Lin

Abstract

The nuclear factor kappaB (NF-κB) is a family of transcription factors that control cell survival, cell proliferation, cell differentiation, inflammatory responses, and innate and adaptive immune responses. Its activation is tightly regulated, and incorrect regulation of NF-κB has been linked to a variety of pathological diseases, including cancer initiation and progression. NF-κB is often constitutively activated in cancer cells to promote cell survival, proliferation, migration, and/or epithelial-to-mesenchymal transition (EMT). Although the mechanism of constitutive NF-κB activation in cancer cells is not fully understood, it has been shown that mutation or aberrant expression of epidermal growth factor receptor (EGFR) contributes to this, and the NF-κB activation, in turn, contributes to cell proliferation, survival, metastasis, and drug resistance in various cancers. Recent study from our lab indicates that CARMA3, similar to the function of CARMA1 in mediating antigen receptor-mediated NF-κB activation, plays an essential role in mediating EGFR-induced NF-κB activation. However, the mechanism on how EGFR induces NF-κB activation is not clearly understood. In this chapter, we describe the methods required to test and characterize the role of a potential signaling component in EGFR-induced NF-κB activation.

Key words NF-κB, EGF, EGFR, IKK, IκBα, p65, CARMA3, BCL10, EMSA

---

1 Introduction

The nuclear factor kappaB (NF-κB) is a family of transcription factors that has been widely recognized and extensively studied for its importance in cell survival, cell proliferation, cell differentiation, inflammatory responses, and innate and adaptive immune responses. In mammalian cells, it contains five structurally conserved family members: RelA (p65), c-Rel, RelB, NF-κB1 (p50; p105), and NF-κB2 (p52; p100). NF-κB signaling is induced by a wide variety of receptors that include receptors for pro-inflammatory cytokines and antigen receptors (AgR) [1]. Upon receptor activation, many signaling intermediates are activated, which leads to the activation of the IκB kinase (IKK) complex [2]. Once activated, the IKK complex induces IκBα phosphorylation, ubiquitination, and degradation [2].

Michael J. May (ed.), *NF-kappa B: Methods and Protocols*, Methods in Molecular Biology, vol. 1280,
DOI 10.1007/978-1-4939-2422-6_6, © Springer Science+Business Media New York 2015

The degradation of IκBα releases NF-κB, which is translocated from the cytoplasm into the nucleus, where it facilitates transcription of its target genes.

NF-κB activation is tightly regulated by positive and negative regulators, and incorrect regulation of NF-κB has been linked to a variety of pathological diseases, including inflammatory and auto-immune diseases, improper responses to microbial infection, aging, and cancer initiation and progression [3–8]. NF-κB is often consti-tutively activated in cancer cells to induce the expression of various proteins involved in the proliferation, survival, migration, and epithelial-to-mesenchymal transition (EMT) of cancer cells and, thus, facilitate the cancer cells to fight against apoptotic stimuli and promote cell proliferation and metastasis [9–11]. Although the mechanism of constitutive NF-κB activation in cancer cells is not fully understood, it has been shown that mutation or aberrant expression of epidermal growth factor receptor (EGFR) contributes to this, and the NF-κB activation, in turn, contributes to cell prolif-eration, survival, metastasis, and drug resistance in various cancers [12–21]. The EGFR family contains four receptor tyrosine kinases designed ErbB1/HER1, ErbB2/HER2, ErbB3/HER3, and ErbB4/HER4. Mutations and aberrant expression of these recep-tors, especially EGFR (HER1) or HER2, are associated with the tumorigenesis of various tissues and drug resistance [22, 23]. EGF can activate EGFR homodimers or EGFR heterodimers with other HER family members. The stimulation of EGFR by EGF induces its intrinsic kinase activity and activates multiple downstream signal-ing cascades that lead to activation of many transcription factors, including NF-κB [24–26].

In efforts to find out the underlying mechanism(s), studies from our lab and others indicate that scaffold proteins CARMA3 and BCL10 play an essential role in EGFR and G protein-coupled receptor (GPCR)-induced NF-κB activation in PKC-dependent manner (Fig. 1) [27–32]. Upon receptor activation, CARMA3, BCL10, and MALT1 form a CBM complex, which leads to activa-tion of IKK and, thus, induces NF-κB activation [27]. CARMA3 and BCL10 are caspase-recruitment domain (CARD)-containing scaffold proteins, while MALT1 is a scaffold protein containing a protease activity. Although it is still not fully determined how CBM complex regulates IKK, CARMA3-dependent and Lys63-linked polyubiquitination of IKK complex may provide a positive signal to activate the IKK complex [33]. However, it remains to be deter-mined how the CBM complex is linked to EGFR signaling pathway and, in a broader view, how these receptors induced NF-κB activa-tion in the context of physiological and pathological conditions.

In order to study the mechanisms of EGFR-induced NF-κB signaling cascade, we describe, in this chapter, the methods required to study a potential molecule that may mediate NF-κB signaling

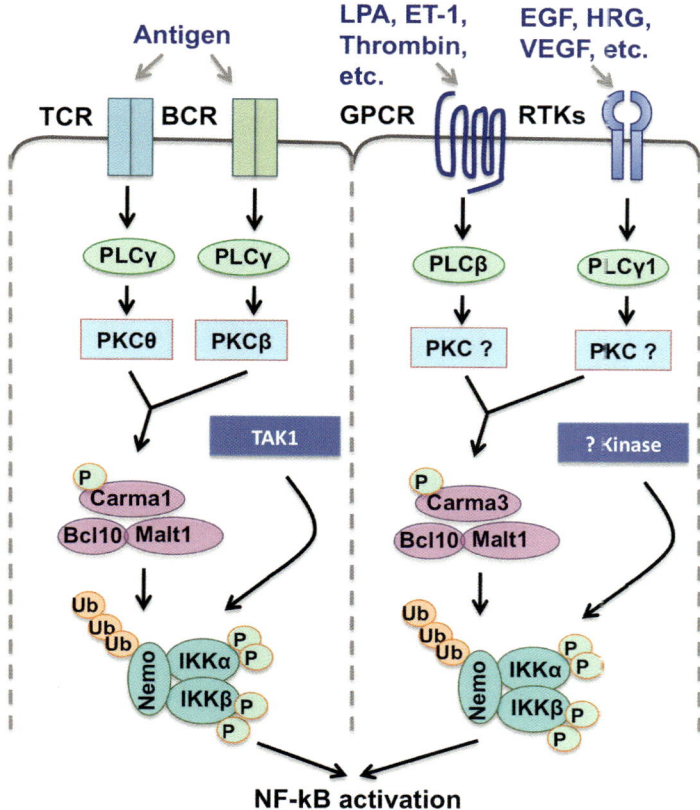

Fig. 1 NF-κB activation induced by CARMA family member-mediated signaling pathways. *TCR* T-cell receptor, *BCR* B-cell receptor, *GPCR* G protein-coupled receptor, *RTK* receptor tyrosine kinase

upon EGFR activation in EGFR overexpressing cancer cell lines. To cover the major known signaling events in EGFR-induced NF-κB activation, we describe how to confirm and characterize the role of a potential molecule in NF-κB signaling by the following methods: (1) Luciferase reporter assay for detecting NF-κB activation when the target protein is overexpressed in HEK293T cells, (2) shRNA transduction to knockdown target gene expression in EGFR overexpressing cancer cell lines, (3) immunoblotting for detecting major NF-κB signaling profiles including IKK phosphorylation and IκBα phosphorylation and degradation, (4) electrophoretic mobility shift assay (EMSA) for detecting DNA binding activity of nuclear NF-κB, (5) quantitative real-time PCR and the enzyme-linked immunosorbent assay (ELISA) for detecting target gene expression induced by NF-κB activation, and (6) immunoprecipitation for detecting the interaction between the target protein and the known NF-κB signaling components.

2 Materials

2.1 Cell Culture and Treatment

All chemicals and reagents are stored at room temperature unless otherwise stated:

1. General rules for cell culture and treatment: (a) all cell culture experiments must be performed in a microbiological safety cabinet using aseptic techniques to ensure sterility, (b) all media, reagents, and supplements must be sterile, and (c) all the equipments, accessories, and supplies must be cleaned by 70 % ethanol before bringing into the microbiological safety cabinet.

2. Dulbecco's Modified Eagle's Medium (DMEM) supplemented with 10 % (v/v) fetal bovine serum (FBS) and 1 % penicillin and streptomycin. Store at 4 °C (*see* **Note 1**).

3. 10× PBS: dissolve 80 g of NaCl, 2.0 g of KCl, 14.4 g of Na_2HPO_4, 2.4 g of KH_2PO_4; adjust pH to 7.4 and adjust volume to 1 l with ddH_2O; sterilize by autoclaving.

4. 1× PBS: 137 M NaCl, 2.7 mM KCl, 10 mM Na_2HPO_4, 1.8 mM K_2HPO_4, pH 7.4. Dilute 10× PBS to 1× PBS with ddH_2O; sterilize by autoclaving.

5. Solution of trypsin (0.05 or 0.25 %) and EDTA (1 mM) (*see* **Note 2**).

6. Serum-free medium: DMEM containing 0.5 % serum (*see* **Note 3**).

7. Human recombinant EGF stored at 1 mg/ml in 0.2 μm-filtered 10 mM acetic acid at −80 °C (*see* **Note 4**).

8. Human TNFα is stored at 10 μg/ml in 0.2 μm-filtered PBS containing 0.1 % (w/v) bovine serum albumin (BSA) at −30 °C.

2.2 Transfection by Calcium Precipitation

1. 2 M $CaCl_2$: dissolve 29.4 g of $CaCl_2$ in 100 ml of ddH_2O and sterilize with a 0.22 μm filter; store at −20 °C.

2. 2× HBSS: dissolve 32.73 g of NaCl, 1.49 g of KCl, 0.43 g Na_2HPO_4, 4.3 g of dextrose, 23.83 g of HEPES in 1.8 l of ddH_2O, and adjust pH to 7.05 and volume to 2 l. Filter with a 0.22 μm filter to sterilize, aliquot, and store at −20 °C (*see* **Note 5**).

3. DNA of interest.

2.3 Luciferase Reporter Assay

1. NF-κB Luciferase reporter plasmid: 5×κB-Luc. Store at −20 °C.

2. *Renilla* Luciferase reporter plasmid: R-Luc. Store at −20 °C.

3. 1× PBS.

4. Lysis buffer: 50 mM HEPES pH 7.4, 150–250 mM NaCl, 1 % NP-40, 1 mM EDTA. Store at 2–8 °C.

5. Lysis buffer with proteinase inhibitors: 50 mM HEPES pH 7.4, 150–250 mM NaCl, 1 % NP-40, 1 mM EDTA, 1 mM PMSF, 1 mM NaF, 1 mM Na_3VO_4, 1 mM DTT, protease inhibitor cocktail (*see* **Note 6**).

6. Dual Luciferase reporter assay kit containing Luciferase Assay Reagent II (LAR II) substrate for Firefly Luciferase, and Stop & Glo® reagent for *Renilla* Luciferase (Promega, Madison, WI, USA). Store at –80 °C.

7. Luminescence microplate reader.

2.4 shRNA Virus Packaging and Gene Knockdown

1. shRNA plasmid DNA and packaging DNA. Store at –20 °C.

2. 2× HBSS. Store at –20 °C.

3. 2 M $CaCl_2$. Store at –20 °C.

4. 50 mM chloroquine: dissolve 2.57 g of chloroquine in 100 ml of ddH_2O and sterilize with a 0.22 μm filter; store at –20 °C.

5. 8 mg/ml Polybrene: dissolve in ddH_2O and sterilize with a 0.22 μm filter; store at –20 °C.

6. 10 mg/ml puromycin or other antibiotics for stable selection. Store at –20 °C.

2.5 Cell Harvest and Sample Preparation for Western Blotting

1. Ice-cold PBS: stay on ice at least 30 min before use.

2. 1 M HEPES stock: dissolve 238.3 g of HEPES (free acid) in 800 ml of ddH_2O, adjust pH to 7.4 with 10 N NaOH, and add ddH_2O to final volume of 1 l.

3. 5 M NaCl: dissolve 292.2 g of NaCl in 800 ml of ddH_2O and add ddH_2O to final volume of 1 l.

4. 0.5 M EDTA, pH 8.0: dissolve 186.1 g of Na_2·EDTA·$2H_2O$ (disodium dehydrate) in about 800 ml of ddH_2O, and adjust pH 8.0 with 18 g NaOH first and then with 10 N NaOH. Add ddH_2O to a final volume of 1 l (*see* **Note 7**).

5. 100 % NP-40.

6. Lysis buffer without protease inhibitor: 50 mM HEPES pH 7.4, 150–250 mM NaCl, 1 % NP-40, 1 mM EDTA. Store at 2–8 °C.

7. Lysis buffer with protease inhibitor: 50 mM HEPES pH 7.4, 150–250 mM NaCl, 1 % NP-40, 1 mM EDTA, 1 mM PMSF, 1 mM NaF, 1 mM Na_3VO_4, 1 mM DTT, protease inhibitor cocktail.

8. 100 mM NaF (100× stock) in water. Store at 2–8 °C.

9. 100 mM Na_3VO_4 (100× stock) in water. Store at 2–8 °C.

10. 100 mM PMSF (100× stock) in 100 % ethanol. Store at 2–8 °C.

11. Complete protease inhibitor cocktail tablets stored at 2–8 °C.

12. Bio-Rad protein assay dye reagent concentrate (5×). Store at 2–8 °C.

13. 1 M Tris–HCl, pH 6.8: dissolve 121.1 g of Tris base in 800 ml of ddH$_2$O, adjust pH to 6.8 with HCl, and add ddH$_2$O to final volume of 1 l.

14. Loading dye (4×): 200 mM Tris–HCl (pH 6.8), 400 mM DTT, 8 % SDS, 0.4 % bromophenol blue, 40 % glycerol. Aliquot and store at –20 °C (*see* **Note 8**).

15. Cell scrapers.

2.6 SDS-PAGE Gel Casting

1. Preparation buffer (4×): 1.5 M Tris–Cl, pH 8.8 at RT. Dissolve 181.65 g Tris base in 800 ml of ddH$_2$O, adjust pH to 8.8 with HCl, and adjust the volume to 1 l with ddH$_2$O.

2. Stacking buffer (4×): 1.5 M Tris–Cl, pH 6.8 at RT. Dissolve 65 g Tris base in 700 ml of ddH$_2$O, adjust pH to 6.8 with HCl, and adjust the volume to 1 l with ddH$_2$O (*see* **Note 9**).

3. 30 % acrylamide/bis solution: store at 4 °C after opening (*see* **Note 10**).

4. 10 % SDS: dissolve 10 g of APS in ddH$_2$O, and adjust the volume to 100 ml.

5. TEMED.

6. 10 % APS: dissolve 10 g of APS in ddH$_2$O and adjust final volume to 100 ml. Store at 4 °C (*see* **Note 11**).

7. Water-saturated butanol solution.

8. 10× running buffer: dissolve 136.4 g of Tris base, 675.6 g of glycine, and 45 g of SDS in 4 l of ddH$_2$O, and adjust final volume to 4.5 l (*see* **Note 12**).

9. 1× running buffer: dilute 10× running buffer to 1× with ddH$_2$O.

10. Protein marker.

2.7 Western Blotting

1. 10× transfer buffer: dissolve 145 of Tris base and 670 g of glycine in 800 ml of ddH$_2$O and adjust the final volume to 4 l (*see* **Note 13**).

2. 1× transfer buffer (pre-cold): dilute 10× transfer buffer to 1× with ddH$_2$O and store at 4 °C (*see* **Note 14**).

3. Methanol (*see* **Note 15**).

4. TBST buffer: dissolve 48.4 of Tris base and 160 g of NaCl in 4 l of ddH$_2$O and adjust pH to 7.6. Add ddH$_2$O to final volume of 20 l. Add 20 ml of Tween 20 and mix well (*see* **Note 16**).

5. Primary antibodies: dilute primary antibodies in TBST containing 1 % BSA and 0.05 % sodium azide. Store at 4 °C (*see* **Note 17**).

6. Horseradish peroxidase (HRP)-labeled secondary antibodies: dilute secondary antibodies in 5 % nonfat dry milk or 5 % BSA in TBST buffer as desired.

7. Enhanced chemiluminescence (ECL) Western blotting substrate (*see* **Note 18**).

8. X-ray film or by digital imaging with a charge-coupled device (CCD) camera-based imager.

9. X-ray film exposure cassette.

2.8 Isolation of Cytoplasmic and Nuclear Factions

1. 1 M HEPES stock: dissolve 238.3 g of HEPES (free acid) in 800 ml of ddH_2O, adjust pH to 7.9 with 10 N NaOH, and add ddH_2O to final volume of 1 l.

2. 1 M KCl stock: dissolve 74 g of KCl in 800 ml of ddH_2O, and adjust the volume to 1 l.

3. EMSA lysis buffer without proteinase inhibitors: 10 mM HEPES pH 7.9, 10 mM KCl, 0.1 mM EDTA, and 0.4 % Nonidet P40. Store at 4 °C.

4. EMSA extraction buffer without proteinase inhibitors: 20 mM HEPES pH 7.9, 0.4 M NaCl, and 1 mM EDTA. Store at 4 °C.

5. Protease inhibitor cocktail: dissolve each tablet in 2 ml of H_2O. Aliquot and keep in –20 °C.

6. EMSA lysis or extraction buffer with proteinase inhibitors: for each 1 ml of lysis buffer or extraction buffer, add 1 μl of 1 M DTT, 5 μl of 100 mM PMSF, 10 μl of 100 mM NaVanadate, and 40 μl of protease inhibitor cocktail.

7. 1× PBS.

8. Cell scraper.

2.9 Probe Labeling for EMSA

1. Complementary consensus oligonucleotide probes for NF-κB and Oct-1. Store at –20 °C (*see* **Note 19**).

2. T4 polynucleotide kinase. Store at –20 °C.

3. 10× incubation buffer: 100 mM HEPES pH 7.9, 400 mM NaCl, 10 mM EDTA, 40 % glycerol. Store at –20 °C.

4. ^{32}P-γ-labeled ATP (10 μCi/μl). Store at 4 °C (*see* **Note 20**).

5. Micro Bio-Spin® 6 chromatography columns. Store at 4 °C.

6. Radioactivity measurement device.

2.10 EMSA Gel Casting

1. 5× TBE buffer (1 l): dissolve 54 g of Tris base and 27.5 g of boric acid in 800 ml of ddH_2O and add 20 ml of 0.5 M EDTA (pH 8.0). Add ddH_2O to final volume of 1 l.

2. 30 % acrylamide/bis solution.

3. TEMED.

4. 10 % APS: dissolve 10 g of APS in ddH_2O and adjust final volume to 100 ml. Store at 4 °C.

2.11 DNA Binding Reaction

1. Poly-d(I–C): dissolve lyophilized poly-d(I–C) (1 mg/ml) in ddH$_2$O. Aliquot ~300 μl/tube. Store at –20 °C.

2. 10× incubation buffer: 100 mM HEPES pH 7.9, 400 mM NaCl, 10 mM EDTA, and 40 % glycerol.

3. 1 M DTT: dissolve 1.55 g of APS in ddH$_2$O and adjust final volume to 100 ml. Store at –20 °C.

4. 10 mM DTT: dilute 1 M DTT to 10 mM with ddH$_2$O. Store at –20 °C.

5. Loading buffer: 50 % glycerol, 0.25 % bromophenol blue, and 0.25 % xylene cyanol.

2.12 EMSA Gel Running

1. EMSA gel running cassette and power supply.

2. ¼× TBE buffer: dilute 5× TBE buffer to ¼× with ddH$_2$O.

3. Long loading tips.

4. Gel drying machine.

5. Whatman filter paper.

6. X-ray film exposure cassette.

2.13 Total RNA Isolation

1. TRIzol: store at 4 °C.

2. Chloroform.

3. 2-Propanol.

4. 70 % ethanol.

5. DEPC-treated water.

2.14 Reverse Transcription

1. RT III reverse transcriptase. Store at –20 °C.

2. Oligo-dT (50 μM) primer or random primers (50 μM). Store at –20 °C.

3. RNase-free water.

4. 5× first-strand buffer. Store at –20 °C.

5. 25 mM dNTP. Store at –20 °C.

6. 100 mM DTT. Store at –20 °C.

7. RNaseOUT ribonuclease inhibitor (Life Technologies, Grand Island, NY, USA). Store at –20 °C.

8. Thermocycler.

2.15 Real-Time PCR

1. Power SYBR® Green PCR Master Mix (Life Technologies). Store at –20 °C.

2. Primers designed according to need and including internal controls, such as GAPDH and HPRT1. Store the primers at –20 °C.

3. cDNA. Store at –20 °C.

4. Real-time thermocycler.

2.16 ELISA

1. ELISA plates.
2. Coating buffer: 1× PBS.
3. 5× assay diluent: 1× PBS containing 5 % BSA. Store at 4 °C.
4. Coating antibodies against target protein such as cytokines. Store at 4 °C.
5. Washing buffer: 1× PBS containing 0.05 % Tween 20.
6. Blocking buffer: 1× assay diluent. Store at 4 °C.
7. Standard for target cytokines. Store at −80 °C.
8. Detection antibodies. Store at 4 °C.
9. TMB reagent (Dako, Carpinteria, CA, USA). Store at 4 °C.
10. Stop solution: mix 8.3 ml of HCl with 91.7 ml of ddH$_2$O.
11. Microplate reader.

2.17 Immunopre-cipitation

1. 1× PBS.
2. Antibodies for immunoprecipitating target proteins and potential interacting proteins (e.g., components of the NF-κB signaling pathways).
3. Protein A or G agarose beads for binding antibodies.
4. Lysis buffer: 50 mM HEPES pH 7.4, 150–250 mM NaCl, 1 % NP-40, 1 mM EDTA.
5. Lysis buffer with proteinase inhibitors: 50 mM HEPES pH 7.4, 150–250 mM NaCl, 1 % NP-40, 1 mM EDTA, 1 mM PMSF, 1 mM NaF, 1 mM Na$_3$VO$_4$, 1 mM DTT, protease inhibitor cocktail.
6. RIPA buffer: 10 mM Tris–Cl (pH 8.0), 1 mM EDTA, 0.5 mM EGTA, 1 % Triton X-100, 0.1 % sodium deoxycholate, 0.1 % SDS, and 140 mM NaCl.
7. 2× loading dye: dilute 4× loading dye to 2× with ddH$_2$O.

3 Methods

3.1 Cell Culture and Treatment

1. Maintain the cells in complete medium (DMEM containing 10 % FBS and 1 % penicillin and streptomycin) in a T75 flask by splitting them 1:4 every other day (*see* **Note 1**).
2. To split the cells, add prewarmed (37 °C) trypsin/EDTA solution to the cells and incubate at 37 °C for a few minutes until complete detachment is verified under microscopy (*Note*: do not let the digestion go overtime. Over-digestion with trypsin may result in poor cell survival.) (*see* **Note 2**).
3. Add an equal volume of complete medium to neutralize the trypsin, followed by spin at 500 × g for 3 min.

4. Remove the supernatant, and resuspend the cells in complete medium.

5. Count the cell concentration and seed $0.5–1 \times 10^6$ cells per well of a six-well plate to obtain 60–90 % confluence one next day.

6. For serum starvation, replace the culture medium with pre-warmed DMEM containing 0.5 % FBS.

7. 16–18 h after serum starvation, stimulate the cells with EGF (100 ng/ml) or TNFα (10 ng/ml) at appropriate concentration for 15, 30, 45, or 60 min.

3.2 Transfection (Fig. 2)

1. Seed 0.3×10^6 of HEK293T cells in a 12-well plate 1 day before transfection.

2. Warm up all the reagents to the room temperature.

3. Dilute 1.6 μg of DNA (*Note*: for luciferase assay, 0.6 μl of 5×κB-Luc (0.1 μg/μl) and 0.6 μl of R-Luc (0.01 μg/μl) should be added for each sample) in water to a final volume of 35 μl.

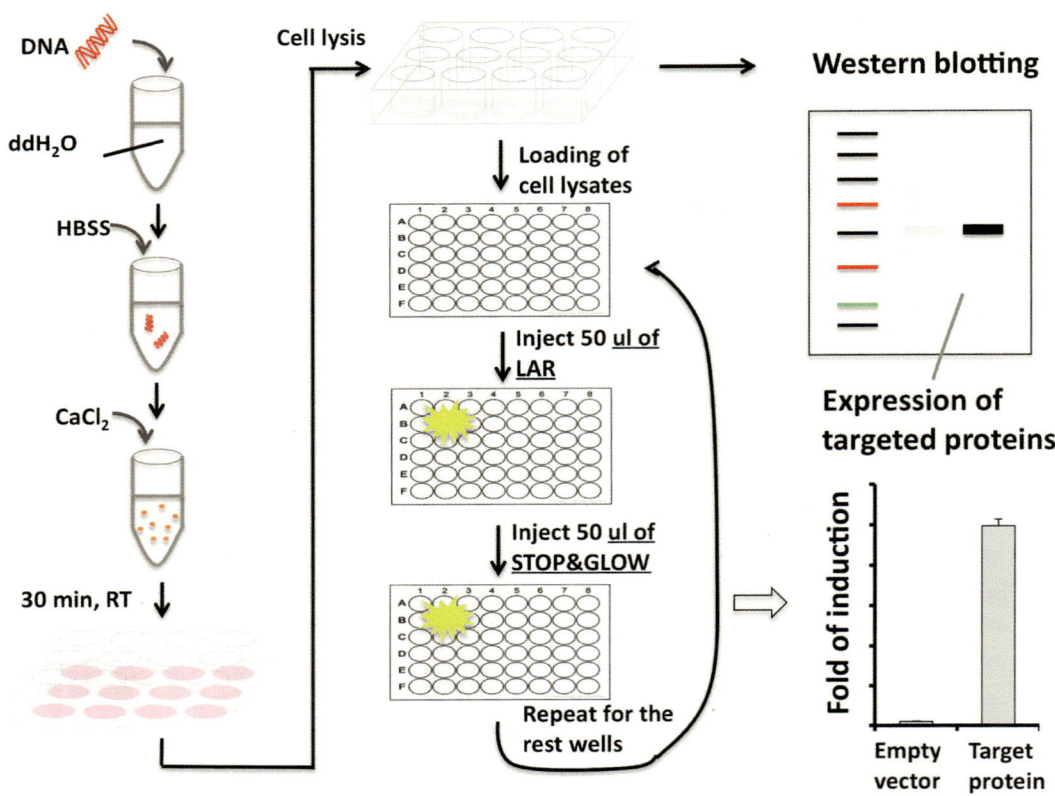

Fig. 2 Flow chart for transfection and luciferase assay. The CaCl₂ method of transfection (*left*) and the luciferase assay (*middle*) are shown along with depictions of results from a Western blot of the expressed protein and increased luciferase activity (*right*)

Table 1
Recommended amounts of materials for transfection of cells grown on separate plate/dish sizes

	Cell/DMEM (million/ml)	DNA/H$_2$0 (µg/µl)	2× HBSS (µl)	2 M CaCl$_2$ (µl)
12-well plate	0.3/0.8	1.6/35	40	5
6-well plate	0.75/2	4/87.5	100	12.5
100 mm dish	4/10	20/437.5	500	62.5

4. Add 40 µl of 2× HBSS to the diluted DNA, and vortex briefly to mix. Add 5 µl of 2 M CaCl$_2$ drop by drop and mix gently by swirling or pipetting the pipette tip.

5. If other sizes of plates or dishes are used, follow Table 1.

6. Incubate the mixture at room temperature for 30 min (*see* **Note 21**).

7. Add the mixture slowly and evenly to the cells (drop by drop). Mix gently.

8. Cells may be harvested 18–24 h after transfection for luciferase assay, or longer, up to 96 h, if desired.

3.3 Luciferase Assay
(Fig. 2)

1. Transfect 60 ng of 5×κB-Luc, 6 ng of R-Luc, and 0.1–1.5 µg of target DNA of interest in triplicates into HEK293T cells seeded in 12-well plates.

2. Harvest the transfected cells (in triplicate) 18–24 h after transfection.

3. Lyse the cells in 60–100 µl of lysis buffer with protease inhibitors on ice for 10–15 min.

4. Spin at 15,000 ×g for 10 min at 4 °C.

5. Transfer the cleared lysates to new tubes and keep them on ice.

6. Add 10 µl of each lysate to the wells of a 96-well assay plate and read in the luminescence microplate reader with automatic injectors using a Dual Luciferase Kit. Alternatively, read the luciferase activity by manual addition of 50 µl of LAR II reagent and measurement of firefly luciferase activity, followed by manual addition of 50 µl of Stop & Glo® reagent and measurement of *Renilla* Luciferase activity. Repeat with the remaining samples.

7 Fold of NF-κB activation can be calculated by normalizing the values of NF-κB-induced firefly luciferase activity to that of EF1-incued *Renilla* Luciferase activity, which serves as internal control, and comparison to vector DNA transfected control.

3.4 shRNA Virus Packaging and Gene Knockdown

1. Transfect HEK293T cells with shRNA and packaging DNA as described in Subheading 3.3. To improve transfection efficiency, add chloroquine to the cells to the final concentration at 25 μM, 5 min prior to transfection (*see* **Note 22**).

2. Replace the medium with fresh complete medium 1-day post transfection.

3. Harvest the virus supernatant once a day on day 3–5.

4. Spin the virus supernatant at $2,000 \times g$ for 5 min to remove cell debris and filter the virus supernatant with a 0.45 μm filter (*see* **Note 23**).

5. Infect your cells of interest with the filtered virus supernatant. To increase infection efficiency, add Polybrene to the cells to the final concentration at 2–10 μg/ml (*see* **Note 24**).

6. Replace the medium with fresh complete medium on day 2.

7. Add appropriate antibiotics to the infected cells on day 3. Allow time for the antibiotics to kill uninfected cells (*see* **Note 25**).

8. To test or confirm knockdown efficiency, perform Western blotting (Fig. 3) or quantitative real-time PCR, if no antibody is available.

3.5 Sample Preparation for Western Blotting (Fig. 4)

1. Place the treated cells on ice, aspirate the medium, and wash the cells (12-well plate) twice with ice-cold PBS.

2. Harvest the cells by scraping and transfer the cells into 1.5 ml tubes.

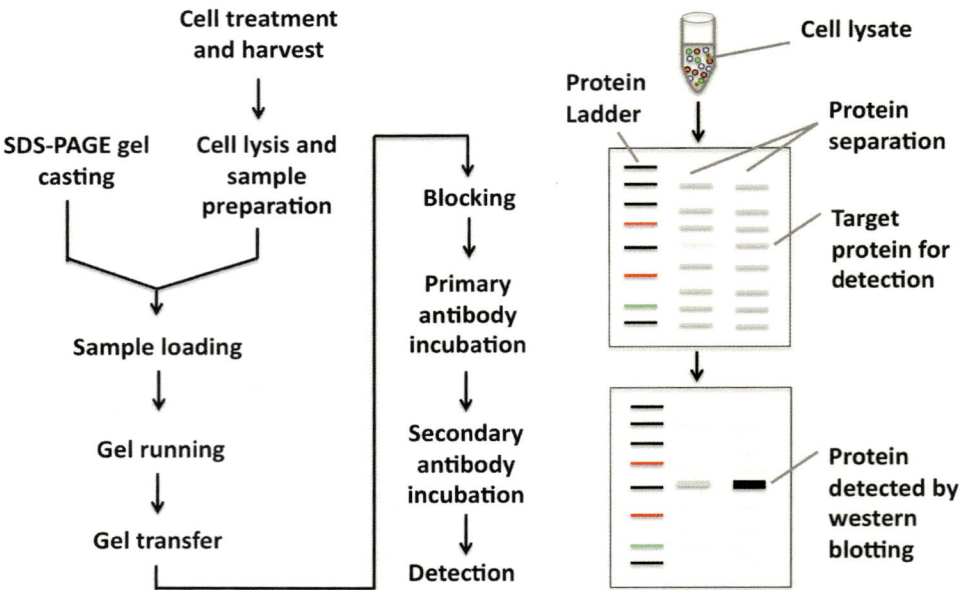

Fig. 3 Flow chart for Western blotting analysis. Depiction of the separated proteins pre- and post blotting is shown on the *right*

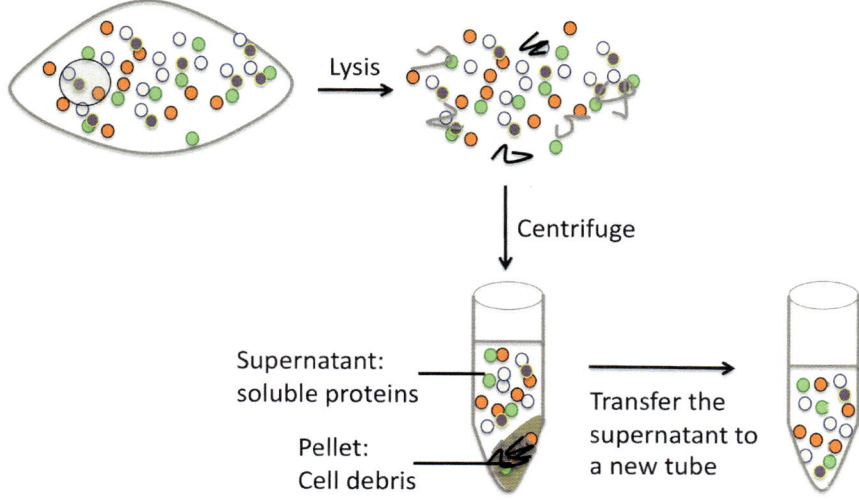

Fig. 4 Sample preparation for Western blotting analysis. Following lysis, centrifugation removes cell debris and DNA. The supernatant is transferred to a new tube and used for analysis

3. Pellet the cells by spinning at $500 \times g$ for 3 min in a refrigerated centrifuge at 4 °C.

4. Aspirate the supernatant and resuspend the cell pellets in 60–100 µl of lysis buffer containing protease inhibitors (*see* **Note 26**).

5. Incubate the lysates on ice for 15 min.

6. Centrifuge at $15,000 \times g$ for 10 min at 4 °C.

7. Transfer the supernatants to new 1.5 ml tubes and keep them on ice.

3.6   Measurement of Protein Concentration

1. Dilute 5× Bio-Rad protein assay dye reagent concentrate to 1× with ddH$_2$O.

2. Aliquot 100 µl to each well of a 96-well assay plate.

3. Set up a protein standard curve by adding 0, 1, 2, 3, 4, and 5 µl of BSA (0.5 µg/µl) to each well in duplicates.

4. Add 0.5–2 µl of the protein lysates above to each, dependent on the protein concentration of the protein lysates (*see* **Note 27**).

5. Mix well and read the plate using a microplate reader.

6. Calculate the protein concentration manually or by appropriate software.

7. Adjust the samples to the same concentration using the lysis buffer.

8. Add 4× loading dye to the samples to the final concentration (1×). Boil the samples at 95 °C for 5–10 min.

9. Spin at $15,000 \times g$ for 1 min. The samples are now ready to be loaded.

3.7 SDS-PAGE Gel Casting (Fig. 5)

1. Clean both sides of long glass plate and short glass plate with water and then 70 % ethanol. Air-dry.

2. Assemble the glass plates in gel-casting cassette.

3. Prepare the separation gel (10 %) by mixing 6.1 ml of water, 5.0 ml of 30 % acrylamide/bis solution, 3.75 ml of 4× separation buffer (pH 8.8), 150 μl of 10 % SDS, 50 μl of 10 % APS, and 10 μl of TEMED.

4. Pour the mixture to the gel cassette until the surface reaches the level about 1.5 cm below the top of the short glass (*see* **Note 28**).

5. Add 100 μl of water-saturated butanol on top of the gel solution.

6. Allow the gel to polymerize at RT for at least 30 min (*see* **Note 29**).

7. Pour off the butanol and rinse twice with water.

8. Prepare the stacking gel (4 %) by mixing 3.6 ml of water, 0.78 ml of 30 % acrylamide/bis solution, 1.5 ml of 4× separation buffer (pH 6.8), 60 μl of 10 % SDS, 30 μl of 10 % APS, and 6 μl of TEMED.

9. Layer the mixture on the top of the separation gel and insert a 15-well comb.

10. Allow the gel to polymerize at room temperature for at least 30 min.

11. Remove the comb and rinse the wells with water (*see* **Note 30**).

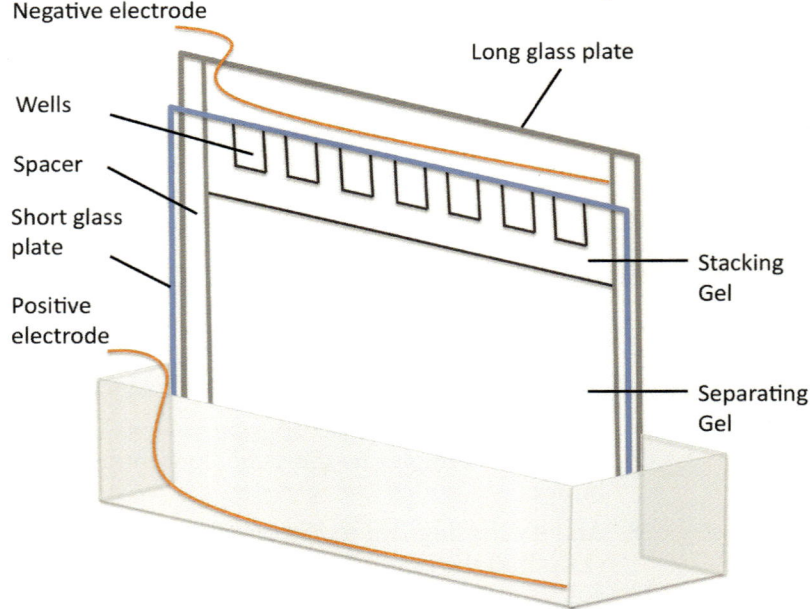

Fig. 5 Western blot gel casting and running. Gels for Western blotting are assembled as shown

3.8 SDS-PAGE Gel Running

1. Assemble the gel running cassette in the tank.
2. Fill the inner and outer chambers with 1× gel running buffer.
3. Load equal amounts of the samples to individual gel wells along with a protein molecular weight standard marker.
4. Connect the gel running cassette to a power supply and run the gel at constant voltage of 90–160 V, RT until the blue dye runs off the bottom of the gel.

3.9 SDS-PAGE Gel Transfer (Fig. 6)

1. Pour the cold transfer buffer in a tray.
2. Wet the transfer cassette, two sponges, two pieces of Whatman filter paper, and one nitrocellulose membrane in the tray.
3. Disassemble the gel from the gel running cassette and cut off the stacking gel.
4. Assemble the gel transfer cassette in the following order: the negative electrode, sponge, filter paper, gel, membrane, filter paper, sponge, and the positive electrode (*see* **Note 31**).
5. Put the cassette to a tank.
6. Fill it with transfer buffer, and insert an ice pack in the tank.
7. Connect to a power supply and run at constant voltage of 90 V for 1 h at 4 °C (*see* **Note 32**).

3.10 Western Blotting (Fig. 3)

1. Assemble the gel transfer cassette and rinse the membrane twice with TBST buffer.
2. Rock the membrane in blocking buffer at RT for 30–60 min.

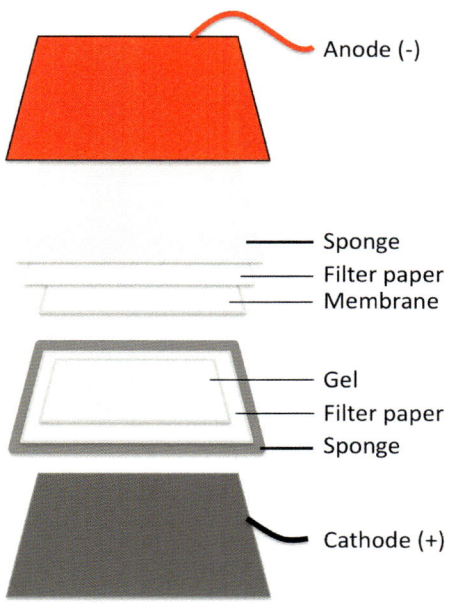

Fig. 6 Wet transfer setup for Western blot gel. Gels and transfer apparatus for Western blotting are assembled as shown

3. Incubate the membrane with primary antibody diluted in TBST containing 1 % BSA at 4 °C overnight on a rocker (*see* **Note 33**).

4. Wash the membrane four times with TBST buffer for 10 min per wash on a rocker at RT.

5. Incubate the membrane with secondary antibody diluted in TBST containing 5 % BSA or milk on a rocker at RT.

6. Wash the membrane four times with TBST buffer.

7. Mix 1 ml of each enhanced chemiluminescence (ECL) reagents for one mini-gel membrane.

8. Pick up the membrane with forceps and place it on a clean flat surface with face up.

9. Pipette the ECL mix on the membrane and incubate at RT for 1 min.

10. Place the membrane in between two plastic sheets with face up in an X-ray film cassette.

11. In a dark room, put one piece of film over the membrane to expose. If necessary, repeat the exposure a few times to obtain optimal results.

3.11   Isolation of Nuclear Extract (Fig. 7)

1. Seed and treat cells as in Subheadings 3.1 and 3.2 (*see* **Note 34**).

2. Place the treated cells on ice, aspirate medium, and wash the cells twice with ice-cold PBS.

3. Harvest the cells by scraping and transfer the cells into 15 ml tubes.

4. Pellet the cells by spinning at $500 \times g$ for 3 min in a refrigerated centrifuge at 4 °C.

5. Aspirate the supernatant and resuspend the cell pellets in 200–400 µl of ice-cold lysis buffer with inhibitors by pipetting up and down without bubbling about ten times.

6. Incubate on ice for 10–15 min.

7. Spin the lysate at $15,000 \times g$ for 1 min at 4 °C. The pellet contains nuclei and should look white and "fluffy" (*see* **Note 35**).

8. Wash the nuclear pellet in 500 µl of lysis buffer without inhibitors. Spin at $15,000 \times g$ for 1 min at 4 °C (*see* **Note 36**).

9. Add 100 µl (or less) of extraction buffer with inhibitors to the pellet and shake vigorously at 4 °C for 15 min on an Eppendorf shaker or vortex at low-medium speed setting (*see* **Note 37**).

10. Spin for 10 min at $15,000 \times g$ for 10 min at 4 °C and transfer the supernatant to a fresh tube and flash freeze aliquots in liquid nitrogen. Store at –70 °C.

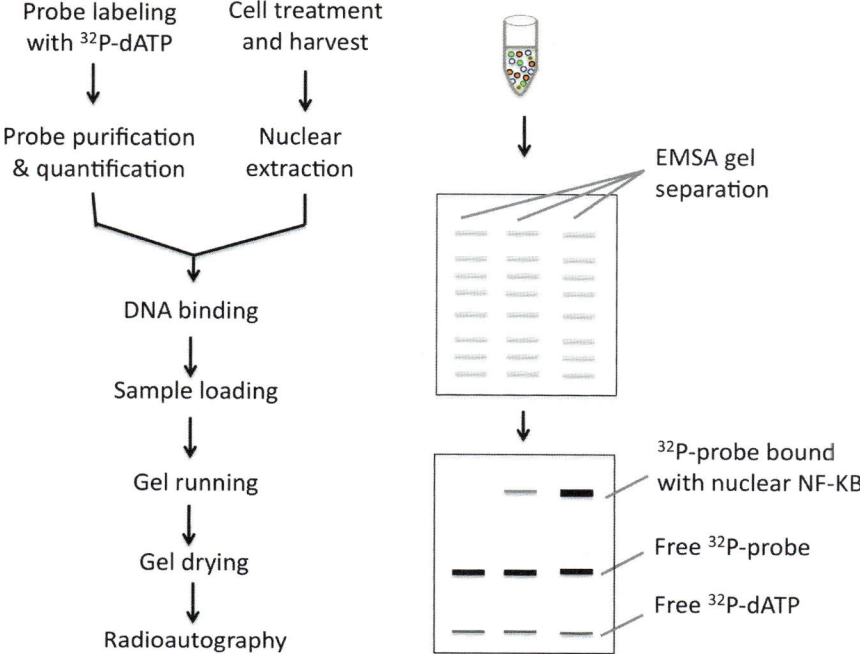

Fig. 7 Flow chart for EMSA. Depiction of the probe bound with proteins, the free probe, and free ^{32}P-dATP is shown on the *right*

3.12 EMSA Probe Labeling (Fig. 7)

1. The probes may be prepared in advance and stored at −20 °C in a radiation-safe box.

2. Mix in the following order: 12 μl of water, 2.5 μl of double-stranded oligonucleotide, 2 μl of 10× incubation buffer, 2.5 μl of ^{32}P-γATP (10 μCi/μl), and 1 μl of T4 polynucleotide kinase.

3. Incubate the mixture at 37 °C for 30 min and purify the probe using a Bio-Rad Spin 6 column.

4. Resuspend the gel in the column and remove any bubbles by inverting the column.

5. Snap off the tip and place the column in a 2 ml tube, remove the cap, and let it drain at room temperature.

6. Discard the drained buffer and remove the remaining packing buffer by spinning at $1,000 \times g$ for 2 min at room temperature.

7. Transfer the column to a new collection tube.

8. Add 25 μl of H_2O to the labeled product mix and transfer it (total 50 μl) to the center of the column.

9. Spin at $1,000 \times g$ for 4 min at room temperature to collect the purified probe.

10. Radioactivity levels of the probe can be assessed using a Geiger counter. The purified probe should be good to use if its activity is not lower than the activity left in the column. Count 1 μl with scintillation fluid. You can get up to $0.2–1 \times 10^6$ cpm/μl activity. A minimum of 2×10^5 cpm/μl should be reached for performance in EMSA analysis.

3.13 EMSA Gel Casting (Fig. 8)

1. Clean both sides of glass plates with water and then 70 % ethanol. Air-dry.

2. Assemble the glass plates with spacer on left and right sides; seal the left, right, and bottom sides with rubber sealer; and hold them together by clips on left and right sides.

3. Prepare the EMSA gel solution by mixing 35 ml of water, 7.5 ml of 30 % acrylamide, 2.25 ml of 5× TBE, 0.35 ml of 10 % APS, and 0.03 ml TEMED (*see* **Note 38**).

4. Pour the mixture into the gel chamber and insert the 20-well comb into the gel.

5. Allow the gel to polymerize at room temperature for 1 h.

6. Set up the gel in the gel running apparatus and fill upper and lower chambers with ¼× TBE buffer.

3.14 DNA Binding of NF-κB and EMSA Gel Running (Fig. 7)

1. Mix 2 μl of 2× incubation buffer, 1 μl of poly-d(I + C) (1.5 mg/ml), 1 μl of DTT (10 mM), and 1 μl of probe (1×10^5 cpm/μl) for a final volume of 5 μl per sample (*see* **Notes 39** and **40**).

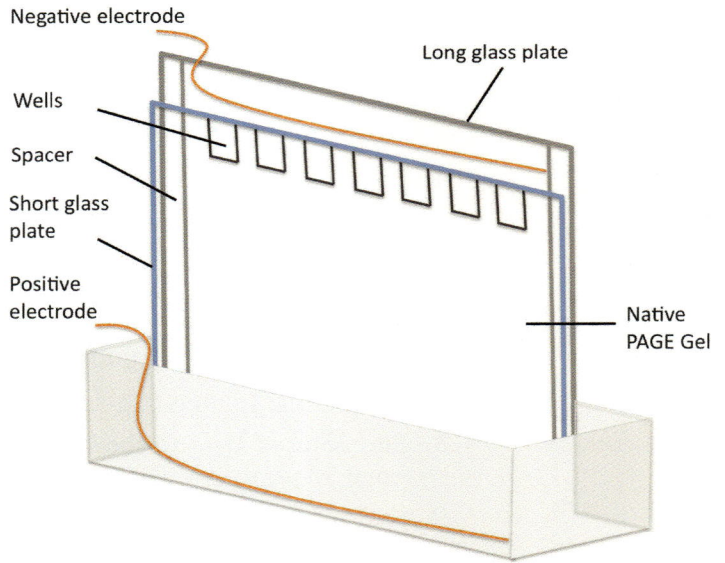

Fig. 8 EMSA gel casting and running. Native PAGE gels for EMSA are assembled as shown

2. Place nuclear extracts on ice before use. Measure protein concentration for each nuclear extract.

3. Mix in the following order: $(15 - x)\,\mu l$ of water, x μl ($x =$ the amount of sample required for 10 μg) of nuclear extract, and 5 μl of the above cocktail.

4. Incubate at room temperature for 15 min.

5. During the incubation, add 2 μl of 10× loading dye on the cap.

6. After incubation, spin down loading dye (quick spin for 3 s) and mix by tapping gently (*see* **Note 41**).

7. Load samples into the EMSA gel.

8. Run on gel at 220 V with constant voltage for no more than 85 min to avoid ^{32}P contamination in the running buffer.

9. Disassemble the gel and blot the gel to one piece of 3 mm Whatman paper.

10. Place the gel to the gel-drying machine. When you dry, put another sheet of Whatman paper on the bottom to avoid ^{32}P contamination in the gel dryer.

11. Dry the gel for 15–30 min at 80 °C with vacuum.

12. Expose to film at –80 °C for 15 min to 2 h.

3.15 Supershift of NF-κB

1. Modify the incubation slightly to preincubate nuclear extract samples with desired antibodies (usually approx. 200 ng Ab/sample) for 30–60 min on ice prior to the EMSA reaction.

2. Add water and MIX and incubate at room temperature for 15 min and proceed as for regular samples (Subheading 3.14).

3.16 Total RNA Isolation (Fig. 9)

1. Seed the cells in a 6-well plate, and treat them if desired.

2. RNA is extremely sensitive to naturally produced RNase found on skin or in bacteria, so gloves should be worn for all procedures.

3. Aspirate the medium and add 1 ml of TRIzol to lyse the cells. Scale up if large amount of cells are needed.

4. Transfer the homogenates to 1.5 ml RNase-free tubes and incubate at RT for 5 min.

5. Add 200 μl of chloroform, shake vigorously (not vortex) for 15 s to mix thoroughly, and incubate at RT for 2–3 min.

6. Centrifuge at $12,000 \times g$ at 4 °C for 15 min.

7. Transfer the upper clear aqueous layer to a fresh 1.5 ml RNase-free tube.

8. Add 0.5 ml of 2-propanol, shake for 15 s to mix well, and incubate at RT for 10 min.

9. Centrifuge samples at $12,000 \times g$ at 4 °C for 10 min to pellet RNA. RNA pellet should appear white fluffy.

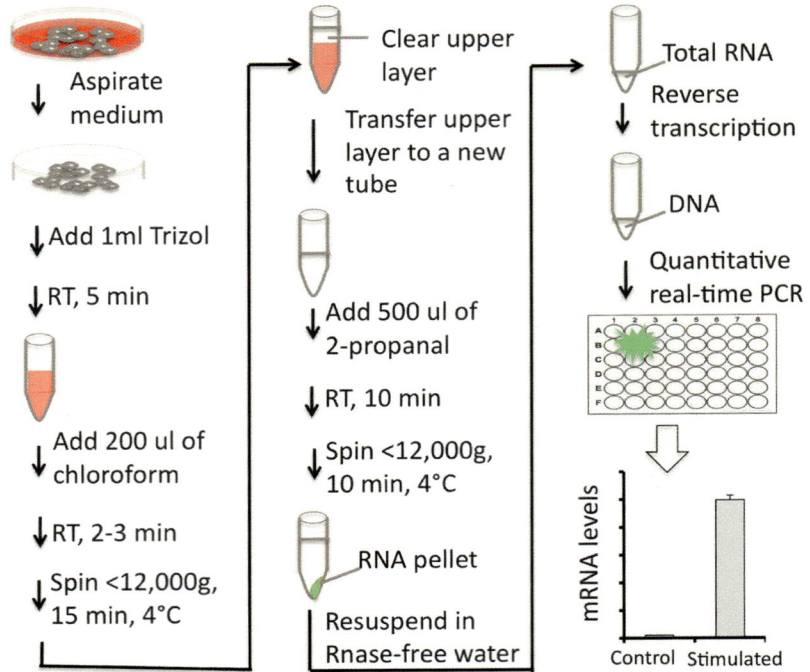

Fig. 9 RNA isolation and quantitation by qPCR. The steps as described in Subheading 3 are shown in the flow chart and a depiction of results of target gene expression from unstimulated and stimulated cells is shown

10. Carefully remove the supernatant, add 1 ml of 75 % DEPC-ethanol, and vortex on low for 5–10 s to wash the pellet.

11. Centrifuge at $7,500 \times g$ at 4 °C for 5 min to pellet RNA again.

12. Carefully remove the supernatant. Air-dry the pellet at RT for 5–10 min (*see* **Note 42**).

13. Add 30–100 μl of RNase-free water, gently pipette to dissolve RNA, and incubate at 55 °C for 5–10 min.

14. Measure RNA concentration using NanoDrop.

3.17 Reverse Transcription (Fig. 9)

1. The following is adapted from the protocol provided with the Invitrogen SuperScript III Reverse Transcriptase enzyme.

2. Mix 1 μl of oligo-dT (50 μM in RNase-free water), 10 pg to 5 μg of total RNA (best to use 1 μg), and adjust the volume to 12.5 μl with RNase-free water in a PCR tube (*see* **Note 43**).

3. Heat at 65 °C for 5 min in a PCR machine; quick chill on ice (at least 1 min).

4. Add 4 μl of 5× first-strand buffer, 0.5 μl of 25 mM dNTP, 1 μl of 0.1 M DTT, 1 μl of RNaseOUT, and 1 μl of SuperScript III RT in order to the RNA-oligo mix.

5. Mix well by gently pipetting up and down, and quick spin. Incubate at 50 °C for 30–60 min, and inactivate the reaction by incubating at 70 °C for 15 min.

6. Use the cDNA immediately for PCR, or store at –20 °C (*see* **Note 44**).

3.18 Quantitative Real-Time PCR (Fig. 9)

1. For Q-RT PCR, we use Power SYBR® Green PCR Master Mix.

2. Design real-time primers and/or probes appropriately.

3. Mix 10 μl of SYBR green mix, 0.45 μl of each primer (20 μM), and 1 μl of cDNA and 8.1 μl of water in a real-time PCR plate, and seal it with a sheet of real-time plate sealer.

4. Quick spin the plate and run the samples by the following protocol: one cycle of 94 °C, 15 s, and 40 cycles of 94 °C, 15 s, and 60 °C, 60 s.

5. Calculate expression of target genes by normalizing to the internal controls, such as GAPDH and HPRT1.

3.19 ELISA (Fig. 10)

1. Seed and treat cells as in Subheadings 3.1 and 3.2 (*see* **Note 45**).

2. Harvest the culture supernatant after cell treatment. Spin it at $500 \times g$ for 5 min to remove the cells/cell debris. The supernatant can be stored at –80 °C

3. Coat an ELISA plate with 100 μl of capture antibodies against the target proteins (cytokines) diluted in coating buffer at RT for 2 h or 4 °C overnight.

4. Wash the plate with 200 μl of washing buffer per well at RT for four times.

5. Dilute the standard(s) in a series within the sensitivity range with 1× assay diluent. Add 100 μl of the diluted standards in duplicates or samples to each well of the coated plate.

6. Incubate at RT for 2 h.

7. Wash the plate with 200 μl of washing buffer per well at RT for four times.

8. Block the plate with blocking buffer at RT for 1 h.

9. Wash the plate with 200 μl of washing buffer per well at RT for four times.

10. Add 100 μl of detecting antibody diluted in 1× assay diluent per well and incubate at RT for 1 h.

11. Wash the plate with 200 μl of washing buffer per well at RT for four times.

12. Add 100 μl of substrate solution to each well and incubate at RT for 5–15 min.

13. Add 50 μl of stop solution to stop the reaction once the desired signal strength is reached.

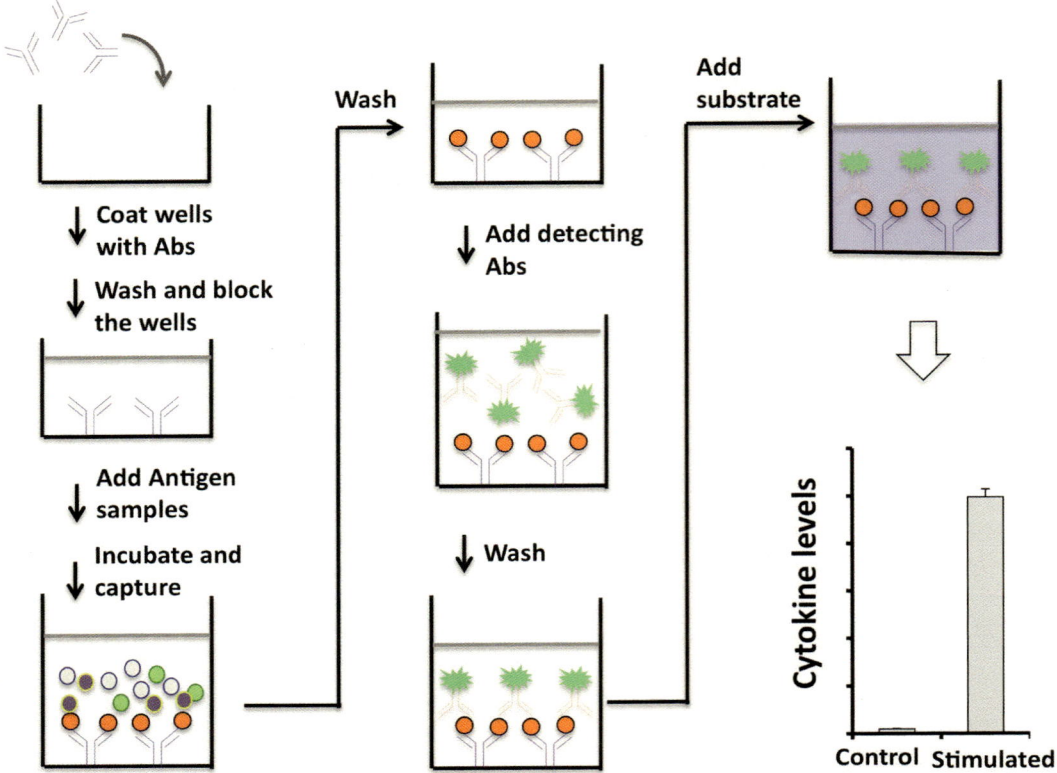

Fig. 10 ELISA flow chart. The steps described in Subheading 3 are shown in the flow chart and a depiction of results of target cytokine protein levels from control and stimulated cells is shown

14. Read the plate at 450 nm. If wavelength subtraction is available, subtract the values of 570 nm from those of 450 nm.

15. Analyze the data and calculate the cytokine concentration according to the standards.

3.20 Immunoprecipitation

1. Seed and treat cells as in Subheadings 3.1 and 3.2.

2. Harvest and lyse the cells in IP lysis buffer with proteinase inhibitors.

3. Incubate the lysate on ice for 10–15 min and spin at $15,000 \times g$ for 15 min.

4. Transfer the clear lysate to a new 1.5 ml tube.

5. Save 30–50 μl lysate as non-IP control and mix with 4× loading buffer for Western blot. Keep the remaining sample on ice.

6. Add 20 μl of protein A or G agarose beads to an IP tube.

7. Equilibrate the beads by washing with 500 μl of IP lysis buffer without proteinase inhibitors.

8. Spin at max speed for 30 s at 4 °C, immediately aspirate supernatant using loading tip, and leave about 20 μl of buffer above the beads to obtain 50 % bead slurry.

9. Add the lysate above to the beads and then add 1–10 μg of immunoprecipitating antibody or a nonspecific matched control antibody. And incubate at 4 °C for 3 h to overnight with gentle rotation.

10. Spin at maximum speed for 30 s at 4 °C, and immediately aspirate supernatant using loading tip. Leave about 10–20 μl supernatant to avoid losing beads.

11. Wash the IPs 4× with 500 μl lysis buffer (without protease inhibitors). Spin and remove supernatant as **step 8**.

12. After the last wash, spin again. Now carefully aspirate the remaining 10–20 μl supernatant immediately after spin.

13. Add 20–40 μl of 4× SDS loading buffer to the beads and boil for 5 min. Quickly spin and mix. Then spin at maximum speed for 30–60 s, and load 10 μl supernatant on the gel to perform Western blot for proteins of interest, or store samples at –20 °C.

4   Notes

1. DMEM or RPMI medium can be used for most cell lines, but do check the culture medium and conditions required for culturing any given cell line before starting to grow it. Some cell lines may require special matrixes such as collagen and gelatin to promote cell attachment or cell growth.

2. Solution of trypsin or trypsin-EDTA ranging from 0.025 to 0.5 % can be used to detach the adherent cells from culture dishes. Incubating cells with trypsin at a very high concentration or for a long time may damage and/or kill the cells. Therefore, it is important to keep the cells exposed to trypsin to a minimum time and dose. If the optimum trypsin digestion conditions for the cells used are unknown, start with trypsin at a low concentration and observe the cell morphology under a microscopy from time to time, to make sure the cells are not over digested.

3. Most cells can be starved at cell culture containing 0.5 % serum. Some cells may require serum at a concentration higher than 0.5 % for them to stay alive and healthy.

4. If treating cells of nonhuman origin, recombinant EGF from the same species as the cells is recommended.

5. The pH of the 2× HBSS is critical for obtaining high transfection efficiency and so prepare three bottles; adjust pH to 7.00, 7.05, or 7.10 and volume to 2 l. Test transfection efficiency to determine which one of the pH is best for transfection in your

system. You can test 2× HBSS by transfecting luciferase reporter plasmid in 293 cells. Save good preps, filter them to sterile, and aliquot. Store at –20 °C.

6. Prepare the lysis buffer with proteinase inhibitors by adding proteinase and phosphatase inhibitors to lysis buffer right before lysing cells.

7. EDTA will not dissolve completely until it reaches pH 8.0 with NaOH. To make 1 l of 0.5 M EDTA, add 18 g NaOH and adjust the pH by slowly adding 10 N NaOH or even NaOH at lower concentration.

8. Loading dye: as an alternative to DTT, β-mercaptoethanol can be added to a final concentration of 400 mM. 4× loading dye without DTT or β-mercaptoethanol can also be made and stored at RT, and DTT or β-mercaptoethanol may be added to this loading dye right before use. SDS is a toxic detergent and so handle SDS, especially SDS powder, with special care, such as gloves and masks.

9. The pH of the separating buffer and stacking buffer is important.

10. Acrylamide/bis solution is extremely toxic. The single unit form of acrylamide is irritant to skin and respiratory tracts and neurotoxic to animals and humans. It may cause cancers in animals and humans. Although the polymerized acrylamide is not toxic, a very small amount of acrylamide may be present in the polymerized acrylamide. Therefore, handle acrylamide very carefully.

11. APS decays slowly in solution, so prepare the solution every 2–3 weeks. If necessary, prepare 10 % APS, aliquot, and store at –20 °C for a few months.

12. Although it is not necessary to adjust pH of 10× and 1× running buffer, check the pH for troubleshooting purpose. It should be ~8.3.

13. Although it is not necessary to adjust pH of 10× transfer buffer, check the pH for troubleshooting purpose. It should be ~8.3.

14. 1× transfer buffer can be reused a few times. Stop using it if the current of transferring is over 0.3 mA under constant voltage of 90 V for one single mini-gel transfer.

15. Methanol may cause toxicity to visual system, so handle it with special care, including using chemical hood.

16. The pH of TBST buffer is critical.

17. Primary antibodies diluted in this buffer can be reused for several times. 0.05 % sodium azide works as a preservative to inhibit the growth of contaminants such as bacteria and fungi in antibody solution.

18. Chemiluminescent substrates may differ in price, ease of use, and sensitivity. Choose the appropriate one for your purpose.

19. There are commercial available complementary consensus oligonucleotide probes for NF-κB and Oct-1. You may also buy sense and antisense consensus oligonucleotides and manually anneal them by the following steps: (1) resuspend the oligos in 500 μm with ddH$_2$O. (2) Add equal amount of both oligos to a tube and add 10× annealing buffer (1,000 mM Tris, pH 8.0, 0.5 M NaCl, 10 mM EDTA) and water to anneal in 1× annealing buffer at 95 °C in a heat block for 10 min and cool down naturally at RT by turning off the power of the heat block. Dilute the annealed probes to 2 pmol with ddH$_2$O, aliquot, and store at –20 °C.

20. A radioisotope safety certificate is required to safely handle radioactive materials.

21. The incubation time may be extended up to 45 min.

22. Chloroquine inhibits autophagy and blocks lysosomal degradation. If added during first hours of transfection, it blocks lysosomal degradation of transfected DNA and thus increases the transfection efficiency. Incubation time of longer than 14 h at the concentration of 25 μm is toxic to cells, so change medium at 6–14 h post transfection.

23. Alternatively, the virus can be concentrated by ultracentrifugation at 50,000×g, 4 °C for 2 h and further purification in sucrose gradient at 20 % by ultracentrifugation at 46,000×g, 4 °C for 2 h. Aliquot and store at –80 °C.

24. Polybrene is a small positive charged polymer. It binds to cell surfaces and neutralizes surface charges, which facilitate the virus particles to bind more efficiently to their receptors. In such a way, Polybrene can greatly enhance the efficiency of virus infection.

25. For a given cell line, the concentration of the specific antibiotic for selection needs to be tittered on parental cells before virus transduction of shRNA viruses.

26. It is important to keep the cells and the cell lysate always on ice to keep protein degradation from proteases and phosphatases, which are active otherwise. This also applies to nuclear extraction, EMSA, and immunoprecipitation experiments.

27. It is important to make sure the amounts of protein lysates added to the wells fall within the detection range 0.5–2.5 μg. If not, either dilute the lysates or add more microliters of your lysates, and remember the dilution factor when you calculate the concentration.

28. The space will be saved for the stacking gel.

29. When the acrylamide becomes polymerized, you can see some refection in the interface of the polymerized gel and the butanol layer.

30. The gel can be used freshly or saved at 4 °C and used within 1 week.

31. Remove any air bubbles between the layers to make sure the even transfer of proteins in the gel to the membrane.

32. Since the transfer buffer can be reused, check the currency when you set up to run. The currency should not exceed 300 mA. If it does, prepare new transfer buffer.

33. Sodium azide may be added to the final concentration of 0.05 % to preserve the diluted antibody for repeated usage.

34. For one EMSA sample, typically start with 10 million for large cells and 20 million for smaller cells. You may use fewer cells by using smaller volume of extraction buffer. Starve and/or stimulate cells, as you desire.

35. If you want to keep the cytoplasmic fraction, transfer supernatant to a fresh tube and spin at $15,000 \times g$ for 15 min at 4 °C. Transfer the supernatant to a new tube. Use it freshly or flash freezes in aliquots in liquid nitrogen for later use.

36. This is a short rinse, so there is no need to disrupt the pellet. You can skip this step. However, to avoid contamination of cytoplasmic proteins in the nuclear extracts, you can resuspend the nuclear pellet in 500 μl of lysis buffer without inhibitors by pipetting for eight times, followed by a spin for 1 min at 4 °C and removal of the supernatant.

37. You can vortex briefly to make sure the pellet is off bottom before shaking at 4 °C for 15 min.

38. EMSA gel may be prepared 1 day before use, wrapped with wet paper towel to keep it from drying, and stored at 4 °C.

39. Every batch of poly-d(I–C) needs to be tested, but generally starts with 2–4 μg/reaction. The more abundant the transcription factor, the less probe you need. For example, for NF-κB, use 10^5 cpm/μl; for SP-1, use 10^4 cpm/μl.

40. Prepare 0.5× more cocktail mix, e.g., if you have 15 samples, make cocktail for 15.5 samples (15 samples + 0.5).

41. For purpose of safety of radioactive materials, it is important to always keep radioactive material on the bottom of tubes.

42. Do not allow the RNA pellet to dry over time, or it will make the RNA very hard to dissolve in water.

43. Remember to use the same techniques as RNA isolation to keep any RNase contamination out of your working system.

44. If using random or gene specific primers, the above protocol is not the optimal one. Please consult the original documented protocol that comes with the Invitrogen enzymes for different primers.

45. For ELISA, typically start with cells seeded in 24-well plates. You may adjust the cell amounts if you need to do ELISA for multiple assays. Starve and/or stimulate cells, as you desire.

Acknowledgments

This work was partially supported by grants from the National Institutes of Health (GM065899 and GM079451) and from the Cancer Prevention Research Institute of Texas (RP120316) to X. Lin.

References

1. Hayden MS, Ghosh S (2004) Signaling to NF-kappaB. Genes Dev 18(18):2195–2224

2. Karin M, Ben-Neriah Y (2000) Phosphorylation meets ubiquitination: the control of NF-[kappa]B activity. Annu Rev Immunol 18:621–663

3. Baud V, Jacque E (2008) The alternative NF-kB activation pathway and cancer: friend or foe? Med Sci (Paris) 24(12):1083–1088

4. Dolcet X et al (2005) NF-kB in development and progression of human cancer. Virchows Arch 446(5):475–482

5. Feinman R, Siegel DS, Berenson J (2004) Regulation of NF-kB in multiple myeloma: therapeutic implications. Clin Adv Hematol Oncol 2(3):162–166

6. Giuliani C et al (2001) Nf-kB transcription factor: role in the pathogenesis of inflammatory, autoimmune, and neoplastic diseases and therapy implications. Clin Ter 152(4):249–253

7. Ivanenkov YA, Balakin KV, Lavrovsky Y (2011) Small molecule inhibitors of NF-kB and JAK/STAT signal transduction pathways as promising anti-inflammatory therapeutics. Mini Rev Med Chem 11(1):55–78

8. Salminen A et al (2008) Activation of innate immunity system during aging: NF-kB signaling is the molecular culprit of inflamm-aging. Ageing Res Rev 7(2):83–105

9. Hideshima T et al (2002) NF-kappa B as a therapeutic target in multiple myeloma. J Biol Chem 277(19):16639–16647

10. Biswas DK et al (2003) Apoptosis caused by chemotherapeutic inhibition of nuclear factor-kappaB activation. Cancer Res 63(2):290–295

11. Biswas DK et al (2005) Crossroads of estrogen receptor and NF-kappaB signaling. Sci STKE 2005(288):pe27

12. Wang Z et al (2006) Epidermal growth factor receptor-related protein inhibits cell growth and invasion in pancreatic cancer. Cancer Res 66(15):7653–7660

13. Ando K et al (2005) Enhancement of sensitivity to tumor necrosis factor alpha in non-small cell lung cancer cells with acquired resistance to gefitinib. Clin Cancer Res 11(24 Pt 1):8872–8879

14. Le Page C et al (2005) EGFR and Her-2 regulate the constitutive activation of NF-kappaB in PC-3 prostate cancer cells. Prostate 65(2):130–140

15. Shimizu M et al (2005) Epigallocatechin gallate and polyphenon E inhibit growth and activation of the epidermal growth factor receptor and human epidermal growth factor receptor-2 signaling pathways in human colon cancer cells. Clin Cancer Res 11(7):2735–2746

16. Marciniak DJ et al (2004) Epidermal growth factor receptor-related peptide inhibits growth of PC-3 prostate cancer cells. Mol Cancer Ther 3(12):1615–1621

17. Wang H et al (2004) Analysis of the activation status of Akt, NFkappaB, and Stat3 in human diffuse gliomas. Lab Invest 84(8):941–951

18. Thornburg NJ, Pathmanathan R, Raab-Traub N (2003) Activation of nuclear factor-kappaB p50 homodimer/Bcl-3 complexes in nasopharyngeal carcinoma. Cancer Res 63(23):8293–8301

19. Bancroft CC et al (2002) Effects of pharmacologic antagonists of epidermal growth factor receptor, PI3K and MEK signal kinases on NF-kappaB and AP-1 activation and IL-8 and

VEGF expression in human head and neck squamous cell carcinoma lines. Int J Cancer 99(4):538–548

20. Bhat-Nakshatri P, Sweeney CJ, Nakshatri H (2002) Identification of signal transduction pathways involved in constitutive NF-kappaB activation in breast cancer cells. Oncogene 21(13):2066–2078

21. Pianetti S et al (2001) Her-2/neu overexpression induces NF-kappaB via a PI3-kinase/Akt pathway involving calpain-mediated degradation of IkappaB-alpha that can be inhibited by the tumor suppressor PTEN. Oncogene 20(11):1287–1299

22. Remon J et al (2014) Acquired resistance to epidermal growth factor receptor tyrosine kinase inhibitors in EGFR-mutant non-small cell lung cancer: a new era begins. Cancer Treat Rev 40(1):93–101

23. Matusan-Ilijas K et al (2013) EGFR expression is linked to osteopontin and Nf-kappaB signaling in clear cell renal cell carcinoma. Clin Transl Oncol 15(1):65–71

24. Biswas DK, Iglehart JD (2006) Linkage between EGFR family receptors and nuclear factor kappaB (NF-kappaB) signaling in breast cancer. J Cell Physiol 209(3):645–652

25. Biswas DK et al (2000) Epidermal growth factor-induced nuclear factor kappa B activation: a major pathway of cell-cycle progression in estrogen-receptor negative breast cancer cells. Proc Natl Acad Sci U S A 97(15):8542–8547

26. Ahmed KM, Cao N, Li JJ (2006) HER-2 and NF-kappaB as the targets for therapy-resistant breast cancer. Anticancer Res 26(6B):4235–4243

27. Blonska M, Lin X (2011) NF-kappaB signaling pathways regulated by CARMA family of scaffold proteins. Cell Res 21(1):55–70

28. Medoff BD et al (2009) CARMA3 mediates lysophosphatidic acid-stimulated cytokine secretion by bronchial epithelial cells. Am J Respir Cell Mol Biol 40(3):286–294

29. Mahanivong C et al (2008) Protein kinase C alpha-CARMA3 signaling axis links Ras to NF-kappa B for lysophosphatidic acid-induced urokinase plasminogen activator expression in ovarian cancer cells. Oncogene 27(9):1273–1280

30. Jiang C, Lin X (2012) Regulation of NF-kappaB by the CARD proteins. Immunol Rev 246(1):141–153

31. Jiang T et al (2011) CARMA3 is crucial for EGFR-induced activation of NF-{kappa}B and tumor progression. Cancer Res 71(6):2183–2192

32. Marsigliante S, Vetrugno C, Muscella A (2013) CCL20 induces migration and proliferation on breast epithelial cells. J Cell Physiol 228(9):1873–1883

33. Grabiner BC et al (2007) CARMA3 deficiency abrogates G-protein-coupled receptor-induced NF-kappaB activation. Genes Dev 21(8):984–986

Chapter 7

Methods to Assess the Activation of the Alternative (Noncanonical) NF-κB Pathway by Non-death TNF Receptors

Caroline Remouchamps and Emmanuel Dejardin

Abstract

The alternative or noncanonical NF-κB pathway regulates the generation of p52-containing NF-κB dimers (e.g., p52/RelB) through a partial degradation (called processing) of the precursor p100 into p52. This pathway is activated by a subset of non-death TNF receptor members, which ultimately activate two kinases: NIK (NF-κB-Inducing Kinase) and IKKα (Inhibitor of κB Kinase alpha). These kinases create a phosphodegron for the E3 ligase SCF-β-TrCP that covalently binds K48-linked polyubiquitin chain onto p100 prior to its proteasomal processing. The resulting p52-containing complexes translocate into the nucleus to activate target genes involved in secondary lymphoid organ development, B cell survival or in osteoclastogenesis.

We describe in this chapter straightforward methods to monitor the activation of the alternative NF-κB pathway. These methods uncover cytosolic and nuclear biochemical modifications of key proteins of the alternative NF-κB pathway required prior to the transcription of NF-κB target genes.

Key words NF-κB, NIK, p100 processing, TRAF, c-IAP1/2 degradation

1 Introduction

The nuclear factor kappa B (NF-κB) is a key transcription factor involved in the transcription of a plethora of genes associated with innate and adaptive immunity, cell survival, or cell proliferation. In human, the NF-κB family contains five proteins named p50, p52, p65, c-Rel, and RelB that form homodimers and heterodimers. In resting cells, NF-κB dimers are kept silent in the cytosol through their binding to members of the Inhibitory kappa B (IκB) family. The latter contains five main proteins that are IκBα, IκBβ, IκBε, p105, and p100. The proteasomal degradation of the IκB inhibitors allows the release of the NF-κB dimers [1]. This process involves the binding of the SCF-β-TrCP E3 ligase to the IκB proteins to build up K48-linked polyubiquitin chains, a posttranslational modification that targets proteins to the proteasome.

Michael J. May (ed.), *NF-kappa B: Methods and Protocols*, Methods in Molecular Biology, vol. 1280,
DOI 10.1007/978-1-4939-2422-6_7, © Springer Science+Business Media New York 2015

The binding of SCF-β-TrCP to IκB proteins requires the phosphorylation of a conserved motif (degron) DS(G/A)ΦXS to create a phosphodegron [2]. The kinases that phosphorylate the degron DS(G/A)ΦXS within IκBs proteins are activated by two main pathways [3]. First, the classical NF-κB pathway activates the IKK complex in which IKKβ is the main kinase that phosphorylates the two phospho-acceptor Serine within the DSGΦXS degron found in IκBα, IκBβ, IκBε, and p105. The subsequent polyubiquitination of these phosphorylated IκB inhibitors triggers their complete degradation through the proteasome (Fig. 1). Second, the alternative NF-κB pathway activates a kinase called NF-κB-Inducing Kinase (NIK) that together with IKKα phosphorylate the two phospho-acceptor Serine within the atypical DSAYXS degron found in p100 [4]. In that case, the polyubiquitination leads to a partial degradation, called processing, of p100 to generate p52 in association with RelB.

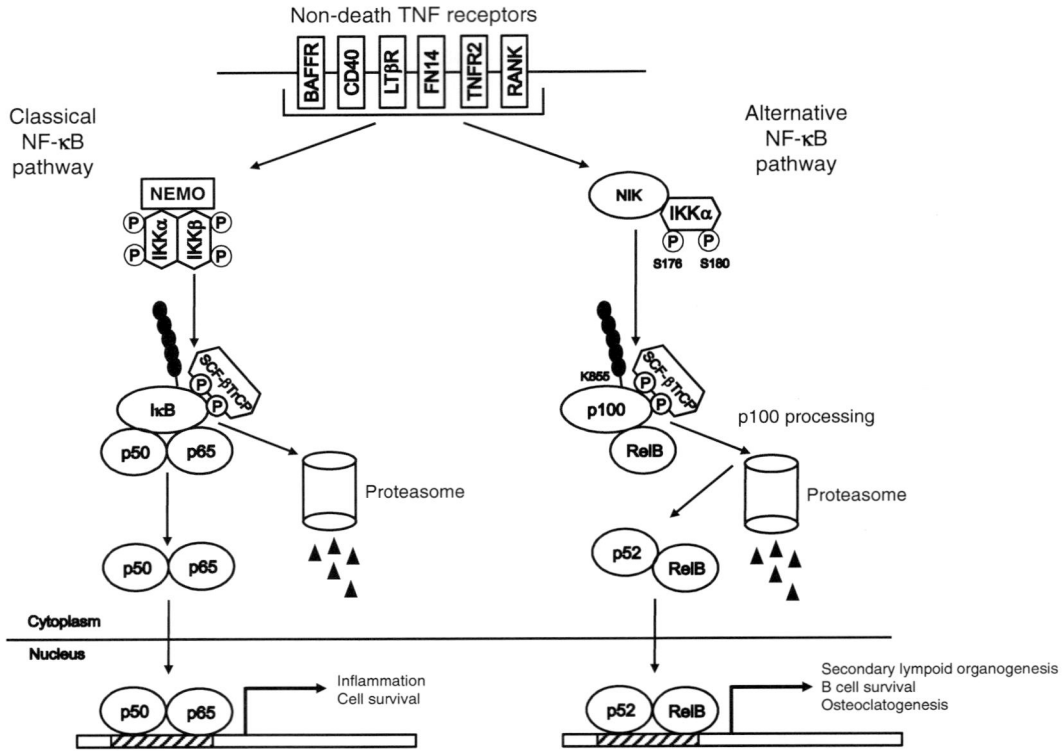

Fig. 1 The classical and the alternative NF-κB signaling pathways are triggered by a subset of TNFR members. The activation of the classical NF-κB pathway relies on the IKK complex containing NEMO/IKKα/IKKβ that phosphorylates IκBs on specific Serine for the binding of the E3 ligase SCF-β-TrCP and the degradation of IκBs by the proteasome leaving p50/p65 free to enter into the nucleus. The alternative NF-κB pathway induces the activity of two kinases, NIK and IKKα that mediate the phosphorylation of p100 prior to its polyubiquitination on K855. These biochemical modifications lead to the processing (partial degradation) of p100/RelB and the release of p52/RelB for its nuclear translocation

The alternative NF-κB pathway is activated by a subset of TNF receptors such as LTβR, BAFFR, CD40, Fn14, TNFR2, and RANK, which altogether activate genes involved in specific immunological processes including secondary lymphoid organ development, B cell survival, and osteoclastogenesis [5–11] (Fig. 1). Once one of these receptors is triggered by its ligand, adaptor proteins such as TRAFs are recruited to the cytoplasmic tail and NIK is activated. NIK, a mitogen-associated protein 3 kinase (MAP3K14), is the central signaling component of the alternative NF-κB pathway. In resting cells, NIK is kept at undetectable levels by a negative regulatory mechanism involving dynamic ubiquitination and proteasomal degradation (Fig. 2a). Indeed, de novo NIK is constantly targeted for degradation by TRAF2/3-c-IAP1/2 complex. TRAF3 physically interacts with NIK and recruits TRAF2, c-IAP1/2. c-IAP1/2 mediates K48-linked polyubiquitination chains onto

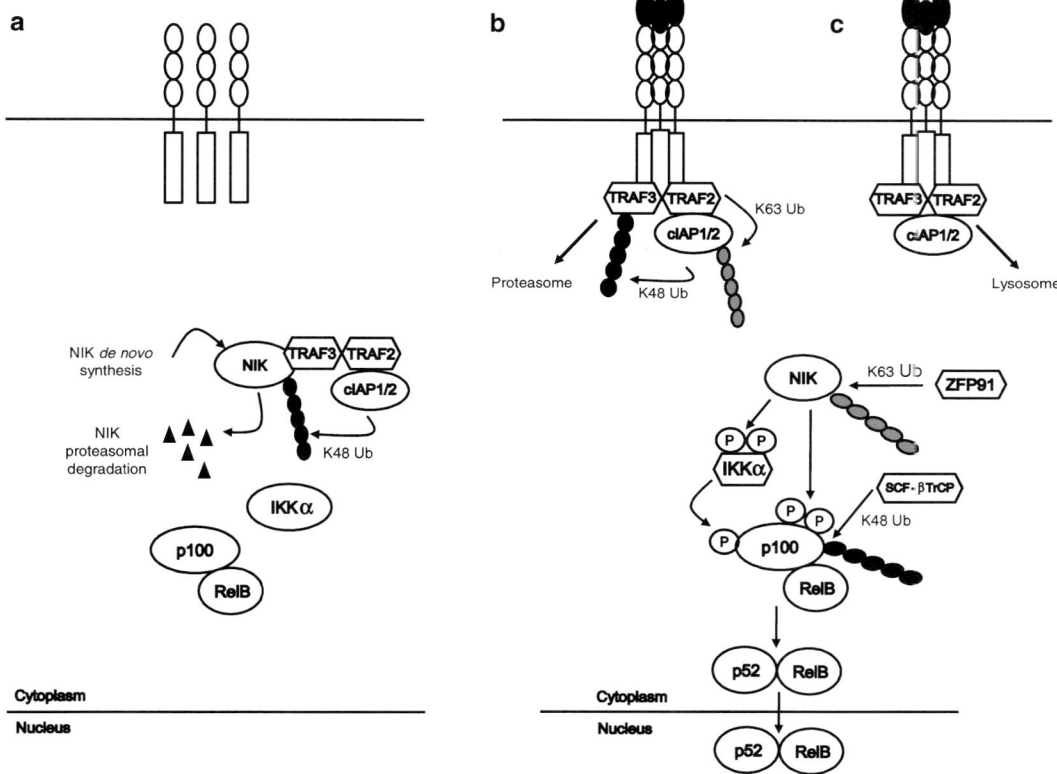

Fig. 2 Model of NIK activation. (**a**) In resting cells, de novo synthesized NIK is constantly targeted by the TRAF2/3-c-IAP1/2 complex for proteasomal degradation. (**b**, **c**) Binding of trimeric ligand to particular TNFR members induce the inhibition of the inhibitory complex TRAF2/3-c-IAP1/2 either through (**b**) proteasomal degradation or (**c**) lysosomal degradation. As a consequence, NIK is ubiquitinated by the E3 ligase ZFP91 for further promoting its stability and participate with IKKα to the phosphorylation of p100. These posttranslational modifications allow the recruitment of the E3 ligase SCF-β-TrCP and the ubiquitination-dependent processing of p100/RelB into p52/RelB

NIK leading to its constant proteasomal degradation [12, 13]. In response to a stimulus such as CD40L, the inhibitory complex is recruited to the receptor where TRAF3 is K48-linked polyubiquitinated by the c-IAP1/2 and degraded by the proteasome (Fig. 2b). Alternatively, some activated receptors like LTβR are internalized, and this recruits the negative regulatory complex TRAF2/3-c-IAP1/2 for its degradation into lysosomes [8] (Fig. 2c).

Thus, TRAF proteins degradation is required for the stabilization of newly synthesized NIK. Yet other posttranslational modifications seem to contribute to the accumulation of NIK. Indeed, a K63-linked polyubiquitination of NIK by the E3 ligase ZFP91 enhances its stability [14]. Recently, two groups crystallized the kinase domain of NIK and found that phosphorylation of the Threonine 559 in the activation loop was not required for its enzymatic activity [15, 16]. NIK structural features are represented in the Fig. 3. Active NIK phosphorylates and activates the kinase IKKα on Ser 176 (*see* Figs. 1 and 4) [17]. However, in resting conditions the processing of p100 is limited due to the presence of a Processing-Inhibitory Domain (PID) and an Ankyrin Repeat Domain (ARD) in its carboxyl-terminal part [4]. However, the inhibitory function of the PID is relieved when Serine 866/870 and Serine 99/108/115/123/872 are phosphorylated by NIK and IKKα, respectively [4, 18]. The phosphorylation of p100 on Serine 866/870 creates the phosphodegron for the E3 ligase

NLS : Nuclear Localization Signal   NES : Nuclear Export Signal

Fig. 3 Regulatory domains of the human NIK protein. Numbers represent the amino acid boundaries of each domain

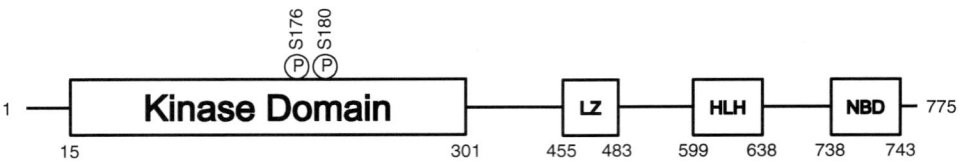

Ⓟ NIK phosphorylation site   LZ : Leucine zipper-like motif   HLH : Helix-loop-helix   NBD : NEMO binding domain

Fig. 4 Regulatory domains of the human IKKα protein. Numbers represent the amino acid boundaries of each domain

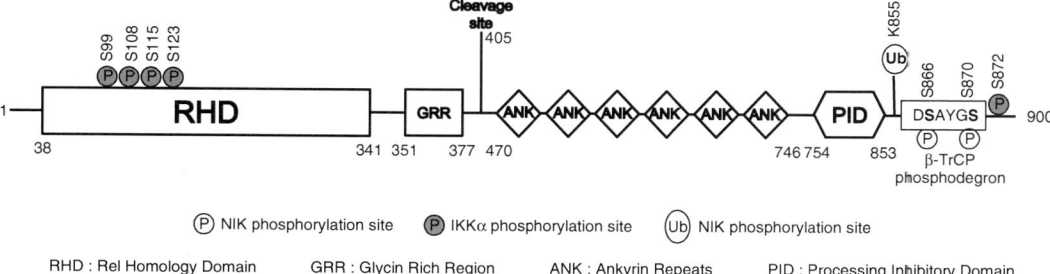

Fig. 5 Regulatory domains of the human p100 protein. Numbers represent the amino acid boundaries of each domain

β-TrCP that promotes the K48-linked polyubiquitination of Lysine 855 (*see* Fig. 5) [19]. The processing of p100 allows the cytoplasmic release of p52 dimerized with RelB.

Once the NF-κB dimers are freed from their IκB inhibitors, they are recognized by transporter proteins called importins, which brings them into the nucleus [20]. The binding of nuclear NF-κB dimers to the chromatin is controlled by many parameters, among which the nucleotide stretch called κB site dictates the affinity towards specific dimers. Whereas the main NF-κB heterodimer of the classical NF-κB pathway (p50/p65) was shown to bind a remarkably loose consensus sequence 5′-GGGRNYYYCC-3′ (R=A or G, Y=T or C) [21], the existence of p52/RelB specific κB sites is still matter of debate. So far, all the κB sites identified for binding p52/ReB displayed comparable affinity for p50/p65 [22]. Interestingly, the X-ray structure of p52/RelB-κB DNA complex revealed that p52/RelB was able to bind a larger spectrum of κB sites than p50/p65 because it was more versatile and less discriminatory in DNA recognition. Arginine 125 of RelB is important to recognize specific κB sites that have more contiguous and centrally located A:T base pairs. Depending on DNA sequence, Arg 125 of RelB allows a switch in the conformation of the dimer in order to interact efficiently with diverse κB sites and thereby activate a broad spectrum of genes involved in inflammation and development [23]. Although much progress is ongoing to refine the specificity of particular NF-kB dimers for the diversity of κB sites in the genome, up to now one can only say that some sites are preferred over others [24, 25].

2   Materials

2.1 Ligands and Agonist Antibodies for TNFR Inducing the Alternative NF-κB Pathway

The ligands and agonist antibodies that we use to activate the alternative pathway are shown in Table 1 along with their species specificities and sources.

Table 1
Ligands and agonist antibodies for TNFR inducing the alternative NF-κB pathway

Name	LTβR		Fn14		CD40		Reference
	Human	Mouse	Human	Mouse	Human	Mouse	
hLTα₁β₂	+	–					Human lymphotoxin alpha 1 beta 2, R&D Systems (Minneapolis, MN, USA)
Anti-hLTbR	+	–					Human Lymphotoxin βR/ TNFRSF3 Antibody, R&D Systems.
Anti-mLTβR	–	+					Mouse Lymphotoxin beta Receptor antibody (clone 3C8), eBioscience (San Diego, CA, USA)
Flag-hLIGHT	+	+					TWEAK (soluble) (human) (recombinant), Enzo Life Science (Farmingdale, NY, USA)
Flag-hTWEAK			+	+			LIGHT (soluble) (human) (recombinant), Enzo Life Science
Flag-hCD40L					+	–	CD40L (soluble) (human) (recombinant), Enzo Life Science

2.2 Antibodies for Western Blotting, Immunoprecipitation, and Supershift Assays

The ligands antibodies that we use for western blotting, immunoprecipitation, and supershift assays to detect activation of the alternative pathway are shown in Table 2 along with their species specificities and sources.

2.3 Whole-Cell Extracts for Immunoblotting

1. Whole-cell lysis (WCE) buffer: 20 mM HEPES (pH 7.9), 350 mM NaCl, glycerol 20 % (v/v), Igepal 1 % (v/v), 1 mM MgCl₂, 0.1 mM EGTA or EDTA with protease and phosphatase inhibitors freshly added (Complete and PhosSTOP, Roche, Madison, WI, USA).

2. SDS lysis buffer: 0.5 % SDS (w/v) with protease and phosphatase inhibitors freshly added (Complete and PhosSTOP, Roche).

3. 10 % SDS-PAGE gel.

2.4 Cell Extracts for Co-immunoprecipitation Experiments

1. TNT buffer: 20 mM Tris–HCl (pH 7.4), 200 mM NaCl, 0.5–1 % (v/v) Triton X-100 with protease and phosphatase inhibitors freshly added (Complete and PhosSTOP, Roche).

2. Protein-A Agarose (Pierce, Waltham, MA, USA).

Table 2
Antibodies for western blotting, immunoprecipitation, and supershift

Name	Human	Mouse	Reference
TRAF2 (H-249)	+	+	Santa Cruz Biotechnology, Santa Cruz, CA, USA
TRAF2 (C-20)	+	+	Santa Cruz Biotechnology
TRAF3 (H-122)	+	–	Santa Cruz Biotechnology
TRAF3 (C-20)	+	+	Santa Cruz Biotechnology
c-IAP1/2	+	+	CycLex, Nagano, Japan
c-IAP1	+	–	R&D Systems
c-IAP2	+	–	R&D Systems
NIK	+	+	Cell Signaling, Danvers, MA, USA
Phospho-NF-kB2 p100 (Ser866/870)	+	+	Cell Signaling
NF-kB p52	+	–	Millipore, Billerica, MA, USA
NF-kB p52 (C-5)	+	+	Santa Cruz Biotechnology
RelB (C-19)	+	+	Santa Cruz Biotechnology
IKKα	+	+	Imgenex, Littleton, CO, USA
IKKα	+	–	BD Pharmingen, Franklin Lakes, NJ, USA
Phospho-IKKα/β (Ser176/180) (16A6)	+	+	Cell Signaling

3. Loading buffer 4×: 6.25 ml of 0.5 M Tris (pH 6.8), 20 ml of glycerol, 4 g of SDS, 10 ml of 1 M β-mercaptoethanol, 2 mg of bromophenol blue.

4. 10 % SDS-PAGE gel.

2.5 Kinase Assays

1. Kinase assay lysis buffer (KLB): 150 mM NaCl, 20 mM Tris–HCl (pH 7.5), 0.2 % (v/v) Igepal, 1 mM EDTA, 10 % (v/v) glycerol with protease and phosphatase inhibitors freshly added (Complete and PhosSTOP, Roche).

2. Kinase assay lysis buffer (KLB) High Salt: 500 mM NaCl, 20 mM Tris–HCl (pH 7.5), 0.2 % (v/v) Igepal, 1 mM EDTA, 10 % (v/v) glycerol.

3. Immunoprecipitation materials: NIK antibody (Cell signaling, Danvers, MA, USA), ChromPure Rabbit Ig, whole molecule (Jackson Immunoresearch, West Grove, PA, USA), Protein-A Agarose beads (Pierce).

4. Kinase assay buffer (KAB): 20 mM HEPES (pH 7.5), 2 mM DTT, 10 mM MgCl$_2$, with protease and phosphatase inhibitors freshly added (Complete and PhosSTOP, Roche).

5. 1.5 mM ATP.

6. ATP [γ^{32}P] 10 mCi/ml - 3,000 Ci/mmol 10 mCi/ml (Perkin Elmer, Waltham, MA, USA).

7. 10 % SDS-PAGE gel.

2.6 GST-Fusion Protein Production

1. BL21 (DE3) pLysS Competent bacteria (Promega, Madison, WI, USA).

2. Luria Broth (LB).

3. Isopropyl β-D-1-thiogalactopyranoside (IPTG).

4. PBS lysis buffer: PBS with 1 % (v/v) Triton X-100 and 1 mM PMSF freshly added.

5. PBS with protease inhibitors (Complete, Roche) freshly added.

6. Glutathione Sepharose 4B (GE Healthcare, Mickleton, NJ, USA).

7. 10 % SDS-PAGE gel.

8. Coomassie blue dye (250 ml): 75 ml of methanol, 25 ml of acetic acid (glacial), 0.25 g of Coomassie blue, 150 ml of H$_2$O.

9. Destaining solution (500 ml): 150 ml of methanol, 50 ml of acetic acid (glacial), 300 ml of H$_2$O.

10. 10 % SDS-PAGE gel.

2.7 Cytoplasmic-Nuclear Extracts

1. Cytoplasmic lysis buffer: 0.2 % (v/v) Igepal (for most cells except for MEFs that are more sensitive and require only 0.05 % (v/v) Igepal), 10 mM HEPES (pH 7.9), 10 mM KCl, 2 mM MgCl$_2$, 0.1 mM EDTA.

2. Nuclear lysis buffer: 20 mM HEPES (pH 7.9), 1.5 mM MgCl$_2$, 0.2 mM EDTA, 420 mM NaCl, 25 % (v/v) glycerol.

3. Washing buffer: 10 mM HEPES (pH 7.9), 20 mM KCl, 2 mM MgCl$_2$, 0.1 mM EDTA.

2.8 Electrophoretic Mobility Shift Assay (EMSA)

1. Probe for NF-κB mobility shift:

    + Strand: 5'- TTGGAGTTGAGGGGACTTTCCCAGG -3'

    – Strand: 5'- TTGGCCTGGGAAAGTCCCCTCAACT -3'

2. 100 mM NaCl.

3. T4 polynucleotide kinase, 3'-phosphatase free (Roche).

4. Quick Spin Columns for radiolabeled DNA purification (Roche).

5. TBE 10× (1 l): 108 g of Tris base, 55 g of boric acid, 7.5 g of EDTA disodium salt, deionized water.

6. 6 % Native PAGE Gel: 12 ml of acrylamide–bisacrylamide 40 % (29:1), 48 ml of deionized water, 20 ml of TBE 1×, 560 μl of ammonium persulfate (APS) 10 % (w/v), 72 μl of *N, N, N'*-tetramethyl-ethylenediamine (TEMED).

7. Buffer D: 20 mM HEPES, 10 mM KCl, 0.2 mM EDTA, 10 % (v/v) glycerol, 5 mM DTT (freshly added).

8. Loading buffer: 30 % (v/v) glycerol, 0.25 % (w/v) bromophenol blue in 20 mM HEPES (neutral pH).

9. Nuclear extracts.

10. 1 μg/μl bovine serum albumin (BSA).

11. Poly (dI-dC) (double strand; Sodium salt).

3 Methods

3.1 Induction of the Alternative NF-κB Pathway

Several receptors of the TNF receptor superfamily can activate both the classical and the alternative pathways. In contrast to the classical pathway, the activation of the alternative pathway is slower and can take up to 2–16 h according to the stimulus. Table 3 shows a range of validated stimuli, doses, and timing required to activate the alternative NF-κB pathway.

3.2 TRAF2/3 Recruitment to the Receptor

TRAF2/3 recruitment can be analyzed by co-immunoprecipitation assays (*see* **Note 1**).

1. Grow cells in 6-well plates and treat with appropriate ligands or agonist antibodies for 30 min to 4 h (*see* Table 3).

2. Add the protease and phosphatase inhibitors to the TNT lysis buffer.

3. Discard the medium and wash the cells with ice-cold PBS.

4. Collect the cells by scraping in ice-cold PBS and transfer into tubes.

Table 3
Stimuli, doses, and timing to activate the alternative NF-κB pathway

Stimuli	Type	Receptor	Dose	Time
hLTα₁β₂	Ligand	hLTβR	1 nM	>2 h
anti-hLTβR	Agonist antibody	hLTβR	0.5–2 μg/ml	>2 h
anti-mLTβR	Agonist antibody	mLTβR	2 μg/ml	>2 h
Flag-hLIGHT	Ligand	h/mLTβR	200 ng/ml	>2 h
Flag-hTWEAK	Ligand	h/mFn14	200 ng/ml	>2 h
Flag-hCD40L	Ligand	hCD40	1 μg/ml	>6 h

5. Pellet the cells by centrifugation at $1,000 \times g$ for 1 min at 4 °C and remove the supernatant.

6. Lyse the cells in 100–200 µl (depending on the size of the pellet) of TNT buffer for 10 min on ice.

7. Centrifuge at maximum speed for 10 min at 4 °C to get rid of the insoluble fraction and transfer the supernatant in a new tube.

8. Keep 50 µl of the extract for input. Adjust the volume of the remaining lysate to 500 µl with TNT 1 % and pre-clear for 1 h with 10 µl of protein-A agarose beads (previously washed three times in TNT 1 %) at 4 °C under rotation. Discard the beads by centrifugation for 1 min at maximum speed. Keep the supernatant.

9. Incubate the supernatant overnight with 1 µg of anti-TRAF2 or anti-TRAF3 antibodies at 4 °C under rotation.

10. Add 10 µl of protein-A agarose beads (previously washed three times in TNT 1 %) for an additional hour at 4 °C under rotation.

11. Centrifuge for 1 min at $6,000 \times g$. Discard the supernatant.

12. Wash the beads three times with 500 µl of TNT 1 %. After the final wash, resuspend to 30 µl of immunoprecipitate and add 10 µl of loading buffer 4×. At this stage the samples can be kept at −20 °C.

13. Boil the immunoprecipitate for 10 min, load on an SDS-PAGE 10 % acrylamide gel, and analyze by immunoblotting with antibody for your TNFR of interest (*see* Table 2).

3.3   c-IAP1/2 Degradation (See Note 2)

1. Grow cells in 6-well plates and treat with specific ligands or agonist antibodies for 30 min up to 6 h (*see* Table 3).

2. Add the protease and phosphatase inhibitors to the WCE lysis buffer.

3. Discard the medium and wash the cells with ice-cold PBS.

4. Collect the cells by scraping in ice-cold PBS and transfer into tubes.

5. Pellet the cells by centrifugation at $1,000 \times g$ for 1 min at 4 °C and remove the supernatant.

6. Lyse the cells in 100–200 µl (depending on the size of the pellet) of WCE buffer for 10 min on ice.

7. Centrifuge at maximum speed for 10 min at 4 °C to get rid of the insoluble fraction and transfer the supernatant to a new tube.

8. Load 50 µg of protein on an SDS-PAGE 10 % acrylamide gel and analyze by immunoblotting with c-IAP1/2 antibody (*see* Table 2).

3.4 NIK Stabilization and Activation (See Note 3)

1. Grow cells in 6-well plates and treat with specific ligands or agonist antibodies for 2–6 h (*see* Table 3).

2. Perform whole-cell extracts as in Subheading 3.3.

3. Load 80 μg of protein on an SDS-PAGE 10 % acrylamide gel and analyze by immunoblotting with a NIK antibody (*see* Table 2).

3.5 NIK Kinase Activity

Another way to assess NIK kinase activity is to determine its ability to autophosphorylate or to phosphorylate a substrate by performing a kinase assay.

3.5.1 GST-p100 Production

A substrate for NIK is a fusion of GST with p100 spanning from amino acids 660–900 (e.g., cloned in the pGex4T1 vector). This substrate is produced as follows:

1. Inoculate one colony of bacterial strain BL-21 (DE3) expressing the construct GST-p100 into individual 5 ml aliquots of LB broth containing suitable antibiotic selection. Grow overnight at 37 °C with shaking.

2. Inoculate 500 ml of LB containing the antibiotic selection with the 5 ml overnight culture from **step 1**.

3. Grow the cultures at 37 °C to reach an OD_{600} of 0.5–1.0 (this should take 3–4 h).

4. Induce the expression of the GST-p100 by adding IPTG to a final concentration of 1 mM.

5. Incubate the cultures for an additional 4 h at 37 °C with shaking.

6. Centrifuge the bacterial culture at $3,500 \times g$ for 10 min at 4 °C and discard the supernatant. At this point, the pellets can be stored frozen at −20 °C if necessary.

7. Resuspend the pellet in 20 ml of PBS lysis buffer and sonicate the bacterial suspension on ice, alternate cycles of 30 s bursts (high frequency) and 30 s of rest on ice. Five cycles of sonication are usually sufficient.

8. Centrifuge the lysate at $12,000 \times g$ for 30 min at 4 °C and transfer the supernatant to a fresh tube.

9. Add 500 μl of a 50:50 slurry solution of glutathione-Sepharose beads equilibrated in PBS lysis buffer to the supernatant of bacterial lysate and incubate for 30 min at room temperature, rotating the tube end over end to ensure mixing.

10. Centrifuge the samples at $750 \times g$ for 1 min at 4 °C to pellet the beads. Remove the supernatant.

11. Wash the beads in 5 ml of ice-cold PBS with protease inhibitors.

12. Centrifuge the samples at $500 \times g$ for 1 min at 4 °C to pellet the beads. Remove the supernatant.

13. Add 5 ml of ice-cold PBS with protease inhibitors. Resuspend the beads by gentle mixing and centrifuge the sample again at $500 \times g$ for 1 min at 4 °C to pellet the beads. Remove the supernatant.

14. Resuspend the beads in 250–500 µl of ice-cold PBS with protease inhibitors and store at −80 °C.

15. Run 5–20 µl of the product as well as a range of concentrations of BSA (100 ng up to 1.5 µg) on an SDS-PAGE 10 % acrylamide gel and stain with Coomassie blue dye for 1 h at room temperature with shaking.

16. After de-staining the gel, estimate the concentration of the GST-p100 produced in regard to the titration of BSA.

3.5.2   Kinase Assay of NIK on GST-p100

1. Grow cells in 10 cm plates and treat with specific ligands or agonist antibodies for 2 up to 6 h (*see* Table 3).

2. Add the protease and phosphatase inhibitors as well as the DTT to KLB.

3. Discard the medium and wash the cells twice with ice-cold PBS.

4. Collect the cells by scraping in ice-cold PBS and transfer into tubes.

5. Pellet the cells by centrifugation at $1,000 \times g$ for 1 min at 4 °C and remove the supernatant.

6. Lyse the cells in 500 µl of KLB for 10 min on ice.

7. Centrifuge at maximum speed for 10 min at 4 °C and transfer the supernatant in a new tube.

8. Keep 1/10 of the extract for input. Pre-clear the remaining extract with 20 µl of Protein-A agarose beads for 1 h at 4 °C under rotation.

9. Centrifuge for 1 min at maximum speed at 4 °C. Transfer the supernatant to a new tube.

10. Immunoprecipitate with 1 µl of NIK antibody or 1 µg of Rabbit irrelevant Ig (negative control) and incubate overnight at 4 °C under rotation.

11. Add 20 µl of protein-A agarose beads for 2 h at 4 °C under rotation.

12. Centrifuge for 1 min at $6,000 \times g$ at 4 °C, discard the supernatant and wash the immunoprecipitate with 500 µl of KLB.

13. Centrifuge again for 1 min at $6,000 \times g$ at 4 °C and discard the supernatant. Wash twice with KLB followed by two washes with KLB High Salt and two further washes with KLB have to be performed. Finally, wash the immunoprecipitate with 500 µl KAB with protease and phosphatase inhibitors freshly added.

14. Keep 1/10 of the immunoprecipitate as an immunoprecipitation control.

15. Resuspend the remaining in 27 µl of KAB+ inhibitors.

16. In a radioactive procedure room, prepare a mixture of cold ATP and ATP γ^{32}P (2 µl of cold ATP 1.5 mM and 1 µl of ATP γ^{32}P 10 µCi/µl per condition). Add 3 µl of the mixture to each tube (cold ATP = 100 µM final).

17. Incubate for 30 min with 1 µg of GST-p100 at 30 °C then boil the samples for 10 min to stop the reaction. At this stage the samples can be stored at –20 °C.

18. Load on an SDS-PAGE 10 % acrylamide gel and analyze NIK autophosphorylation and GST-p100 phosphorylation by autoradiography.

3.6 Detecting IKKα Activation

1. Grow cells in 6-well plates and treat with appropriate ligands or agonist antibodies for 2 up to 4 h (see Table 3).

2. Perform whole-cell extracts as described in Subheading 3.3.

3. Load 50 µg of protein on an SDS-PAGE 10 % acrylamide gel and analyze by immunoblotting with anti-phospho IKKα Ser 176/180 antibody (see Table 2).

3.7 Detecting p100 Phosphorylation on Ser866/870

1. Grow cells in 6-well plates and treat with ligands or agonist antibodies for up to 4 h (see Table 3).

2. Perform whole-cell extracts as described in Subheading 3.3.

3. Load 50 µg of protein on an SDS-PAGE 10 % acrylamide gel and analyze by immunoblotting with anti-phospho-NF-κB p100 (Ser866/870) antibody (see Table 2).

3.8 Detecting p100 Processing to p52

1. Grow cells in 6-well plates and treat with specific ligands or agonist antibodies for at least 2 h (see Table 3).

2. Add the protease and phosphatase inhibitors to the SDS lysis buffer (see **Note 4**).

3. Discard the medium and wash the cells with ice-cold PBS.

4. Collect the cells by scraping in ice-cold PBS and transfer into tubes.

5. Pellet the cells by centrifugation at $1,000 \times g$ for 1 min at 4 °C and remove the supernatant.

6. Lyse the cells in 100–200 µl (depending on the size of the pellet) of SDS 0.5 %. The extract will probably be viscous due to the extraction of genomic DNA.

7. To render the extract less viscous, either boil it for 10 min at 95 °C, or incubate with DNAse I at 37 °C prior to boiling.

8. Load 30 µg of protein on an SDS-PAGE 10 % acrylamide gel and analyze by immunoblotting with anti-NF-κB p100/p52 antibody (see Table 2) (see **Note 5**).

3.9 Detecting p52/ RelB Translocation into the Nucleus

To analyze p52 and RelB translocation, cytoplasmic and nuclear extracts have to be made (*see* **Note 6**).

1. Grow cells in 10 cm dishes and treat with appropriate ligands or agonist antibodies for at least 2 h (*see* Table 3).

2. Discard the medium and wash the cells twice with ice-cold PBS.

3. Collect the cells by scraping in ice-cold PBS and transfer into tubes.

4. Pellet the cells by centrifugation at $100 \times g$ for 1 min at 4 °C and remove the supernatant.

5. Lyse the cells in 400 μl of cytoplasmic lysis buffer for 10 min on ice.

6. Centrifuge for 5 min at $100 \times g$. Transfer the supernatant (the cytoplasmic fraction) into a new tube.

7. Wash the pellet (nuclei) with 400 μl of washing buffer without resuspending the pellet.

8. Centrifuge at $100 \times g$ for 1 min at 4 °C and discard the supernatant. Repeat the wash one more time.

9. Add two volumes of nuclear lysis buffer to the pellet. Resuspend the pellet and incubate for 30 min at 4 °C under rotation.

10. Centrifuge at maximum speed for 30 min. Transfer the supernatant (the nuclear fraction) into a new tube.

11. Load 30 μg of protein of cytoplasmic and nuclear fractions on an SDS-PAGE 10 % acrylamide gel and analyze by immunoblotting with anti-NF-κB p52 and anti-RelB antibodies (*see* Table 2).

3.10 Detecting p52/ RelB Binding to κB Sites

The activity of binding can be measured by EMSA experiment on nuclear extracts of cells treated with an inducer of the alternative NF-κB pathway (*see* **Note 7**).

3.10.1 Nuclear Extracts

Nuclear extracts for EMSA are made as described in Subheading 3.9, **steps 1–10**

3.10.2 Probe Radiolabeling

1. In a tube, add 2 μl of annealed oligonucleotides (100 ng), 1 μl of T4 polynucleotide kinase PNK 10 U/μl (10 units), 2 μl of buffer PNK 10×, 4 μl of ATP γ^{32}P 10 μCi/μl, and 11 μl of water (final volume = 20 μl).

2. Incubate for 1 h at 37 °C.

3. Purify the radiolabeled probe using the Quick spin columns for radiolabeled DNA purification.

4. Measure the activity of the radiolabeled probe in a scintillation counter.

EMSA

1. Make a 6 % Native PAGE gel.

2. Switch on the cooler and pre-run the gel for 20 min at 300 V.

3. Make up the following reaction mixture for each sample: 5 μg of nuclear protein, 1 μl of Poly(dI-dC) (1 μg/μl), 1 μl of BSA (1 μg/μl), 8 μl of buffer D (add DTT just before use), and radiolabeled probe (100,000 cpm). Adjust the volume with water up to 17 μl.

4. For supershift assays make the following reaction mixture: 5 μg of nuclear proteins, 1 μl of Poly(dI-dC) (1 μg/μl), 1 μl of BSA (1 μg/μl), 8 μl of buffer D (with DTT freshly added), and 1 μg of antibody. Incubate on ice for 20 min then add the radiolabeled probe (100,000 cpm). Adjust the volume with water to 17 μl.

5. Incubate for 30 min at room temperature.

6. Add 3 μl of blue Loading Buffer for EMSA.

7. Load the samples on the gel and run for 2–3 h until the blue dye front reaches ¾ of the way down the gel.

8. Dry the gel for 45 min at 80 °C in a gel dryer.

9. Put the dried gel in an autoradiography cassette.

10. Expose overnight to X-ray film then develop the next day.

4   Notes

1. Once the receptor binds its ligand, the adaptor TRAF proteins are recruited through specific binding sites. Previously, using LTβR as a prototype, we showed that internalization of the receptor and binding of TRAF2 and TRAF3 are a prerequisite for the activation of p100 processing [8].

2. Ligands of the TNF family such as LTα₁β₂, CD40L, or Tweak were shown to induce c-IAP1/2 depletion through proteasomal and/or lysosomal degradative pathways [8, 26]. Synthetic small molecules mimicking the protein Smac (Smac mimetics) sensitize some cancer cell lines to TNFR1-dependent cell death and can be used as tools to induce c-IAP1/2 depletion [27–29].

3. In resting conditions, NIK is kept at a very low level through a proteasomal-dependent degradation mediated by the TRA2/3-c-IAP1/2 complex. When cells are stimulated by an inducer of the alternative NF-κB pathway, TRAFs are recruited to the activate receptor and c-IAP1/2 are depleted. NIK is no longer targeted for negative regulation and can accumulate in its active form.

4. Although p100 processing is the ultimate step to generate cytoplasmic p52, the latter translocates into the nucleus to

bind enhancer and promoter regions. To analyze correctly the ratio p100/p52, reflecting the degree of activation of the pathway, it is therefore necessary to extract properly the whole p52 content, including p52 associated to the chromatin fraction. A classical lysis buffer with a detergent such as Igepal or Triton X-100 will not be sufficient to extract entirely all the proteins associated to the chromatin. A more stringent extraction can be performed using SDS as detergent in the lysis buffer.

5. The ratio p100/p52 can be quantified with software such as Image J.

6. p52 preferentially binds RelB to form a transcriptionally active dimer (the transactivation domain in this dimer is in RelB) which translocates into the nucleus.

7. No κB sites are exclusively recognized by p52/RelB. Nevertheless, sequences have been described to bind p50/p65 as well as p52/RelB. Nuclear extracts from untreated HT29 cells contained only low levels of NF-κB DNA binding activity. After 6 h of treatment with an agonist anti-LTβR antibody, the binding is increased. The band observed on the gel is supershifted with antibodies against p52, RelA (p65), or RelB [7].

Acknowledgments

We thank the Centre Anticancéreux (CAC) from the University of Liège (Belgium), the Fédération belge Contre le Cancer (FCC) (Belgium), and the Interuniversity Attraction Poles (IAP7/32) (Belgium) for their funding. C.R. is supported by a fellowship from the Télévie (Belgium), and E.D. is a Research Associate at the FNRS (Belgium).

References

1. Vallabhapurapu S, Karin M (2009) Regulation and function of NF-kappaB transcription factors in the immune system. Annu Rev Immunol 27:693–733

2. Kanarek N, Ben-Neriah Y (2012) Regulation of NF-kappaB by ubiquitination and degradation of the IkappaBs. Immunol Rev 246(1):77–94

3. Dejardin E (2006) The alternative NF-kappaB pathway from biochemistry to biology: pitfalls and promises for future drug development. Biochem Pharmacol 72(9):1161–1179

4. Xiao G, Harhaj EW, Sun SC (2001) NF-kappaB-inducing kinase regulates the processing of NF-kappaB2 p100. Mol Cell 7(2):401–409

5. Claudio E et al (2002) BAFF-induced NEMO-independent processing of NF-kappa B2 in maturing B cells. Nat Immunol 3(10):958–965

6. Coope HJ et al (2002) CD40 regulates the processing of NF-kappaB2 p100 to p52. EMBO J 21(20):5375–5385

7. Dejardin E et al (2002) The lymphotoxin-beta receptor induces different patterns of gene expression via two NF-kappaB pathways. Immunity 17(4):525–535

8. Ganeff C et al (2011) Induction of the alternative NF-{kappa}B pathway by lymphotoxin {alpha}{beta} (LT{alpha}{beta}) relies on

internalization of LT{beta} receptor. Mol Cell Biol 31(21):4319–4334

9. Munroe ME, Bishop GA (2004) Role of tumor necrosis factor (TNF) receptor-associated factor 2 (TRAF2) in distinct and overlapping CD40 and TNF receptor 2/CD120b-mediated B lymphocyte activation. J Biol Chem 279(51): 53222–53231

10. Novack DV et al (2003) The IkappaB function of NF-kappaB2 p100 controls stimulated osteoclastogenesis. J Exp Med 198(5): 771–781

11. Saitoh T et al (2003) TWEAK induces NF-kappaB2 p100 processing and long lasting NF-kappaB activation. J Biol Chem 278(38): 36005–36012

12. Vallabhapurapu S et al (2008) Nonredundant and complementary functions of TRAF2 and TRAF3 in a ubiquitination cascade that activates NIK-dependent alternative NF-kappaB signaling. Nat Immunol 9(12):1364–1370

13. Zarnegar BJ et al (2008) Noncanonical NF-kappaB activation requires coordinated assembly of a regulatory complex of the adaptors cIAP1, cIAP2, TRAF2 and TRAF3 and the kinase NIK. Nat Immunol 9(12):1371–1378

14. Jin X et al (2010) An atypical E3 ligase zinc finger protein 91 stabilizes and activates NF-kappaB-inducing kinase via Lys63-linked ubiquitination. J Biol Chem 285(40): 30539–30547

15. de Leon-Boenig G et al (2012) The crystal structure of the catalytic domain of the NF-kappaB inducing kinase reveals a narrow but flexible active site. Structure 20(10): 1704–1714

16. Liu J et al (2012) Structure of the nuclear factor kappaB-inducing kinase (NIK) kinase domain reveals a constitutively active conformation. J Biol Chem 287(33):27326–27334

17. Ling L, Cao Z, Goeddel DV (1998) NF-kappaB-inducing kinase activates IKK-alpha by phosphorylation of Ser-176. Proc Natl Acad Sci U S A 95(7):3792–3797

18. Xiao G, Fong A, Sun SC (2004) Induction of p100 processing by NF-kappaB-inducing kinase involves docking IkappaB kinase alpha (IKKalpha) to p100 and IKKalpha-mediated phosphorylation. J Biol Chem 279(29): 30099–30105

19. Amir RE et al (2004) Mechanism of processing of the NF-kappa B2 p100 precursor: identification of the specific polyubiquitin chain-anchoring lysine residue and analysis of the role of NEDD8-modification on the SCF(beta-TrCP) ubiquitin ligase. Oncogene 23(14):2540–2547

20. Fagerlund R et al (2008) NF-kappaB p52, RelB and c-Rel are transported into the nucleus via a subset of importin alpha molecules. Cell Signal 20(8):1442–1451

21. Chen FE et al (1998) Crystal structure of p50/p50 heterodimer of transcription factor NF-κB bound to DNA. Nature 391:410–413

22. Britanova L, Makeev V, Kuprash D (2008) In vitro selection of optimal RelB/p52 DNA-binding motifs. Biochem Biophys Res Commun 365(3):583–588

23. Fusco AJ et al (2009) NF-kappaB p52:RelB heterodimer recognizes two classes of kappaB sites with two distinct modes. EMBO Rep 10(2):152–159

24. Siggers T et al (2012) Principles of dimer-specific gene regulation revealed by a comprehensive characterization of NF-kappaB family DNA binding. Nat Immunol 13(1):95–102

25. Wang VY et al (2012) The transcriptional specificity of NF-kappaB dimers is coded within the kappaB DNA response elements. Cell Rep 2(4):824–839

26. Varfolomeev E et al (2012) Cellular inhibitors of apoptosis are global regulators of NF-kappaB and MAPK activation by members of the TNF family of receptors. Sci Signal 5(216):ra22

27. Bertrand MJ et al (2008) cIAP1 and cIAP2 facilitate cancer cell survival by functioning as E3 ligases that promote RIP1 ubiquitination. Mol Cell 30(6):689–700

28. Varfolomeev E et al (2007) IAP antagonists induce autoubiquitination of c-IAPs, NF-kappaB activation, and TNFalpha-dependent apoptosis. Cell 131(4):669–681

29. Vince JE et al (2007) IAP antagonists target cIAP1 to induce TNFalpha-dependent apoptosis. Cell 131(4):682–693

Chapter 8

Systematic Detection of Noncanonical NF-κB Activation

Zhaoxia Qu and Gutian Xiao

Abstract

In unstimulated cells, NF-κB dimers usually exist as latent complexes in the cytoplasm with the IκB (inhibitor of NF-κB) proteins or IκB-like protein p100, the precursor of NF-κB2 mature form p52. Accordingly, there are two major mechanisms leading to NF-κB activation: inducible degradation of IκBs and processing of p100 to generate p52 (selective degradation of the C-terminal IκB-like sequence of p100), which are termed the canonical and noncanonical NF-κB pathways, respectively. While activation of the canonical NF-κB pathway plays critical roles in a wide range of biological processes, the noncanonical NF-κB pathway has important but more restricted roles in both normal and pathological processes. Systematic detection of the noncanonical NF-κB pathway activation is very important for understanding the physiological role of this pathway in biological processes, and for the diagnosis, prevention, and treatment of related diseases. We describe here the methods we employ to detect noncanonical NF-κB activation in cells and tissues. These methods are immunoblotting, co-immunoprecipitation, immunofluorescence, immunohistochemistry, chromatin immunoprecipitation (ChIP) analysis, and electrophoretic mobility shift assay (EMSA). Noncanonical NF-κB-induced gene expression changes can be determined by gene array analysis and quantitative real-time PCR.

Key words NF-κB, NF-κB2, p100, p52, NF-κB-inducing kinase, Immunoblotting (IB) assay, Immunoprecipitation (IP) assay, Electrophoretic mobility shift assay (EMSA), Chromatin immunoprecipitation (ChIP) assay

1 Introduction

The mammalian NF-κB family consists of five closely related DNA binding proteins: RelA (p65), RelB, c-Rel, NF-κB1/p50, and NF-κB2/p52. They can form various homodimers and heterodimers. All five NF-κB members share a highly conserved 300-amino-acid-long N-terminal Rel homology domain (RHD), which is responsible for DNA binding, dimerization, and nuclear translocation. In resting cells, NF-κB dimers are sequestered in the cytoplasm as latent complexes through binding to inhibitors of κB (IκB), a family of ankyrin repeat domain (ARD)-containing proteins that interact with the RHD of NF-κB. Signal induced IκB degradation liberates NF-κB dimers, allowing them to translocate into

Michael J. May (ed.), *NF-kappa B: Methods and Protocols*, Methods in Molecular Biology, vol. 1280,
DOI 10.1007/978-1-4939-2422-6_8, © Springer Science+Business Media New York 2015

the nucleus to regulate gene expression. This IκB degradation-based mechanism of NF-κB activation is called the canonical NF-κB pathway [1, 2].

Among the five mammalian NF-κB members, RelA (p65), RelB, and c-Rel are directly synthesized as mature proteins, and each contains a transactivation domain at their C-termini, which is responsible for induction of gene expression. Unlike RelA (p65), RelB, and c-Rel, the other two NF-κB members, NF-κB1/p50 and NF-κB2/p52, do not contain a transactivation domain, and are synthesized indirectly as larger precursors, p105 and p100, respectively. Both p105 and p100 contain characteristic ARD of IκB at their C termini, and function as IκB-like NF-κB inhibitors. Unlike the complete degradation of IκB by proteasome, proteasome-mediated processing of p105 and p100 selectively degrades their ARD-containing C-terminal regions, and leaves intact their N-terminal parts, which become mature forms of NF-κB1/p50 and NF-κB2/p52, respectively. Thus, processing of p105 and p100 not only serves to generate mature NF-κB members but also disrupts of the IκB-like function of these precursor proteins, resulting in release of p100/p105-binding NF-κB dimers and their subsequent nuclear translocation. While the processing of p105 is constitutive, p100 processing is tightly controlled and highly inducible. The p100 processing-mediated NF-κB activation is termed the noncanonical NF-κB pathway [2].

Unlike the canonical pathway, which can be activated by a plethora of stimuli, the noncanonical NF-κB pathway is induced by only a handful of stimuli, including B-cell activating factor (BAFF), CD40 ligand, lymphotoxin β (LTβ), receptor activator of NF-κB ligand (RANKL), tumor necrosis factor (TNF)-like weak inducer of apoptosis (TWEAK) [3–8]. In addition, while the canonical NF-κB pathway activation is rapid, activation of the noncanonical NF-κB pathway is slow and depends on de novo protein synthesis of NF-κB-inducing kinase (NIK), a serine/threonine kinase in the MAP3K family [9]. Although its mRNA expression is relatively abundant, the protein level of NIK is normally very low because of its constitutive degradation via a tumor necrosis factor receptor-associated factor 3 (TRAF3)-dependent mechanism [10]. TRAF3 functions as a scaffold between NIK and TRAF2, which in turn recruits cellular inhibitors of apoptosis 1 and 2 (c-IAP1/2) into the NIK complex [10–12]. Within the complex, c-IAP1 or c-IAP2 acts as the E3 ubiquitin ligase to mediate NIK polyubiquitination and proteolysis, thereby keeping its abundance below the threshold required for its function. In response to noncanonical NF-κB stimuli, either TRAF2 and TRAF3 or c-IAP1 and c-IAP2 are degraded by the proteasome, resulting in stabilization and accumulation of the newly synthesized NIK, thereby allowing NIK proteins to form oligomers and cross-phosphorylate each other for their activation [12–14]. Self-activated NIK in turn activates the IκB kinase (IKK) complex and specifically recruits IKKα, also known as IKK1, into the p100 complex to phosphorylate p100, leading to p100 ubiquitination by the β-transducin repeat-containing protein (β-TrCP) E3 ubiquitin ligase and processing by the proteasome to generate p52 [8, 15–17]. The processed product p52 together with its NF-κB binding partner translocates into the nucleus to regulate gene expression.

The noncanonical NF-κB signaling pathway plays important roles in immune responses and lymphoid organogenesis, as well as cell proliferation and survival. Germ-line knockout of *nfkb2* gene without expression of both p100 and p52 results in severe defects in B-cell function and impairment in the formation of proper architecture in peripheral lymphoid organs [18, 19]. Similarly, mice deficient in key signaling molecules in activating the noncanonical NF-κB pathway, such as NIK-deficient, NIK-aly mutation, and IKKα mutation knock-in mice, also displayed abnormalities in lymphoid tissue development and B cell functions [12, 20–22]. In addition, disruption of different receptors involved in the noncanonical NF-κB activation results in distinct deficiencies in the immune system. For example, LTβR knockout mice show deficiencies in the development of secondary lymphoid organs such as peripheral lymph nodes, Peyer's patches, as well as alterations of the splenic microarchitecture [23]. Mice with BAFF-R deficiency exhibit reduced survival of peripheral B cells and failure of T cell-dependent antibody formation [24–26]. CD40-null mice present with impaired immunoglobulin class switching and germinal center formation [27]. RANK-deficient mice manifest profound osteopetrosis resulting from an apparent block in osteoclast differentiation, and a marked deficiency of B cells in the spleen. Although they retain mucosal-associated lymphoid tissues including Peyer's patches, these mice exhibit complete absence of all other peripheral lymph nodes [28].

Given the critical roles of the noncanonical NF-κB signaling in controlling immune system organogenesis and responses, as well as cell proliferation and survival, it is plausible that aberrant activation of this pathway can cause immune disorders and cancers [1, 2]. Deletion of p100 C-terminus with constitutive p52 production leads to abnormal lymphocyte proliferation and cytokine production, causes enlargement of lymph nodes, histopathological alterations of other hematopoietic tissues, and marked gastric hyperplasia, resulting in early postnatal death in mice [29]. Mice overexpressing BAFF, the major physiological inducer of p100 processing in B-cells, also show similar lymphocytic disorders, and additional autoimmune features [30, 31]. Additionally, B-lymphocyte specific deletion of the negative regulator TRAF3, which causes hyper-activation of the noncanonical NF-κB signaling in B lymphocytes, leads to development of B-cell lymphoma in mice [32]. Consistent with the mouse studies, aberrantly persistent activation of this noncanonical NF-κB signaling pathway in human is associated with various autoimmune diseases and tumors, particularly lymphoid malignancies.

In humans, genetic mutations or epigenetic alterations of noncanonical NF-κB signaling regulators, which causes positive regulator hyper-activation or negative regulator inactivation, leading to constitutive activation of this signaling pathway, has frequently been found in hematologic cancers, such as multiple myeloma, diffuse large B-cell lymphoma, Waldenstrom's macroglobulinemia, and mucosal associated lymphoid tissue lymphomas [33–41]. Examples include mutations causing over-activation of receptors, such as CD40, BAFF-R, RANK and LTβR [35–39]. In addition, inactivating mutations or epigenetic alterations of TRAF-cIAP complex members, cIAP1/2, TRAF2, and TRAF3, which negatively regulates NIK stability, have also been found in many cases [36–41]. Furthermore, in some cases NIK, the key regulator of the noncanonical NF-κB signaling, has been found to be mutated, amplified, or cleaved by API2-MALT1 fusion oncoprotein created by chromosome translocation to generate a C-terminal fragment that retains kinase activity and is resistant to proteasomal degradation, leading to constitutive p100 processing [33, 34, 36, 38, 39].

Constitutive processing of p100, caused by the loss of its C-terminal processing inhibitory domain (PID) due to *nfκb2* gene rearrangements [8], is associated with the development of various lymphomas/leukemia such as cutaneous T-cell lymphomas, B-cell non-Hodgkin lymphomas, chronic lymphocytic leukemia, and myelomas in humans [42, 43]. In contrast to the inducible processing of p100, which mainly occurs in the cytoplasm, the constitutive processing of these p100 mutants (p100ΔCs) occurs in the nucleus, and requires their DNA binding. After binding to κB promoters, the p100ΔCs recruit the proteasome to form a stable proteasome/p100ΔC/ κB promoter complex. This complex then mediates the p100ΔC processing by endoproteolytic cleavage, and subsequent degradation of the cleaved C-terminal portion [44].

Aberrant persistent activation of noncanonical NF-κB signaling is also associated with T-cell transformation by the human T-cell leukemia virus type I (HTLV-I), an etiological agent of an acute and fatal T-cell malignancy, adult T cell leukemia (ATL) [45]. The pathogenic processing of p100 in HTLV-I-transformed cells is induced by the virus-encoded oncoprotein Tax. Tax is largely responsible for the oncogenic ability of HTLV-1, and potently activates both canonical and noncanonical NF-κB pathways. In activation of the canonical NF-κB pathway, Tax physically interacts with IKKγ, recruiting Tax to the IKK catalytic subunits, IKKα and IKKβ, leading to IKK activation, IκB phosphorylation by the activated IKK, and subsequent ubiquitination and degradation of IκB proteins [46]. In activation of the noncanonical NF-κB pathway, Tax directly binds to p100 and IKKγ, specifically recruiting

IKKα, but not IKKβ, to p100, triggering p100 phosphorylation, ubiquitination and ultimate processing [47]. Besides the β-TrCP-dependent ubiquitination, Tax-induced p100 processing also employs an alternative β-TrCP-independent mechanism, which involves p100 nuclear translocation [48]. Tax-induced canonical NF-κB activation induces expression of p100, facilitating noncanonical NF-κB activation. Additionally, Tax-induced noncanonical NF-κB activation suppresses expression of WW domain–containing oxidoreductase (WWOX), a tumor suppressor that specifically inhibits canonical NF-κB [49]. This positive feed-forward loop potentiates Tax-induced both canonical and noncanonical NF-κB activation.

Besides HTLV-1, other oncogenic viruses, such as Kaposi's sarcoma-associated herpesvirus/human herpesvirus 8 (KSHV/HHV-8) and Epstein-Barr virus (EBV) also employ NF-κB for their oncogenic action [50]. KSHV is associated with the development of Kaposi's sarcoma, primary effusion lymphoma (PEL), and multicentric Castleman's disease (MCD). EBV infection is etiologically associated with several malignancies, such as Burkitt's lymphoma, gastric carcinomas, nasopharyngeal carcinomas, and Hodgkin's disease. Both the canonical and noncanonical NF-κB pathways were found to be constitutively activated in cancer cells infected with these viruses. In addition, aberrant activation of the noncanonical NF-κB pathway has also been observed in several other tumors such as melanoma, breast cancer, and pancreatic cancer [51–53].

In addition to cancer, deregulated noncanonical NF-κB signaling has been found in certain autoimmune diseases. Increased plasma concentration of BAFF, is often seen in patients with autoimmune diseases, such as systemic lupus erythemetosus, rheumatoid arthritis, and Sjögren's syndrome [54–56]. BAFF primarily activates the noncanonical NF-κB pathway in B cells. Overproduction of BAFF may induce aberrant noncanonical NF-κB activation in autoreactive B-cells, leading to their survival instead of normal elimination, which could potentially contribute to the development of autoimmune diseases.

In noncanonical NF-κB activation, p100 is processed to generate p52. The processed product p52 together with its binding partner translocate into the nucleus, bind to its target DNA, and induce target genes expression changes. To detect noncanonical NF-κB activation, we first need to examine whether processing of p100 to generate p52 is increased compared to control. This can be found out by immunoblotting using anti-p100 N-terminal antibody. Next, we can investigate nuclear translocation of the processed product p52 by immunoblotting with separation of cytosolic and nuclear fractions, as well as immunofluorescence and/or immunohistochemistry. The association of p52 containing NF-κB dimers with target DNA sequences can be revealed by chromatin immunoprecipitation (ChIP) analysis and electrophoretic mobility shift assay (EMSA). The noncanonical NF-κB-induced gene expression changes can be determined by gene array analysis and quantitative real-time PCR.

To investigate the upstream events leading to noncanonical NF-κB activation, we need to inspect signaling molecules upstream of p100 processing. We can examine protein levels of NIK, IKKα, and their association by immunoblotting, co-immunoprecipitation, immunofluorescence, and/or immunohistochemistry. We can also examine protein levels of TRAF2, TRAF3, cIAP1/2 and their association with each other by immunoblotting, co-immunoprecipitation, immunofluorescence, and/or immunohistochemistry. As noncanonical NF-κB activation is known to be induced by BAFF, CD40 ligand, LTβ, RANKL, TWEAK, we can also examine the levels of these stimuli and their receptors. In the following sections we will describe methods used for detection of noncanonical NF-κB activation.

2   Materials

2.1   Immunoblotting (IB)

1. 1.5 M Tris–HCl (pH 8.8): To prepare 1 l, dissolve 181.65 g Tris base in ~800 ml of diH$_2$O. Adjust the pH to 8.8 with concentrated HCl. Bring up the volume to 1 l with diH$_2$O. Make sure to let the solution cool down to room temperature before making the final pH adjustment.

2. 10 % (w/v) ammonium persulfate (APS): Dissolve 1 g ammonium persulfate in 10 ml of H$_2$O and store as 1 ml aliquots at –20 °C.

3. *N, N, N′*-tetramethyl-ethylenediamine (TEMED).

4. Water-saturated n-butanol: Combine 50 ml of n-butanol with 5 ml of diH$_2$O in one bottle and shake. Use the top phase only. Store at room temperature.

5. Phosphate buffered saline (PBS): 137 mM NaCl, 2.7 mM KCl, 10 mM Na$_2$HPO$_4$, 1.8 mM KH$_2$PO$_4$. Prepare 1× PBS by diluting 10× PBS 1:10 with diH$_2$O and sterilize. To prepare 1 l of 10× PBS stock, dissolve 80 g of NaCl, 2 g of KCl, 14.4 g of Na$_2$HPO$_4$, 2.4 g of KH$_2$PO$_4$, in 800 ml of diH$_2$O. Adjust the pH to 7.4 with HCl or NaOH, and then add diH$_2$O to 1 l. Dispense the solution into aliquots and sterilize by autoclaving for 20 min at 15 psi (1.05 kg/cm^2) on liquid cycle or by filter sterilization. Store at room temperature.

6. Tris Buffered Saline (TBS): 20 mM Tris, 150 mM NaCl. Prepare 1× TBS by diluting 10× TBS 1:10 with diH$_2$O and sterilize. To prepare 1 l of 10× TBS stock, dissolve 24.2 g of Tris, 87.6 g of NaCl in 800 ml of H$_2$O. Adjust pH to 7.6 with

concentrated HCl and then add H_2O to 1 l. Autoclave and store at room temperature.

7. TBST: 20 mM Tris, 150 mM NaCl, 0.1 % (v/v) Tween-20.

8. 1 M Tris–HCl: To prepare 1 l of 1 M Tris–HCl, dissolve 121.1 g of Tris base in 800 ml of H_2O. Adjust to the required pH (e.g., pH 7.4, 7.6, 8.0) by adding concentrated HCl.

   Allow the solution to cool to room temperature before making final adjustments to the pH. Adjust the volume of the solution to 1 l with H_2O. Dispense into aliquots and sterilize by autoclaving (*see* **Note 1**).

9. 0.5 M Ethylenediaminetetraacetic acid (EDTA) (pH 8.0): To prepare 1 l, add 186.1 g of disodium EDTA · $2H_2O$ to 800 ml of H_2O. Stir vigorously on a magnetic stirrer. Adjust the pH to 8.0 with NaOH (~20 g of NaOH pellets). Bring the volume to 1 l with diH_2O. Autoclave to sterilize. The disodium salt of EDTA will not go into solution until the pH of the solution is adjusted to ~8.0 by the addition of NaOH.

10. 1 M Dithiothreitol (DTT): Dissolve 1.5 g of DTT in 10 ml diH_2O. Store as 1 ml aliquots in dark at –20 °C. Do not autoclave DTT-containing solutions.

11. 100 mM phenylmethylsulfonyl fluoride (PMSF): To prepare a 100 mM solution, add 17.4 mg of PMSF per milliliter of isopropanol. Store at –20 °C (*see* **Note 2**).

12. Radio Immuno Precipitation Assay (RIPA) Cell lysis buffer: 50 mM Tris–HCl (pH 7.4), 150 mM NaCl, 1 mM EDTA, 0.25 % (w/v) sodium deoxycholate, 1 % (v/v) Nonidet P-40, 10 mM sodium fluoride (NaF), 1 mM sodium orthovanadate, 1 mM DTT, and 1 mM PMSF. To prepare 1 l, add 50 ml of 1 M Tris–HCl (pH 7.4), 30 ml of 5 M NaCl, 2 ml of 0.5 M EDTA (pH 8.0), 2.5 g of sodium deoxycholate, 10 ml of Nonidet P-40, bring volume to 1 l with diH_2O (*see* **Note 3**).

13. Buffer B for subcellular fractionation: 10 mM HEPES, pH 7.9, 10 mM KCl, 0.4 % Nonidet P-40, 0.1 mM EDTA, 0.1 mM EGTA, 1 mM dithiothreitol, 1 mM phenylmethylsulfonyl fluoride (*see* **Note 3**).

14. Buffer C for subcellular fractionation: 20 mM HEPES, pH 7.9, 0.4 M NaCl, 0.1 mM EDTA, 1 mM dithiothreitol, 1 mM phenylmethylsulfonyl fluoride (*see* **Note 3**).

15. 0.5 M Tris–HCl pH 6.8: To prepare 100 ml, add 6 06 g of Tris base to ~60 ml of diH_2O. Adjust to pH 6.8 with concentrated HCl. Bring the volume to 100 ml with diH_2O. Store at 4 °C.

16. 10 % SDS (sodium dodecyl sulfate) (w/v) in diH_2O.

17. 0.1 % bromophenol blue (w/v) in diH_2O.

18. 2× Laemmli buffer: 125 mM Tris–HCl, 4 % SDS, 20 % glycerol, 0.005 % bromophenol blue. To prepare 10 ml, mix

together 2.5 ml of 0.5 M Tris–HCl (pH 6.8), 4 ml of 10 % SDS, 2 ml of glycerol, 0.5 ml of 0.1 % bromophenol blue, and 1 ml of diH$_2$O. Check the pH and adjust to pH 6.8 if necessary. Store the buffer as 1 ml aliquots at −80 °C. Add 50 µl of 2-mercaptoethanol per ml of buffer immediately prior to use.

19. Running buffer: 25 mM Tris, 192 mM glycine, 0.1 % (w/v) SDS. Prepare 1× running buffer by diluting 10× running buffer stock 1:10 with diH$_2$O. To prepare a 10× running buffer stock, dissolve 30.3 g of Tris base, 144 g of glycine, 10 g of SDS in 1 l H$_2$O. Check the pH and adjust to 8.3 if necessary. Store at room temperature.

20. Transfer buffer: 25 mM Tris, 192 mM glycine, 20 % (v/v) methanol. To prepare 1 l of 1× transfer buffer: mix together 100 ml of 10× transfer buffer, 200 ml of methanol and 700 ml dH$_2$O. Store at 4 °C. To prepare 1 l of 10× transfer buffer, dissolve 30.3 g of Tris, 144 g of glycine in diH$_2$O, bring volume up to 1 l with diH$_2$O.

21. Blocking buffer: 5 % w/v nonfat dry milk in 1× TBST. For 20 ml, add 1 g of nonfat dry milk to 20 ml of 1× TBST and mix well.

22. Primary antibody dilution buffer: 1× TBST with 5 % BSA or 5 % nonfat dry milk as indicated on primary antibody datasheet; for 20 ml, add 1.0 g of BSA or nonfat dry milk to 20 ml of 1× TBST and mix well.

23. Prestained molecular weight marker such as Bio-Rad precision plus protein all blue standards (Bio-Rad, Hercules, CA, USA).

2.2 Co-immunoprecipitation (Co-IP)

1. PBS.
2. Protein A/G-agarose beads.
3. Primary antibodies against noncanonical NF-κB signaling proteins including p100/p52, RelB, NIK, TRAF3, c-IAP1/2 and IKKα.
4. RIPA cell lysis buffer.
5. 2× Laemmli buffer.

2.3 Immunohistochemistry (IHC)

1. PBS.
2. O.C.T (optimal cutting temperature).
3. Acetone.
4. Methanol.
5. Neutral buffered formalin (3.7 % formaldehyde): to prepare 1 l, dissolve 6.5 g of dibasic sodium phosphate, anhydrous (Na$_2$HPO$_4$), and 4 g of monobasic sodium phosphate, monohydrate (NaH$_2$PO$_4$·H$_2$O), in 900 ml of diH$_2$O. Adjust pH to

7.4 if necessary with 1 M NaOH or 1 M HCl. Then add 100 ml of 37 % formaldehyde solution (formalin) and mix.

6. Xylene.

7. Ethanol.

8. Hematoxylin.

9. TBST.

10. 10 mM sodium citrate buffer pH 6.0: to prepare 1 l, add 2.94 g of sodium citrate trisodium salt dihydrate to 1 l diH$_2$O. Adjust pH to 6.0.

11. 1 mM EDTA pH 8.0.

12. TE buffer pH 9.0: 10 mM Tris–HCl pH 9.0; 1 mM EDTA. To prepare 1 l, add 1.21 g of Tris base and 0.372 g of EDTA to 950 ml of diH$_2$O. Adjust pH to 9.0, and then adjust final volume to 1,000 ml with diH$_2$O.

13. Pepsin working solution: 1 mg/ml in Tris–HCl pH 2.0.

14. Trypsin Stock Solution (0.5 %): Dissolve 50 mg of trypsin in 10 ml of diH$_2$O. Store as 1 ml aliquots at –20 °C.

15. Calcium chloride stock solution (1 %): Dissolve 0.1 g of calcium chloride in 10 ml of diH$_2$O. Store at 4 °C.

16. Trypsin working solution: to prepare 10 ml, add 1 ml of trypsin stock solution (0.5 %), 1 ml of calcium chloride stock solution (1 %) to 7 ml of diH$_2$O. Adjust pH to 7.8 with 1 N NaOH and bring volume to 10 ml with diH$_2$O. Store as aliquots at –20 °C.

17. Blocking solution: TBST/5 % normal serum from the species in which the secondary antibody was raised. Add 250 μl normal sera to 5 ml of 1× TBST.

18. 3 % hydrogen peroxide: to prepare 50 ml, add 5 ml of 30 % H$_2$O$_2$ to 45 ml of diH$_2$O.

19. Biotinylated secondary antibody.

20. Labeled streptavidin biotin (LSAB)2 system HRP (horseradish peroxidase) for the qualitative detection of antigens by light microscopy (Dako, Carpinteria, CA, USA).

21. 3,3′-diaminobenzidine (DAB) reagent or another suitable substrate for detection of HRP in tissue sections.

2.4 Immuno-fluorescence

1. PBS.

2. TBST.

3. Normal serum from the species in which the secondary antibody was raised.

4. Blocking solution: TBST containing 5 % normal serum from the species in which the secondary antibody was raised. To 5 ml of 1× TBST add 250 μl of normal serum.

5. Primary antibodies against noncanonical NF-κB signaling proteins including p100/p52, RelB, NIK, TRAF3, c-IAP1/2, and IKKα.

6. Biotinylated secondary antibodies.

7. Streptavidin–fluorescein.

8. Fluorochrome-conjugated secondary antibody.

9. Hoechst or DAPI DNA Dyes.

2.5 Chromatin Immunoprecipitation (ChIP)

1. PBS+ Protease Inhibitor Cocktail (PIC: Cell Signaling Technology, Danvers, MA, USA): Prepare PBS + PIC by diluting concentrated PIC to 1× in cold PBS. Prepare 3 ml of PBS + PIC per 25 mg of tissue to be processed and keep on ice.

2. 1.25 M glycine (10× stock): To prepare 100 ml, dissolve 9.4 g of glycine in diH$_2$O to 100 ml.

3. Lysis buffer 1: 50 mM KCl; 0.5 % (v/v) IGEPAL CA-630 (Sigma-Aldrich, St. Louis, MO, USA); 25 mM HEPES (pH 7.9); 10 μg/ml Leupeptin; 20 μg/ml Aprotinin; 125 μM DTT; 1 mM PMSF (*see* **Note 4**).

4. Wash buffer 2: 50 mM KCl; 25 mM HEPES (pH 7.9); 10 μg/ml Leupeptin; 20 μg/ml Aprotinin; 125 μM DTT; 1 mM PMSF (*see* **Note 4**).

5. SDS lysis buffer 3: 1 % SDS; 10 mM EDTA; 50 mM Tris–HCl (pH 8.0); 10 μg/ml Leupeptin; 20 μg/ml Aprotinin; 125 μM DTT; 1 mM PMSF (*see* **Note 4**).

6. Buffer 4: 1 % Triton X-100; 2 mM EDTA; 150 mM NaCl; 20 mM Tris–HCl (pH 8.0).

7. Buffer 5 (for bead preparation): 1 mM EDTA; 10 mM Tris–HCl (pH 8.0).

8. Buffer 6 (Bead washing buffer 1): 0.1 % SDS; 1 % Triton X-100; 2 mM EDTA; 20 mM Tris–HCl (pH 8.0); 150 mM NaCl.

9. Buffer 7 (Bead washing buffer 2): 0.1 % SDS; 1 % Triton X-100; 2 mM EDTA; 20 mM Tris–HCl (pH 8.0); 500 mM NaCl.

10. Buffer 8 (bead washing buffer 3): 250 mM LiCl; 1 % NP-40; 1 % sodium deoxycholate; 1 mM EDTA; 10 mM Tris–HCl (pH 8.0) (*see* **Note 5**).

11. Buffer 9 (Extract buffer): 1 % SDS; 0.1 M NaHCO$_3$.

12. TE buffer: 1 mM EDTA; 10 mM Tris–HCl (pH 8.0).

2.6 Quantitative Real-Time Polymerase Chain Reaction (PCR)

1. Nuclease-free H$_2$O.

2. Trizol (Life Technologies, Carlsbad, CA, USA) or RNeasy purification kit (Qiagen, Valencia, CA, USA) for RNA extraction.

3. Primers for genes of interest.

4. Template DNA generated by reverse transcription using a suitable reverse transcriptase enzyme (e.g., from Life Technologies).

5. 2× SYBR green PCR master mix (Life Technologies).

2.7 Gene Array Analysis

1. RNeasy purification kit (Qiagen).

2. Agilent platform (2100 Bioanalyzer; Agilent Technologies, Palo Alto, CA, USA).

3. Gene array analysis software.

2.8 Electrophoretic Mobility Shift Assay (EMSA)

1. PBS.

2. Hypotonic lysis buffer A: 10 mM HEPES (pH 7.9); 1.5 mM $MgCl_2$; 10 mM KCl; 0.5 mM DTT; 0.2 mM PMSF; Protease inhibitor cocktail (*see* **Note 3**).

3. Hypertonic buffer B: 20 mM HEPES (pH 7.9); 1.5 mM $MgCl_2$; 420 mM NaCl; 0.2 mM EDTA pH 8.0; 25 % (v/v) Glycerol; 0.5 mM DTT; 0.2 mM PMSF; Protease inhibitor cocktail (*see* **Note 3**).

4. T4 Polynucleotide Kinase 10× Buffer: 700 mM Tris–HCl (pH 7.6), 100 mM $MgCl_2$, 50 mM DTT.

5. 5× TBE buffer: 450 mM Tris base; 450 mM Boric acid; 10 mM EDTA (pH 8.0).

6. 0.5× TBE running buffer: Prepare by diluting 5× TBE buffer 1:10 with diH_2O.

7. 5× gel shift binding buffer: 50 mM Tris–HCl (pH 7.5); 5 mM $MgCl_2$; 250 mM KCl; 2.5 mM EDTA pH 8.0; 15 % (v/v) Glycerol; 2.5 mM DTT.

8. 10× gel loading buffer: 250 mM Tris–HCl pH 7.5; 0.02 % bromophenol blue; 40 % glycerol.

9. Poly dI-dC (Sigma-Aldrich) to prevent nonspecific DNA binding.

10. DNA 3′ Biotin end labeling kit such as the Pierce kit available from Thermo Scientific (Waltham, MA, USA).

11. Positively charged nylon membrane (e.g., Biodyne™ B Membrane, Thermo Scientific).

12. Chemiluminescent nucleic acid detection module kit such as the Pierce kit available from Thermo Scientific.

---

3 Methods

3.1 Immunoblotting

3.1.1 Gel Casting

1. Locate all the materials needed for the SDS PAGE electrophoresis system: tank, lid with power cables, electrodes and power supply, cell buffer dam, casting stands, casting frames, combs, glass plates, and spacers.

Table 1
Percentage of SDS-PAGE gels recommended for resolution of specific target protein M.W. ranges

Acrylamide %	M.W. range (kDa)
7	50–500
10	20–300
12	10–200
15	3–100

Table 2
Volumes of solutions required to make resolving gels of different percentages

Acrylamide percentage	6 %	8 %	10 %	12 %	15 %
H_2O	5.2 ml	4.6 ml	3.8 ml	3.2 ml	2.2 ml
Acrylamide–Bis-acrylamide (30 %/0.8 % w/v)	2 ml	2.6 ml	3.4 ml	4 ml	5 ml
1.5 M Tris–HCl (pH 8.8)	2.6 ml	2.6 ml	2.6 ml	2.6 ml	2.6 ml
10 % (w/v) SDS	100 µl	100 µl	100 µl	100 µl	100 µl
10 % (w/v) ammonium persulfate (APS)	100 µl	100 µl	100 µl	100 µl	100 µl
TEMED	10 µl	10 µl	10 µl	10 µl	10 µl

2. For each SDS-PAGE gel, assemble one small and one large glass plate separated by spacers on both sides with casting frame on casting stand.

3. Prepare the resolving gel solution following the volumes in Tables 1 and 2. Swirl the solution gently but thoroughly. Choose resolving gel acrylamide percentage according to the size of the target protein in the sample (Table 1). Volumes of the gel solutions differ according to the size, thickness, and number of gel casting. Adjust the volume proportionally. For a 10 ml resolving gel (*see* **Note 6**) follow the volumes provided in Table 2.

4. Pipet appropriate amount of resolving gel solution into the gap between the glass plates.

5. To make the top of the resolving gel be horizontal and prevent drying, carefully layer 1 ml of water-saturated butanol on top of the gel solution.

6. Wait for 20–30 min to let the gel solidify.

7. Remove layer of liquid on top of gel.

8. Prepare stacking gel as shown in Table 3. Adjust the volume proportionally as needed.

Table 3
Volumes of solutions required to make 5 ml of stacking gel

H₂O	2.975 ml
0.5 M Tris–HCl (pH 6.8)	1.25 ml
Acrylamide–Bis-acrylamide (30 %/0.8 % w/v)	0.67 ml
10 % (w/v) SDS	50 μl
10 % (w/v) ammonium persulfate (APS)	50 μl
TEMED	5 μl

9. Pipet stacking gel solution on top of the resolving gel until an overflow.

10. Insert the well-forming comb without trapping air under the teeth. Wait for 20–30 min to let the gel solidify.

11. Make sure of complete solidification of the stacking gel and take out the comb. Take the glass plates out of the casting frame and set them in the cell buffer dam. Pour 1× running buffer into the inner chamber and keep pouring after overflow until the buffer surface reaches the required level in the outer chamber.

12. Remove gel and holder from casting cell and mount onto electrode assembly.

13. Insert electrode assembly into tank.

14. Remove combs and fill tank with 1× running buffer.

3.1.2 Sample Preparation from Suspension Cell Culture

1. Pellet the cells by centrifugation, and then aspirate the supernatant and wash cells once with ice-cold PBS.

2. Lyse cells by adding ice-cold RIPA buffer supplemented with a protease inhibitor mixture. Keep on ice for 30 min.

3. Centrifuge the cell lysate at $12,000 \times g$ for 20 min at 4 °C.

4. Transfer the supernatant to a fresh tube. Keep on ice.

3.1.3 Sample Preparation from Adherent Cell Culture

1. Place the cell culture on ice. Aspirate the medium from the cultures and wash the cells with ice-cold PBS.

2. Lyse cells by adding ice-cold RIPA buffer supplemented with a protease inhibitor mixture. Immediately scrape the cells off the plate and transfer the extract to a microcentrifuge tube. Keep on ice for 30 min.

3. Centrifuge the cell lysate at $12,000 \times g$ for 20 min at 4 °C.

4. Transfer the supernatant to a fresh tube and keep on ice.

3.1.4 Sample Preparation from Tissues

1. Break a big piece of frozen tissue into small pieces while frozen.
2. Take no more than 50–100 mg of frozen tissue, add 5× volume of RIPA buffer supplemented with a protease inhibitor mixture, homogenize the tissue using a mortar–pestle or sonication and centrifuge the tissue lysate at $12,000 \times g$ for 15 min at 4 °C.
3. Transfer the supernatant to a fresh tube.
4. Centrifuge the supernatant again at $12,000 \times g$ for 20 min at 4 °C.
5. Transfer the supernatant to a fresh tube and keep on ice.

3.1.5 Subcellular Fractionation of Samples

1. Incubate cells in Buffer B supplemented with protease inhibitor mixture.
2. Let the cells swell on ice for 15 min, then centrifuge at $12,000 \times g$ for 5 min at 4 °C.
3. Transfer the supernatant, which is the cytoplasmic extract, to a new tube.
4. Wash the pellet (nucleus) three times with Buffer B.
5. Suspend the pellet in Buffer C supplemented with protease inhibitor mixture, and rotate at 4 °C for 15 min, then centrifuge at $12,000 \times g$ for 5 min at 4 °C. The resultant supernatant is the nuclear extract (*see* **Note 7**).

3.1.6 SDS-PAGE and Transfer

1. Measure the protein concentration in each sample using NanoDrop spectrophotometer, protein staining by Bradford reagent or another protein assay system such as BCA.
2. Take the same protein amount from each sample (e.g., 10–20 µg of protein), add RIPA buffer to a final volume of 10 µl then add 10 µl 2× Laemmli buffer.
3. Heat each sample to 95–100 °C for 5 min; cool on ice.
4. Microcentrifuge for 5 min.
5. Load each sample onto SDS-PAGE gel. In addition, load a prestained molecular weight marker to determine molecular weights and verify the electro-transfer.
6. Set an appropriate voltage (e.g., 100 V) and run the electrophoresis (*see* **Note 8**).
7. Electro-transfer to nitrocellulose or PVDF (polyvinylidene fluoride) membrane in 1× transfer buffer (*see* **Note 9**).

3.1.7 Immunostaining of the Membrane

1. (Optional) After transfer, wash membrane with TBS for 5 min at room temperature.
2. Incubate membrane in blocking buffer for 1 h at room temperature.
3. Wash three times for 5 min each with TBST.

4. Dilute primary antibody in primary antibody dilution buffer as recommended in the product datasheet. Incubate membrane in primary antibody solution with gentle agitation overnight at 4 °C.

5. Wash three times for 5 min each with TBST.

6. Dilute HRP-conjugated species-appropriate secondary antibody in blocking buffer. Incubate membrane with the secondary antibody solution with gentle agitation for 1 h at room temperature.

7. Wash three times for 5 min each with TBST.

8. Incubate the blot with the Chemiluminescent-HRP substrate made up according to the particular manufacturer's instructions.

9. Expose the blot to X-ray film for the appropriate time period that yields desired results.

3.2 Co-immuno-precipitation (Co-IP)

1. Prepare sample lysates as described in the immunoblotting protocol (Subheading 3.1).

2. Wash protein A/G-agarose beads twice with PBS and make a 50 % protein A/G agarose working solution (in PBS).

3. (Optional pre-clearing) Add 100 μl of 50 % protein A/G agarose per ml of sample solution. Incubate with gentle rocking for 30–60 min at 4 °C. Centrifuge at $14,000 \times g$ for 15 min at 4 °C, transfer the supernatant to new tubes, and discard protein A/G-agarose beads.

4. Quantify total protein using BCA or another method.

5. Dilute the total protein to 1 μg/μl with PBS to reduce the concentrations of detergents. If the concentration of the target protein is low, dilute the total protein to 10 μg/μl.

6. Add the appropriate amount of primary antibody recommended for immunoprecipitation (typically 1–10 μg). Incubate with gentle rocking overnight at 4 °C.

7. Add 100 μl of 50 % protein A/G agarose beads per ml of sample solution. Incubate with gentle rocking for 1–3 h at 4 °C.

8. Microcentrifuge $14,000 \times g$ for 1 min at 4 °C. Wash the pellet five times with ice-cold cell lysis buffer. Keep on ice during washes.

9. Resuspend the pellet with 2× Laemmli buffer. Vortex, then microcentrifuge for 1 min at $14,000 \times g$.

10. Heat the sample to 95–100 °C for 2–5 min and microcentrifuge for 1 min at $14,000 \times g$.

11. Collect the supernatant. Proceed to SDS-PAGE, transfer to membrane, and analyze by immunoblotting (Subheading 3.1) (*see* **Note 10**).

3.3 Immun-ohistochemistry (IHC)

3.3.1 Cell Culture, Cytology Smears, and Cytospin Slide Preparation

1. Grow cultured cells on sterile glass cover slips or slides, or prepare cytospin or cytology smears. Wash briefly with PBS and fix cells by either: (a) 5 min in $-10\ ^{\circ}$ C methanol, air-dry (recommended method); (b) 2 min in cold acetone, air-dry; (c) 15 min in 10 % formalin in PBS (keep wet).

2. Wash in three changes of PBS.

*3.3.2 Frozen Tissue Section Slide Preparation (See **Note 11**)*

1. Freeze tissue in cryo-embedding medium (e.g., O.C.T.) on top of a block mostly immersed in liquid nitrogen.

2. Cut 4–8-µm thick cryostat sections. Mount sections on slides at room temperature. Slides may be stored at $-70\ ^{\circ}$ C.

3. Thaw slides at room temperature prior to fixing and staining.

4. Fix slides in cold acetone for 10 min and keep refrigerated (other fixation procedures may also be used).

5. Wash in three changes of PBS.

3.3.3 Formalin-Fixed, Paraffin-Embedded Tissue Section Slide Preparation

1. Fix tissues in neutral-buffered formalin and embed in paraffin blocks.

2. Cut 4–6 µm thick tissue sections, and apply to slides. Air-dry the slides.

3. Incubate the slides at 56–58 °C for 2 h.

4. Deparaffinize the sections 3×5 min with xylene (*see* **Note 12**).

5. Hydrate the sections gradually through graded alcohols: 2×10 min in 100 % ethanol, then 2×10 min in 95 % ethanol.

6. Wash sections 2×5 min in diH_2O.

3.3.4 Heat-Induced Antigen Retrieval

Certain antigenic determinants are masked by formalin fixation and paraffin embedding and may be retrieved by this heat-induced method or an enzymatic method (Subheading 3.3.5) (*see* **Note 13**).

1. Preheat to 95–100 °C steamer or water bath with staining dish containing either 10 mM sodium citrate buffer pH 6.0. 1 mM EDTA pH 8.0 or TE Buffer pH 9.0 (*see* **Note 14**).

2. Immerse slides in the staining dish. Place the lid loosely on the staining dish and incubate for 15–30 min (*see* **Note 15**).

3. Turn off steamer or water bath and remove the staining dish to room temperature and allow the slides to cool for 30 min.

4. Wash the sections 3×5 min with TBST.

3.3.5 Enzymatic Antigen Retrieval

1. Cover pre-warmed sections with pre-warmed pepsin working solution or trypsin working solution.

2. Incubate for 10–20 min at 37 °C in a humidified chamber (*see* **Note 16**).

3. Wash the sections 3×5 min with TBST.

3.3.6 Permeabilization	This step is only required for intracellular antigen staining (*see* **Note 17**).

1. Incubate the samples for 10 min with PBS containing 0.25 % Triton X-100 (or 100 μM digitonin or 0.5 % saponin) (*see* **Note 18**).

2. Wash the samples 3×5 min with PBS.

3.3.7 Staining

1. Block each section with blocking solution for 1 h at room temperature.

2. Remove the blocking solution and add to each section primary antibody diluted in recommended antibody diluent or blocking solution (*see* **Note 19**). Incubate overnight at 4 °C.

3. Remove the antibody solution and wash sections 3×5 min with TBST.

4. If using an HRP conjugate for detection, incubate sections in 3 % hydrogen peroxide for 15 min. Wash the sections 3×5 min with TBST.

5. Add to each section biotinylated secondary antibody, diluted in TBST. Incubate for 30 min at room temperature.

6. Remove the secondary antibody solution and wash sections 3×5 min with TBST.

7. Apply LSAB2 Streptavidin-HRP to each section and incubate for 30 min at room temperature.

8. Remove LSAB2 Streptavidin-HRP and wash sections 3×5 min with TBST.

9. Apply DAB or suitable substrate to each section and monitor staining closely.

10. As soon as the sections develop immerse the slides in diH$_2$O.

11. If desired, counterstain the sections in hematoxylin.

12. Wash the sections 2×5 min with diH$_2$O.

13. Dehydrate the sections as follows: 2×1 min in 95 % ethanol, then 2×1 min in 100 % ethanol, then 2×1 min in Xylene.

14. Mount coverslips on sections in mounting medium. Examine the slides under light microscope after drying.

3.4 Immuno-fluorescence

Prepare slides as described in immunohistochemistry protocol (Subheading 3.3).

3.4.1 Slide Preparation

3.4.2 Immuno-fluorescence Staining

1. Block each sample with blocking solution for 1 h at room temperature.

2. Remove blocking solution and add primary antibody diluted in recommended antibody diluent or blocking solution to each sample.

Incubate in a humidified chamber for 1 h at room temperature or overnight at 4 °C (*see* **Note 20**).

3. Remove the antibody solution and wash samples 3×5 min with TBST.

4. Add biotinylated secondary antibody or fluorochrome-conjugated secondary antibody to each sample, diluted in PBS with 1.5–3 % normal blocking serum. Incubate for 45 min at room temperature.

5. Remove the secondary antibody solution and wash the samples 3×5 min with TBST. If fluorochrome-conjugated secondary antibody is used, incubate in a dark chamber and omit the next step.

6. Incubate with fluorescent streptavidin conjugate for 15 min in a dark chamber (*see* **Note 21**). Wash extensively with PBS.

7. Counterstain with 0.1–1 μg/ml Hoechst or DAPI (DNA stain) for 1 min. Rinse with PBS.

8. Mount the coverslip with aqueous mounting medium.

9. Seal the coverslip with nail polish to prevent drying and movement under the microscope.

10. Examine the slides under fluorescence microscope with appropriate filters. Store slides in the dark at 4 °C.

3.4.3 Double Immunofluorescence Staining

In order to examine the co-distribution of two or more different antigens in the same sample, a double (or more) immunofluorescence procedure can be carried out.

1. Block each sample with blocking solution for 1 h at room temperature.

2. Remove the blocking solution and add to each sample the mixture of two primary antibodies diluted in blocking solution (*see* **Note 22**).

3. Incubate in a humidified chamber for 1 h at room temperature or overnight at 4 °C.

4. Remove the primary antibody mixture and wash the samples 3×5 min with PBS.

5. Incubate the samples with the mixture of two secondary antibodies diluted in blocking solution for 1 h at room temperature in dark (*see* **Note 23**).

6. Remove the secondary antibody mixture and wash the samples 3×5 min with PBS in dark.

7. Counter stain with 0.1–1 μg/ml Hoechst or DAPI (DNA stain) for 1 min. Rinse with PBS.

8. Mount the coverslip with aqueous mounting medium.

9. Seal the coverslip with nail polish to prevent drying and movement under microscope.

10. Examine the slides under a fluorescence microscope with appropriate filters. Store the slides in the dark at 4 °C.

3.5 Chromatin Immunoprecipitation

3.5.1 Sample Preparation from Tissues

1. (*See* **Notes 24**) Weigh the fresh or frozen tissue sample. Use 25 mg of tissue for each IP to be performed. Keep the tissue cold to avoid protein degradation.

2. Place the tissue sample in a 60 or 100 mm dish on ice and finely mince using a clean scalpel or razor blade.

3. Transfer the minced tissue to a 15 ml conical tube.

4. Add 1 ml of PBS + PIC (1× protease inhibitor cocktail in PBS) per 25 mg of tissue to the conical tube.

5. To cross-link proteins to DNA, add 37 % formaldehyde to a final concentration of 1.5 % (45 μl per 1 ml of tissue in PBS + PIC) and rock at room temp for 20 min.

6. Stop cross-linking by adding 100 μl of 10× glycine per 1 ml of tissue and rock for 5 min at room temperature.

7. Centrifuge the tissue at 800 × g for 5 min at 4 °C.

8. Remove the supernatant and wash one time with 1 ml of PBS + PIC per 25 mg of tissue.

9. Repeat centrifugation at 800 × g for 5 min at 4 °C.

10. Remove supernatant and resuspend the tissue in 1 ml of PBS + PIC per 25 mg of tissue and store on ice.

11. Transfer the tissue resuspended in PBS + PIC to a Dounce homogenizer. Disaggregate the tissue pieces with 20–25 strokes. Check for single-cell suspension by microscope (optional).

12. Transfer the cell suspension to a 15 ml conical tube and centrifuge at 1,000 × g for 5 min at 4 °C. Remove supernatant from cells and immediately continue with chromatin preparation and immunoprecipitation.

3.5.2 Sample Preparation from Cultured Cells

1. Start with 1×10^7 to 5×10^7 adherent or suspension cells in culture. Cross-link proteins to DNA by adding formaldehyde drop-wise directly to the culture medium to a final concentration of 1 % (for example, add 270 μl of 37 % formaldehyde into 10 ml of culture medium). Incubate with shaking for 10 min at room temperature.

2. Stop the cross-linking reaction by adding 10× glycine to a final concentration of 1× (125 mM) to the medium and incubate with shaking for 5 min at room temperature.

3. Wash cells with ice-cold PBS. Keep the samples cold from this step forward.

4. For adherent cells, aspirate the medium and rinse twice with PBS. Add 5 ml of cold PBS, and scrape the cells into a 15 ml conical tube. Add 3 ml of PBS to dishes and transfer the remainder of the cells to the 15 ml tube. Centrifuge at $800 \times g$ for 5 min at 4 °C. Carefully aspirate off the supernatant (*see* **Note 25**).

5. For suspension cells, wash by centrifuging $800 \times g$ for 5 min at 4 °C, remove the supernatant carefully, and resuspend the pellet in PBS. Repeat this wash for three times (*see* **Note 25**).

3.5.3 Chromatin Preparation and Immunoprecipitation

1. Resuspend the cell pellets in Lysis Buffer 1 (1 ml per 1.5×10^7 cells). Keep on ice for 10 min to break the cell membrane.

2. Centrifuge at $1,500 \times g$ for 5 min at 4 °C to separate nuclei from the cytoplasmic component. Remove the supernatant.

3. (Optional step) Wash the nuclear pellet twice with 1 ml of Wash Buffer 2. Centrifuge at $1,500 \times g$ for 5 min at 4 °C to pellet the nuclei.

4. Resuspend the nuclear pellets in 1 ml of SDS Lysis Buffer 3.

5. Sonicate the nuclear lysates on ice to shear the DNA to lengths of between 500 and 1,500 base pairs (*see* **Note 26**).

6. Centrifuge at $12,000 \times g$ for 10 min at 4 °C then transfer the supernatant to a new tube.

7. Dilute the supernatants 1:10 in Buffer 4. Save 20 μl of the diluted supernatant as 1 % of input DNA control to evaluate the efficiency of ChIP and PCR.

8. For each ChIP, use 2 ml of diluted supernatant. Perform immunoclearing with 40 μl of pretreated protein G-agarose (50 % slurry in 10 mM Tris–HCl, 1 mM EDTA, pH 8.0) and 1 μl of IgG for 2 h at 4 °C. Centrifuge at $1,500 \times g$ for 5 min at 4 °C then transfer the supernatant to a new tube.

9. Perform immunoprecipitation for 6 h or overnight at 4 °C with 2 ml of immunocleared sample, 1–20 μg of specific antibodies and 60 μl of pretreated 50 % protein G agarose slurry (*see* **Note 27**).

10. For a negative control, perform an immunoprecipitation with IgG by incubating 2 ml of supernatant with 60 μl of pretreated protein G agarose (50 % Slurry) for 6 h or overnight at 4 °C with rotation.

11. After immunoprecipitation, wash the beads sequentially for 10 min each in 1 ml of Buffer 6, Buffer 7, and Buffer 8 (*see* **Note 28**). Wash the precipitates three times with 1 ml of TE buffer for 5 min each.

12. The immunoprecipitates can now be analyzed by immunoblotting. Add 25 μl of 1× Laemmli buffer per sample and boil for 10 min.

Load 20 µl per lane and perform immunoblot procedure as described in Subheading 3.1.

13. To analyze the immunoprecipitated DNA, extract the samples two times with Buffer 9. Extract with 300 µl Buffer 9 for 10 min first, and repeat with 200 µl buffer 9 for 10 min (*see* **Note 29**).

14. Purify immunoprecipitated DNA from the extracts using a DNA purification kit according to kit manufacturer's protocol, or by following **steps 15–18** below.

15. Add 20 µl of 5 M NaCl and 2 µl of 10 µg/µl RNase A to the combined eluates (500 µl) and reverse protein–DNA cross-links by heating at 65 °C for 4 h or overnight (*see* **Note 30**).

16. Add 5 µl of 20 µg/µl proteinase K to digest protein at 55 °C for 2 h.

17. Recover DNA by phenol–chloroform extraction (600 µl/sample) and ethanol precipitation (1 ml/sample plus 20 µg glycogen). Wash pellets with 70 % ethanol and air-dry.

18. Resuspend the pellets in 30 µl of TE buffer. The purified DNA now can be used for semiquantitative PCR or quantitative real-time PCR analysis (*see* **Note 31**).

3.5.4 Semiquantitative Polymerase Chain Reaction (PCR) Analysis

1. Gently vortex and briefly centrifuge all solutions after thawing.

2. Set up PCR reactions as shown in Table 4 (*see* **Notes 32–35**).

3. Gently vortex the samples and spin down to collect drops.

4. Place the tubes in a thermocycler (*see* **Note 36**) and perform PCR using the recommended thermal cycling conditions in Table 5 (*see* **Note 37**).

Table 4
Volumes of solutions and components required to set up a semiquantitative PCR reaction

Stock solution	Amount to add in 50 µl	Final concentration
10× Taq buffer	5 µl	1×
10 mM each dNTP Mix	1 µl	0.2 mM
10 µM forward primer	0.5–5 µl	0.1–1 µM
10 µM reverse primer	0.5–5 µl	0.1–1 µM
25 mM MgCl$_2$	2–8 µl	1–4 mM
Template DNA		10 pg–1 µg
Taq DNA polymerase	1.25 U	
Nuclease-free water	to 50 µl	

Table 5
Recommended temperatures and cycles for a semiquantitative PCR reaction

Step	Temperature (°C)	Time (min)	Number of cycles
Initial denaturation	95	5	1
Denaturation	95	0.5	15–32
Annealing	Tm - 5	0.5	
Extension	72	1 min/kb	
Final extension	72	5–10	1
Hold	4		

Table 6
Volumes of solutions and components required to set up a QRT-PCR reaction

Stock solution	Amount to add in 25 µl	Final concentration
10 µM forward primer	0.25–2.5 µl	0.01–1 µM
10 µM reverse primer	0.25–2.5 µl	0.01–1 µM
Template DNA		10 pg–1 µg
2× SYBR Green PCR master mix	12.5 µl	
Water, nuclease-free	to 25 µl	

5. After PCR, load the same volume of reaction products into wells of an agarose gel, and run electrophoreses. Images of stained DNA (PCR products) are then obtained and analyzed by visual comparison or with computer software.

3.6 Quantitative Real-Time Polymerase Chain Reaction (PCR)

Quantitative real-time PCR (QRT-PCR) can be used to determine the levels of known noncanonical NF-κB-regulated genes in cell or tissue samples.

1. Design primers for target sequence (*see* **Notes 38–40**). Synthesize designed primer oligos.

2. For gene quantification from mRNA, extract RNA from samples using Trizol or the RNeasy purification kit. Reverse transcribe RNA into cDNA following the manufacturer's protocol provided with the reverse transcriptase enzyme of choice.

3. Gently vortex and briefly centrifuge all solutions after thawing.

4. Set up PCR reactions as shown in Table 6 (*see* **Notes 33, 34, 41**, and **42**).

5. Gently vortex the samples and spin down to collect drops.

Table 7
Recommended temperatures and cycles for a QRT-PCR reaction

Step	Temperature	Time	Number of cycles
Initial denaturation	95 °C	10	1
Denaturation	95 °C	15 s	40
Annealing/extension	60 °C	1 min	
Denaturation	95 °C	15 s	1
Annealing/extension	60 °C	1 min	1
Dissociation (melt curve)	0.3 °C increment until 95 °C		

6. Place the tubes in a real-time PCR machine and perform PCR using the recommended thermal cycling conditions shown in Table 7.

7. Analyze the results using the compatible software.

3.7 Gene Array Analysis

Gene array analysis can be used to determine the expression of known or novel noncanonical NF-κB-regulated genes in cell or tissue samples.

1. Isolate RNA from samples using RNeasy purification kit (Qiagen).

2. Examine the integrity of the RNAs using the Agilent platform (2100 Bioanalyzer; Agilent Technologies, Palo Alto, CA) (*see* **Note 43**).

3. Send out RNA samples for gene array analysis to a gene array analysis facility (*see* **Note 44**).

4. Analyze the results using gene array analysis software.

3.8 Electrophoretic Mobility Shift Assay (EMSA)

For EMSA analysis the DNA probes can be either labeled using ^{32}P-ATP or by nonradioactive labeling using 3′ end labeling with Biotin.

3.8.1 DNA Probe Labeling with ^{32}P-ATP

1. Mix the reagents listed in Table 8, adding the "hot" ATP last (*see* **Note 45**). Incubate at 37 °C for 30 min.

2. Stop the reaction by adding 1 μl of 0.5 M EDTA or heat inactivation at 65 °C for 30 min.

3. Clean up the probe using Qiagen PCR purification kit.

3.8.2 DNA 3′ Biotin End Labeling

1. Prepare the labeling reactions by adding components from the DNA 3′ end labeling kit in the order listed in Table 9 (*see* **Notes 46** and **47**).

Table 8
Reaction mixture to label DNA probe for EMSA using ^{32}P-ATP

Reagents	Amount to add in 10 µl	Final concentration
DNA probe (1.75 pmol/µl)	2 µl	0.35 pmol/µl
T4 polynucleotide kinase 10× buffer	1 µl	1×
[γ-^{32}P]ATP (3,000 Ci/mmol at 10 mCi/ml)	1–2.2 µl	0.33–0.7 pmol/µl
T4 polynucleotide kinase (5–10 U/µl)	1 µl	0.5–1 U/µl
Nuclease-free water	to 10 µl	

Table 9
Reaction mixture to label DNA probe for EMSA using Biotin 3′ end DNA labeling

Reagents	Amount to add in 50 µl	Final concentration
diH$_2$O	25 µl	
5× TdT reaction buffer	10 µl	1×
DNA probe (1 µM)	5 µl	100 nM
Biotin-11-UTP (5 µM)	5 µl	0.5 µM
TdT (2 U/µl)	5 µl	0.2 U/µl

2. Incubate reactions at 37 °C for 30 min.

3. Add 2.5 µl 0.2 M EDTA to stop each reaction.

4. Add 50 µl chloroform–isoamyl alcohol to each reaction to extract the TdT. Vortex the mixture briefly, then centrifuge for 1–2 min at high speed in a microcentrifuge to separate the phases. Transfer the upper aqueous phase to a new tube.

3.8.3 Nuclear Extract Preparation

Keep all the solutions cold and perform all steps at 4 °C or on ice.

1. Harvest about 1–5×10^7 cells. For adherent cells, scrape off with 1 ml of cold PBS and transfer to a 1.5 ml tube. For suspension cells, centrifuge at $800 \times g$ for 5 min, then resuspend the cell pellet in 1 ml of cold PBS, and transfer to a 1.5 ml tube.

2. Pellet the cells by centrifugation at $800 \times g$ for 5 min at 4 °C.

3. Resuspend cell pellet in 300 µl of Hypotonic Lysis Buffer A and incubate for 20–30 min on ice.

4. Centrifuge at $1,500 \times g$ for 5 min at 4 °C to separate nuclei from cytoplasmic component. Remove the supernatant.

Table 10
Volumes of solutions to make native gels of specific percentages for EMSA

Acrylamide percentage	4 %	5 %	6 %
diH$_2$O	37.8 ml	36.2 ml	34.5 ml
Acrylamide–Bis-acrylamide (30 %/0.8 % w/v)	6.7 ml	8.3 ml	10 ml
5× TBE buffer	5 ml	5 ml	5 ml
10 % (w/v) ammonium persulfate (APS)	500 µl	500 µl	500 µl
TEMED	40 µl	40 µl	40 µl

5. Resuspend the pellet in 1–1.5 ml of Buffer A, and incubate for 15–20 min. Centrifuge at $1,500 \times g$ for 10 min, and discard the supernatant. Repeat this step once.

6. Resuspend the pellet in 50 µl Buffer B, and incubate for 20 min. Centrifuge at $12,000 \times g$ for 10 min and collect the supernatant as nuclear extract.

7. Determine the protein concentration of the extracts using NanoDrop spectrophotometer, protein staining by Bradford, or another assay.

8. Dilute the nuclear extract to 1 µg/µl protein concentration with Buffer A. Store extracts that will not be used as aliquots at −80 °C to avoid repeated thawing of the samples if the EMSA needs to be repeated.

3.8.4 Gel Preparation and Pre-run

1. Prepare a native polyacrylamide gel in 0.5× TBE or use a pre-cast DNA retardation gel. The appropriate polyacrylamide percent depends on the size of the target DNA and the binding protein. Most systems use a 4–6 % polyacrylamide gel in 0.5× TBE. The formulation for a 50 ml gel solution is shown in Table 10, which can be scaled up or down as needed.

2. Pre-run the gel. Place the gel in the electrophoresis unit and clamp it to obtain a seal. Fill the inner chamber with 0.5× TBE to a height several millimeters above the top of the wells. Fill the outside of the tank with 0.5× TBE to just above the bottom of the wells, which reduces heat during electrophoresis. Flush wells and pre-electrophorese the gel for 30–60 min. Apply 100 V for an $8 \times 8 \times 0.1$ cm gel.

3.8.5 Binding Reactions

1. Thaw all the components and bring to room temperature before mixing. Prepare the reactions in the order shown in Table 11. Before adding the labeled DNA probe, incubate the reactions at room temperature for 10–20 min. Generally, unlabeled competitors are at 200–300-fold molar excess of the labeled probe in the binding reaction.

Table 11
Volumes of solutions and components for binding reactions for EMSA

Components	Negative control (μl)	Test sample (μl)	Specific competitor (μl)	Nonspecific competitor (μl)	Super shift (μl)
Nuclease-free H₂O	to 20	to 20	to 20	to 20	to 20
5× gel shift binding buffer	4	4	4	4	4
1 μg/ μl Poly dI-dC	1	1	1	1	1
Nuclear extract		1–5	1–5	1–5	1–5
Unlabeled specific competitor			2		
Unlabeled nonspecific competitor				2	
Antibody					2
Labeled DNA probe	2	2	2	2	2

2. If a supershift is being performed to detect specific NF-κB proteins, antibodies (e.g., anti-p52, anti-RelB) should be added to the reaction as shown in Table 11.

3. Incubate binding reactions at room temperature for 20–30 min.

4. Add 2 μl of 10× loading buffer to each 20 μl binding reaction, pipetting up and down several times to mix (*see* **Note 48**). Do not vortex or mix vigorously.

3.8.6 Electrophoreses of Binding Reactions

1. Switch off current to the electrophoresis gel after pre-run.

2. Flush the wells and then load 20 μl of each sample onto the polyacrylamide gel.

3. Switch on current and electrophorese samples until the bromophenol blue dye has migrated approximately 2/3 to 3/4 down the length of the gel. Maintain a gel temperature of <30 °C to prevent overheating that can lead to loss of protein–DNA interactions.

3.8.7 Detection of Radiolabeled DNA

1. Open the gel plates and place the gel on a sheet of Whatman filter paper.

2. Cover with plastic wrap and dry on a gel dryer.

3. Expose the gel to X-ray film for 1 h to overnight at −70 °C with an intensifying screen. Alternatively, analyze the gel using phosphorimaging instrumentation.

3.8.8 Detection of Biotin-Labeled DNA

1. Soak nylon membrane in 0.5× TBE for at least 10 min.

2. Sandwich the gel, nylon membrane, and blotting paper in a clean electrophoretic transfer unit (*see* **Note 49**). Use 0.5× TBE cooled to ~10 °C with a circulating water bath. Use very clean forceps and powder-free gloves, and handle the membrane only at the corners.

3. Transfer at 380 mA (~100 V) for 30 min. Typical transfer times are 30–60 min at 380 mA using a standard tank transfer apparatus for mini gels (e.g., $8 \times 8 \times 0.1$ cm).

4. When the transfer is complete, place the membrane with the bromophenol blue side up on a dry paper towel. There should be no dye remaining in the gel.

5. Allow buffer on the membrane surface to absorb into the membrane. This will only take a minute. Do not let the membrane dry.

6. The transferred DNA is then cross-linked to the membrane by one of the following methods (any one of **steps 7–9**).

7. Cross-link Method 1: Using a commercial UV-light cross-linking instrument equipped with 254 nm bulbs, cross-link at 120 mJ/cm^2 (45–60 s exposure using the auto-cross-link function).

8. Cross-link Method 2: Cross-link at a distance of approximately 0.5 cm from the membrane for 5–10 min with a handheld UV lamp equipped with 254 nm bulbs.

9. Cross-link Method 3: Cross-link for 10–15 min with the membrane face down on a transilluminator equipped with 312 nm bulbs.

10. After cross-linking using any of the methods in **steps 7–9**, the membrane may be stored dry at room temperature for several days. Do not allow the membrane to get wet again until ready to proceed with detection by chemiluminescence.

11. To detect biotin-labeled DNA cross-linked on the membrane, use the chemiluminescent nucleic acid detection module kit.

4   Notes

1. If the 1 M solution has a yellow color, discard it and obtain Tris of better quality. The pH of Tris solutions is temperature-dependent and decreases ~0.03 pH units for each 1 °C increase in temperature. For example, a 0.05 M solution has pH values of 9.5, 8.9, and 8.6 at 5, 25, and 37 °C, respectively.

2. PMSF is inactivated in aqueous solutions. The rate of inactivation increases with increasing pH and is faster at 25 °C than at 4 °C.

The half-life of a 20 mM aqueous solution of PMSF is ~35 min at pH 8.0. This short half-life means that aqueous solutions of PMSF can be safely discarded after they have been rendered alkaline (pH >8.6) and stored for several hours at room temperature.

3. DTT, PMSF and protease inhibitor mixture supplement should be added immediately prior to use.

4. DTT, PMSF, leupeptin, and aprotinin should be added immediately before use.

5. The 10 % sodium deoxycholate stock solution must be protected from light.

6. 10 % APS and TEMED must be added right before each use.

7. The purity of the obtained fractions can be confirmed by immunoblotting using antibodies against Hsp90 (cytoplasm), or α-tubulin (cytoplasm), and histone H3 (nucleus).

8. The running time can vary depending on the voltage used, the gel concentration and the target protein size.

9. For PVDF it is essential to pre-wet the membrane in methanol prior to transfer.

10. In immunoblotting, for proteins with molecular weights of 50 kDa, it is recommended to use light-chain specific or conformation specific (recognizing only native but not denatured IgG) anti-IgG as secondary antibody to minimize masking produced by denatured heavy chains (50 kDa). For proteins with molecular weights of 25 kDa, it is recommended to use Fc fragment specific or conformation specific anti-IgG to minimize masking produced by denatured light chains (25 kDa).

11. For tissues containing high levels of endogenous biotin (which may result in higher background staining), we recommend following the IHC-P (paraffin) protocol, as endogenous biotin is normally destroyed in paraffin-embedded tissue.

12. Do not allow slides to dry at any time hereafter. Drying out will cause nonspecific antibody binding and therefore high background staining.

13. Consult antibody data sheet for specific recommendations on antigen retrieval. If the antigen retrieval method is not stated on the antibody datasheet, then the optimal method must be tested out experimentally.

14. In our experience, TE buffer (pH 9.0) generally gives a stronger staining than 10 mM sodium citrate buffer (pH 6.0).

15. Optimal incubation time should be determined empirically. The efficiency of heat-induced antigen retrieval is largely a function of temperature, time, and pH and chemical composition

of the buffer. Too short incubation may lead to insufficient antigen retrieval, whereas prolonged heating may cause increased background staining, as well as damage to the morphology.

16. Optimal incubation time should be determined empirically. The enzymatic digestion time is closely related to the fixation time. If the tissue has been shortly fixed, the enzymatic digestion should be diminished; otherwise the target proteins may be digested away. On the other hand, if the fixation time has been prolonged, the digestion time may need to be increased to achieve sufficient antigen retrieval. Keep in mind that prolonged enzymatic retrieval can sometimes damage the morphology of the section.

17. If the target antigen is located intracellularly, it is very important to permeabilize the cells to allow antibody to gain access to the inside of the cells to detect the protein. These include intracellular proteins and transmembrane proteins whose epitopes are in the cytoplasmic region. Since acetone fixation also permeabilizes cells, acetone fixed samples do not require permeabilization.

18. Triton X-100 is the most popular detergent for improving the penetration of the antibody. However, it is not appropriate for the use of membrane-associated antigens since it destroys membranes.

19. Optimal antibody concentration should be determined by titration. High antibody concentration tends to increase non-specific staining, whereas low antibody concentration tends to lower staining intensity.

20. The staining is largely a function of antibody concentration, and incubation time. With longer incubation time, the antibody concentration can be lowered. Optimal antibody concentration should be determined by titration.

21. Optimal streptavidin conjugate concentration for a given application should be determined by titration; recommended range is 10–20 μg/ml in PBS.

22. Primary antibodies specific for different antigens, raised in different species (e.g., rabbit against human antigen-1 and mouse against human antigen-2, if the antigens are human proteins) can be used either in parallel (in a mixture) or in a sequential way.

23. The two secondary antibodies are specific for the two different primary antibodies used and conjugated with two different fluorochromes (e.g., Texas Red-conjugated against rabbit and FITC-conjugated against mouse). They can be used either in parallel (in a mixture) or in a sequential way.

24. When harvesting tissue, remove unwanted material such as fat and necrotic material from the sample. Tissue can then be processed and cross-linked immediately or snap frozen in liquid nitrogen and stored at –80 °C for processing later. Generally, use 25 mg of tissue for each immunoprecipitation to be performed. The chromatin yield does vary between tissue types, and some tissues such as brain may require more than 25 mg for each immunoprecipitation.

25. Cells can be used immediately for ChIP assay or snap frozen in liquid nitrogen and stored in liquid nitrogen or –80 °C freezer for later use.

26. Sonication conditions must be determined empirically for each cell or tissue type, and sonicator model. Avoid foaming by keeping the tip end of the sonicator near the bottom of the sample tubes, but not touching. Placing the tip end near the surface induces foaming, which can result in unequal shearing of DNA samples.

27. Different antibodies have different affinity to target proteins, and for each target the amount of antibody needs to be tested case by case. Generally, polyclonal antibodies are much better than monoclonal antibodies.

28. Because different antibodies have different affinities to target proteins, for some cases, the Buffer 8 washing step can be omitted.

29. Extraction at 65 °C can enhance the efficiency.

30. Generally overnight is better. Reverse cross-linking overnight if time permitting. Reverse cross-linking for the input control samples similarly.

31. As a positive control for PCR, DNA prepared from samples prior to immunoprecipitation can be used as total or input DNA. For additional controls, use primers designed to anneal to promoters known not to be regulated by the protein of interest, or primers designed to anneal to regions of the gene known not to bind to the protein of interest.

32. For each ChIP DNA PCR reaction, a control PCR using purified DNA from input (no ChIP) control and a no antibody control PCR can be performed. An additional control PCR can also be performed using a primer set to a non-target DNA sequence.

33. The PCR reaction volumes can be scaled up or down as long as the final concentrations of the reaction components remain the same.

34. To decrease pipetting load and pipetting variations between reactions, a master mix can be prepared, distributed into each PCR tube, followed by addition of the rest component(s).

35. Some 10× Taq buffer contains $MgCl_2$, it may not be necessary to add $MgCl_2$ in these cases.

36. When using thermal cyclers without a heated lid, overlay the reaction mixture with 25 μl of mineral oil.

37. For gene quantitation, the PCR reactions must be within the linear range of amplification. Optimal cycles of PCR for gene quantitation need to be determined empirically.

38. The primer pair has to be specific to the target DNA sequence, i.e., the primer pair does not amplify pseudogenes or other related genes/DNA sequences.

39. Primers that yield 50- to 150-bp amplicons work the best. Primers that yield slightly larger amplicons will also work fine. Try to design primers with GC content in the 30–80 % range, Tm close to 60 °C, no more than two G and/or C bases in the 3′ end five nucleotides. Try to avoid runs of identical nucleotide, potential secondary structure formation, or primer dimer formation.

40. For gene expression quantification, try to design primers with amplicons spanning intron(s) totaling at least 1,000 bp in size to avoid amplification of the target gene in genomic DNA. If poly dT-primed cDNA is used, it would be better to use primers near the 3′ end of the gene.

41. For each test PCR reaction, perform a control PCR reaction using a primer set to a DNA sequence that should not be changed throughout the experimental conditions.

42. For each primer set, perform a non-template control PCR.

43. Each sample is assigned an RNA integrity number (RIN) score: 10 corresponds to a pure, un-degraded total RNA sample while a RIN of 1 corresponds to a completely degraded sample. It is recommended to only use samples with RIN's greater than 7.

44. At the gene array analysis facility, the RNA samples are converted to labeled cDNA via reverse transcription. Then the labeled cDNAs are hybridized to identical membrane or glass slide arrays. After removing the unhybridized cDNA, the hybridized cDNAs are detected and quantitated.

45. Aliquot out reagents from kit to use in the radiation area. DO NOT use the stocks directly.

46. Mix reactions gently. Do not vortex.

47. Complementary oligos should be end-labeled separately and then annealed before use. Anneal oligos by mixing together equal amounts of labeled complementary oligos and incubating the mixture for 1 h at room temperature. Oligonucleotides with high melting temperatures or secondary structure may require denaturation and slow cooling for optimal annealing

(e.g., denature at 90 °C for 1 min, then slowly cool and incubate at the melting temperature for 30 min). Freeze annealed oligos that will not be used immediately, and thaw on ice for use. Removal of the unincorporated Biotin-11-UTP is not necessary for use in EMSA.

48. The dyes in the gel-loading buffer sometimes interfere with the proteins DNA binding. If this occurs, we recommend that the gel loading buffer be added only to the negative control.

49. Use clean transfer sponges. Avoid using sponges that have been used in Western blots.

References

1. Xiao G, Fu J (2011) NF-kappaB and cancer: a paradigm of Yin-Yang. Am J Cancer Res 1(2): 192–221

2. Xiao G et al (2006) Alternative pathways of NF-kappaB activation: a double-edged sword in health and disease. Cytokine Growth Factor Rev 17(4):281–293

3. Claudio E et al (2002) BAFF-induced NEMO-independent processing of NF-kappa B2 in maturing B cells. Nat Immunol 3(10):958–965

4. Coope HJ et al (2002) CD40 regulates the processing of NF-kappaB2 p100 to p52. EMBO J 21(20):5375–5385

5. Dejardin E et al (2002) The lymphotoxin-beta receptor induces different patterns of gene expression via two NF-kappaB pathways. Immunity 17(4):525–535

6. Novack DV et al (2003) The IkappaB function of NF-kappaB2 p100 controls stimulated osteoclastogenesis. J Exp Med 198(5):771–781

7. Saitoh T et al (2003) TWEAK induces NF-kappaB2 p100 processing and long lasting NF-kappaB activation. J Biol Chem 278(38): 36005–36012

8. Xiao G, Harhaj EW, Sun SC (2001) NF-kappaB-inducing kinase regulates the processing of NF-kappaB2 p100. Mol Cell 7(2):401–409

9. Qing G, Qu Z, Xiao G (2005) Stabilization of basally translated NF-kappaB-inducing kinase (NIK) protein functions as a molecular switch of processing of NF-kappaB2 p100. J Biol Chem 280(49):40578–40582

10. Liao G et al (2004) Regulation of the NF-kappaB-inducing kinase by tumor necrosis factor receptor-associated factor 3-induced degradation. J Biol Chem 279(25): 26243–26250

11. Grech AP et al (2004) TRAF2 differentially regulates the canonical and noncanonical pathways of NF-kappaB activation in mature B cells. Immunity 21(5):629–642

12. Senftleben U et al (2001) Activation by IKKalpha of a second, evolutionary conserved, NF-kappa B signaling pathway. Science 293(5534):1495–1499

13. Vallabhapurapu S et al (2008) Nonredundant and complementary functions of TRAF2 and TRAF3 in a ubiquitination cascade that activates NIK-dependent alternative NF-kappaB signaling. Nat Immunol 9(12):1364–1370

14. Zarnegar BJ et al (2008) Noncanonical NF-kappaB activation requires coordinated assembly of a regulatory complex of the adaptors cIAP1, cIAP2, TRAF2 and TRAF3 and the kinase NIK. Nat Immunol 9(12):1371–1378

15. Fong A, Sun SC (2002) Genetic evidence for the essential role of beta-transducin repeat-containing protein in the inducible processing of NF-kappa B2/p100. J Biol Chem 277(25):22111–22114

16. Xiao G, Fong A, Sun SC (2004) Induction of p100 processing by NF-kappaB-inducing kinase involves docking IkappaB kinase alpha (IKKalpha) to p100 and IKKalpha-mediated phosphorylation. J Biol Chem 279(29): 30099–30105

17. Ling L, Cao Z, Goeddel DV (1998) NF-kappaB-inducing kinase activates IKK-alpha by phosphorylation of Ser-176. Proc Natl Acad Sci U S A 95(7):3792–3797

18. Franzoso G et al (1998) Mice deficient in nuclear factor (NF)-kappa B/p52 present with defects in humoral responses, germinal center reactions, and splenic microarchitecture. J Exp Med 187(2):147–159

19. Caamano JH et al (1998) Nuclear factor (NF)-kappa B2 (p100/p52) is required for normal splenic microarchitecture and B cell-mediated immune responses. J Exp Med 187(2): 185–196

20. Yin L et al (2001) Defective lymphotoxin-beta receptor-induced NF-kappaB transcriptional

activity in NIK-deficient mice. Science 291(5511):2162–2165

21. Shinkura R et al (1999) Alymphoplasia is caused by a point mutation in the mouse gene encoding Nf-kappa b-inducing kinase. Nat Genet 22(1):74–77

22. Miyawaki S et al (1994) A new mutation, aly, that induces a generalized lack of lymph nodes accompanied by immunodeficiency in mice. Eur J Immunol 24(2):429–434

23. Futterer A et al (1998) The lymphotoxin beta receptor controls organogenesis and affinity maturation in peripheral lymphoid tissues. Immunity 9(1):59–70

24. Thompson JS et al (2001) BAFF-R, a newly identified TNF receptor that specifically interacts with BAFF. Science 293(5537):2108–2111

25. Shulga-Morskaya S et al (2004) B cell-activating factor belonging to the TNF family acts through separate receptors to support B cell survival and T cell-independent antibody formation. J Immunol 173(4):2331–2341

26. Sasaki Y et al (2004) TNF family member B cell-activating factor (BAFF) receptor-dependent and -independent roles for BAFF in B cell physiology. J Immunol 173(4):2245–2252

27. Kawabe T et al (1994) The immune responses in CD40-deficient mice: impaired immunoglobulin class switching and germinal center formation. Immunity 1(3):167–178

28. Dougall WC et al (1999) RANK is essential for osteoclast and lymph node development. Genes Dev 13(18):2412–2424

29. Ishikawa H et al (1997) Gastric hyperplasia and increased proliferative responses of lymphocytes in mice lacking the COOH-terminal ankyrin domain of NF-kappaB2. J Exp Med 186(7):999–1014

30. Mackay F et al (1999) Mice transgenic for BAFF develop lymphocytic disorders along with autoimmune manifestations. J Exp Med 190(11):1697–1710

31. Khare SD et al (2000) Severe B cell hyperplasia and autoimmune disease in TALL-1 transgenic mice. Proc Natl Acad Sci U S A 97(7): 3370–3375

32. Moore CR et al (2012) Specific deletion of TRAF3 in B lymphocytes leads to B-lymphoma development in mice. Leukemia 26(5): 1122–1127

33. Rosebeck S et al (2011) Cleavage of NIK by the API2-MALT1 fusion oncoprotein leads to noncanonical NF-kappaB activation. Science 331(6016):468–472

34. Rosebeck S, Lucas PC, McAllister-Lucas LM (2011) Protease activity of the API2-MALT1 fusion oncoprotein in MALT lymphoma development and treatment. Future Oncol 7(5): 613–617

35. Pham LV et al (2011) Constitutive BR3 receptor signaling in diffuse, large B-cell lymphomas stabilizes nuclear factor-kappaB-inducing kinase while activating both canonical and alternative nuclear factor-kappaB pathways. Blood 117(1):200–210

36. Demchenko YN et al (2010) Classical and/or alternative NF-kappaB pathway activation in multiple myeloma. Blood 115(17): 3541–3552

37. Compagno M et al (2009) Mutations of multiple genes cause deregulation of NF-kappaB in diffuse large B-cell lymphoma. Nature 459(7247):717–721

38. Keats JJ et al (2007) Promiscuous mutations activate the noncanonical NF-kappaB pathway in multiple myeloma. Cancer Cell 12(2): 131–144

39. Annunziata CM et al (2007) Frequent engagement of the classical and alternative NF-kappaB pathways by diverse genetic abnormalities in multiple myeloma. Cancer Cell 12(2):115–130

40. Nagel I et al (2009) Biallelic inactivation of TRAF3 in a subset of B-cell lymphomas with interstitial del(14)(q24.1q32.33). Leukemia 23(11):2153–2155

41. Braggio E et al (2009) Identification of copy number abnormalities and inactivating mutations in two negative regulators of nuclear factor-kappaB signaling pathways in Waldenstrom's macroglobulinemia. Cancer Res 69(8):3579–3588

42. Rayet B, Gelinas C (1999) Aberrant rel/nfkb genes and activity in human cancer. Oncogene 18(49):6938–6947

43. Sun SC, Xiao G (2003) Deregulation of NF-kappaB and its upstream kinases in cancer. Cancer Metastasis Rev 22(4):405–422

44. Qing G, Qu Z, Xiao G (2007) Endoproteolytic processing of C-terminally truncated NF-kappaB2 precursors at kappaB-containing promoters. Proc Natl Acad Sci U S A 104(13): 5324–5329

45. Qu Z, Xiao G (2011) Human T-cell lymphotropic virus: a model of NF-kappaB-associated tumorigenesis. Viruses 3(6):714–749

46. Xiao G, Sun SC (2000) Activation of IKKalpha and IKKbeta through their fusion with HTLV-I tax protein. Oncogene 19(45):5198–5203

47. Xiao G et al (2001) Retroviral oncoprotein Tax induces processing of NF-kappaB2/p100 in T cells: evidence for the involvement of IKKalpha. EMBO J 20(23):6805–6815

48. Qu Z et al (2004) Tax deregulation of NF-kappaB2 p100 processing involves both beta-TrCP-dependent and -independent mechanisms. J Biol Chem 279(43):44563–44572

49. Fu J et al (2011) The tumor suppressor gene WWOX links the canonical and noncanonical NF-kappaB pathways in HTLV-I Tax-mediated tumorigenesis. Blood 117(5):1652–1661

50. de Oliveira DE, Ballon G, Cesarman E (2010) NF-kappaB signaling modulation by EBV and KSHV. Trends Microbiol 18(6):248–257

51. Wu JT, Kral JG (2005) The NF-kappaB/IkappaB signaling system: a molecular target in breast cancer therapy. J Surg Res 123(1):158–169

52. Storz P (2013) Targeting the alternative NF-kappaB pathway in pancreatic cancer: a new direction for therapy? Expert Rev Anticancer Ther 13(5):501–504

53. Ueda Y, Richmond A (2006) NF-kappaB activation in melanoma. Pigment Cell Res 19(2):112–124

54. Groom J et al (2002) Association of BAFF/BLyS overexpression and altered B cell differentiation with Sjogren's syndrome. J Clin Invest 109(1):59–68

55. Zhang J et al (2001) Cutting edge: a role for B lymphocyte stimulator in systemic lupus erythematosus. J Immunol 166(1):6–10

56. Cheema GS et al (2001) Elevated serum B lymphocyte stimulator levels in patients with systemic immune-based rheumatic diseases. Arthritis Rheum 44(6):1313–1319

Chapter 9

Noncanonical NF-κB Activation and SDF-1 Expression in Human Endothelial Cells

Kelly A. McCorkell and Michael J. May

Abstract

NF-κB is a family of transcription factors regulated through two distinct signaling cascades, the classical and the Noncanonical NF-κB pathways. Noncanonical NF-κB plays important roles in the immune system, as it is necessary for lymphoid organogenesis and B-cell survival and differentiation, as well as osteoclastogenesis. In the last few years, there has been an increased number of studies focusing on both identifying the upstream events that regulate the noncanonical NF-κB pathway as well as determining the physiological roles of noncanonical NF-κB in normal and disease pathologies, such as cancer and autoimmune diseases. Dysregulation of noncanonical NF-κB has now been associated with the pathogenesis of several types of lymphomas and autoimmune diseases and is believed to contribute to chronic inflammatory diseases, including ulcerative colitis. These studies suggest that targeting the Noncanonical pathway, similar to classical NF-κB, may have some therapeutic potential in the future; however, there is still quite a bit about the regulation of the noncanonical signaling that remains to be defined. In this chapter we describe the use of HUVEC, as an in vitro model for examining noncanonical NF-κB signaling in response to different stimuli. We demonstrate two different methods to measure noncanonical NF-κB activation: the processing of p100 to p52, and noncanonical NF-κB-dependent gene expression of CXCL12. The first method examines a key regulatory requirement for noncanonical NF-κB activation, by which p100 undergoes proteolytic cleavage to relieve the inhibition of NF-κB dimers for nuclear translocation and activation of gene transcription. The latter demonstrates the downstream effects of activated noncanonical NF-κB in response to stimuli.

Key words Noncanonical NF-κB, Lymphotoxin-beta receptor (LTβR), CD40, p100 processing, CXCL12, Stromal derived factor-1 (SDF-1), Human umbilical cord vein endothelial cells (HUVEC)

1 Introduction

The NF-κB signaling pathway is a major regulator of gene expression necessary for many physiological functions, especially inflammatory and immune responses. Activation of NF-κB can occur through two different signaling paradigms, the classical NF-κB or noncanonical NF-κB pathways. Classical NF-κB is activated by a large variety of stimuli and is the pathway used for the majority of known NF-κB functions, such as cell survival and differentiation, inflammation and

Michael J. May (ed.), *NF-kappa B: Methods and Protocols*, Methods in Molecular Biology, vol. 1280,
DOI 10.1007/978-1-4939-2422-6_9, © Springer Science+Business Media New York 2015

innate and adaptive immunity [1]. In contrast, noncanonical NF-κB responds to a smaller subset of activators, including Lymphotoxin-β receptor (LTβR), CD40, B-cell activating factor receptor (BAFF-R), receptor activator of NF-κB (RANK), among a few others and regulates BAFF and several lymphoid chemokines, such as CXCL13 (BLC), CCL19 (ELC), CCL21 (SLC), and CXCL12 (SDF). Similar to classical NF-κB, the noncanonical NF-κB pathway also plays important functions in immunity, as the noncanonical NF-κB pathway is required for secondary lymphoid organ development, maintenance of splenic architecture, thymic organogenesis and self-tolerance, B-cell survival, differentiation, and maturation as well as osteoclastogenesis [2–6].

Dysregulated NF-κB results in a diverse number of disease states, such as cancer, chronic inflammation, arthritis, atherosclerosis, and metabolic disorders, however these have been attributed to the classical NF-κB signaling paradigm [7–10]. Until recently, the role for noncanonical NF-κB in disease had been poorly understood, however aberrant noncanonical NF-κB activity has now been associated with ulcerative colitis and several autoimmune diseases, including systemic lupus erythematosus, rheumatoid arthritis, and Sjögren's syndrome. Additionally, dysregulated stability of NF-κB inducing kinase (NIK), a key upstream kinase of noncanonical NF-κB activation, has been implicated in several hematological cancers, such as multiple myeloma, diffuse large B-cell lymphoma, and mucosal-associated lymphoid tissue lymphomas [11–13]. As research efforts for studying the regulation of NF-κB for drug development in therapeutic treatment of inflammatory diseases and cancer expand, so will our understanding of the functions associated with the noncanonical pathway.

Much of the research performed in our laboratory focuses on noncanonical NF-κB's role in regulating the phenotype and function of vascular endothelial cells (EC) at sites of acute and chronic inflammation, particularly because endothelial cell dysfunction is a major contributor to the pathogenesis of chronic inflammatory diseases, such as atherosclerosis and many autoimmune diseases [14]. In this chapter, we describe the use of Human Umbilical Vein Endothelial Cells (HUVEC) as a model system for studying the regulation of noncanonical NF-κB activation and chemokine upregulation in vitro. The start of this chapter describes the derivation, culturing and treatment of primary HUVEC from umbilical cord tissue. We then describe two methods used to demonstrate the activation of noncanonical NF-κB in response to stimuli. The first involves immunoblotting to examine the processing of p100 to p52 (Fig. 1), a key event in noncanonical signaling that allows for the nuclear translocation of NF-κB dimers [15]. Lastly, we describe the methods necessary to perform quantitative real-time PCR to determine the effects of different stimuli on NF-κB-dependent gene expression of CXCL12, a known noncanonical NF-κB gene target in EC (Fig. 2).

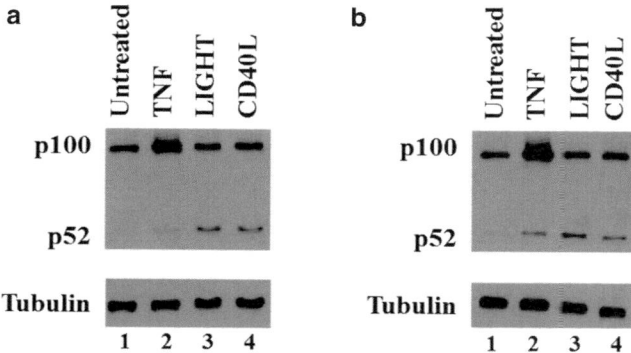

Fig. 1 Noncanonical NF-κB activation as measured by p100 processing to p52 in HUVEC. Immunoblots of protein extracted from HUVEC left untreated (*lane 1*) or treated for 8 h (**a**) or 24 h (**b**) with 10 ng/ml TNF (*lane 2*), 100 ng/ml LIGHT (*lane 3*), or 100 ng/ml CD40L (*lane 4*) were examined for p100 cleavage and p52 generation (*top panel*). Tubulin was used as a loading control (*bottom panel*)

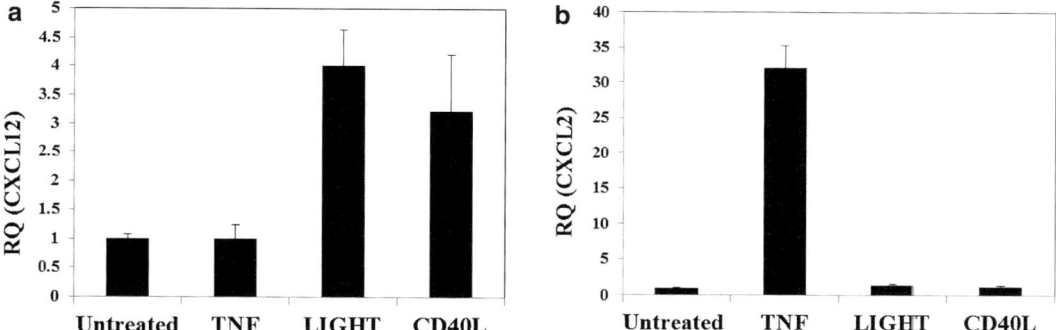

Fig. 2 Noncanonical NF-κB-dependent gene expression of CXCL12 (SDF-1) in HUVEC. HUVEC were either untreated or incubated with TNF (10 ng/ml), LIGHT (100 ng/ml), and CD40L (100 ng/ml) for 24 h were harvested for mRNA extraction and cDNA generation. The cDNA was subjected to quantitative Real-time PCR to examine expression levels of CXCL12 (**a**), a noncanonical NF-κB gene target and CXCL2 (**b**), a classical NF-κB gene target. Gene expression values were calculated relative to the expression of the β-Actin endogenous control

To demonstrate the activation of noncanonical activation in HUVEC, we examined protein levels of p100 and p52 in response to different stimuli. The NF-κB transcription factor activated by noncanonical signaling is a dimer comprised of NF-κB2 (p100/p52) and RelB. In untreated cells, NF-κB2 exists as its precursor protein form, p100. The p100 protein contains an inhibitory sequence at its C-terminus, which acts to sequester p100/RelB in the cytoplasm. Stimulation of HUVEC with noncanonical stimuli, LIGHT and CD40L activates the upstream NIK and IKKα kinases to trigger the proteolytic cleavage of the C-terminus of p100, processing it into its active p52 form. This results in an accumulation of p52 protein compared to untreated cells. This is specific to noncanonical activation, as treatment with a classical NF-κB cytokine, TNF, shows no increase in p52 relative to p100. Rather TNF results in the accumulation of p100 and basal p52 as a result of p100 being a classical NF-κB gene target (Fig. 1).

To demonstrate the activation of noncanonical NF-κB-dependent gene expression in vascular endothelial cells, we examined the gene expression levels of a noncanonical gene target, CXCL12 (SDF1). Treatment of HUVEC with either LIGHT or CD40L results in a three to four-fold increase in CXCL12 expression compared to untreated EC. TNF stimulated EC results in the upregulation of CXCL2, a classical NF-κB gene target and but has no effect on CXCL12 expression (Fig. 2).

2   Materials

All reagents should be prepared using deionized water and should be stored at room temperature unless otherwise noted. All methods involving the isolation, culturing, and stimulations of HUVEC should be conducted using sterile reagents and in a vertical flow hood. All waste should be considered as potentially biohazardous and should follow institutional regulations for proper waste disposal.

2.1 Isolation of HUVEC

1. Umbilical Cord (*see* **Note 1**).

2. Serrated Hemostatic Clamps: two sterile clamps for each umbilical cord.

3. Serrated Forceps: One pair of 8 in. (203.2 mm) serrated straight or curved tipped forceps.

4. Phosphate Buffered Saline (PBS): Sterile or sterile-filtered PBS free of calcium and magnesium.

5. Vacutainer Blood Collection Set: Blood collection set that includes a winged 21 G needle with tubing connected with a 0.75 in. (19 mm) Luer Adapter for syringe attachment (*see* **Note 2**).

6. Syringes: Sterile wrapped 10 ml syringes without attached needles.

7. Syringe Filters: 28 mm Diameter syringe filters with a 0.45 μm pore SFCA membrane.

8. Collagenase Type I: Dissolve 10 mg of collagenase I in 10 ml of sterile PBS per cord and sterile filter using a 10 ml syringe and syringe filter (*see* **Note 3**).

9. Absorbent Bench Underpads.

10. Small metal tray: an appropriate size to hold umbilical cords in the CO_2 incubator.

11. Clear plastic laboratory wrap.

12. Personal Protective Equipment (PPE): Laboratory coats and double pairs of gloves are necessary for protection from blood.

13. Tissue Culture Flask: one T25 flask per cord.

2.2 Culturing and Passaging of HUVEC

1. Gelatin from Porcine Skin: Prepare 0.1 % (w/v) solution of gelatin in warm ultra-pure water and autoclave. Store at 4 °C (*see* **Note 4**).

2. VascuLife® Basal medium (Lifeline Cell Technology, Frederick MD, USA).

3. VascuLife® VEGF-Mv LifeFactors Kit (Lifeline Cell Technology): Contains recombinant human fibroblast growth factor (rhFGF, 5 ng/ml), recombinant human insulin growth factor-1 (rh IGF-1, 15 ng/ml), recombinant human epidermal growth factor (rh EGF, 5 ng/ml), recombinant human vascular endothelial growth factor (rh VEGF, 5 ng/ml), Ascorbic Acid (50ug/ml), hydrocortisone hemisuccinate (1.0 μg/ml), L-glutamine (10 mM), heparin sulfate (0.75 U/ml), and fetal bovine serum (FBS, 5 % (v/v)). Contents are stored at –20 °C and thawed overnight at 4 °C before being added to the VascuLife® Basal medium.

4. Penicillin–Streptomycin.

5. Complete endothelial cell medium: One 500 ml bottle of VascuLife® Basal Medium supplemented with a package of VascuLife® VEGF-Mv LifeFactors Kit, penicillin, and streptomycin under the laminar flow hood. Stored at 4 °C for up to 1 month (*see* **Note 5**).

6. Sterile PBS without calcium and magnesium. Stored at room temperature.

7. Trypsin–EDTA solution: 0.25 % trypsin and 1 mM ethylenediaminetetraacetic acid (EDTA) warmed to room temperature (*see* **Note 6**).

8. Tissue culture flasks: One T75 flask per cord.

9. Tissue culture dishes: several 6-well and 12-well dishes.

2.3 Cytokine Preparation

1. Recombinant Human TNF: Reconstituted at 10 μg/ml in sterile PBS containing 0.1 % (w/v) bovine serum albumin (BSA). Stored in aliquots of 5–20 μl at –80 °C.

2. Recombinant Human LIGHT/TNFSF14: Reconstituted at 50 μg/ml in sterile PBS containing 0.1 % (w/v) BSA. Stored in aliquots of 8–20 μl at –80 °C.

3. Recombinant Human sCD40-Ligand: Reconstituted in 10 mM Sodium Phosphate, pH 7.5 to a concentration of 20 μg/ml. Stored in aliquots of 10–20 μl at –20 °C.

4. Sodium Phosphate Buffer (0.1 M), pH 7.5: Mix 16.0 ml of Solution A (0.2 M monobasic sodium phosphate) and 84.0 ml of Solution B (0.2 M dibasic sodium phosphate) and dilute to a total volume of 200 ml.

5. Monobasic sodium phosphate (0.2 M): Dissolve 27.6 g $NaH_2PO_4 \cdot H_2O$ per liter.

6. Dibasic sodium phosphate (0.2 M): Dissolve 53.65 g $Na_2HPO_4 \cdot 7H_2O$ per liter.

2.4 Protein Sample Preparation for Immunoblotting

1. TNT lysis buffer: 50 mM Tris–HCl (pH 6.8), 150 mM NaCl, 1 % (v/v) Triton X-100 prechilled on ice (*see* **Note 7**).

2. Complete protease inhibitor cocktail tablets (Roche Applied Science, Indianapolis, IN, USA). Stored at 4 °C (*see* **Note 8**).

3. Phosphatase Inhibitors: 1 mM sodium fluoride (NaF), 1 mM β-glycerolphosphate (*see* **Note 9**).

4. Rubber policeman (*see* **Note 10**).

5. PBS in a 25 ml beaker for rinsing the rubber policeman (*see* **Note 11**).

6. Coomassie Plus protein assay reagent kit containing BSA (2 mg/ml) protein standard (Pierce, Rockford, IL, USA). Reagent stored at 4 °C and BSA stored at –20 °C once aliquoted (*see* **Note 12**).

7. Sample Buffer (5×): 0.3 M Tris–HCl (pH 6.8), 5 % (w/v) sodium dodecyl sulfate (SDS), 50 % (v/v) glycerol, 100 mM DTT, 0.03 % (w/v) bromophenol blue. Stored at –20 °C in 0.5 ml aliquots (*see* **Note 13**).

2.5 SDS-Polyacrylamide Gel Electrophoresis (SDS-PAGE)

1. Separating buffer (4×): 1.5 M Tris–HCl, pH 8.8, 0.4 % (w/v) SDS. Stored at room temperature.

2. Stacking buffer (4×): 0.5 M Tris–HCl, pH 6.8, 0.4 % (w/v), SDS. Store at room temperature.

3. Thirty percent acrylamide–bis solution (37.5:1 with 2.6 % C) stored at 4 °C (*see* **Note 14**).

4. N,N,N,N′-tetramethyl ethylenediamine (TEMED): Stored at room temperature.

5. Ammonium persulfate (APS): Prepare as 10 % (w/v) and store at 4 °C (*see* **Note 15**).

6. Methanol.

7. Running buffer: 25 mM Tris, 250 mM glycine, 0.1 % (w/v) SDS. Prepare a 5× stock stored at room temperature (*see* **Note 16**).

8. Prestained protein molecular weight standard markers.

2.6 Immunoblotting for p100/p52

1. Transfer buffer: 25 mM Tris, 192 mM glycine, 20 % (v/v) methanol (*see* **Note 17**).

2. Methanol.

3. Polyvinylidene fluoride (PVDF) transfer membrane.

4. Thick cellulose chromatography filter paper.

5. Tris-buffered saline with Tween (TBS-T): 25 mM Tris–HCl, pH 8.0, 140 mM NaCl, 3 mM KCl, 0.05 % Tween-30 (*see* **Note 18**).

6. Blocking buffer: 5 % (w/v) nonfat dried milk in TBS-T.

7. Primary antibodies: Dilute mouse anti-p100/p52 (EMD Millipore, Billerica, MA, USA) at 1:1,000 in blocking buffer and mouse anti-Tubulin (Sigma-Aldrich, St. Louis, MO) at 1:5,000 in blocking buffer (*see* **Note 19**).

8. Secondary antibodies—Dilute horseradish peroxidase-conjugated donkey anti-mouse IgG in blocking buffer at 1:5,000 immediately before use (*see* **Note 20**).

9. Western blotting luminal reagent kit. Stored at 4 °C.

10. Autoradiography film.

2.7 RNA Preparation for Reverse Transcription

1. RNeasy® Mini Kit: RNeasy Mini Kit (Qiagen, Valencia, CA, USA) contains RLT lysis buffer, RW1 buffer, RPE buffer, RNase-free water, and RNeasy Mini Spin Columns (*see* **Note 21**).

2. β-mercaptoethanol (14.3 M).

3. 100 % ethanol.

4. 70 % ethanol.

5. QIAshredder homogenizer kit (Qiagen): Kit contains spin tubes for lysate homogenization (*see* **Note 22**).

6. RNase-Free DNase Set (Qiagen): Kit contains RDD buffer, RNase-free water, and DNase I enzyme (*see* **Note 23**).

2.8 Reverse Transcription and cDNA Synthesis

1. High-Capacity cDNA Reverse Transcription Kit: cDNA Reverse Transcription kit (Applied Biosystems, Carlsbad, CA, USA) contents include nuclease-free H_2O, 10× Reverse Transcriptase Buffer, 10× Reverse Transcriptase Random Primers, 100 mM dNTP mix, and 50 U/μl of MultiScribe Reverse Transcriptase (*see* **Note 24**).

2. Total RNA: samples prepared in advanced as template (*see* Subheading 3.6).

3. Thermal Cycler (*see* **Note 25**).

4. Spectrophotometer.

2.9 Quantitative Real-Time PCR

1. TaqMan® Gene Expression Master Mix (Applied Biosystems) (2×) stored at 4 °C.

2. TaqMan® Gene Expression Assays for CXCL12 (SDF1) and CXCL2 (MIP2α): One tube for each assay (20×) contains two unlabeled primers (900 nM per primer, final concentration),

one 6-FAM™ dye-labeled TaqMan® MGB probe (250 nM, final concentration). Stored at –20 °C and protected from light (*see* **Note 26**).

3. TaqMan® Human ACTB (BETA-ACTIN) Endogenous Control (VIC®/MGB Probe): One tube (20×) contains two unlabeled primers (150 nM per primer, final concentration), one 6-VIC® dye-labeled TaqMan® MGB probe (250 nM, final concentration). Stored at –20 °C and protected from light (*see* **Note 26**).

4. cDNA samples: Dilute cDNA to 25 ng/μl in nuclease-free water.

5. Reaction plate and optical adhesive film.

6. Real-time PCR instrument: ABI Biosystems instrument (*see* **Note 27**).

3 Methods

3.1 *Isolation of HUVEC*

1. Prepare the laminar flood hood prior to starting to work with the cords by removing any objects from inside the laminar flow hood to avoid blood contamination.

2. Place an absorbent bench pad into the laminar flow hood making sure to cover the entire working surface of the hood.

3. Place two sterilized 500 ml beakers in the hood. Fill one with 100 ml of PBS and the second beaker with 90 ml of PBS per umbilical cord.

4. Prepare the 10 % (w/v) Collagenase type I solution and syringe filter the solution in the laminar flow hood into a sterile 50 ml tube.

5. Place enough syringes, blood collection sets and clamps for each cord in the hood and unwrap all items from their outer wrappers or sterile packaging. Leave the blood collection needles and tubing in the inner plastic wrapping at this point. Make sure Kimwipes and paper towels are within reaching distance of the hood and are easily accessible.

6. Put all PPE including a lab coat and put on two pairs of gloves on each hand.

7. Place umbilical cords in the 500 ml beaker containing 100 ml of PBS. Once the cords are removed, this beaker will continue to be used as a waste beaker for collection of PBS washes of the umbilical cord vein.

8. Swirl the cords around in the PBS using 8-in. serrated straight forceps to loosen any excess blood (*see* **Note 28**).

9. Take out one umbilical cord at a time, using a Kimwipes wipe the outer surfaces of umbilical cord to remove exterior blood

and place on the bench pad. Examine outer walls of the umbilical cord for any holes or tears (*see* **Note 29**).

10. Massage the umbilical cord by sliding the pointer and middle fingers along the length of the cord while applying pressure to break up blood in lumen of the vein. Use a little bit of extra pressure around areas of blood clots by rolling the umbilical between the forefinger and thumb.

11. Starting at the center of the umbilical cord, using the palm of a hand slide the hand towards one end pushing out any blood left inside the vein and arteries in the cord. Repeat this several times in one direction.

12. Starting from the center, reverse the direction to remove blood from other section of the umbilical cord.

13. Check both ends of the umbilical cord to decide which end to cannulate. The easiest end of the cord to cannulate will have a clear unobstructed vein, does not have any obvious blood clots within 2.5 cm of the end of the cord and have a vein that is not stretched out more than 0.5 mm (*see* **Note 30**).

14. Remove the Vacutainer blood collection kit from the inner plastic wrapper by holding the winged part of the needle while avoiding the bottom of the needle that will be placed into the cord.

15. Slide the needle into the vein of the umbilical cord, twirling the needle back and forth as its being pushing in so that needle slides in easily without damaging the walls of the vein (*see* **Note 31**).

16. Once in, hold the needle in the vein and secure in place by closing a hemostatic clamp around the cord about 1.5 cm down from the end of the cord.

17. Fill a 10 ml syringe with sterile PBS from the second beaker of PBS and screw the syringe onto the luer adapter.

18. Holding the umbilical cord over the waste beaker by the clamp-push the 10 ml of PBS from the syringe through the vein to wash out remaining blood.

19. Refill the syringe with 10 ml of PBS and repeat wash two times (*see* **Note 32**).

20. Clamp the uncannulated end of the cord closed by placing a second hemostatic clamp around the cord about 1.5 cm from the end of the cord.

21. Fill the 10 ml syringe with 10 ml of the collagenase solution prepared earlier and fill the umbilical cord until the cord feels firm. The volumes of collagenase solution necessary to fill the vein of the umbilical cord will vary based on cord size. Smaller cords may only take 5 ml of collagenase and larger cords may take all 10 ml (*see* **Note 33**).

22. Place the filled cords, with attached clamps and needles into a laboratory wrapped metal tray and cover with the excess laboratory wrap and place in a 37 °C CO_2 incubator for 20 min.

23. Following incubation, cords are removed from the tray and laid out on the bench pad.

24. Keeping the syringe attached, massage the cords again by applying gentle pressure of pointer and middle fingers helping to loosen the enzymatically digested endothelial cells from the vein wall.

25. Sliding the palm of the hand along the cord, push the collagenase solution containing cells back into the attached syringe.

26. Transfer the collagenase solution containing endothelial cells from the syringe to a sterile 50 ml tube.

27. Fill the syringe with 10 ml of PBS and fill the vein of the umbilical cord. Massage the cord again and push the fluid back into the syringe.

28. Transfer the PBS from the syringe into the 50 ml tube containing the collagenase solution and repeat the PBS washes three more times pooling the washes and original collagenase solution from that cord.

29. Remove the hemostatic clamps and place into the metal tray to be bleached and washed. Remove the blood collection needle, tubing, and syringe and place in biohazard sharps containers. Discard the cord into double bagged biohazard bags.

30. Pellet endothelial cells by centrifugation at $300 \times g$ for 5 min at room temperature.

31. Pour off supernatant into the waste beaker and resuspend the pellet in 4 ml of complete endothelial cell medium. Transfer the resuspended cells into a sterile T25 flask (*see* **Note 34**).

32. Place the T25 flask of HUVEC at 37 °C in 5 % CO_2 incubator overnight with the cap left loose for air exchange.

33. Wash each T25 flask three times with PBS to wash off cell debris and cells that did not adhere. Replace with fresh complete endothelial cell medium and incubate cells till they reach about 80 % confluence (*see* **Note 35**).

3.2 Culturing of HUVEC and Treatments

1. Once HUVEC reach about 80 % confluence in the T25 flasks, cell need to be expanded and transferred to a T75 flask.

2. Gelatin coat one T75 flask, for each T25 to be expanded, by pipetting 3 ml of 0.1 % (w/v) gelatin solution into the flask. Swirl the gelatin around the flask so the gelatin coats the entire bottom surface of the T75. Aspirate excess gelatin and allow flask to dry for 30 min in the laminar flow hood (*see* **Note 36**).

3. Aspirate off medium from the HUVEC in the T25 flasks and pipette 5 ml of PBS into flask to wash cells.

4. Aspirate the PBS from the flask and add 0.5 ml of 0.25 % Trypsin–EDTA solution. Place the flask at 37 °C in the 5 % CO_2 incubator for 5 min for enzymatic cell dissociation.

5. Following the 5 min incubation with trypsin/EDTA solution, tap the flask against the palm of a hand to ensure cells are detached (*see* **Note 37**).

6. Resuspend supernatant in 12 ml complete endothelial cell medium and transfer the entire cell solution to a gelatin coated T75 flask.

7. Incubate cells at 37 °C in 5 % CO_2 incubator with the cap left loose for air exchange.

8. Complete endothelial cell medium is replaced every second day until cells reach 80–90 % confluence for their final passage.

9. When HUVEC reach 80–90 % confluence in a T75, cells will be passaged for the last time and seeded for experimental treatments. One T75 of HUVEC can seeded at a maximum of 36 ml of the medium containing cells, so experiments must be planned around the total use of 36 ml per T75 flask.

10. Gelatin coat three 12-well dishes by adding 0.5 ml of 0.1 % gelatin solution to each well. Tap dishes to ensure even coating of the entire growth area of the well with gelatin Aspirate the excess gelatin from each well and leave the dish opened to dry in the laminar flow hood for 30 min (*see* **Note 38**).

11. Once gelatin coated dishes have dried, aspirate medium from the T75 flask and wash cells with 5 ml of PBS.

12. Add 1 ml of trypsin/EDTA solution and incubate for 5 min at 37 °C in 5 % CO_2 incubator.

13. Following incubation with trypsin–EDTA, tap flask against palm of hand to aid cell detachment (*see* **Note 37**).

14. Resuspend cells in 36 ml complete endothelial cell medium, making sure cells are evenly resuspended and there are no visual cell clumps.

15. Seed the cell/medium solution by pipetting 1 ml of cell suspension into each well of the gelatin coated 12-well dishes and incubate dishes at 37 °C in 5 % CO_2 incubator.

16. Cells are rested for a minimum of 18 h following seeding for cells to return to steady state prior to treatment (*see* **Note 39**).

17. HUVEC are stimulated in wells by addition of 1 μl of TNF per ml of medium for a final concentration of 10 ng/ml TNF, 2 μl of LIGHT per ml of medium for a final concentration of 100 ng/ml or 5 μl of CD40L per ml of medium for a final concentration of 100 ng/ml.

18. Treated cells are kept at 37 °C in 5 % CO_2 incubator for their appropriate incubation times (*see* **Note 40**).

3.3 Sample Preparation for Immunoblotting

1. Aspirate medium from wells and wash each well with 1 ml of PBS.

2. Place plate on ice and add 100 μl of TNT lysis buffer containing complete protease inhibitor cocktail and phosphatase inhibitors to each well.

3. Incubate the plate on ice for 15 min and then scrape the cells in each well with a rubber policeman.

4. Tilt the plate on a 45° angle against the ice so the lysates will collect at the bottom of their wells.

5. Transfer the lysates to new 1.5 ml microfuge tubes and vortex at max for 10 s.

6. Spin tubes for 10 min at $16,000 \times g$ in a refrigerated bench top microfuge set to 4 °C (*see* **Note 41**).

7. Remove supernatant to a new 1.5 ml tube and place on ice while determining protein concentration (*see* **Note 12**).

8. Set up protein standard curve by dilution from the stock of BSA (2 mg/ml) to achieve concentrations of 0.5, 1, 2.5, 5, 10, and 20 μg/ml in water. Pipette duplicate aliquots of 150 μl of water (blank) and the protein standards into a 96-well plate.

9. Dilute 2 μl of each HUVEC lysate in 500 μl of water and vortex.

10. Pipette 150 μl of each diluted sample in duplicate into new wells of the 96-well plate containing the blanks and standards.

11. Add 150 μl of the Coomassie Plus reagent to each well and read the plate using a microplate reader. Calculate the protein concentration in each sample manually by constructing a graph using the standard curve or by using software specific for the plate reader used (*see* **Note 12**).

12. Prepare samples for SDS-PAGE (*see* Subheading 3.4) in sample buffer with each containing 20 μg of protein (*see* **Note 42**).

13. The remaining lysates should be frozen for short time frames at –20 °C for a future immunoblot. For longer-term storage, remaining lysates should be stored at –80 °C.

3.4 SDS-Polyacrylamide Gel Electrophoresis (SDS-PAGE)

1. These directions are for casting and running 10 % SDS gels using the Bio-Rad Mini Protean Electrophoresis apparatus (Bio-Rad, Hercules, CA, USA). Directions for other systems are similar and a potential alternative to making gels would be purchasing precast 10 % gels (*see* **Note 43**).

2. Clean the back spacer plates and front short plates before use. Plates should be wiped down with 70 % ethanol to avoid gel sticking unevenly to the plates.

3. Assemble the glass plates in the gel-casting apparatus placing a short front plate with a back spacer plate lined up evenly.

4. In a 50 ml conical tube, prepare a 10 % resolving gel by mixing 6.25 ml of water, 3.75 ml of 4× separating buffer, 5 ml of 30 %

acrylamide–bis solution, 50 µl of APS solution, and 20 µl of TEMED (*see* **Note 44**). Mix by inversion.

5. Pipette the solution into the gel chamber leaving space for the stacking gel. Layer 1 ml of methanol onto the resolving gel while the gel solidifies.

6. Check the left over resolving gel mixture in the 50 ml tube for polymerization. Polymerization can take between 5 and 10 min.

7. Begin preparation of the stacking gel in a 50 ml conical with exception to the addition of TEMED, by mixing 3.05 ml of water, 1.25 ml of 4× stacking buffer, 0.65 ml of 30 % acrylamide–bis solution, and 25 µl of APS solution.

8. Once the gel is polymerized, place a piece of paper towel along the top of each gel chamber and invert the casting mold to absorb the methanol out of the gel chamber in preparation of the stacking gel.

9. Add 10 µl of TEMED to the stacking gel solution and mix by inversion.

10. Pour stacking gel on top of the resolving gel and insert a 10-well comb (*see* **Note 45**). The stacking will polymerize in 5–10 min.

11. While waiting for the stacking gel to polymerize, prepare 1 l of 1× running buffer from the 5× stock.

12. Assemble the electrophoresis cassette in the tank when gel has fully polymerized. Two gels will be required to create an inner chamber, using a dam plate if only one gel is required. Fill inner chamber entirely and 1/3 of the outer chamber with 1× running buffer.

13. Remove the comb by pulling straight up on both sides of the comb to keep wells intact.

14. Place the samples (*see* **step 12** of Subheading 3.3) in a heat block (95 °C) for 5 min and then spin at top speed in a bench top microfuge for 2 min.

15. Using a gel-loading tip, add 10 µl of protein molecular weight standard to the first well. Continue the addition of samples to subsequent wells using gel-loading tips.

16. Connect the power supply and run at 150–180 V until the samples have separated over the entire length of the gel (visualized by watching the separation of the prestained markers, *see* **Note 46**). The gel is run at room temperature.

3.5 Immunoblotting for p100/p52

1. These directions for protein transfer are for the Bio-Rad Mini-Trans-Blot® Electrophoretic Transfer Cell wet transfer apparatus.

2. Fill the transfer chamber with prefilled complete transfer buffer (containing 20 % methanol) and place an ice pack in the outer chamber.

3. Cut two pieces of chromatography filter paper and a single piece of PVDF membrane large enough to cover the gel (*see* **Note 47**).

4. Saturate the PVDF membrane with 100 % methanol and immediately immerse membrane in transfer buffer. Do not allow membrane to dry.

5. Fill a dish (large enough to hold the transfer cassette) with complete transfer buffer. Immerse the membrane, filter paper, and foam pads in transfer buffer.

6. Remove the gel from the electrophoresis apparatus and trim away the stacking gel using a clean razor blade.

7. Assemble the gel transfer "sandwich" (keeping all materials immersed in transfer buffer) in the following order: sponge, filter paper, gel, PVDF membrane, filter paper, and sponge.

8. Remove any air bubbles from the assembly using a roller (*see* **Note 48**).

9. Place the sandwich in a cassette and insert the cassette into the inner chamber of the transfer apparatus with the gel side of the sandwich closest to the negative (black) electrode and the membrane side of the sandwich closest to the positive (red) electrode (*see* **Note 49**).

10. Run the transfer at 125 V for 1 h 20 min at 4 °C (*see* **Note 50**).

11. When transfer is complete the markers should be visible on the membrane. Disassemble the cassette and sandwich starting at one corner and place the PVDF membrane in blocking buffer on a rocker for 30 min at room temperature.

12. Place the membrane in a bag containing p100/p52 primary antibody and seal the bag making sure to remove all air bubbles. Rock the membranes overnight at 4 °C (*see* **Note 51**).

13. The next day, remove membranes from primary antibody and wash with TBS-T four times for 5 min. Place dishes containing membranes on a rocker for each wash changing buffer between the washes (*see* **Note 52**).

14. Dilute the appropriate HRP-conjugated secondary antibody in blocking buffer and incubate with the membrane for 1 h on a rocker at room temperature.

15. Remove the membranes from secondary antibody and wash with TBS-T four times for 5 min, changing buffer between washes.

16. During the final wash, mix 1 ml of each of the luminal reagents for chemiluminescent detection. Immediately after discarding the TBS-T from the final wash, transfer the blots using forceps to a piece of saran or laboratory wrap on a flattened surface placing blots protein side up. Carefully pipette the mixed luminal reagent onto the blot. Incubate at room temperature for 3 min (*see* **Note 53**).

17. Pick up the membrane using forceps and gently dab off the luminal using Kimwipes. Place the membrane between two sheets of acetate (*see* **Note 54**) and plate into an X-ray film cassette.

18. Place a piece of film over membrane and test a range of exposure times to obtain optimal band intensity. A typical result is shown in Fig. 1.

19. Following exposure for optimal band intensity, membrane can be removed from the film cassette and incubated in TBS-T for a 10 min wash.

20. Following the TBS-T wash, place membrane in a bag containing anti-Tubulin primary antibody and seal the bag making sure to remove all air bubbles. Rock the membranes overnight at 4 °C (*see* **Note 55**).

21. Repeat **steps 12–18** (*see* Subheading 3.5) to wash membrane, incubate with the appropriate HRP-conjugated secondary antibody and prepare for chemiluminescent detection of Tubulin.

3.6 Preparation of RNA

1. These directions are specific to the use of the Qiagen RNeasy Mini Kit for purifying RNA from cultured cells. There are many alternative kits that follow similar steps in regards to the lysis of cells, specific membranes for RNA isolation and wash steps, however each kit comes with its own set up proprietary buffers.

2. HUVEC cells should be passaged and treated as described in Subheading 3.2.

3. Prior to the end of cytokine treatment endpoint prepare buffers from RNeasy kit for RNA extraction adding four volumes of 100 % ethanol to the concentrated RPE buffer.

4. Prepare RLT buffer for cell lysis by transferring 350 μl/well of RLT buffer to a 15 ml tube and adding 10 μl of β-mercaptoethanol for each ml of RLT buffer (*see* **Note 56**).

5. At the endpoint of cytokine incubation, aspirate medium from cells and wash each well with 1 ml of PBS.

6. Add 350 μl of RLT buffer containing β-mercaptoethanol to each well and disrupt cells by pipetting (*see* **Note 57**).

7. Transfer cell lysate to a QIAshredder spin column and spin at full speed for 2 min to homogenize the cell lysate.

8. Discard the spin column and mix one volume (350 μl) of 70 % ethanol with the homogenized lysate by pipetting.

9. Transfer the 700 μl cell lysate to an RNeasy spin column and spin at $8,000 \times g$ for 15 s. Discard the flow-through.

10. Add 350 μl of RW1 buffer to the RNeasy spin column and spin at $8,000 \times g$ for 15 s. Discard the flow-through.

11. For each sample, prepare a master mix of DNase digestion buffer by adding 10 μl DNase I stock solution to 70 μl of RDD buffer for each sample to be treated and mix by inversion.

12. Add 80 μl of DNase master mix directly to each RNeasy column membrane and incubate for 15 min at room temperature (*see* **Note 58**).

13. Add 350 μl of Buffer RW1 to RNeasy spin column and centrifuge for 15 s at 8,000×*g*. Discard the flow-through.

14. Add 500 μl of RPE buffer to the RNeasy spin column and centrifuge at 8,000×*g* for 15 s. Discard the flow-through.

15. Add 500 μl of RPE buffer to the RNeasy spin column and centrifuge for 2 min at 8,000×*g*.

16. Place the RNeasy spin column into a new 2 ml collection tube (provided in the kit) and spin at 16,000×*g* for 1 min to remove any residual RPE buffer.

17. Place the RNeasy spin column into a 1.5 ml collection tube. Add 40 μl of RNase-free water directly to the spin column and incubate for 1 min at room temperature.

18. Centrifuge the RNeasy spin column for 1 min at 8,000×*g* to elute the RNA (*see* **Note 59**).

19. Use the RNA sample directly for cDNA synthesis (*see* Subheading 3.7) or store at −80 °C for a later time.

3.7 Preparation of cDNA

1. These directions are for the cDNA synthesis from RNA samples using the High-Capacity cDNA Reverse Transcription kit (*see* **Note 24**).

2. Prepare a master mix of reverse transcriptase reaction solution in a 1.5 ml tube by adding the following 4.2 μl of H_2O, 2 μl of 10× RT buffer, 2 μl of Random Primer buffer, 0.8 μl of dNTPs, and 1 μl of reverse transcriptase enzyme for each RNA sample collected in Subheading 3.6.

3. Pipet 10 μl of the reverse transcriptase reaction mixture into each PCR tube.

4. Pipet 10 μl of each RNA sample into the reverse transcriptase reaction and mix by pipetting.

5. Load the tubes into a thermal cycler and subject the reaction mixtures to the following cycling conditions: 10 min at 25 °C, 120 min at 37 °C, 5 min at 85 °C, and a hold step at 4 °C (*see* **Note 25**).

6. Prepare 200 μl samples of each cDNA diluted at 25 ng/μl in nuclease-free water for direct use in the quantitative real-time PCR reaction. The remaining cDNA samples can be stored at 4 °C for future use.

3.8 Performing Quantitative Real-Time PCR

1. The following methods for the quantitative real-time PCR reaction setup and for running the plate are specific to the use of the ABI Biosystems TaqMan® gene expression assays and on an ABI Biosystems Fast Real-time PCR instrument. A less expensive alternative to the TaqMan® gene expression assays is the use of SYBR Green dye reaction mixes which can be set up similarly, however the instructions for running those reactions on an ABI PCR instrument or other Real-time PCR instruments may vary greatly from the directions listed here.

2. Thaw TaqMan® gene expression primer/probe sets on ice.

3. Prepare master mixes of the individual gene assays for each sample in quadruplicate, by adding 5 μl of the 2× TaqMan Gene Expression Master Mix, 0.5 μl nuclease-free water, 0.5 μl of the 20× TaqMan Gene Expression primer/probe assay per sample in a 1.5 ml tube.

4. Vortex the TaqMan master mixes and centrifuge briefly to bring the liquid down to the bottom of the tube.

5. Pipet 6 μl of each master mix into the appropriate wells of the reaction plate.

6. Pipet 4 μl of each cDNA sample (25 ng/μl) directly into the reaction mixture in quadruplicate (*see* **Note 60**).

7. Seal the reaction plate with an optical adhesive film using a hard plastic identification card or credit card to smooth the cover over the plate and push out any air bubbles (*see* **Note 61**).

8. Centrifuge the plate briefly to bring down the reactions to the bottom of the wells and to remove air bubbles.

9. Load the plate into the real-time thermal cycler and open the software for the PCR instrument.

10. Create a new experiment and select the Relative ($\Delta\Delta C_T$) quantification plate as the assay.

11. Select the TaqMan® reagents mode or the FAST mode for the ramp speed for TaqMan assays.

12. Select the appropriate detectors for the CXCL12 and CXCL2 with the FAM fluorophore and the BETA-ACTIN (β-Actin) detector with the VIC fluorophore. If these are unavailable, create these new detectors.

13. Assign your reactions containing the β-Actin primer/probe as the endogenous control samples. Assign all other samples as reference or target samples.

14. Under the plate layout tab, highlight the quadruplicate wells of each sample and input the sample name and the reaction assay loaded in those locations on the plate (*see* **Note 62**).

15. Under the instrument tab set the reaction volume to 10 μl/well.

16. The PCR program should be automatically filled in based on the FAST mode program. That program involves a single heating step of 95 °C for 20 s, followed by 40 cycles of 95 °C for 3 s and 60 °C for 30 s.

17. Save file and load the plate into the machine.

18. Click start.

19. Once the plate has finished, close the experiment.

20. For analysis, create a new experiment and select the Relative ($\Delta\Delta C_T$) quantification study as the assay. Add the plate of the data that was just run.

21. From this assay page, look at the raw data under the amplification curve tab.

22. Ensure the settings for the baseline and threshold are correct (*see* **Note 63**).

23. From the dropbox on top, choose the untreated sample as your control sample.

24. Scan through the individual well Ct values to ensure the quadruplicates look similar and omit any outliers (*see* **Note 64**).

25. Go to File and then Export to export your .csv file which can be opened in Excel.

26. Quantifications of the gene expression in each sample for a gene assay are plotted using the RQ values determined in the ABI software. The positive error bar is the difference between the RQmax and the RQ value. The negative error bar is the difference between the RQ value and the RQmin.

4   Notes

1. We consistently isolate HUVEC from discarded umbilical cord tissue obtained directly from the Labor and Delivery departments at the Hospitals within in our Health system following a protocol approved by our institution's Internal Review Board. This is a much cheaper alternative to purchasing HUVEC and allows us to work with the cells as early as passage 2. HUVEC can be obtained through many commercial sources, including ATCC (Manassas, VA, USA), Life Technologies and Lifeline Cell Technology among several others. We have successfully used HUVEC from these sources.

2. Although there are many different kinds of blood collection sets commercially available, the blood collection set is not used for collecting cord blood in this protocol but rather to cannulate the vein for pushing/pulling fluids through the cords and so sets with attached tubes should be avoided. The best collection set

for these purposes will have winged (butterflied) 21 G needles that have 12 in. (305 mm) tubing with an attached luer adapter for separate syringe attachments.

3. Although collagenase solutions can be prepared as a larger stock and aliquoted in smaller working volumes to avoid multiple freeze–thaws, we always prepare collagenase solutions fresh, an hour prior to beginning the isolation since the protease activity is sensitive to the freeze–thaw process.

4. A 0.1 % stock of gelatin should be stored at 4 °C but must be warmed to room temperature before use to coat flasks and dishes.

5. Endothelial cell isolation and maintenance requires a complete growth medium that is low in serum amounts and supplemented with growth factors. There are many commercially available medium kits that provide kits of growth factors and basal medium, but we find that our endothelial cells isolated and maintained in VascuLife® VEGF-Mv cell culture medium more rapid proliferation following isolation thereby shortening the delay between isolation and treatments. Additionally isolated endothelial cells maintain healthy cell morphology a passage longer compared to previous isolations using alternate endothelial cell medium.

6. Unopened bottles of trypsin can be stored at –20 °C. Once a bottle of trypsin solution has been thawed and opened, we store it at 4 °C and warm it to room temperature before adding it to cells.

7. TNT lysis buffer can be stored at room temperature or at 4 °C however after several weeks the buffer becomes cloudy and should not be used for cell lysis. We often prepare fresh TNT after one month of storage regardless of its appearance.

8. Dissolve one tablet of a mini-complete protease inhibitor cocktail dissolved in 400 μl of water to make a 25× stock. This can be stored at –20 °C in aliquots if not used immediately. Frozen aliquots should be not be used with repeated freeze–thaws. Protease inhibitors should not be stored in TNT lysis buffer and should be added fresh the same day as performing cell lysis.

9. Phosphatase inhibitors must be added fresh to TNT lysis buffer the same day as performing cell lysis and should not be stored in TNT lysis buffer. Both NaF and β-glycerolphosphate can be prepared as 1 M stocks and stored in small aliquots at –20 °C. Aliquots should only be thawed once and discarded; therefore, we make 10 μl aliquots.

10. We use the rubber end of a plunger from a 1 ml syringe as a rubber policeman to scrape cells instead of cell scrapers because

they are a better fit in small wells for scraping HUVEC cells which do not detach easily.

11. Swirl the rubber policeman in PBS and wipe with Kimwipes in between wells to avoid contamination between the different samples.

12. Using the Coomassie Plus Protein Reagent, protein concentrations can be determined using a microplate reader equipped with a 595-nm filter. We use a Bio-Rad Model 680 microplate reader utilizing Microplate Manager-III software.

13. A 1 M Tris–HCl pH 6.8 stock cannot be used during the preparation of Sample Buffer (5×) because the required volume of 1 M Tris will be higher than volume availability. We use a 2 M Tris–HCl pH 6.8 stock for preparation of Sample Buffer.

14. Unpolymerized acrylamide is a potentially harmful neurotoxin. Extreme care should be taken when handling acrylamide including the wearing of gloves and safety glasses.

15. 10 % stocks of APS solution can be stored at 4 °C for up 2 weeks without affecting polymerization.

16. We make a large stock of 20 l of 5× running buffer and can be stored at room temperature for up to 3 months.

17. We make a 20 l stock of Transfer buffer containing Tris and Glycine but without the Methanol. On the day of the immunoblotting, we make 1 l of complete Transfer buffer using 800 ml of the transfer buffer stock and 200 ml methanol and prechill at 4 °C at least 1 h prior to setting up the transfer. This 1 l will be sufficient to fill the transfer apparatus and have enough for setting up the transfer "sandwich."

18. For convenience during the repeated use of TBS-T, we make a 20 l stock of 1× TBS-T which can be stored for up to 3 months at room temperature.

19. Primary antibodies diluted in blocking buffer containing 0.1 % (w/v) sodium azide can be reused up to three times.

20. Sodium azide inhibits the enzymatic activity of HRP and should not be used for the preparation of secondary antibodies in blocking buffer. Secondary antibodies should be prepared freshly and are not kept for multiple uses.

21. There are multiple methods for preparation of RNA from cells, including alternative commercially available kits for RNA isolation; however we find the Qiagen kit allows for quick and reproducible RNA concentrations for further cDNA synthesis and quantitative reverse-transcriptase experiments. The cheapest alternative is the use organic extraction protocols; however, the steps are cumbersome and more time consuming.

22. The RNeasy Mini Kit requires a step to homogenize the lysate and provides three options, two of which are much cheaper than buying an additional kit. We use the QIAshredder spin kit, as it is the quickest method, thereby reducing the total time the RNA is being kept at room temperature. Additionally, this kit provides a consistent lysate homogenate between the samples.

23. The on-column DNase digestion method is optional using the RNeasy Mini Kit and a separate treatment of RNA samples can be performed using DNase I treatment, however we perform the on-column DNase digestion step for consistency between the samples. Additionally, a separate DNase I treatment will leave extra buffers in the digestion which may not be compatible with our cDNA synthesis protocol and requires a heat inactivation step to stop the DNase I which could result in some RNA sample degradation.

24. There are several cDNA synthesis kits commercially available. We choose to use the Applied Biosystems High Capacity cDNA synthesis kit because this kit is most compatible with the Gene Expression Assays and Real-Time PCR systems which are from Applied Biosystems.

25. We use a thermal cycler for ease; however, the reaction of synthesizing cDNA can be easily performed using a few water baths or heat blocks.

26. There are several different ways Real-time PCR can be run for the quantification of gene expression. We use the TaqMan® Gene Expression Assays specifically for the detection of CXCL12 in HUVEC because it has provided the most reliable and consistent detection of CXCL12 between experimental repeats performed in HUVEC isolated from various umbilical cord tissues. In our laboratory, we do also perform quantitative Real-time PCR protocols using SYBR® Green Dye with synthesized oligonucleotides for the quantification of expression of other gene targets. This would provide a cheaper alternative to using the TaqMan® reagents described in this chapter, but requires reliable and target specific primers for quantification purposes.

27. Since ABI Biosystems markets the TaqMan® Gene Expression Assays, their Real-time PCR instruments contain the appropriate filters required for the detection and quantification of a number of dyes including the FAM and VIC dyes used in this protocol. Real-time PCR instruments from other sources may be suitable for use with these methods if they contain the appropriate filters however we have never tested this.

28. Use gentle force when moving umbilical cords using forceps. Keep just enough pressure to have a grip on the cord without

squeezing into the umbilical cord wall. This could damage the interior vessels. We often try to grab cords near their end so that any damaged tissue can be cut away while leaving the remaining portion of the cord intact for isolation.

29. If there are any tears or holes near the end of the umbilical cord, the cord can be cut to remove the damaged ends. If any damage had occurred to the outer walls near the center of the cord, the cord should be discarded and not used for isolation.

30. If neither end of the umbilical cord matches the description of one that might be good for cannulation, we will cut off 2 cm portions of the end furthest from any blood clots until we find a clear vein that will be easy to slide the needle into.

31. The fit of the needle into the vein should be snug. If the needle is very easy to slide into the vein because the vein opening is much larger than the needle, there will problems with leakage of your cells and solutions at later steps. If there is a lot of resistance to the needle sliding in, there may be a clot near the end. When this occurs, we cut away a small piece from the end and retry with a new sterile needle or alternatively cannulate the other end of the cord.

32. By this third wash, the PBS should be close to clear. If the third PBS wash still contains a lot of blood, it is best to go back and massage the cord against the hood surface to help break up any remaining blood clots that were missed and perform a few extra PBS washes before filling the cord with collagenase solution.

33. When the cord is full, there will be resistance while pushing the plunger into the syringe. Sometimes when the cord is over-filled, some of the solution will be pushed back into the syringe. Do not continue to push collagenase into the cord past this resistance, even if the cord feels like it is not full, as this may cause damage to the vein wall and compromise the rest of the isolation.

34. If any of the umbilical cords were longer than 10–11 cm and the cell pellet looks large, we might resuspend the pellet in 8 ml of complete endothelial cell medium and split the cells between two T25 flasks.

35. HUVEC isolated from fresh umbilical cords following this protocol, typically reach 80 % confluence and are ready for expansion by 2 to 3-days following the isolation. If the cells are not confluent by this time and still remain sparse, we would usually not continue to use them for experiments since they may have started to lose their endothelial cell phenotype.

36. We prop the T75 flasks up at a 45° angle against a tip box while it is drying in the hood so any excess gelatin will collect at bottom. This can be aspirated before adding the cell suspension.

37. Complete detachment of cells should be verified microscopically before resuspending the cells for transfer.

38. Experiments can be done in other tissue culture plates depending on the needs of the experiment, such as the number of treatments and time-points required. Any combination of dishes can be used as long as they can be appropriately seeded using only a total of 36 ml of the cells suspension. We find that using 12-well dishes allows us the opportunity to have a large number of test conditions (36 different conditions) while at the same time providing a sufficient number of cells per condition to make protein and RNA lysates from. This minimizes our need to pool cells from different umbilical cords to cover our treatment requirements.

39. The proteolytic activity of trypsin can cleavage and internalization of cell surface receptors, so cells must be allowed time to resynthesize these receptors.

40. We often stagger our stimulations times so that we can collect cells at the same endpoint prior to lysing cells for protein or RNA. To do this, we begin the stimulations of HUVEC for our 24 h incubations the night before the cells will be collected. We will then perform our 8 h stimulations exactly 8 h prior to the end of the 24 h incubation. This provides consistency between our sample lysates since they are processed at the same time using the same freshly prepared buffers.

41. If a refrigerated microfuge is unavailable, a standard bench top microfuge can be prechilled for 2 h in a cold room or large refrigerator.

42. Protein concentrations can be increased if the signal is too weak as long as it fits within in the volume allowed by the well thickness. Protein concentrations can also be decreased if the signal is too strong. We have found that 10–20 µg of protein from HUVEC is sufficient for obtaining a strong signal for examining p100 processing to p52.

43. Precast gels can be obtained from several commercial sources. These are an expensive alternative to casting gels but provide reproducible high-quality results.

44. These volumes are sufficient for making two 0.75 mm thick gels.

45. Once TEMED is added to the stacking gel, the polymerization reaction begins, so the combs should be placed into the gel apparatus immediately following the pouring of the stacking gel.

46. Unless we are looking for proteins of smaller sizes, we often run our gels long enough that any proteins lower than the 36 kD protein band in our protein molecular weight standard have run off the gel. For the detection of both the p52 protein and Tubulin (57 kD) on the same blot, we run these gels so that the 36 kD protein molecular weight standard is at the very

bottom of the gel. This will provide enough distance between the 50 and 60 kD proteins on a 10 % SDS-PAGE gel that p52 and Tubulin will not be too close for detection.

47. Chromatography filter paper cut to pieces of 3" × 4" provide sufficient coverage of the gel and fit in the transfer cassette without overhanging. A membrane size of 2.5" × 3.5" will cover the separating portion of the gel.

48. A cut down 5 or 10 ml plastic pipette or a 15 ml tube can be used as a roller.

49. If using the Bio-Rad apparatus, the cassette has a black side and an opaque side. The black side can be used as a convenient guide to assemble the sandwich and align the cassette with the negative electrode in the transfer tank.

50. For convenience, we set up the apparatus in a chromatography refrigerator to transfer at 4 °C. Alternatively a 4 °C cold room can be used.

51. We have found the Impulse Sealer from Hualian Packaging Machinery Co. Ltd. (Wenzhou City, China) to be ideal for immunoblotting purposes.

52. Lids from pipette tip boxes work well as dishes for incubating membranes with TBS-T and secondary antibody solutions.

53. It is not necessary to perform this step in a dark room. However, the subsequent exposure to film should be performed under a safety light in a dark room.

54. Plastic sheet protectors available from most stationary stores can be cut to size and used for holding membranes for exposure instead of acetate sheets.

55. We re-probe membranes for detection of multiple proteins without performing stripping procedures when proteins of interest are different enough in size that we can distinguish between them. In this case, we run the gel longer allowing the small proteins to run off giving us enough room to distinguish between the p52 and Tubulin proteins following chemiluminescent detection. If detecting proteins of similar size or if your antibody detects extra nonspecific bands, membranes must be stripped using stripping buffers prior to placing membranes into a different primary antibody.

56. Volumes stated in these methods are specific to the use of a 12-well dish containing HUVEC. Volumes need to be adjusted for other cell culture dishes based on cell number and the guidelines provided by the manufacturer should be reviewed.

57. Gloves should be worn during the rest of the steps to prevent contamination of the RNA samples by RNases on skin.

58. It is important to pipette the DNase mixture onto the center of the column membrane and not along the sides of the tube. The volume is too small for it to cover the entire bottom and will not come in contact with the RNA lysates on the membrane.

59. The caps of the provided 1.5 ml collection tube often snap off when too many tubes are centrifuged at one time. We have found it best to separate the tubes to leave empty holes in between. We have also had low incidence of caps breaking when the tubes are spun with the hinge of the spin column facing towards the center of the rotor.

60. When placing the quadruplicate samples into wells, keep them next to each other. This will make it easier to input sample names into the plate layout on the instrument software when preparing to run the plate and for the analysis of the raw data.

61. Wear gloves when placing the optical adhesive film onto the plate to avoid getting any skin oils onto the optical cover while trying to smooth the film out. Oils or fingerprints can provide background noise for fluorescent readings.

62. Make sure when you assign the name of the sample or treatment that this name is the same for all three reactions assays. The name must appear exactly the same between those quadruplicates in the CXCL12 reaction as in the β-ACTIN reaction and the CXCL2. The software will automatically set the quantification of your target genes to the endogenous control, but will not provide RQ values if it does not recognize the samples as being the same.

63. The baseline should not need adjusting unless the water in the reaction has some contamination. Rather than adjusting the baseline, we would set up a new reaction and plate. There is an option on the analysis screen to automatically set Threshold. We usually let the software automatically set the Threshold by clicking this option. If you are looking to compare raw data from one reaction plate to another reaction plate, this threshold must be the same between the two plates. In those cases, we will manually set the threshold so that it is the same between the two different analysis files.

64. We only manually omit a sample reading, if one of the C_T values is obviously different from the other three replicates and missed by the software. The software does automatically omit samples from the quadruplicate readings if it is deemed significantly different from the other replicates and does not include it as part of the RQ determination. We never omit more than one of the quadruplicate data points for a sample so that the RQ, RQmin, and RQmax values are always determined from three C_T values.

References

1. Hayden MS, Ghosh S (2004) Signaling to NF-kappaB. Genes Dev 18:2195–2224

2. Dejardin E, Droin NM, Delhase M et al (2002) The lymphotoxin-β receptor induces different patterns of gene expression via two NF-kappaB pathways. Immunity 17:525–535

3. Coope HJ, Atkinson PG, Huhse B et al (2002) CD40 regulates the processing of NF-kappaB2 p100 to p52. EMBO J 15:5375–5385

4. Claudio E, Brown K, Park S et al (2002) BAFF-induced NEMO-independent processing of NF-kappaB2 in maturing B cells. Nat Immunol 3:958–965

5. Novack DV, Yin L, Hagen-Stapleton A et al (2003) The IκB function of NF-κB2 p100 controls stimulated osteoclastogenesis. J Exp Med 198:771–781

6. Sun S-C (2011) Non-canonical NF-κB signaling pathway. Cell Res 21:71–85

7. Courtois G, Gilmore TD (2006) Mutations in the NF-κB signaling pathway: implications for human disease. Oncogene 25:6831–6843

8. Karin M (2006) Nuclear factor-κB in cancer development and progression. Nature 441:431–436

9. Tak PP, Firestein GS (2001) NF-κB: a key role in inflammatory disease. J Clin Invest 107:7–11

10. Baker RG, Hayden MS, Ghosh S (2011) NF-κB, inflammation and metabolic disease. Cell Metab 13:11–22

11. Dejardin E (2006) The alternative NF-κB pathway from biochemistry to biology: pitfalls and promises for future drug development. Biochem Pharmacol 72:1161–1179

12. Ranzani B, Reichardt AD, Cheng G (2011) Non-canonical NF-κB signaling activation and regulation: principles and perspectives. Immunol Rev 244:44–54

13. Brown KD, Claudio E, Siebenlist U (2008) The roles of the classical and alternative nuclear factor-κB pathway: potential implications for autoimmunity and rheumatoid arthritis. Arthritis Res Ther 10:212

14. Sitia S, Tomasoni L, Atzeni F et al (2010) From endothelial dysfunction to atherosclerosis. Autoimmun Rev 9:830–834

15. Solan NJ, Miyoshi H, Carmona EM et al (2002) RelB cellular regulation and transcriptional activity are regulated by p100. J Biol Chem 277:1405–1418

Chapter 10

Stable Reconstitution of IKK-Deficient Mouse Embryonic Fibroblasts

Carolyn M. Gray and Michael J. May

Abstract

Retroviral transduction is an invaluable technique in molecular biology used to express proteins encoded by nonviral genes in mammalian cells. A key feature of this technique is the ability to create cell lines that stably express the protein of interest and can be cultured long term. Here we describe a retroviral transduction procedure for mouse embryonic fibroblasts (MEFs) that uses Platinum-E cells to rapidly package high-titer, helper-free retrovirus. This technique is useful to study the role of key signaling kinases in the NF-κB signal transduction pathway.

Key words Retroviral transduction, IKK, Mouse embryonic fibroblast, Cloning, Western blot, Luciferase

1 Introduction

The rapid generation of high-titer retrovirus for in vitro and in vivo transduction has been a landmark technology in the field of gene transfer [1]. This technique involves the transient transfection of a gene of interest along with the viral packaging genes *gag*, *pol*, and *env* into a packaging cell line to create a retroviral supernatant. This supernatant can then be used to stably transduce target cells to study the gene of interest [4]. Virions bind to host-specific receptors on the target cell surface, facilitating viral fusion and infection of the viral genome. Early work by Pear et al. employed the MuLV LTR to drive the expression of the viral structural proteins in Bosc23 cells [1]. The development of the Platinum-E (Plat-E) cell line [2] improved this technique in two main ways: (1) the more potent EF1α promoter was used to drive gene expression and (2) bicistronic constructs contained viral structural genes (*gag-pol* or *env*) followed by an IRES sequence and a selectable gene cassette, thereby ensuring selection of cells with the necessary virion packaging components. Additionally, Plat-E cells have been shown to produce viral titers of 1×10^7 U/mL [2].

Michael J. May (ed.), *NF-kappa B: Methods and Protocols*, Methods in Molecular Biology, vol. 1280,
DOI 10.1007/978-1-4939-2422-6_10, © Springer Science+Business Media New York 2015

Reconstitution of key signaling components in knockout murine embryonic fibroblasts (MEFs) has been an invaluable approach employed to understand the critical regulatory role of the inhibitor of kappa B kinase (IKK) complex in NF-κB signaling [3]. Reintroduction of a mutant version of IKKα or IKKβ, for example, into a knockout cell lines has yielded insight into the structural requirements for classical and noncanonical NF-κB signal transduction [4, 5]. In this chapter we outline a protocol to reintroduce IKKα into IKKα-deficient mouse embryonic fibroblasts (MEFs). The resulting cell line will be useful to investigate the role of IKKα in noncanonical NF-κB signaling or to explore IKK-independent roles of nuclear IKKα [5–9]. We will transiently transfect Plat-E cells with the GFP-expressing MigR1 retroviral construct [10] containing FLAG-IKKα. The FLAG tag serves as a molecular handle to more easily access IKKα-containing complexes (not shown). Retroviral supernatants are used to transduce IKKα-deficient MEFs. Cells are then FACS sorted based on GFP expression to create stable cell lines expressing the retroviral vector. Protein expression will be monitored by immunoblot, and IKKα function will be assessed through activation of the noncanonical NF-κB signaling pathway. We show that the stable reconstitution of IKKα-deficient cells with a FLAG-tagged version of the protein restores signaling capacity to these cells. This procedure can then be used to assess the role of IKKα in NF-κB signaling pathways by making truncated versions of the protein, for example [4, 5], or to study the role of other key components of the NF-κB signaling cascade.

2   Materials

2.1   General Cell Culture (See Note 1)

1. 37 °C Cell Culture Incubator with 5 % CO_2.

2. 100 mm cell culture plates.

3. Dulbecco's Modified Eagle's Medium (DMEM) supplemented with 10 % (v/v) Fetal Bovine Serum (FBS), 1 % Penicillin/Streptomycin, and 1 % L-Glutamine (*see* **Note 2**).

4. IKKα−/− MEFs (kindly made available to the community by Dr. Inder Verma, Division of Biological Sciences, UCSD; or Dr. Michael Karin, Department of Pharmacology, UCSD).

2.2   Retrovirus Preparation and Transfection

1. MigR1 Retroviral Vector (http://www.addgene.org/27490/) with or without IKKα cDNA inserted into the Xho1 restriction site (*see* **Note 3**).

2. Platinum-E ecotropic packaging cell line.

3. Opti-MEM reduced Serum-free medium (Life Technologies, Carlsbad, CA, USA), warmed to room temperature.

4. FuGene 6 Transfection Reagent, stored at 4 °C.

5. Polybrene (8 mg/mL stock in sterile PBS, 50 μL aliquots stored at –20 °C).

2.3 Cell Lysis and Preparation for Immunoblotting

1. PBS (100 mL) placed on ice at least 30 min before use.

2. TNT lysis buffer containing 50 mM Tris–HCl pH 6.8, 150 mM NaCl, and 1 % (v/v) Triton X–100 placed on ice at least 30 min before use (*see* **Note 4**).

3. Complete protease inhibitor cocktail tablets (Roche Applied Science, Indianapolis, IN, USA) to be added to TNT lysis buffer immediately before use (*see* **Note 5**). Tablets are stored at 4 °C.

4. Coomassie Plus protein assay reagent stored at 4 °C and a solution of BSA (2 mg/mL) for generating a standard protein curve stored in aliquots at –20 °C (*see* **Note 6**).

5. Sample buffer (5×): 0.3 M Tris–HCl pH 6.8, 5 % (w/v) sodium dodecyl sulfate (SDS), 50 % (v/v) glycerol, 100 mM dithiothreitol (DTT), and 0.03 % (w/v) bromophenol blue. Stored in aliquots at –20 °C.

6. Plunger from a 1-mL syringe for scraping cells (*see* **Note 7**).

2.4 SDS-Polyacrylamide Gel Electrophoresis (SDS-PAGE)

1. Separating buffer (4×): 1.5 M Tris–HCl, pH 8.8, 0.4 % (w/v) SDS. Stored at room temperature.

2. Stacking buffer (4×): 0.5 M Tris–HCl, pH 6.8, 0.4 % (w/v) SDS. Stored at room temperature.

3. Thirty percent acrylamide/bis solution (37.5:1 with 2.6 % C) stored at 4 °C (Bio-Rad, Hercules, CA, USA) (*see* **Note 8**).

4. *N*,*N*,*N*,*N*′-tetramethyl ethylenediamine (TEMED): Stored at room temperature.

5. Ammonium persulfate (APS) 10 % (w/v) solution in ultrapure water prepared immediately before use.

6. Methanol stored at room temperature.

7. Running buffer: 25 mM Tris, 250 mM glycine, 0.1 % (w/v) SDS prepared from 5× stock stored at room temperature (*see* **Note 9**).

8. Prestained protein molecular weight standard markers (*see* Blue Plus2; Invitrogen, Carlsbad, CA, USA).

2.5 Immunoblotting

1. Transfer buffer: 25 mM Tris, 192 mM glycine, 20 % (v/v) methanol (*see* **Note 10**).

2. Methanol at room temperature.

3. Immobilon-P polyvinylidene fluoride (PVDF) transfer membrane and Whatman filter paper.

4. Tris-buffered saline with Tween (TBS-T) 25 mM Tris–HCl, pH 8.0, 140 mM NaCl, 3 mM KCl, 0.05 % Tween-20 (*see* **Note 11**).

5. Blocking buffer: 5 % (w/v) nonfat dried milk in TBS-T. Made fresh for each experiment.

6. Primary antibodies: rabbit anti-IKKα (Novus Biologicals, Littleton, CO, USA), rabbit anti-p100 (Cell Signaling Technology, Danvers, MA, USA), and mouse anti-tubulin (Sigma Aldrich, St. Louis, MO, USA) diluted to 1:1,000, 1:1,000, and 1:10,000, respectively, in blocking buffer (*see* **Note 12**).

7. Secondary antibodies: horseradish peroxidase-conjugated donkey anti-rabbit IgG and donkey anti-mouse IgG diluted to 1:10,000 in blocking buffer immediately before use (*see* **Note 13**).

8. ECL Solution 1: 200 μL 250 mM luminol (in DMSO), 88 μL of 90 mM p-coumaric acid (in DMSO), 2 mL of 1 M Tris–HCl pH 8.5, 17.7 mL of dH$_2$O.

9. ECL Solution 2: 12 μL of 30 % H$_2$O$_2$, 2 mL of Tris–HCl pH 8.5, 18 mL of dH$_2$O.

2.6 Noncanonical NF-κB Activation

1. Six-well tissue culture dish.

2. Rat anti-LTβR agonistic antibody (clone 5G11) (Abcam, Cambridge, MA, USA).

3. Rubber policeman (*see* **Note 7**).

4. Ice-cold PBS.

5. TNT Lysis buffer (*see* **Note 4**).

3 Methods

3.1 Fugene Transfection for Retrovirus Production

1. The day before transfection, trypsinize Plat-E cells to create a single cell suspension. For a 10-cm culture dish, remove culture medium and wash cells in 5 mL of sterile PBS. Add 1 mL of 0.25 % trypsin and place dish at 37 °C. Incubate cells in trypsin for 3–5 min, or until they slide off the plate when tapped.

2. Resuspend cells in 5 mL of DMEM and move to a 15-mL conical tube to create a single cell suspension by pipetting cells up and down along the side of the tube.

3. Plate cells to a final density of 6×10^6 cells into two 10-cm culture dishes in a final volume of 10 mL of medium (6×10^5 cells/mL) (*see* **Note 14**). Return plate to incubator overnight.

4. The next day, cells should be 70–80 % confluent. If cells are too dense, replate cells at lower density for the following day. If cells are less than 50 % confluent, wait 12–24 h before continuing with transfection.

5. Calculate the amount of Fugene required for transfection using a ratio of 3 µL Fugene per 1 µg of DNA according to the manufacturer's instructions. For a 10-cm culture dish, dilute 18 µL of Fugene in 500 µL of prewarmed serum-free Opti-MEM (*see* **Note 15**) in a 50-mL conical tube. Use a different conical tube per condition. Tap the tube gently to mix and incubate for 3 min at room temperature.

6. Pipette 6 µg of DNA (MigR1 vector or MigR1-IKKα) into each Fugene/Opti-MEM mixture. Tap the tube gently to mix and incubate at room temperature for a minimum of 45 min (*see* **Note 16**).

7. Change the culture medium on Plat-E cells to 10 mL of pre-warmed DMEM. Return cells to incubator.

8. Add transfection mix to Plat-E cells drop-wise, using a p1000 pipette. Do not swirl the plate. Return plate to 37 °C incubator overnight. Cells can stay in their culture medium (*see* **Note 17** and Subheading 3.2).

9. Allow 12–18 h for transfection. After 12 h, change medium to 7 mL of fresh DMEM for retroviral collection.

10. 24–36 h after transfection, collect retroviral supernatant into a 50-mL conical tube. Add 7 mL of fresh DMEM to plate (*see* **step 13**). Spin collected supernatant for 5 min at $2,500 \times g$ to pellet any cells.

11. Sterile filter retroviral supernatant into a new 15-mL conical tube using a 0.45 µm syringe-top filter and 10 mL syringe.

12. Virus may be stored on ice and used for transduction immediately (*see* Subheading 3.2) or may be snap frozen on dry ice with ethanol and stored at −80 °C.

13. Retroviral supernatant may continue to be collected every 12–18 h, repeating **steps 10–12**, until 72 h posttransfection or until Plat-E cells have reached maximum confluency.

3.2 Retroviral Transduction of IKKα−/− MEFs

1. Twenty-four hours before transduction, seed IKKα−/− MEFs into 10-cm culture dishes so they are 60–70 % confluent at the start of transduction.

2. Warm retrovirus at room temperature or 37 °C, just until warm (*see* **Note 18**).

3. Add polybrene (1:1,000) to virus to a final concentration of 8 µg/mL (*see* **Note 19**). Vortex to mix.

4. Replace medium on IKKα−/− culture dishes with MigR1 or MigR1-IKKα retrovirus. Return to incubator.

5. Twelve to twenty-four hours later, replace conditioned retroviral medium with 10 mL of complete DMEM for 2 h. Return to incubator.

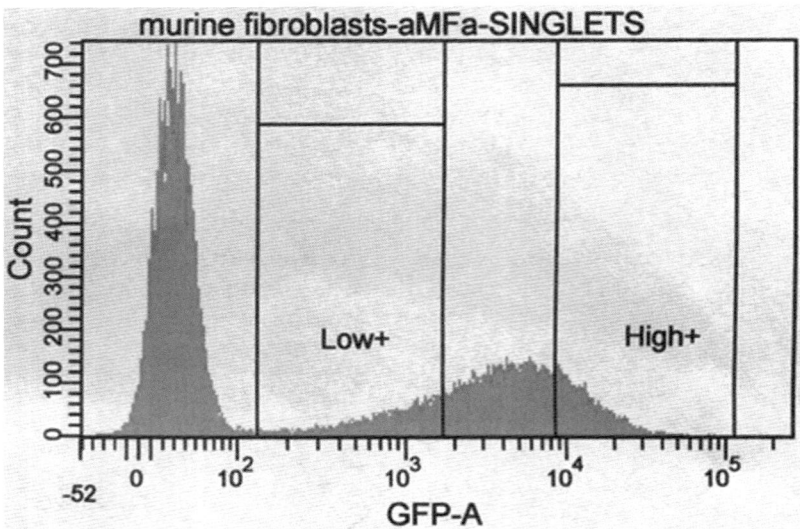

Fig. 1 Example of FACS sort on low and high GFP+ MEF populations. Cells were transduced with two rounds of MigR1-IKKα retrovirus over 48 h. Cells were trypsinized and prepared for FACS sort. Sorting was performed on a FACS DiVa by the Perelman School of Medicine Cell Sorting Facility at the University of Pennsylvania

6. Repeat transduction **steps 2–4** with fresh (or freshly thawed) virus.

7. Monitor transduction efficiency by GFP expression by live fluorescent microscopy (*see* **Note 20**).

8. Passage transduced MEFs at high density, typically a 1:3 split into three different plates. Use 1 mL of single cell suspension to check for IKKα expression by immunoblot (*see* Subheading 3.3).

9. FACS sort GFP+ MEFs to select for cells expressing the viral vector (*see* **Note 21**; an example of GFP expression is shown in Fig. 1).

3.3 Preparing Cells for Immunoblot

1. Collect cells in a 1.5 mL Eppendorf tube and pellet cells by centrifugation for 5 min at $300 \times g$ in a tabletop microcentrifuge.

2. Aspirate off medium and wash each tube with ice-cold PBS, spinning to pellet cells in between washes.

3. Place the tube on ice and add 200 μL of TNT lysis containing complete protease inhibitor cocktail to each well.

4. Incubate the tube on ice for 30 min and then vortex each tube for 10 s.

5. Spin for 20 min at $16,000 \times g$. This should be performed in a refrigerated benchtop microcentrifuge set to 4 °C. If a refrigerated centrifuge is not available, a standard benchtop microfuge should be chilled in a cold room or large refrigerator for at least 2 h prior to beginning the lysis.

6. Remove supernatant to a new 1.5-mL tube and place sample on ice while determining protein content (*see* **Note 22**).

7. Set up a protein standard curve by dilution from the stock of BSA (2 mg/mL) to achieve concentrations of 0.5, 1, 2.5, 5, 10, and 20 µg/mL in water. Pipette duplicate aliquots of 150 µL of water (blank) and the protein standards into a 96-well plate.

8. Dilute 2 µL of each lysate in water. Typically, 1/10 cell suspension from a confluent 10 cm dish lysed in 200 µL of TNT will produce a sample that can be diluted 1:250 to achieve a concentration within the range of the standard curve. It may also be necessary to adjust the dilution accordingly. Pipette 150 µL of each diluted sample in duplicate into the 96-well plate containing the blanks and standards.

9. Add 150 µL of the Coomassie Plus protein reagent to each well and read the plate using a microplate reader. Calculate the protein concentration in each sample manually by constructing a graph using the standard curve or by using software specific for the plate reader used (*see* **Note 22**).

10. Prepare samples for SDS-PAGE (*see* Subheading 3.3) in sample buffer with each containing 15 µg of protein.

11. The remaining lysates should be snap frozen on dry ice containing ethanol and then stored at −80 °C for future use.

3.4 SDS-PAGE

1. The following directions are for casting and running 10 % SDS gels using the Bio-Rad Mini-Protean Electrophoresis apparatus. Other systems have similar directions available from the manufacturer and an alternative to making gels would be purchasing precast 10 % gels (*see* **Note 23**).

2. Clean the 0.75 mm back and short front glass plates with dH₂O before use. Leave the washed plates at room temperature until completely dry before use.

3. Assemble the glass plates in the gel-casting apparatus.

4. Pipette water into the gel-casting chamber to check for leaks. If water leaks out, clean the bumper sponges and adjust the apparatus.

5. Prepare a 10 % resolving gel solution (*see* **Note 24**) in a 50 mL beaker by mixing 6.25 mL of water, 3.75 mL of 4× separating buffer, 5 mL of 30 % acrylamide/bis solution, 100 µL of 10 % APS solution, and 20 µL of TEMED. Gently mix by pipetting up and down (*see* **Note 25**).

6. Immediately pour the solution into the gel chamber leaving space for the stacking gel (about 2 mm below the height of the comb). Layer 100 µL of 100 % methanol onto the gel to remove any bubbles and flatten the surface. Save the remaining gel solution in the beaker.

7. After 5–10 min check for polymerization in the remaining gel solution in the beaker. While waiting for the gel to polymerize, prepare the stacking gel in a beaker, omitting the APS and TEMED. To prepare the stacking gel, mix 3.05 mL of water, 1.25 mL of 4× stacking buffer, and 0.65 mL of 30 % acrylamide/bis solution.

8. Pour off the methanol from the top of the polymerized gel using a paper towel to catch the liquid.

9. Add 50 μL of 10 % APS solution and 10 μL of TEMED to the stacking gel solution and gently pipette up and down to mix.

10. Immediately pour stacking gel on top of the resolving gel and insert a 0.75 mm ten-well comb. Save the remaining stacking gel in the beaker to check for polymerization. Wait 5–10 min for the stacking gel to polymerize.

11. During this period, prepare 1 L of running buffer from the 5× stock. Store at room temperature.

12. When the gel has fully polymerized, assemble the electrophoresis cassette in the gel tank and pour running buffer into the inner chamber.

13. Wait to see if the runner buffer leaks out of the inner chamber. If leakage occurs, reassemble the gel apparatus and refill the inner chamber with running buffer.

14. Remove the comb and flood the wells with running buffer.

15. Fill the outer chamber ½ full with running buffer (*see* **Note 26**).

16. Place the samples (*see* Subheading 3.2) in a heat block (95 °C) for 10 min and then spin at top speed in a benchtop microcentrifuge for 1 min.

17. Using a gel-loading tip, add 5 μL of protein molecular weight standard to the first well. Load samples into the next wells with gel-loading tips.

18. Connect the power supply and run at 120–160 V at room temperature until the samples have separated over the entire length of the gel (visualized by watching the separation of the prestained markers; *see* **Note 27**).

3.5 Immunoblotting for FLAG-IKKα

1. These directions are for the Bio-Rad Mini-Trans-Blot® Electrophoretic Transfer Cell wet transfer apparatus.

2. Prechill 1 L transfer buffer to 4 °C at least 1 h before use.

3. Cut two pieces of Whatman filter paper and a single piece of Immobilon-P large enough to cover the gel.

4. Fill a pyrex dish large enough to hold the transfer cassette with transfer buffer. Place the open cassette black side down in transfer buffer. Immerse the foam pads and filter paper in transfer buffer.

5. Remove the gel from the electrophoresis apparatus and trim away the stacking gel and any excess gel using a clean razor blade.

6. Saturate the Immobilon-P with 100 % methanol and immediately immerse the membrane in transfer buffer. Do not allow the membrane to dry.

7. Keeping all materials immersed in transfer buffer, arrange the gel transfer "sandwich" in the following order: black side of cassette, sponge, filter paper, gel, membrane, filter paper, sponge, and white side of the cassette.

8. Before closing the cassette, remove any air bubbles from the assembled sandwich using a roller (*see* **Note 28**).

9. Place the cassette in a transfer tank, matching up the black side of the cassette with the black (negative) electrode and the white side of the cassette with the red (positive) electrode.

10. Place an ice pack in the open side of the transfer tank and fill the tank with prechilled transfer buffer.

11. Run the transfer at 125 V for 1.5 h at 4 °C (*see* **Note 29**).

12. When transfer is complete the markers should be visible on the membrane. Disassemble the apparatus, trimming away any excess membrane with clean scissors and place the membrane in blocking buffer. Incubate on a rocker at room temperature for 60 min.

13. To simultaneously probe for IKKα (or FLAG) and tubulin (*see* **Note 30**), carefully cut the membrane in half below the 64-kDa marker. Place the upper half of the gel in a bag containing anti-IKKα (or anti-FLAG) and the lower half in a bag containing anti-tubulin (*see* Subheading 2.4). Seal the bags making sure to remove all air bubbles and rock the membranes overnight at 4 °C (*see* **Note 31**).

14. Remove membranes from primary antibodies and wash four times with TBS-T for 10 min each. All washes should be performed on a rocker at room temperature.

15. Dilute the appropriate HRP-conjugated secondary antibody 1:10,000 in blocking buffer (*see* Subheading 2.4) and incubate with the membrane (rocking) for 1 h at room temperature.

16. Remove the membranes from secondary antibody and wash four times with TBS-T for 10 min each change of buffer.

17. During the final wash, prepare ECL solutions 1 and 2 in separate 50-mL conical tubes as follows: For solution 1, mix 17.7 mL of deionized H_2O, 2 mL of Tris–HCl pH 8.5, 88 µL of p-coumaric acid, and 200 µL of luminol. For solution 2, mix 18 mL of deionized H_2O, 2 mL of Tris–HCl pH 8.5, and 12 µL of 30 % hydrogen peroxide (*see* **Note 32**). Do not mix solutions until ready for chemiluminescence detection.

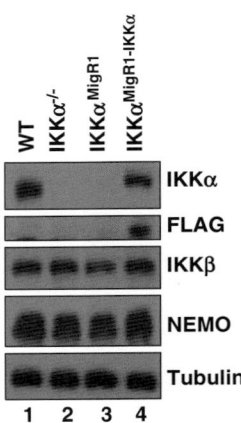

Fig. 2 Immunoblot of IKK expression. IKKα−/− MEFs were transduced with MigR1 (*lane 3*) or MigR1-IKKα (*lane 4*). Lysates were immunoblotted for IKKα and FLAG to confirm protein levels as compared to WT MEFs (*lane 1*). IKKβ and NEMO are shown to confirm that transduction does not interfere with other components of the IKK complex. Tubulin serves as a control for protein loading. IKKα in *lane 4* runs with slower electrophoretic mobility due to the FLAG tag

18. After the final wash, mix together equal parts (i.e., 10 + 10 mL) of ECL solutions 1 and 2 and place in a small dish (lid/wash container). Using forceps, transfer the blots to the bucket containing the ECL, placing the blots protein side up. Incubate at room temperature for 1 min with gentle rocking (*see* **Note 33**).

19. Pick up the membrane using forceps and gently shake off any excess ECL. Place the membrane between a clear sheet protector (*see* **Note 34**) and place into an X-ray film cassette. Use a roller to remove any bubbles and collect any excess liquid with a paper towel or Kim wipe.

20. In a secure dark room, place a piece of film over membrane and test a range of exposure times to obtain optimal band intensity. A typical result is shown in Fig. 2.

3.6 Noncanonical NF-κB Activation

1. The day before, seed 2 wells each: WT, IKKα$^{\text{MigR1}}$, and IKKα^{WT} MEFs in a six-well culture dish. Cells should be split 1:10 from a semi-confluent stock plate, and 2 mL of single cell suspension should be seeded into each well. It may take 2 days for cells to reach confluence for stimulation.

2. When cells are 90 % confluent, change medium to fresh pre-warmed DMEM.

3. Treat one well of each cell type with 3 μL/mL anti-LTβR cross-linking Ab (6 μL per well in 2 mL of DMEM).

4. Incubate at 37 °C for 8 h.

5. Before incubation period is over, prepare 1.5 mL of fresh TNT lysis buffer and chill PBS on ice.

6. To stop incubation, remove medium from cells by aspiration and wash each well twice in 1 mL of ice-cold PBS.

7. Add 200 μL ice-cold TNT lysis buffer to cells and incubate for 30 min on ice.

8. To collect lysates, scrape cells with rubber policeman stopper, rinsing stopper in ice between each sample.

9. Collect cell lysates into labeled 1.5 mL Eppendorf tubes and vortex.

10. Spin cells for 20 min at maximum speed in a tabletop microcentrifuge at 4 °C to pellet cell debris.

11. Snap freeze cell lysates or proceed to determine protein concentration and prepare samples for SDS-PAGE (*see* Subheading 3.3, **step 6**).

12. To determine noncanonical NF-κB activity, immunoblot for p100/p52 using anti-p100/p52 antibody from cell signaling diluted 1:1,000 in blocking buffer, according to the protocol outlined in Subheadings 3.4–3.5. A typical result is shown in Fig. 3.

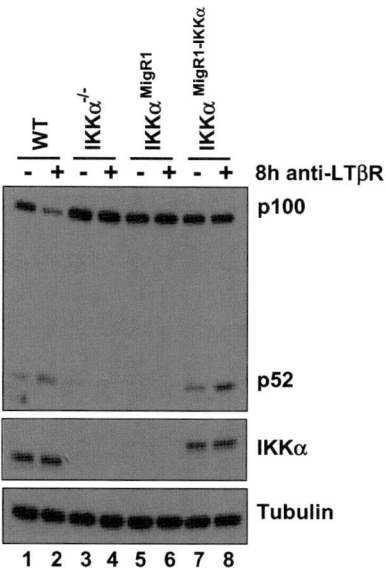

Fig. 3 Noncanonical NF-κB activation. Cells were either untreated (−) or treated (+) with an agonistic antibody against the Lymphotoxin beta receptor (LTbR) for 8 h. Whole cell lysates were immunoblotted for p100/p52 to determine the extent of noncanonical NF-κB activity. IKKα and Tubulin are shown as controls

4   Notes

1. All cell culture should be performed in a tissue culture hood with sterile techniques. All medium and culture bottles should be sprayed with 70 % ethanol and completely wiped down with paper towels before use in the tissue culture hood.

2. Both Plat-E cells and MEFs are cultured in DMEM. DMEM is stored at 4 °C and should be warmed to 37 °C in a water bath for approximately 30 min immediately before use.

3. The MigR1 retroviral vector with expanded multiple cloning sites [10] was developed and kindly provided to the community by Warren Pear (Department of Pathology and Lab Medicine, The University of Pennsylvania). The vector was modified in our laboratory to contain a Kozak sequence followed by the FLAG sequence. IKKα cDNA was received from Michael Karin (UCSD), amplified by PCR and cloned into the XhoI restriction site of the MigR1 retroviral vector to be in frame with the FLAG tag.

4. TNT may become cloudy in solution. To avoid this we create a stock of 50 mM Tris–HCl pH 6.8 and 150 mM NaCl (TN), which is stored at room temperature. 1 % (v/v) Triton X–100 is added fresh for each experiment.

5. One complete protease inhibitor tablet dissolved in 400 μL ultrapure water makes a 25× stock solution. Any solution not used immediately may be aliquoted and stored at –20 °C.

6. The Coomassie Plus protein assay from Roche comes with BSA ampoules. Extreme care should be taken when opening glass ampoules. We cover the bottleneck with paper towel before cracking along the indicated seam. BSA should be stored at –20 °C in aliquots suitable for your standard curve.

7. We use the plunger from a 1 mL syringe as a "rubber policeman" to scrape adherent cells from tissue culture dishes. To avoid contamination, the plunger should be completely washed and dried between samples.

8. Unpolymerized acrylamide is a particularly potent neurotoxin. Extreme care should be taken when working with liquid acrylamide, including gloves and safety goggles.

9. A large stock of 5× running buffer (i.e., 10 L) may be prepared and stored at room temperature for up to 3 months. When diluting the stock solution, add concentrated 5× buffer (200 mL) to 800 mL of water to avoid bubbles in the 1× solution.

10. For convenience, a large stock (10–20 L) of transfer buffer *without* methanol (25 mM Tris, 192 mM glycine) may be stored at room temperature, but to preserve transfer efficiency,

methanol (20 % v/v) should be added fresh to this base solution before each experiment.

11. A large stock of 1× TBS-T (10–20 L) may be stored at room temperature for up to 3 months.

12. Primary antibodies may be reused up to five times if sodium azide (0.002 %) is added to the diluted antibody solution. Sodium azide is highly toxic and should only be handled in a fume hood. Gloves must be worn when measuring sodium azide or when working with a stock solution.

13. Sodium azide interferes with HRP activity and should never be added to secondary antibodies. Secondary antibodies should be made fresh for each experiment.

14. All subsequent transfection steps should be performed in replicate for control (MigR1) and experimental (MigR1-IKKα) conditions. If multiple rounds of retroviral transduction are to be performed, this procedure may be scaled up to multiple plates per condition for retroviral collection.

15. These instructions are for Fugene-6 transfection reagent (Roche). Be extremely cautious not to let undiluted Fugene touch the sides of the conical tube or any plastic surface other than the pipette tip. This interferes greatly with the transfection efficiency. Alternative transfection procedures may be performed according to the manufacturer's instructions.

16. The transfection mixture must be incubated at room temperature for at least 45 min, but should be used within 3 h of incubation.

17. If retrovirus is to be used fresh, target MEFs should be seeded for transduction the same day Plat-E cells are transfected. A subconfluent plate of IKKα–/– MEFs split 1:10 should be ready for transduction the following day.

18. Retroviral half-life is reported to be 3–6 h at 37 °C [11]. Therefore, retrovirus should be incubated at 37 °C for no longer than 30 min before transduction. Virus may be stored on ice for 1–2 h before use, or it should be frozen for future use.

19. Polybrene is a cationic polymer added to enhance virion fusion with the target cell membrane. Polybrene amounts may need to be titrated for different cell lines; however we find 8 µg/mL to be suitable for the transduction of MEFs. Polybrene should be reconstituted in sterile PBS (8 mg/mL) and stored in aliquots at –20 °C.

20. Live cell microscopy or FACS may be used to monitor GFP expression. We prefer microscopy so cells may be monitored during the transduction procedure without additional passages.

21. Unsorted cells should be frozen on liquid nitrogen to preserve cell line stocks. Our cells are sorted on a FACS DiVa sorter by

the Flow Cytometry and Cell Sorting core facility at the University of Pennsylvania. We routinely collect populations expressing the highest and lowest 2–5 % GFP. These individual lines are monitored for protein expression by Western blot to choose the expression level closest to the physiological level of expression in untransduced WT MEFs.

22. Protein concentration is determined using a microplate reader (such as Bio-Rad Model 680) with a 595 nm filter. We use Microplate Manager III software to set up a standard curve protocol for use with the Coomassie Plus Reagent.

23. Precast gels are an expensive alternative to casting your own gels.

24. This recipe is for 3–0.75 mm gels. It can be scaled up for four to six gels, if necessary.

25. The resolving gel recipe may be prepared up to 30 min ahead of time if the APS and TEMED are excluded. This provides time for the solution to degas while plates are being cleaned and assembled. The polymerization reaction occurs when APS and TEMED are added to the solution.

26. If the inner chamber continues to leak, fill the entire outer chamber with running buffer.

27. The gel may be run until the dye front reaches the bottom of the tank and begins to emerge into the runner buffer.

28. A piece of 5 or 10 mL pipette makes a useful roller.

29. The ice pack will keep the transfer buffer cool, but we run the entire transfer in a chilled chromatography refrigerator. If a refrigerator is not available, the transfer tank can be surrounded by ice.

30. Tubulin is used as a loading control. The molecular weights of tubulin and IKKα are 55 and 87 kD, respectively. Therefore to simultaneously probe for tubulin and IKKα, the membrane may be cut at the 64 kD marker (or similar) using a clean razor blade and ruler for guidance.

31. We have found the Impulse Sealer from Hualian Packaging Machinery Co. Ltd. (Wenzhou City, China) and Kapak SealPAK pouches (VWR, Arlington Heights, IL, USA) ideal for this procedure.

32. 30 % hydrogen peroxide is corrosive and should be handled with care. Aqueous solution should be stored at 4 °C in an opaque container.

33. This step does not need to be done in a dark room. Subsequent steps should be performed in a dark room under a safety light.

34. Sheet protectors from any office supply store are suitable for chemiluminescence detection.

References

1. Pear WS, Nolan GP, Scott ML et al (1993) Production of high-titer helper-free retroviruses by transient transfection. Proc Natl Acad Sci U S A 90:8392–8396

2. Morita S, Kojima T, Kitamura T (2000) Plat-E: an efficient and stable system for transient packaging of retroviruses. Gene Ther 7:1063–1066

3. Hinz M, Scheidereit C (2014) The IkappaB kinase complex in NF-kappaB regulation and beyond. EMBO Rep 15:46–61

4. Solt LA, Madge LA, May MJ (2009) NEMO-binding domains of both IKKalpha and IKKbeta regulate IkappaB kinase complex assembly and classical NF-kappaB activation. J Biol chem 284:27596–27608

5. Gray CM, Remouchamps C, McCorkell KA et al (2014) Noncanonical NF-kappaB signaling is limited by classical NF-kappaB activity. Sci Signal 7:ra13

6. Senftleben U, Cao Y, Xiao G et al (2001) Activation by IKKalpha of a second, evolutionary conserved, NF-kappa B signaling pathway. Science 293:1495–1499

7. Dejardin E, Droin NM, Delhase M et al (2002) The lymphotoxin-beta receptor induces different patterns of gene expression via two NF-kappaB pathways. Immunity 17:525–535

8. Yamamoto Y, Verma UN, Prajapati S et al (2003) Histone H3 phosphorylation by IKK-alpha is critical for cytokine-induced gene expression. Nature 423:655–659

9. Razani B, Zarnegar B, Ytterberg AJ et al (2010) Negative feedback in noncanonical NF-kappaB signaling modulates NIK stability through IKKalpha-mediated phosphorylation. Sci Signal 3:ra41

10. Pear WS, Miller JP, Xu L et al (1998) Efficient and rapid induction of a chronic myelogenous leukemia-like myeloproliferative disease in mice receiving P210 bcr/abl-transduced bone marrow. Blood 92:3780–3792

11. Pear W (2001) Transient transfection methods for preparation of high-titer retroviral supernatants. Curr Protoc Mol Biol Chapter 9: Unit 9.11

Chapter 11

Dissecting NF-κB Signaling Induced by Genotoxic Agents via Genetic Complementation of NEMO-Deficient 1.3E2 Cells

Shawn S. Jackson and Shigeki Miyamoto

Abstract

The transcription factor NF-κB regulates expression of a diverse set of genes to modulate multiple biological and pathological processes. Among these, NF-κB activation in response to genotoxic agents has received considerable attention due to its role in regulating cancer cell resistance to chemo- and radiation therapy. Furthermore, induction of this pathway by endogenous damage is further implicated in normal developmental processes, such as B cell development, and premature aging, among others. This pathway also serves as a signaling model in which nuclear initiated signals (DNA damage) are communicated to a cytoplasmic target (IκB kinase and NF-κB). Several of the critical molecular events of this nuclear to cytoplasmic NF-κB signaling cascade were discovered, in part, by genetic complementation analyses of the NEMO-deficient 1.3E2 mouse pre-B cell line. This chapter describes methods used to generate and analyze such reconstitution cell systems and certain caveats that are critical for proper interpretation of NEMO mutant defects.

Key words NF-κB, NEMO, DNA damage, EMSA, 1.3E2, Genetic complementation, Radiation, Etoposide, Camptothecin, Doxorubicin

1 Introduction

When cells are exposed to various genotoxic agents, they mount a cellular response collectively known as the DNA damage response (DDR) [1]. DDR is critical for proper recognition and repair of damaged DNA, activation of signaling events that control cell cycle checkpoint activation, and induction of senescence or apoptosis pathways, among others [2, 3]. Cellular responses to DNA damage are often initiated by the PI3K-related protein kinases, ATM (ataxia telangiectasia mutated) [4, 5], ATR [6], and DNA-PK [7] through phosphorylation of downstream target proteins. Some of these responses involve transcriptional activation of specific genes. For example, the best studied transcription factor in the DDR is p53, a tumor suppressor which can be stabilized by an ATM-dependent mechanism to induce transcription of genes such as p21 cell

Michael J. May (ed.), *NF-kappa B: Methods and Protocols*, Methods in Molecular Biology, vol. 1280,
DOI 10.1007/978-1-4939-2422-6_11, © Springer Science+Business Media New York 2015

cycle-dependent kinase inhibitor [8] or Bax, a cell death inducer [9]. The transcription factor NF-κB can also be activated by a variety of DNA damaging agents to initiate transcription of genes critical for cell proliferation as well as survival and inflammatory responses [10, 11]. As such, NF-κB activation by genotoxic agents is implicated in the development of resistance to chemotherapeutic agents and ionizing radiation (IR) in multiple cancer cell types [12]. Thus, there is considerable interest in understanding the mechanisms involved in NF-κB signaling induced by genotoxic agents [13–15].

Previous studies have demonstrated the critical role of ATM in mediating NF-κB signaling in response to IR [16], a topoisomerase I poison, camptothecin (CPT) [11, 17], and the topoisomerase II poisons, etoposide (VP16) and doxorubicin [13, 14]. The role of IκB kinases (IKKα and IKKβ) had also been described in IR [18] and CPT-induced signaling pathway [10]. However, the details of the molecular link between nuclear-localized ATM and cytoplasmic-localized IKK complexes remained unclear. Our group used a genetic reconstitution strategy to dissect this signaling pathway by employing the NEMO-deficient 1.3E2 mouse pre-B cell line. 1.3E2 cells are derived from the 70Z/3 line through immunoselection as cells that failed to activate NF-κB after treatment with bacterial lipopolysaccharide (LPS) or phorbol esters [19]. The original cloning and functional analysis of NEMO (also known as IKKγ) in several cell systems, including 1.3E2 cells by genetic complementation, defined the essential requirement of NEMO for NF-κB activation by these stimuli [20, 21]. In a similar strategy, we introduced various NEMO deletion and point mutants into 1.3E2 cells and isolated stable pools and individual clones to characterize the associated defects in NF-κB signaling induced by CPT and LPS. This analysis led to the initial discovery of the critical role of specific posttranslational modifications (SUMOylation, phosphorylation, and ubiquitination) of NEMO in NF-κB activation in response to CPT and a variety of other genotoxic agents, including IR, VP16, and doxorubicin (reviewed in ref. 22). While SUMOylation of NEMO was found to be ATM independent, subsequent phosphorylation and ubiquitination of NEMO were determined to be ATM dependent [13, 14]. Thus, this particular cell system was instrumental in defining a set of conserved signaling events induced by different genotoxic agents to mediate ATM-dependent NF-κB activation.

There are several advantages of the 1.3E2 cell reconstitution system to study NF-κB signaling induced by genotoxic agents. First, activation of NF-κB by multiple genotoxic agents as measured by electrophoretic mobility shift assay (EMSA) in wildtype NEMO-reconstituted cells (and the parent 70Z/3 cells) is very robust, often approaching the maximal activation induced by a strong NF-κB inducer, LPS. This provides a large range of NF-κB activation defects that can be detected in association with different NEMO mutants.

For comparison, for an unknown reason, mouse embryo fibroblasts (MEFs) often show very limited (or undetectable) NF-κB activation as measured by EMSA in response to genotoxic agents, thereby severely limiting dissection of this signaling pathway in NEMO-deficient MEF lines. Only when exceedingly high doses of genotoxic agents (except for CPT) are utilized does one observe a modest increase in NF-κB activation could be detected despite activation of ATM at much lower doses. Second, the growth of 1.3E2 cells is quite rapid. The doubling time is in the order of 12 h, which allows for the growth of individual clones in a reasonable amount of time (less than 2 weeks from a single cell to mass culture with millions of cells). Third, clonogenic survival of these cells is also reasonably high so that ~50 % of individual clones will consistently grow out as mass culture. Finally, these cells are sensitive to apoptosis induced by genotoxic agents and NF-κB activation improves their survival. As such, biochemical analyses of individual clones can be directly linked to changes in functional outcomes (NF-κB-dependent gene expression and cell survival). A major weakness of this cell system is the difficulty of transiently transfecting various constructs and siRNAs, thereby preventing convenient gain-of-function and loss-of-function analyses of other signaling components. Another weakness is the difficulty of cell imaging to evaluate subcellular localization of NEMO and other proteins due to their spherical shape, suspension growth, and a relatively large nuclear to cytoplasm volume ratio.

Of all the techniques used to measure NF-κB activity, EMSA (colloquially known as a "gel shift") is one of the most robust and reproducible. While the emergence of qRT-PCR and phospho-specific antibodies has decreased the reliance on gel shifts in the field, it still remains the most direct way of measuring one of the early biochemical events associated with NF-κB activation. First described in 1981 [23], the EMSA was quickly incorporated into the field of NF-κB signaling and was the first widely used technique to measure NF-κB activity [24].

Our lab continues to use the EMSA as the mainstay of measuring NF-κB activation for several reasons. First, it is a rapid assay, capable of measuring the activity from frozen cell pellets to film (or phosphor-image cassette) exposure in approximately 4 h. Second, it is relatively easy to master. Even the least experienced members of a research group can learn the gel shift technique relatively quickly with minimal training. Third, the assay is quantitative and reproducible. Finally, unlike phospho-specific antibodies or immune-kinase assay, whose data quality frequently depends on the specific batches of antibodies used, EMSA analysis is not subject to batch-to-batch variation. Typically, with EMSA, the entire process of cell treatment to data analysis may be accomplished within a working day (if activation is robust) or following overnight exposure (if activation is relatively weak), thereby minimizing the amount of time required for planning of subsequent experiments.

While our group continues to utilize ^{32}P labeling of double-stranded oligo probes, others have begun using fluorescent, infrared, and biotinylated oligonucleotide probes. In our limited testing of some of these alternative nonradioactive methods, we found the sensitivity of EMSA assay is higher with ^{32}P-labeled probes. In addition, we use size mis-matched plus-strand template and minus-strand primers in extension reactions, followed by column purification to generate uniformly double-stranded DNA probe populations. We found that this protocol gives a better quality EMSA result than annealing plus-strand and minus-strand oligos of the same length to generate the probe. Accordingly, this chapter presents detailed protocols for generation of 1.3E2 cell clones reconstituted with different NEMO constructs and analysis of their activation potentials by EMSA, along with caveats and pitfalls associated with such analyses.

2 Materials

2.1 Culture of 1.3E2 Cells

1. 1.3E2 Growth Medium: RPMI 1640 with 10 % FBS (v/v) and containing β-mercaptoethanol to a final concentration of 0.05 mM. Add penicillin (100 IU/ml) and streptomycin (100 µg/ml).

2. Dimethylsulfoxide (DMSO).

3. G418 (Geneticin) or other appropriate selection agent.

4. Plasmid containing gene of interest (*see* **Note 1**).

5. 5 % CO_2 incubator.

6. Tissue culture materials (96-well plate; 60, 100, and 150 mm dishes; 15 and 50 ml spin tubes).

2.2 Electroporation of Plasmid DNA

1. Plasmid DNA—Expression of plasmid DNA may be based on pCDNA3.1 (+) or retroviral vectors. We have had success with pCDNA3 and pBABE-based vectors.

2. Cell electroporator such as the Bio-Rad Gene Pulser (Bio-Rad, Hercules, CA, USA) or equivalent.

3. Mammalian electroporation cuvettes.

4. Centrifuge suitable for tissue culture samples.

2.3 Selection and Isolation of Clonal Populations

1. Selection agent, as determined by the selection cassette (e.g., G418, Puromycin).

2. Culture materials as listed in Subheading 2.1.

2.4 Removal of Nonviable Cells

1. Lymphocyte separation medium (Cellgro, Manassas, VA, USA).

2. Culture materials as listed in Subheading 2.1.

2.5 Isolation of Stable Clones	1. Tissue culture microscope suitable for visualizing individual cells and isolating them by micropipette. 2. Culture materials as listed in Subheading 2.1.
2.6 Analysis and Treatment of Clones	1. Freezing medium: 10 % DMSO in FBS. 2. 2× SDS sample buffer (Life Technologies, Carlsbad, CA, USA). 3. PAGE materials and equipment. 4. NEMO antibody (FL-419, sc-8330) (Santa Cruz Biotechnology, Santa Cruz, CA,USA). 5. Bradford reagent. 6. Spectrophotometer capable of measuring OD_{595}. 7. Lipopolysaccharide. 8. Etoposide (VP-16). 9. Camptothecin (CPT). 10. IL-1β, mouse.
2.7 General Stock Reagents for Electrophoretic Mobility Shift Assay (EMSA)	1. 4 M NaCl. 2. 3 M NaOAc ($C_2H_3NaO_2$)—pH 5.2. 3. 1 M $MgCl_2$. 4. 1 M Tris—pH 8. 5. 0.5 M EDTA—pH 8. 6. 0.5 M EGTA—pH 8. 7. 0.1 M DTT (*see* **Note 2**).
2.8 Annealing the Single-Stranded Oligonucleotides	1. Plus-strand and minus-strand oligonucleotides for NF-κB (Table 1; *see* **Note 3**). 2. Klenow 10× Buffer (Promega, Madison, WI, USA). 3. Hot plate.
2.9 End-Filling of the Double-Stranded Oligonucleotide	1. DNA Polymerase I Large Fragment (Klenow) (Promega). 2. Klenow 10× Buffer. 3. dNTP (10 mM each, 40 mM total). 4. 37 °C heat block or water bath.

Table 1
Oligonucleotides used for EMSA

Igκ κB plus-strand oligo	5′-CTC AAC AGA **GGG GAC TTT CC**G AGA GGC CAT-3′
Igκ κB minus-strand oligo	5′-ATG GCC TCT C-3′
Oct1 consensus double-stranded oligo	5′-TGT CGA ATG CAA ATC ACT AGA A-3′ 3′-ACA GCT TAC GTT TAG TGA TCT T-5′

NF-κB consensus sequence: 5′-GGGRNNYYCC-3′

2.10   Oligo Cleanup

1. Sephadex G50 (pre-prepared columns also available; GE Healthcare, Mickleton, NJ, USA).

2. Columns: we use 1 ml syringes utilizing cotton wool frits loaded with G50 fine beads, washed with 1× TE (10 mM Tris pH 8, 1 mM EDTA pH 8). These are typically placed in 15 ml conical tubes for centrifugation.

3. 3 M NaOAc ($C_2H_3NaO_2$)—pH 5.2.

4. Ethanol.

5. –80 °C Freezer.

6. Centrifuge capable of handling 15 ml conical tubes.

2.11 ^{32}P-γ-ATP Labeling

1. Radiation specific equipment—dedicated bench space, protective boxes, shields, Geiger counter, scintillation counter.

2. Oct1 double-stranded oligo (Promega).

3. Igκ κB oligo (generated in Subheading 3.2).

4. T4 Polynucleotide Kinase (Promega).

5. ^{32}P-γ-ATP (3,000 Ci/mmol, 10 mCi/ml).

6. Sephadex G50.

7. 0.1 M DTT.

2.12   Casting Native PAGE Gel

1. Gel electrophoresis system (we use 20 × 20 cm plates for higher resolution).

2. EMSA Acrylamide Stock Solution: 30 % w/v acrylamide, 2 % w/v bis-acrylamide. Heat slightly (to 37 °C) and mix to dissolve. Sterilize by flowing through a 0.45 μm filter. Stable for up to 6 months at 4 °C (see **Note 4**).

3. 10× TBE: 108 g Tris base, 55 g boric acid, 40 ml of 0.5 M EDTA pH 8 in a total volume of 1 l.

4. 10 % w/v Ammonium persulfate (APS).

5. N,N,N',N'-tetramethylethylenediamine (TEMED).

2.13   Generating Total Cell Extracts

1. TOTEX Buffer: 20 mM HEPES pH 7.9, 350 mM NaCl, 1 mM $MgCl_2$, 0.5 mM EDTA, 0.1 mM EGTA, 20 % glycerol, 1 % NP-40. Can be stored for up to 6 months at 4 °C. Prior to use, a small volume of TOTEX should be aliquoted. Fresh DTT should be added to a final concentration of 1 mM and protease inhibitors should be added to a final working concentration of 1×.

2. 100× Halt™ Protease Inhibitor Cocktail (Thermo, Waltham, MA, USA) or other suitable inhibitor cocktail.

3. 0.1 M DTT.

2.14 Preparation of Samples for EMSA	1. Bradford reagent. 2. Spectrophotometer capable of reading at OD_{595}. 3. TOTEX Buffer (*see* Subheading 2.13).
2.15 Assembling the DNA Binding Reaction	1. 5× EMSA Binding Buffer: 375 mM NaCl, 75 mM Tris pH 7.5, 7.5 mM EDTA, 20 % glycerol, 100 µg/mL BSA. Can be stored for up to 6 months at 4 °C. 2. Poly(dI-dC) (Sigma-Aldrich, St. Louis, MO, USA): 25U of powdered Poly(dI-dC) in 1 mL of 200 mM NaCl. Heat in a 1 L water bath at 45 °C for 5 min and allow the water bath to cool passively to room temperature. Aliquot in 100 µL volumes. Stable at –20 °C for approximately 1 year. 3. Labeled double-stranded oligo as generated in Subheading 3.2.
2.16 Native Gel Electrophoresis	1. EMSA running buffer: 0.25× TBE. 2. DNA loading buffer containing bromophenol blue.
2.17 Drying the Gel and Exposing to Autoradiography Film or PhosphorImager Screen	1. Gel dryer with vacuum pump. 2. 3 mm thick 35×45 cm chromatography paper. 3. Plastic cling wrap. 4. Film cassette and intensifier screen or a PhosphorImager screen and cassette setup.

---

3 Methods

3.1 Generating 1.3E2 Stable Cell Lines	We recommend that when one attempts to perform this reconstitution study, one begins with only NEMO-WT (positive), empty vector (negative), and mock (no selection) controls to ensure the quality of the 1.3E2 batch used. When one becomes proficient, multiple mutants may be analyzed in parallel. To ensure the phenotypes of each NEMO mutant analyzed are indeed linked to that specific mutant, we also recommend that the NEMO mutant status be revalidated by sequencing NEMO cDNA generated by RT-PCR from individual 1.3E2 stable clones. This is to guard against potential contamination of different clones (especially wild-type or unintended mutants of NEMO) that might have occurred during extensive culturing of individual clones.
3.1.1 Culture of 1.3E2 Cells	1. Grow 1.3E2 cells for at least three passages (from cryopreservation) in 1.3E2 Growth Medium (*see* **Note 5**). 2. Do not allow cells to exceed $1–1.2 \times 10^6$ cells/ml.

3.1.2 Electroporation of Plasmid DNA

1. Concentrate 1.3E2 cells (from $\leq 1 \times 10^6$ cells/ml culture) by centrifugation ($400 \times g$ for 5 min) to a final concentration of $7-9 \times 10^6$ cells/ml.

2. Place the concentrated mixture on ice and aliquot 750 µl of concentrated cells (total of $\sim 5-7 \times 10^6$ cells) into the electroporation cuvette.

3. Add 40 µg plasmid DNA and incubate on ice for 15 min (*see* **Note 6**).

4. Electroporate the cells after having thoroughly dried the outside of each cuvette (*see* **Notes 7–9**).

5. Remove the white/pink layer of froth that appears at the top of the cuvette. Transfer the remaining cells into a 100 mm tissue culture dishes containing 10 ml of RPMI medium (*see* **Note 10**).

3.1.3 Selection of Clonal Populations

1. Begin selection for resistant cells at approximately 24 h at a dose that has been shown to be efficacious under your conditions. We routinely use ~ 500 µg/ml G418.

2. Continue to expand the volume of medium spiking in additional selection medium as needed (*see* **Note 11**).

3.1.4 Removal of Nonviable Cells

1. After approximately 3–5 days it will become necessary to remove the majority of the dead and nonviable cells present in culture.

2. Centrifuge cells at $400 \times g$ for 5 min.

3. Resuspend cells in 4 ml of fresh medium.

4. Place 3 ml of lymphocyte separation medium in a new 15 ml conical tube.

5. Carefully layer 4 ml of culture medium containing the lymphocytes on top of the 3 ml of lymphocyte separation medium (*see* **Note 12**).

6. Centrifuge cells at $400 \times g$ for 30 min.

7. Discard approximately 2.5–3 ml of medium from the very top of the tube.

8. Harvest viable cells, located at the interface between the culture medium and the lymphocyte separation medium. As additional selection will be performed, err on the side of taking more volume than less volume. Dead cells or cell debris will form a pellet at the bottom of the tube (*see* **Note 13**).

9. Dilute the mixture containing the viable cells into 10 ml of fresh medium and centrifuge at $400 \times g$ for 5 min to remove lymphocyte separation medium.

10. Resuspend cell pellet in fresh medium, transfer to a tissue culture dish. Continue to select as appropriate (monitor mock control).

3.1.5   Isolation of Stable Clones

1. After complete selection has taken place determined by lack of viable cells in the mock control, select single cells to generate individual stable clones (*see* **Note 14**).

2. Dilute cell culture such that one can identify no more than one or a few viable cell(s) per microscope field at a total magnification 100×.

3. Using a 10 μL pipette, begin aspirating single cells and dispensing into a 96-well dish containing medium with selection. The concentration of G418 is kept relatively low at 200 μg/ml to allow for growth of individual stable clones (*see* **Notes 15** and **16**).

4. We recommend generating over a dozen viable cell lines for analysis by western blot. For most purposes, we pick 12–18 cells from the A, B, and C plates (*see* Subheading 3.1.2 and **Note 10**) with the assumption that only 50 % of the clones will grow out under ideal conditions (*see* **Note 17**).

3.1.6   Treatment of Clones (See Notes 18 and 19)

1. To test stable cell lines for NF-κB activation as measured by EMSA, treat genetically reconstituted 1.3E2 cells with the following: Lipopolysaccharide (10 μg/ml), IL-1β (10 ng/ml), and Etoposide (VP-16) (10 μM) (*see* **Note 20**).

2. We typically employ treatment times of 30 min for canonical NF-κB stimuli and 120 min for DNA damaging agents. We find that time course and dose response experiments are extremely valuable when testing out new cell lines and reagents.

3. Centrifuge cells as above to pellet them.

4. Remove medium and transfer cells to 1.5 ml Eppendorf tube using 1 ml of ice-cold PBS.

5. Centrifuge cells in microfuge to pellet them. Remove PBS (*see* **Note 21**).

3.2   Electrophoretic Mobility Shift Assay (EMSA)

3.2.1   Annealing the Single-Stranded Oligonucleotides

1. To generate double-stranded oligonucleotides suitable for end labeling and NF-κB—DNA binding, we use complimentary plus-strand and minus-strand oligonucleotides (Table 1) and assemble the reaction as shown in Table 2.

2. Place reaction mixture in a 1 L beaker containing approximately 800 mL of near boiling water (90–100 °C).

3. Heat near boiling (90–100 °C) for 10 min.

4. Allow the water to passively cool to room temperature. This should take approximately 2–3 h.

3.2.2   End-Filling of the Double-Stranded Oligonucleotide

1. Assemble the reaction as shown in Table 3.

2. Incubate at 37 °C for 2 h.

Table 2
Volumes for annealing single-stranded oligonucleotides

62.5 µl	Sterile H$_2$O
12.5 µl	Plus-strand 30-mer oligo (430 ng/µl)
12.5 µl	Minus-strand 10-mer oligo (140 ng/µl)
10 µl	10× DNA polymerase buffer (Klenow Buffer)
100 µl	Total

Table 3
Volumes for end filling double-stranded oligonucleotides

100 µl	Annealing reaction (from Table 2)
81 µl	Sterile H$_2$O
4 µl	dNTP (10 mM each, 40 mM total)
10 µl	10× DNA polymerase buffer (Klenow Buffer)
5 µl	DNA polymerase I large fragment (Klenow)
200 µl	Total

3.2.3 Oligo Cleanup

1. Prepare two Sephadex G50 columns per oligo to be purified.
2. Load the 200 µl reaction from Subheading 3.2.2 on the G50 column.
3. Centrifuge $400 \times g$ for 3 min.
4. Load on second G50 column and centrifuge $400 \times g$ for 3 min.
5. Transfer to 1.5 ml microcentrifuge tube.
6. Add 20 µl 3 M Sodium Acetate (pH 5.2).
7. Add 500 µl Ethanol.
8. Store at –80 °C for 30 min.
9. Spin at maximum speed in a microcentrifuge at 4 °C for 10 min.
10. Carefully remove supernatant (pellet may be difficult to visualize).
11. Wash 1× with 200 µl 70 % Ethanol.
12. Air-dry for 5 min.
13. Resuspend in 400 µl of TE to a desired final concentration of ~20 ng/µl.
14. Aliquot double-stranded κB oligo in labeled screw-top tubes for probe labeling (*see* **Notes 22** and **23**).

3.2.4 *³²P-γ-ATP Labeling*

1. Using the κB oligo generated above or the commercial Oct1 oligo, assemble the labeling reaction shown in Table 4 in screw-top tubes. This reaction can be scaled up or down to fit the needs of the lab.

2. Incubate at 37 °C on the dry heat block for 1 h.

3. Add 100 µL of TE to the reaction.

4. Clean up and remove unincorporated ³²P by first loading the reaction on a G50 column.

5. Centrifuge at $400 \times g$ for 3 min, ensuring that unincorporated ³²P-γ-ATP is removed from the eluent (frequently associated with green dye).

6. Transfer to a new screw cap tube and label.

7. Place in plastic radiation box at −20 °C (*see* **Note 24**).

3.2.5 *Casting Native PAGE Gel*

1. Spray glass plates with 70 % ethanol, clean with benchtop paper wipes (*see* **Note 25**).

2. Assemble glass plates with 1.5 mm spacers.

3. Assemble polyacrylamide gel mixture as shown in Table 5.

Table 4
Volumes for ³²P-γ-ATP end labeling of double-stranded oligonucleotides

κB/Oct Oligo (~20 ng/µl)	3 µl
H₂O	32 µl
10× T4 PNK buffer	5 µl
0.1 M DTT	3 µl
T4 PNK	3 µl
³²P-γ-ATP	6 µl
Total	**50 µl**

Table 5
Volumes for casting a native PAGE gel

50 ml	Sterile H₂O
8 ml	EMSA acrylamide
1.5 ml	10× TBE
600 µl	10 % APS
60 µl	TEMED
~60 ml	Total

4. Slowly pour gel to avoid bubbles and place wells (typically 15-well comb) at a depth of 5 mm.

5. Allow gel to polymerize at room temperature until the samples are ready from the step below. This will typically be over 1 h, but if one wants to pour during the sample incubation below, let the gel polymerize at least 15 min.

6. Remove the comb and rinse wells gently with water.

7. Remove the bottom spacer.

8. Place the cast gel in the electrophoresis box. Add TBE running buffer (0.25×) to the electrophoresis apparatus (*see* **Note 26**).

3.2.6 Generating Total Cell Extracts

1. Treated samples should be pelleted (and optionally stored at −80 °C until ready for processing). If samples have been frozen, briefly thaw them on ice.

2. Remove any residual PBS buffer by quick spin followed by careful removal by pipet.

3. Resuspend the cell pellet in TOTEX buffer with 0.1 mM DTT and protease inhibitors freshly added, at a ratio of approximately 3:1, TOTEX to pellet volume (*see* **Note 27**).

4. Lyse for 30 min on ice, gently tapping the microcentrifuge tube every 10 min.

5. Spin in a 4 °C tabletop microcentrifuge at maximum speed (~18,000×g) for 10 min.

6. Transfer the supernatant to a new tube and discard the pelleted debris.

3.2.7 Preparation of Samples for EMSA

1. Determine the protein concentration for each sample using your preferred method. We routinely use Bradford reagent measured at OD_{595}.

2. Equalize protein concentrations by adding additional TOTEX buffer to samples that are more concentrated (*see* **Notes 28** and **29**).

3.2.8 Assembling the DNA Binding Reaction

1. On ice, assemble the binding reaction shown in Table 6 (*see* **Note 30**).

2. Incubate on ice for 20 min.

3. Add 1 μl of radiolabeled probe. Optionally, 50× cold unlabeled probe can be added to demonstrate specificity. Alternatively, add 1 μl of antibodies for each of the NF-κB family members to perform "supershift" assay.

4. Mix gently by tapping the tube.

5. Incubate at room temperature for 20 min.

Table 6
Binding reaction for EMSA

– μl	5–10 μg total protein in TOTEX
2 μl	5× binding buffer
1 μl	Poly(dI-dC)
0.1 μl	0.1 M DTT
– μl	Sterile H₂O
9 μl	Total

3.2.9 Native Gel Electrophoresis

1. Load the DNA binding reaction on to the prepared polyacrylamide gel (*see* **Note 31**).

2. Load a single lane on the edge of the gel (not containing a sample) with DNA loading dye with bromophenol blue. This approximates free (unbound) probe as it migrates through the gel.

3. Electrophorese at constant voltage at room temperature. A 20×20 cm gel run at 200 V will finish in approximately 80 min (*see* **Note 32**).

3.2.10 Drying the Gel and Exposing to Autoradiography Film

1. After the gel electrophoresis has completed, carefully separate the glass plates. The polyacrylamide gel should remain on the bottom plate.

2. Apply a double-thickness (folded) 20×20 cm piece of Whatman paper to the gel. This will stick to the gel and allow for removal from the glass plate.

3. Place two sheets of Whatman paper on the bottom of the gel dryer.

4. Place the gel (on the double-sided Whatman paper), gel side up.

5. Cover the polyacrylamide gel with plastic wrap (*see* **Note 33**).

6. Place another piece of Whatman paper on top of the plastic wrap.

7. Dry gel, following manufacturer instructions.

8. Remove the dried gel from the gel dryer. Discard the plastic wrap and Whatman paper, except for that which is adhered to the dried gel.

9. If necessary, cut the excess Whatman paper to dried gel size and wrap with new plastic wrap and secure with tape. Make sure to cover the entire front and back of the gel-Whatman paper to reduce the risk of contaminating cassettes.

10. Expose to film. Optionally if required, an intensifying screen can be used and placed at –80 °C.

11. Develop the film after an exposure from 2 h to overnight. The gel may be exposed to a PhosphorImager screen and developed on a Typhoon, Storm, or equivalent PhosphorImager device for quantitation of signal intensities.

12. By running Oct-1 control binding in parallel with each sample, changes in specific binding to NF-κB probe can be evaluated. Typically, the Oct-1 binding is constant across the lanes. If this varies from sample to sample, the quality of data obtained should be viewed with caution.

4   Notes

1. As an optional step the plasmid can be nicked using a restriction site away from the gene of interest/selection cassette to linearize the DNA.

2. DTT needs to be relatively fresh otherwise NF-κB DNA binding will be severely reduced.

3. The κB site in Table 1 is the Igκ-κB site. The κB site (bolded and underlined) may be mutated to generate mutant probe to show specificity. The major NF-κB complex activated by genotoxic agents and LPS stimulation is p50-p65 heterodimer. A minor p50-cRel heterodimer may also be seen depending on the duration of cell stimulation. A p50-p50 homodimer is also often observed with this probe. Other NF-κB binding sites could also be used. A similar design may be used with different plus-strand oligos with complementary minus-strand 10-mers. Depending on the specific κB sites used, different amounts of NF-κB dimers might be detected upon cell stimulation.

4. Acrylamide and bis-acrylamide are neurotoxic. Use of a mask and gloves are required.

5. If cells are used immediately after thawing from cryopreservation, their growth and phenotypes tend not to be reliable. Thus, we routinely passage 3 or more times before using them for experiments. If these cells are allowed to grow higher than ~1.2×10^6 cells/ml for multiple passages, their health tends to be compromised (cell shape may appear irregular, rather than "shiny spheres") and cells may undergo apoptosis more readily. Consequently, their phenotypes may become less reproducible. Cryopreserve early passage cells in multiple vials and when cells appear irregular, do not use them for reconstitution experiments. Moreover, check for potential mycoplasma contamination, avoiding the use of mycoplasma-infected cell lines during experiments.

6. Include a mock DNA control (without selection) to monitor when appropriate selection has occurred.

7. We recommend the following settings as a starting point: Voltage (kv), 0.25 kV; Capacitance, 950 µF.

8. Plasmid DNA may be restriction digested to generate linear DNA to minimize the random breakage of NEMO and selection cassette in the plasmid DNA during genomic integration. However, our experience indicates that this procedure is unnecessary for successful isolation of numerous independent clones. Alternatively, retroviral infection may be used instead of electroporation to generate stable clones. However, 1.3E2 cells require the use of protamine sulfate (at a concentration of 4 µg/ml) instead of polybrene, as we found that polybrene can be particularly toxic in this cell line.

9. We had success in functionally reconstituting 1.3E2 cells with murine and human NEMO cDNA. There are several splice variants of murine NEMO. We had functional reconstitution with mouse NEMO cDNA cloned from the parental 70Z/3 cells as well as murine NEMO obtained from Dr. Alain Israel's group who originally cloned NEMO. Tagged or untagged versions of NEMO (both mouse and human) can be used successfully. We were able to reconstitute functional NF-κB activation by genotoxic agents and LPS using the following N-terminal tags: 6xMyc, 2xHA, eGFP, and His6. However, researchers need to be mindful that these tags add different charge and size properties to NEMO, which might modulate certain aspects of cell signaling.

10. For each distinct cell line you wish to generate, we recommend immediately splitting the initial electroporated cells into three separate dishes (labeled A, B, and C). This guarantees at least three absolutely distinct clones for downstream analysis. Otherwise, as these cells grow rapidly, it is relatively easy to isolate multiple clones derived from the same founder clone.

11. Transiently transfected cells often survive for 3 or 4 days, and therefore a high dose of G418 coupled with selection for greater than 4 days will ensure the survival of only stable transformants.

12. Great care must be taken to avoid mixing these two layers. Alternatively, cell suspension may be placed in a tube and then lymphocyte separation medium may be placed at the bottom of the cell with a pipet (underlaying rather than overlaying).

13. A distinct viable cell layer *may or may not* be present at the interface between the two solutions. Ensure that the majority of this layer is harvested whether the cell layer is visible or not.

14. Stable pools can be cryopreserved at −80 °C (10 % DMSO + 90 % FBS). However, the number of cells is typically low and viability of cells upon thawing may become compromised.

15. Use pipets with barrier tips to prevent cross-contamination of stable clones. Once becoming proficient, one can pick a large number of clones in a few hours. To reduce potential contamination, it is recommended that this task is done when traffic in the area is minimal.

16. We have observed that 1.3E2 cell growth is negatively affected by the presence of dead cells, which occurs during cloning by limited dilution. As such, we prefer to pick individual clones by hand. This operation may be performed inside a biological safety cabinet, although we have not had any contamination issues performing this outside of a hood. Alternatively, it is possible to employ FACS to deposit individual clones in 96 wells. Regardless of the method employed, researchers should practice isolating lymphocytes before attempting to do this with newly electroporated cell populations.

17. Stable cell lines can be frozen down at −80 °C in cryopreservation solution while analysis of the clones is performed to avoid unnecessarily culturing each cell line. Because different clones grow out at different rates, cryopreservation may be necessary to be able to handle the heavy culturing load.

18. Clones should be analyzed for the expression of NEMO protein through western blotting. In each gel, run a positive control sample (we recommend running 70Z/3 cell extracts as a control for "physiological" expression). It is typical that each clone will express varying steady-state levels of NEMO, some with little to no expression, some near the 70Z/3 level, and some much higher than 70Z/3 level. We typically analyze at least three independent clones that express relatively similar level as 70Z/3 (one from each of the A, B, and C plates to ensure clonality). Other methods, such as immunofluorescence, may also be used to test the uniformity of expression of exogenous NEMO.

19. Because of the variability in the levels of expression in individual cells within a given pool of stable cells, patterns of NF-κB activation may not accurately represent the phenotype of a given mutation. Moreover, the activation phenotypes may drift over several passages as the certain clones outgrow others. As such, we generally do not analyze stable pools. We also recommend that wildtype NEMO is included in each experiment as a positive control. This is especially important if the phenotype of a NEMO mutant results in partial loss of function.

20. For routine screening purposes, cells are treated in 60 mm dishes in 3 ml of medium. 3×10^6 cells can yield a sufficient

amount of total proteins for NF-κB, Oct1, and Western blot analysis. However, until proficient, we recommend that researchers use 1×10^7 cells in 100 mm dish per sample.

21. Cell pellets can be frozen at −80 °C for several weeks or more prior to analysis by EMSA.

22. Oligos not to be immediately end-labeled can be aliquoted and stored at −20 °C. It may be helpful to aliquot volumes that will be used during each labeling reactions (1–3 μl depending on scale) in screw-top tubes so as not to freeze-thaw the double-stranded oligo batch multiple times.

23. The double-stranded cold κB oligos may be used for competition experiments to demonstrate the specificity of NF-κB complexes detected in EMSA analysis.

24. Oligos may be stored frozen at −20 °C; however, users must keep in mind that ^{32}P has a relatively short half-life of 14.29 days. As the radioactivity decay, the concentration of cold probe will increase, which may function as a competitor for NF-κB binding to radioactive probe thereby compromising the binding assay. As such, fresh probe should be made every 2 weeks or less.

25. Optionally apply a thin layer of SigmaCote (Sigma-Aldrich) or other siliconizing reagents to prevent gel adhesion to one of the glass plates. This will help to remove one plate off the gel during the later disassembling step.

26. Some EMSA gel mixtures include glycerol to stabilize the complexes during electrophoresis. However, we found that the NF-κB complexes form tighter banding patterns if glycerol is left out of the gel. One may use thinner spacers; however, since the gel is only ~4 %, it is very easy to tear or distort. One may also run "mini-EMSA" gels. However, the quality of NF-κB banding patterns may suffer.

27. This step will require some optimization by the user, but ultimately samples should be prepared such that their protein concentration is 3–10 μg/μl. We found that reasonably high-quality data could be obtained using total cell extracts, rather than nuclear extracts. However, nuclear extracts may be used instead. The quality of nuclear extracts may vary from sample to sample in each experiment set. If different volumes of nuclear extracts are used in different samples within an experimental set, the quality of the EMSA suffers significantly (e.g., due to different amounts of salt and contaminating genomic DNA). As such the quality of nuclear extracts in each sample needs to be validated.

28. Samples should be equalized such that the protein concentration is consistent between all samples and somewhere between 3 and 10 μg/μl (dictated by the final volume of TOTEX buffer to

be added to the reaction mixture). Samples less concentrated than this are not suitable for EMSA.

29. Because the interaction between NF-κB and DNA is highly salt sensitive, the same final amount of TOTEX should be present in each of the samples and do not add more than 3 μl of TOTEX buffer per 10 μl reaction mixture.

30. We frequently use a master-mix template to minimize pipetting error. This template is available at http://miyamotolab.wisc.edu/EMSA.xls.

31. As the samples do not contain loading dye, care must be taken when loading the gel. Tangential lighting is helpful to visualize the glycerol containing reaction as it sinks to the bottom of the well.

32. The gel may be run in a cold room. This will increase the running time. We found that this is unnecessary for NF-κB or Oct1 EMSA.

33. The plastic wrap is required to prevent the gel from adhering to another piece of Whatman paper.

Acknowledgements

The authors would like to thank the many Miyamoto lab members, past and present, for their help in formulating this chapter. We would like to particularly thank Shelly Wuerzberger-Davis who has trained generations of lab members and who made significant contributions to this protocol. This work was funded by F30CA171840 (S.S.J.), NIH R01CA077474 (S.M.), NIH R01GM083681 (S.M.).

References

1. Harper JW, Elledge SJ (2007) The DNA damage response: ten years after. Mol Cell 28:739–745

2. Matsuoka S, Huang M, Elledge SJ (1998) Linkage of ATM to cell cycle regulation by the Chk2 protein kinase. Science 282:1893–1897

3. Elledge SJ (1996) Cell cycle checkpoints: preventing an identity crisis. Science 274:1664–1672

4. Kastan MB et al (1992) A mammalian cell cycle checkpoint pathway utilizing p53 and GADD45 is defective in ataxia-telangiectasia. Cell 71:587–597

5. Canman CE (1998) Activation of the ATM kinase by Ionizing radiation and phosphorylation of p53. Science 281:1677–1679

6. Cimprich KA, Shin TB, Keith CT, Schreiber SL (1996) cDNA cloning and gene mapping of a candidate human cell cycle checkpoint protein. Proc Natl Acad Sci U S A 93:2850–2855

7. Carter T, Vancurová I, Sun I, Lou W, DeLeon S (1990) A DNA-activated protein kinase from HeLa cell nuclei. Mol Cell Biol 10:6460–6471

8. Macleod KF et al (1995) p53-dependent and independent expression of p21 during cell growth, differentiation, and DNA damage. Genes Dev 9:935–944

9. Miyashita T et al (1994) Tumor suppressor p53 is a regulator of bcl-2 and bax gene expression in vitro and in vivo. Oncogene 9:1799–1805

10. Huang TT et al (2000) NF-kappaB activation by camptothecin. A linkage between nuclear DNA damage and cytoplasmic signaling events. J Biol Chem 275:9501–9509

11. Piret B, Piette J (1996) Topoisomerase poisons activate the transcription factor NF-kappaB in ACH-2 and CEM cells. Nucleic Acids Res 24:4242–4248

12. Karin M, Cao Y, Greten FR, Li Z-W (2002) NF-kappaB in cancer: from innocent bystander to major culprit. Nat Rev Cancer 2:301–310

13. Huang TT, Wuerzberger-Davis SM, Wu Z-H, Miyamoto S (2003) Sequential modification of NEMO/IKKγ by SUMO-1 and ubiquitin mediates NF-κB activation by genotoxic stress. Cell 115:565–576

14. Wu Z-H, Shi Y, Tibbetts RS, Miyamoto S (2006) Molecular linkage between the kinase ATM and NF-kappaB signaling in response to genotoxic stimuli. Science 311:1141–1146

15. Stilmann M et al (2009) A nuclear poly(ADP-ribose)-dependent signalosome confers DNA damage-induced IkappaB kinase activation. Mol Cell 36:365–378

16. Li N et al (2001) ATM is required for IkappaB kinase (IKK) activation in response to DNA double strand breaks. J Biol Chem 276:8898–8903

17. Piret B, Schoonbroodt S, Piette J (1999) The ATM protein is required for sustained activation of NF-kappaB following DNA damage. Oncogene 18:2261–2271

18. Li N, Karin M (1998) Ionizing radiation and short wavelength UV activate NF-kappaB through two distinct mechanisms. Proc Natl Acad Sci U S A 95:13012–13017

19. Rooney JW, Emery DW, Sibley CH (1990) 1.3E2, a variant of the B lymphoma 70Z/3, defective in activation of NF-kappa B and OTF-2. Immunogenetics 31:73–78

20. Yamaoka S et al (1998) Complementation cloning of NEMO, a component of the IkappaB kinase complex essential for NF-kappaB activation. Cell 93:1231–1240

21. Rothwarf DM, Zandi E, Natoli G, Karin M (1998) IKK-gamma is an essential regulatory subunit of the IkappaB kinase complex. Nature 395:297–300

22. McCool KW, Miyamoto S (2012) DNA damage-dependent NF-κB activation: NEMO turns nuclear signaling inside out. Immunol Rev 246:311–326

23. Garner MM, Revzin A (1981) A gel electrophoresis method for quantifying the binding of proteins to specific DNA regions: application to components of the Escherichia coli lactose operon regulatory system. Nucleic Acids Res 9:3047–3060

24. Sen R, Baltimore D (1986) Multiple nuclear factors interact with the immunoglobulin enhancer sequences. Cell 46:705–716

Part III

Methods to Study the Control of NF-κB Signaling

Chapter 12

Visualizing TCR-Induced POLKADOTS Formation and NF-κB Activation in the D10 T-Cell Clone and Mouse Primary Effector T Cells

Suman Paul and Brian C. Schaefer

Abstract

T cells are an immune cell lineage that play a central role in protection against pathogen infection. Antigen, in the form of pathogen-derived peptides, stimulates the T-cell receptor (TCR), leading to activation of the transcription factor, nuclear factor kappa B (NF-κB). The subsequent NF-κB-dependent gene expression program drives expansion and effector differentiation of antigen-specific T cells, leading to the adaptive anti-pathogen immune response. The cell surface TCR transmits activating signals to cytosolic NF-κB by a complex signaling cascade, in which the adapter protein Bcl10 plays a key role. We have previously demonstrated that TCR engagement leads to the formation of cytosolic Bcl10 clusters, called POLKADOTS, that provide a platform for the assembly of the terminal signaling complex that ultimately mediates NF-κB activation. In this chapter, we describe the methods utilized to visualize the formation of TCR-induced POLKADOTS and to study the temporal association between POLKADOTS formation and nuclear translocation of the NF-κB subunit, RelA/p65.

Key words T cells, TCR, Bcl10, POLKADOTS, NF-κB, RelA/p65, Confocal microscopy

1 Introduction

The human body is constantly exposed to pathogens, many of which can cause severe morbidity and mortality in the absence of an effective immune response. T cells are white blood cells that play a key role in detecting the presence of pathogens and executing the anti-pathogen immune response. These T-cell functions require recognition of pathogen components by the cell surface TCR, leading to intracellular signaling and initiation of transcriptional programs.

Upon entering the body, intact pathogens and pathogen components are captured and digested by a heterogeneous group of cells collectively called "antigen-presenting cells" (APCs). Peptides from digested pathogen proteins are then displayed on the APC surface in a complex manner [1, 2], which will not be discussed in detail here. T cells utilize the cell surface TCR to scan the APC

Michael J. May (ed.), *NF-kappa B: Methods and Protocols*, Methods in Molecular Biology, vol. 1280,
DOI 10.1007/978-1-4939-2422-6_12, © Springer Science+Business Media New York 2015

surface in order to identify the presence of foreign antigens. When a TCR specifically binds to a pathogen peptide displayed by the APC, the TCR initiates a signaling cascade in the T-cell cytosol. Signals transduced by the TCR ultimately trigger activation of several key transcription factors, including NF-κB, which control T-cell activation, proliferation, and differentiation [3–5]. The expanded populations of differentiated, pathogen-specific T cells (called "effector" T cells), directly or in combination with other immune cells, are responsible for limiting the spread of invading microorganisms. Transmission of TCR activation signals to NF-κB is therefore of central importance in the successful initiation of a T-cell-mediated immune response.

TCR activation of NF-κB is dependent on a complex signaling cascade (Fig. 1). Following antigen engagement of the TCR, early membrane-proximal signaling events lead to activation of a specific protein kinase C (PKC) isoform, PKCθ. Through phosphorylation of the large adaptor protein, Carma1, activated PKCθ drives the association of Carma1 with a preexisting complex of the small adaptor, Bcl10, and the protease, Malt1, thereby forming the "CBM" complex. Data suggest that the CBM complex recruits ubiquitin ligases, particularly TNF receptor-associated factor 6 (Traf6). Traf6-mediated ubiquitination of CBM complex components is believed to recruit both the IκB kinase complex (IKK) and its activating kinase, Tak1. The consequence of bringing IKK into close proximity to Traf6 and Tak1 is the activation of IKK via Tak1-mediated phosphorylation and Traf6-mediated polyubiquitination [6]. IKK phosphorylates the inhibitor of κBα (IκBα), resulting in proteasomal IκBα degradation and release of NF-κB from cytosolic sequestration. The freed NF-κB then translocates to the nucleus, where it initiates transcription of genes required for T-cell proliferation, differentiation, and effector function [7].

Imaging T-cell interaction with APCs enables visualization of key protein translocation events that are intimately associated with this signaling cascade. In unstimulated T cells, PKCθ and RelA have a distinctly cytosolic distribution, while Bcl10 is present diffusely throughout the T-cell cytosol and nucleus. Immediately after TCR ligation, PKCθ rapidly translocates to the immunological synapse (IS), which is the zone of contact between the APC and T cell [8, 9]. Within minutes following PKCθ IS translocation, Bcl10 and Malt1 are recruited to preexisting aggregates ("speckles") of the multi-domain adapter protein, p62, forming prominent cytosolic clusters, called POLKADOTS [10–12]. POLKADOTS also contain the phosphorylated, activated form of IKK, and they are transiently associated with phosphorylated IκBα and NF-κB [13], immediately prior to NF-κB nuclear translocation. These data suggest that POLKADOTS represent a cytosolic signalosome that directs terminal activation of NF-κB [13]. The formation of Bcl10-Malt1 cytosolic POLKADOTS and the temporal association of

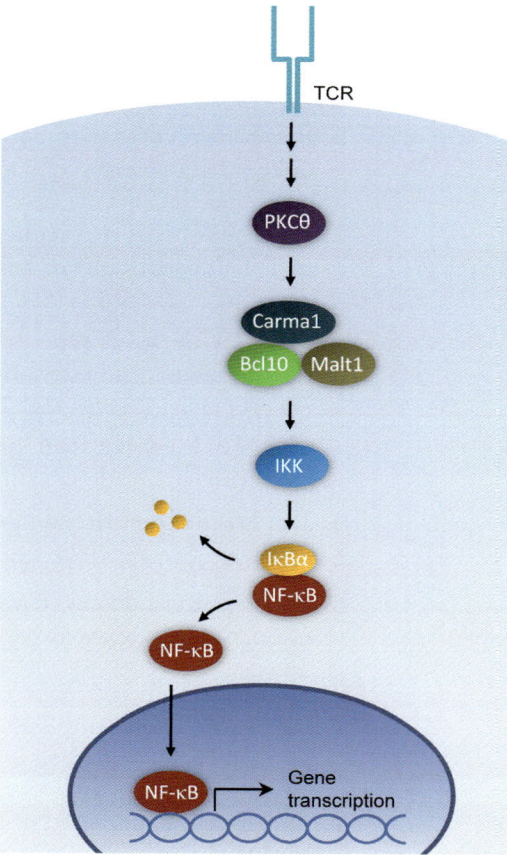

Fig. 1 The TCR-to-NF-κB pathway. In T lymphocytes, the TCR activation signal is transmitted through several upstream signaling proteins to activate PKCθ. Activated PKCθ drives formation of the Carma1-Bcl10-Malt1 (CBM) complex. The CBM complex, assisted by ubiquitin ligase TRAF6 (not shown), recruits and activates IKK. IKK-mediated phosphorylation of IκB leads to IκB degradation, allowing free NF-κB to translocate to the nucleus

POLKADOTS formation with NF-κB activation can be observed using both ectopic expression of fluorescent protein-tagged Bcl10 and by antibody staining of endogenous Bcl10. We have obtained equivalent data using both the D10 T-cell clone and in vitro differentiated primary murine effector T cells [10]. In this chapter, we provide detailed guidance regarding the generation of stable Bcl10-GFP-expressing D10 T-cell lines, stimulation of D10 T cells with anti-CD3 antibody or antigen-loaded APCs, staining for the endogenous NF-κB subunit RelA/p65, and microscopic imaging to reveal the POLKADOTS formation and RelA activation. We also provide protocols for microscopic imaging of endogenous POLKADOTS via staining for endogenous Bcl10 following anti-CD3+ anti-CD28 or APC+ antigen stimulation of primary effector T cells.

2   Materials

2.1   Cloning of Bcl10-GFP into a Retroviral Vector

1. Bcl10 with a C-terminal fluorescent fusion protein.
2. Retroviral vector with an internal ribosome entry site (IRES)-linked drug-selectable marker downstream of the cloning site.

2.2   HEK293T Cell Transfection for Generation of Retrovirus

1. HEK293T cells maintained in DMEM (*see* below) and grown in T-175 tissue culture flasks.
2. Dulbecco's modified Eagle's medium (DMEM) supplemented with 10 % (v/v) fetal bovine serum, glutamine, penicillin, streptomycin, and gentamicin.
3. IMDM (Iscove's Modified Dulbecco's medium) supplemented with 10 % (v/v) fetal bovine serum, glutamine, penicillin, streptomycin, and gentamicin.
4. 1× phosphate buffered saline (1× PBS).
5. Solution of 0.05 % trypsin with 0.2 g/L EDTA.
6. Tissue culture dishes (6-well plates).
7. Sterile, endotoxin-free water.
8. 2.5 M calcium chloride.
9. 2× HEPES solution: 140 mM NaCl, 1.5 mM NaPhosphate, 50 mM HEPES (free acid), pH 7.05.

2.3   D10 T-Cell Retroviral Infection and Antibiotic Selection

1. D10 T cells maintained in EHAA medium with IL-2 (*see* below) and grown in T-75 tissue culture flasks (*see* **Note 1**).
2. EHAA medium supplemented with 10 % (v/v) heat-inactivated fetal bovine serum, 4.6 mM HEPES pH 7.4, 50 µM 2-mercaptoethanol, glutamine, penicillin, streptomycin and gentamicin, and murine IL-2 (1–2 ng/mL final conc.).
3. 1,000× Polybrene (10 mg/mL).
4. Tissue culture dish (24-well plate).
5. Temperature-controlled centrifuge capable of accepting 24-well plates and spinning at 1,200×g.
6. Selective agent: e.g., G418, Zeocin, or hygromycin.

2.4   T-Cell Stimulation with Anti-CD3 or Anti-CD3+ Anti-CD28

1. Coverslips (square, #1.5 thickness, 18×18 mm).
2. Coating solution 1: 1 M HCl in 70 % ethanol.
3. Coating solution 2: 100 µg/mL poly-D-lysine hydrobromide (MW 70,000–150,000).
4. Anti-CD3 antibody (clone 145-2C11), 100 µg/mL (final concentration).
5. Anti-CD28 antibody (clone 37.51), 1 µg/mL (final concentration).

2.5 T-Cell Stimulation with APC⁺ Antigen

1. Antigen-presenting cells (APCs): We use CH12 B cells [14] for conalbumin stimulation of D10 T cells. For SEB (staphylococcal enterotoxin B) stimulation of C57BL/6 primary T cells, we use CHb B cells as APCs [14]. Both cell lines are maintained in EHAA medium supplemented with ingredients listed in Subheading 2.3, **item 2**, with the exception of IL-2 (*see* **Note 2**).

2. Antigen: Conalbumin 10 mg/mL stock in 1× Hanks' balanced salt solution (HBSS) or 1 mg/mL SEB (staphylococcal enterotoxin B).

3. Poly-D-lysine-coated coverslips (#1.5 thickness). Incubate coverslips for 6 h at ambient temperature in a bacterial petri dish filled with solution of 3 mg/mL poly-D-lysine hydrobromide (MW 70,000–150,000) dissolved in water. Coverslips are stored frozen at –20 °C until use. On the day of the experiment, wash the required number of coverslips in a bacterial petri dish filled with distilled water. Add coverslips, swirl petri dish, discard water, and replace. Repeat 5×. Place coverslip in well of 6-well tissue culture dish (or 35 mm dish). Aspirate surface of coverslip to remove residual water.

2.6 Antibody Staining of D10 T-Cell Clones or CD4⁺ T$_H$2 Cells and Slide Preparation

1. Fix solution, 3 % paraformaldehyde, 3 % sucrose, 1× PBS: In a chemical fume hood, heat ~800 mL of ddH$_2$O to exactly 56 °C with continuous mixing. Add 30 g paraformaldehyde, and then add a few drops of 10 N NaOH, until the majority of paraformaldehyde has entered solution (there is usually a little insoluble material). Add 30 g sucrose and 100 mL of 10× PBS. Adjust pH to about 7.5 (use pH paper for this measurement). Transfer to a 1 L graduated cylinder, and add ddH$_2$O to a final volume of 1 L. Filter through a 0.22 or 0.45 micron filter to remove insoluble material, dust, etc. Aliquot solution to 50 mL tubes, and store at –20 °C. Aliquots should be thawed at ≤37 °C to improve longevity of fix solution through multiple freeze-thaw cycles. Mix well after thawing; paraformaldehyde and sucrose will settle to bottom when thawed.

2. Permeabilization solution: 0.2 % Triton X-100 and 0.1 % sodium azide dissolved in 1× PBS (*see* **Note 3**).

3. Blocking solution: 10 % FBS and 0.1 % sodium azide in DMEM (*see* **Note 3**).

4. Mounting medium: We use 90 % glycerol with p-phenylenediamine (PPD) as mounting medium (for instructions regarding making PPD mounting medium, *see* **Note 4**). Alternatively, any commercially available mounting medium with a photobleaching inhibitor will also work.

5. Microscopy slides.

2.7 CD4+ T_H2 Effector Cell Differentiation

1. Mice as source of CD4+ T cells (we use C57BL/6 mice).

2. Anti-CD3 antibody, 100 μg/mL final concentration.

3. Anti-CD28 antibody, 1 μg/mL final concentration.

4. Hanks' balanced salt solution (HBSS).

5. T_H2 working solution: EHAA medium (with supplements as described in Subheading 2.3, **item 2**, but note that higher IL-2 concentration is specified here) plus anti-IL-12 antibody (20 μg/mL), anti-IFN-γ antibody (20 μg/mL), IL-2 (10 ng/mL), and IL-4 (4 ng/mL).

6. CD4+ T-cell negative selection immunomagnetic bead purification kit (we use a negative selection kit from Invitrogen) and an appropriate magnet.

7. EHAA medium (*see* Subheading 2.3, **item 2**).

8. 24- and 6-well tissue culture plates, T-25 and T-75 tissue culture flasks.

9. 70 μm cell strainer, 50 mL conical tubes for collecting and washing cells.

10. Centrifuge capable of spinning 50 mL conical tubes at $213 \times g$.

3   Methods

Using D10 T cells or primary CD4+ effector T cells, we have shown that TCR stimulation by a specific antigen-loaded APC results in translocation of PKCθ to the IS and cytosolic clustering of Bcl10 (Fig. 2a). Although stimulation with immobilized antibody does not generally result in efficient translocation of PKCθ to the plasma membrane, robust Bcl10 clustering in the cytosol is induced by anti-CD3 (for D10 T cells) or by anti-CD3+ anti-CD28 (for primary Th2 CD4 cells). The phenomenon of TCR-induced cytosolic Bcl10 clustering/POLKADOTS formation occurs at similar time points post-stimulation in both the D10 T-cell clone (Fig. 2) and in mouse primary T_H2 cells (Fig. 3). Antigen-induced PKCθ enrichment at the IS can be detected within seconds of stimulation. Bcl10 cytosolic aggregations are initiated at about 10 min post-stimulation and are well formed by approximately 20 min after initial stimulation. In unstimulated T cells, the NF-κB subunit, RelA (p65), is bound to IκB and has a diffuse cytosolic distribution (Figs. 2 and 3). At 10 min post-stimulation, RelA transiently associates with Bcl10 cytosolic clusters (Fig. 2a). By 20 min after stimulation, as a result of IκB degradation, RelA has translocated to the nucleus, indicating successful activation of NF-κB (Figs. 2 and 3).

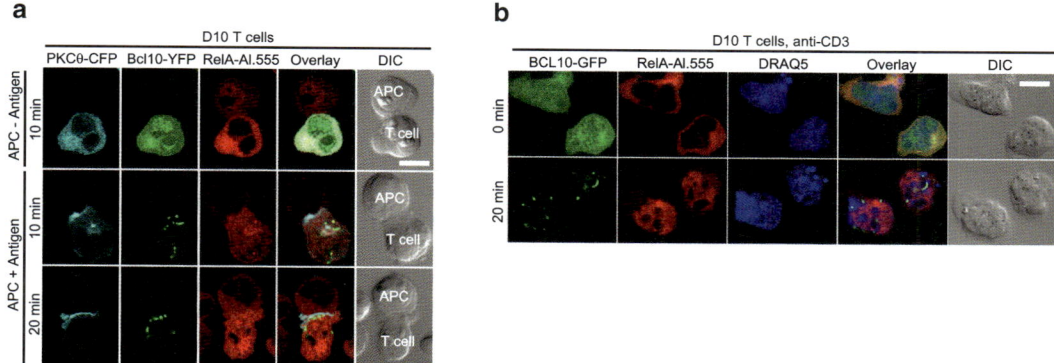

Fig. 2 POLKADOTS formation and RelA activation in D10 T-cell clones overexpressing fluorescent protein fusions of Bcl10. (a) D10 T cells expressing PKCθ-CFP and Bcl10-YFP stimulated with APCs loaded with or without antigen (conalbumin) and fixed at 10 or 20 min post-stimulation. Conjugates were stained with a rabbit anti-RelA primary and anti-rabbit Alexa-555 secondary antibody. (b) D10 T cells expressing Bcl10-GFP stimulated with anti-CD3-coated coverslips and fixed at 0 or 20 min post-stimulation. Cells were stained with a rabbit anti-RelA antibody and anti-rabbit Alexa-555 secondary antibody. DRAQ5 was used to visualize the nucleus. Scale bar, 10 μm

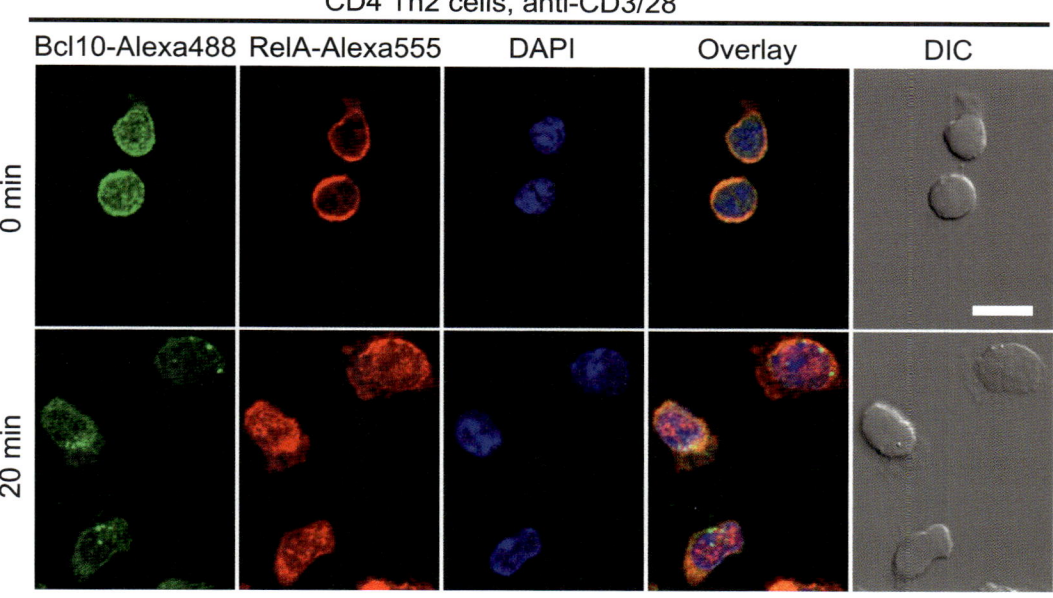

Fig. 3 Antibody staining of endogenous Bcl10 and RelA to visualize POLKADOTS formation and NF-κB activation in CD4+ T$_H$2 cells. Mouse primary T$_H$2 effector cells were stimulated on anti-CD3/CD28-coated coverslips and fixed at 0 and 20 min post-stimulation. Cells were stained with a mouse anti-Bcl10 and rabbit anti-RelA primary antibodies and followed by anti-mouse Alexa-488 and anti-rabbit Alexa-555 secondary antibodies. DAPI was used to visualize the nucleus. Scale bar, 10 μm

We provide a detailed protocol for stimulating T cells using antigen-loaded APCs or coverslips coated with anti-CD3+ anti-CD28. We also provide instruction for visualizing post-stimulation Bcl10 clustering/POLKADOTS formation and for tracking the movement of NF-κB from the cytosol to POLKADOTS and ultimately into the nucleus.

Both endogenous and overexpressed Bcl10 proteins form POLKADOTS. We describe the methods used to generate stable cell lines expressing fluorescent protein-tagged Bcl10 protein, and we provide a protocol for staining endogenous Bcl10 in primary T_H2 cells. We also provide methodology for detecting terminal activation of NF-κB using antibodies directed toward RelA in combination with stains for nuclear DNA, such as DAPI (Fig. 3) or DRAQ5 (Fig. 2b).

3.1 Cloning Bcl10-GFP into a Retroviral Vector

1. Employing standard molecular biology techniques, directly fuse GFP (or an alternative fluorescent protein) to murine Bcl10. We have found that fluorescent protein fusions to the C-terminus of Bcl10 redistribute and signal to NF-κB in a manner indistinguishable from endogenous Bcl10. By contrast, we have observed that fusions to the Bcl10 N-terminus exhibit a different redistribution pattern (e.g., such constructs are inefficiently recruited to POLKADOTS) (unpublished data). A caveat to placing fluorescent proteins at the Bcl10 C-terminus is that the C-terminally cleaved form of Bcl10 [15] will lack the fluorescent tag.

2. Clone the Bcl10-GFP fusion into a retroviral vector which includes a drug-selectable marker (allowing selection by G418, hygromycin, Zeocin, or alternative drug). To ensure expression of Bcl10-GFP in every cell that survives selection, the selectable marker should be cloned downstream of Bcl10-GFP as an IRES cassette, rather than being driven by an independent promoter. We have previously described a set of vectors, called pE vectors [16], that work well for this purpose. Similar vectors are commercially available.

3.2 Transfecting HEK293T Cells to Generate Retrovirus

This method describes the production of retrovirus coding for a Bcl10-GFP fusion protein. This virus can be used to infect D10 T cells or primary T cells. The transfection method is adapted from [17] and results in near 100 % transfection efficiency and production of high-titer retrovirus. Other transfection methods can be substituted, if empirically determined to be highly efficient:

1. HEK293T cells are grown in DMEM containing FBS (10 %) in T-175 culture flask (see **Note 5**). When approaching confluence, passage cells by gently washing in 1× PBS followed by PBS aspiration. Incubate cell monolayer with 3 mL of trypsin solution for 2 min (detachment verified by microscopy).

Add at least 3 mL of DMEM (with serum) to inactivate trypsin. Transfer cells to a 50 mL conical tube. Create a single-cell suspension by repeated pipetting (at least 10×). Count cells using a hemocytometer or alternative method.

2. *Day* 1. Coat each well of a 6-well culture plate with 0.5–1 mL of poly-D-lysine (100 µg/mL) and incubate for 5–10 min at ambient temperature. Aspirate the poly-D-lysine and wash 6× with 1× PBS (*see* **Note 6**). Plate 6×10^5 HEK293T cells in 1 mL of medium per well of a 6-well plate. Incubate overnight at 37 °C, 5 % CO_2.

3. *Day* 2. Replace DMEM with 1.5 mL of pre-warmed (37 °C) IMDM.

4. Assemble transfection master mixture in an Eppendorf tube: For each well, mix 2.4 µg Bcl10-GFP retroviral construct, 0.6 µg pCL-Eco helper plasmid [18], 7.5 µL of 2.5 M calcium chloride, and endotoxin-free water to a final volume of 75 µL. Add 75 µL of 2× HEPES, gently pipetting up and down 3–4× to ensure proper mixing. After exactly 1 min at ambient temperature, add master mixture dropwise onto the HEK293T cells in the 6-well plate (*see* **Note 7**).

5. *Day* 3. After 24 h, replace medium with 2 mL of pre-warmed (37 °C) DMEM.

6. *Day* 4. At approximately 48 h post-transfection, collect supernatant (which contains retrovirus) and keep supernatant at 4 °C (for storage up to 2 weeks) or –80 °C (for long-term storage) (*see* **Note 8**).

3.3 Retroviral Infection of D10 T Cells to Produce Stable Cell Lines

This protocol describes the generation of stable D10 T cells expressing Bcl10-GFP with the use of retrovirus produced in Subheading 3.2:

1. *Day* 1. Transfer 1×10^6 D10 T cells to a 15 mL conical tube. Centrifuge cells at $500 \times g$ for 10 min. Aspirate EHAA medium.

2. Resuspend D10 T-cell pellet in 2 mL of retroviral supernatant (generated in Subheading 3.2).

3. Add murine IL-2 (2 ng/mL final concentration) and 2 µL of 1,000× Polybrene.

4. Transfer the T-cell-virus suspension into one well of a 24-well plate and place plate inside a zipper seal bag. Seal bag, removing excess air, and tape overhanging edges. Transfer plate into centrifuge adaptor. Ideally, a covered plate adaptor should be used, so that any loose edges of the bag do not create drag as the plate is spun.

5. Spin the plate at $1,200 \times g$ for 2 h at 32 °C. It is critical that this step is *NOT* performed at low temperature (i.e., below 15 °C), which would inhibit virus fusion with target cells.

6. After centrifugation, transfer the D10 T cells to a 15 mL conical tube. Spin the 15 mL conical tube at $500 \times g$ for 10 min to collect the D10 T cells.

7. Aspirate medium, resuspend pellet in fresh EHAA medium, and transfer to T-25 tissue culture flask. Place flask inside a 37 °C incubator.

8. *Day 2.* 24 h post-retroviral transduction, add the appropriate antibiotic for selection. For pENeo-Bcl10-GFP, add G418 to a final concentration of 1 mg/mL. If a virus with a different selectable marker is used, we have determined the following optimal concentrations for D10 T cells: Zeocin, 25 μg/mL; hygromycin, 0.3 mg/mL.

9. Select cells for ≥3–5 days for G418 and ≥7–9 days for hygromycin or Zeocin. Set up a parallel control/uninfected D10 T-cell selection to empirically determine the time point at which uninfected cells are all dead. If the cells reach a density greater than 4×10^5/mL, it will be necessary to dilute the cells (in the appropriate concentration of antibiotic-supplemented complete medium) to 1×10^5/mL. Even with high-titer virus, the infection frequency of D10 T cells is often quite low. It may thus take in excess of 2 weeks for selected cells to grow out in numbers sufficient for passaging and freezing. We recommend keeping cells under selection until frozen. Also, maintaining under continuous selection may be required to keep expression of Bcl10-GFP to desired levels.

3.4 APC⁺ Antigen Stimulation of D10 T Cells and Antibody Staining of D10 T Cells to Observe POLKADOTS Formation and NF-κB Activation

This method describes stimulation of D10 T cells with APC to induce POLKADOTS formation in D10 cells:

1. *Day 1.* CH12 B cells are used as APC to stimulate D10 T cells. CH27 cells or any other I-A^k-positive cell type can also be substituted. CH12 cells are maintained in EHAA medium (without IL-2) at a concentration of $1–5 \times 10^5$ cells/mL. Load CH12 cells with conalbumin, by resuspending 5×10^5 cells in 5 mL of EHAA medium in a T-25 tissue culture flask, and add conalbumin to a final concentration of 100–250 μg/mL. Incubate overnight (12–24 h) in a 37 °C tissue culture incubator. It is also important to include a CH12 control with no conalbumin.

2. *Day 2.* For each experimental sample, 1×10^5 CH12 B cells and 1×10^5 D10 T cells will be required. Remember to prepare enough D10 T cells for both the CH12+ conalbumin and CH12 control conditions. Harvest the appropriate number of cells by centrifugation at $500 \times g$ in a 15 mL conical tube. Resuspend both CH12 B cells and D10 T cells to a final concentration of 1×10^5 cells/100 μL. In a 1.5 mL tube, add 100 μL of CH12 B cells to 100 μL of D10 T cells (*see* **Note 9**).

3. Spin CH12 and D10 mixture at ≥500×*g* for 30 s to initiate CH12 D10 cell contact and conjugate formation.

4. Place Eppendorf tube in 37 °C water bath/bead bath for 10 min.

5. About 12 min prior to end of incubation period, pipette cell mixture up and down 5 times to dissociate large cell clumps (T-cell-B-cell conjugates will not be disrupted by this procedure).

6. Pipette the 200 μL of cell mixture onto a poly-D-lysine-coated coverslip (prepared as described in Subheading 2.5), placed inside one well of a 6-well plate or 35 mm dish.

7. Cover plate and place it inside a 37 °C incubator for 7–10 min (*see* **Note 9**).

8. Terminate T-cell stimulation by adding 2 mL of fix solution, and incubate for 10 min. This step and all subsequent steps are performed on the benchtop at room temperature. *It is important that the cells are not allowed to dry at any time during the remaining steps.*

9. Aspirate fixation solution and wash 3× with 1×PBS. If no antibody staining is desired, skip to **step 21**. For staining intracellular proteins with fluorescent antibodies, continue to **step 10**.

10. Add 1–2 mL permeabilization solution and incubate for 5 min at ambient temperature.

11. Carefully aspirate permeabilization solution and wash 3× with ~2 mL of 1× PBS. Add PBS to dish using a pipette or squirt bottle directed to the edge of the dish. (Applying directly to the coverslip will dislodge cells).

12. Carefully aspirate excess PBS from the dish, without touching the coverslip (*see* **Note 10**).

13. Carefully add 200 μL of blocking solution on top of the coverslip and incubate for 20 min at ambient temperature (*see* **Note 11**).

14. Dilute primary antibody (anti-Bcl10, anti-RelA, etc.) at required concentration in blocking solution in a 1.5 mL tube. Centrifuge at 10,000×*g* for 10 min to precipitate any particulate matter.

15. Carefully aspirate blocking solution from coverslip (*see* **Note 12**).

16. Add 200 μL of primary antibody diluted in blocking solution to top of coverslip and incubate overnight at ambient temperature (*see* **Note 13**). Avoid disturbing the particulate pellet at the bottom of the tube.

17. *Day 3*. Wash 3× with 1×PBS, taking care not to dislodge cells from coverslip.

18. Dilute secondary antibody (appropriate fluorescent-tagged antibody for species/isotype of primary antibody) at required concentration in blocking solution (*see* **Note 13**). Centrifuge in 1.5 mL tube at 10,000×*g* for 10 min to precipitate any particulate matter.

19. Aspirate excess PBS (as in **step 12**).

20. Place 200 μL of secondary antibody diluted in blocking solution on coverslip and incubate for 1 h.

21. Wash 3× with 1×PBS.

22. Place 10 μL of mounting medium on slide.

23. Using fine-tipped forceps, carefully lift coverslip from well of a 6-well plate (keep surface with cells facing up). Wipe the bottom (non-cell side) with a Kimwipe to remove PBS. While holding the coverslip in a vertical position, gently touch the lowest corner to a Kimwipe to remove excess PBS from the cell side. Place coverslip with cell side facing *down* on top of mounting medium on slide (*see* **Note 14**). If necessary, gently push on top of coverslip to eliminate bubbles that are present in the mounting medium between the slide and coverslip.

24. Using blotting paper strips, absorb excess mounting medium and residual PBS from the sides of coverslip.

25. Seal edges of coverslip by applying clear nail polish on each side of coverslip.

26. Let nail polish dry at room temperature for 10–15 min, and store slides at –20 °C (*see* **Note 15**).

3.5 Stimulation of D10 T Cells on Anti-CD3-Coated Coverslips to Visualize POLKADOTS Formation and NF-κB Activation

Steps 1–13 describe the process of coating coverslips with anti-CD3 antibody, which should ideally be done a day before the planned D10 T-cell stimulation. The remainder of the protocol provides detailed instructions for use of anti-CD3-coated coverslips to stimulate D10 T cells:

1. *Day 1.* Place the required number of coverslips into wells of a 6-well tissue culture plate.

2. Add 1–2 mL of *Solution 1* (1 M HCl in 70 % ethanol) to each well to completely cover the coverslips.

3. Incubate at ambient temperature for 15 min.

4. Aspirate and discard Solution 1.

5. Heat 6-well plate with coverslips on a heat block/bead bath at 55–60 °C for 30 min.

6. Add 1–2 mL of poly-D-lysine (100 μg/mL) to each well to cover the coverslip.

7. Incubate at room temp for 15 min.

8. Aspirate poly-D-lysine (if recovered using a sterile method, this poly-D-lysine can be reused several times).

9. Heat 6-well plate with coverslips on a heat block/bead bath at 55–60 °C for 30 min.

10. Allow plate to cool to ambient temperature and carefully pipette 150–200 μL of anti-CD3 antibody on top of each coverslip.

11. Insert plate with coverslips and coating solution inside a zipper seal bag (taking care not to displace coating solution). Incubate at 4 °C overnight.

12. *Day 2.* After 16–24 h, collect anti-CD3 antibody and store for reuse (antibody can be reused several times).

13. Wash coverslip 6× with 1× PBS. Leave final PBS wash on coverslip to prevent drying of coverslip surface (which can denature the antibody). Aspirate the last wash immediately before adding D10 T cells (**step 16**).

14. Harvest D10 T cells from a culture growing at optimal density (2–5×10^5 cells/mL) in EHAA medium (with supplementation described in Subheading 2.3).

15. Centrifuge D10 T cells at $500 \times g$ for 5–10 min at ambient temperature and resuspend cells at a final concentration of 1×10^5 cells/100 µL.

16. Pipette 200 µL of the D10 T-cell suspension onto an anti-CD3-coated coverslip (from **step 13**).

17. Place 6-well plate with coverslips inside a 37 °C tissue culture incubator for 20 min (optimal for visualization of POLKADOTS and nuclear RelA. If the goal is to visualize early POLKADOTS and/or POLKADOTS-RelA association, the incubation time should be reduced to 10 min).

18. Remove plate from incubator and fix cells on the coverslip via gentle pipetting of 1–2 mL of fix solution at the edge of the well (i.e., not directly on the coverslip). For fixation, permeabilization, antibody staining, and mounting on slides, follow Subheading 3.4, **steps 9–26**.

3.6 Confocal Microscopy to Image T Cells

We use a 40× 1.4 NA oil objective and a Zeiss LSM 710 NLO confocal microscope to acquire the fluorescent images. The Zeiss Zen software in multitrack mode was used to operate the LSM 710 NLO. Other confocal microscopes of similar sensitivity can be substituted. Less sensitive confocal systems or epifluorescence systems can be used to detect Bcl10-GFP, but successful detection of endogenous proteins (particularly endogenous Bcl10) would be challenging. The selection of excitation light source, filters, etc., will depend on the exact combination of fluorescent proteins and/or fluorescent antibodies used in a given experiment.

3.7 In Vitro Differentiation of Murine T_H2 Cells

This is a modification of the effector cell differentiation protocol published in [19]. All procedures must be performed inside a tissue culture hood, unless otherwise specified. **Steps 1–3** describe the process of coating 24-well plates with anti-CD3 and anti-CD28 antibody, which must be performed a day prior to initiating the differentiation protocol. **Step 3** onward describes the process of collecting, stimulating, and culturing naïve CD4+ T cells to generate T_H2 cells (*see* **Note 16**):

1. *Day* 1. Pipette 200 µL of anti-CD3 antibody (100 µg/mL) and 10 µL of anti-CD28 antibody (1 µg/mL) into one well of a 24-well plate. Gently swirl or tap the plate so that the antibody solution completely covers the bottom of the well.

2. Insert 24-well plate into a zipper-lock bag and incubate at 4 °C overnight.

3. *Day* 2. Collect the anti-CD3/anti-CD28 antibody mixture (keep mixture in a separate container; this can be reused several times) and wash well 3–6× with sterile 1× PBS.

4. Dissect lymph nodes from desired strain of mice (we use C57BL/6) and collect in a sterile, disposable tube or dish filled with sterile 1× HBSS.

5. Collect lymphocytes from lymph nodes by forcing lymph nodes through a 70 µm cell strainer, followed by pipetting at least 5 mL of 1× HBSS through the strainer mesh.

6. Isolate CD4+ T cells from suspension of mixed lymphocytes using negative selection immunomagnetic beads or a similar technique. Cell purity (≥90 % CD4+ T cells) should be verified by flow cytometry.

7. Count CD4+ T cells and resuspend at final concentration of $1 \times 10^6/250$ µL of EHAA medium.

8. Pipette 250 µL of CD4+ T-cell suspension into an anti-CD3+ anti-CD28-coated well of 24-well plate.

9. Prepare 3 mL of T_H2 solution, and add 500–750 µL of T_H2 solution to the 250 µL of CD4+ cell suspension in the 24-well plate (*see* **Note 17**).

10. Place 24-well plate inside a 37 °C incubator.

11. *Days* 3–5. Observe cells daily for development of blast morphology (i.e., cells should gradually get larger). If the cells are growing well and appear to be using up the medium (cell culture medium will appear cloudy and/or more yellow in color), replace with fresh T_H2 solution, without dislodging the cells at the bottom of the wells. We do not recommend disturbing the cells for counting at this point in the differentiation process.

12. *Day* 6 *or* 7. Five to six days after initial stimulation, collect cells into a 50 mL tube. Fill tube with sterile 1× HBSS and centrifuge at $220 \times g$ for 10 min at 4 °C. Resuspend cell pellet in 2–5 mL of sterile 1× HBSS and count cells.

13. Fill tube with sterile 1× HBSS and centrifuge as in **step 12** to pellet cells. Resuspend pellet at 2×10^5 cells/mL in EHAA medium (supplemented as in Subheading 2.3) and transfer cells to a 6-well plate (for volumes of 4 mL or less) or a T-25 tissue culture flask (for volumes >4 mL).

14. Incubate cells at 37 °C for additional 2–3 days to further expand the Th2 cells to yield the number of cells required for experiments. Count cells daily and maintain at a density in the range of 1–5×10^5 cells/mL by adding fresh EHAA medium (supplemented as described in Subheading 2.3) to support the growing population of cells. As the cell population expands, cells can be transferred to larger volume tissue culture flasks (*see* **Note 18**).

3.8 Anti-CD3⁺ Anti-CD28 Stimulation of CD4⁺ T_H2 Cells to Visualize POLKADOTS Formation and NF-κB Activation

T_H2 cells behave similarly to D10 T cells, in terms of post-stimulation Bcl10 clustering/POLKADOTS formation and NF-κB activation. Coverslips coated with anti-CD3/anti-CD28 antibody can be utilized to stimulate T_H2 cells and initiate T-cell activation. On day 1, prepare anti-CD3/anti-CD28-coated coverslips by following Subheading 3.5, **steps 1–13**, with the following exception: for **step 10**, pipette 150 µL of anti-CD3 antibody (100 µg/mL) *and* 10 µL of anti-CD28 antibody (1 µg/mL) onto the coverslip. Wash and recover antibodies as detailed in Subheading 3.5. On day 2, collect T_H2 cells growing at optimal density (2–5×10^5 cells/mL). Stimulate T_H2 cells by following Subheading 3.5, **steps 14–18**.

4 Notes

1. D10 cells (D10.G4) are a clone of helper T cells produced from AKR/J mice immunized with conalbumin [20]. The original D10 clone requires cycles of antigen restimulation and rest, like most T-cell clones. We use a D10 subclone that proliferates continuously in IL-2 (and requires 1–2 ng/mL IL-2 for survival). This subclone, sometimes referred to as "D10-IL2," retains sensitivity to TCR stimulation, but does not require antigen stimulation for in vitro proliferation. These T cells grow well when kept at optimum density of 1–5×10^5 cells/mL. Cells should be split to 1×10^5 cells/mL the day prior to retroviral infection or use in stimulation experiments, yielding ~2×10^5 cells/mL on the day of the experiment. D10 T cells allowed to overgrow will neither infect efficiently nor stimulate efficiently. We have found that the behavior of this clone can become heterogeneous after many weeks of continuous culture (e.g., sensitivity to TCR stimulation can be reduced). For consistent results, discard D10 cell cultures after 2–3 months of continuous growth, and thaw out a fresh batch.

2. CH12 and CHb cells grow well when maintained at densities similar to D10 cells. Do not add IL-2 to CH12 or CHb culture medium, as IL-2 is toxic to these cell lines. Note that other I-A^k- or I-A^b-expressing APCs can be substituted for CH12 and CHb cells, respectively.

3. Permeabilization and blocking solutions can be prepared in advance and stored for months at 4 °C.

4. Prepare PPD and mounting medium using the following protocols. *Important: PPD is toxic, avoid contact or inhalation. It is also light sensitive. Wrap the vial in aluminum foil to extend the shelf life.* After obtaining a new bottle of PPD, dissolve the entire contents in 10× PBS to a final concentration of 20 mg/mL. Aliquot in 1.0 mL aliquots, and freeze at −80 °C. This solution will be purple, and it requires ~1 h for PPD to solubilize (covered in foil on a rotator). PPD precipitates from solution following thawing. Thus, place thawed PPD in a foil-covered tube on a rotator until it all goes back into solution (~20 min). For preparing the mounting solution, aliquot 17 mL of glycerol (>99 % pure) in a 50 mL conical tube with a clean stir bar. Add 1.0 mL of 20 mg/mL of PPD (in 10× PBS) to the vial. Put the mounting solution on a rotator (slow speed) to mix (a stir bar facilitates mixing); cover with foil and mix for 5–10 min. Add 1.0 mL of carbonate-bicarbonate buffer (*see* below) to the mounting solution. Mix on the rotator. Add Buffer A (*see* below) 100 µL at a time (mix well!), until the pH reaches about 8.5. *Important:* To check pH, mix 100 µL of mounting medium with about 1 mL of ddH$_2$O in a 1.5 mL tube, and check the pH with appropriately scaled pH paper. Undiluted mounting medium isn't absorbed well by the paper, making accurate pH determination difficult. The final color of the mounting solution will be pale orange. At neutral pH, PPD does not prevent bleaching, and there will be substantial fading of fluorescence of your stained samples. Aliquot into 1.5 mL tubes and store in −70 °C freezer in a light-tight box. Store working aliquots at −20 °C. Warm *briefly* in a 37 °C water bath immediately before use. PPD will rapidly degrade if kept at ambient or higher temperatures for extended lengths of time. When PPD becomes dark orange in color, replace with a new aliquot. Buffer A: 0.2 M anhydrous sodium carbonate (2.12 g/100 mL). Buffer B: 0.2 M sodium bicarbonate (1.68 g/100 mL). For carbonate-bicarbonate buffer, mix 4 mL of Buffer A with 46 mL of Buffer B and bring volume to 200 mL with ddH$_2$O. pH will be ~9.

5. Cells should be ~50–60 % confluent on the day the cells are passaged into 6-well plates. One T-175 flask at this density yields ~2×10^7 total cells. Splitting HEK293T cells for 2 days in a row prior to day 1 can further increase transfection efficiency. Cells should not be allowed to reach confluence during the days prior to transfection, as this will greatly reduce transfection efficiency. A simple method to determine transfection efficiency is to transfect HEK293T cells with a GFP expression plasmid under control of the CMV immediate early promoter

(e.g., pcDNA3-GFP), followed by flow cytometry analysis for GFP fluorescence. GFP-expressing cells should be compared to control cells transfected with empty vector.

6. Collected poly-D-lysine can be reused several times without significant loss of activity.

7. We find it easier to make a master mixture of a total volume of 1,050 µL (150 µL per well for a total of 6 wells of a 6-well plate, plus an extra 150 µL to account for pipetting error).

8. At 4 °C, MMLV ecotropic retroviral supernatants have a half-life of approximately 2 weeks. Note that viral supernatants used immediately after harvest or within the first 72 h of harvest will likely contain viable HEK293T cells, which can survive the antibiotic selection. For this reason, viral supernatants should not be used until they have been stored for >72 h at 4 °C (or frozen and thawed). For freezing retrovirus, snap-freeze small aliquots (~1.5 mL) in 2 mL cryotubes, using pulverized dry ice. Virus remains viable with minimal loss of titer when frozen in this manner and stored at –80 °C. One freeze-thaw cycle reduces titer by twofold. Multiple freeze-thaw cycles should not be performed, as this vastly diminishes titer. To double the amount of virus supernatant recovered, add 2 mL of pre-warmed (37 °C) DMEM to transfected HEK293T cells immediately after the first harvest and collect the supernatant 24 h later. This second harvest generally has a viral titer similar to the first harvest (within a factor of 2). To empirically determine a viral titer, the following method can be used with viruses encoding an insert that expresses a fluorescent protein fusion: Plate NIH/3T3 cells at 1×10^5 cells/well in a 6-well plate (with 2 mL complete DMEM per well). For each virus to be titered, use 1–3 wells. Include two extra wells for obtaining cell counts (expect $2–3 \times 10^5$). Approximately 24 h later, trypsinize the two extra wells and count. Record the cell number. For the remaining wells of NIH/3T3 cells, add Polybrene to a 1× final concentration (10 µg/mL). To each well, add one viral supernatant to a final dilution of 1:500 or 1:1,000. Return cells to incubator for ~40 h. Harvest cells via trypsinization (as described for HEK293T cells in Subheading 3.2). Determine the percentage of GFP-expressing cells in the population via flow cytometry. Calculate titer by multiplying the average number of NIH/3T3 cells in the two counted wells × viral supernatant dilution × percentage of GFP-positive cells. For example, if there were 2.5×10^5 cells per well, the viral supernatant was diluted 1:500, and the final percentage of positive cells by FACS was 4 %, the titer would be $(2.5 \times 10^5) \times (500) \times (0.04) = 5 \times 10^6$ infectious particles/mL. A good retroviral titer using the described methods is 1×10^7 particles/mL. Note that the effective titer on D10 T cells and primary T cells is much lower than on NIH/3T3 cells.

9. This protocol outlines steps required to observe POLKADOTS formation at 20 min post-TCR stimulation. For longer periods of antigen stimulation (>1 h), it is advisable to raise the medium volume to 1 mL for the water bath incubation to promote cell viability (this "extra" medium is then aspirated at the end of the incubation period to leave 200 μL to plate on the coverslip). When increasing incubation times, note that time of plating on poly-D-lysine coverslips is always ~10 min prior to the end of the stimulation period. Incubating T-cell-B-cell conjugates on poly-D-lysine coverslips for >10 min tends to result in cells crawling across coverslips, leaving membrane fragments that produce considerable background fluorescence. If the goal is to visualize early POLKADOTS and/or POLKADOTS-RelA association, cells must be fixed at 10 min. Thus, cells are immediately plated on coverslips after centrifugation to form T-cell-APC conjugates and fixed following 10 min in the 37 °C incubator.

10. This procedure creates a hydrophobic surface that helps minimize the volume of blocking and antibody solutions that are added on top of coverslip in the next steps. Alternatively, paraffin-coated strips can be placed at the bottom of the 6-well plate below the coverslip. Paraffin has hydrophobic properties and helps prevent leakage of blocking and antibody solutions from the top of the coverslip to the area surrounding the coverslip.

11. We find that 200 μL is the optimal volume of blocking or antibody solution to place onto the coverslip to minimize the chance of leakage off the coverslip surface (which reduces the efficiency of antibody labeling). Alternatively, a 150–180 μL volume can be used, if difficulty is encountered in keeping the solution on top of the coverslip.

12. Gently aspirate the solution without touching the base of the coverslip, as that can dislodge the attached cells.

13. The optimal concentration of primary and secondary antibodies will have to be empirically determined. Also, for multiple antibody labeling, the use of primary antibodies from different species (e.g., one mouse, one rabbit) or isotypes (e.g., one mouse IgG_1, one IgG_{2a}) is required. For such experiments, it is critical to purchase polyclonal secondary antibodies that have been cross-adsorbed against antibodies from other species (e.g., in the example above, the anti-mouse antibody should be cross-adsorbed against antibodies from other species, including rabbit, and the rabbit antibody should be cross-adsorbed against antibodies from other species, including mouse).

14. We find it easier to lift the coverslip from the bottom of a 6-well plate if we leave the final PBS wash in the well. This procedure will also keep the cells from drying out when

multiple coverslips are being mounted (drying destroys cell morphology and produces autofluorescence).

15. Slides can be stored for weeks to months at −20 °C with minimum loss of fluorescence.

16. Mouse naïve CD4⁺ T cells do not form POLKADOTS in response to TCR stimulation. Therefore, it is important to differentiate naïve CD4⁺ T cells into effector cells for studies of Bcl10 redistribution into POLKADOTS.

17. The T_H2 solution must be made fresh each time on the day of the experiment.

18. To ensure proper T_H2 differentiation, check for production of T_H2 cytokines (e.g., by performing an intracellular IL-4 stain) after restimulation of the differentiated cells.

Acknowledgments

Supported by grants from the US National Institutes of Health (Al057481 to B.C.S.), Center for Neuroscience and Regenerative Medicine (CNRM) (to B.C.S.), and predoctoral fellowships (to S.P.) from the American Heart Association (10PRE3150039) and the Henry M. Jackson Foundation. The views expressed are those of the authors and do not necessarily reflect those of the Uniformed Services University or the Department of Defense.

References

1. Blum JS, Wearsch PA, Cresswell P (2013) Pathways of antigen processing. Annu Rev Immunol 31:443–473

2. Neefjes J, Jongsma ML, Paul P et al (2011) Towards a systems understanding of MHC class I and MHC class II antigen presentation. Nat Rev Immunol 11:823–836

3. Paul S, Schaefer BC (2013) A new look at T cell receptor signaling to nuclear factor-kappaB. Trends Immunol 34:269–281

4. Smith-Garvin JE, Koretzky GA, Jordan MS (2009) T cell activation. Annu Rev Immunol 27:591–619

5. Thome M, Charton JE, Pelzer C et al (2010) Antigen receptor signaling to NF-kappaB via CARMA1, BCL10, and MALT1. Cold Spring Harb Perspect Biol 2:a003004

6. Sun L, Deng L, Ea CK et al (2004) The TRAF6 ubiquitin ligase and TAK1 kinase mediate IKK activation by BCL10 and MALT1 in T lymphocytes. Mol Cell 14:289–301

7. Vallabhapurapu S, Karin M (2009) Regulation and function of NF-kappaB transcription factors in the immune system. Annu Rev Immunol 27:693–733

8. Dustin ML, Chakraborty AK, Shaw AS (2010) Understanding the structure and function of the immunological synapse. Cold Spring Harb Perspect Biol 2:a002311

9. Monks CR, Freiberg BA, Kupfer H et al (1998) Three-dimensional segregation of supramolecular activation clusters in T cells. Nature 395:82–86

10. Paul S, Kashyap AK, Jia W et al (2012) Selective autophagy of the adaptor protein Bcl10 modulates T cell receptor activation of NF-kappaB. Immunity 36:947–958

11. Rossman JS, Stoicheva NG, Langel FD et al (2006) POLKADOTS are foci of functional interactions in T-Cell receptor-mediated signaling to NF-kappaB. Mol Biol Cell 17:2166–2176

12. Schaefer BC, Kappler JW, Kupfer A et al (2004) Complex and dynamic redistribution of NF-kappaB signaling intermediates in response to T cell receptor stimulation. Proc Natl Acad Sci U S A 101:1004–1009

13. Paul SKA, Traver MK, Kashyap AK et al (2014) TCR signals to NF-κB are transmitted by cytosolic p62-Bcl10-Malt1 signalosome. Sci Signal 13:ra45

14. Haughton G, Arnold LW, Bishop GA et al (1986) The CH series of murine B cell lymphomas: neoplastic analogues of Ly-1+ normal B cells. Immunol Rev 93:35–51

15. Rebeaud F, Hailfinger S, Posevitz-Fejfar A et al (2008) The proteolytic activity of the paracaspase MALT1 is key in T cell activation. Nat Immunol 9:272–281

16. Schaefer BC, Mitchell TC, Kappler JW et al (2001) A novel family of retroviral vectors for the rapid production of complex stable cell lines. Anal Biochem 297:86–93

17. Jordan M, Schallhorn A, Wurm FM (1996) Transfecting mammalian cells: optimization of critical parameters affecting calcium-phosphate precipitate formation. Nucleic Acids Res 24:596–601

18. Naviaux RK, Costanzi E, Haas M et al (1996) The pCL vector system: rapid production of helper-free, high-titer, recombinant retroviruses. J Virol 70:5701–5705

19. Fitch FW, Gajewski TF, Hu-Li J (2006) Production of TH1 and TH2 cell lines and clones. Curr Protoc Immunol Chapter 3:Unit 3.13

20. Kaye J, Porcelli S, Tite J, Jones B, Janeway CA Jr (1983) Both a monoclonal antibody and antisera specific for determinants unique to individual cloned helper T cell lines can substitute for antigen and antigen-presenting cells in the activation of T cells. J Exp Med 158:836–856

Chapter 13

Detection of Recombinant and Cellular MALT1 Paracaspase Activity

Daniel Nagel and Daniel Krappmann

Abstract

MALT1 (mucosa-associated lymphoid tissue protein 1) is a key regulator of antigen-induced NF-κB activation in the adaptive immune response. Activation of proteolytic activity of the MALT1 paracaspase was shown to boost the immune response. Additionally, MALT1 proteolytic activity is essential for the survival of MALT1-dependent lymphoma, such as the activated B-cell type (ABC) of diffuse large B-cell lymphoma (DLBCL) or MALT lymphoma. The functional impact of MALT1 paracaspase on T-cell activation and lymphomagenesis suggests that MALT1 is a promising therapeutic target for the treatment of autoimmune diseases and distinct lymphoma entities. To evaluate the requirement of MALT1 in further detail, direct measurement of its activity status is of great importance. We have established a fluorogenic cleavage assay which can be used to measure activity of recombinant and cellular MALT1. Here we describe the basis of the cleavage assay and include a detailed protocol for recombinant production of MALT1 and also the cellular immunoprecipitation of endogenous MALT1 to determine its proteolytic activity.

Key words MALT1, Cysteine protease, Activity assay, Autoimmunity, Inflammation, Lymphoma

1 Introduction

MALT1 is an essential mediator of antigen-dependent lymphocyte activation. As an adaptor protein, it bridges the CARMA1-BCL10-MALT1 (CBM) complex to canonical IκB kinase (IKK)/NF-κB signaling [1]. The central caspase-like domain confers activity of a cysteine-type protease that is activated upon antigen stimulation in T-cells [2, 3]. Due to its high homology to mammalian caspases, MALT1 was termed "paracaspase" [4], but the substrate specificity and mechanism of activation are quite distinct from caspases [5]. MALT1 paracaspase activity is critical for an optimal T-cell activation [2, 3, 6]. Congruent with this, MALT1 is critical for disease onset in a murine model of multiple sclerosis (EAE, experimental autoimmune encephalomyelitis) [7, 8]. In addition the proteolytic activity of MALT1 is essential for the survival of MALT1-dependent lymphoma like ABC-DLBCL or MALT lymphoma that expresses

Michael J. May (ed.), *NF-kappa B: Methods and Protocols*, Methods in Molecular Biology, vol. 1280,
DOI 10.1007/978-1-4939-2422-6_13, © Springer Science+Business Media New York 2015

the oncogenic fusion protein API2-MALT1 [9–11]. Interference with MALT1 proteolytic activity therefore represents novel promising approach for precision therapy in MALT1-dependent lymphoma and to balance an overshooting immune reaction in autoimmune diseases such as multiple sclerosis or rheumatoid arthritis.

To characterize the pathophysiological role of MALT1 and to better evaluate the possibility of inhibiting MALT1 in certain diseases, it is of importance to be able to measure the activity status of MALT1 in cells. Here we describe a MALT1 cleavage assay that is applicable for high-throughput screenings and has been used to identify first small molecule MALT1 inhibitors [12]. The cleavage reaction is stable and robust and easily applied to measure the activity of recombinant MALT1 and allows the direct evaluation of different inhibiting or activating influences on MALT1 activity. In addition, we have also adapted the system to measure cellular MALT1 activity after enrichment by immunoprecipitation (IP) from cell extracts. In this respect, the assay is reminiscent to the detection of protein kinase activities after IP (e.g., IKKα/β activity), but it does not require the use of radioactivity. This allows the direct evaluation of cellular MALT1 activity in response to genetic or pharmaceutical manipulations of the cells in an isolated milieu independent of individual substrates. Transfection of cells with MALT1 expressing constructs and isolation of the exogenous proteins also allows to analyze the functional impact of MALT1 mutations.

2 Materials

2.1 MALT1 Protein Expression and Purification

1. Prokaryotic expression vector system pGEX-6P-1.
2. *E. coli* strain BL21-CodonPlus (DE3) RIPL (Agilent, Santa Clara, CA, USA).
3. LB medium with 100 μg/ml ampicillin prepared immediately before use.
4. Isopropyl-β-D-thiogalactopyranoside (IPTG) stock solution (1 M) stored at –20 °C.
5. Bacterial lysis buffer: 50 mM HEPES pH 7.5, 150 mM NaCl, 2 mM $MgCl_2$, 10 % glycerin (v/v), 0.1 % Triton X-100 (v/v), 1 mM dithiothreitol (DTT), complete protease inhibitor cocktail tablets (Roche, Madison, WI, USA) stored at 4 °C. Addition of DTT immediately before use.
6. Purification buffer: 50 mM Tris–HCL pH 8.0, 150 mM NaCl.
7. Glutathione elution buffer: 50 mM Tris–HCL pH 8.0, 150 mM NaCl, 15 mM reduced glutathione. Prepared immediately before use.

8. Storage buffer: 20 mM Tris–HCL pH 8.0, 500 mM NaCl, 20 mM β-mercaptoethanol, 2.5 mM EDTA, 10 % saccharose (w/v), 0.1 % CHAPS (w/v). Purified proteins are stored at –80 °C.

9. GSTrap HP columns (GE Healthcare, Mickleton, NJ, USA).

10. Amicon Ultra-4 centrifugal filter unit (Millipore, Billerica, MA, USA) with an exclusion size of 10 kDa.

11. Colloidal Coomassie for detection of purified proteins in an SDS gel.

12. ÄKTA fast protein liquid chromatography (FPLC) system (GE Healthcare).

2.2 Recombinant MALT1 Cleavage Reaction

1. Black nonbinding 384-well microplates.

2. Cleavage buffer: 50 mM MES pH 7.0, 150 mM NaCl, 10 % saccharose (w/v), 0.1 % CHAPS (w/v), 1 M sodium or ammonium citrate, 10 mM DTT. Buffer is stored at RT and DTT is added immediately before use.

3. Mepazine (Chembridge, San Diego, CA, USA) dissolved in DMSO (50 mM) and stored at –20 °C.

4. Ac-LRSR-AMC fluorogenic MALT1 substrate peptide (Peptides International, Louisville, KY, USA) dissolved in DMSO (20 mM) and stored at –80 °C.

2.3 Cell Culture for Analysis of Cellular MALT1 Activity

1. RPMI 1640 medium supplemented with 10 % FBS (v/v) and 100 U/ml penicillin/streptomycin.

2. Mepazine dissolved in DMSO (50 mM) and stored at –20 °C.

3. Phorbol 12-myristate 13-acetate (PMA) and ionomycin stored at –20 °C as 200 ng/ml and 300 ng/ml stock solution in H_2O, respectively.

4. Anti-CD3/CD28 antibodies (Becton Dickinson, Franklin Lakes, NJ, USA) to stimulate the cells via TCR cross-linking.

5. Cellular lysis buffer: 50 mM HEPES pH 7.5, 150 mM NaCl, 2 mM $MgCl_2$, 10 % glycerin (v/v), 0.1 % Triton X-100 (v/v), 1 mM dithiothreitol (DTT), complete protease inhibitor cocktail tablets. DTT is added immediately before use.

2.4 Immuno-precipitation of Endogenous MALT1 and Cellular Cleavage Reaction

1. Anti-MALT1 antibody (Santa Cruz Biotechnologies, Santa Cruz, CA, USA).

2. Protein-G Sepharose (GE Healthcare).

3. Phosphate-buffered saline (PBS) precooled at 4 °C.

4. 26 G syringe for purification of Protein-G Sepharose beads.

5. Black nonbinding 384-well microplates.

6. Cleavage buffer: 50 mM MES pH 7.0, 150 mM NaCl, 10 % saccharose (w/v), 0.1 % CHAPS (w/v), 1 M sodium or

ammonium citrate, 10 mM DTT. Buffer is stored at RT and DTT is added immediately before use.

7. Ac-LRSR-AMC dissolved in DMSO (20 mM) and stored at −80 °C.

---

3 Methods

3.1 Recombinant MALT1 Cleavage Reaction

To measure the activity of recombinant MALT1, we use either the full-length or a truncated protein that harbors at least the complete paracaspase, and C-terminal Ig3 domain as deletion of the Ig-like domain may lead to inactivity of the protein [13]. Mutation of the active site in MALT1 C464A renders the enzyme completely inactive [6].

3.1.1 Protein Expression and Purification

1. Transfer 20 ml of BL21-CodonPlus (DE3) RIPL pre-culture containing a MALT1 expression vector into 1 L LB-amp medium and incubate at 37 °C to an OD_{600} of 0.7–0.8.

2. Cool the culture to ~21 °C and induce the expression of GST-MALT1 by addition of IPTG (50 μM). Incubate the culture overnight at 18 °C (*see* **Note 1**).

3. Harvest the bacteria by centrifugation at $3,000 \times g$ and resuspend the pellet in lysis buffer (7 ml of lysis buffer for each original 1 L of culture) on ice.

4. Sonicate the bacterial suspension on ice for 10× 20 s (*see* **Note 2**).

5. Centrifuge at $20,000 \times g$ (4 °C) for 2× 30 min to clarify the lysate from debris.

6. To bind GST-MALT1 proteins to GSTrap HP columns, apply clarified bacterial lysates via a 10 ml Superloop to an ÄKTA liquid chromatography system.

7. Wash with 7 column volumes to get rid of bacterial proteins and then elute GST-MALT1 proteins with the glutathione elution buffer.

8. Concentrate the proteins with an Amicon Ultra-4 spin column into storage buffer to a final concentration of 2–6 μg/μl (*see* **Note 3**).

9. Check purity by SDS-PAGE and colloidal Coomassie staining.

3.1.2 Cleavage Reaction

1. GST-MALT1 WT or C464A mutant expressed and purified as described above.

2. Mix 100 nM GST-MALT1 (in 18.5 μl of cleavage buffer) with 1 μM mepazine or DMSO (0.5 μl) and add 50 μM Ac-LRSR-AMC (1 μl in cleavage buffer) to a final volume of 20 μl in a 384-well microplate (*see* **Note 4**).

3. Measure the cleavage activity for 2 h at 30 °C in a microplate reader with 360 nm excitation and 460 nm emission (Fig. 1).

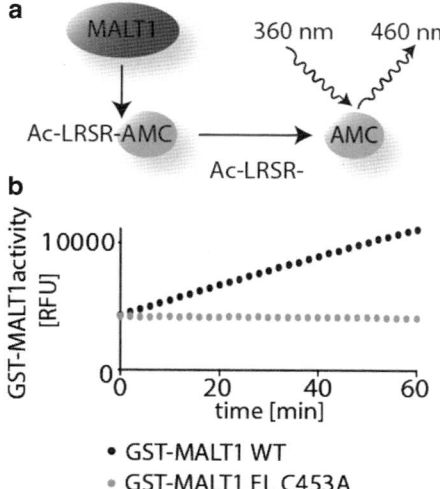

Fig. 1 Cleavage reaction of recombinant MALT1. (**a**) Scheme of the cleavage reaction. (**b**) Result of a 60 min cleavage reaction. MALT1 WT protein leads to a stable fluorescence release, while the catalytic inactive MALT1 C153A mutant (isoform B) is unable to cleave the fluorogenic substrate

3.2 Measurement of Cellular MALT1 Activity

The cleavage reaction can be used to analyze the activity of endogenous MALT1 using cell lines and primary cells. In addition the transfection of cells and expression of exogenous MALT1 may represent a good method to analyze the activity of MALT1 mutants in a cellular context. In this protocol we describe the measurement of immunoprecipitated MALT1 from stimulated Jurkat T-cells.

1. 5×10^6 Jurkat T-cells are left untreated and stimulated with 0.4 PMA and 0.6 μl of ionomycin or anti-CD3/CD28 for at least 30 min to activate MALT1. Cells with constitutive MALT1 activity (e.g., ABC-DLBCL lymphoma cells) can be stimulated to increase MALT1 activity.

2. MALT1 inhibitors such as mepazine can be added 3 h prior to stimulation and cell lysis to monitor their cellular effects.

3. Pellet the cells via centrifugation at $300 \times g$ and lyse with cellular lysis buffer by rotating for 20 min at 4 °C.

4. Centrifuge the cell lysate for 10 min at $20{,}000 \times g$ to get rid of cell debris.

5. Incubate the cell lysate with 700 ng of a MALT1 antibody overnight at 4 °C.

6. Incubate the lysate with 15 μl of protein-G beads for 60 min at 4 °C (*see* **Note 5**).

7. Pellet the beads via centrifugation for 1 min at $400 \times g$ and 4 °C and wash three times with precooled PBS at $400 \times g$ for 3 min.

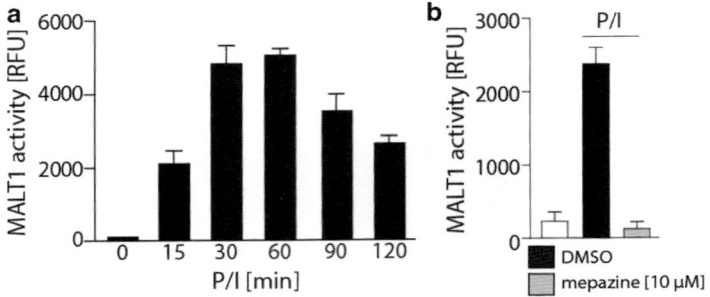

Fig. 2 Measurement of cellular MALT1 activity. (**a**) Kinetic of MALT1 activity after IP of the protein from PMA/ionomycin (P/I)-stimulated Jurkat T-cells. MALT1 activity is induced 15 min after stimulation of the cells and has a maximum of 60 min. For the quantification of the results, the release of fluorescence of a 90 min cleavage reaction was determined and depicted as relative fluorescence units (RFU) in a bar diagram. *Bars* show the mean of at least three independent experiments with SD indicated. (**b**) Inhibition of cellular MALT1 activity with mepazine. Jurkat T-cells were incubated with DMSO or 10 μM mepazine for 3 h and subsequently stimulated with P/I for 30 min. Treatment with mepazine led to a total decline of MALT1 activity compared to DMSO-incubated control cells. *Bars* are showing the mean of at least three independent experiments with SD indicated

8. After the last washing step is completed, completely discard the PBS using a syringe.

9. Gently add 45 μl of cleavage buffer to the beads and pellet the beads with a brief centrifugation step at $400 \times g$.

10. Transfer 49 μl of the beads to a 384-well microplate and add 1 μl of Ac-LRSR-AMC to a final concentration of 20 μM.

11. Measure the cleavage activity for 2 h at 30 °C in a microplate reader with 360 nm excitation and 460 nm emission (Fig. 2).

4   Notes

1. Recombinant MALT1 tends to be insoluble or inactive and sometimes high amounts of truncated proteins are produced. The reason could be the high abundance of leucine in the N-terminus of the protein (~25 % within the first 50 amino acids), which may result in a premature termination of protein translation. To avoid these problems, always produce at low temperature and use low concentrations of IPTG. This should slow the metabolism of the bacteria and subsequently the translation of MALT1.

2. Gently sonicate at the upper part of the suspension but avoid the formation of foam as this will result in a lower amount of soluble protein fraction. It is also important to cool the suspension as it tends to get warm due to the sonication. Therefore it is highly advisable that the whole procedure takes place on ice. In addition it should be done in intervals, e.g., a break of 30 s between each 20 s sonication step. This may be most comfortably achieved by using 2 or more suspension vials. Do not increase sonication steps as this will limit the amount of soluble proteins.

3. It is not overly important to further purify MALT1 via gel filtration as the glutathione in the elution buffer is not interfering with MALT1 activity. Nevertheless it is important to change the buffer during the concentration step. To this end add storage buffer to the glutathione elution buffer to a maximum of 4 ml and concentrate the protein. After the first concentration to 200–300 µl, add another volume of storage buffer up to 4 ml and concentrate MALT1 for the second time. After the final step, make sure to properly flush the centrifugal filter with 100–200 µl of storage buffer to gather residual amounts of the protein.

4. It is important to dilute the protein and the substrate in cleavage buffer. This will allow a proper reaction and also limit the amount of DMSO as MALT1 activity will be impaired with more than 1 % of DMSO in the reaction.

5. For the first application, protein-G sometimes needs to be washed with PBS to get rid of residual ethanol. The final suspension of protein-G we use is diluted 1:3 in PBS.

References

1. Thome M et al (2010) Antigen receptor signaling to NF-kappaB via CARMA1, BCL10, and MALT1. Cold Spring Harb Perspect Biol 2(9): a003004

2. Rebeaud F et al (2008) The proteolytic activity of the paracaspase MALT1 is key in T cell activation. Nat Immunol 9(3):272–281

3. Coornaert B et al (2008) T cell antigen receptor stimulation induces MALT1 paracaspase-mediated cleavage of the NF-kappaB inhibitor A20. Nat Immunol 9(3): 263–271

4. Uren AG et al (2000) Identification of paracaspases and metacaspases: two ancient families of caspase-like proteins, one of which plays a key role in MALT lymphoma. Mol Cell 6(4):961–967

5. Wiesmann C et al (2012) Structural determinants of MALT1 protease activity. J Mol Biol 419:4–21

6. Duwel M et al (2009) A20 negatively regulates T cell receptor signaling to NF-kappaB by cleaving Malt1 ubiquitin chains. J Immunol 182(12):7718–7728

7. Brustle A et al (2012) The NF-kappaB regulator MALT1 determines the encephalitogenic potential of Th17 cells. J Clin Invest 122(12): 4698–4709

8. Mc Guire C et al (2013) Paracaspase MALT1 deficiency protects mice from autoimmune-mediated demyelination. J Immunol 190(6): 2896–2903

9. Hailfinger S et al (2009) Essential role of MALT1 protease activity in activated B cell-like diffuse large B-cell lymphoma. Proc Natl Acad Sci U S A 106(47):19946–19951

10. Ferch U et al (2009) Inhibition of MALT1 protease activity is selectively toxic for activated B cell-like diffuse large B cell lymphoma cells. J Exp Med 206(11):2313–2320

11. Rosebeck S et al (2011) Cleavage of NIK by the API2-MALT1 fusion oncoprotein leads to noncanonical NF-kappaB activation. Science 331(6016):468–472

12. Nagel D et al (2012) Pharmacologic inhibition of MALT1 protease by phenothiazines as a therapeutic approach for the treatment of aggressive ABC-DLBCL. Cancer Cell 22(6): 825–837

13. Hachmann J et al (2012) Mechanism and specificity of the human paracaspase MALT1. Biochem J 443(1):287–295

Chapter 14

TRAF Protein Function in Noncanonical NF-κB Signaling

Anna D. Reichardt, Jose Pindado, Shivam A. Zaver, and Genhong Cheng

Abstract

Nuclear factor-κB (NF-κB) signaling is classified into the canonical and noncanonical pathways. We describe in this chapter the methods used to study the noncanonical pathway, including derivation of primary cells, pathway stimulation, and immunoblotting.

Key words Noncanonical NF-κB, NIK, TRAF3, TRAF2, IKKα, p100

1 Introduction

NF-κB signaling is classified into the canonical and noncanonical pathways [1]. Only a small subset of the tumor necrosis factor receptor superfamily (TNFRSF) is able to activate the noncanonical NF-κB pathway through the release of the p52/RelB dimer. This subset of TNFRSFs includes cluster of differentiation 40 (CD40), lymphotoxin β-receptor (LTβR), B-cell-activating factor receptor (BAFFR), receptor activator of NF-κB (RANK), fibroblast growth factor-inducible 14 (Fn14), and cluster of differentiation 27 (CD27) [2].

The control of NF-κB-inducing kinase (NIK) posttranslational stability is the key locus of noncanonical signaling. NIK is continuously synthesized, but at baseline, its expression level is maintained at low to undetectable levels via proteasome-mediated degradation. Basal NIK degradation is mediated by a TNF receptor-associated factor (TRAF)—cellular inhibitor of apoptosis proteins (cIAP) E3 ubiquitin ligase complex, composed of the proteins TRAF2, TRAF3, and cIAP1/2. TRAF3 acts as the substrate-binding subunit and TRAF2 as a linker, connecting TRAF3/NIK to cIAP1/2, which in turn mediate the K48-linked polyubiquitination of NIK that targets it for degradation (Fig. 1).

Michael J. May (ed.), *NF-kappa B: Methods and Protocols*, Methods in Molecular Biology, vol. 1280,
DOI 10.1007/978-1-4939-2422-6_14, © Springer Science+Business Media New York 2015

Upon receptor ligation, though, this process of constitutive NIK degradation is blocked. Receptor ligation triggers the recruitment of the NIK-containing complex to the receptor, where TRAF2-mediated K63-linked polyubiquitination of cIAP1/2 redirects the K48-ubiquitin ligase activity of cIAP1/2 toward TRAF3 instead of NIK. This induced degradation of TRAF3 releases NIK from the TRAF-cIAP1/2 complex, allowing accumulation of de novo synthesized NIK (Fig. 1) [6, 7].

Once allowed to accumulate, NIK undergoes phosphorylation at key residues in its activation loop. NIK then phosphorylates IKKα, which phosphorylates p100, marking p100 for partial proteasomal digestion and releases of the p100 N-terminus, p52, in complex with RelB [3–9].

In addition to its role propagating noncanonical NF-κB activation, IKKα also functions as a negative feedback regulator. Activated IKKα directly phosphorylates NIK; this phosphorylation results in NIK destabilization and serves to inhibit excessive accumulation of NIK following receptor ligation (Fig. 1) [10]. This IKKα-dependent negative regulation of NIK stability functions independently of the TRAF-cIAP complex that regulates NIK levels in unstimulated cells.

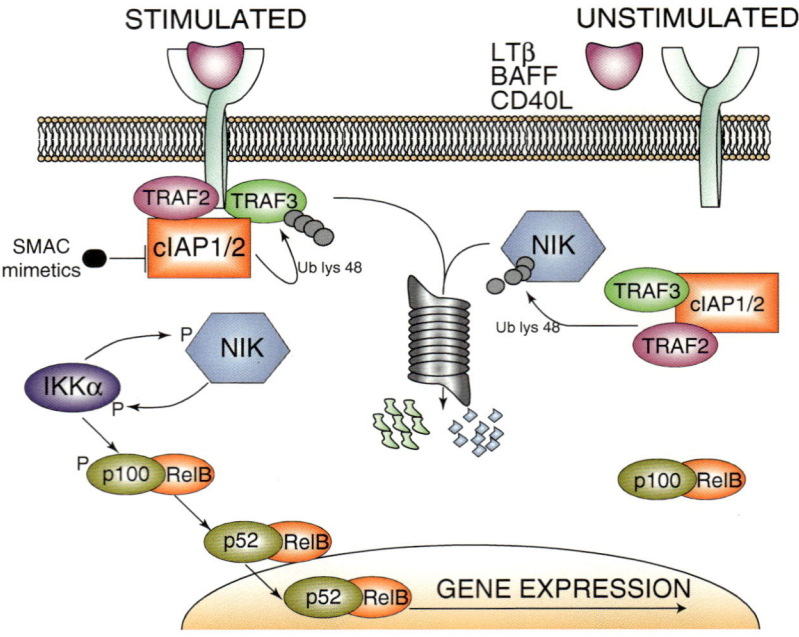

Fig. 1 Schematic of the noncanonical NF-κB pathway. The unstimulated condition is shown on the *right*: the TRAF/cIAP ubiquitin ligase complex tags NIK with K48-linked polyubiquitin, NIK is degraded, and p100 is not processed. The stimulated condition is shown on the *left*: the TRAF/cIAP ubiquitin ligase complex is recruited to the activated receptor, cIAP1/2 switch their K48 polyubiquitin ligase activity toward TRAF3 and TRAF3 is degraded, newly synthesized NIK accumulates and activates IKKα, IKKα phosphorylates p100, and p100 is partially degraded into p52, releasing the p52/RelB dimer

Physiological functions of the noncanonical pathway include secondary lymphoid organogenesis and architecture organization, thymic epithelial cell differentiation, B-cell maturation and survival, dendritic cell (DC) maturation, and bone metabolism [11–15].

The hyperactivation of noncanonical signaling has been linked to the etiopathogenesis of certain autoimmune pathologies (including systemic lupus erythematosus) and hematologic cancers (such as multiple myeloma) [16, 17]. In particular, the current understanding of negative feedback regulation is quite limited, and it is sure that new components and regulatory mechanisms of the noncanonical NF-κB pathway will be discovered. It is hoped that an improved understanding of the pathway's activation and feedback regulation will someday permit the modulation of noncanonical NF-κB signaling for therapeutic purposes. We describe in this chapter the methods used to study the noncanonical pathway, including derivation of primary cells from mice, pathway stimulation, and immunoblotting to detect activation of the pathway.

2   Materials

2.1   Generation of Murine Embryonic Fibroblasts (MEFs)

1. Isoflurane.
2. 70 % ethanol.
3. Sterile PBS.
4. 0.25 % Trypsin-EDTA (Life Technologies, Carlsbad, CA, USA).
5. Dulbecco's Modified Eagle's Medium (DMEM).
6. Fetal bovine serum (FBS).
7. Penicillin and streptomycin (Pen/Strep) (100 U/ml each).
8. MEF culture medium: DMEM, supplemented with 10 % FBS and 1 % Pen/Strep, filter-sterilized.

2.2   B-Cell Isolation

1. 30- and 70-μm cell strainers.
2. ACK Lysing Buffer (Life Technologies).
3. β-Mercaptoethanol (β-ME).
4. MACS (Magnetic Cell Sorting) column and separator (Miltenyi Biotec, San Diego, CA, USA).
5. An antibody cocktail for immune depletion consisting of biotinylated anti-CD43, anti-CD4, and anti-Ter-119 (Miltenyi Biotec).
6. Anti-biotin microbeads, beads conjugated with a monoclonal anti-biotin antibody (Miltenyi Biotec).
7. MACS Buffer: PBS without Ca^{2+} or Mg^{2+} supplemented with 5 % FBS and 2 mM EDTA; filter-sterilized.
8. B-cell culture medium: RPMI 1640 supplemented with 10 % FBS, 50 μM β-ME, and 1 % Pen/Strep. Filter-sterilized.

2.3 Bone Marrow-Derived Macrophage (BMDM) Isolation and Culture

1. Wash medium: RPMI 1640, supplemented with 5 % FBS.

2. BMDM culture medium: DMEM, supplemented with 10 % FBS and 20 % L929-conditioned medium; filter-sterilized.

3. L929 culture medium: DMEM supplemented with 10 % FBS; filter-sterilized.

4. L929-conditioned medium generated as described in Subheading 3.3.2. The L929-conditioned medium provides a source of M-CSF required for BMDM growth.

2.4 Cell Stimulation

1. Anti-LTβR antibody (AdipoGen, San Diego, CA, USA).

2. Proteasome inhibitor MG132 (Calbiochem, Billerica, MA, USA).

3. Anti-CD40 antibody (Enzo Life Sciences, Farmingdale, NY, USA).

4. Recombinant Mouse CD40 ligand (R&D Systems, Minneapolis, MN, USA).

5. SMAC mimetic compound. We use LBW242 generously provided by Novartis Pharmaceuticals.

6. Recombinant human BAFF (R&D Systems).

7. Recombinant Mouse RANK ligand (R&D Systems).

8. MEF stimulation medium: DMEM.

9. B-cell stimulation medium: RPMI, supplemented with 2 % FBS + 50 μM β-ME.

10. BMDM stimulation medium: DMEM, supplemented with 2 % FBS.

2.5 Western Blotting Reagents

1. Protein assay kit. We use the Pierce Bicinchoninic Acid (BCA) Protein Assay Kit (Thermo Scientific, Waltham, MA, USA).

2. SDS-PAGE handcast system or a commercially available precast SDS-PAGE gel.

3. Vertical electrophoresis system.

4. Tank-style gel transfer system.

5. Polyvinylidene difluoride (PVDF) membranes.

6. TBST: Tris-buffered saline (TBS: 150 mM Tris–HCl, pH 7.5; 150 mM NaCl) containing 0.1 % Tween-20.

7. Nonfat dry milk or an alternative blocking agent, such as a BSA protein solution.

8. Chemiluminescent substrate. We use SuperSignal West Pico Chemiluminescent Substrate (Thermo Scientific).

9. Anti-NIK rabbit antibody (Cell Signaling Technology, Danvers, MA, USA).

10. Anti-NF-κB2 p100/p52 antibody (Cell Signaling Technology).

11. Anti-RelB antibody (Cell Signaling Technology).

12. Anti-TRAF3 antibody (Santa Cruz Biotechnology, Santa Cruz, CA, USA).

2.6 Buffers and Solutions

1. Blocking solution: 5 % milk in TBST.

2. RIPA Buffer: 50 mM Tris–HCl pH 7.5, 150 mM NaCl, 1 % NP40, 0.5 % deoxycholic acid, 0.1 % SDS.

3. Buffer A: 10 mM HEPES, 1.5 mM MgCl$_2$, 10 mM KCl, 0.5 mM DTT, 0.05 % NP40, pH 7.9.

4. Buffer B: 5 mM HEPES, 1.5 mM MgCl$_2$, 0.2 mM EDTA, 0.5 mM DTT, 26 % glycerol (v/v), 0.4 M NaCl, pH 7.0.

5. SDS loading buffer, 2× stock: 100 mM Tris–Cl (pH 6.8), 4 % (w/v) SDS, 20 % (v/v) glycerol, 0.2 % (w/v) bromophenol blue, 200 mM β-ME (add β-ME directly before use). Store stock at room temperature; after adding β-ME, you may prepare and store aliquots at –20 °C.

2.7 Protease and Phosphatase Inhibitors

For protease and phosphatase inhibitors, add stocks to a final concentration of 1 mM PMSF, 1 mM Na-orthovanadate, 10 mM NaF, and 1 mM DTT. For each 10 ml of buffer B, add one protease inhibitor tablet. Crush tablet and mix well so that tablet dissolves completely. Store at 4 °C. Buffer may be reused for up to 2 weeks after preparation (without adding additional protease/phosphatase inhibitors); with the caveat, that fresh PMSF must be added each time the buffer is reused. After 2 weeks, prepare fresh buffer with new protease/phosphatase inhibitors:

1. Complete Mini Protease Inhibitor (PI) Cocktail Tablets (Roche Diagnostics, Madison, WI, USA).

2. 100 µM phenylmethylsulfonyl fluoride (PMSF) stock; prepare in ethanol and store at –20 °C.

    Keep solution wrapped in aluminum foil as PMSF is light sensitive.

3. 1 M dithiothreitol (DTT) stock prepared in ddH$_2$O and stored in aliquots at –20 °C.

4. 0.5 M Na-fluoride (NaF) stock prepared in ddH$_2$O, and store at 4 °C. Keep wrapped in aluminum foil as NaF is light sensitive.

5. 100 mM Na-orthovanadate stock prepared in a fume hood in ddH$_2$O. Use HCl to adjust pH to 9.0, boil until colorless, and then cool to room temperature. Repeat cycle until solution remains at pH 9.0 after boiling and cooling. Use ddH$_2$O to bring up to initial volume. Store in aliquots at –20 °C.

3   Methods

3.1   Generation Murine Embryonic Fibroblasts (MEFs)

3.1.1   Primary MEF Isolation (See **Notes 1** *and* **2***)*

1. Set up timed pregnancies in mice, and harvest fetuses between gestation days 14 and 15 (*see* **Note 3**).

2. Euthanize mice using an approved institutional euthanization protocol (*see* **Note 4**).

3. Sterilize the entire mouse with 70 % ethanol and then cut open the peritoneal cavity (*see* **Notes 5** and **6**).

4. Dissect out the uterus and then cut through the uterine wall to expose individual embryos.

5. Dissect fetuses from the uterus, removing extrauterine/placental tissue. Place each individual fetus into a 60-mm culture dish containing 3 ml of PBS.

6. Remove the heart, liver, and head from each fetus (*see* **Note 7**). Wash thoroughly with PBS to remove as much blood as possible.

7. Transfer individual carcasses to a fresh 60-mm culture dish containing ~2–3 ml of 0.25 % Trypsin/EDTA. Use dissection scissors/scalpel to cut/shred embryos into small pieces.

8. Incubate for 3 h, rocking dishes every 35–60 min (*see* **Note 8**).

9. Add 10 ml of MEF culture medium to each dish; transfer suspension into a 15-ml tube. Break up digested tissues into suspension by vigorous pipetting.

10. Incubate for 15 min, allowing large pieces to sediment. Directly plate the supernatant into one 10-cm dish per embryo. Add an additional 5 ml of MEF culture medium to the tube (containing sedimentation); break up any clumps by pipetting.

11. Repeat the 15-min sedimentation and combine the supernatant into the original 10-cm dish. These are primary MEF cells labeled passage 0 (p0).

3.1.2   Primary MEF Cell Culture

1. Incubate freshly isolated MEFs for 24 h. Cells are now ready to be maintained and expanded in culture or to be frozen for later use. This protocol is to maintain and expand primary MEF cells in culture. Volumes listed are for 10-cm plates. These volumes can be scaled up or down as needed.

2. Aspirate the medium. Use 10 ml of PBS to wash and remove any non-adherent cells and then repeat this wash.

3. Add 10 ml of fresh MEF culture medium to plate, and culture the cells to near confluence. Cells are now designated p1.

4. Aspirate the medium and then wash twice with 10 ml of PBS. Add 3 ml of 0.05 % Trypsin-EDTA; incubate for 5 min at 37 °C.

5. Add 5 ml of MEF culture medium to resuspend the cells, and transfer to a 15-ml tube.

6. Centrifuge the cells for 5 min at $400 \times g$ at room temperature (RT). Aspirate the supernatant and resuspend cells into MEF culture medium. Split into new 10-cm plates at a ratio of 1:5.

7. Culture the cells to near confluence and then split again following **steps 2–6** (*see* **Note 9**).

3.2 Splenic B-Cell Isolation

3.2.1 Preparation of Splenic Single-Cell Suspension (See **Notes 1** and **10**)

1. For each mouse/spleen that will be collected, prepare a separate 10-cm plate, adding to each plate a 70-μm cell strainer and 8–10 ml of B-cell medium (*see* **Note 11**).

2. Use 6-week- to 6-month-old mice. Euthanize mice with isoflurane (*see* **Note 4**).

3. Sterilize mouse with 70 % ethanol. Cut 1-in. incision into the peritoneal cavity along the left side of the mouse, halfway between the front and back legs (approximate).

4. Grasp spleen with forceps, and pull it free from connective tissue; place spleen into cell strainer.

5. Make several cuts into spleen. With plunger of 6-ml syringe, use circular motions to press the spleen pieces against the bottom of the plate (through the cell strainer). Macerate spleen until mostly fibrous tissue remains. Rinse cell strainer at regular intervals with B-cell medium.

6. Transfer suspended cells into a 15-ml tube (*see* **Note 12**). Pellet cells for 10 min at $300 \times g$ at 4 °C. Aspirate the supernatant; resuspend cells into 5 ml of ACK lysis buffer, and incubate for 5 min at room temperature.

7. Add 10 ml of B-cell medium. Pellet cells for 10 min at $300 \times g$ at 4 °C. Aspirate the supernatant, resuspend the pellet into 20 ml of B-cell medium, and pellet cells 10 min at $300 \times g$ at 4 °C.

8. Aspirate the supernatant, and resuspend the pellet into 3 ml of B-cell medium. Cells are ready to count (*see* **Note 13**).

3.2.2 Magnetic Labeling and Separation

We use the Miltenyi Biotech B-cell isolation kit. In this protocol, non-B cells are depleted from single-cell suspension to yield an enriched B-cell population. Non-B cells (such as T cells, NK cells, dendritic cells, and macrophages) are labeled using a cocktail of biotin-conjugated antibodies against CD43, CD4, and Ter-119. These non-B cells will be retained within the column that contains beads that are pre-conjugated to anti-biotin antibodies, whereas the B cells will pass through the column:

1. Start with a single-cell suspension prepared from the spleen, as described above.

2. Pellet for 10 min at $200 \times g$ at 4 °C. Aspirate the supernatant, and resuspend the pellet into MACS buffer (*see* **Note 14**). Pass the cells through 30-μm filter and then rinse the filter with additional 1 ml of MACS buffer (*see* **Notes 15** and **16**).

3. Pellet the cells for 10 min at $200 \times g$ at 4 °C. Aspirate the supernatant, and resuspend the pellet into MACS buffer with 40 μl of buffer per 10^7 total cells.

4. Add 10 μl of the biotin-antibody cocktail per 10^7 total cells. Mix well and incubate for 5 min in a refrigerator (2–8 °C) (*see* **Note 17**).

5. Add another 30 μl of MACS buffer per 10^7 total cells. Add 20 μl of anti-biotin microbeads per 10^7 total cells. Mix well and incubate for 10 min in a refrigerator (2–8 °C).

6. Add MACS buffer to a minimum total volume of 500 μl. Cells are now ready for magnetic separation. Based on total cell count, choose the appropriately sized column (MS, LS, or XS) and proper MACS separator. This protocol is written specifically for use with the MS column; Subheading 4 contains specific details regarding the use of the LS column (*see* **Note 18**).

7. Place MS column into the magnetic field of the MACS separator. Rinse the column with 500 μl of MACS buffer (*see* **Note 19**). Wait for the column reservoir to empty.

8. Apply the cell suspension onto the column; collect the flow-through that are the unlabeled cells and represent the enriched B-cell population (*see* **Note 20**). Wait for the column reservoir to empty.

9. Wash the column with 500 μl of MACS buffer, collect the flow-through, and combine with the flow-through collected in the previous step (*see* **Note 19**).

10. Pellet the combined flow-through for 10 min at $200 \times g$ at 4 °C. Aspirate that supernatant and resuspend the cells into B-cell medium. These are purified B cells and are ready to use for stimulation (*see* **Note 21**). Count the cells and save a minimum of two million for later FACS analysis.

3.3 BMDM Cell Isolation

3.3.1 BMDM Isolation and Culture (See **Note 1***)*

1. Euthanize a 4–12-week-old mouse (*see* **Note 4**).

2. Sterilize the abdomen and hind legs with 70 % ethanol. Make an incision along the midline of the abdomen, pulling outward to expose the hind legs. Peel the skin from the top of each hind leg down over the foot. Cut off and discard the foot. Cut each hind leg off at hip joint. Be sure to leave the femur intact.

3. Collect the femur and tibia from each hind leg. Remove as much muscle tissue as possible, and sever the leg bones proximal to each joint.

4. Insert a 27-G needle into each bone marrow cavity. Fill a 10-ml syringe with wash medium, and flush out the bone marrow into a 50-ml tube on ice. Continue to flush with wash medium (approximately 2–5 ml total) until the bone marrow cavity appears white.

5. Pipette up and down several times to bring the cells into suspension. Pellet the cells for 10 min at $400 \times g$ at room temperature (RT). Aspirate the supernatant; resuspend the pellet in 1 ml of ACK buffer, and incubate for 5 min at RT (*see* **Note 22**).

6. Pellet for 10 min at $400 \times g$ at RT. Aspirate the supernatant and resuspend the cells in 10 ml of wash medium.

7. Repeat **step 6**.

8. Pellet for 10 min at $400 \times g$ at RT and then aspirate the supernatant.

9. Resuspend the cells in culture medium and then count and plate into 6-well dishes at a concentration of approximately 1×10^6 cells/ml (*see* **Note 23**).

10. Incubate for 48 h in a humidified 37 °C, 10 % CO_2 incubator. Wash cells twice with PBS and add fresh BMDM culture medium (*see* **Note 24**).

11. Incubate for 48 h in a humidified 37 °C, 10 % CO_2 incubator. Wash cells twice with PBS and then add fresh BMDM culture medium.

12. Incubate in a humidified 37 °C, 10 % CO_2 incubator. Approximately 5–6 days after isolation, the dish will contain purified adherent macrophages (*see* **Note 25**).

3.3.2 Preparation of L929-Conditioned Medium

1. Culture L929 cells in L929 medium. When almost confluent, passage cells 1:3.

2. Split cells (1:3) into 15-cm plates and then add 25 ml of L929 medium per plate.

3. Incubate for up to 7 days in a humidified 37 °C, 10 % CO_2 incubator. Maintain culture without passaging or changing medium even after cells reach confluence. Supernatant will be ready to harvest 3–4 days after the cells reach confluence (*see* **Note 26**).

4. Collect and filter the supernatant through 0.45-μm filter.

5. Aliquot into 50-ml conical tubes, and store at –20 °C. This is now L929-conditioned medium (*see* **Note 27**).

3.4 Activation of Noncanonical NF-κB in MEFs (See **Notes 1** *and* **28***)*

We use two ligands to activate the noncanonical NF-κB pathway in MEF cells: an agonist antibody against LTβR (used at 2 μg/ml) and TWEAK (used at 500 ng/ml). The same stimulation protocols and time courses are used for both α-LTβR and TWEAK. We also frequently use second mitochondria-derived activator of caspases (SMAC) mimetics to activate the noncanonical NF-κB pathway. SMAC mimetics are small molecules that induce proteasomal degradation of cIAP1/2; in the absence of cIAP1/2, NIK is not targeted for K48-linked polyubiquitination. A number of SMAC mimetic compounds are available, but we use the SMAC mimetic LBW242 provided by Novartis at 1 µM. We use the same stimulation protocol and time courses that we use for α-LTβR and TWEAK. The proteasome inhibitor MG132 can be used to induce the accumulation of NIK by preventing its basal degradation by the proteasome. MG132 does not activate the noncanonical pathway as it blocks the processing of p100 to p52. A shorter time course should be used when incubating cells with MG132 as prolonged treatment causes cell death. We treat MEFs with MG132 at 10 µM, for 3 h. Finally, the loss of genes encoding proteins required for basal noncanonical NF-κB pathway regulation such as TRAF2, TRAF3, and cIAP1/2 causes basal activation of noncanonical NF-κB. TRAF3$^{-/-}$, TRAF2$^{-/-}$, and cIAP1/2$^{-/-}$ mice have been generated and characterized, and we have used these mice to generate knockout MEF cell lines. TRAF3$^{-/-}$, TRAF2$^{-/-}$, and cIAP1/2$^{-/-}$ MEF cells display basal activation of the noncanonical NF-κB pathway in the absence of stimulation:

1. Plate MEFs into 6-well tissue culture plates at a density of approximately 70 % confluence. Incubate cells in complete MEF cell culture medium (*see* **Note 29**).

2. Incubate for at least 12 h in a humidified 37 °C, 10 % CO_2 incubator, to allow cells to adhere firmly to the culture plate.

3. Aspirate the complete cell medium, and replace with MEF stimulation medium. We recommend using 1 ml of stimulation medium for each well of the 6-well plate (*see* **Notes 30** and **31**).

4. Stimulate MEFs using the ligands or stimuli described above. To detect noncanonical NF-κB activation, we recommend using a stimulation time course of multiple time points that extends for at least 6 h. During the stimulation time course, incubate in a humidified 37 °C, 10 % CO_2 incubator (*see* **Notes 32** and **33**).

5. When the stimulation time course is complete, place the cell culture plate onto ice. Aspirate the stimulation medium and wash cells twice with 1 ml of PBS (4 °C) (*see* **Notes 34** and **35**).

6. Add 1 ml of cold PBS to each 6-well plate. Keeping the plates on ice, harvest cells by mechanical scraping (*see* **Note 36**). Using a p1000 pipette aid, transfer the cells to a prechilled 15-ml tube.

7. Pellet for 4 min at $300 \times g$ at 4 °C, then aspirate the supernatant, and resuspend the pellet in 1 ml of cold PBS.

8. Transfer the cells to prechilled 1.7-ml Eppendorf tubes. Pellet for 4 min in a tabletop centrifuge at maximum speed at 4 °C. Aspirate the supernatant. The cell pellet is now ready to be lysed, or it can be stored at −80 °C if the experiment requires only the analysis of whole-cell extracts (*see* **Notes 37** and **38**).

9. Resuspend the pellet in 50 µl of prechilled RIPA buffer containing protease and phosphatase inhibitors (*see* **Note 39**). Incubate for 10 min on ice vortexing every 2 min. Incubate an additional 10 min on ice.

10. Spin for 15 min in tabletop centrifuge at maximum speed at 4 °C. Transfer the supernatant into fresh Eppendorf tubes. This is the whole-cell lysate (WCL). Proceed to the immunoblotting protocol described in Subheading 3.7.1.

3.5 Activation of Noncanonical NF-κB in Primary B Cells (See Note 1)

We use either BAFF (100 ng/ml), an agonist antibody against CD40 (5 µg/ml) or the SMAC mimetic LBW242 (1 µM) to activate the noncanonical NF-κB pathway in primary B cells. We use the same stimulation protocol and time course for these as described for MEFs above (Subheading 3.4). As with MEF cells, we also use MG132 at 10 µM, for 4 h to stabilize NIK. In addition to primary MEFS, it is also possible to activate the noncanonical pathway in B-cell lines: for example, we stimulate CH12.LX B cells using BAFF at 100 ng/ml, and we stimulate A20 cells with the agonist antibody against CD40 (5 µg/ml):

1. Transfer the cell culture into 50-ml tubes, and pellet for 5 min at $300 \times g$ at RT.

2. Aspirate the supernatant and resuspend the cells in 1 ml of B-cell stimulation medium. Pipette up and down to remove any cell clumps. Add a sufficient volume of additional B-cell stimulation medium such that cells are at a countable density. Gently mix the cells and then remove an aliquot to count (*see* **Note 40**).

3. Adjust the concentration of B cells with stimulation medium to a target of 2×10^6 B cells/ml. You will need 2×10^6 B cells for each time point (*see* **Note 41**).

4. For each time point, add 1 ml of the B-cell suspension to a 12-well tissue culture dish.

5. Stimulate the cells using the ligands or stimuli described above. As with MEFs (Subheading 3.4) to detect noncanonical NF-κB activation, we recommend using a relatively long stimulation time course that extends to a minimum of 6 h. During cell stimulation, incubate the cells in a humidified 37 °C, 10 % CO_2 incubator.

6. When time course is complete, place the cell culture plate onto ice. Transfer the cells to a prechilled 1.7-ml Eppendorf tube on ice pipetting up and down to ensure that any cells that settled during the stimulation time course are now resuspended (*see* **Note 34**).

7. Pellet for 4 min at $500 \times g$ at 4 °C. Aspirate the supernatant and resuspend the pellet in 1 ml of ice cold PBS.

8. Repeat **step 7**

9. Pellet for 4 min at $500 \times g$ at 4 °C and aspirate the supernatant. The cell pellet is now ready for lysis or can be stored at −80 °C if you plan to generate whole-cell lysates (*see* **Notes 37** and **38**).

10. Resuspend the pellet into 50 µl of prechilled RIPA buffer containing protease and phosphatase inhibitors (*see* **Note 39**). Incubate for 10 min on ice, vortexing every 2 min.

11. Incubate for an additional 10 min on ice. Pellet for 15 min in a tabletop centrifuge at maximum speed at 4 °C. Transfer the supernatant into a fresh Eppendorf tube. This is the WCL. Proceed to the immunoblotting protocol (Subheading 3.7.1).

3.6 Activation of Noncanonical NF-κB in BMDMs (See Note 1)

1. Aspirate the complete cell medium and replace with stimulation medium. We recommend using 1 ml of stimulation medium for each well of a 6-well plate (*see* **Notes 23** and **42**).

2. Stimulate the BMDMs using mRANK ligand (100 ng/ml) using a relatively long stimulation time course that extends at least 6 h. During cell stimulation, incubate the cells in a humidified 37 °C, 10 % CO_2 incubator (*see* **Notes 32** and **33**).

3. When the stimulation time course is complete, place the cell culture plate onto ice. Aspirate the stimulation medium, and wash cells twice with 1 ml of PBS (4 °C) (*see* **Notes 34** and **35**).

4. Add 1 ml of cold PBS to each 6-well plate. Keeping plates on ice, harvest the cells by mechanical scraping (*see* **Note 36**). Transfer the cells to 15-ml tubes. Pellet for 4 min at $300 \times g$ at 4 °C.

5. Aspirate the supernatant and resuspend the pellet into 1 ml of cold PBS and then transfer to prechilled 1.7-ml Eppendorf tubes. Pellet for 4 min in a tabletop centrifuge at maximum speed at 4 °C.

6. Aspirate the supernatant. The cell pellet is ready for lysis, or it can be stored at −80 °C if it is to be used to generate a whole-cell lysate (*see* **Notes 37** and **38**).

7. Resuspend the pellet in 50 µl of prechilled RIPA buffer containing protease and phosphatase inhibitors (*see* **Note 39**). Incubate for 10 min on ice, vortexing every 2 min.

8. Incubate for an additional 10 min on ice. Pellet for 15 min in a tabletop centrifuge at maximum speed at 4 °C. Transfer the supernatant to a fresh Eppendorf tube, and proceed to immunoblotting (Subheading 3.7.1).

3.7 Detection of Noncanonical NF-κB Activation

There is a significant overlap between the genes targeted by the p50/RelA and p50/c-Rel dimers (representing activation of the canonical NF-κB pathway) and genes targeted by the p52/RelB dimer (representing activation of the noncanonical NF-κB pathway).

This presents a problem if you want to look specifically at the effects of noncanonical NF-κB activation, and differentiating whether a certain gene activation profile is the result of activation of the canonical or noncanonical NF-κB pathway is difficult. Therefore, our lab uses immunoblotting to monitor the activation of the noncanonical NF-κB pathway. We monitor the pathway at several steps, including the accumulation of de novo synthesized NIK after stimulation, the processing of p100 into the p52 fragment, the degradation of TRAF3 required to liberate NIK from the TRAF-cIAP ubiquitin ligase complex, and the translocation of the p52/RelB dimer from the cytoplasmic into the nuclear compartment (Fig. 2).

3.7.1 Immunoblotting Protocol (See Note 43)

1. Start with either WCLs (as described above for each cell type) or cytoplasmic/nuclear extracts (Subheading 3.8). Quantify protein concentration using the Pierce BCA protein assay kit.

2. Add SDS loading dye (our stock is 2×); incubate for 5 min at 95 °C (*see* **Note 44**).

3. Spin for 30 s at maximum speed in a tabletop centrifuge at RT. The samples are ready to proceed with immunoblotting, or if needed, samples can now be stored at –20 °C.

4. Load equal amounts of cell lysates to the wells of an 8.5 % SDS-PAGE gel; separate proteins as usual by SDS-PAGE (*see* **Notes 45** and **46**).

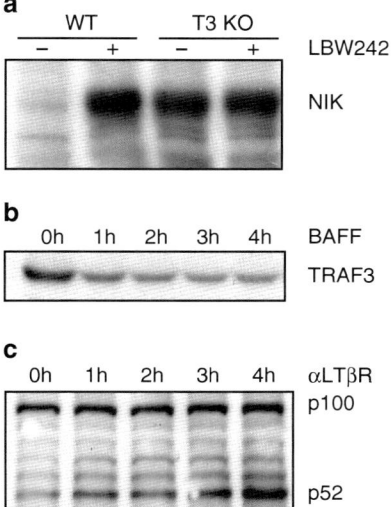

Fig. 2 Detection of noncanonical NF-κB activation by immunoblot. (**a**) MEFs derived from wild-type (WT) or TRAF3$^{-/-}$ mice were treated with the SMAC mimetic LBW242 (1 μM) for 8 h, and WCL was probed using anti-NIK. (**b**) Primary splenic B cells from WT mice were stimulated with BAFF (100 ng/ml); WCL was probed for TRAF3. (**c**) WT MEFs were treated with an agonist antibody against LTβR (2 μg/ml); WCL was probed using anti-p100/p52

5. Transfer to PVDF membrane by tank transfer (*see* **Notes 47–50**). Mark the position of the standards after transfer is complete.

6. Block the membrane using either 5 % nonfat dry milk in TBST or a 2–5 % BSA solution for at least 30 min or overnight at 4 °C (*see* **Note 51**).

7. Incubate the membrane with primary antibody diluted in the same buffer as used above.

8. Wash the membrane with TBST at least 3 times for 5 min each.

9. Incubate the membrane with secondary antibody that is either anti-rabbit or anti-mouse antisera conjugated to horseradish peroxidase diluted in TBST as per the manufacturer's recommendations for 30–60 min.

10. Wash the membrane with TBST at least 4 times with 30 min for each wash.

11. Visualize using enhanced chemiluminescence (ECL). For ECL, first wash the membrane once with TBS without Tween-20 for 5 min. Then, drain off the buffer and add a small volume of the ECL reagents. Incubate for 1 min and then remove the membrane. Remove any excess reagent from the membrane by gently blotting with a kimwipe. Develop the blot.

12. Rinse the membrane with TBST. Either re-blot the membrane for another protein of interest, or dry and save the membrane for later use such as probing for loading controls (*see* **Note 52**).

3.7.2 NIK Accumulation

1. Start with cell pellets generated from stimulation time course described above for each cell type. Quantify the protein concentration, and add SDS loading dye.

2. Perform the immunoblotting protocol, loading 100 µg of total protein/lane (*see* **Notes 53** and **54**).

3. Block the membrane using 5 % nonfat dry milk in TBST (*see* **Note 55**).

4. Incubate the membrane with anti-NIK primary antibody overnight at 4 °C with shaking, using 1:1,000 concentration diluted in TBST with milk.

5. Incubate with anti-rabbit secondary antibody for 1 h, using 1:5,000 concentration diluted in TBST with milk.

6. When the blot is developed, the NIK band will be visible at approximately 130 kDa (Fig. 2).

3.7.3 p100/p52 Processing

1. Start with cell pellets generated from stimulation time course described above for each cell type. Quantify the protein concentration and add SDS loading dye.

2. Perform the immunoblotting protocol, loading approximately 20-µg total protein/lane (*see* **Note 56**).

3. Block the membrane with 5 % nonfat dry milk in TBST.

4. Incubate the membrane with anti-p100/p52 primary antibody overnight at 4 °C with shaking using 1:1,000 concentration diluted in TBST with milk.

5. Incubate with anti-rabbit secondary antibody for 1 h, using 1:5,000 concentration diluted in TBST with milk. An example of a developed blot is shown in Fig. 2.

3.7.4 TRAF3 Degradation

1. Start with cell pellets generated from stimulation time course described above for each cell type. Quantify the protein concentration and add SDS loading dye.

2. Perform the immunoblotting protocol, loading approximately 40-µg total protein/lane.

3. Block the membrane with 5 % nonfat dry milk in TBST.

4. Incubate the membrane with anti-TRAF3 primary antibody overnight at 4 °C with shaking using 1:1,000 concentration diluted in TBST with milk.

5. Incubate with anti-rabbit secondary antibody for 1 h, using 1:5,000 concentration diluted in TBST with milk. An example of a developed blot is shown in Fig. 2 (*see* **Note 57**).

3.8 Nuclear Translocation of p52/ RelB Complexes

To observe the translocation of the p52/RelB dimer from the cytoplasm to the nucleus, it is first necessary to perform a nuclear/ cytoplasmic fractionation. This protocol must be performed on cells immediately after stimulation and cannot be performed on cell pellets that have first been frozen. Perform all steps on ice and all centrifugations at 4 °C.

3.8.1 Nuclear Fractionation Protocol

1. Start with cell pellets generated from stimulation time course described above for each cell type. The pellets must be kept on ice during this protocol (*see* **Note 58**).

2. Prepare an aliquot of buffer A with protease inhibitors for lysing the cell pellet. For washes, buffer A without additional protease inhibitors can be used.

3. Resuspend the pellet into buffer A. For a confluent 60-mm plate of MEF cells, we resuspend the pellet in 200 µl of buffer A. Incubate on ice for 10 min.

4. Centrifuge in a tabletop centrifuge at 4 °C at $1,000 \times g$ for 10 min.

5. Remove the supernatant to fresh Eppendorf tubes: this is the cytoplasmic extract. Quantitate protein concentration using a BCA assay kit. This extract can be stored at −80 °C, but avoid repeated freeze/thaw cycles.

6. Wash the pellet 3 times using 1 ml of buffer A each time (this does not need to contain protease inhibitors) (*see* **Note 59**).

7. Resuspend the pellet into buffer B with protease inhibitors. When starting with a confluent 60-mm plate, we resuspend the nuclear pellet into 30 μl.

8. Pipette up and down without introducing bubbles and leave on ice for 30 min (*see* **Note 60**).

9. Spin for 20 min in tabletop centrifuge at maximum speed at 4 °C. Collect the supernatant and quantitate protein concentration. This nuclear extract can be stored at –80 °C (avoid repeated freeze/thaw cycles).

10. Perform immunoblotting protocol of nuclear and cytoplasmic extracts, as described above for WCLs (Subheading 3.7.1).

4   Notes

1. All solutions/equipment coming into contact with cells must be sterile; use aseptic technique.

2. All incubations are performed in a humidified 37 °C, 10 % CO_2 incubator unless specified.

3. If null mutant embryos are lethal before E14, then sacrifice mice earlier.

4. We use isoflurane to euthanize mice, but there are alternate methods including carbon dioxide inhalation and cervical dislocation. All procedures involving mice must be performed following institutionally approved lab animal protocols.

5. The entire mouse should be soaked with the 70 % ethanol solution.

6. Be careful to avoid puncturing intestines as this increases the possibility of bacterial contamination.

7. Save the fetus heads as they can be used later for genotyping.

8. Do not leave the tissue in trypsin for too long as this will cause the cells to die.

9. During early passages (approximately passages 1–6), the cells will grow to confluence and need to be split every 3–4 days. Proliferation will dramatically slow after passage 6. Then, around passage 10, cells will undergo a growth crisis and will stop any apparent proliferation. Continue splitting cells and changing medium as needed. If you maintain these cells in culture, they will eventually begin to proliferate again, and these will now be considered 3T3 cells. We recommend performing any experiments that require primary MEF cells at an early passage, such as passage 3.

10. Unless otherwise stated, tissues and cell suspensions should be kept on ice at all times and all steps performed at 4 °C.

11. We collect the spleens into complete RPMI medium. You may also collect into MACS buffer.

12. If needed, you may further disperse any clumps in the cell suspension, using a 6-ml syringe with a 19-G needle to draw up and expel the suspension several times.

13. If immediate processing of the single-cell suspension is not possible, then it is best to pellet the cell suspension and keep this pelleted suspension on ice. This will help maintain viability and reduce any loss of cells that may stick to the plastic walls of the conical tubes. Simply resuspend the pellet when you are ready to process the samples.

14. We present our recipe for MACS buffer here. There are many variations. You may use from 2 % to 5 % FBS. The FBS can be replaced with BSA (0.1–0.5 %). The 2-mM EDTA can be replaced with 0.5 % sodium citrate.

15. Before using, rinse the 30-μm filter with MACS buffer.

16. Volumes given in this protocol are meant to be used for up to 1×10^7 total cells. If increased cell numbers are used, simply scale up all reagent volumes.

17. Miltenyi Biotec recommends an incubation temperature of 2–8 °C for incubation with the biotin-antibody cocktail and the anti-biotin microbeads; if incubating on ice, you may require longer incubation times. Higher incubation temperatures and/or longer incubation periods during these steps can lead to increased nonspecific cell labeling.

18. Miltenyi Biotec recommends the MS column for a maximum total cell count of 2×10^8 cells, the LS column for a maximum of 2×10^9 cells, and the XS column for a maximum of 2×10^{10} cells. Choose the appropriate column based on the cell count obtained after preparation of the single-cell suspension. Different MACS separators can hold different column sizes—make sure that whatever column you choose can fit in whichever MACS separator you use.

19. Rinse the LS column with 3 ml of MACS buffer.

20. When loading the cell suspension onto the column, avoid formation of any air bubbles.

21. If you wish to collect the magnetically labeled cells that are bound by the column, then simply remove the column from the magnetic MACS separator, load the column with MACS buffer (500 μl for the MS column, 3 ml for the LS column), and then push a plunger into the column in order to flush out the labeled cells. This fraction is the non-B-cell fraction. FACS analysis can be performed to determine the pre- and post-purification.

22. The ACK-lysis step, which removes red blood cells, is optional.

23. BMDM stimulations are best when they are performed in the same plate in which the cells are derived. Therefore, it is best to perform this initial plating and derivation in the same plate that you plan to later use for stimulations. It is possible, if necessary, to replate the cells after derivation. In this case, cells should be incubated for 10 min with a solution of 5-mM EDTA in PBS, then scraped, collected, and replated in the same medium (DMEM with 10 % FBS) at a density of 0.2–0.3 million cells/ml medium. Allow at least 12 h for the cells to adhere to the plate surface before beginning any stimulation.

24. Macrophage progenitors will adhere to the dish and will not be washed away with PBS. The PBS washes remove any contaminating, non-adherent cells.

25. Bone marrow monocyte/macrophage progenitor cells will proliferate and differentiate into mature BMMs when grown in the presence of M-CSF. Our lab uses L929-conditioned medium as a source of M-CSF (L929 cells secrete M-CSF into the culture medium). It is also possible to directly add purchased M-CSF instead.

26. The L929 medium will not turn yellow even though the cells reach confluence; this is normal.

27. The percentage of L929-cell-conditioned medium to be used varies depending on batch. FACS analysis of the BMDM cells can be used to quantitate how much of a particular batch of L929-conditioned medium provides optimal growth.

28. After prolonged passage number, cells begin to exhibit basal levels of p100/p52 processing; we recommend using early passage cells that do not show basal levels of p100/p52 processing.

29. Different MEF cell lines will vary in cell size, which in turn will affect how many cells should be plated to achieve approximately 70 % confluence. A good starting point is 5×10^5 cells/6 wells; optimize this number for the specific MEF cell line used.

30. We stimulate MEF cells in medium that does not contain FBS, but stimulation can be performed in lower-percentage medium (approximately 2–3 % FBS).

31. The volumes we give here are specifically optimized for individual 6 wells; simply scale up or down depending on the size of the tissue culture plate being used.

32. When performing any stimulation, you want to plan ahead so that you will collect all the different time points together, and so you will stimulate the various samples at different times. However, do make sure that all sample time points including the "0" time point are switched into stimulation medium at the beginning of the time course instead of the complete culture

medium that contains FBS. During a stimulation time course, you may have to remove plates from the humidified incubator (e.g., to add ligand to another well of a 6-well plate). When doing this, work quickly and minimize any disturbances to the cells.

33. In our hands, we find that to see NIK accumulation in MEF cells requires a minimum of 5 h for most stimulus conditions. We find that p100/p52 processing can be observed earlier. This is partly because NIK is harder to detect by immunoblotting and then p100/p52. Regardless, plan your stimulation time course depending on exactly what readouts you plan to use to detect activation of the noncanonical NF-κB pathway.

34. Before beginning to harvest cells after a stimulation, make sure to have a bucket of ice ready to put samples on, cold PBS, and a tabletop centrifuge that has been precooled to 4 °C. When collecting the samples, work quickly and keep the cells at 4 °C. This cold temperature will help minimize any signaling changes that may occur during the harvesting process versus those that occur during the actual signaling time course.

35. When washing, add PBS along the side walls of culture plates to minimize any disturbance to cells.

36. When harvesting cells by scraping, the focus should be on covering the entire surface of the plate instead of pushing down on the plate with the cell scraper. We use a grid-type movement when scraping cells—first covering the surface vertically and then covering it again horizontally.

37. Before beginning cell lysis, make sure that the RIPA buffer is ready. It should be prechilled to 4 °C, and all protease and phosphatase inhibitors should have already been added.

38. During lysis, keep the samples on ice to minimize protein degradation.

39. We recommend resuspending the cell pellet into a small volume of RIPA buffer. Especially if you plan to immunoblot for NIK accumulation, you will need to load a large amount of protein onto the SDS-PAGE gel, and so you will require concentrated samples. If you only plan to look at p100/p52 processing, then this is not as important as less protein is needed, and so even dilute samples are acceptable. *See* Subheading 3.7 for details.

40. You will plate the B cells at a concentration of 2×10^6 cells/ml. Therefore, when resuspending the B cells into the stimulation medium, you do not want to add too much medium and over-dilute the cells. If you do accidentally over-dilute the cells, though, simply spin the cells down again, resuspend them in a smaller volume, and recount them.

41. The volumes and cell numbers we give are optimized for a 12-well tissue culture plate size. For other cell culture sizes, simply scale up or down the volumes and cell numbers.

42. The volumes we give are optimized for individual 6 wells; scale up or down for your plate size.

43. If you are planning to examine WCLs, then you may freeze cell pellets at −80 °C after stimulation, and proceed with lysis and the rest of the immunoblotting protocol at a later time. An initial freeze will in fact aid lysis of the cells. However, avoid repeated freeze/thaw cycles of cell pellets, since this will lead to sample degradation. If you are planning to examine nuclear-cytoplasmic fractionation extracts, though, then you cannot freeze the cell pellets before performing the fractionation as freezing will disrupt cellular compartments.

44. If necessary, WCLs can be stored at −80 °C. However, because each freeze/thaw cycle leads to protein degeneration, we prefer to determine protein concentration and add SDS loading dye immediately after cell lysis. These samples can then be stored at −20 °C without fear of sample degeneration. You must determine protein concentration before adding loading dye!

45. We cast our own SDS-PAGE gels; however, you may buy your own precast SDS-PAGE gels.

46. We recommend 8.5 % SDS-PAGE for resolution; you may vary this depending on which proteins you wish to detect.

47. We transfer to PVDF, but it is also possible to use nitrocellulose membrane.

48. We use a wet tank-style transfer system, but a semidry transfer system can also be used.

49. Be careful not to introduce any bubbles when setting up the transfer, as bubbles will block the transfer of proteins onto the membrane from the gel.

50. Use forceps to handle the membrane.

51. We have found best results using milk as a blocking reagent; however, in certain cases including phosphoprotein detection, milk is not an option and we use BSA (2–5 %) instead.

52. As loading controls for WCLs, we use either anti-β-actin (1:1,000) or anti-HSP90 (1:5,000). There are many alternative antibodies available to use for as loading controls. For nuclear-cytoplasmic fractionation, we use different loading controls both to confirm equal loading of protein and to confirm good separation of the cellular compartments. There are again many different options available. As a cytoplasmic marker, we use either anti-HSP90 (1:5,000) or anti-tubulin (1:1,000), and as a nuclear marker we use anti-Oct1 (1:1,000).

53. To blot for NIK, we load up to 200 µg/sample for western blot; 100 µg is a good starting point.

54. When blotting for NIK, we use lower-percentage gels to improve resolution of NIK (which runs at just under 130 kDa) from nonspecific bands that appear slightly higher than 100 kDa.

55. We have achieved best results using milk when blotting for NIK to block membranes and during primary and secondary antibody incubations. We have found that BSA and other commercial blocking reagents do not work as well as milk.

56. To blot for p100/p52, we load approximately 20 µg/sample for western blots. We have found using the antibody from Cell Signaling that loading too much protein makes it difficult to detect changes in p100/p52 processing after stimulation.

57. Depending on the cell type, the TRAF3 antibody may show single or double bands. B cells typically show a single band, whereas 3T3 and MEF cells show double bands. In all cases, the band corresponding to TRAF3 appears at approximately 70 kDa.

58. When performing the nuclear-cytoplasmic fractionation, it is necessary to begin with more cells than would be used for WCL.

59. The washes are very important to remove excess cytoplasmic proteins from the nuclear fraction. We recommend generous washing of the nuclear pellet.

60. Avoid introducing bubbles during this step. If the pellet is small making it difficult to pipette up and down without introducing bubbles, then let the pellet rotate overnight to aid lysis.

References

1. Hayden MS, Ghosh S (2004) Signaling to NF-kappaB. Genes Dev 18(18):2195–2224

2. Razani B, Reichardt AD, Cheng G (2011) Non-canonical NF-kappaB signaling activation and regulation: principles and perspectives. Immunol Rev 244(1):44–54

3. Matsushima A, Kaisho T, Rennert PD et al (2001) Essential role of nuclear factor (NF)-kappaB-inducing kinase and inhibitor of kappaB (IkappaB) kinase alpha in NF-kappaB activation through lymphotoxin beta receptor, but not through tumor necrosis factor receptor I. J Exp Med 193(5):631–636

4. Senftleben U, Cao Y, Xiao G et al (2001) Activation by IKKalpha of a second, evolutionary conserved, NF-kappa B signaling pathway. Science 293(5534):1495–1499

5. Yin L, Wu L, Wesche H et al (2001) Defective lymphotoxin-beta receptor-induced NF-kappaB transcriptional activity in NIK-deficient mice. Science 291(5511):2162–2165

6. Vallabhapurapu S, Matsuzawa A, Zhang W et al (2008) Nonredundant and complementary functions of TRAF2 and TRAF3 in a ubiquitination cascade that activates NIK-dependent alternative NF-kappaB signaling. Nat Immunol 9(12):1364–1370

7. Zarnegar BJ, Wang Y, Mahoney DJ et al (2008) Noncanonical NF-kappaB activation requires coordinated assembly of a regulatory complex

of the adaptors cIAP1, cIAP2, TRAF2 and TRAF3 and the kinase NIK. Nat Immunol 9(12):1371–1378

8. Xiao G, Fong A, Sun SC (2004) Induction of p100 processing by NF-kappaB-inducing kinase involves docking IkappaB kinase alpha (IKKalpha) to p100 and IKKalpha-mediated phosphorylation. J Biol Chem 279(29): 30099–30105

9. Solan NJ, Miyoshi H, Carmona EM et al (2002) RelB cellular regulation and transcriptional activity are regulated by p100. J Biol Chem 277(2):1405–1418

10. Razani B, Zarnegar B, Ytterberg AJ et al (2010) Negative feedback in noncanonical NF-kappaB signaling modulates NIK stability through IKKalpha-mediated phosphorylation. Sci Signal 3(123):ra41

11. Novack DV, Yin L, Hagen-Stapleton A et al (2003) The IkappaB function of NF-kappaB2 p100 controls stimulated osteoclastogenesis. J Exp Med 198(5):771–781

12. Mackay F, Figgett WA, Saulep D et al (2010) B-cell stage and context-dependent requirements for survival signals from BAFF and the B-cell receptor. Immunol Rev 237(1): 205–225

13. Hostager BS, Bishop GA (2013) CD40-mediated activation of the NF-kappaB2 pathway. Front Immunol 4:376

14. Ma DY, Clark EA (2009) The role of CD40 and CD154/CD40L in dendritic cells. Semin Immunol 21(5):265–272

15. Jin J, Xiao Y, Chang JH et al (2012) The kinase TBK1 controls IgA class switching by negatively regulating noncanonical NF-kappaB signaling. Nat Immunol 13(11):1101–1109

16. Enzler T, Bonizzi G, Silverman GJ et al (2006) Alternative and classical NF-kappa B signaling retain autoreactive B cells in the splenic marginal zone and result in lupus-like disease. Immunity 25(3):403–415

17. Pham LV, Fu L, Tamayo AT et al (2011) Constitutive BR3 receptor signaling in diffuse, large B-cell lymphomas stabilizes nuclear factor-kappaB-inducing kinase while activating both canonical and alternative nuclear factor-kappaB pathways. Blood 117(1):200–210

Chapter 15

Roles of c-IAP Proteins in TNF Receptor Family Activation of NF-κB Signaling

Eugene Varfolomeev, Tatiana Goncharov, and Domagoj Vucic

Abstract

Precise regulation of survival and signaling pathways is essential for proper maintenance of organismal homeostasis, development, and immune defense. Inhibitor of apoptosis (IAP) proteins are evolutionarily conserved regulators of cell death and immune signaling that impact numerous cellular processes. Initially characterized as inhibitors of apoptosis, the ubiquitin ligase activity of IAP proteins is critical for modulating various signaling pathways (e.g., NF-κB, MAPK) and cellular fate. Cellular IAP1 and IAP2 regulate the pro-survival canonical NF-κB pathway by ubiquitinating RIP1 and enabling recruitment of kinase (IKK) and E3 ligase (LUBAC) complexes. On the other hand, c-IAP1 and c-IAP2 are negative regulators of noncanonical NF-κB signaling by promoting ubiquitination and consequent degradation of the NF-κB-inducing kinase NIK. In this article, we describe the involvement of c-IAP1 and c-IAP2 in NF-κB signaling and provide detailed methodology for examining how c-IAPs exert their functional roles.

Key words IAP, Inhibitor of apoptosis, NF-κB, IAP antagonist, TNF, c-IAP, RING domain, Ubiquitin, RIP1, NIK, TRAF2, Proteasomal degradation

1 Introduction

1.1 Regulation of NF-κB Signaling by c-IAP Proteins

The evolutionarily conserved family of inhibitor of apoptosis (IAP) proteins encompasses structurally related regulators of many critical cellular processes [1]. Among the human IAP proteins, cellular IAP1 and IAP12 (c-IAP1 and c-IAP2) and X chromosome-linked IAP (XIAP) are probably the most studied, although other IAP proteins (NAIP, ML-IAP, survivin, ILP2, and Apollon) also play important roles in cell survival, cell cycle, inflammation, and overall homeostasis. IAP proteins contain one to three copies of a signature baculovirus IAP repeat (BIR) domain that regulates protein–protein interactions that are necessary for IAP function [2]. XIAP, c-IAP1, c-IAP2, ML-IAP, and ILP2 also contain a carboxy-terminal really interesting new gene (RING) domain, which imparts them with ubiquitin ligase activity [3]. Ubiquitination involves covalent modification of target proteins with the 76-amino acid protein

Michael J. May (ed.), *NF-kappa B: Methods and Protocols*, Methods in Molecular Biology, vol. 1280,
DOI 10.1007/978-1-4939-2422-6_15, © Springer Science+Business Media New York 2015

ubiquitin and requires the enzymatic activity of an ubiquitin-activating enzyme (E1), an ubiquitin-conjugating enzyme (E2), and an ubiquitin ligase (E3) [4]. The attachment of a single ubiquitin molecule to a lysine residue of the substrate protein yields monoubiquitination [4]. However, since ubiquitin contains seven lysine residues and a free amino-terminus, polyubiquitin chains can be synthesized through eight different isopeptide linkages [5]. Lysine 48-linked polyubiquitin chains predominantly target proteins for proteasomal degradation, whereas lysine 63-, amino-terminal methionine-, and, in some cases, lysine11-linked chains provide a scaffolding platform for the assembly of signaling complexes [4, 6].

Although IAP proteins were characterized initially as antiapoptotic factors, the regulation of cell survival and homeostasis by IAP proteins is not limited to cell death pathways. A number of studies conducted in recent years have established IAP proteins as important regulators of MAPK and in particular NF-κB signaling pathways [7–10]. The NF-κB family of transcription factors regulates the expression of a vast number of genes involved in cell survival, inflammation, and immunity [11]. NF-κB family transcription factors NF-κB1 (p105/p50), NF-κB2 (p100/p52), RelA (p65), RelB, and c-Rel act as homodimers or heterodimers whose activation is regulated by phosphorylation and ubiquitination [12]. In canonical NF-κB signaling, the inhibitor of κB (IκB) binds NF-κB proteins RelA and p50 in the cytoplasm to block them from entering the nucleus in unstimulated cells [12]. The binding of TNF to TNFR1 triggers the recruitment of the proximal receptor-associated complex that includes TRADD, RIP1, TRAF2, and TRAF2-associated c-IAP1 and c-IAP2 proteins [9, 13]. Aggregation at the receptor complex results in c-IAP-mediated ubiquitination of RIP1, TRAF2, and the c-IAPs themselves, and this promotes the recruitment of the IκB kinase (IKK) complex, transforming growth factor β-activating kinase 1, the TAK1–TAB2/3 (TAK1-binding protein 2/3) complex, and the linear ubiquitin chain assembly complex, LUBAC [14].

IKK-γ or NEMO (NF-κB essential modifier) in the IKK complex and TAB2/3 in the TAK1–TAB2/3 complex bind to polyubiquitin chains on RIP1, which brings the kinase TAK1 into proximity with IKK-β, thus allowing phosphorylation of IKK-β. Subsequent phosphorylation of IκB by IKK-β is recognized by the E3 ligase complex SCF–β-TRCP, which promotes IκB ubiquitination and degradation. Autoubiquitination of c-IAP1/2 proteins enables the recruitment of LUBAC, which assembles linear polyubiquitin chains on NEMO and RIP1 [14, 15]. LUBAC is comprised of two regulatory components: the heme-oxidized IRP2 ubiquitin ligase 1 homolog (HOIL-1L) and SHANK-associated RH domain interactor (SHARPIN) and the E3 ligase HOIL-1-interacting protein (HOIP) [16–18]. Linear polyubiquitin chains stabilize TNFR-associated signaling complexes, whereas a reduction in the amount of LUBAC diminishes NF-κB signaling [19]. Besides the TNFR1-associated complex, the ubiquitin ligase activity of c-IAP1 and c-IAP2 is critical for linking a number of TNF family receptors to distal kinase and E3 ligase complexes IKK, TAK1/TAB2/3, and LUBAC [9]. DR3, FN14, LT-βR, CD40, and CD30 all rely on the adaptor protein TRAF2 and the E3 ligases c-IAP1/c-IAP2 to stimulate canonical NF-κB signaling [7–10, 20].

Although positive regulators of the canonical NF-κB pathway, c-IAP proteins are crucial negative regulators of noncanonical NF-κB signaling [10]. The NF-κB-inducing kinase, NIK, initiates noncanonical signaling by phosphorylating IKK-α, which leads to phosphorylation and proteasomal processing of p100 to p52 [21]. NIK is not abundant in unstimulated cells because it is ubiquitinated constitutively by c-IAP1 and c-IAP2 proteins. This posttranslational modification targets NIK for proteasomal degradation [22]. A cytoplasmic complex comprising the adaptor proteins TRAF2 and TRAF3 links the E3 ligases c-IAP1/c-IAP2 to NIK [9, 23]. The activation of a number of TNF family receptors (including FN14, LT-βR, CD40) by their respective ligands or agonistic antibodies leads to the recruitment of TRAF2, TRAF3, and c-IAP1/c-IAP2 to the receptor complex [9, 23]. This membrane-associated aggregation causes dimerization of the c-IAP proteins and stimulation of their E3 ligase activity. The ubiquitination of TRAF2, TRAF3, and the c-IAPs themselves results in the degradation of these proteins [9, 20]. The absence of TRAF2, TRAF3, or the c-IAP proteins allows NIK to accumulate in cells and activate noncanonical NF-κB signaling. Thus, c-IAP proteins are critical for keeping this signaling pathway suppressed so that harmful induction of cytokine expression and inflammation is avoided.

The expression of IAP proteins is elevated in many tumor types. Combined with their functional importance for the regulation of survival and signaling pathways, this makes IAP proteins attractive targets for therapeutic intervention. Among several strategies used for targeting IAP proteins, the most advanced and attractive approach involves SMAC-mimicking small-molecule antagonists [24, 25]. IAP antagonists, such as BV6, bind to select BIR domains of IAP proteins and promote rapid proteasomal degradation of c-IAP1 and c-IAP2 proteins [2, 22, 26]. This chemically induced depletion of c-IAPs can be used to study the role of c-IAP1/c-IAP2 in NF-κB and other signaling pathways.

2   Materials

2.1   Equipment

1. Power supply. We use the Bio-Rad (Hercules, CA, USA) PowerPac™ HC High-Current Power Supply.

2. PAGE gel running apparatus. We use the Criterion™ cell running chamber (Bio-Rad).

3. Gel transfer apparatus such as the Criterion™ Blotter With Plate Electrodes (Bio-Rad).

4. Blotting-Grade Blocker (nonfat dried milk).

5. Benchtop centrifuges for large and small volumes such as the Allegra 6 (Beckman Coulter, Pasadena, CA, USA) and 5417R (Eppendorf, Hamburg, Germany) centrifuges, respectively.

6. Protein assay kit such as the Pierce BCA Protein Assay Kit (Thermo Scientific, Rockford, IL, USA).

7. Cell scrapers.

8. Sonicator. We use a Branson Sonifier 450 (Fisher Scientific, Waltham, MA, USA).

2.2 Cells and Reagents

1. Cells: HT1080 human fibrosarcoma cells, HT29 and SW620 human colorectal adenocarcinoma cells, Ku812F human chronic myelogenous leukemia cells, and Daudi and Ramos human Burkitt's lymphoma cells can be obtained from ATCC (Manassas, VA, USA). Ku812F cells are suitable for studying TL1A signaling; Daudi and Ramos are responsive to CD40L; HT1080, HT29, and SW620 respond to TWEAK and LIGHT, whereas all these cell lines are suited to studying TNF-α signaling.

2. Adherent cell lines are grown in 50:50 Dulbecco's modified Eagle's and FK12 medium.

3. RPMI medium supplemented with 10 % FBS, penicillin, streptomycin, and 2 mM L-glutamine for growth of suspension cell lines.

4. Anti-CD40 antibody (R&D Systems, Minneapolis, MN, USA) to trigger CD40 signaling pathways and to immunoprecipitate the CD40 receptor signaling complex.

Flag-tagged recombinant TNF-α (Enzo Life Sciences, Farmingdale, NY, USA) for stimulation of TNFR1 signaling pathways and to immunoprecipitate the TNFR1 receptor signaling complex.

5. Anti-DR3 antibody (R&D Systems) to immunoprecipitate the DR3 receptor signaling complexes.

6. The following are recombinant cytokines and proteins (R&D Systems) used to stimulate cells: TNF-α; TL1A; TWEAK; LIGHT; CD40L.

7. Transfection reagents to transfect siRNA. We routinely use Lipofectamine RNAiMAX (Invitrogen, Carlsbad, CA, USA), in Opti-MEM (Life Technologies, Carlsbad, CA, USA).

8. IAP antagonist BV6 [22] dissolved in DMSO at 20 mM.

9. Proteasome inhibitor MG-132 (Millipore, Billerica, MA, USA), prepared in DMSO at 20 mM (1,000×).

10. Lysosome inhibitor CA-074Me (Millipore), prepared in DMSO at 20 mM (1,000×).

11. 15- and 50-ml test tubes, 10- or 15-cm dishes, 175-cm² flasks, and 6- and 24-well plates can be purchased from Corning.

12. Stacked cell culture chambers. We use 10-Stack Corning CellSTACK Cell Culture Chambers (Sigma-Aldrich, St. Louis, MO, USA).

13. Triton lysis buffer (TLB): 20 mM Tris–HCl, pH 7.5, 150 mM NaCl, 4 mM EDTA, 1 % Triton X-100 supplemented with a protease and phosphatase inhibitor cocktail.

14. SDS lysis buffer (SLB): TLB supplemented with 1 % SDS.

15. Ubiquitin binding buffer (UBB): 20 mM Tris–HCl (pH 7.5), 135 mM NaCl, 1.5 mM MgCl₂, 1 mM EGTA, 1 % Triton X-100, 20 μM MG132, 4 mM NEM, 20 mM iodoacetamide, and protease inhibitor cocktail. For solubilization, this buffer is made up containing 6 M urea.

16. Cell Dissociation Buffer, enzyme-free (Life Technologies).

17. The following are sense siRNA oligos for *c-IAP1* and/or *c-IAP2* transfection experiments at 20 μM stock concentrations: *c-IAP1*— <hcIAP112S> *UCGCAAUGAUGAUGUCA-AAtt*; <hcIAP113S> *GAAUGAAAGGCCAAGAGUUtt*; *c-IAP2*— <hc25S> *UCTAACACAAGAUCAUUGAtt*; <hc29S> *AUU CGGUACAGUUCACAUGtt*.

18. Complete Mini EDTA-free protease inhibitors (Roche, Madison, WI, USA).

19. Glutathione agarose (Thermo Scientific).

20. Protein A/G beads (Thermo Scientific).

21. Lysis buffer 1: 0.5 % Triton X-100, 100 mM NaCl, 40 mM Tris–HCl pH 7.5, 1 mM CaCl₂, 1 mM MgCl₂, complete EDTA-free protease inhibitor cocktail.

22. Lysis buffer 2: 1 % Triton X-100, 0.1 % SDS, 100 mM NaCl, 40 mM Tris–HCl pH 7.5, 1 mM CaCl₂, 1 mM MgCl₂, complete EDTA-free protease inhibitor cocktail.

2.3 Materials for Western Blotting

The materials we use for Western blotting (listed 1–6 below) are from Bio-Rad. Other systems, reagents, and gels can be used:

1. 4× XT sample buffer.

2. 20× XT reducing agent.

3. 20× XT MOPS running buffer.

4. Criterion XT Bis-Tris Gel 18 wells.

5. Criterion XT Bis-Tris Gel 26 wells.

6. Nitrocellulose/Filter Paper Sandwiches.

7. Transfer buffer. We use Novex NuPAGE Transfer Buffer (20×) (Invitrogen).

8. Protein marker such as SeeBlue Plus2 (Invitrogen).

9. Enhanced chemiluminescence (ECL) reagents. Several kits are available, but we use the Western Lightning Plus-ECL kit (PerkinElmer, Waltham, MA, USA).

10. PBS-T buffer: PBS supplemented with TWEEN 20 (0.05 % final concentration).

11. BLOTTO (membrane-blocking solution): PBS-T with blocker reagent (5 % final concentration—nonfat dried milk).

12. PBS-T BSA membrane-blocking solution for the detection of phosphorylated proteins: PBS-T with bovine serum albumin (BSA; 5 % final concentration).

2.4 Antibodies

The following antibodies are required for Western blotting (1:1,000 dilutions if not indicated otherwise). The antibodies from specific providers that we have used successfully are listed:

1. Anti-phospho-IκB, anti-IκB, anti-phospho-p38, anti-p38, anti-JNK, anti-pan-cadherin, anti-HSP90, anti-SP1, anti-FN14, anti-tubulin (Cell Signaling, Danvers, MA, USA).

2. Anti-phospho-JNK, anti-RIPK1, anti-TRADD (1:250), anti-TRAF2, anti-NEMO (BD, East Rutherford, NJ, USA).

3. Anti-TRAF3 (Invitrogen).

4. Anti-NF-κB p100/p52 (1:2,000); anti-TRAF6 (Millipore, Billerica, MA, USA).

5. Anti-c-IAP1, anti-IKK2, anti-TAK1, anti-TNFR1 (1:500), anti-lymphotoxin β receptor (1:500), anti-CD40 (1:500), anti-DR3 (1:500) (R&D Systems).

6. Anti-c-IAP2, anti-HOIP (Novus Biologicals, Littleton, CO, USA).

7. Anti-actin (1:5,000) (Sigma-Aldrich).

8. K11-, K48-, K63-ubiquitin chain-specific antibodies [27, 28] (Genentech, South San Francisco, CA, USA).

9. The following are HRP-conjugated antibodies and streptavidin (Jackson ImmunoResearch, West Grove, PA, USA), reconstituted with 0.5 ml of PBS, mixed with 0.5 ml of 100 % glycerol, and used as secondary antibodies: anti-IgG1 (1:10,000); anti-IgG2a (1:5,000); anti-rabbit IgG (1:5,000), anti-goat IgG (1:10,000), anti-rat (1:10,000), streptavidin (1:1,000).

3 Methods

3.1 Western Blot-Based Assessment of the Activation of MAPKs and NF-κB Signaling Pathways by TNF Ligands

3.1.1 siRNA Transfection

1. Rinse the cells once with warm trypsin and trypsinize them for 3–5 min. Collect the cells and neutralize the trypsin with 20 ml of growth medium without antibiotics. Count the cells and plate an appropriate number for transfection (*see* **Note 1**).

2. On the next morning, check the confluence of the seeded cells (*see* **Note 1**). If the cells are too sparse, then wait for 8–24 h before proceeding with transfection. However, if the cells are more than 80 % confluent or grown in clumps, repeat the seeding with fewer cells.

3. Place 1 ml of Opti-MEM in each marked 15-ml tube. Add 30 μl of stock siRNA duplexes.

4. Mix RNAiMAX reagent with Opti-MEM in a separate tube; use 30 μl of reagent and 1 ml of medium for each transfection.

5. Add 1 ml of diluted RNAiMAX to each siRNA and gently mix.

6. Incubate tubes for 10 min at room temperature and then add to the cells in 10-cm dishes. Swirl the dishes gently.

7. Grow transfected cells for 48 h before treating them with TNF ligands.

3.1.2 Cell Growth and BV6 Treatment

1. Grow an appropriate number of cells to semi-confluence (*see* **Note 2**).

2. Treat the cells with 1 μM of BV6 for 4–12 h or DMSO as a control. It is important to use BV6 diluted in DMSO to 0.2–2 μM.

3.1.3 Treatment of Cells with TNF Ligands or Agonistic Anti-TNF Receptor Antibodies

1. For adherent cells, rinse the cells once with warm PBS and then apply warm nonenzymatic detachment buffer (*see* **Note 3**). Incubate the cells at 37 °C for 5–10 min, periodically monitoring cell detachment under the microscope. For cells grown in suspension, go directly to **step 3**.

2. Transfer the cells into 50-ml conical tubes. Wash the dishes once with 10 ml of growth medium and add it to the collected cells.

3. Centrifuge the cells for 5 min at $1,200 \times g$.

4. Resuspend the cells in the appropriate growth medium and seed them into 24- or 6-well dishes (*see* **Note 4**).

5. Prepare treatment reagents in serum-free medium (*see* **Note 5**). The final concentration of TNF-α should be 20 ng/ml, 100 ng/ml of other cytokines and 200 ng/ml of anti-CD40 agonistic antibodies. Before addition to the medium, CD40L, LIGHT, and TWEAK should be cross-linked with anti-His

antibody. Anti-CD40 antibody should be cross-linked with IgG2b. Perform cross-linking at room temperature for 5–10 min in a minimal volume (i.e., 20–100 μl). Mix the ligands with 2× the amount of anti-His or IgG2b antibodies.

6. Treat the cells for 5, 15, 30, and 60 min or 12–24 h in the case of noncanonical NF-κB activation (*see* **Note 6**).

7. Detach the cells by pipetting (in the case of a short treatment) or with a cell scraper (e.g., if adherent cells were treated for several hours).

8. Transfer the cells into 1.5-ml test tubes, add 0.5 ml of cold PBS to the wells, and then transfer the remaining cells.

9. Spin cells at $1,200 \times g$ for 3–5 min at 4 °C and discard all liquid.

3.1.4 Cell Lysis and Protein Sample Preparation

1. Lyse the cells in TLB (*see* **Note 7**). Incubate the lysates for 20–30 min on ice.

2. Spin the lysates at $14,000 \times g$ for 10 min at 4 °C.

3. Transfer supernatants to new 1.5-ml tubes. Determine the protein concentration using BCA Protein Assay Reagent following the instruction manual.

4. Prepare protein samples for SDS-PAGE: in a new 1.5-ml tube, combine 1/20 of the final volume XT reducing agent, 1/4 XT sample buffer, appropriate amounts of protein lysates, and water (*see* **Note 8**).

5. Heat the samples for 5 min at 90–95 °C and then microfuge for a short time (10–20 s) to collect the samples at the bottom of the tube.

3.1.5 SDS-PAGE and Transfer to Membranes

These methods are for the Bio-Rad Criterion precast gel system:

1. Assemble gel units.

2. Prepare 1× MOPS XT running buffer (about 800 ml for each full gel unit).

3. Use 26-well gels if the samples contain 1.5 μg/μl protein or less. Otherwise, use 18-well gels because they typically give better sample resolution.

4. Load 10 μl of protein standard and 12–15 μl of the protein samples on the gel.

5. Run the samples at 200 V for 1 h.

6. Prepare 1 L per unit of 1× transfer buffer supplemented with methanol (20 % final concentration) (*see* **Note 9**).

7. Assemble transfer chamber.

8. Perform the protein transfer at 4 °C 120 V for 90 min.

3.1.6 Immunoblotting

1. When the transfer is complete, place the membranes in plastic boxes (*see* **Note 10**). One may combine several membranes for blocking.

2. Incubate membranes with 50–100 ml of BLOTTO or PBS-T BSA at RT for 60 min or at 4 °C overnight on a rotator (60 rpm).

3. Once the membranes are blocked, add primary antibodies in BLOTTO or PBS-T BSA reagents diluted fivefold with PBS-T to achieve 1 % blocking reagent (*see* **Note 11**). Take into consideration that one may utilize the same membrane to detect two different proteins (*see* **Note 12**).

4. Incubate membranes with primary antibodies for 8 h or longer at 4 °C on a rotator (60 rpm).

5. Wash the membrane for 15–30 min with several (2–4) changes of PBS-T, and then incubate them for 1 h at RT on a shaker (60 rpm) with secondary antibodies prepared in 1 % BLOTTO.

6. Wash membranes as described above for 30 min.

7. Place the membranes on Saran wrap, remove excess liquid with paper towels, and then incubate the membranes with ECL reagent for 1 min at RT.

8. Remove excess liquid with paper towels, cover membranes with Saran wrap, and place the membrane into X-ray cassettes.

9. Proceed with X filmography performing several exposures starting from the longest (3–4 min) to obtain optimal intensity of the band.

3.2 Subcellular Fractionation of Proteins

Grow and treat cells as described in Subheading 3.1 with the following modifications: (1) use four million adherents and eight million suspension cells per sample, and (2) when detecting and analyzing proteins that are degraded by the proteasome and/or lysosome, pretreat the cells with MG-132 and/or CA-074Me for 30 min prior to TNF ligand treatment:

1. Lyse the cells in TLB.

2. Centrifuge the lysate at $1,200 \times g$ and then dissolve the pellet in SLB (*see* **Note 13**).

3. Sonicate the samples using a microtip with output power 2, 80 % of duty cycle for 3–4 short, 3–5 s bursts (*see* **Note 14**).

4. Prepare the samples for SDS-PAGE, taking care to proceed to sample boiling as soon as possible after sonication.

5. Proceed with Western blot detection of proteins as described above.

3.3 Immunoprecipitation of Endogenous Receptor-Associated Protein Signaling Complexes

1. For immunoprecipitation of endogenous receptor-associated protein complexes, grow cells ($1–3 \times 10^8$ cells per time point) in one or two 10-layer stackers (*see* **Note 15**).

2. Treat half of the cells with 1 µM BV6 (as described in Subheading 3.1.2) and the other half with DMSO. Leave the cells in a 37 °C incubator overnight.

3. Wash cells once with PBS at room temperature and then detach using Cell Dissociation Buffer (*see* **Note 16**).

4. Collect the cells in a 500-ml bottle and centrifuge for 5 min at 4 °C at 1,200×g.

5. Resuspend the cells in warm medium and distribute evenly into 50-ml conical tubes for treatment.

6. Leave cells untreated or treat them with 1 µg/ml of TL1A, anti-CD40 antibody, or anti-Flag or anti-His antibody-cross-linked ligands (Flag-TNFα, His-TWEAK, His-LIGHT) for 5 or 30 min.

7. Wash cells with cold PBS and then lyse them in 1 % Triton X-100 buffer (10 pellet volumes of lysis buffer).

8. After incubation on ice for 30 min, centrifuge the lysates at 14,000×g for 10 min.

9. Collect the supernatants and discard the pellets unless additional lysis procedures are planned. Reserve 100 µl of soluble lysates to determine protein concentration (as described in Subheading 3.1.4), and run in SDS-PAGE for the analysis of protein levels by Western blotting as described in Subheadings 3.1.5 and 3.1.6.

10. Incubate cell lysates with GST-linked agarose beads in 15- or 50-ml conical tubes depending on the sample volume for 1 h at 4 °C with continuous rotation.

11. Precipitate the mixture of lysates and beads by centrifugation at 1,200×g for 5 min, and collect the supernatants; precipitated beads can be discarded.

12. Incubate the supernatants with anti-DR3 antibody for 2 h and add protein A/G beads for another 3 h. For other tagged ligands and antibody combinations, add protein A/G beads for 3 h.

13. Centrifuge immunoprecipitation mixtures for 5 min at 1,200×g, remove supernatants, and transfer the pelleted beads to 1.5-ml Eppendorf tubes.

14. Wash the beads with 1 ml of lysis buffer five times with centrifugation at 1,200×g for 2 min between washes.

15. After the last wash, remove the supernatants, add 4× LDS loading dye supplemented with reducing agent, and heat for 5–7 min at 90–95 °C along with cleared lysates prepared at **step 9**.

16. Proceed to SDS-PAGE, membrane transfer, and Western blotting as described in Subheadings 3.1.5 and 3.1.6.

3.4 Immuno-precipitation of Membrane Receptor-Associated Protein Signaling Complexes

1. For BR3-associated protein signaling complex, grow and treat cells as described in the immunoprecipitation protocol in Subheading 3.3.

2. Treat cells with anti-Flag antibody-cross-linked Flag-BAFF (2 μg/ml) for 5 and 30 min.

3. Following PBS washes, resuspend the cells in lysis buffer 1 and keep for 20 min on ice.

4. Triton-soluble and Triton-insoluble fractions should be separated by centrifugation at $14,000 \times g$ for 10 min at 4 °C.

5. The Triton-insoluble pellets should be resolubilized in lysis buffer 2 followed by sonication as described in Subheading 3.2.

6. Clear the lysates by centrifugation at $14,000 \times g$ for 10 min to remove the remaining insoluble material, and collect supernatants/cleared lysates.

7. Set aside 50 μl of lysates to determine protein concentration (as described in Subheading 3.1.4), and run in SDS-PAGE for the analysis of protein levels by Western blotting as described in Subheading 3.1.5 and 3.1.6.

8. Incubate cellular lysates with GST-agarose beads and immunoprecipitate with protein A/G beads as described above.

9. Wash immunoprecipitated protein complexes, resolve on SDS-PAGE, and immunoblot with the indicated antibodies as described in Subheadings 3.1.5 and 3.1.6.

3.5 Secondary Immunoprecipitation of Ubiquitinated Proteins in Receptor Signaling Complexes

1. After **step 14** (Subheading 3.3) of the primary immunoprecipitation, disrupt the immunoprecipitated endogenous receptor complexes by incubation in UBB buffer containing 6 M urea for 20 min at room temperature with constant soft rocking.

2. After incubation, pellet disrupted immunoprecipitates by centrifugation at $1,200 \times g$ for 5 min and collect supernatants.

3. Dilute supernatants twofold in UBB buffer.

4. Incubate diluted supernatants with K11-, K48-, or K63-linkage-specific anti-ubiquitin antibodies overnight at 4 °C with rotation.

5. On the next day, add 55–75 μl of protein A/G beads and rotate at 4 °C for another 3 h.

6. Wash the immunoprecipitated proteins five times as described in Subheading 3.3.

7. Following washes, add 4× loading dye and resolve immunoprecipitates and lysates on SDS-PAGE and immunoblot as described in Subheadings 3.1.5 and 3.1.6.

4   Notes

1. Use 0.5–0.8 million cells per 10-cm dish. It is important to get a 40–70 % confluent dish the next morning. One 10-cm dish allows the inclusion of 3–5 experimental points. However, the appropriate number of plates to transfect may vary between different cell lines and depend on protein contents of specific cell lines. Aim to get at least 2 µg/ml of protein lysates.

2. Use 1.5–3 million adherent cells and 3–6 million suspension cells per one experimental point. The number of experimental points defines the total number of cells needed for the experiment.

3. Use 5 ml of detachment buffer for a 10-cm dish or 75-cm^2 flask and 15 ml for a 15-cm dish or 175-cm^2 flask.

4. Use 24-well plates for short (i.e., up to 1.5 h) treatment periods and 6-well dishes for longer times. Seed cells in 1 ml or 3 ml of growth medium per well into 24- or 6-well dishes, respectively.

5. Use 0.1 ml of serum-free medium for the treatment of 1 well. The appropriate amounts of treatment reagents should be calculated based on the final volume (1.1 or 3.1 ml per well).

6. Use a 5 min time point for testing the phosphorylation status of IκB-α; 15 or 30 min corresponds to the peak of JNK and p38 activation for most of the cells; 30 min corresponds to the detection of the lowest amounts of total IκB-α for most of the cells, and 60 min is the time of expected accumulation of the newly synthesized IκB-α.

7. Use 30–50 µl of TLB for small pellets (10 µl), 70 µl of TLB for medium pellets (>10, <25 µl), and 100–140 µl for pellets of large size (25–50 µl)—approximately 3–5× volume of the lysis buffer relative to the pellet volume. Visually compare the cell pellet size with defined volume of the liquid in 1.5-ml test tube.

8. The final volume of protein sample to run can be defined on the basis of the number of loadings, loading volume (typically 15 µl), concentration, and total amount of obtained protein sample. A concentration of the protein below 2 µg/µl is considered to be low; 3–15 µg/µl is normal. Try to avoid running the samples with extremely low (less than 0.5 µg/µl) or high (15 µg/µl) protein. In general, the final concentration of the protein in the loading sample should be between 3 and 4 µg/µl.

9. Prepare transfer buffer in advance—transfer buffer should not be warm (not higher than 20 °C) since warm liquids can damage nitrocellulose membranes.

10. Use a box in which size does not significantly exceed the membrane sizes, but is deep enough to accommodate at least 50 ml of liquid.

11. One may incubate several membranes together (up to 5) with an appropriate amount of antibody solution that can be defined based on the number and sizes of membranes used (10–20 ml per up to three membranes, 15–25 ml per five membranes). It is also important to ensure that all of the membranes are shaking and immersed in liquid during incubation.

12. The expression of two proteins that run at different molecular weights on SDS-PAGE (e.g., 40 and 60 kDa) can be detected simultaneously using the same membrane. In this case, the membrane can be cut into lower and upper parts. It is also possible to cut the membrane prior to blocking if one of the parts of the membrane will be used for the detection of phosphorylated protein.

13. Use at least 100 μl of SLB per sample (200 μl is the optimal amount). Lower volumes will lead to significant sample loss during sonication, and volumes higher than 200 μl can cause significant dilutions of the samples.

14. Try to avoid touching tube walls since it may result in bubbling of the samples. Avoid heating of the sample, and if additional sonications are needed, cool the sample on ice.

15. For most adherent cell lines, one needs 15–18 confluent 225-mm3 flasks to seed two 10 stackers with cells in 1.5 l per stacker. Recommend to seed a single layer or 1 stack of cells per condition as this way cells can be observed under the microscope.

16. Cells within 10-layer stackers need to be washed with warm PBS and incubated with 500 ml of Cell Dissociation Buffer for 20–25 min in the TC incubator. To make sure all the cells are removed from the stacker, the stackers should be tapped by hand or against the hard surface. After the cells are removed, stackers could be rinsed with additional 250 ml of PBS to collect remaining cells.

Acknowledgments

We thank the researchers at Genentech who helped with the suggestions, reagents, and comments. The authors are employees of Genentech, Inc.

References

1. Varfolomeev E, Vucic D (2011) Inhibitor of apoptosis proteins: fascinating biology leads to attractive tumor therapeutic targets. Future Oncol 7(5):633–648

2. Ndubaku C, Cohen F, Varfolomeev E, Vucic D (2009) Targeting inhibitor of apoptosis (IAP) proteins for therapeutic intervention. Future Med Chem 1(8):1509–1525 [review]

3. Vaux DL, Silke J (2005) IAPs, RINGs and ubiquitylation. Nat Rev Mol Cell Biol 6(4):287–297

4. Hershko A, Ciechanover A (1998) The ubiquitin system. Annu Rev Biochem 67:425–479

5. Pickart CM, Fushman D (2004) Polyubiquitin chains: polymeric protein signals. Curr Opin Chem Biol 8(6):610–616

6. Vucic D, Dixit VM, Wertz IE (2011) Ubiquitylation in apoptosis: a post-translational modification at the edge of life and death. Nat Rev Mol Cell Biol 12(7):439–452

7. Bertrand MJ, Milutinovic S, Dickson KM, Ho WC, Boudreault A, Durkin J et al (2008) cIAP1 and cIAP2 facilitate cancer cell survival by functioning as E3 ligases that promote RIP1 ubiquitination. Mol Cell 30(6):689–700

8. Silke J, Brink R (2010) Regulation of TNFRSF and innate immune signalling complexes by TRAFs and cIAPs. Cell Death Differ 17(1): 35–45

9. Varfolomeev E, Goncharov T, Maecker H, Zobel K, Komuves LG, Deshayes K et al (2012) Cellular inhibitors of apoptosis are global regulators of NF-kappaB and MAPK activation by members of the TNF family of receptors. Sci Signal 5(216):ra22

10. Varfolomeev E, Vucic D (2008) (Un)expected roles of c-IAPs in apoptotic and NF-κB signaling pathways. Cell Cycle 7(11):1511–1521

11. Hayden MS, Ghosh S (2004) Signaling to NF-kappaB. Genes Dev 18(18):2195–2224

12. Scheidereit C (2006) IκB kinase complexes: gateways to NF-κB activation and transcription. Oncogene 25(51):6685–6705

13. Micheau O, Tschopp J (2003) Induction of TNF receptor I-mediated apoptosis via two sequential signaling complexes. Cell 114(2):181–190

14. Gentle IE, Silke J (2011) New perspectives in TNF-R1-induced NF-kappaB signaling. Adv Exp Med Biol 691:79–88

15. Haas TL, Emmerich CH, Gerlach B, Schmukle AC, Cordier SM, Rieser E et al (2009) Recruitment of the linear ubiquitin chain assembly complex stabilizes the TNF-R1 signaling complex and is required for TNF-mediated gene induction. Mol Cell 36(5):831–844

16. Ikeda F, Deribe YL, Skanland SS, Stieglitz B, Grabbe C, Franz-Wachtel M et al (2011) SHARPIN forms a linear ubiquitin ligase complex regulating NF-kappaB activity and apoptosis. Nature 471(7340):637–641

17. Tokunaga F, Nakagawa T, Nakahara M, Saeki Y, Taniguchi M, Sakata S et al (2011) SHARPIN is a component of the NF-kappaB-activating linear ubiquitin chain assembly complex. Nature 471(7340):633–636

18. Gerlach B, Cordier SM, Schmukle AC, Emmerich CH, Rieser E, Haas TL et al (2011) Linear ubiquitination prevents inflammation and regulates immune signalling. Nature 471(7340):591–596

19. Walczak H (2011) TNF and ubiquitin at the crossroads of gene activation, cell death, inflammation, and cancer. Immunol Rev 244(1):9–28 [Research Support, Non-U.S. Gov't Review]

20. Vince JE, Chau D, Callus B, Wong WW, Hawkins CJ, Schneider P et al (2008) TWEAK-FN14 signaling induces lysosomal degradation of a cIAP1-TRAF2 complex to sensitize tumor cells to TNFalpha. J Cell Biol 182(1):171–184

21. Dejardin E (2006) The alternative NF-kappaB pathway from biochemistry to biology: pitfalls and promises for future drug development. Biochem Pharmacol 72(9):1161–1179

22. Varfolomeev E, Blankenship JW, Wayson SM, Fedorova AV, Kayagaki N, Garg P et al (2007) IAP antagonists induce autoubiquitination of c-IAPs, NF-κB activation, and TNFα-dependent apoptosis. Cell 131(4):669–681

23. Vallabhapurapu S, Matsuzawa A, Zhang W, Tseng PH, Keats JJ, Wang H et al (2008) Nonredundant and complementary functions of TRAF2 and TRAF3 in a ubiquitination cascade that activates NIK-dependent alternative NF-kappaB signaling. Nat Immunol 9(12): 1364–1370

24. LaCasse EC, Mahoney DJ, Cheung HH, Plenchette S, Baird S, Korneluk RG (2008) IAP-targeted therapies for cancer. Oncogene 27(48):6252–6275

25. Fulda S, Vucic D (2012) Targeting IAP proteins for therapeutic intervention in cancer. Nat Rev Drug Discov 11(2):109–124

26. Vince JE, Wong WW, Khan N, Feltham R, Chau D, Ahmed AU et al (2007) IAP antagonists target cIAP1 to induce TNFalpha-dependent apoptosis. Cell 131(4):682–693

27. Matsumoto ML, Wickliffe KE, Dong KC, Yu C, Bosanac I, Bustos D et al (2010) K11-linked polyubiquitination in cell cycle control revealed by a K11 linkage-specific antibody. Mol Cell 39(3):477–484

28. Newton K, Matsumoto ML, Wertz IE, Kirkpatrick DS, Lill JR, Tan J et al (2008) Ubiquitin chain editing revealed by polyubiquitin linkage-specific antibodies. Cell 134(4): 668–678

Chapter 16

Elucidating Dynamic Protein–Protein Interactions and Ubiquitination in NF-κB Signaling Pathways

Noula Shembade and Edward W. Harhaj

Abstract

The Nuclear factor-kappaB (NF-κB) family of transcription factors plays critical roles in inflammatory responses and host defense; however, uncontrolled NF-κB activation can be deleterious by promoting autoimmune diseases and cancers. Lysine K63 (K63)-linked polyubiquitination has emerged as an important regulatory mechanism in NF-κB signaling by regulating dynamic protein–protein interactions that trigger NF-κB signaling. RIP1 and TRAF6 serve as key substrates of K63-linked polyubiquitin chains in tumor necrosis factor receptor (TNFR) and interleukin-1 receptor (IL-1R) pathways respectively as a mechanism to recruit TAK1 and IKK kinases by associated ubiquitin-binding adaptor molecules. Activation of IKKβ by TAK1 induces IκBα phosphorylation, degradation, and downstream NF-κB activation. The ubiquitin-editing enzyme A20 maintains transient NF-κB activation by opposing the K63-linked polyubiquitination of RIP1 and TRAF6. A20 inducibly interacts with the adaptor molecule TAX1BP1 and the E3 ligases Itch and RNF11 to form an A20 ubiquitin-editing enzyme complex. Notably, loss-of-function somatic mutations or polymorphisms in human A20 are associated with B-cell lymphomas or a variety of autoimmune diseases as a result of dysregulated NF-κB activation. In this chapter, we summarize the protocols routinely used in our laboratories to examine ubiquitination and NF-κB signaling.

Key words NF-κB, MEFs, Ubiquitination assay, Co-immunoprecipitation, Western blot

1 Introduction

Inflammation constitutes a protective response by the host to harmful stimuli such as injury and infection. Immune cells recruited to sites of inflammation produce proinflammatory cytokines including tumor necrosis factor (TNF) or interleukin-1β (IL-1β) that trigger activation of the NF-κB family of transcription factors [1]. NF-κB induces the expression of hundreds of genes that together coordinate the inflammatory response that aids in resolving infections or repairing tissue damage. NF-κB activity is tightly controlled by numerous mechanisms since constitutive activation of NF-κB promotes chronic inflammation which can lead to inflammation-induced tissue damage and malignancies [2].

Michael J. May (ed.), *NF-kappa B: Methods and Protocols*, Methods in Molecular Biology, vol. 1280,
DOI 10.1007/978-1-4939-2422-6_16, © Springer Science+Business Media New York 2015

The NF-κB family is composed of five members, all bearing amino (N)-terminal Rel-homology domains: NF-κB1 (p50/p105), NF-κB2 (p52/p100), p65 (RelA), RelB, and c-Rel. Each of these proteins can form homodimers and heterodimers [3]. The prototypical p65/p50 dimer is sequestered as an inactive latent transcription factor in the cytoplasm by members of the IκB family, all containing a series of ankyrin repeat domains [4, 5]. In response to diverse stress stimuli or infection with harmful microbes, cytoplasmic NF-κB dimers are rapidly mobilized to the nucleus to regulate gene expression. All NF-κB activating stimuli converge at the IκB kinase (IKK) complex containing catalytic kinase subunits IKKα and IKKβ and the regulatory subunit NEMO (also known as IKKγ) [6]. Two distinct NF-κB pathways have been described: the classical (canonical) or alternative (noncanonical) pathways [7]. In the classical NF-κB pathway, IKKβ phosphorylates IκBα on two serine residues in a NEMO-dependent manner which then triggers proteolysis of IκBα through the ubiquitin/proteasome pathway, thus facilitating NF-κB nuclear import and gene induction [8].

In the alternative NF-κB pathway, IKKα is activated in a NEMO-independent manner by the NF-κB inducing kinase (NIK) in response to specific ligands of the TNF superfamily, including BAFF, lymphotoxin-β, and CD40L [9, 10]. Activated IKKα phosphorylates p100 that results in ubiquitination and partial degradation of p100 by the proteasome to generate p52. RelB/p52 heterodimers translocate into the nucleus to regulate genes involved in cell survival and lymphoid organogenesis [11]. NIK is mainly regulated post-translationally by a TRAF2/TRAF3/cIAP1/cIAP2 complex that triggers NIK degradation [12]. TNF superfamily ligands that activate noncanonical NF-κB signaling promote TRAF3 degradation that results in the stabilization and activation of NIK [13].

Ubiquitin is a 76 amino acid polypeptide that plays diverse roles in cells ranging from protein degradation to receptor trafficking and DNA damage repair. Ubiquitin is covalently attached to lysine residues on protein substrates in an enzymatic cascade catalyzed by three classes of proteins: E1 (ubiquitin-activating), E2 (ubiquitin-conjugating), and E3 (ubiquitin ligase) enzymes [14]. The E1 enzyme initially activates ubiquitin in an ATP-dependent manner and forms a thioester linkage between the catalytic cysteine of the E1 and the carboxyl (C)-terminal glycine of ubiquitin. Activated ubiquitin is then transferred to a specific E2 ubiquitin-conjugating enzyme to form an E2 ubiquitin thioester. The E3 ubiquitin ligase then conjugates ubiquitin to a lysine residue in the substrate. Monoubiquitination refers to a single ubiquitin moiety conjugated to a lysine residue on a substrate. Polyubiquitination consists of chains of ubiquitin monomers linked to one another via one of seven lysine residues or the initiating methionine (Met1).

Although all types of polyubiquitin chains (Met1, K7, K11, K27, K29, K33, K48, K63) have been identified in mammalian cells by mass spectrometry [15], K48 and K63-linked polyubiquitin chains appear to be most abundant and have been most intensively studied. Polyubiquitin chains linked by lysine 48 (K48) generally target the substrate for degradation by the 26S proteasome. Conversely, K63-linked polyubiquitin chains are usually not linked to protein degradation, but rather regulate signal transduction, receptor trafficking, and the DNA damage response [16]. Indeed, K63-linked polyubiquitin chains, as well as Met1-linked linear ubiquitin chains, play a central role in the regulation of IKK and NF-κB [17].

The TNF-NF-κB signaling cascade illustrates the key role of ubiquitin in the regulation of dynamic stimulus-dependent protein–protein interactions. TNF binding to TNF receptor 1 (TNFR1) results in receptor trimerization and the recruitment of multiple signaling proteins to form a receptor proximal signaling complex. Initially, the adaptor protein TRADD is recruited to TNFR1, which subsequently recruits RIP1, the E2 ubiquitin-conjugating enzyme UBCH5, E3 ligases TNF receptor-associated factor 2 (TRAF2) and cIAP1/2, and the LUBAC E3 complex that specifically synthesizes linear ubiquitin chains [18–21]. RIP1 is targeted for K63-linked and linear polyubiquitination by cIAP1/2 and LUBAC respectively, which mediates the recruitment of ubiquitin-binding domain protein complexes, including TAB2/TAB3 and NEMO, and activation of TAK1 and IKK kinases [22–24]. This signaling paradigm appears to be conserved in multiple NF-κB pathways since TAK1 and IKK are recruited and activated by the E3 ligase TRAF6 after it is auto-ubiquitinated in the Toll-like receptor 4 (TLR4)/IL-1R signaling pathways. The E2 enzymes Ubc13 and UbcH5c are important for the K63-linked polyubiquitination of TRAF6 [17, 25, 26].

Given the role of NF-κB as a central mediator of inflammatory processes, tight regulation and homeostatic control is paramount to prevent tissue damage and systemic inflammation. Indeed, NF-κB is negatively regulated by numerous mechanisms, including NF-κB-dependent induction of IκBα in a negative feedback loop [27]. In addition, there are a number of well-described inhibitors of NF-κB, most notably the deubiquitinases A20 (also known as TNFAIP3) and CYLD [28]. Mice lacking A20 develop lethal multi-organ inflammation and cachexia due to uncontrolled NF-κB signaling [29]. A20 also functions as a tumor suppressor in B cells and loss-of-function mutations have been identified in various subtypes of B-cell lymphomas [30]. A20 contains an N-terminal ovarian tumor (OTU) family deubiquitinase domain and seven zinc finger domains [31]. A20 zinc finger four (ZF4) interacts with K63-linked polyubiquitin chains [32], whereas zinc finger 7 (ZF7) binds to linear polyubiquitin chains to suppress LUBAC and TNFR signaling [33, 34]. Early mechanistic studies have shown that A20 functions as a ubiquitin-editing enzyme in the TNFR pathway by removing K63-linked polyubiquitin chains from RIP1 and via intrinsic E3 ligase activity, A20 zinc finger 4 (ZF4) conjugates K48-linked polyubiquitin chains to elicit RIP1 degradation to terminate signaling [35]. Given that the in vitro specificity of the A20 DUB domain was directed for K48-linked, but not K63-linked polyubiquitin chains [36], it was plausible that A20 required essential contributions from other cellular proteins. Indeed, Tax1 binding protein 1 (TAX1BP1), previously shown to interact with TRAF6 and A20 in yeast two-hybrid screens [37, 38], was found to function as an essential ubiquitin-binding adaptor molecule for A20 [39, 40]. In addition, the E3 ligases Itch and RNF11 were also demonstrated to suppress NF-κB signaling by regulating A20 function [41, 42]. Together, A20, TAX1BP1, Itch, and RNF11 inducibly interact in response to TNF or IL-1 stimulation to form an A20 ubiquitin-editing complex that is triggered by TAX1BP1 phosphorylation by IKKα [43]. A20 also terminates NF-κB signaling by disrupting the interactions between E2 ubiquitin-conjugating enzymes and E3 ligases. A20 inhibits TRAF6, TRAF2, and cIAP1 E3 ligase activation by blocking interactions with Ubc13 and UbcH5c resulting in suppression of K63-linked polyubiquitination of RIP1 and TRAF6 in TNFR and TLR4/IL-1R signaling pathways [44].

In this chapter, we focus on the techniques commonly used in our laboratories to elucidate the mechanisms of negative regulation of NF-κB activation. Specifically, protocols on how to generate murine embryonic fibroblasts (MEFs) and examine stimulus-dependent ubiquitination and protein–protein interactions (co-immunoprecipitations) are described.

2   Materials

2.1   Reagents Needed to Make MEFs

MEFs are commonly used to examine the functional effects of gene deletion on NF-κB signaling events. Primary MEFs may be used for experiments; in addition, cells can be easily immortalized after transfection of an SV40 large T antigen plasmid. All reagents should be sterilized to minimize the risk of cell contamination.

1. Pregnant mouse.
2. Dissection instruments: scissors, forceps, razor blade, and scalpel.
3. 1× phosphate buffered saline (PBS).
4. 100 mm diameter tissue culture dishes.

5. Complete DMEM medium containing 20 % fetal bovine serum, heat inactivated, sterile-filtered; L-glutamine; 1× penicillin–streptomycin; and 0.2 % 0.1 M beta mercaptoethanol.

6. 50 ml conical tubes.

7. 6 cc syringe and 18 gauge needle.

8. T75 tissue culture flasks with vented caps.

9. 0.05 and 0.25 % Trypsin/EDTA (1×).

10. SV40 large T-antigen plasmid (kindly provided by Dr. David Ron from New York University).

11. FuGENE® HD transfection reagent (Roche Applied Science).

12. Laminar flow hood.

2.2 Reagents Needed for NF-κB Activation of MEFs, Co-immuno-precipitations, and Ubiquitination Assays

1. TNF.

2. IL-1β.

3. 100 mm diameter tissue culture dishes or 6-well plates.

4. 1.5 ml microcentrifuge tubes.

5. Complete Protease Inhibitor (PI) EDTA-free cocktail tablets (Roche Applied Science; Indianapolis, IN) for inhibition of serine and cysteine proteases.

6. RIPA lysis buffer: 50 mM Tris base pH 6.8, 150 mM NaCl, 1 % Igepal, 0.5 % deoxycholic acid, 0.1 % SDS, 10 mM NaF, 10 mM DTT, 0.2 mM Na_3VO_4, one PI tablet per 10 ml. of buffer, 1 mM phenylmethylsulfonyl fluoride (PMSF) (add freshly in buffer directly before use) (*see* **Note 1**).

7. Bio-Rad protein assay kit (Bio-Rad; Hercules, CA).

8. Protein A agarose beads.

9. Ubiquitination wash buffer: 50 mM Tris base pH 6.8, 150 mM NaCl, 1 % Igepal, 0.5 % deoxycholic acid, 1 M urea, 1 mM N-ethylmaleimide (NEM).

2.3 Reagents Needed for SDS-PAGE and Western Blotting

1. Mini-gel apparatus and power supply.

2. Acrylamide gels. We typically use 8.75 % acrylamide gels to examine RIP1 and TRAF6 polyubiquitination or endogenous protein–protein interactions. The following are needed to make acrylamide gels:

   (a) Lower gel Tris buffer: 1.5 M Tris–HCl pH 8.8, 0.4 % SDS.

   (b) Upper gel Tris buffer: 0.5 M Tris–HCl pH 6.8, 0.4 % SDS.

   (c) 30 % Acrylamide and bis-acrylamide solution (30:0.8), 10 % SDS, 10 % ammonium persulfate (APS), *N,N,N,'*-tetramethyl-ethylenediamine (TEMED) (Sigma Chemical Company, St. Louis, MO, USA).

For 10 ml solution of 8.75 % lower gel add: 4.5 ml of H_2O, 2.5 ml of lower gel Tris buffer, 2.9 ml of acrylamide solution, 100 µl of 10 % SDS, 100 µl of 0 % APS, and 5 µl of TEMED. Wait approximately 60 min for gel to polymerize. For 10 ml solution of 4 % upper gel add: 6 ml of H_2O, 2.5 ml of upper gel Tris buffer, 1.5 ml of acrylamide solution, 100 µl of 10 % SDS, 100 µl of 10 % APS, and 10 µl of TEMED.

3. Molecular weight marker.

4. SDS gel running buffer: 25 mM Tris Base, 192 mM glycine, 0.1 % SDS.

5. 2× sample buffer: 4 % SDS, 62.5 mM Tris pH 6.8, 0.004 % Bromophenol Blue, 10 % Glycerol, 5 % β-mercaptoethanol, 8 M urea.

6. Whatman 3 mm chromatography paper.

7. Western blot transfer buffer: 26.9 mM Tris Base, 194 mM glycine, 20 % methanol.

8. Western blot transfer system.

9. Membrane Wash Buffer (PBST): add 0.25 % Tween® 20 (Sigma) to 1× PBS.

10. Membrane blocking buffer: PBST containing 5 % (w/v) non-fat dry milk powder.

11. BioTrace™ Nitrocellulose Transfer Membrane (Pall Life Sciences; Port Washington, NY).

12. Primary antibodies. For IP and western blotting the following are the sources of the NF-κB pathway and ubiquitin related antibodies we commonly use: Anti-TRAF6 (Santa Cruz Biotechnology, Santa Cruz, CA, USA), anti-TAX1BP1 (Abcam, Cambridge, MA, USA), anti-Itch (BD Biosciences Pharmingen, San Jose, CA, USA), anti-A20 (BD Biosciences Pharmingen), anti-A20 (Santa Cruz Biotechnology, anti-RIP1 (BD Biosciences Pharmingen), anti-Ubc13 (Invitrogen, Grand Island, NY, USA), anti-Ubc13 (Abcam), anti-K63 linkage-specific ubiquitin (Biomol, Farmingdale, NY, USA), anti-K63 linkage specific polyubiquitin (D7A11) rabbit mAb (Cell Signaling Technologies, Danvers, MA, USA), and anti-ubiquitin (Assay Designs, Farmingdale, NY, USA) (*see* **Note 2**).

13. Secondary antibody. Anti-mouse or anti-rabbit IgG horseradish peroxidase (HRP) (GE Healthcare; Pittsburgh, PA, USA).

14. ECL detection reagent-Western Lightning Plus ECL (Perkin Elmer; Waltham, MA, USA).

15. Dry milk powder.

16. Bovine serum albumin (BSA), fraction V (Sigma).

3 Methods

3.1 Generation of MEFs

1. Set up timed breeding and check and record the date of the plug to establish the age of the embryos.

2. When embryos are E13.5 days old, euthanize pregnant mouse by CO_2 asphyxiation and lay the mouse on its back on clean bench paper.

3. Wipe the abdominal area with 70 % ethanol and use scissors to make an incision across the belly. Grasp the skin above and below the cut and tear the skin apart to expose the viscera of the gut.

4. Dissect out the uterus using sterile scissors and forceps and place embryos in a 100 mm tissue culture dish containing sterile 1× PBS in a laminar flow hood.

5. Isolate individual embryos, transfer to a new dish containing 1× PBS, and remove the head and liver using a scalpel and forceps. Place remainder of embryo in a dish containing 5 ml of 0.25 % trypsin/EDTA and mince the embryo with a razor blade into small pieces. Make sure to use a sterile blade for each embryo.

6. Pipet up and down several times to disrupt tissues and place in a tissue culture incubator at 37 °C and 5 % CO_2 for 5 min (*see* **Note 3**).

7. Add 5 ml of complete DMEM, pipet several times, transfer the cell suspension (avoid pieces of tissue and debris) to a 50 ml conical tube containing 20 ml of prewarmed complete DMEM, centrifuge at $100 \times g$ in a tabletop centrifuge for 5 min., resuspend the pellet in 20 ml of fresh DMEM and transfer the cells to a T75 flask.

8. Put the flask back in the incubator at 37 °C. The fibroblasts should grow and divide in 1–2 days. Change medium as necessary.

9. When cells reach confluence (typically after 3–4 days), wash cells gently with 1× PBS, add 2 ml of 0.05 % trypsin/EDTA, and split at a ratio of 1:5 with fresh DMEM.

10. After the second passage of MEFs, the cells can be frozen (3×10^6 cells/vial), seeded in plates or dishes for experiments, or immortalized by transfecting SV40 large T-antigen plasmid using FuGene® HD.

11. For transfection, cells are seeded in 6 well plates (1×10^6) and transfected with 2 μg of the Large T Ag plasmid following the manufacturer's instructions. The cells will become immortalized after 4–6 weeks.

3.2 Co-immuno-precipitation and Ubiquitination Assays

1. Seed MEFs in 100 mm tissue culture dishes (minimum of 2×10^7 cells per sample) at ~80 % confluence and the next day treat with either TNF (10 ng/ml) or IL-1β (10 ng/ml). To examine endogenous RIP1 or TRAF6 K63-linked polyubiquitination, treat cells with TNF or IL-1β for 30 or 60 min.

2. Wash cells with ice cold 1× PBS, remove the PBS by spinning down $(500 \times g)$ for 10 s at high speed and lyse the pellet in 750 µl of RIPA buffer. Complete lysis of the cell pellet is important for detection of endogenous protein–protein interactions (*see* **Note 1**).

3. Spin down the lysate at high speed $(10,000 \times g)$ for 10 min at 4 °C and transfer the lysate supernatant to a new prechilled 1.5 ml tube. Quantitate protein amounts and use equal amounts of total protein (generally ~600–700 µg) for IPs.

4. For the pre-clearing step, add 25 µl of protein A agarose beads and rotate in the cold room for 90 min.

5. Centrifuge for 15 s at $10,000 \times g$ and transfer the supernatants to fresh prechilled tubes.

6. Add 0.5 µg of primary antibody (e.g., TRAF6 or RIP1) to immunoprecipitate the protein of interest and rotate in the cold room overnight. *See* Fig. 1 for an example of IL-1β-inducible protein interactions with TRAF6.

7. The following day, add 30–40 µl of protein A agarose beads and rotate for an additional 2–3 h.

8. Centrifuge, aspirate the supernatants, and wash the beads three times with ~1 ml of RIPA buffer (*see* **Note 4**).

9. Add 30–40 µl of 2× sample buffer and boil for 5 min at 95 °C before loading the gel.

3.3 SDS-PAGE and Western Blotting

1. Prepare acrylamide gels, assemble the mini-gel apparatus, and add SDS running buffer.

2. Load MW protein marker (in a separate lane) and samples on the gel and electrophorese at 100 V until the blue dye front exits the gel.

3. Disassemble the glass plates and set up the gel transfer using pre-cut chromatography paper and nitrocellulose membrane. Transfer the gel in the cold room overnight at 30 V using a western blot transfer system.

4. For ubiquitination assays of endogenous proteins, autoclave the membrane for 30 min (put the membrane in ddH$_2$O) and allow the membrane to cool down to room temperature. *See* Fig. 2 for representative endogenous RIP1 and TRAF6 K63-linked polyubiquitination assays.

5. Block the membrane in 5 % nonfat milk prepared in 1× PBST for 1 h at room temperature with gentle shaking before adding

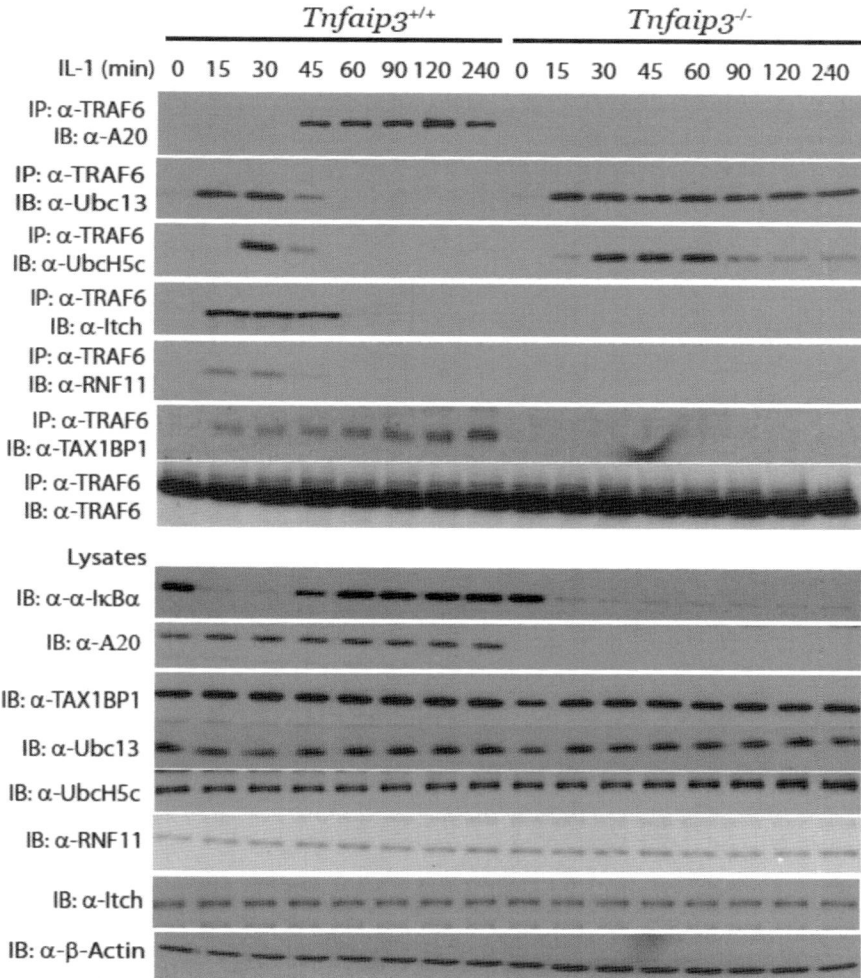

Fig. 1 Disruption of interactions between E2 and E3 enzymes in the IL-1R pathway by A20 and TAX1BP1. Kinetics of TRAF6, Ubc13, UbcH5c, Itch, RNF11, A20, and TAX1BP1 interactions in control and A20-deficient MEFs. *A20*⁺/⁺ and *A20*⁻/⁻ MEFs were stimulated with IL-1 for the indicated times. Proteins from lysates were immunoprecipitated with TRAF6 antibody and detected by immunoblotting with antibodies to A20, Ubc13, UbcH5c, Itch, RNF11, TAX1BP1, or TRAF6. Lysates were subjected to immunoblotting with anti-IκBα, A20, TAX1BP1, Ubc13, UbcH5c, RNF11, Itch, and β-actin. Figure reproduced from ref. 44

primary antibody (0.2 μg/ml for Santa Cruz antibodies, 0.05 μg/ml for others) in 1× PBST (*see* **Note 5**).

6. Shake gently overnight in the cold room.

7. The following day, wash the membrane three times (10 min each) at room temperature with 1× PBST.

8. Add the secondary antibody (1:7,000) and shake at room temperature for 1 h.

9. Wash the membrane five times (10 min each) at room temperature with 1× PBST.

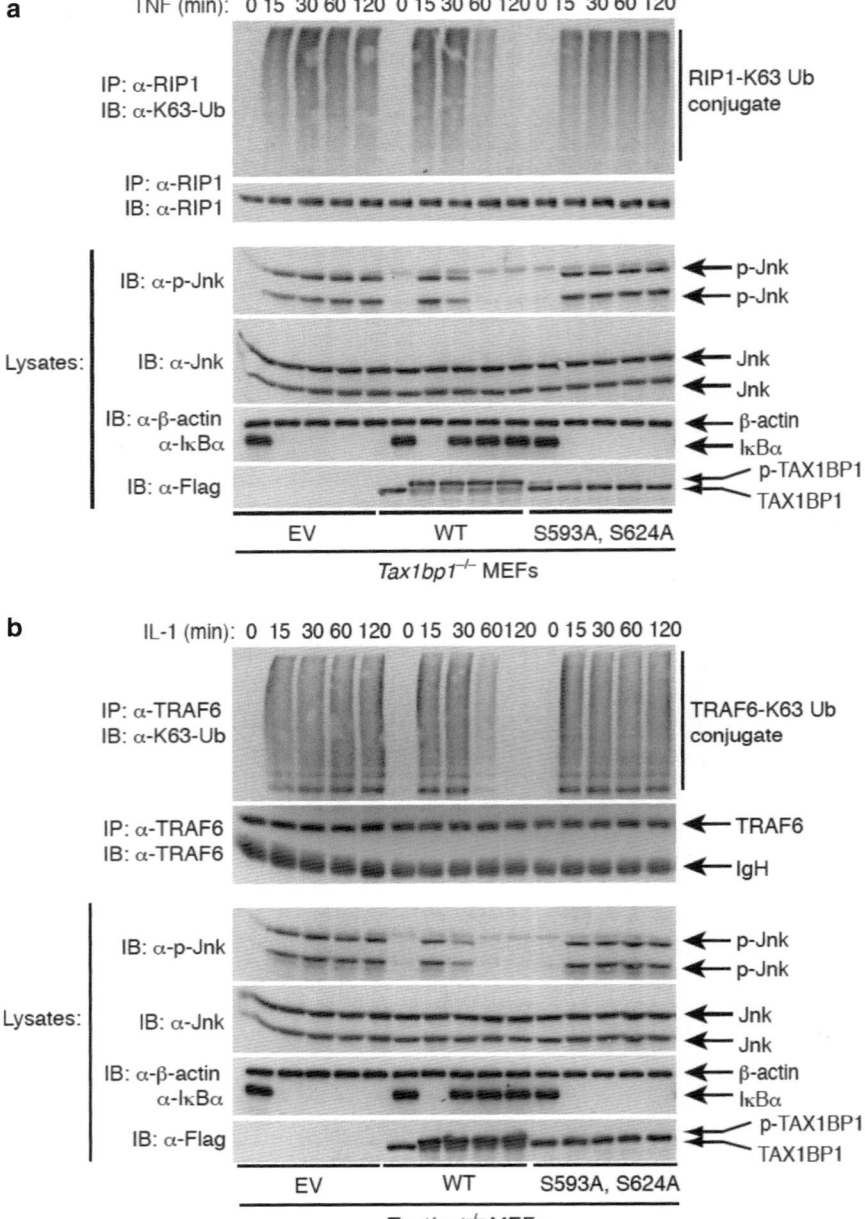

Fig. 2 TAX1BP1 phosphorylation is essential for the termination of NF-κB signaling, JNK phosphorylation and RIP1 ubiquitination. (**a**, **b**) Immunoassay of lysates of *Tax1bp1*⁻/⁻ MEFs transfected with empty vector, Flag-TAX1BP1 or Flag-TAX1BP1 S593A, S624A, then treated with TNF (**a**) or IL-1 (**b**) for the indicated times. Immunoprecipitation was performed with anti-RIP1 and immunoblot with anti-K63-specific ubiquitin (K63-Ub) and anti-RIP1 (**a**). Immunoprecipitation was also performed with anti-TRAF6 and immunoblot with anti-K63-specific ubiquitin (K63-Ub) and anti-TRAF6 (**b**). Immunoblot was also performed with anti-JNK, anti-phospho-JNK, anti-IκBα, anti-β-actin, and anti-Flag (**a**, **b**). Figure reproduced from ref. 43

10. Add ECL reagents to the membrane and incubate for 1 min. Remove the ECL reagents, dry the membrane and wrap in plastic wrap.

11. Expose the membrane to film and develop in a dark room.

4 Notes

1. For ubiquitination assays also supplement the lysis buffer with 1 mM NEM to inhibit deubiquitinases in the lysates.

2. When conducting IPs for endogenous proteins, it is advisable to use a negative control antibody with the same isotype as the primary antibody used for IP.

3. At this step, the embryo tissue can also be passed through a sterile 6 cc syringe with an 18 gauge needle to remove pieces of tissue and debris prior to incubation at 37 °C.

4. For ubiquitination assays include an extra wash with ubiquitination wash buffer (RIPA buffer containing 1 M urea and 1 mM NEM) to reduce background from ubiquitinated proteins that may have precipitated with the protein of interest.

5. If blotting with phospho-specific antibodies from Cell Signaling Technologies, incubate membrane overnight with diluted antibody in 5 % (w/v) BSA, 1× TBS, and 0.1 % Tween® 20 in the cold room.

Acknowledgements

This work was supported by National Institutes of Health grants (RO1GM083143 and RO1CA135362 to E.W.H.) and a Stanley J. Glaser Research Award and American Cancer Society Institutional Grant award to N.S. We thank Drs. Alfonso Lavorgna and Soratree Charoenthongtrakul for critical reading of the manuscript.

References

1. Vallabhapurapu S, Karin M (2009) Regulation and function of NF-kappaB transcription factors in the immune system. Annu Rev Immunol 27:693–733

2. Karin M, Greten FR (2005) NF-kappaB: linking inflammation and immunity to cancer development and progression. Nat Rev Immunol 5:749–759

3. Baldwin AS Jr (1996) The NF-kappa B and I kappa B proteins: new discoveries and insights. Annu Rev Immunol 14:649–683

4. Perkins ND (2007) Integrating cell-signalling pathways with NF-kappaB and IKK function. Nat Rev Mol Cell Biol 8:49–62

5. Ghosh S, Hayden MS (2008) New regulators of NF-kappaB in inflammation. Nat Rev Immunol 8:837–848

6. Hacker H, Karin M (2006) Regulation and function of IKK and IKK-related kinases. Sci STKE 2006(357):re13

7. Pomerantz JL, Baltimore D (2002) Two pathways to NF-kappaB. Mol Cell 10:693–695

8. Karin M, Ben-Neriah Y (2000) Phosphorylation meets ubiquitination: the control of NF-kappaB activity. Annu Rev Immunol 18:621–663

9. Xiao G, Harhaj EW, Sun SC (2001) NF-kappaB-inducing kinase regulates the processing of NF-kappaB2 p100. Mol Cell 7:401–409

10. Sun SC (2011) Non-canonical NF-kappaB signaling pathway. Cell Res 21:71–85

11. Dejardin E, Droin NM, Delhase M, Haas E, Cao Y, Makris C, Li ZW, Karin M, Ware CF, Green DR (2002) The lymphotoxin-beta receptor induces different patterns of gene expression via two NF-kappaB pathways. Immunity 17:525–535

12. Zarnegar BJ, Wang Y, Mahoney DJ, Dempsey PW, Cheung HH, He J, Shiba T, Yang X, Yeh WC, Mak TW, Korneluk RG, Cheng G (2008) Noncanonical NF-kappaB activation requires coordinated assembly of a regulatory complex of the adaptors cIAP1, cIAP2, TRAF2 and TRAF3 and the kinase NIK. Nat Immunol 9: 1371–1378

13. Vallabhapurapu S, Matsuzawa A, Zhang W, Tseng PH, Keats JJ, Wang H, Vignali DA, Bergsagel PL, Karin M (2008) Nonredundant and complementary functions of TRAF2 and TRAF3 in a ubiquitination cascade that activates NIK-dependent alternative NF-kappaB signaling. Nat Immunol 9:1364–1370

14. Hershko A, Ciechanover A (1998) The ubiquitin system. Annu Rev Biochem 67:425–479

15. Meierhofer D, Wang X, Huang L, Kaiser P (2008) Quantitative analysis of global ubiquitination in HeLa cells by mass spectrometry. J Proteome Res 7:4566–4576

16. Chen ZJ, Sun LJ (2009) Nonproteolytic functions of ubiquitin in cell signaling. Mol Cell 33:275–286

17. Chen ZJ (2005) Ubiquitin signalling in the NF-kappaB pathway. Nat Cell Biol 7:758–765

18. Tada K, Okazaki T, Sakon S, Kobarai T, Kurosawa K, Yamaoka S, Hashimoto H, Mak TW, Yagita H, Okumura K, Yeh WC, Nakano H (2001) Critical roles of TRAF2 and TRAF5 in tumor necrosis factor-induced NF-kappa B activation and protection from cell death. J Biol Chem 276:36530–36534

19. Wertz IE, Dixit VM (2008) Ubiquitin-mediated regulation of TNFR1 signaling. Cytokine Growth Factor Rev 19:313–324

20. Micheau O, Tschopp J (2003) Induction of TNF receptor I-mediated apoptosis via two sequential signaling complexes. Cell 114: 181–190

21. Haas TL, Emmerich CH, Gerlach B, Schmukle AC, Cordier SM, Rieser E, Feltham R, Vince J, Warnken U, Wenger T, Koschny R, Komander D, Silke J, Walczak H (2009) Recruitment of the linear ubiquitin chain assembly complex stabilizes the TNF-R1 signaling complex and is required for TNF-mediated gene induction. Mol Cell 36:831–844

22. Bertrand MJ, Milutinovic S, Dickson KM, Ho WC, Boudreault A, Durkin J, Gillard JW, Jaquith JB, Morris SJ, Barker PA (2008) cIAP1 and cIAP2 facilitate cancer cell survival by functioning as E3 ligases that promote RIP1 ubiquitination. Mol Cell 30:689–700

23. Mahoney DJ, Cheung HH, Mrad RL, Plenchette S, Simard C, Enwere E, Arora V, Mak TW, Lacasse EC, Waring J, Korneluk RG (2008) Both cIAP1 and cIAP2 regulate TNFalpha-mediated NF-kappaB activation. Proc Natl Acad Sci U S A 105:11778–11783

24. Ea CK, Deng L, Xia ZP, Pineda G, Chen ZJ (2006) Activation of IKK by TNFalpha requires site-specific ubiquitination of RIP1 and polyubiquitin binding by NEMO. Mol Cell 22: 245–257

25. Xia ZP, Sun L, Chen X, Pineda G, Jiang X, Adhikari A, Zeng W, Chen ZJ (2009) Direct activation of protein kinases by unanchored polyubiquitin chains. Nature 461:114–119

26. Yamamoto M, Okamoto T, Takeda K, Sato S, Sanjo H, Uematsu S, Saitoh T, Yamamoto N, Sakurai H, Ishii KJ, Yamaoka S, Kawai T, Matsuura Y, Takeuchi O, Akira S (2006) Key function for the Ubc13 E2 ubiquitin-conjugating enzyme in immune receptor signaling. Nat Immunol 7:962–970

27. Sun SC, Ganchi PA, Ballard DW, Greene WC (1993) NF-kappa B controls expression of inhibitor I kappa B alpha: evidence for an inducible autoregulatory pathway. Science 259:1912–1915

28. Sun SC (2008) Deubiquitylation and regulation of the immune response. Nat Rev Immunol 8:501–511

29. Lee EG, Boone DL, Chai S, Libby SL, Chien M, Lodolce JP, Ma A (2000) Failure to regulate TNF-induced NF-kappaB and cell death responses in A20-deficient mice. Science 289:2350–2354

30. Kato M, Sanada M, Kato I, Sato Y, Takita J, Takeuchi K, Niwa A, Chen Y, Nakazaki K, Nomoto J, Asakura Y, Muto S, Tamura A, Iio M, Akatsuka Y, Hayashi Y, Mori H, Igarashi T, Kurokawa M, Chiba S, Mori S, Ishikawa Y, Okamoto K, Tobinai K, Nakagama H, Nakahata T, Yoshino T, Kobayashi Y, Ogawa S (2009) Frequent inactivation of A20 in B-cell lymphomas. Nature 459:712–716

31. Harhaj EW, Dixit VM (2012) Regulation of NF-kappaB by deubiquitinases. Immunol Rev 246:107–124

32. Bosanac I, Wertz IE, Pan B, Yu C, Kusam S, Lam C, Phu L, Phung Q, Maurer B, Arnott D, Kirkpatrick DS, Dixit VM, Hymowitz SG (2010) Ubiquitin binding to A20 ZnF4 is required for modulation of NF-kappaB signaling. Mol Cell 40:548–557

33. Verhelst K, Carpentier I, Kreike M, Meloni L, Verstrepen L, Kensche T, Dikic I, Beyaert R (2012) A20 inhibits LUBAC-mediated NF-kappaB activation by binding linear polyubiquitin chains via its zinc finger 7. EMBO J 31:3845–3855

34. Tokunaga F, Nishimasu H, Ishitani R, Goto E, Noguchi T, Mio K, Kamei K, Ma A, Iwai K, Nureki O (2012) Specific recognition of linear polyubiquitin by A20 zinc finger 7 is involved in NF-kappaB regulation. EMBO J 31:3856–3870

35. Wertz IE, O'Rourke KM, Zhou H, Eby M, Aravind L, Seshagiri S, Wu P, Wiesmann C, Baker R, Boone DL, Ma A, Koonin EV, Dixit VM (2004) De-ubiquitination and ubiquitin ligase domains of A20 downregulate NF-kappaB signalling. Nature 430:694–699

36. Komander D, Reyes-Turcu F, Licchesi JD, Odenwaelder P, Wilkinson KD, Barford D (2009) Molecular discrimination of structurally equivalent Lys 63-linked and linear polyubiquitin chains. EMBO Rep 10:466–473

37. De Valck D, Jin DY, Heyninck K, Van de Craen M, Contreras R, Fiers W, Jeang KT, Beyaert R (1999) The zinc finger protein A20 interacts with a novel anti-apoptotic protein which is cleaved by specific caspases. Oncogene 18: 4182–4190

38. Ling L, Goeddel DV (2000) T6BP, a TRAF6-interacting protein involved in IL-1 signaling. Proc Natl Acad Sci U S A 97:9567–9572

39. Shembade N, Harhaj NS, Liebl DJ, Harhaj EW (2007) Essential role for TAX1BP1 in the termination of TNF-alpha-, IL-1- and LPS-mediated NF-kappaB and JNK signaling. EMBO J 26:3910–3922

40. Iha H, Peloponese JM, Verstrepen L, Zapart G, Ikeda F, Smith CD, Starost MF, Yedavalli V, Heyninck K, Dikic I, Beyaert R, Jeang KT (2008) Inflammatory cardiac valvulitis in TAX1BP1-deficient mice through selective NF-kappaB activation. EMBO J 27:629–641

41. Shembade N, Harhaj NS, Parvatiyar K, Copeland NG, Jenkins NA, Matesic LE, Harhaj EW (2008) The E3 ligase Itch negatively regulates inflammatory signaling pathways by controlling the function of the ubiquitin-editing enzyme A20. Nat Immunol 9:254–262

42. Shembade N, Parvatiyar K, Harhaj NS, Harhaj EW (2009) The ubiquitin-editing enzyme A20 requires RNF11 to downregulate NF-kappaB signalling. EMBO J 28:513–522

43. Shembade N, Pujari R, Harhaj NS, Abbott DW, Harhaj EW (2011) The kinase IKKalpha inhibits activation of the transcription factor NF-kappaB by phosphorylating the regulatory molecule TAX1BP1. Nat Immunol 12: 834–843

44. Shembade N, Ma A, Harhaj EW (2010) Inhibition of NF-kappaB signaling by A20 through disruption of ubiquitin enzyme complexes. Science 327:1135–1139

Chapter 17

Immunoblot Analysis of Linear Polyubiquitination of NEMO

Yoshiteru Sasaki, Hiroaki Fujita, Misa Nakai, and Kazuhiro Iwai

Abstract

Stimulation with inflammatory cytokines such as TNF-α and IL-1 activates the canonical NF-κB pathway through the activation of the IKK complex. The mechanism underlying IKK activation has been extensively studied and the involvement of the ubiquitin system has been well documented. We have recently reported that a novel ubiquitin ligase complex, LUBAC is involved in the activation of the IKK complex. LUBAC consists of one catalytic subunit, HOIP and two accessory molecules, HOIL-1L and SHARPIN and activates the IKK complex by conjugating the linear polyubiquitin chains to NEMO (IKKγ), the regulatory subunit of IKK complex. In this chapter, we describe the protocol for the detection of the linear polyubiquitination of NEMO by the immunoblotting using anti-linear ubiquitin antibody.

Key words LUBAC, Linear polyubiquitin chain, NEMO (IKKγ), IKK complex

1 Introduction

The NF-κB transcription factor controls the expression of genes involved in inflammation, immune responses and cell survival [1]. Five NF-κB proteins have been described in mammalian cells: RelA, c-Rel, RelB, p50/NF-κB1, and p52/NF-κB2. p50/NF-κB1 and p52/NF-κB2 are generated by the processing of their precursor forms, p105 and p100, respectively. NF-κB forms a heterodimer or homodimer of Rel family proteins and resides in the cytoplasm through binding to its inhibitor proteins named inhibitor of κB (IκB) in resting cells. NF-κB activation is mediated through either the canonical or alternative pathways [2]. In the canonical NF-κB pathway, stimulation of various receptors such as tumor necrosis factor (TNF) receptor, Toll-like receptors (TLRs), CD40, and lymphocyte antigen receptor activates the IκB kinase (IKK) complex consisting of IKK1(α) and IKK2(β), and NEMO (also called IKKγ), which phosphorylates IκB proteins and leads to their degradation by the 26S proteasome. Stimulation of BAFF-R, CD40 and lymphotoxin-β receptor activates the alternative NF-κB pathway in which p100/NF-κB2 is processed to p52/NF-κB2.

Michael J. May (ed.), *NF-kappa B: Methods and Protocols*, Methods in Molecular Biology, vol. 1280, DOI 10.1007/978-1-4939-2422-6_17, © Springer Science+Business Media New York 2015

The ubiquitin system is known to be involved in the activation of NF-κB [3]. Ubiquitination is one of a number of posttranslational modifications of cellular proteins. In most cases, the ubiquitin system modifies and controls the function of proteins through the attachment of polyubiquitin chains [4, 5]. The polyubiquitin chains are generated via one of seven Lys (K) residues of ubiquitin via a cascade of reactions catalyzed by three enzymes: ubiquitin-activating enzyme (E1), ubiquitin-conjugating enzyme (E2), and ubiquitin-protein ligase (E3) (Fig. 1). Each type of polyubiquitin chain regulates the conjugated protein differently [6]. For example, K48-linked polyubiquitin works as a signal for proteasome-mediated degradation and plays an important role in the degradation of IκB proteins. K63-linked polyubiquitin chains are involved in signal transduction and DNA repair and are proposed to be involved in the activation of the IKK complex [7]. However cells lacking Ubc13, an E2, which specifically generates K63 chains showed almost normal activation of NF-κB by TNF-α, IL-1, and TLR stimulation [8] and K63-linked chains were found to be dispensable for TNF-α-mediated NF-κB activation [9]. We recently identified a novel ubiquitin ligase complex named linear ubiquitin chain assembly complex (LUBAC), which specifically generates N-terminally linked linear polyubiquitin chains [10, 11]. LUBAC is composed of HOIP, HOIL-1L, and SHARPIN and the RING-IBR-RING domain of HOIP mediates the ubiquitin

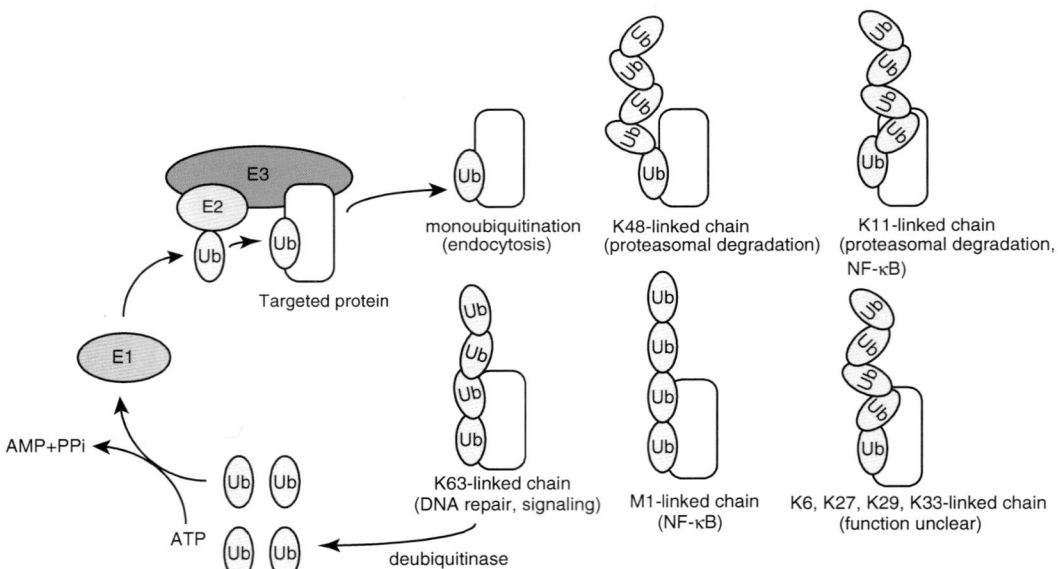

Fig. 1 The ubiquitin system and its cellular functions. Ubiquitin is conjugated to targeted proteins by the cooperative reaction of three enzymes, E1, E2, and E3. How ubiquitination of the protein regulates its cellular function depends on the type of ubiquitin chains. Ubiquitin is recycled by the activity of deubiquitinating enzymes

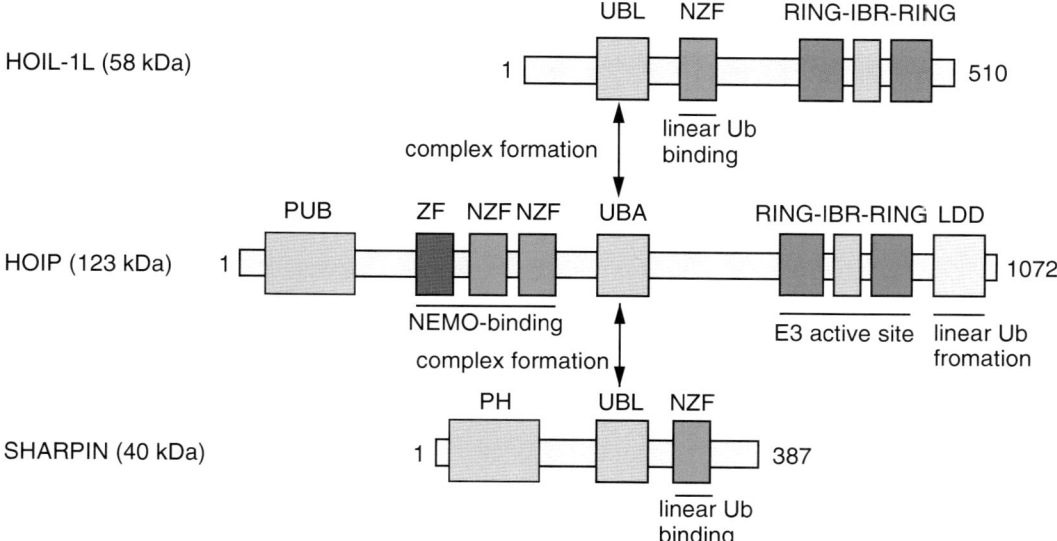

Fig. 2 Schematic representation of LUBAC. LUBAC is composed of HOIL-1L, HOIP, and SHARPIN. The RING-IBR-RING domain of HOIP is the E3 active site. *UBL* ubiquitin-like, *NZF* NPl4-type zinc finger, *RING* really interesting new gene, *IBR* in-between RING, *ZF* zinc finger, *UBA* ubiquitin-associated domains, *PUB* PNGase/UBA or UBX, *PH* Pleckstrin homology domain

ligase activity [11, 12] (Fig. 2). We and other groups showed that TNF-α-mediated activation of canonical NF-κB pathway was impaired in cells lacking either HOIL-1L or SHARPIN and LUBAC conjugated linear polyubiquitin chain to NEMO and activated the IKK complex [11, 12]. It is reported that the UBAN domain of NEMO preferentially binds to the linear polyubiquitin chain [13, 14]. Moreover, we recently demonstrated that linearly polyubiquitinated NEMO activates IKK more potently than unanchored linear chains and linear chain-mediated activation of IKK2 involved homotypic interaction of the IKK2 kinase domain [15]. Thus, we propose the following model for the function of the linear polyubiquitin chain in the activation of the IKK complex. The linear polyubiquitin chains conjugated to NEMO are recognized by the UBAN domain of NEMO in another IKK complex, inducing the multimerization of the IKK complex. Upon multimerization, IKK2 could dimerize and be phosphorylated by transphosphorylation (Fig. 3).

Because the importance of the linear polyubiquitin chain in the regulation of NF-κB system is becoming clear, the development of a convenient method to examine the linear polyubiquitination of cellular proteins has been awaited. In order to detect the linear polyubiquitination of proteins easily, we tried and succeeded to generate a monoclonal antibody which specifically recognizes the linear polyubiquitin chain [16]. In this chapter we describe the

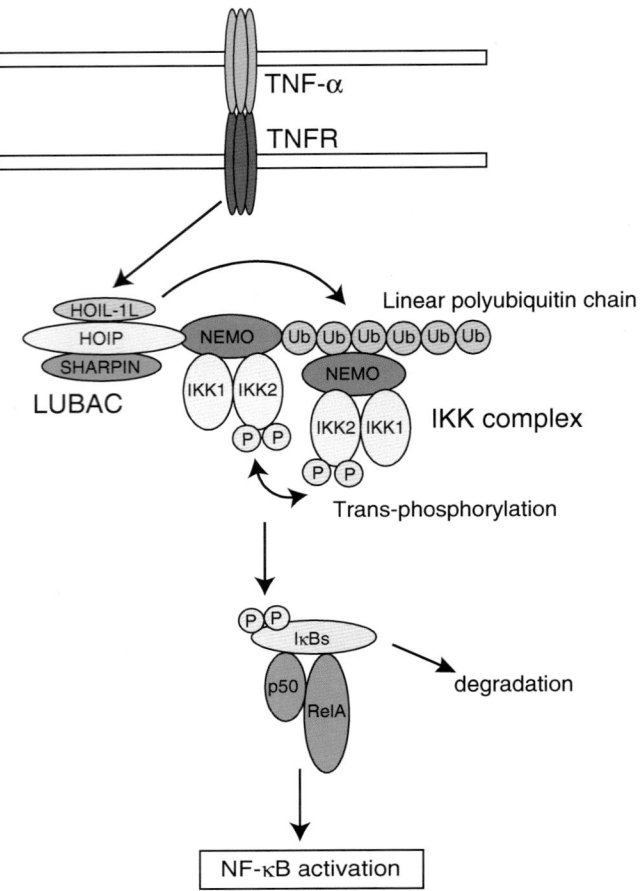

Fig. 3 Schematic representation of the LUBAC-mediated activation of the canonical NF-κB pathway. Upon TNF-α stimulation, LUBAC conjugates linear polyubiquitin chains to NEMO. Linear polyubiquitin chains conjugated to NEMO could be recognized by NEMO in another IKK complex and induce multimerization of the IKK complex. Upon multimerization, IKK2 dimerizes and is phosphorylated by trans-autophosphorylation and then activated. The activated IKK complex phosphorylates IκB protein and induces the degradation of IκB protein by the proteasome

protocol for the detection of the linear polyubiquitination of protein by immunoblotting using this anti-linear ubiquitin antibody. We show the linear polyubiquitination of NEMO induced by TNF-α stimulation in human Jurkat T cell line as an example. We also introduce the hot lysis method by which we can detect linear ubiquitin chains covalently attached to the protein and show the linear polyubiquitination of NEMO by the overexpression of LUBAC in HEK293T cells.

2 Materials

2.1 Cell Culture

1. Jurkat T cells and HEK293T human embryonic kidney cells (American Type Culture Collection, Manassas, VA, USA).

2. RPMI 1640 medium supplemented with 10 % fetal bovine serum (FBS), and penicillin–streptomycin solution for Jurkat cells.

3. Dulbecco's Modified Eagle's Medium (DMEM) supplemented with 10 % FBS, and penicillin–streptomycin solution for HEK293T cells.

4. Human recombinant TNF-α dissolved at 0.1 mg/ml in sterile water.

5. Phosphate buffered-saline (PBS, pH 7.4).

6. 10 cm culture dishes.

7. 10 cm culture dishes coated with Collagen type 1.

8. 15 ml centrifuge tubes.

2.2 Transfection

1. Lipofectamine 2000 transfection reagent (Invitrogen, Carlsbad, CA, USA).

2. Opti-MEM reduced serum medium (Life Technologies, Carlsbad, CA, USA).

3. High expression level Mammalian Expression plasmids (e.g., pcDNA3.1) containing subcloned cDNAs encoding HA-tagged HOIP, MYC-tagged HOIL-1L, T7-tagged SHARPIN, and FLAG-tagged NEMO.

2.3 Cell Lysis

1. 1 % Triton X-100 lysis buffer: 1 % (w/v) Triton X-100, 20 mM Tris–HCl (pH 7.5), 150 mM NaCl.

2. 1 % sodium dodecyl sulfate (SDS) in PBS.

3. 200 mM phenylmethylsulfonyl fluoride (PMSF): Dissolve 1.7149 g of PMSF in 50 ml of ethanol. Store at –20 °C.

4. Protease inhibitor cocktail (Sigma-Aldrich, St. Louis, MO, USA).

5. Phosphatase inhibitor cocktail (Nacalai Tesque, Kyoto, Japan).

6. 1.5 ml microcentrifuge tubes.

7. 5 ml microcentrifuge tubes.

8. 1 ml syringes.

9. 25 gauge needles.

10. Bradford protein assay solution. We use a commercially available Bradford protein assay solution (Protein assay CBB solution) (Nacalai Tesque).

11. Bovine Serum Albumin (BSA).

12. Spectrophotometer.

2.4 Immuno-precipitation

1. Protein A Sepharose (GE Healthcare, Mickleton, NJ, USA).

2. Immunoprecipitating antibodies: anti-NEMO (Santa Cruz Biotechnology, Santa Cruz, CA, USA), Rabbit anti-FLAG polyclonal antibody (Sigma-Aldrich, St. Louis, MO, USA).

2.5 Sodium Dodecyl Sulfate–Polyacrylamide Gel Electrophoresis (SDS-PAGE)

1. 29:1 acrylamide–*bis*-acrylamide solution: 29 g acrylamide and 1 g *bis*-acrylamide in 100 ml dH$_2$O.

2. Lower Buffer: 0.5 M Tris–HCl (pH 8.8), 0.4 % SDS (w/v).

3. Upper Buffer: 1.5 M Tris–HCl, (pH 6.8), 0.4 % SDS (w/v).

4. Ammonium persulfate: 10 % solution in water.

5. *N*,*N*,*N*,'-Tetramethyl-ethylenediamine (TEMED).

6. Electrophoresis Buffer: 25 mM Tris, 192 mM glycine, 0.1 % SDS.

7. 2× Laemmli Sample Buffer: 125 mM Tris–HCl (pH 6.8), 10 % 2-mercaptoethanol, 4 % SDS, 10 % glycerol, 0.004 % bromophenol blue.

8. PAGE gel casting apparatus. We use the AE-6500 Dual Mini Slab system (ATTO, Tokyo, Japan).

2.6 Immunoblotting

1. Polyvinylidene Difluoride (PVDF) transfer membrane (0.45 μm).

2. Whatman 3M Filter paper.

3. Transfer Buffer: 100 mM Tris, 192 mM glycine, 20 % (v/v) methanol.

4. Electroblotting transfer apparatus. We use the Bio-Rad Trans-blot cell system (Bio-Rad, Hercules, CA, USA).

2.7 Detection of Linear Polyubiquitin Chains

1. TBST: 20 mM Tris–HCl (pH 7.6), 138 mM NaCl, 0.1 % Tween-20.

2. Blocking Buffer: 5 % nonfat dry milk in TBST.

3. Primary antibodies: mouse anti-linear ubiquitin monoclonal antibody (LUB9), mouse anti-phospho-IκBα (Ser32/36) monoclonal antibody (Cell Signaling Technology, Danvers, MA, USA), rabbit anti-IκBα antibody (Santa Cruz Biotechnology), rabbit anti-NEMO antibody (Santa Cruz Biotechnology), mouse anti-Tubulin monoclonal antibody (Santa Cruz Biotechnology), and mouse anti-FLAG monoclonal antibody (M2) (Sigma-Aldrich).

4. Secondary antibodies: HRP-conjugated horse anti-mouse IgG (Cell Signaling Technology), HRP-conjugated donkey anti-rabbit IgG (GE Healthcare).

5. Chemiluminescent Substrates. For detection of immunoblotted proteins we use SuperSignal West Pico Chemiluminescent Substrate (Thermo, Waltham, MA, USA) and SuperSignal West Dura Extended Duration Substrate (Thermo).

6. Imaging System such as ImageQuant LAS4000 (GE Healthcare) or the ChemiDoc MP System (Bio-Rad).

7. Antibody Stripping Buffer: 62.5 mM Tris–HCl (pH 6.7), 0.2 % SDS, 100 mM 2-mercaptoethanol (*see* **Note 1**).

3   Methods

In this section, we describe the protocol for immunoblot detection of linear polyubiquitination of NEMO induced by TNF-α stimulation in human Jurkat T cell line. However, this protocol can be readily used for other cell types with some modifications. Results from a representative experiment are shown in Fig. 4.

3.1 Cell Culture

1. Grow Jurkat cells in RPMI 1640 medium supplemented with 10 % FCS, and penicillin–streptomycin solution.

2. Prepare the required number of 15 ml tubes.

3. Count cell number and suspend 1.5×10^7 cells in 2 ml culture medium in a 15 ml tube.

4. Pre-warm the cells in 37 °C water bath for 30 min.

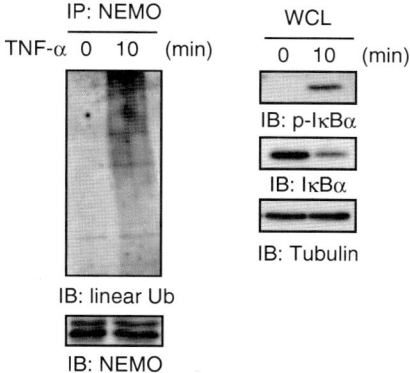

Fig. 4 Jurkat cells were stimulated with 10 ng/ml of human TNF-α for the indicated times. Whole-cell lysates (WCL) were prepared and immunoprecipitated with an anti-NEMO antibody and then immunoblotted with an anti-linear ubiquitin antibody (LUB9) (*left panel*). WCL were also immunoblotted with anti-phospho-IκBα and anti-IκBα antibody (*right panel*) to confirm the activation of the canonical NF-κB pathway. Tubulin was used as a loading control

5. Stimulate the cells with 10 ng/ml human TNF-α for the desired times.

6. Stop the stimulation quickly by adding 10 ml of ice-cold PBS.

7. Spin down the cells at $1,600 \times g$ at 4 °C for 1 min and carefully aspirate the supernatant and remove as much medium as possible (*see* **Note 2**).

3.2 Preparation of Whole-Cell Lysates

1. Suspend the pellets in 250 μl of 1 % Triton X-100 lysis buffer and transfer to a prechilled 1.5 ml microcentrifuge tube (*see* **Note 3**). Incubate on ice for 10 min.

2. Centrifuge the samples for 10 min at $20,000 \times g$ at 4 °C.

3. Transfer the supernatants to new prechilled 1.5 ml microcentrifuge tubes.

3.3 Measuring the Protein Concentration of Whole-Cell Lysates

1. Prepare CBB reagent by diluting one part of Protein assay CBB solution with three parts of distilled, deionized water. Prepare five dilutions of BSA (0, 4, 8, 16, 20 mg/ml) as a standard. Add 1 μl of standard solution and sample solution to 1 ml CBB reagent. Incubate at room temperature for 10 min.

2. Measure absorbance at 595 nm and calculate the protein concentration.

3. Place 20 μg of whole-cell lysate into 1.5 ml microcentrifuge tubes and add the same volume of 2× Laemmli sample buffer and mix well.

4. Incubate the samples at 95 °C for 5 min. These whole-cell lysate samples will be used to determine whether the canonical NF-κB pathway is activated normally by TNF-α stimulation.

3.4 Immunoprecipitation of NEMO

1. Place 1,600 μg of whole-cell lysates in 1.5 ml microcentrifuge tubes and add 1 μg of anti-NEMO antibody to each sample. Incubate samples at 4 °C for 60 min while rotating the tubes on a rotator.

2. Add 20 μl of protein A sepharose and incubate at 4 °C for 60 min with rotation.

3. Pellet the sepharose beads by centrifugation ($9,000 \times g$ at 4 °C for 1 min). Carefully aspirate the supernatants.

4. Add 1 ml of ice-cold 1 % Triton X-100 lysis buffer to the tubes and wash the sepharose beads by inverting the tubes several times.

5. Pellet the sepharose beads by centrifugation ($9,000 \times g$ at 4 °C for 1 min). Carefully aspirate the supernatants. Repeat this washing step for four times.

6. Remove the final wash as completely as possible. Add 20 μl of 2× Laemmli sample buffer to the tubes and heat them at 95 °C for 5 min.

3.5 SDS-PAGE

1. Prepare a 7.5 % acrylamide SDS-PAGE gel as follows: For the running gel mix together in a beaker 3.75 ml of 29:1 acryl-amide–*bis*-acrylamide solution, 3.75 ml Lower Buffer, 7.5 ml dH$_2$O, 50 µl of 10 % APS, 10 µl TEMED. Pour into gel apparatus and allow the solution to polymerize. Make up the stacking gel by mixing together 0.65 ml of 29:1 acrylamide–*bis*-acrylamide solution, 1.25 ml of Upper Buffer, 3.05 ml of dH$_2$O, 25 µl of 10 % APS, and 5 µl of TEMED. When the running gel has polymerized add the stacking gel above the running gel and place an appropriately sized comb in the stacking gel. Allow the stacking gel to polymerize.

2. Load the samples and pre-stained molecular marker to the wells of the SDS-PAGE gel.

3. Run the samples at constant current until the bromophenol blue (BPB) in the sample buffer gets to the bottom of the gel.

4. Disassemble the electrophoresis apparatus and take out the gel.

3.6 Western Blotting

1. Place the gel in transfer buffer for 1 min.

2. Cut a piece of PVDF membrane and four pieces of Whatman 3M paper to be slightly larger than the gel.

3. Soak the PVDF membrane in 100 % methanol for 1 min and the place it in transfer buffer.

4. Place two Whatman 3M papers presoaked in transfer buffer on the Bio-Rad Trans-Blot Semidry apparatus.

5. Place the PVDF membrane in top of the blot papers and then put the gel on the membrane.

6. Place two Whatman 3M papers presoaked in transfer buffer on the gel.

7. Carefully place the upper electrode on the top of the stack.

8. Run for 50 min at 200 mA.

9. Carefully disassemble the apparatus and take out the membrane (*see* **Note 4**).

3.7 Immuno-detection of Linear Polyubiquitin Chains

1. Block the membrane with TBST containing 5 % nonfat dry milk for 2 h at room temperature.

2. Wash the membrane briefly two times with TBST.

3. Incubate the membrane with anti-linear polyubiquitin antibody diluted with TBST containing 5 % nonfat dry milk at 4 °C overnight (*see* **Note 5**).

4. Wash the membrane three times with TBST for 10 min with shacking.

5. Incubate the membrane with HRP-conjugated anti-mouse IgG antibody diluted with TBST containing 5 % nonfat dry milk for 1 h at room temperature.

6. Wash the membrane three times with TBST for 10 min with shacking.

7. Immerse the membrane in chemiluminescent substrates solution.

8. Cover the membrane with transparent wrapping film.

9. Capture an image of the bands on the membrane using the imaging system.

3.8 Stripping the Anti-polyubiquitin Antibody and Re-probing with Anti-NEMO

1. Incubate the membrane in antibody stripping buffer at 50 °C for 10–15 min with occasional agitation (*see* **Note 6**).

2. Wash the membrane briefly two times with TBST.

3. Wash the membrane two times with TBST for 10 min with shaking.

4. Repeat the immunodetection protocol (Subheading 3.7) using anti-NEMO as the primary antibody and HRP-conjugated anti-rabbit IgG as the secondary.

3.9 The Hot Lysis Method

NEMO contains a UBAN domain that recognizes linear ubiquitin chains. Thus, both linear ubiquitin chains covalently and non-covalently attached to NEMO will be detected by the immunoblotting protocol described above. To detect linear ubiquitin chains covalently attached to NEMO, the following hot lysis method must be performed in order to eliminate any non-covalently associated proteins. Here, we describe the protocol for the hot lysis method for the detection of the linear polyubiquitination of NEMO induced by over-expression of LUBAC in HEK293T (Fig. 5).

3.9.1 Cell Culture and Transfection

1. Grow HEK293T cells in DMEM medium supplemented with 10 % FCS, and penicillin–streptomycin solution.

2. Count cells and seed 2×10^6 cells in 10 ml culture medium in 10 cm collagen-coated dishes.

3. Incubate cells overnight in CO_2 incubator (37 °C/5 % CO_2).

4. Change the medium to fresh DMEM supplemented with 10 % FCS without penicillin-streptomycin solution.

5. Add 1.5 μg of pcDNA3.1-HA-HOIP, 0.5 μg of pcDNA3.1-MYC-HOIL-1 L, 0.5 μg of pcDNA3.1-T7-SHARPIN, and 0.2 μg of pcDNA3.1-FLAG-NEMO to 500 μl of Opti-MEM medium.

6. Dilute 5.4 μl of Lipofectamine 2000 in 500 μl of Opti-MEM medium and incubate for 5 min at room temperature.

7. Add diluted Lipofectamine 2000 to Opti-MEM medium containing plasmids and incubate for 20 min at room temperature.

IP: FLAG-
NEMO

FLAG-NEMO +   +
LUBAC −   +

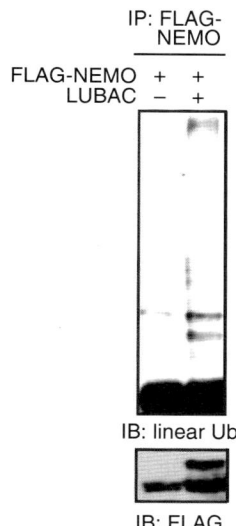

IB: linear Ub

IB: FLAG

Fig. 5 HEK293T cells were transfected with the expression plasmid for FLAG-NEMO with or without the expression plasmids for LUBAC. After 48 h whole-cell lysates were prepared with the hot lysis method and immunoprecipitated with rabbit anti-FLAG polyclonal antibody and then immunoblotted with an anti-linear ubiquitin antibody (LUB9) (*upper panel*). The membrane was re-blotted with mouse anti-FLAG monoclonal antibody to confirm that the equal amounts of the NEMO protein were immunoprecipitated (*lower panel*)

8. Add Lipofectamine-DNA mixture to cells.

9. Incubate cells for 48 h in a CO_2 incubator (37 °C/5 % CO_2).

3.9.2 Preparation of Whole-Cell Lysates

1. Aspirate the supernatant and wash the cells with 4 ml of ice-cold PBS.

2. Lyse cells with 0.5 ml of 1 % SDS in PBS and transfer to a prechilled 1.5 ml microcentrifuge tube.

3. Boil for 10 min at 95 °C.

4. Shear cell lysates by drawing several times through a 25-gauge needle attached to a 1 ml syringe.

5. Centrifuge the samples for 5 min at $20,000 \times g$ at room temperature.

6. Transfer the supernatants to 5 ml microcentrifuge tubes.

7. Dilute the samples with lysis buffer to decrease the concentration of SDS to 0.1 %.

8. Measure the protein concentration of whole-cell lysates as described in Subheading 3.3.

3.9.3 Immuno-
precipitation
of FLAG-NEMO

1. Place 2.2 mg of whole-cell lysates in 5 ml microcentrifuge tubes and add 2 μg of rabbit anti-FLAG polyclonal antibody to each sample. Incubate samples at 4 °C for 90 min while rotating the tubes on a rotator.

2. Add 30 μl of protein A sepharose and incubate at 4 °C for 60 min with rotation.

3. Pellet the sepharose beads by centrifuging at $800 \times g$ at 4 °C for 2 min. Carefully aspirate the supernatants.

4. Add 1 ml of ice-cold 1 % Triton X-100 lysis buffer to the tubes and transfer the sepharose to 1.5 ml microcentrifuge tube.

5. Wash the sepharose beads with inversion of the tubes several times.

6. Pellet the sepharose beads by centrifugation ($9,000 \times g$ at 4 °C for 1 min).

7. Carefully aspirate the supernatants and repeat this washing step four times.

8. Remove the final wash as completely as possible. Add 20 μl of 2× Laemmli sample buffer to the tubes and heat them at 95 °C for 5 min.

9. Analyze samples by immunoblotting as described in Subheading 3.7 and shown in Fig. 5.

4 Notes

1. This antibody stripping protocol is based on the protocol of ECL Western Blotting Detection System (GE Healthcare). 2 % SDS is used in the original protocol but the treatment with 2 % SDS sometimes reduces the signal strength of re-probing. Thus, we use 0.2 % SDS instead of 2 % SDS.

2. Do not freeze the cells for lysis. Cells should be lysed directly with 1 % Triton X-100 lysis buffer. Freezing cells makes it difficult to detect the linear polyubiquitination of NEMO.

3. Add 1/100 volume of 200 mM PMSF, protease inhibitor cocktail, and phosphatase inhibitor cocktail to 1 % Triton X-100 lysis buffer just before use.

4. Divide the membrane into two pieces so that one contains the immunoprecipitation samples and the other contains the samples from the whole-cell lysates.

5. At the same time the phosphorylation and the degradation of IκBα should be examined by using the membrane of the whole-cell lysates in order to check whether the TNF-α stimulation has worked or not.

6. The incubation time of the original protocol is 30 min. But the stripping antibody for 30 min reduces the signal strength of re-probing. Pilot studies should be performed to determine the optimal incubation time at which signal from the initial reaction is absent but signal from re-probing is detectable.

Acknowledgments

This work is partly supported by the Targeted Proteins Research Program (TPRP) and grants from the Ministry of Education, Culture, Sports, Science, and Technology of Japan to K.I. and/or Y.S.

References

1. Vallabhapurapu S, Karin M (2009) Regulation and function of NF-kappaB transcription factors in the immune system. Annu Rev Immunol 27:693–733

2. Hayden MS, Ghosh S (2004) Signaling to NF-kappaB. Genes Dev 18(18):2195–2224

3. Bhoj VG, Chen ZJ (2009) Ubiquitylation in innate and adaptive immunity. Nature 458(7237):430–437

4. Hershko A, Ciechanover A (1998) The ubiquitin system. Annu Rev Biochem 67:425–479

5. Glickman MH, Ciechanover A (2002) The ubiquitin-proteasome proteolytic pathway: destruction for the sake of construction. Physiol Rev 82(2):373–428

6. Kulathu Y, Komander D (2012) Atypical ubiquitylation—the unexplored world of polyubiquitin beyond Lys48 and Lys63 linkages. Nat Rev Mol Cell Biol 13(8):508–523

7. Deng L, Wang C, Spencer E, Yang L, Braun A, You J, Slaughter C, Pickart C, Chen ZJ (2000) Activation of the IkappaB kinase complex by TRAF6 requires a dimeric ubiquitin-conjugating enzyme complex and a unique polyubiquitin chain. Cell 103(2):351–361

8. Yamamoto M, Okamoto T, Takeda K, Sato S, Sanjo H, Uematsu S, Saitoh T, Yamamoto N, Sakurai H, Ishii KJ, Yamaoka S, Kawai T, Matsuura Y, Takeuchi O, Akira S (2006) Key function for the Ubc13 E2 ubiquitin-conjugating enzyme in immune receptor signaling. Nat Immunol 7(9):962–970

9. Xu M, Skaug B, Zeng W, Chen ZJ (2009) A ubiquitin replacement strategy in human cells reveals distinct mechanisms of IKK activation by TNFalpha and IL-1beta. Mol Cell 36(2):302–314

10. Kirisako T, Kamei K, Murata S, Kato M, Fukumoto H, Kanie M, Sano S, Tokunaga F, Tanaka K, Iwai K (2006) A ubiquitin ligase complex assembles linear polyubiquitin chains. EMBO J 25(20):4877–4887

11. Tokunaga F, Sakata S, Saeki Y, Satomi Y, Kirisako T, Kamei K, Nakagawa T, Kato M, Murata S, Yamaoka S, Yamamoto M, Akira S, Takao T, Tanaka K, Iwai K (2009) Involvement of linear polyubiquitylation of NEMO in NF-kappaB activation. Nat Cell Biol 11(2):123–132

12. Tokunaga F, Nakagawa T, Nakahara M, Saeki Y, Taniguchi M, Sakata S, Tanaka K, Nakano H, Iwai K (2011) SHARPIN is a component of the NF-kappaB-activating linear ubiquitin chain assembly complex. Nature 471(7340):633–636

13. Lo YC, Lin SC, Rospigliosi CC, Conze DB, Wu CJ, Ashwell JD, Eliezer D, Wu H (2009) Structural basis for recognition of diubiquitins by NEMO. Mol Cell 33(5):602–615

14. Rahighi S, Ikeda F, Kawasaki M, Akutsu M, Suzuki N, Kato R, Kensche T, Uejima T, Bloor S, Komander D, Randow F, Wakatsuki S, Dikic I (2009) Specific recognition of linear ubiquitin chains by NEMO is important for NF-kappaB activation. Cell 136(6):1098–1109

15. Fujita H, Rahighi S, Akita M, Kato R, Sasaki Y, Wakatsuki S, Iwai K (2014) Mechanism underlying IkappaB kinase activation mediated by the linear ubiquitin chain assembly complex. Mol Cell Biol 34(7):1322–1335

16. Sasaki Y, Sano S, Nakahara M, Murata S, Kometani K, Aiba Y, Sakamoto S, Watanabe Y, Tanaka K, Kurosaki T, Iwai K (2013) Defective immune responses in mice lacking LUBAC-mediated linear ubiquitination in B cells. EMBO J 32(18):2463–2476

In Vitro Detection of NEMO–Ubiquitin Binding Using DELFIA and Microscale Thermophoresis Assays

Michelle Vincendeau, Daniel Krappmann, and Kamyar Hadian

Abstract

Canonical NF-κB signaling in response to various stimuli converges at the level of the IκB kinase (IKK) complex to ultimately activate NF-κB. To achieve this, the IKK complex uses one of its regulatory subunit (IKKγ/NEMO) to sense ubiquitin chains formed by upstream complexes. Various studies have shown that different Ubiquitin chains are involved in the binding of NEMO and thereby the activation of NF-κB. We have utilized two distinct biochemical methods, i.e., Dissociation-Enhanced Lanthanide Fluorescence Immunoassay (DELFIA) and Microscale Thermophoresis (MST), to detect the interaction of NEMO to linear and K63-linked Ubiquitin chains, respectively. Here, we describe the brief basis of the methods and a detailed underlying protocol.

Key words NF-κB, IKK, NEMO, Ubiquitin, Linear, K63, DELFIA, Microscale thermophoresis

1   Introduction

Besides phosphorylation and ubiquitination, NF-κB activation from various stimuli makes use of protein–protein interactions (PPI) to allow signal progression [1]. Here, a major PPI step is the interaction of IKKγ/NEMO with poly-Ubiquitin, which connects the upstream receptor proximal complex with downstream NF-κB activation, i.e., IκBα phosphorylation and degradation through the active IKK complex. NEMO strongly interacts with linear Met1-linked Ubiquitin chains [2, 3], but it can also bind to other linkage types like K63 or K11-linked chains [4–8]. Several efforts have been done to disclose the nature of these interactions in vitro and in cell systems [3, 9].

Popular strategies to identify and biophysically characterize PPI are isothermal titration calorimetry (ITC) and surface plasmon resonance (SPR) [10, 11]. These assays are extensively employed to provide key parameters for their respective targets, such as their dissociation constant (K_d), stoichiometry, and the kinetics of binding (k_{on} and k_{off}). Besides these advantages, these assays have

Michael J. May (ed.), *NF-kappa B: Methods and Protocols*, Methods in Molecular Biology, vol. 1280,
DOI 10.1007/978-1-4939-2422-6_18, © Springer Science+Business Media New York 2015

also disadvantages such as the high amount of protein consumption in the case of ITC. Therefore, this assay is not suitable for investigation of K63-linked Ubiquitin chains.

Here we describe two recently established techniques to detect NEMO–Ubiquitin interaction [3]. The first technique relies on an ELISA-based PPI interaction assay, DELFIA system, using the time-resolved fluorescence readout to visualize the binding. The second method is a novel biophysical assay, Microscale Thermophoresis [12], that detects binding by analyzing protein movement in a thermal gradient. Both techniques are very distinct from each other as the DELFIA assay is a more rigid method that utilizes immobilized proteins similar to SPR, while MST is an in solution assay reminiscent to ITC. The big advantage of both assays is that they are performed with very little amount of protein (nM–µM range). This low consumption of material is in particular favorable for analyzing ubiquitin chains generated through a ligation reaction (e.g., K63 or K11-linked chains), as their amount is usually limited. Comparing results from these methods with different preferences can provide more insight into the nature of the NEMO–Ubiquitin binding.

2   Materials

2.1 Components for Protein Expression and Purification

1. Prokaryotic expression vector system pASK-IBA3plus (IBA GmbH, Germany).

2. E. coli strain BL21-CodonPlus (DE3) RIPL (Agilent, San Clara, CA, USA).

3. LB Medium with 100 µg/ml ampicillin.

4. Isopropyl-β-D-thiogalactopyranoside (IPTG) stock solution (1 M).

5. Anhydrotetracycline stock solution (2 mg/ml in DMF).

6. Lysis buffer for StrepTagII purification: 100 mM Tris–HCl pH 8.0, 150 mM NaCl, 0.5 mM DTT, 0.5 mg/ml lysozyme and Roche (Madison, WI, USA) protease inhibitor tablet.

7. Lysis buffer for His-Tag purification: 75 mM phosphate buffer pH 7.4, 400 mM NaCl, 10 mM imidazole, 0.5 mg/ml lysozyme, and Roche EDTA-free protease inhibitor tablet.

8. StrepTagII purification buffer: 100 mM Tris–HCl, pH 8.0, 150 mM NaCl, 0.5 mM DTT.

9. StrepTagII elution buffer: 100 mM Tris–HCl, pH 8.0, 150 mM NaCl, 0.5 mM DTT, 2.5 mM D-desthiobiotin.

10. His-Tag purification buffer: 20 mM phosphate buffer pH 7.4, 500 mM NaCl, 30 mM imidazole.

11. His-Tag elution buffer: 20 mM phosphate buffer pH 7.4, 500 mM NaCl, 500 mM imidazole.

12. Desalting buffer: 20 mM Tris–HCl, pH 8.0.

13. StrepTrap HP columns (GE Healthcare, Mickleton, NJ, USA).

14. HisTrap Fast Flow columns (GE Healthcare).

15. HiTrap Desalting columns (GE Healthcare).

16. Amicon Ultra-4 centrifugal filter unit (Millipore, Billerica, MA, USA) with an exclusion size of 3 kDa.

17. ÄKTA fast protein liquid chromatography (FPLC) system (GE Healthcare).

18. Superdex-75 gel filtration column (GE Healthcare).

2.2 Ubiquitin Linkage Reaction Components

1. Ubiquitin Conjugation Reaction buffer kit (Boston Biochem, Boston, USA).

2. E1 enzyme UBE1 (Boston Biochem).

3. Ubc13/Uev1a (Boston Biochem).

4. Ubi-D77 (Enzo Lifescience, Farmingdale, NY, USA).

5. PBS for subsequent gel filtration with a Superdex-75 column.

2.3 Material for DELFIA Assays

1. StrepTactin coated 96-well plates (IBA GmbH, Germany).

2. DELFIA assay buffer (PerkinElmer, Waltham, MA, USA).

3. DELFIA washing buffer (PerkinElmer).

4. DELFIA enhancer solution (PerkinElmer).

5. Anti-HisTag DELFIA Europium-N1 (PerkinElmer).

2.4 Material for Microscale Thermophoresis (MST)

1. MST NT-647 labeling Kit (NanoTemper, Munich, Germany) containing: NT-647 NHS dye, Buffer Labeling NHS, Column A (Buffer Exchange), Column B (Purification).

2. Standard MST capillaries (NanoTemper).

3. DMSO.

4. MST labeling elution buffer: 25 mM Tris–HCl pH 8.0, 100 mM NaCl.

5. MST assay buffer: 25 mM Tris–HCl pH 8.0, 100 mM NaCl, 0.1 % BSA, 0.1 % Tween 20, 0.5 mM DTT.

3   Methods

3.1 Protein Expression and Purification Procedures

3.1.1 Protein Expression

1. Transfer 20 ml of BL21-CodonPlus (DE3) RIPL pre-culture containing the expression vector of interest into 1 L LB-Amp and incubate at 37 °C.

2. Let the culture grow until OD_{600} value of 0.8–1.0.

3. Induce the culture with 1 mM IPTG and 200 ng/ml anhydro-tetracycline and incubate the induced culture o/n at 21 °C.

4. On the next day, harvest the bacteria by centrifugation at $3,000 \times g$.

5. Resuspend the pellets in either StrepTagII purification lysis buffer or His-Tag purification lysis buffer on ice (*see* **Note 1**).

6. After 20 min at RT, sonicate the bacterial suspension on ice (*see* **Notes 2** and **3**).

7. Centrifuge at $20,000 \times g$ (4 °C) for 20 min to clarify the lysate from debris.

8. Centrifuge the supernatant again at $20,000 \times g$ (4 °C) for 20 min for final clarification.

3.1.2 StrepTagII Purification of NEMO Proteins Using an ÄKTA Purifier Station

1. Apply bacterial lysates containing the StrepII-tagged proteins of interest on the StrepTrap columns using an ÄKTA Superloop and the StrepTagII purification buffer.

2. Wash bacterial proteins away using eight column volumes of StrepTagII washing buffer.

3. Elute target proteins with the help of the StrepTagII elution buffer.

4. Subsequently desalt eluted proteins using a Desalting column and the Desalting buffer to remove the D-desthiobiotin.

5. After desalting, immediately add NaCl to a final concentration of 100 µM (i.e., 50 µl of a 2 M stock NaCl to 1 ml of desalted protein) (*see* **Note 4**).

6. Take an Amicon Ultra-4 spin column to concentrate the proteins of interest if needed (*see* **Note 5**).

7. Check purity by SDS-PAGE and Coomassie staining (Fig. 1).

3.1.3 HisTag Purification Procedure (Ubiquitin Proteins) Using an ÄKTA Purifier Station

1. Apply bacterial lysates containing the His-tagged proteins of interest on the HisTrap columns using a Superloop and the HisTag purification buffer.

2. Wash bacterial proteins away using eight column volumes of HisTag washing buffer.

3. Elute His-tagged proteins using a gradient of His-Tag elution buffer.

4. Subsequently desalt eluted proteins using a Desalting column and the Desalting buffer to remove the imidazole.

5. After desalting, immediately add NaCl to reach a final concentration of 100 µM (i.e., 50 µl of a 2 M stock NaCl to 1 ml of desalted protein) (*see* **Note 4**).

6. Use an Amicon Ultra-4 spin column to concentrate the proteins of interest if needed.

7. Check purity by SDS-PAGE and Coomassie Staining (Fig. 1).

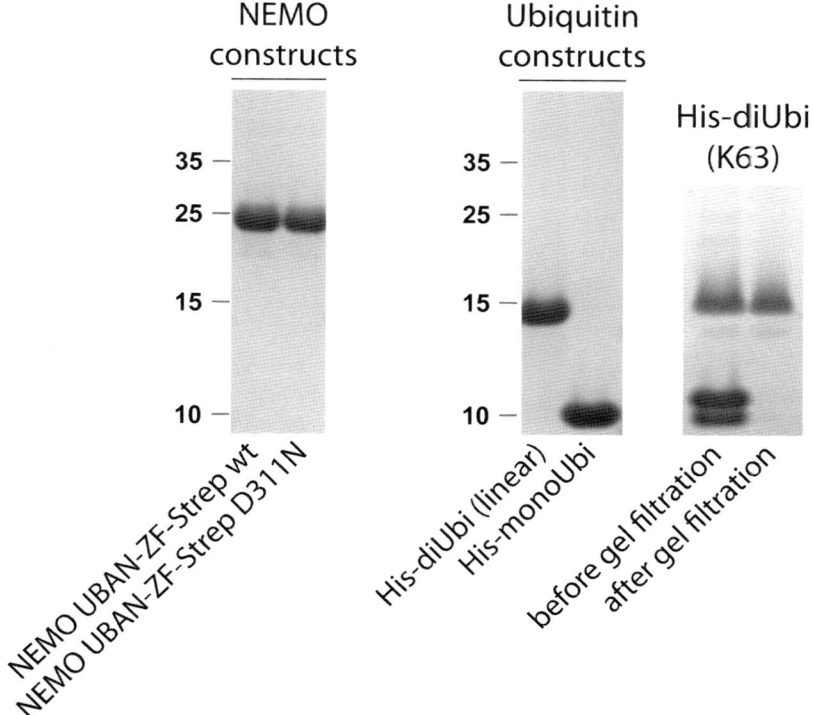

Fig. 1 Purification of NEMO and ubiquitin constructs. 2 μg of the purified proteins was loaded on a 15 % SDS gel. The SDS gel was stained with Colloidal Coomassie to prove the purity of NEMO and Ubiquitin recombinant proteins

3.2 Ubiquitin Linkage Reaction

1. Express and purify the His-Ubiquitin K63R protein as described above.

2. Mix 1 mg of His-Ubiquitin K63R, 1 mg of Ubi-D77, 0.1 μM E1 enzyme UBE1, and 6.5 μM K63-specific E2 complex Ubc13/Uev1a all in ligation buffer (from Ubiquitin Conjugation Reaction buffer Kit) reaching a final volume of 90 μl. Add 10 μl of 10× Mg-ATP to start the reaction.

3. Incubate the reaction for 4 h at 37 °C.

4. Add 10 μl of 10× E1 Stop buffer to stop the reaction.

5. Use the ÄKTA purifier station and a Superdex-75 gel filtration column to separate excess of non-ligated mono-ubiquitin from the diUbiquitin moieties (*see* **Note 6**).

6. After gel filtration analyze the individual elution fractions by SDS PAGE and Western blotting to identify those fractions that only contain the ligated diUbiquitin.

7. Pool the positive fractions and concentrate the proteins using an Amicon Ultra-4 spin column.

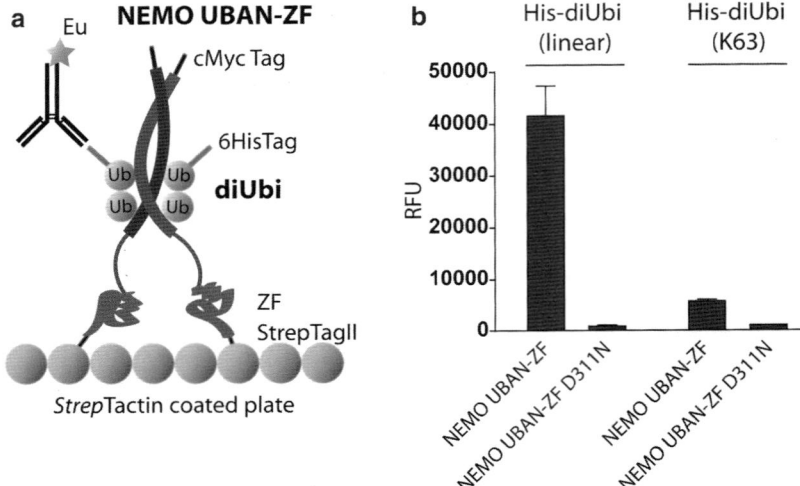

Fig. 2 DELFIA assay. (**a**) The experimental setup of the DELFIA assay is depicted. Recombinant c-Myc-NEMO-UBAN-ZF-StrepTagII dimer is bound to a StrepTactin-coated plate. Next, His-labeled diUbiquitin binds to NEMO and can subsequently be detected by an Europium-labeled anti-His antibody. (**b**) NEMO-UBAN-ZF wild type shows a strong and robust signal when incubated with linear His-diUbiquitin, while no signal is detected after incubation of linear His-diUbiquitin with the NEMO UBAN-ZF D311N binding mutant. K63-linked His-diUbiquitin chains also bind to NEMO-UBAN-ZF wild type. However, this binding is much weaker than the NEMO UBAN-ZF binding to linear diUbiquitin

3.3 Dissociation-Enhanced Lanthanide Fluorescent Immunoassay (DELFIA)

This assay system is an ELISA-based protein-protein interaction assay that uses the time-resolved fluorescence technology through Europium fluorescence. Here, a delay time between excitation and emission guarantees a significant reduction in background fluorescence. Moreover, the primary detection antibody is loaded with Europium so that no secondary antibody is needed (Fig. 2a).

1. Add 20 pmol of StrepII-tagged NEMO in 100 μl of DELFIA assay buffer to StrepTactin coated 96-well plates.

2. Incubate for 2 h allowing the NEMO protein to bind to the StrepTactin plate (*see* **Note 7**).

3. Wash unbound fraction away using DELFIA washing buffer (*see* **Note 8**).

4. Add 250 pmol of His-tagged Ubiquitin protein in 100 μl of DELFIA assay buffer to the bound fraction of NEMO proteins.

5. Again incubate for 2 h so that NEMO proteins and Ubiquitin proteins can interact (*see* **Note 7**).

6. Wash excess of Ubiquitin proteins away using DELFIA washing buffer (*see* **Note 8**).

7. Add 500 ng/ml Europium (Eu)-labeled anti-HisTag antibody in DELFIA assay buffer.

8. Incubate the reaction for 1 h (*see* **Note 7**).

9. Wash away excess of anti-His antibody with DELFIA washing buffer (*see* **Note 8**).

10. Add 100 µl of DELFIA enhancer solution to activate Eu-fluorescence (*see* **Note 9**).

11. Measure the time-resolved signal in an appropriate plate reader (Ex: 340 nm—Em: 615 nm) (Fig. 2b).

3.4 Microscale Thermophoresis (MST) Assay

The labeling reaction is performed with a reactive dye (NT-647) that uses the *N*-hydroxy succinimide (NHS)-ester chemistry to efficiently react with primary amines of proteins forming highly stable dye–protein conjugates.

3.4.1 Protein Labeling for MST Assays

1. Re-buffer 100 µl of NEMO protein at a concentration of 20 µM into the labeling NHS buffer provided with the labeling Kit using column A.

2. Prepare a 50 µM NT-647 NHS dye solution in dimethyl sulfoxide (DMSO).

3. Mix 100 µl of NEMO protein with 100 µl of NT-647 dye and incubate this labeling reaction for 30 min.

4. Meanwhile, equilibrate column B with 8 ml of MST labeling elution buffer (not provided with the Kit) (*see* **Note 11**).

5. After 30 min, let the 200 µl of labeling reaction enter the column B (*see* **Notes 10** and **11**).

6. Add 300 µl of labeling elution buffer to the column (*see* **Note 10**). At this stage the labeled protein and the free dye will separate on the column (*see* **Note 11**).

7. Add further 600 µl of MST labeling elution buffer and collect the labeled protein fraction in a fresh tube (*see* **Notes 10** and **11**).

3.4.2 MST Assays and Data Evaluation

The biophysical method of thermophoresis makes use of a thermal gradient, where proteins can move depending on their charge, size, and hydration shell. These preferences change in the process of protein–protein interactions. However, this method can also be used to detect protein binding to nucleic acid, small molecules, etc. To measure MST assays the NanoTemper Monolith NT.115 device is needed.

1. To measure the binding of NEMO to different Ubiquitin chains (i.e., linear vs. K63 linked chains), the NEMO protein needs to be titrated in a 12-point serial dilution starting at 200 µM in MST assay buffer. To do so, use a 96-well V-shaped plate and transfer 10 µl of MST assay buffer into wells 2–12. Subsequently add 20 µl of 200 µM NEMO solution to well 1 and perform serial dilution by transferring 10 µl from well to well. Discard 10 µl from last well.

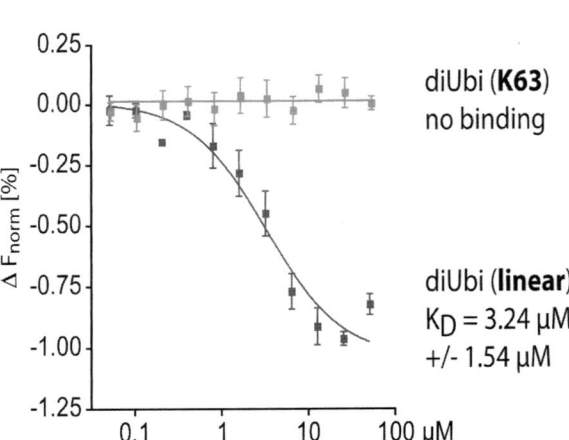

Fig. 3 The graph displays the % changes of normalized fluorescence from MST assays. The NEMO UBAN-ZF binds to linear diUbiquitin, whereas K63-linked diUbiquitin does not

2. In a second step add 10 µl labeled Ubiquitin protein to each well. The stock solution of labeled Ubiquitin protein should be adjusted in a way that it reaches approx. 600–800 fluorescence counts in the NanoTemper Monolith NT.115 instrument. After the Ubiquitin addition the highest NEMO concentration is 100 µM and the Ubiquitin has a final fluorescence count of approx. 300–400.

3. Incubate reaction for 30 min.

4. Soak each dilution step into a standard capillary and place the capillaries in the corresponding tray of the NanoTemper Monolith NT.115 device (*see* **Notes 12** and **13**).

5. Measure MST signal in the NanoTemper Monolith NT.115 device (instrument settings: red LED, LED power: 20 %, laser power: 23 %, temperature: 25 °C) (*see* **Notes 14** and **15**).

6. For data analysis, plot NEMO concentration (logarithmic scale) against % changes of normalized fluorescence (ΔF_{norm} [%]) (Fig. 3). Use standard software for curve fitting (nonlinear regression) and for determination of K_D values (EC$_{50}$ values).

4 Notes

1. For His-Tag purification it is most important to have an EDTA free system, i.e., no EDTA in any buffer and the usage of EDTA-free protease inhibitor tablets.

2. During the sonification step it is important that the lysate is not getting too warm. The complete procedure needs to take

place on ice and it should be done in intervals, i.e., 10–15 20-s cycles. Between the cycles the lysates should chill for approx. 30 s on ice.

3. It is also very important to avoid foam formation during sonification. Otherwise sonification will not work properly and the output of the soluble protein fraction will decrease noticeably. There is always a small foam layer on top of the lysate. Always first pass this layer and enter the liquid area and then turn on the sonicator. This will help to avoid further foam formation.

4. After the desalting procedure, the NaCl should be added immediately.

5. NEMO proteins should not be concentrated higher than 1–2 mg/ml to avoid precipitation of the proteins.

6. During gel filtration collect small elution fractions (~200 µl). This will help to better discriminate between fractions that contain either the ligated diUbiquitin or the monoUbiquitin or both. Apply all fractions around the peaks to a 15 % SDS gel and identify the fraction only containing the diUbiquitin. Unify these fractions and concentrate.

7. During the incubation steps let the plate shake carefully (~300 rpm).

8. Use 5–6 cycles of washing procedure for each step. Here, add 100 µl of wash buffer for each cycle and invert the plate to discard the wash solution. Tap on a paper towel to remove the remaining wash solution.

9. It is beneficial but not mandatory to incubate the plates with the Enhancement solution for 3–5 min. During this incubation step keep the plate in dark.

10. It is important to carry out the purification of labeled proteins as described. This will best separate the labeled protein from unlabeled dye.

11. If BSA is needed in the final buffer, add it after the labeling and purification step! DO NOT add before (even not to the MST elution buffer) as this will lead to co-labeling of BSA.

12. Be careful to not touch the capillary in the middle where it will be measured. Always touch it at the ends.

13. Be sure that there are no bubbles in the middle of the capillary after soaking of the sample.

14. The final fluorescence unit is optimally in the range of 300–400. This region is a good agreement for good sensitivity with protein saving. However, also regions of 200–1,000 are OK. Fluorescence below 200 will affect quality of data.

15. Before MST measurement, the instrument always performs a capillary scan. Here it is important that the top of the peaks look round. Any spike depicts aggregation of the protein and the measurement needs to be discarded.

Acknowledgements

This research was originally published in The Journal of Biological Chemistry. Kamyar Hadian, Richard A. Griesbach, Scarlett Dornauer, Tim M. Wanger, Daniel Nagel, Moritz Metlitzky, Wolfgang Beisker, Marc Schmidt-Supprian, and Daniel Krappmann. NF-κB Essential Modulator (NEMO) Interaction with Linear and Lys-63 Ubiquitin Chains Contributes to NF-κB Activation. Journal of Biological Chemistry. 2011; 286:26107–26117. © The American Society for Biochemistry and Molecular Biology.

References

1. Bauch A, Superti-Furga G (2006) Charting protein complexes, signaling pathways, and networks in the immune system. Immunol Rev 210:187–207

2. Rahighi S, Ikeda F, Kawasaki M et al (2009) Specific recognition of linear ubiquitin chains by NEMO is important for NF-kappaB activation. Cell 136:1098–1109

3. Hadian K, Griesbach RA, Dornauer S et al (2011) NF-kappaB essential modulator (NEMO) interaction with linear and lys-63 ubiquitin chains contributes to NF-kappaB activation. J Biol Chem 286:26107–26117

4. Lo YC, Lin SC, Rospigliosi CC et al (2009) Structural basis for recognition of diubiquitins by NEMO. Mol Cell 33:602–615

5. Wu CJ, Conze DB, Li T et al (2006) Sensing of Lys 63-linked polyubiquitination by NEMO is a key event in NF-kappaB activation [corrected]. Nat Cell Biol 8:398–406

6. Yoshikawa A, Sato Y, Yamashita M et al (2009) Crystal structure of the NEMO ubiquitin-binding domain in complex with Lys 63-linked di-ubiquitin. FEBS Lett 583:3317–3322

7. Laplantine E, Fontan E, Chiaravalli J et al (2009) NEMO specifically recognizes K63-linked polyubiquitin chains through a new bipartite ubiquitin-binding domain. EMBO J 28:2885–2895

8. Dynek JN, Goncharov T, Dueber EC et al (2010) c-IAP1 and UbcH5 promote K11-linked polyubiquitination of RIP1 in TNF signalling. EMBO J 29:4198–4209

9. Kensche T, Tokunaga F, Ikeda F et al (2012) Analysis of nuclear factor-kappaB (NF-kappaB) essential modulator (NEMO) binding to linear and lysine-linked ubiquitin chains and its role in the activation of NF-kappaB. J Biol Chem 287:23626–23634

10. Ghai R, Falconer RJ, Collins BM (2012) Applications of isothermal titration calorimetry in pure and applied research–survey of the literature from 2010. J Mol Recognit 25:32–52

11. Willander M, Al-Hilli S (2009) Analysis of biomolecules using surface plasmons. Methods Mol Biol 544:201–229

12. Jerabek-Willemsen M, Wienken CJ, Braun D et al (2011) Molecular interaction studies using microscale thermophoresis. Assay Drug Dev Technol 9:342–353

Chapter 19

Use of Fluorescence Spectroscopy for Quantitative Investigations of Ubiquitin Interactions with the Ubiquitin-Binding Domains of NEMO

Virginie Dubosclard, Elisabeth Fontan, and Fabrice Agou

Abstract

Ubiquitin serves as a signal for a variety of cellular processes and its specific interaction with ubiquitin-binding domain (UBD) regulates key cellular events including protein degradation, cell-cycle control, DNA repair, and kinase activation. Several binding mechanisms for isolated UBDs have been reported in recent years. However, little is known about the mechanism through which proteins containing multiple-UBDs achieve specificity for a particular oligomer of polyUb. The NF-κB essential modulator (NEMO, also known IKKγ), which plays a key role in the NF-κB signaling pathway, belongs to the latter family of proteins since it contains two distal NOA (also known UBAN/CC2-LZ/NUB) and ZF UBDs, separated by an unstructured proline-rich linker of about 40 residues in length. Here, we show a new procedure for fast purification of this bipartite domain. We also describe the use of intrinsic fluorescence spectroscopy for quantitative investigations of ubiquitin interactions between two distal ubiquitin-binding domains of NEMO (NOA and ZF). This spectroscopic method has many advantages over other techniques like GST pulldown and Biacore's SPR for monitoring avid interactions between two UBDs, especially when UBDs are located at significant distance from each other within the protein.

Key words NF-κB, Polyubiquitin chains, NEMO, Ubiquitin-binding domains, Intrinsic fluorescence spectroscopy, Protein purification

1 Introduction

Interactions with ubiquitin (Ub) play a key role in biology and are increasingly being identified as important for almost all cell processes. This small protein can form eight different types of chain, through specific linkage (M1, K6, K11, K27, K29, K33, K48, and K63-linked chains). These chains are recognized by ubiquitin-binding domains (UBDs) and are classified into several families on the basis of their structure and function [1]. Ub–UBD interactions constitute a powerful signaling system regulating mechanisms of protein degradation, cell-cycle control, DNA repair, and kinase activation. Ub–UBD interactions have been characterized in detail

Michael J. May (ed.), *NF-kappa B: Methods and Protocols*, Methods in Molecular Biology, vol. 1280,
DOI 10.1007/978-1-4939-2422-6_19, © Springer Science+Business Media New York 2015

for isolated UBDs, but little is known about the mechanism by which proteins containing multiple-UBDs achieve linkage specificity for a particular oligomer of polyUb.

The NF-κB essential modular protein (NEMO) is a key component of the canonical NF-κB pathway [2]. It contains two distal NOA (also known UBAN/CC2-LZ/NUB) and ZF UBDs, separated by an unstructured proline-rich linker of about 40 residues in length (Fig. 1). The affinities of these isolated UBDs for different types of ubiquitin chains have been determined with several biochemical and biophysical methods [3–5]. However, little is known about the way in which polyubiquitin chains specifically interact with the bipartite UBD of NEMO. We have shown that specificity for the K63 poly-ubiquitin chain requires both the NOA and ZF domains of NEMO [6]. However, the structural basis of polyubiquitin recognition by NEMO remains to be elucidated.

Many techniques for quantifying interactions between proteins have been described, including GST pulldown, native gel electrophoresis, isothermal titration calorimetry (ITC), surface plasmon resonance (SPR), biolayer interferometry, and microscale thermophoresis [7]. ITC can be used to determine binding constants and thermodynamics (n, K, ΔH, and ΔS), but this approach

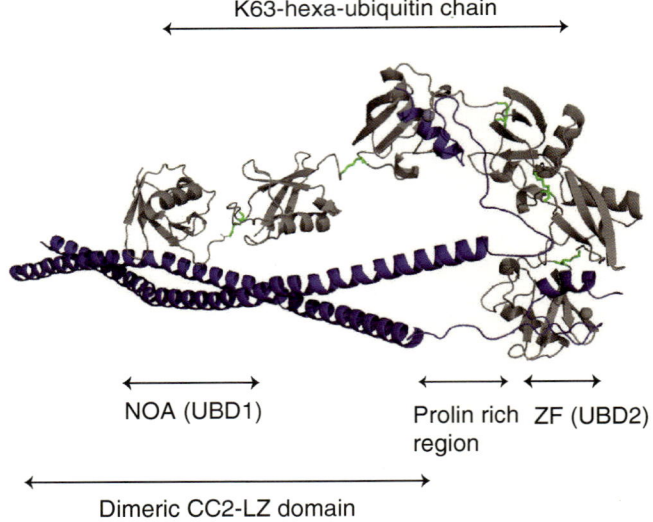

Fig. 1 Structural model of one K63-hexaubiquitin chain (*gray*) bound to the bipartite domain of NEMO (NOAZ in *blue*). The two UBDs in tandem of each NEMO monomer (NOA/UBAN/NUB in the CC2-LZ domain and the ZF domain) are separated by an unstructurated proline-rich linker of about 40 residues in length. In this model, showing the NOAZ domain in its dimeric state, avid interactions result from the binding between one NOA UBD and two ZF UBDs per dimer. The residues colored in *green* are the lysine residues involved in the linkage of the K63-hexaubiquitin chain, highlighting the high rotational flexibility and spacing of each ubiquitin unit resulting from this particular K63 linkage

often requires large amounts of material. Moreover, weak-affinity allosteric behavior is generally difficult to detect because a classical (rectangular) binding curve devoid of cooperativity invariably gives rise to a sigmoidal shape. Biacore and biolayer interferometry techniques require the attachment of the protein to a surface via a tag or bivalent antibody. This may result in an artificial conformation and multivalency of the immobilized protein, potentially increasing or decreasing binding. A recently developed sensitive fluorescence-based technique, microscale thermophoresis, is based on the labeling of the protein with a reactive dye [8]. This labeling may alter the structure and function of the protein and, thus, its interaction with its partner. Finally, specific transient protein–protein interactions, which are usually governed by rapid off-rate constants, cannot be investigated by GST pulldown and non-denaturing electrophoresis because these techniques operate under non-equilibrium conditions.

Intrinsic fluorescence spectroscopy has several advantages over the techniques described above: (1) the interactions between proteins occur in solution and often require very small amounts of biological material, and (2) binding events occur under ideal equilibrium conditions. Furthermore, many mutagenesis tools are now available for the simple introduction of a single tryptophan residue into a particular domain of the protein of interest (*see* **Note 1**). This natural aromatic residue can serve as a sensitive, local probe for monitoring positive or negative binding cooperativity between two binding sites, even if they are located at considerable distance from each other within the protein.

We describe here the use of fluorescence spectroscopy for quantitative investigations of ubiquitin interactions between distal ubiquitin-binding domains of NEMO (NOA and ZF). We generated a mutant with a F395W single point mutation resulting in a substitution within the second ZF UBD. This mutation has no effect on NEMO function [5], and can be used as a local fluorescent probe for monitoring ubiquitin binding to the ZF UBD, with or without the contribution of the other distal NOA UBD (*see* **Note 2**). As a control, we generated a similar fluorescent mutant (F395W) that also had a double mutation (Y308A/D311N) affecting the first NOA UBD. These mutations have been reported to result in the total abolition of interactions between the isolated NOA UBD and linear or K63 polyubiquitin chains [3, 9, 10]. We describe an atypical purification method for the bipartite NOAZ domain of NEMO, based on successive denaturation/renaturation steps, and show, by fluorescence spectroscopy, that the human bipartite domain of NEMO interacts with a hexaubiquitin chain rather than a tetraubiquitin chain, highlighting the importance of chain length for synergic interactions between the two UBDs and polyubiquitin chains. This complex was not observed with a double mutant form of the first NOA UBD (Y308A/D311N) unable to bind ubiquitin, indicating that positive cooperation between the NOA and ZF UBDs is required for efficient interaction with long ubiquitin oligomers, such as hexaubiquitin chains.

---

2 Materials

2.1 Protein Purification

1. Extraction buffer: 20 mM Tris–HCl pH 8, 20 mM KCl, 5 % glycerol, and protease inhibitor cocktail (Complete EDTA-free: Roche Diagnostic GmbH, Mannheim, Germany) added to the extraction buffer immediately before use.

2. French press.

3. Sonicator.

4. Urea in powder form (Sigma-Aldrich, St. Louis, MO, USA).

5. Denaturation buffer: 50 mM Tris–HCl pH 8, 1 M NaCl, 6 M urea. Stored at room temperature (*see* **Note 3**).

6. Renaturation buffer: 50 mM Tris–HCl pH 8, 1 M NaCl.

7. Elution buffer: 50 mM Tris–HCl pH 8, 1 M NaCl, 1 M imidazole.

8. Column packed with nickel-charged resin (Ni-NTA Superflow: Qiagen, Valencia, CA, USA).

9. Native equilibration buffer of the Ni-NTA column: 50 mM Tris–HCl pH 8, 1 M NaCl, 10 mM imidazole.

10. Buffer: 50 mM Tris–HCl pH 8, 150 mM NaCl, 1 mM TCEP (tris(2-carboxyethyl)phosphine).

11. Akta Prime (GE Healthcare, Bio-Sciences, AB, Uppsala, Sweden).

12. Centrifugal filter devices with a 10 kDa cutoff (Amicon Ultra 10, Merck Millipore, Cork, Ireland).

13. 12 % SDS-PAGE gel (Bio-Rad, Hercules, CA, USA).

14. Coomassie blue stain for gel.

15. SDS-PAGE running buffer: MOPS buffer (Bio-Rad).

2.2 Quality Control and Folding of Recombinant Proteins: Zinc Titration

1. Proteinase K (Merck, Whitehouse Station, NJ, USA): 10 mg/ml solution in water.

2. Thermostatically controlled water bath.

3. 4-(2-pyridylazo)resorcinol monosodium salt (PAR, Sigma): 2.5 mM working solution in water.

4. Zinc solution: 1 M $ZnCl_2$ (Sigma) working solution in protein buffer: 50 mM Tris–HCl, pH 8, 150 mM NaCl.

5. EDTA: 0.5 M solution in water, pH 8 (*see* **Note 4**).

6. Costar 96-well flat-bottomed cell culture plates.

7. Microplate reader (TECAN Infinite 500: TECAN, Morrisville, NC, USA).

2.3   Quality Control and Folding of Recombinant Proteins: Circular Dichroism Spectroscopy

1. Hellma QS 121.000 quartz Suprasil cuvette with a 0.2 mm light path (Hellma Analytics, Mulheim, Germany).

2. Aviv 215 circular dichroism spectrometer (Aviv Biomedical, Lakewood, NJ, USA).

3. Peltier temperature controller.

4. Buffer: 50 mM Tris–HCl pH 8, 150 mM NaCl, 1 mM TCEP (tris(2-carboxyethyl)phosphine).

2.4   K63-Polyubiquitin Synthesis and Purification

1. Monoubiquitin (Boston Biochem, Cambridge, MA, USA). Stored at 4 °C.

2. Purified yeast recombinant E1 (Boston Biochem). Stored at −80 °C.

3. Yeast MmS2/UbC13 recombinant E2 ubiquitin-conjugating enzymes, both purified as described by Hofmann [11]. Stored at −80 °C.

4. Ligation buffer for ubiquitin synthesis: 40 mM Tris–HCl, pH 7.5, 10 mM $MgCl_2$, 10 mM ATP.

5. Resource S prepacked column (GE Healthcare Bio-Sciences AB, Uppsala, Sweden).

6. Buffer A: 50 mM ammonium acetate, pH 4.5.

7. Buffer B: Buffer A supplemented with 500 mM NaCl.

8. Storage buffer: 50 mM Tris–HCl, pH 7.6.

9. Sample concentrator with a 10 kDa cutoff (Amicon Ultra 10, Merck Millipore, Cork, Ireland).

2.5   Linear GST-Ub$_4$ Purification

1. The linear-Ub$_4$ gene was inserted into the pGEX 4T2 vector such that the encoded protein was fused to the C-terminus of GST.

2. Extraction buffer: 20 mM Tris–HCl, pH 8, 30 mM KCl, 1 mM TCEP.

3. Protease inhibitor cocktail tablets (Complete EDTA-free).

4. French press.

5. Sonicator.

6. Glutathione Sepharose 4B Fast Flow (GE Healthcare).

7. Column equilibration buffer: 0.1 % (v/v) Triton X-100 in PBS, pH 7.5.

8. Washing buffer: 50 mM Tris–HCl, pH 8.

9. Elution buffer: 50 mM Tris–HCl, pH 8, 20 mM reduced glutathione (Sigma-Aldrich) (see **Note 5**).

10. Storage buffer: 50 mM Tris–HCl, pH 7.5, 150 mM NaCl, 0.4 mM DTE (dithioerythritol).

11. Human plasma thrombin immobilized on agarose beads, 0.3 mg/ml (Calbiochem, San Diego, CA, USA), stored at 4 °C.

2.6 Fluorescence Spectroscopy

1. Binding Buffer: 50 mM Tris–HCl pH 8, 150 mM NaCl, 1 mM TCEP.

2. QuantaMaster QM4CW high-sensitivity multimodular fluorometer (Photon Technology International, Edison, NJ, USA).

3. Peltier temperature controller.

4. Fluorometer cuvette type 23-3.45/Q/3 (Starna, Atascadero, CA, USA).

5. Adaptor type FCA 3 (Starna).

3 Methods

For the protocols and experiments described below, the cDNA sequences corresponding to residues 259–419 of the human NEMO protein were mutated at position 395 to generate the WT fluorescent NOAZ F395W, and at position 308, 311 and 395 to generate the fluorescent NOAZ Y308A/D311N F395W mutant. The mutations were performed by the standard overlap extension PCR method. The PCR fragments were then inserted into the pET28b vector and the constructs were checked by DNA sequencing.

3.1 Purification of the Human Bipartite UBD Domain of NEMO (Residues 259–419)

The following procedure is an atypical purification method including successive denaturation and renaturation steps that was developed for the NOAZ domain of NEMO, despite the recombinant protein being expressed in a soluble form in *E. coli*. This may seem surprising, because such procedures are generally recommended only for recombinant proteins forming inclusion bodies in *E. coli* [12]. However, in the particular case of the human NOAZ domain of NEMO, which contains many lysine residues exposed at the surface of the protein, this method has the advantage of rapidly removing any *E. coli* chaperone proteins (DnaJ, DnaK and GrpE) and nucleic acids present, because these interactions cannot occur in a buffer containing 6 M urea. Traces of nucleic acids and chaperone proteins are often present when the protein is purified under native conditions, and the presence of these contaminants might interfere with ubiquitin binding in fluorescence spectroscopy experiments.

3.1.1 Resuspension of the Bacterial Pellet and Cell Lysis

1. Resuspend the bacterial pellet in extraction buffer at a ratio of 5.5 ml of buffer per gram of dry pellet.

2. Lyse the resuspended bacteria by two passages through a French press followed by four bursts of sonication on ice for 15 s each, with an interval of 30 s between sonications (*see* **Notes 6–8**).

3.1.2 Cell Lysate Clarification and Denaturation

1. Clarify the cell lysate by centrifugation for 30 min at $10,000 \times g$ and 4 °C.

2. Denature the proteins by slowly adding urea powder to the supernatant until a final urea concentration of 6 M is reached.

3. Stir for 30 min at room temperature (*see* **Notes 3** and **9**).

*3.1.3 Protein Renaturation and Purification on Nickel-Charged Resin (See **Note 10**)*

1. Load the denatured supernatant onto a Ni-NTA column at a flow rate of 2 ml/min.

2. Nonspecific interactions of the Ni-NTA matrix with nucleic acids and proteins from *E. coli* are prevented by thoroughly washing the column with a urea-containing denaturation buffer until absorbance returned to baseline values.

3. Renature the proteins directly on the column, by applying a reverse, linear urea gradient (6 to 0 M urea) at a low flow rate of 1 ml/min, in six column volumes of renaturation buffer, and then equilibrate the column with three column volumes of this buffer (*see* **Note 11**).

4. Elute the protein from the column at a flow rate of 1 ml/min with a linear imidazole gradient, from 0.01 to 1 M, in five column volumes of elution buffer.

5. After this first purification step, some protein contaminants eluted with the protein of interest could still be detectable on an SDS-PAGE gel. A second purification is therefore performed as follows.

6. Pool the protein fractions and dialyze them against the native equilibration buffer of the Ni-NTA column.

7. Load the protein onto another Ni-NTA column at a flow rate of 2 ml/min and elute with a linear imidazole gradient, from 0 to 1 M in five column volumes, at a flow rate of 1 ml/min.

8. Pool the protein fractions and concentrate with a 10 kDa Amicon concentrator. Exchange the buffer by carrying out several washes with Buffer. This will eliminate the imidazole and decrease the NaCl concentration gradually to 150 mM.

9. At this step, the final concentration of the protein will be about 200 μM, based on absorbance at 280 nm. Analyze the purity of the protein by SDS-PAGE on a 12 % polyacrylamide gel followed by Coomassie blue staining and compare with a protein purified under native conditions. As shown in the inset of Fig. 2a, the two proteins are highly homogeneous, with similar purity levels.

3.2 Zinc Titration

Determine the zinc concentration in the refolded NOAZ domain and the binding stoichiometry of Zn^{2+} per protein subunit in a PAR colorimetric assay [13]. This method is based on the use of PAR, which forms a high-affinity complex with Zn^{2+}, modifying its absorbance spectrum.

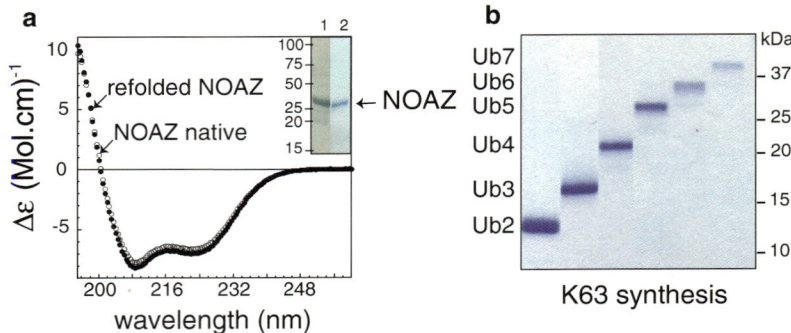

Fig. 2 (**a**) Circular dichroism spectra recorded at 20 °C and 1 mg/ml for the NOAZ domain purified in non-denaturing conditions (*open circle*) or by a procedure involving successive steps of urea-induced denaturation and renaturation (*filled circle*) (*see* Subheading 3.1 for details). *Inset*, SDS-PAGE analysis followed by Coomassie blue staining of the NOAZ domain purified by a denaturation/renaturation procedure (*lane 1*) or in non-denaturing conditions (*lane 2*). (**b**) SDS-PAGE gel stained with Coomassie blue, showing the various K63-polybiquitin chains produced by enzymatic synthesis

1. First release the zinc ion by digestion of the NOAZ domain in the presence of 10 μg/ml proteinase K for 30 min at 60 °C (*see* **Note 12**).

2. Before quantification of the zinc-PAR complex by the measurement of absorbance at 492 nm, treat 96-well culture plates with 0.5 M EDTA and thoroughly rinsed with water to decrease the background signal (*see* **Note 13**).

3. Set up and plot a calibration curve with a range of known concentrations of $ZnCl_2$, from 0 to 25 μM in protein buffer.

4. Measure absorbance at 492 nm in a TECAN Infinite 500 microplate reader.

5. As a positive control, determine the amount of Zn^{2+} bound to NOAZ with the NOAZ domain purified in non-denaturing conditions (Table 1). When purified in denaturing conditions, the NOAZ domain has a binding stoichiometry close to 1:1, similar to that of the NOAZ domain purified in non-denaturing conditions. Thus, purification through successive unfolding/refolding steps does not affect the binding of the protein to the Zn^{2+} ion and results in a native-like protein (*see* **Note 14**).

3.3   Circular Dichroism (CD) Spectroscopy

Assess whether the recombinant NOAZ domain purified in denaturing conditions has been correctly refolded, by comparing its CD spectrum with that of the same NOAZ domain purified in non-denaturing conditions [14] (*see* **Note 15**).

1. Extensively dialyze the proteins against Buffer. Check the protein concentrations by determining absorbance at 280 nm with the extinction coefficient of the protein concerned (*see* **Note 16**), and dilute the protein solutions to 1 mg/ml with dialysis buffer.

Table 1
Zinc titration in a 4-(2-pyridylazo)resorcinol (PAR) colorimetric assay for NOAZ domains purified in non-denaturing conditions or in denaturation/renaturation conditions

	Protein concentration (µM)	Zinc estimated concentration (µM)	Zinc concentration/ protein concentration
NEMO NOAZ purified in native conditions	8	9.6	1.2 ± 0.1
	4	4.8	1.2 ± 0.1
NEMO NOAZ purified in unfolding–refolding conditions	13	12.5	1.0 ± 0.1
	4	4.8	1.2 ± 0.1

Zn^{2+} titration of NOA ΔZF, lacking the ZF domain was used as a negative control, and the murine NOAZ domain used as a positive control (the presence of one zinc atom per NOAZ subunit was previously established by flame atomic absorption spectroscopy, as described by Vinolo [15])

2. Obtain each CD spectrum as the average of three scans recorded from 195 to 260 nm at 20 °C, with a bandwidth of 1 nm. Subtract the signal of the blank (buffer alone) from all CD data, which are then converted from units of ellipticity (degrees) to the molar differential extinction coefficient $\Delta \varepsilon$ (M^{-1} cm^{-1}) (*see* **Note 17**).

3. No significant spectral difference is observed between the two NOAZ domains purified by unfolding/refolding or in non-denaturing conditions (Fig. 2b). Moreover, a similar $A_{222/208 \, nm}$ ratio of 1.06, greater than 1, is obtained for both proteins, indicating the presence of a coiled-coil structure. Thus, the experimental conditions for purification described here allowed the correct refolding of the NOAZ domain after denaturation in urea. NOAZ has a three-dimensional structure with a high α-helix content, similar to that of NOAZ domain purified under non-denaturing conditions (*see* **Note 18**).

3.4 Synthesis and Purification of Polyubiquitin Chains

3.4.1 K63-Linked Chains

1. Generate K63-linked ubiquitin chains from 8 to 10 mg/ml untagged monoubiquitin in ligation buffer supplemented with 2 mM DTE (*see* **Notes 19** and **20**).

2. Initiate the conjugation reaction by adding 0.1 µM yeast E1 and 4 µM yeast MmS2/UbC13 complex and incubating for 6 h at 37 °C (*see* **Note 21**).

3. At the end of the incubation period, stop the reaction by adding 1 mM DTE (freshly prepared) and 1 mM EDTA. The sample can be stored at −20 °C.

4. Add 0.03 volumes of 2 N acetic acid to the sample before loading onto the column. Check that the pH was approximately 4.5.

5. Load the acidified sample reaction (1 ml) onto a 1 ml Resource S column pre-equilibrated with buffer A.

6. Wash the column with approximately 10 column volumes of buffer and elute the protein with a linear gradient of NaCl (0–0.3 M) in 40 column volumes.

7. Pool peak fractions, dialyze, concentrate and store at −80 °C in storage buffer (*see* **Note 22**).

8. Analyze each peak by running on a 12 % SDS-PAGE gel and staining with Coomassie blue (Fig. 2b).

3.4.2 M1 (Linear) Tetraubiquitin Chains

1. Mix GST-Ub4 (8 mg) with 200 μl of thrombin-settled agarose gel, and incubate the mixture for 6 h at room temperature with shaking.

2. Remove the thrombin-agarose gel by centrifugation at 2,000 × *g* for 10 min at 4 °C.

3. Load the supernatant onto a Glutathione Sepharose 4 Fast Flow (20 ml) column to remove the free GST released by thrombin hydrolysis.

4. Linear Ub4 protein, eluted in the column flowthrough, should be concentrated, dialyzed against storage buffer and stored at −80 °C (*see* **Note 22**).

5. Assess the purity of the linear Ub4 by running on a 12 % SDS-PAGE gel and staining with Coomassie blue.

3.5 Measurements of Polyubiquitin Interactions with the UBD Domain of NEMO by Intrinsic Fluorescence Spectroscopy

The following intrinsic fluorescence spectroscopy experiment to measure polyubiquitin interactions with the UBD of NEMO was performed using the NEMO domains and polyubiquitin generated as described above.

One day before the experiment, all NOAZ domains and ubiquitin chains were extensively dialyzed against Binding Buffer. Protein concentrations were then determined with an extinction coefficient of 13,200 M^{-1} cm^{-1} for the WT NOAZ (F395W), 11,710 M^{-1} cm^{-1} for the mutant Y308A/D331N NOAZ (F395W), 5,100 M^{-1} cm^{-1} for tetraubiquitin chains, and 10,200 M^{-1} cm^{-1} for hexaubiquitin chains. NEMO samples were then diluted to a concentration of 5 μM with buffer and placed on ice. Fluorescence titration experiments were performed at 25 °C, in a cuvette with an optical pathlength of 3 mm. The wavelength at which emission was maximal was determined for the WT and the defective NOA UBD (Y308A/D311A) mutant, by an emission scan from 310 to 450 nm, with an excitation wavelength of 297 nm. The excitation wavelength was set to 297 nm (the red edge of the tryptophan absorption spectrum), to reduce the contribution of the tyrosyl residues to total fluorescence (the NOAZ domain and hexaubiquitin contain five and six tyrosyl residues, respectively). The excitation and emission bandwidths were set to 2 nm and 4 nm, respectively. For each NEMO protein, emission was maximal at 345 nm. This emission wavelength was therefore used for subsequent experiments (*see* **Note 23**).

We carried out titration experiments on the bipartite domain of NEMO with linear tetraubiquitin (M1-Ub$_4$), K63-linked tetraubiquitin or hexaubiquitin, chains as follows:

1. Each sample was titrated at 25 °C by successively adding aliquots of 2–10 µl of 40 µM polyubiquitin chain to 120 µl of NEMO sample (*see* **Note 24**). Depending on the sample, an interval of 2–10 min was left between additions of polyubiquitin chains, to allow the sample to reach equilibrium.

2. For determination of the background signal, we carried out the same titration process with polyubiquitin chains in buffer alone, recovered from dialysis, using the same instrumental setup as for protein titration (*see* **Note 25**).

3. For each titration point, the fluorescence intensity at 345 nm, F_i, was corrected for dilution, background and inner filter effect (*see* **Note 26**), according to the following formula: $F_{icorr} = (F_i - B_i)(V_i/V_0)10^{0.5\ l\ (Aiex + Aiem)}$, where F_{icorr} is the corrected fluorescence intensity at a given point of titration i, F_i is the experimentally measured fluorescence intensity, B_i is the background signal corresponding to the titration of polyubiquitin chains with the buffer alone, V_i is the volume of the sample at a given titration point, V_0 is the initial volume of the sample, l, is the total length of the cuvette in cm, A_{iex} and A_{iem} are the absorbances of the sample at the excitation and emission wavelengths, respectively.

4. Corrected fluorescence intensity was plotted against polyubiquitin concentration (chain concentration). The experimental binding curve was then fitted by nonlinear regression, with the most appropriate binding model used to calculate the binding parameters (dissociation constant K_D, stoichiometry n, and cooperativity m for a sigmoidal binding curve).

The addition of increasing concentrations of hexa-polyubiquitin chains to the NOAZ WT (F395W), but not to the Y308A/D311N NOAZ mutant with a defective NOA UBD, led to an increase in fluorescence intensity, which reached a plateau at about 2–4 µM (Fig. 3). This reflects the formation of a specific complex between the ZF UBD of NOAZ and hexa-polyubiquitin chains, which occurs only when the NOA UBD is functional. No significant interaction with tetraubiquitin chains was observed in the same conditions, indicating that an optimal length of at least six ubiquitin residues is required for optimal interaction with K63 chains (inset of Fig. 3). In conclusion, intrinsic fluorescence spectroscopy is a powerful and sensitive method that can highlight avid interactions between two distal UBDs and a long oligomer of K63 polyubiquitin chains (*see* **Notes 27–29**).

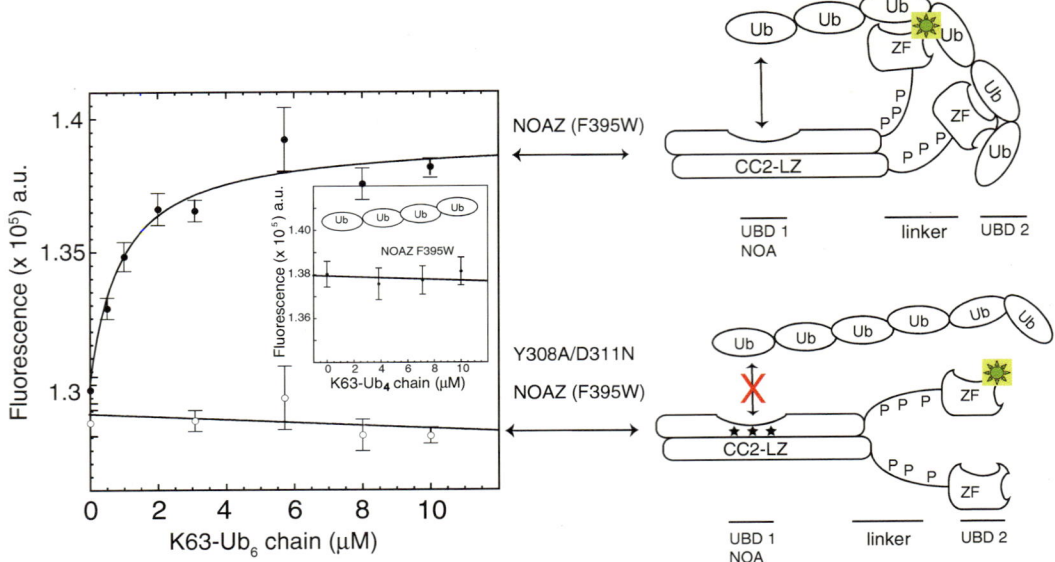

Fig. 3 Fluorescence titrations with K63-hexaubiquitin chains of the WT (F395W) NOAZ (*filled circle*) and the Y308A/D311N mutant (*open circle*) defective for ubiquitin binding by the distal NOA UBD. *Inset*: fluorescence titration of the WT (F395W) NOAZ with K63 tetraubiquitin chains. *Right panel*: schematic diagram showing the binding events observed in fluorescence spectroscopy with the fluorescent WT NOAZ, NOAZ (F395W), and the fluorescent double mutant defective in ubiquitin binding within the NOA binding site, Y308A/D311N NOAZ (F395W). The mutations introduced in the NOA (UBD1) mutant are indicated by an *asterisk*, whereas the tryptophan fluorescence probe located in the ZF domain (UBD 2) is indicated by a *green ideogram*

4 Notes

1. The three aromatic amino acids present in proteins (tryptophan, phenylalanine, and tyrosine) are potential probes for fluorescence spectroscopy. However, protein fluorescence is generally due to tryptophan residues, which have a higher absorbance and quantum yield of emission than tyrosine and phenylalanine. For these reasons, tryptophan is the fluorescence probe most frequently used for the monitoring of conformational changes in proteins.

2. Introducing a point mutation (F395W) could alter any function of the protein. In NEMO, this mutation has no deleterious effect because the F395W mutant fully restored TNF-α-induced NF-κB activation, to WT levels, in a genetic complementation assay [5].

3. A freshly prepared urea solution should be used because urea solutions slowly break down to release cyanate and ammonium ions. The cyanate ions can modify proteins by reacting with their amino groups. Furthermore, urea is not sufficiently soluble at 4 °C to generate a 6 M urea solution in water at this temperature.

All purification steps involving urea-containing buffers should therefore be performed at room temperature.

4. The dissolution of EDTA in water can be facilitated by adjusting the pH of the solution to 8.

5. The pH should be checked because reduced glutathione acidifies the buffer.

6. If a French press is not available, the bacteria can be lysed by incubation with lysozyme (0.5 mg/ml) for 20 min at room temperature followed by treatment with DNase (20 μg/ml) in presence of 10 mM $MgCl_2$ for 30 min at room temperature.

7. Cell lysate sonication leads to the shearing of genomic DNA, resulting in a lower viscosity.

8. It is generally advisable to keep the temperature low during cell lysis (homogenization and sonication).

9. Precautions should be taken when dissolving the urea, because this significantly increases the volume of cell lysate.

10. If no chromatographic system is available, batch resin systems may be used.

11. To ensure that all the urea is removed from the Ni-NTA column, the refraction index of the buffer leaving the column can be measured at any time.

12. Incubating the protein solution at temperatures between 50 and 60 °C significantly increases the catalytic efficiency of proteinase K.

13. Incubating plates with EDTA solution and then washing thoroughly with water eliminates any bivalent ion contaminants present and significantly increases assay accuracy. After EDTA treatment, a failure to rinse thoroughly with water may result in EDTA chelating of the released zinc ions and competing for ZnII binding with the PAR reactant.

14. In some cases, the ZnII/protein ratio may exceed 1, due to the presence of trace amounts of nickel ions, which also react with the PAR reactant. These Ni^{2+} ions generally originate from the Ni-NTA resin used for protein purification.

15. After each CD or fluorescence titration experiment, the cuvette should be thoroughly cleaned with 2 % Hellmanex solution, extensively rinsed with distilled water and dried with ethanol. The washing quality of the cuvette can be checked each time by simply recording the fluorescence signal with the buffer alone.

16. Most buffers and salts absorb in the far-UV region, from 180 to 200 nm. To reduce the background signal, it is advisable (1) to increase the protein concentration, (2) to decrease the length of the light path, (3) to degas the buffer extemporaneously to remove the dissolved air, and (4) to use buffers and

salts that are relatively transparent in the far-UV region, such as phosphate/borate/cacodylate and sodium fluoride.

17. The CD signal is converted to the molar differential extinction coefficient $\Delta\varepsilon$ (M^{-1} cm^{-1}) as follows: $\Delta\varepsilon_{norm} = \theta_{obs} / (L_P \times 3,298 \times 10 \times L_C \times C)$, where $\Delta\varepsilon_{norm}$ is the normalized CD signal, θ_{obs} is the CD signal measured in degrees, L_P is the number of peptide bonds within the protein, L_C is the pathlength of the cuvette, and C is the molar concentration of protein.

18. We want to draw the attention on the fact that the measurement of the molecular mass of a protein by gel filtration can be erroneous because of the elongated shape of this protein. Indeed, the NOAZ domain elutes as an apparent tetramer according to the calibration curve of the column using standard globular proteins. However, sedimentation velocity experiments showed that it forms a full dimer with a frictional coefficient typical of an elongated protein (1.95), confirming that the aberrant chromatographic behavior in gel filtration was due to the elongated shape of the NOAZ domain. Combined with CD experiments, these data also supported the view that the NOAZ domain is properly folded.

19. Monoubiquitins fused at their terminus to any tag should not be used as a substrate for the MmS2/UbC13 heterodimeric complex, due to inefficient conjugation [11].

20. The pH of the ATP solution should be adjusted to 7 with a weak base, such as 1 M histidine, to extend the half-life of ATP, and the neutralized ATP can be stored at –20 °C. ATP is instable in acidic conditions, in which it may be irreversibly converted into its diphosphate form.

21. The E1 activating enzyme of mammalian origin can also be used for K63 synthesis with the heterodimeric yeast MmS2/Ubc13 E2 complex.

22. Fractions were frozen in liquid nitrogen.

23. When the fluorescence spectrum of a protein is recorded, a weak shoulder may appear on the high-energy side of the maximum fluorescence wavelength. This is due to the inelastic Raman scatter of the water. The wavelength of Raman scattering, λ_{RA}, can be calculated as follows: $1/\lambda_{RA} = 1/\lambda_{Ex} - 0.00033$. For instance, if the total fluorescence of a sample is not very high, a Raman scatter peak will be observed at $\lambda_{RA} = 329$ nm when the sample is irradiated at 297 nm. For correction of the fluorescence spectrum, the spectrum of the blank (buffer alone) should be recorded with the same instrumental setup as used for the protein and the spectrum obtained should be subtracted from the crude spectral data.

24. In titration experiments, it is often recommended that the final volume of the added binding partner at the end of the titration should not exceed about 20 % of the initial total volume of the protein sample.

25. As a control, it is crucial to measure the fluorescence emission signal for each type of polyubiquitin chain in the buffer alone, although ubiquitin carries no tryptophan residues. Indeed, UV contaminants displaying Trp-like fluorescence spectra were often observed, particularly following the isolation of M1-(linear) polyubiquitin chains from fusions with the GST-tag. An example of such an experiment with the linear M1-tetraubiquitin chain is shown in Fig. 4. The polyubiquitin chain was produced in *E. coli* and purified from its GST-fusion tag by a standard procedure according to the manufacturer instruction described elsewhere. Note that no significant trace of protein contaminants was detectable by SDS-PAGE followed with Coomassie blue staining (not shown). The incremental addition of a polyubiquitin chain induces a similar increase in fluorescence signal as the buffer alone and the NOAZ F395W mutant. This increase in fluorescence, which varies linearly with polyubiquitin concentration, reflects the presence of UV contaminants in the polyubiquitin preparation. This contamination, which may be due to

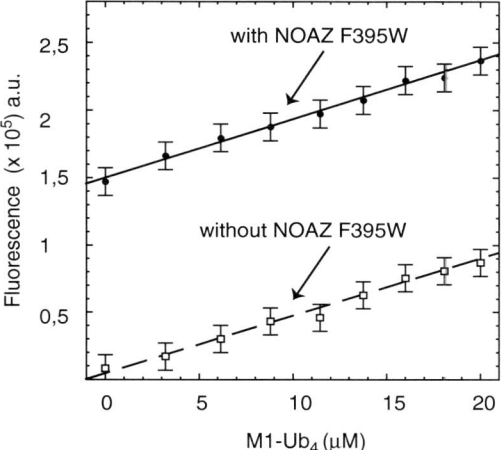

Fig. 4 Control titration experiments showing the presence of fluorescent impurities in the purified M_1-tetraubiquitin sample. Independent titrations of the buffer alone (without NOAZ) with M1-tetraubiquitin were performed with the same aliquots, 1–20 μl of M1-Ub$_4$ and the same experimental setup as for titration with the fluorescent WT (F395W) NOAZ. The addition of M_1-Ub$_4$ induced a change in fluorescence similar to that observed with the NOAZ domain and buffer alone, indicating the presence of fluorescent impurities in the M1-Ub$_4$ sample purified from *E. coli*. This high background signal makes impossible to monitor M_1-Ub$_4$ binding under these experimental conditions

the presence of GST and *E. coli* protein traces, considerably decreases the sensitivity of the assay, preventing observations of binding with a linear tetraubiquitin chain.

26. The inner filter effect can introduce an error and non-linearity into the relationship between fluorescence emission and absorbance. This effect is due to the solution at the front of the cuvette being exposed to a higher intensity of excitation light than the sample nearest the opposite side of the cuvette. In general, it is important to correct the fluorescence intensity for each titration point when sample absorption exceeds 0.01 at the excitation and emission wavelengths.

27. To get the optimal high signal/background ratio, the user must be aware of the lifetime of the xenon lamp, which is usually around 2,000 h. The state of the xenon lamp can be checked by periodically recording the Raman spectrum of water with an excitation wavelength of 350 nm and an emission scan from 360 to 450 nm, in 1 nm increments, with both excitation and emission bandwidths set to 5 nm.

28. Before measurement, the sample must reach thermal equilibrium, to minimize signal drift.

29. In some cases, protein samples must be centrifuged for 30 min at $10,0000 \times g$ before starting the experiment, and the protein concentration in the supernatant is again checked by measuring absorbance at 280 nm. This is an important point in fluorescence spectroscopy, because sample turbidity increases light scattering due to the presence of large particles in solution resulting from the aggregation or denaturation of proteins.

Acknowledgments

The authors thank Samuel Levy, MD, for his critical reading of the manuscript, and Prof. Alain Israël for fruitful discussions and continuous support. This work was supported, in whole or in part, by the BNP-Paribas Foundation and Institut de Recherches SERVIER (Croissy sur Seine, France).

References

1. Behrends C, Harper JW (2011) Constructing and decoding unconventional ubiquitin chains. Nat Struct Mol Biol 18:520–528

2. Yamaoka S, Courtois G, Bessia C et al (1998) Complementation cloning of NEMO, a component of the IkappaB kinase complex essential for NF-kappaB activation. Cell 93:1231–1240

3. Rahighi S, Ikeda F, Kawasaki M et al (2009) Specific recognition of linear ubiquitin chains by NEMO is important for NF-kappaB activation. Cell 136:1098–1109

4. Dynek JN, Goncharov T, Dueber EC et al (2010) c-IAP1 and UbcH5 promote K11-linked polyubiquitination of RIP1 in TNF signalling. EMBO J 29:4198–4209

5. Ngadjeua F, Chiaravalli J, Traincard F et al (2013) Two-sided ubiquitin binding of NF-kappaB essential modulator (NEMO) zinc finger unveiled by a mutation associated with anhidrotic ectodermal dysplasia with immunodeficiency syndrome. J Biol Chem 288: 33722–33737

6. Laplantine E, Fontan E, Chiaravalli J et al (2009) NEMO specifically recognizes K63-linked poly-ubiquitin chains through a new bipartite ubiquitin-binding domain. EMBO J 28:2885–2895

7. Kastritis PL, Bonvin AM (2013) On the binding affinity of macromolecular interactions: daring to ask why proteins interact. J R Soc Interface 10:20120835

8. Wienken CJ, Baaske P, Rothbauer U et al (2010) Protein-binding assays in biological liquids using microscale thermophoresis. Nat Commun 1:100–104

9. Hubeau M, Ngadjeua F, Puel A et al (2011) New mechanism of X-linked anhidrotic ectodermal dysplasia with immunodeficiency: impairment of ubiquitin binding despite normal folding of NEMO protein. Blood 118:926–935

10. Lo YC, Lin SC, Rospigliosi CC et al (2009) Structural basis for recognition of diubiquitins by NEMO. Mol Cell 33:602–615

11. Hofmann RM, Pickart CM (2001) In vitro assembly and recognition of Lys-63 polyubiquitin chains. J Biol Chem 276:27936–27943

12. Burgess RR (2009) Refolding solubilized inclusion body proteins. Methods Enzymol 463:259–282

13. Hartwig A, Schwerdtle T, Bal W (2010) Biophysical analysis of the interaction of toxic metal ions and oxidants with the zinc finger domain of XPA. Methods Mol Biol 649: 399–410

14. Clarke DT (2011) Circular dichroism and its use in protein-folding studies. Methods Mol Biol 752:59–72

15. Vinolo E, Sebban H, Chaffotte A et al (2006) A point mutation in NEMO associated with anhidrotic ectodermal dysplasia with immunodeficiency pathology results in destabilization of the oligomer and reduces lipopolysaccharide- and tumor necrosis factor-mediated NF-kappa B activation. J Biol Chem 281: 6334–6348

Chapter 20

Generation of a Proteolytic Signal: E3/E2-Mediated Polyubiquitination of IκBα

Robert A. Chong, Kenneth Wu, Jordan Kovacev, and Zhen-Qiang Pan

Abstract

A key regulatory node in NF-κB signaling is the removal of the IκBα inhibitor, whose levels are tightly controlled by the ubiquitin–proteasome system. In response to signal activation and transmission, ubiquitin E1, E2, and E3 enzymes are employed to generate a lysine 48-linked ubiquitin chain that triggers degradation of IκBα by the proteasome. In this chapter we describe an in vitro biochemical approach to reconstitute the ubiquitination system. To do so, we detail methods for the preparation of the relevant enzymes and substrate, as well as for the execution of the reaction with high efficiency. This sensitive and highly reproducible readout can be applied to the study of proteins, small molecules, and other factors that modulate IκBα ubiquitination, thereby producing outcomes that impact NF-κB signaling to advance the course of improving human health.

Key words NF-κB, IκBα ubiquitin, E3 ubiquitin ligase, SCF, Nedd8, Cdc34, UbcH5c, Polyubiquitination

1 Introduction

Nuclear factor kappa enhancer binding protein (NF-κB) is an essential regulator of diverse processes fundamental to the cell including survival, immunity, and inflammation. Canonical NF-κB signaling can be initiated through a number of ligands, such as tumor necrosis factor (TNF)-α, interleukin-1β (IL-1β), or toll-like receptor (TLR) ligands, which activate TNFR, IL-1R, and TLRs, respectively. Through a variety of intermediate steps, these signaling cascades ultimately converge on NF-κB, which is freed to translocate to the nucleus to induce the expression of many target genes. Importantly, regulation of this signaling pathway in the absence of stimulation is achieved through inhibitor of NF-κB (IκB) family members, predominantly IκBα, which binds directly to NF-κB and inactivates it via sequestration in the cytoplasm. It has been well documented that the cytokine-triggered rapid translocation of NF-κB

Michael J. May (ed.), *NF-kappa B: Methods and Protocols*, Methods in Molecular Biology, vol. 1280,
DOI 10.1007/978-1-4939-2422-6_20, © Springer Science+Business Media New York 2015

from the cytoplasm to the nucleus requires ubiquitin (Ub)-dependent proteasomal degradation of IκBα [1].

Subsequent biochemical identification and characterization of the IκBα degradation signal/determinant (degron) has elucidated the molecular details required for the translation of ligand/cytokine responses to proteolytic signals. As illustrated in Fig. 1, the IκBα N terminus features a phosphodegron, composed of the DSG box (DS^{32}GΦXS^{36}) [2]. In response to external cues, the IκB kinase (IKK) complex (composed of IKKα, IKKβ, and IKKγ) is activated, resulting in the deposition of phosphates onto IκB Serine 32 and Serine 36 (Fig. 1). This phosphorylation triggers the interaction of an E3 ubiquitin ligase (*SCF^{β}TrCP*) with IκBα, which directs the assembly of multiple Ub moieties onto lysine 21 and lysine 22 (K21, K22) of IκBα [3–5]. As a consequence, Ub-modified IκBα is degraded by the 26S proteasome, thereby liberating NF-κB that enters the nucleus to execute its transcription activity. Incredibly, IκBα is degraded within 3–5 min in cells after exposure to the requisite ligands/cytokines, indicating a need for the generation of proteolytic signal in a rapid fashion.

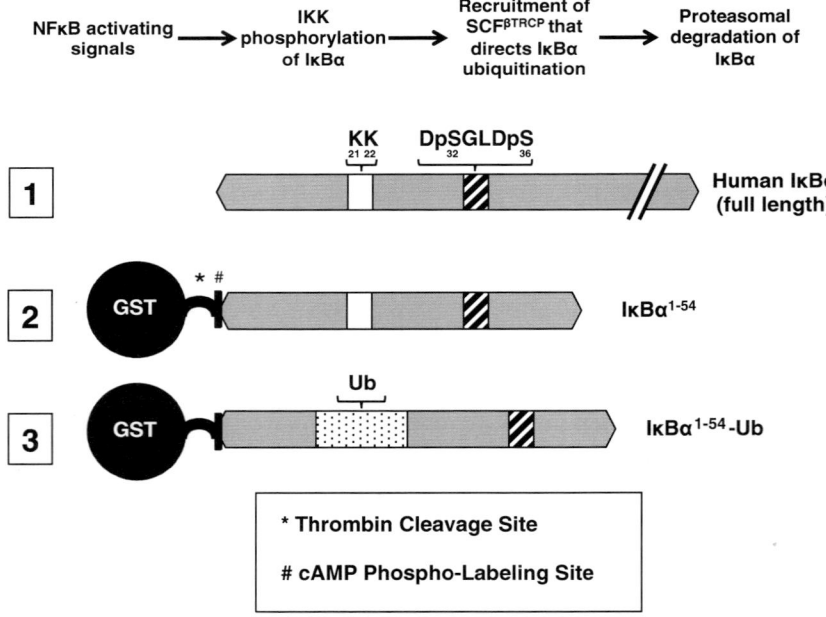

Fig. 1 Ubiquitin enzymes generate a K48 linked polyubiquitin chain on IκBα. The methods provided here detail a procedure to recapitulate the degradation signal on IκBα [1] in vitro by means of model substrates [2] and [3]. IκBα$^{1-54}$ [2] contains an intact K21 and K22 to accept ubiquitins. IκBα$^{1-54}$-Ub [3] contains an Ub in place of K21 and K22, which mimics a monoubiquitinated substrate and thus only requires Cdc34 (see text for details; **Note 12**). Both model substrates contain a cleavable N terminal GST tag, a thrombin cleavage site (*asterisk*), as well as a cAMP consensus site (*hash*) that is required for ^{32}P labeling

This chapter summarizes recent advances on biochemical reconstitution of the polyubiquitination of IκBα and provides protocols for the generation of such proteolytic signal with rapid kinetics. In brief, the polyubiquitination of IκBα requires a cascade of ubiquitination enzymes (E1 activating, E2 conjugating, and E3 ligase). The Ub E1 activates an Ub moiety, which is transferred to an E2 conjugating enzyme to form a charged thiol ester (E2 ~ Ub). In the case of IκBα, the E3 Ub ligase $SCF^\beta TrCP$ is a four protein complex composed of Skp1 (adaptor), cullin 1 (CUL1; scaffold), ROC1 (E2 recruitment, also called Rbx1), and βTrCP (substrate targeting), in which CUL1 must be modified by a small ubiquitin-like protein, Nedd8, for interaction with the charged E2 [6, 7]. Once phosphorylated by the activated IKK, IκBα is bound to the Nedd8-modified form of $SCF^\beta TrCP$, resulting in the ROC1-mediated recruitment of the Ub-charged E2. It has been shown that the efficient assembly of Ub chains is dependent on two E2s, with UbcH5 initiating the reaction by forming predominantly substrate-Ub linkages, and Cdc34 extending the chain by forming K48 Ub-Ub linkages [8]. It should be noted that non-degradative ubiquitin signals such as K63 or linear chains formed by other Ub E2/E3 pairings also play an important role in the transmission of the NF-κB signal [1, 9].

The ability to reconstitute the IκBα degradation signal has resulted in the discovery of many key determinants that govern ubiquitin and NF-κB biology. Given that overactive NF-κB signaling is a key contributor to tumorigenesis and inflammation, more recent efforts have been directed toward blocking NF-κB activation [10]. In the context of ubiquitination, new and exciting therapeutic efforts are designed to specifically block specific components of the ubiquitin–proteasome pathway [11–13]. The method detailed here provides a model for accurately modeling Ub chain formation, and can be used as a platform to discover and characterize molecular agents capable of perturbing IκBα ubiquitination.

2 Materials (*See* Note 1)

2.1 Preparation of FLAG-IKKβ $^{S177E/ S181E}$

1. pFAST-FLAG-IKKβ$^{S177E/S181E}$ expression bacmid [14].

2. Standard tissue culture plates and flasks.

3. Grace's Insect Medium, supplemented with 10 % FBS (Life Technologies, Carlsbad, CA, USA).

4. Grace's Insect Medium, unsupplemented (Life Technologies).

5. Cellfectin II Transfection Reagent (Life Technologies).

6. PBS (phosphate-buffered saline), chilled at 4 °C before using.

7. Lysis buffer 1: 20 mM HEPES-NaOH, pH 7.5, 5 mM KCl, 1.5 mM $MgCl_2$, 1 mM DTT, 1 mM phenylmethylsulfonyl fluoride (PMSF), and 0.2 μg/ml of antipain and leupeptin.

8. Dounce Homogenizer.

9. Buffer A: 25 mM Tris–HCl (pH 7.5), 1 mM EDTA, 0.01 % Nonidet P-40 (NP-40), 10 % glycerol, 1 mM dithiothreitol (DTT), 0.1 mM PMSF, and 0.2 µg/ml of antipain and leupeptin.

10. Economy Flex-Column (Kimble Chase, Rockwood, TN, USA).

11. Anti-FLAG M2 Beads (Sigma-Aldrich, St. Louis, MO, USA). Stored at –20 °C.

12. 3× FLAG Peptide (Sigma-Aldrich), dissolved to 5 mg/ml in TBS (50 mM Tris–HCl (pH 7.4), with 150 mM NaCl).

13. Amicon Ultra-15 and Ultra-50 Centrifugation Filters (Millipore, Billerica, MA, USA).

14. BaculoTiter™ Assay kit (Invitrogen, Carlsbad, CA, USA).

2.2 Preparation of Ub E2 Conjugating Enzymes (Cdc34 and UbcH5c)

1. pET-3a-HIS-HA-Cdc34 plasmid [14], and pET-3a-HIS-UbcH5c [15].

2. BL21 Competent Cells, stored at –80 °C (New England Biolabs, Ipswich, MA, USA).

3. Luria Broth (LB) Base, diluted to 15.5 g/L, autoclaved, and stored at 4 °C (Sigma-Aldrich).

4. Carbenicillin, stock is 100 mg/ml in H_2O, and stored at –20 °C (Sigma-Aldrich).

5. LB Agar Plates + carbenicillin (100 µg/ml) (Sigma-Aldrich).

6. Isopropyl β-D-1-thiogalactopyranoside (IPTG), stock is 1 M in H_2O.

7. Lysis buffer 2: 20 mM Tris–HCl, pH 8.0, 1 % Triton X-100, 0.5 M NaCl, 2 mM phenylmethylsulfonyl fluoride, 0.4 µg/ml antipain, and 0.2 µg/ml leupeptin (*see* **Note 2**).

8. Imidazole wash buffer: Buffer A + 50 mM NaCl + 5 mM imidazole.

9. Imidazole elution buffer: Buffer A + 50 mM NaCl + 250 mM imidazole.

10. Ni-nitrilotriacetic acid-agarose (Ni-NTA) beads (Qiagen, Valencia, CA, USA).

11. 4× SDS loading buffer: 40 % glycerol, 240 mM Tris–HCl (pH 6.8), 8 % SDS, and 0.04 % bromophenol blue dissolved in H_2O.

12. Coomassie Brilliant Blue solution: Coomassie R-250 (Bio-Rad, Hercules, CA, USA), methanol (30 % v/v), glacial acetic acid (10 % v/v).

13. Destaining solution: methanol (10 % v/v), glacial acetic acid (10 % v/v).

14. SnakeSkin Pleated Dialysis Tubing, 3,500 MWCO (Pierce, Rockford, IL, USA).

15. Q-Sepharose column (GE Healthcare, Mickleton, NJ, USA).

16. Amicon Ultra-15 Centrifugation Filters, 10 K (Millipore).

17. Sephadex-75 gel filtration column (GE Healthcare).

18. AKTA FPLC (GE Healthcare).

2.3 Preparation of SCF$^{\beta TRCP}$ Ub E3 Ligase

1. GST-βTRCP-Skp1 and GST-Skp1-βTRCP expressing baculoviruses [16, 17].

2. Glutathione elution buffer: Buffer A + 50 mM NaCl + glutathione (10 mM).

3. Glutathione Sepharose 4B (GE Healthcare).

4. Biotinylated Thrombin, Thrombin Dilution Buffer, and Thrombin cleavage buffer (Millipore).

5. Streptavidin Sepharose (GE Healthcare).

2.4 Preparation of Radioactive Substrates

1. GST-IκBα$^{1-54}$ and GST-IκBα-Ub E. coli expression plasmids [8].

2. γ-[^{32}P]-ATP 6,000 Ci/mmol 150 mCi/ml Lead, 5 mCi (PerkinElmer).

3. cAMP kinase (Promega, Madison, WI, USA).

4. Thermal mixer.

5. Compact mini-column (Affymetrix, Santa Clara, CA, USA).

2.5 In Vitro Reconstitution Reaction with IκBα$^{1-54}$ Peptide

1. Nedd8 protein, stored at −80 °C (Boston Biochem, Cambridge, MA, USA).

2. Nedd8 E1 enzyme (APP-BP1/Uba3), stored at −80 °C (Boston Biochem).

3. Ubc12, stored at −80 °C (Boston Biochem).

4. UbcH5c, stored at −80 °C.

5. Human Ub, stored at −20 or −80 °C (Boston Biochem).

6. Na-glutamate, stock is 1 M.

2.6 In Vitro Reconstitution Reaction with IκBα-Ub Fusion

1. Nedd8 protein, stored at −80 °C (Boston Biochem).

2. Nedd8 E1 enzyme (APP-BP1/Uba3), stored at −80 °C (Boston Biochem).

3. Ubc12, stored at −80 °C (Boston Biochem).

4. UbcH5c, stored at −80 °C.

5. Human Ub, stored at −20 or −80 °C (Boston Biochem).

6. Na-glutamate, stock is 1 M.

2.7 Running the Gel	1. NOVEX 4–20 % Bis-Tris Protein Gel (12–15 well, 1.0 mm).
	2. 10× Tris/Glycine/SDS Running Buffer, diluted to 1× with distilled H_2O (Bio-Rad).
	3. Protein marker (Bio-Rad).
	4. Power supply.
2.8 Drying the Gel	1. 3 mm Whatman Chromatography Paper (GE Healthcare).
	2. Fixing solution: isopropanol (25 % v/v), acetic acid (10 % v/v) in water.
	3. Gel dryer.
2.9 Autoradiography and Exposure	1. Autoradiography Film.
	2. Film developer.
	3. Phospho-imager.

3   Methods

3.1   Preparation of FLAG-IKKβ$^{S177E/S181E}$ (See* Note 3*)

1. For routine maintenance of Sf9 cells, culture them in Grace's Insect Medium supplemented with FBS. Ensure that they are >95 % viable for optimal transfection. For baculovirus preparation, seed cells in Grace's Insect Medium, unsupplemented.

2. To prepare baculovirus-expressing FLAG-IKKβ$^{S177E/S181E}$, transfect 9×10^5 Sf9 cells in 1 well of a 6-well plate with 2 μg recombinant bacmid DNA (pFast-FLAG-IKKβ$^{S177E/S181E}$) using 8 μl of Cellfectin II reagent.

3. Collect the P1 viral stock. To generate a high-titer viral stock, perform two additional rounds of reinfection to obtain a P3 viral stock.

4. Determine the titer of the virus using the BaculoTiter™ Assay kit.

5. For large-scale purification of FLAG-IKKβ$^{S177E/S181E}$, infect 3×10^7 Sf9 cells at a multiplicity of infection (MOI) of 10 with baculovirus expressing Flag-IKK$^{S177E/S181E}$ for 60 h at 27 °C.

6. Harvest the cells, wash 1× with cold PBS, and resuspend in two cell pellet volumes of lysis buffer 1.

7. Incubate 10 min on ice.

8. Lyse cells using a prechilled dounce homogenizer (20 strokes).

9. Adjust the cell extract to 0.2 M NaCl with the addition of 5 M NaCl.

10. Sonicate the lysate (4× 20 s) and centrifuge it at $14,000 \times g$ for 45 min at 4 °C to clear (*see* **Note 4**).

11. Mix soluble extracts with 0.6 mL of M2 anti-FLAG beads (pre-equilibrated with lysis buffer 1 + 0.2 M NaCl) for 4 h at 4 °C.

12. Pack beads into a Flex-Column and wash with 20 mL of lysis buffer 1 + 0.2 M NaCl and 10 mL of buffer A + 50 mM NaCl.

13. Elute protein with 1 ml of 3× FLAG peptide (1 mg/ml) in buffer A + 50 mM NaCl.

14. Repeat an additional four times to ensure that the bound protein is eluted.

15. Add protein to the centrifugation filter and concentrate via repeated centrifugation at $4,000 \times g$.

16. The expected recovery is approximately 1 mg FLAG-IKKβS177E/S181E out of 3×10^7 infected Sf9 cells.

3.2 Preparation of Ub E2 Conjugating Enzymes (Cdc34 and UbcH5c)

1. Transform pET-3a-HIS-HA-Cdc34 plasmid into BL21 cells. Incubate cells on ice for 30 min, heat shock at 37 °C for 30 s, and return to ice for 2 min. Plate on LB agar plates with carbenicillin.

2. Pick a single colony and inoculate in a starter culture overnight in LB +100 μg/ml carbenicillin (*see* **Note 5**).

3. Add 5 mL of the starter bacterial colony to 500 mL (1:100 dilution) of LB with 0.4 % glucose in the presence of carbenicillin. Grow at 37 °C until the optical density measured at 600 reaches 0.5 (~3 to 3.5 h).

4. Induce culture with IPTG at a final concentration of 1 mM overnight at 25 °C.

5. Pellet bacterial cells by centrifugation at $5,000 \times g$ for 15 min at 4 °C.

6. Resuspend the cells in 1/25 of the culture volume of lysis buffer 2.

7. Sonicate the lysate (4× 20 s) and centrifuge it at $10,000 \times g$ for 30 min.

8. Mix 10 mL of soluble extracts with 3 mL of Ni-nitrilotriacetic acid-agarose beads (Ni-NTA) beads and rotate the mixture for 2 h at 4 °C.

9. Pack beads into a Flex-Column and wash with 30 mL of buffer A + 50 mM NaCl, and 10 mL of imidazole wash buffer.

10. Elute protein with a 40 mL of linear gradient of 5–250 mM imidazole in buffer A + 50 mM NaCl, collecting 1 mL fractions.

11. Determine peak fractions by subjecting aliquots of the fractions (1 μl) to SDS-PAGE (Subheading 3.7). Stain the gel for

10 min in Coomassie solution, rinse with distilled water, and destain in destaining solution.

12. Pool peak fractions, fill dialysis tubing with them, and properly seal the ends. Dialyze overnight by placing the tubing in 10–20× the volume of buffer A + 50 mM NaCl.

13. Load material onto a Q-sepharose column. Wash column with 30 mL of buffer A + 50 mM NaCl.

14. Elute the protein with a 40 mL linear gradient of 50–500 mM NaCl in buffer A. Analyze by SDS-PAGE as in **step 11** and pool the peak fractions.

15. Concentrate the peak fractions using centrifugal filters and purify them using FPLC using a Sephadex-75 gel filtration column. Pool the Cdc34 peak fractions.

16. The yield is approximately 0.5 mg of purified Cdc34 (from 0.5 L culture) (*see* **Note 6**).

17. To purify UbcH5c, repeat **steps 1–9** with pET-3a-HIS-UbcH5c.

18. Elute three times with 5 mL of imidazole elution buffer and check the concentration of each fraction on a Coomassie gel.

19. Dialyze the UbcH5c containing eluates overnight by placing the tubing in 10–20× the volume of buffer A + 50 mM NaCl.

20. Concentrate the UbcH5c using Amicon centrifugal filter and determine the concentration. The expected recovery is approximately 8 mg of UbcH5c (from 0.5 L of culture).

3.3 Preparation of SCF^βTRCP E3 Ligase (See Note 7)

1. As in Subheading 3.1, **steps 1–10**, generate baculovirus expressing GST-βTRCP-Skp1.

2. Pack a column with 2 mL of glutathione sepharose.

3. Pass the soluble extract (20 mL) through the column.

4. Wash the column with 20 mL of lysis buffer 1 + 0.2 M NaCl.

5. Wash the column with 15 mL of buffer A + 50 mM NaCl.

6. Elute with glutathione elution buffer three times, 5 mL each.

7. Check for protein by SDS-PAGE and staining by Coomassie.

8. To remove glutathione and concentrate the eluted protein, add the protein to a centrifugation filter and wash with 5 mL of buffer A + 50 mM NaCl twice. This should yield approximately 0.8 mg (0.8 mg/ml).

9. To remove the GST, digest purified GST-βTRCP-Skp1 in a reaction mixture (1 mL) containing 10U biotinylated thrombin and 1× cleavage buffer for overnight at 4 °C.

10. Mix the thrombin digested mixture with pre-equilibrated streptavidin beads (200 μl) for 4 h at 4 °C, shaking on a rotator.

11. To remove streptavidin beads containing biotinylated thrombin, pass the mixture through a compact column containing a filter and collect the flow-through.

12. The final yield is approximately 0.15 mg of βTRCP-Skp1 out of 10^7 infected Sf9 cells.

13. Repeat **steps 1–12** with GST-ROC1-CUL1.

14. The final yield is approximately 0.4 mg of ROC1-CUL1 out of 10^7 infected Sf9 cells.

3.4 Preparation of Radioactive Substrates (See Note 8)

1. To generate the substrate using *E. coli* expression, repeat **steps 1–7** in Subheading 3.2, except with a plasmid expressing GST-IκBα$^{1-54}$.

2. Repeat **steps 4–13** in Subheading 3.3 to isolate the substrate.

3. To phosphorylate substrates, incubate ~25 μg of purified GST-IκBα$^{1-54}$ with 0.1 μg of FLAG-IKKβ$^{S177E/S181E}$ in a reaction (100 μl) containing 50 mM Tris–HCl (pH 7.4), 10 mM MgCl$_2$, 1 mM ATP, and 0.1 mg/ml BSA. Incubate the reaction mixture at 37 °C for 15 min.

4. Add a second aliquot (0.1 μg) of FLAG-IKKβ$^{S177E/S181E}$ and incubate the reaction mixture at 37 °C for an additional 15 min. *See* Fig. 2 for an example of the phosphorylation reaction, which results in slowed migration of GST-IκBα$^{1-54}$ on denaturing gel (compare lane 1 and lanes 2–5). For further comments, *see* **Note 9**.

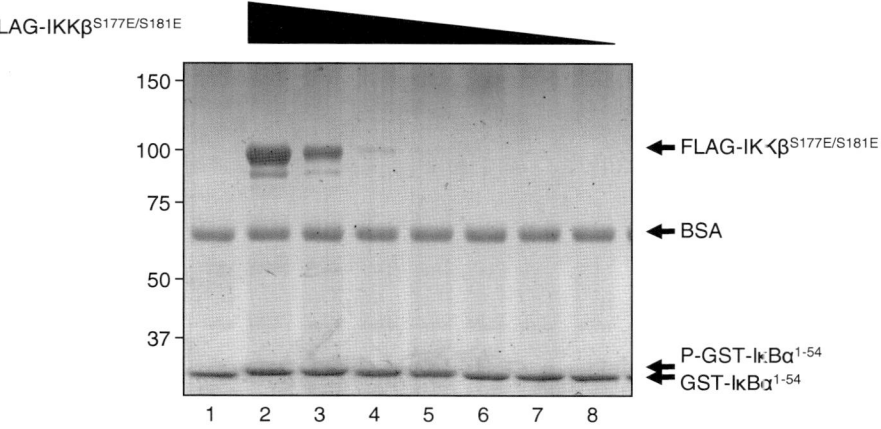

Fig. 2 Analysis of FLAG-IKK$^{S177E/S181E}$ activity. Phosphorylation of S32 and S36 on IκBα is an essential step in creating a competent phosphodegron that supports ubiquitination. To check the activity of the purified FLAG-IKKS177E/^{S181E}, an in vitro kinase reaction was performed. 1 μg of GST-IκBα$^{1-54}$ was incubated with varying amounts of purified FLAG-IKK$^{S177E/S181E}$ (0.04–4.0 μg) and analyzed by SDS-PAGE followed by Coomassie staining. In comparison to the untreated (*lane 1*), substrates treated with sufficient amounts of FLAG-IKK$^{S177E/S181E}$ (*lanes 2–5*) resulted in slowed gel mobility exhibited by GST-IκBα$^{1-54}$

5. To inactivate FLAG-IKKβ$^{S177E/S181E}$, add 2 μl of 0.5 M EDTA for a final concentration of 10 mM.

6. To have the IKKβ$^{S177E/S181E}$-phosphorylated GST-IκBα$^{1-54}$ re-bound to glutathione beads, add 10 μl of pre-equilibrated glutathione beads to the reaction, and mix for 60 min at 4 °C.

7. Wash 3× with lysis buffer 1 + 0.2 M NaCl, and 2× with buffer A + 50 mM NaCl. Remove the supernatant, careful not to disturb the beads.

8. From this point forward, take care to wear proper protective equipment when using ^{32}P and ^{32}P-labeled substrates, and ensure proper disposal of all materials (*see* **Note 10**).

9. To the beads, add 20 mM Tris–HCl (pH 7.4), 12 mM MgCl$_2$, 50 mM NaCl, 25 μM ATP, 2 mM NaF, cAMP kinase (2.5 U), and γ-[^{32}P]-ATP (1 μl) to a final volume of 100 μl.

10. On a thermal mixer, incubate the reaction mixture at 37 °C for 30 min, with shaking at 1,300 rpm.

11. Wash 3× with lysis buffer 1 + 0.2 M NaCl, and 2× with buffer A + 50 mM NaCl. Remove the supernatant, careful not to disturb the beads.

12. To cleave GST, add 20 mU of biotinylated thrombin (30 μl) to the mixture and incubate at room temperature on a thermal mixer for 60 min (*see* **Note 11**).

13. To re-purify ^{32}P-labeled, GST-free IκBα$^{1-54}$, add 10 μl of pre-equilibrated streptavidin beads and incubate at 4 °C for 60 min, shaking on a thermal mixer.

14. Centrifuge the mixture to pellet the beads. Collect the ^{32}P-IκBα$^{1-54}$ containing supernatant.

15. Wash the beads 2× with 30 μl buffer A + 50 mM NaCl. Pool the supernatant with the two washes, filter the pool using the compact mini-column, and freeze it at –20 °C until needed.

16. Alternatively, **steps 1–14** in this section can be performed with a GST-IκBα-Ub fusion, which contains a single Ub moiety fused in place of K21 and K22 in IκBα$^{1-54}$ [8] (*see* **Note 12**).

3.5 In Vitro Reconstitution Reaction with IκBα$^{1-54}$ Peptide

1. Thaw all reagents on a metal block on ice (*see* **Note 13**).

2. To generate the active SCF^{βTRCP} E3 ligase, CUL1 must be neddylated. In a 3 μl reaction mixture, incubate 1.5 pmol of purified ROC1-CUL1 with 50 mM Tris–HCl, pH 7.4, 5 mM MgCl$_2$, 2 mM NaF, 10 nM okadaic acid, 2 mM ATP, 0.5 mM DTT, 0.1 mg/ml BSA, Nedd8 (20 μM), APP-BP1/Uba3 (83 nM), and Ubc12 (15 μM). These amounts are for a single reaction, and should be scaled up accordingly to generate a master mix. Incubate the reaction at room temperature for 10 min (*see* **Note 14**). This should result in approximately 50 % neddylated CUL1.

3. To assemble the quaternary SCF$^{\beta TRCP}$ complex, add 3 pmol of purified βTrCP1-Skp1 to the reaction mixture and continue incubation in the presence of 0.1 M Na-glutamate (*see* **Note 15**) for 10 min at room temperature.

4. In order to assemble the E3-substrate complex, incubate the neddylated SCF$^{\beta TRCP}$ (0.15 μM) with ^{32}P-IκBα$^{1-54}$ (0.15 μM). The total reaction volume from **steps 2–4** should be brought up to 5 μl. This now contains active SCF$^{\beta TRCP}$ in complex with the ^{32}P-IκBα$^{1-54}$ substrate, and is referred to as Mix I.

5. In a separate tube, generate the E2 charging reaction, which is referred to as Mix II. Combine Ub E1 (0.1 μM), UbcH5c (17 μM), Cdc34 (17 μM), Ub (50 μM), 50 mM Tris–HCl, pH 7.4, 5 mM MgCl$_2$, 2 mM NaF, 10 nM okadaic acid, 2 mM ATP, 0.5 mM DTT, and 0.1 mg/ml BSA, in a volume of 5 μl. Incubate the reaction for 5 min at 37 °C. At this point, the E2s have been charged with Ub to generate E2 ~ Ub (*see* **Note 16**).

6. Initiate the ubiquitination reaction by combining Mix I (5 μl) and Mix II (5 μl) and incubating at 37 °C for 15 min.

7. Quench the reaction with the addition of SDS loading dye.

8. Figure 3 provides an example of the ubiquitination, showing that the combination of UbcH5c and Cdc34 efficiently converted the substrate to lengthy ubiquitination products.

3.6 In Vitro Reconstitution Reaction with IκBα-Ub Fusion

1. As an alternative to the 2E2 reaction described in Subheading 3.5, it is also possible to use the GST-IκBα-Ub fusion substrate (*see* **Note 17**).

2. Repeat the steps in Subheading 3.5, except with IκBα-Ub fusion and H$_2$O in place of UbcH5.

3. Figure 4 provides an example of the ubiquitination, showing quantitative conversion of the input substrate into extensive ubiquitination species.

3.7 Running the Gel

1. Remove the gel from the wrapper, remove the comb, and flush the wells with water before placing in the gel apparatus.

2. Pour running buffer into the chamber and ensure that it covers the wells of the gel.

3. Boil the quenched reactions for 5 min at 95 °C and centrifuge briefly afterward to collect the sample (*see* **Note 18**).

4. Load 2–5 μl of protein standard, along with your samples, using gel-loading tips.

5. Connect the power supply and run at 150 V until the loading dye approaches the bottom of the gel (or until the 10 kDa marker approaches the bottom).

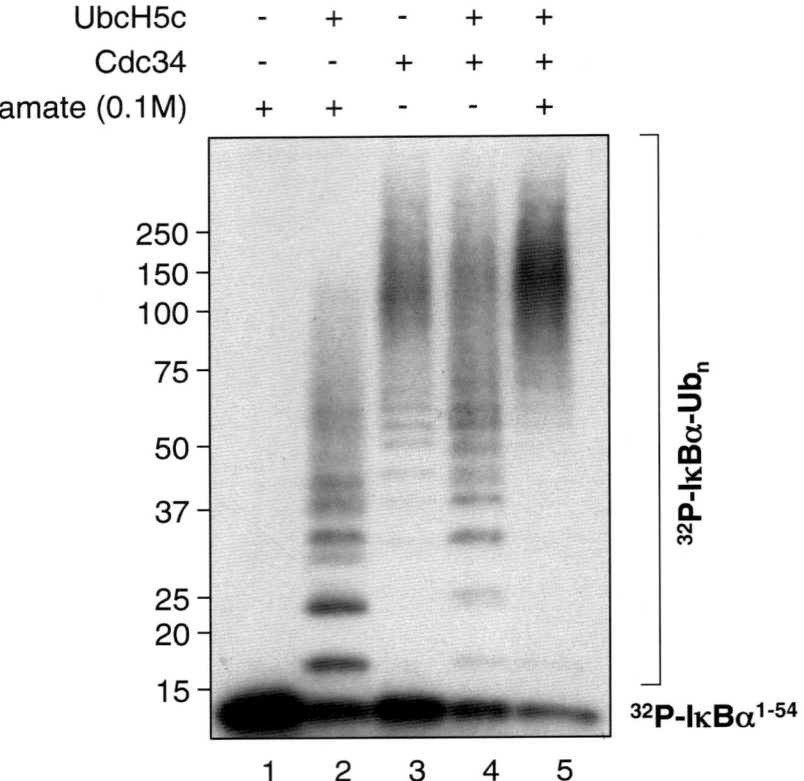

Fig. 3 Generation of polyubiquitin signal on ^{32}P-IκBα^{1-54}. To generate polyubiquitin chains, 0.5 pmol of ^{32}P-IκBα^{1-54} (phosphorylated and cleaved) was incubated with 1.5 pmol of SCF$^{\beta TRCP}$ for 10 min at room temperature to generate a substrate-E3 complex (Mix I). Na-Glutamate was added to Mix I, as indicated. Separately, UbcH5c and/or Cdc34 as indicated was charged in the presence of Ub and E1 at 37 °C for 5 min to generate Mix II. The reaction was initiated by combining Mix I and Mix II for 15 min at 37 °C, and quenched by the addition of 4× SDS loading dye. The reactions were analyzed by SDS-PAGE and autoradiography

3.8 Drying the Gel

1. When the gel is finished running, remove it from the cassette, cut off the wells and bottom of the gel, and place it into fixing solution for 10 min at room temperature.

2. Cut four sheets of Whatman paper as well as a sheet of clear plastic lab wrap, all slightly larger than the size of the gel.

3. Assemble the sandwich by placing two layers of Whatman paper first on the bottom. Wet these with approximately 2 mL of fixing solution in the area where the gel will go.

4. Using a glass plate, transfer the gel to the pre-wetted Whatman paper, taking care not to introduce waves into the gel.

5. Place the clear plastic wrap over the gel.

6. Place the remaining two layers of Whatman paper over the plastic wrap.

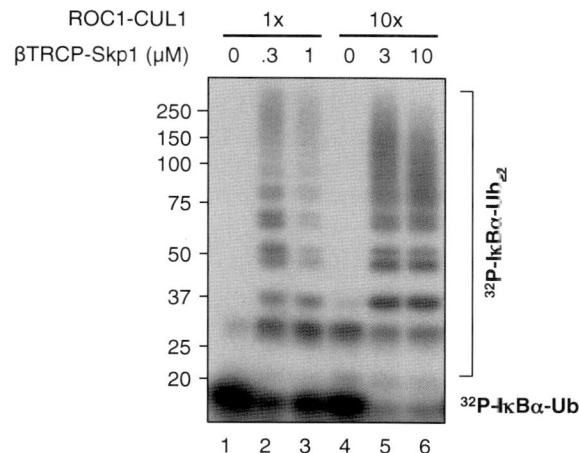

Fig. 4 Generation of a polyubiquitin signal on ^{32}P-IκBα$^{1-54}$-Ub. To establish proper conditions for assembling the quaternary E3 ligase complex, either 0.47 μM (1×) or 4.7 μM (10×) of purified ROC1-CUL1 was neddylated and then incubated with increasing amounts of βTRCP-Skp1 for 10 min, as indicated. These were then incubated with the ^{32}P-IκBα$^{1-54}$-Ub fusion substrate to generate a substrate-E3 complex, or Mix I. The E2 charging reaction, or Mix II, was generated by combing Ub E1, Ub, and Cdc34 (13 μM) for 5 min at 37 °C. Mix I and Mix II were then combined and incubated for 15 min at 37 °C, quenched with 4× SDS loading dye, and analyzed by SDS-PAGE followed by autoradiography. It was determined that 1.5 pmol of SCF^{βTRCP} (*lane 5*) represented the optimal concentration of E3. Notably, Cdc34 alone was sufficient to generate the degradation signal in this experiment because we employed a fusion substrate (*see* Fig. 1)

7. Dry the gel in a drying apparatus for 40 min at 80 °C, or until the gel becomes dry and fixed tightly to the sheet of Whatman paper below it (*see* **Note 19**).

3.9 Autoradiography and Exposure

1. Using lab tape, affix the dried gel to an autoradiography cassette.

2. In a dark room, expose the gel with film anywhere from minutes to days, depending on how fresh the ^{32}P signal is. The signal can be increased with the addition of an autoradiography screen.

3. Develop the film.

4. If desired, quantification of the reaction can also be performed by phosphorimaging.

4   Notes

1. The IKKβ$^{S177E/S181E}$ kinase, the SCF^{βTrCP} E3 ligase complex and the IκBα radiolabeled substrate are not commercially available. While E1 activating enzyme for Ub or Nedd8, as well as E2

enzymes UbcH5, Cdc34 and Ubc12 can be purchased from Boston Biochem, we provide the protocols for the preparation of UbcH5c and Cdc34 for cost-saving.

2. For purifications not involving nickel pull-down, 5 mM EDTA and 5 mM EGTA can be added to the lysis buffer to inhibit cellular proteases.

3. IKKβ$^{S177E/S181E}$ bears serine to glutamate substitutions at amino acid positions 177 and 181 and is constitutively active [18, 19].

4. If the mixture is still too viscous, pass the lysate through a syringe with a 21G syringe several times.

5. Glycerol stocks can be made and stored long term at –80 °C.

6. Cdc34 is stored at –80 °C. Multiple freeze–thaw cycles can decrease the activity of the enzyme, so it is best stored in small aliquots. All other enzymes should also be stored at –80 °C.

7. SCF^{βTRCP} can also be affinity purified from FCHT293 cells, a HEK293-based cell line that constitutively expresses HA-βTRCP2 and FLAG-CUL1 [6]. In this case, a single affinity step can be used to prepare highly active SCF^{βTRCP}, with ~50 % of CUL1 in Nedd8-modified form. However, the yield of the E3 complex from this procedure is very low.

8. It has been our experience that the use of ^{32}P-labeled substrate provides visualization and quantification of ubiquitination products at levels far more effective than those seen with standard immunoblot analysis. In this case, GST-IκBα$^{1-54}$ contains the first 54 amino acids of IκBα, which include K21 and K22 (the physiological sites of ubiquitination) and S32 and S36 (sites of IKKβ phosphorylation). Additionally, it contains an N terminal cAMP consensus site for ^{32}P-labeling.

9. Efficient phosphorylation of the substrate is paramount to reaction efficiency. It is suggested to monitor GST-IκBα$^{1-54}$ on the phosphorylation-induced alteration of gel mobility as shown in Fig. 2.

10. Work in the field suggests that the use of fluorescence may replace ^{32}P labeling in the future. However, at present, the best method to track the IκBα substrate is through radioactive labeling of the cAMP consensus site. We find that the amount of signal is directly proportional to input and is therefore the most reliable.

11. For quick assays, re-purification of the substrate can be avoided and the mixture can be used or frozen directly at this point. However, repeated freeze–thaw cycles (more than three) will markedly reduce reaction efficiency.

12. GST-IκBα-Ub fusion uses the fused Ub as a receptor for Ub chain assembly by Cdc34 and therefore, bypasses the requirement of UbcH5 to initiate ubiquitination. However, the reaction still depends on the interaction between SCF^{βTRCP} and

IκBα DSG degron (Fig. 1; ref. 8) and thus faithfully measures the elongation phase of the polyubiquitination of IκBα.

13. It is important to only thaw enzymes immediately before use, and put them away immediately afterwards.

14. It is difficult to accurately pipet less than 0.2 μl of a volume. A larger reaction mix may be used.

15. Glutamate (0.1 M) stimulates the ability of UbcH5 to initiate ubiquitination of IκBα by SCF$^{\beta TRCP}$ [20].

16. Longer incubation times at 37 °C may result in auto-ubiquitination of E2, which should be minimized.

17. UbcH5 is needed to prime the reaction, and is best suited for generating a limited number of Ubs on a substrate via substrate-Ub ligation. Cdc34, on the other hand, is efficient in Ub-Ub ligation and is needed to extend a growing chain. Both are required for rapid generation of a long Ub chain. The IκBα-Ub fusion mimics a UbcH5 primed substrate and therefore obviates the need for the addition of UbcH5 to the reaction.

18. The reactions can be stored at −20 °C at this point, and can be run at a later time.

19. Ensure that the drying is complete, as the film will stick to the gel if it is still wet.

Acknowledgements

We thank J. Nayak and other members of the Pan lab for their assistance with protocols and careful reading of the method. We are grateful to J. Hurwitz and I. Tappin for assistance with baculovirus preparation. R.A.C. was supported by NIH fellowship 1F30DK095572-01. Z.-Q.P. is the receipt of the 2013 Jiangsu special medical expert award. This work was supported by Public Health Service grants GM61051 and CA095634 to Z.-Q. P.

References

1. Skaug B, Jiang X, Chen ZJ (2009) The role of ubiquitin in NF-kappaB regulatory pathways. Annu Rev Biochem 78:769–796

2. Chen ZJ, Parent L, Maniatis T (1996) Site-specific phosphorylation of IkappaBalpha by a novel ubiquitination-dependent protein kinase activity. Cell 84:853–862

3. Scherer DC, Brockman JA, Chen Z et al (1995) Signal-induced degradation of I kappa B alpha requires site-specific ubiquitination. Proc Natl Acad Sci U S A 92:11259–11263

4. Chen Z, Hagler J, Palombella VJ et al (1995) Signal-induced site-specific phosphorylation targets I kappa B alpha to the ubiquitin-proteasome pathway. Genes Dev 9:1586–1597

5. Tan P, Fuchs SY, Chen A et al (1999) Recruitment of a ROC1-CUL1 ubiquitin ligase by Skp1 and HOS to catalyze the ubiquitination of I kappa B alpha. Mol Cell 3:527–533

6. Yamoah K, Oashi T, Sarikas A et al (2008) Autoinhibitory regulation of SCF-mediated

ubiquitination by human cullin 1's C-terminal tail. Proc Natl Acad Sci U S A 105:12230–12235

7. Duda DM, Borg LA, Scott DC et al (2008) Structural insights into NEDD8 activation of cullin-RING ligases: conformational control of conjugation. Cell 134:995–1006

8. Wu K, Kovacev J, Pan Z-Q (2010) Priming and extending: a UbcH5/Cdc34 E2 handoff mechanism for polyubiquitination on a SCF substrate. Mol Cell 37:784–796

9. Iwai K (2012) Diverse ubiquitin signaling in NF-κB activation. Trends Cell Biol 22:355–364

10. Xu S, Patel P, Abbasian M et al (2005) In vitro SCFβ-Trcp1-mediated IκBα ubiquitination assay for high-throughput screen. Methods Enzymol 399:729–740

11. Soucy TA, Smith PG, Milhollen MA et al (2009) An inhibitor of NEDD8-activating enzyme as a new approach to treat cancer. Nature 458:732–736

12. Ceccarelli DF, Tang X, Pelletier B et al (2011) An allosteric inhibitor of the human Cdc34 ubiquitin-conjugating enzyme. Cell 145:1075–1087

13. Shen M, Schmitt S, Buac D et al (2013) Targeting the ubiquitin-proteasome system for cancer therapy. Expert Opin Ther Targets 17(9):1091–1108

14. Gazdoiu S, Yamoah K, Wu K et al (2007) Human Cdc34 employs distinct sites to coordinate attachment of ubiquitin to a substrate and assembly of polyubiquitin chains. Mol Cell Biol 27:7041–7052

15. Ohta T, Michel JJ, Schottelius AJ et al (1999) ROC1, a homolog of APC11, represents a family of cullin partners with an associated ubiquitin ligase activity. Mol Cell 3:535–541

16. Zheng N, Schulman B, Song L et al (2002) Structure of the Cul 1–Rbx 1–Skp 1–F boxSkp 2 SCF ubiquitin ligase complex. Nature 416:703–709

17. Li T, Pavletich NP, Schulman BA et al (2005) High-level expression and purification of recombinant SCF ubiquitin ligases. Methods Enzymol 398:125–142

18. Mercurio F (1997) IKK-1 and IKK-2: cytokine-activated IB kinases essential for NF-B activation. Science (New York, NY) 278:860–866

19. Delhase M, Hayakawa M, Chen Y et al (1999) Positive and negative regulation of IkappaB kinase activity through IKKbeta subunit phosphorylation. Science (New York, NY) 284:309–313

20. Kovacev J, Wu K, Spratt DE et al (2014) A snapshot at ubiquitin chain elongation: lysine 48-tetra-ubiquitin slows down ubiquitination. J Biol Chem 289(10):7068–7081

Chapter 21

Control of NF-κB Subunits by Ubiquitination

Patricia E. Collins, Amy Colleran, and Ruaidhrí J. Carmody

Abstract

NF-κB is an essential regulator of inflammation and is also required for normal immune development and homeostasis. The inducible activation of NF-κB by a wide range of immuno-receptors such as the toll-like receptors (TLR), Tumour Necrosis Factor receptor (TNFR), and antigen T cell and B cell receptors requires the ubiquitin-triggered proteasomal degradation of IκBα to promote the nuclear translocation and transcriptional activity of NF-κB dimers. More recently, an additional role for ubiquitination and proteasomal degradation in the control of NF-κB activity has been uncovered. In this case, it is the ubiquitination and proteasomal degradation of the NF-κB subunits that play a critical role in the termination of the NF-κB-dependent transcriptional response induced by receptor activation. The primary trigger of NF-κB ubiquitination is DNA binding by NF-κB dimers and is further controlled by specific phosphorylation events which regulate the interaction of NF-κB with the E3 ligase complex and the deubiquitinase enzyme USP7. It is the balance between ubiquitination and deubiquitination that shapes the NF-κB-mediated transcriptional response. This chapter describes methods for the analysis of NF-κB ubiquitination.

Key words NF-κB, Ubiquitin, Immunoprecipitation, Immunoblotting

1 Introduction

Ubiquitination is a multistep process that results in the covalent addition of one or more ubiquitin molecules to a substrate protein [1, 2]. The first step requires the activation of ubiquitin, an 8 kDa protein, via a high-energy thioester linkage with the E1 enzyme. Activated ubiquitin is then transferred to the active site of an E2 ubiquitin-conjugating enzyme which functions together with an E3 ubiquitin-protein ligase to conjugate ubiquitin to the substrate protein, usually at a lysine group [3]. Substrate specificity is determined by the E3 ligase of which there are hundreds in mammalian cells and which are broadly divisible into two groups: the HECT family [4] and the RING family [5]. Many substrates are monoubiquitinated, but most undergo multiple rounds of ubiquitination to form polyubiquitin chains at lysine residues. The best known role for ubiquitination is in protein turnover by the proteasome, but intensive research in the past decade has identified additional

Michael J. May (ed.), *NF-kappa B: Methods and Protocols*, Methods in Molecular Biology, vol. 1280,
DOI 10.1007/978-1-4939-2422-6_21, © Springer Science+Business Media New York 2015

roles for ubiquitination in the regulation of protein localization and activity [6]. These different functions of ubiquitin can be distinguished by the nature of the linkages formed; thus, proteasomal degradation requires the polyubiquitination of substrate proteins in which ubiquitin groups are linked through lysine 48 (K48) of ubiquitin, whereas linkages such as lysine 63 (K63) lead to alternative non-proteolytic outcomes [7]. Other non-proteolytic linkages include K6, K11, K27, K29, K33, as well as linear ubiquitin chains resulting from the attachment of the C-terminal of one ubiquitin to the N-terminal methionine of another [2, 7].

Ubiquitination plays a prominent role in the activation of NF-κB and is required for the inducible proteasomal degradation of IκBα, an essential step in initiating the nuclear translocation of NF-κB dimers in the classical NF-κB pathway [8]. The ubiquitination of IκBα is dependent on the phosphorylation of IκBα by the inhibitor of kappaB kinase (IKK) complex. Phosphorylation by IKK promotes the binding of the E3 ubiquitin ligase β-TrCP to the phosphorylation sites on IκBα and triggers K48-linked polyubiquitination which in turn leads to degradation by the 26S proteasome [9]. In addition to the inducible proteasomal degradation of IκBα, non-proteolytic ubiquitination is also involved in the activation of the IKK complex by a wide range of inflammatory cytokines. K63 polyubiquitination mediated by TRAF6 is important in the activation of the IKK complex by the TAK1 kinase complex [10], and the binding of linear ubiquitin chains by the IKK complex scaffold protein NEMO is also essential for IKK activation [11]. Ubiquitination also plays a pivotal role in the activation of the noncanonical NF-κB pathway which requires the signal-induced ubiquitination and proteasomal processing of p100 to release RelB- and p52-containing dimers into the nucleus [8, 12].

More recently, it has emerged that the ubiquitination of NF-κB subunits themselves is an important regulatory mechanism controlling NF-κB transcriptional activity. The ubiquitination of NF-κB occurs on the promoters of target genes and is dependent on DNA binding [13–15]. The predominant form of NF-κB ubiquitination appears to be K48-linked chains which trigger the proteasomal degradation of NF-κB dimers [13, 14]. The relationship between promoter-bound NF-κB and ubiquitin-mediated degradation appears to be a major limiting factor in the transcription of NF-κB target genes, and it is essential for the prompt termination of transcriptional responses [13, 14]. Supressor of Cytokine Signalling 1 (SOCS1) [16] has been identified as an E3 ligase for the p65/RelA NF-κB subunit, and a complex composed of COMMD1 [17], cullin-2 [17], and GCN5 [18] is required in order to catalyze ubiquitination. Phosphorylation at serine 468 influences the ubiquitination of p65/RelA [15] as does the acetylation of lysine residues of p65/RelA that are also targeted for ubiquitination [19]. The identification of USP7 as a deubiquitinase for p65/RelA revealed that the ubiquitination of NF-κB occurs to a greater degree than initially appreciated [20]. Thus, the balance between the ubiquitination and deubiquitination of NF-κB profoundly influences the strength of transcriptional response.

The ubiquitination of NF-κB subunits is a key mechanism for the control of transcriptional activity and provides a therapeutic target of significant potential. However, much remains to be understood of the pathways and mechanisms that control NF-κB ubiquitination and proteasomal degradation. Here, we describe how to study NF-κB ubiquitination using methods that assess endogenous NF-κB ubiquitination, ubiquitination of transiently transfected NF-κB, and the ubiquitination of recombinant NF-κB in vitro.

2 Materials

2.1 Cell Culture and Treatment with TNFα and LPS

1. Dulbecco's Modified Eagle's Medium (DMEM) supplemented with 10 % (v/v) fetal calf serum, glutamine (2 mM), penicillin (10 U/ml), and streptomycin (10 μg/ml) (see Note 1).

2. Human or mouse TNFα stored in 10 μg/ml aliquots in cell culture medium and stored at −80 °C.

3. Lipopolysaccharides (LPS) stored in 100 ng/ml aliquots in cell culture medium and stored at −20 °C.

4. Trypsin–EDTA (0.05 %).

5. 6 and 10 cm tissue culture dishes.

6. 1.5 ml microfuge tube.

2.2 Transfection

1. TurboFect Transfection Reagent (Thermo Fisher, Waltham, MA, USA). Stored at 4 °C.

2. DMEM without serum or antibiotics.

3. HA-tagged ubiquitin expression vector (see Notes 2 and 3).

4. Epitope-tagged NF-κB subunit expression vector.

5. 1.5 ml microfuge tubes.

2.3 Immunoprecipitation

1. Protein G-conjugated agarose beads (50 % w/v) (Millipore, Billerica, MA, USA).

2. 1 % (w/v) sodium dodecyl sulfate (SDS) solution (see Note 4).

3. Radioimmunoprecipitation assay (RIPA) lysis buffer: 50 mM Tris–Cl pH 7.4, 150 mM NaCl, 1 mM EDTA, 1 % Nonidet P-40, and 0.25 % sodium deoxycholate, placed on ice at least 30 min before use (see Note 4).

4. Protease inhibitors pepstatin (1 mg/ml), leupeptin (2 mg/ml), aprotinin (2 mg/ml), and phenylmethanesulfonyl fluoride (PMSF) (100 mM) (see Note 4).

5. Phosphatase inhibitors Na$_3$VO$_4$ (100 mM) and NaF (100 mM) (*see* **Note 4**).

6. 1 M *N*-ethylmaleimide (*NEM*) in isopropanol (*see* **Note 4**).

7. Immunoprecipitating antibody: rabbit anti-p65 (Santa Cruz Biotechnology, Santa Cruz, CA).

8. Phosphate buffered saline (PBS) containing 0.05 % (v/v) Tween-20, stored at 4 °C.

9. Sample buffer (2×): 100 mM Tris–HCl pH 6.8, 4 % (w/v) SDS, 20 % (v/v) glycerol, 0.2 % (w/v) bromophenol blue, 200 mM β-mercaptoethanol. Stored without β-mercapto-ethanol at room temperature (*see* **Notes 5** and **6**).

10. 1.5 ml microfuge tube and 15 or 50 ml centrifuge tubes.

11. Cell scraper.

2.4 In Vitro Ubiquitination Assay

The ubiquitination reagents used for in vitro ubiquitination assays are all from Boston Biochem (Boston, MA, USA).

1. Recombinant p65.

2. Energy regeneration solution (ERS).

3. Ubiquitin activating enzyme (UBE1).

4. UbcH5a.

5. Ubiquitin-FLAG (*see* **Note 7**).

6. HeLa S-100 fraction.

7. Ubiquitin aldehyde.

8. Lactacystin.

9. Glutathione agarose.

2.5 SDS–Polyacrylamide Gel Electrophoresis (SDS-PAGE)

1. 1.5 M Tris–HCl pH 8.8.

2. 1 M Tris–HCl pH 6.8.

3. 10 % (w/v) sodium dodecyl sulfate (SDS) (*see* **Note 4**).

4. 10 % (w/v) ammonium persulfate (APS). Stored in aliquots at −20 °C (*see* **Note 4**).

5. 30 % acrylamide/bis-acrylamide solution (29:1) stored at 4 °C (*see* **Note 8**).

6. *N,N,N,N′*-tetramethylethylenediamine (TEMED).

7. Isopropanol.

8. Tris–glycine electrophoresis buffer containing: 25 mM Tris, 250 mM glycine, 0.1 % (w/v) SDS. Prepare from 5× stock (*see* **Note 9**).

9. Prestained protein molecular weight standards.

10. Hamilton syringe or gel-loading tips.

2.6 Immunoblotting

1. Transfer buffer: 48 mM Tris, 39 mM glycine, 0.0375 % SDS, 20 % methanol. Prepare from 5× stock (*see* **Note 10**).

2. Amersham Hybond ECL nitrocellulose membrane (GE Healthcare, Mickleton, NJ, USA) and chromatography filter paper.

3. Ponceau S solution 0.1 % (w/v) in 5 % acetic acid.

4. PBS with 0.05 % (v/v) Tween-20 (PBS-T).

5. Blocking buffer: 5 % (w/v) nonfat dried milk in PBS-T. Made immediately prior to use, and stored at 4 °C for 1–2 days.

6. Primary antibodies: mouse anti-HA (Roche, Madison, WI, USA), mouse anti-ubiquitin (Tebu-Bio, Le Perray-en-Yvelines, France), rabbit anti-p65 (Santa Cruz Biotechnology, Santa Cruz, CA, USA), and mouse anti-FLAG M2 (Sigma-Aldrich, St. Louis, MO, USA), diluted to 1:1,000, 1:1,000, 1:10,000, and 1:5,000, respectively, in blocking buffer.

7. Secondary antibodies: horseradish peroxidase-conjugated anti-mouse IgG and anti-rabbit IgG (GE Healthcare) diluted to 1:1,000 in blocking buffer before use.

8. Chemiluminescent Western blotting detection reagent (Advansta, Menlo Park, CA, USA) stored at 4 °C (*see* **Note 11**).

9. Chemiluminescence detection CL-XPosure film (Thermo Scientific).

10. Stripping Buffer (Thermo Scientific).

2.7 Special Equipment

1. Tissue culture hood.

2. Refrigerated centrifuge.

3. Sonicator.

4. End-to-end rotator.

5. Electrophoresis and wet transfer systems.

6. Rocker.

3 Methods

The upstream events during the activation of NF-κB require the ubiquitin-triggered proteasomal degradation of IκBα in the cytoplasm [8]. Following translocation to the nucleus, the ubiquitination and proteasomal degradation of NF-κB itself plays an important role in the termination of the transcriptional response. The ubiquitination and proteasomal degradation of NF-κB appears to occur only to a small pool of nuclear protein, and so the investigation of specific regulatory events in the control of ubiquitination typically employs the transient transfection of the NF-κB subunit of interest along with an epitope-tagged ubiquitin expression vector as described in the first method. This method takes advantage of the availability of high-affinity antibodies to epitope tags that overcome the sensitivity problems associated with the available antibodies against ubiquitin. The poor sensitivity of many commercially available antibodies against ubiquitin has often presented difficulties in the detection of endogenous ubiquitination. However, the inclusion of certain steps, as described in the second method, can overcome the limitations of the antibodies available.

In both cell-based protocols, the denaturing lysis of cells ensures that the ubiquitination of the protein of interest and not an interacting partner protein is measured. The first protocol described (Subheading 3.1) allows the role of specific posttranslational modifications, or interacting proteins of NF-κB, in the ubiquitination of NF-κB subunits to be assessed. This technique has been used to establish the regulation of NF-κB ubiquitination by the deubiquitinase USP7 [20] (Fig. 1). The second technique (Subheading 3.6) has been employed to show the ubiquitination of NF-κB in the context of an inducible NF-κB response such as that following the activation of TLRs (Fig. 2). The third method described (Subheading 3.7) enables the analysis of NF-κB ubiquitination using recombinant proteins in an in vitro setting (Fig. 3). This technique has been used to identify and characterize components of the E3 ligase complex [16].

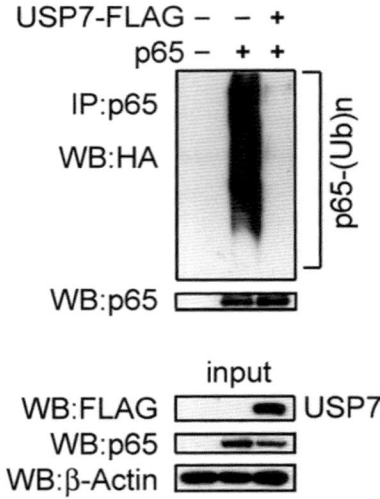

Fig. 1 The deubiquitinase USP7 inhibits NF-κB p65 ubiquitination. HEK293T cells were transiently transfected with plasmids encoding HA-ubiquitin, as well as p65 and FLAG-USP7 as indicated. 24 h following transfection, cells were lysed and p65 immunoprecipitated using anti-p65 antibody. Immunoprecipitates were resolved by SDS-PAGE, and ubiquitinated p65 was detected by Western blot using anti-HA antibody. Immunoprecipitation and expression of transfected plasmids were confirmed by Western blot analysis of input lysates and antibodies against p65, FLAG, and β-actin

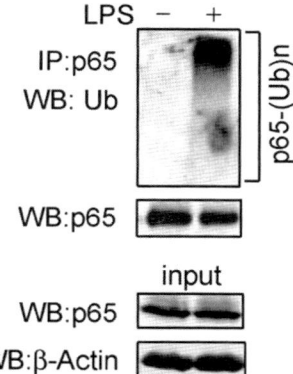

Fig. 2 LPS induces NF-κB p65 ubiquitination in RAW 264.7 macrophage cells. Cells were treated with LPS (100 ng/ml) for 1 h. 20 μM of the proteasome inhibitor MG132 was included for the final 30 min of LPS treatment. Cells were then lysed and immunoprecipitated with anti-p65 antibody. Immunoprecipitates were resolved by SDS-PAGE, and ubiquitinated p65 was detected by Western blot using anti-ubiquitin antibody. Immunoprecipitates and input lysates were analyzed by Western blot using anti-p65 and anti-β-actin antibodies

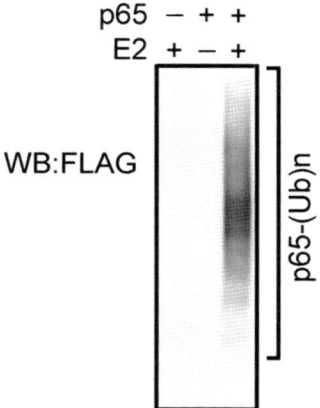

Fig. 3 In vitro ubiquitination of recombinant GST-NF-κB p65. In vitro ubiquitination reactions were assembled which included recombinant FLAG-ubiquitin and incubated for 2 h at 37 °C. p65 was affinity purified using GSH-agarose beads and resolved by SDS-PAGE. Ubiquitinated GST-p65 was detected by Western blot using anti-FLAG antibody. The omission of recombinant p65 and E2 ligase from reactions served as negative controls

3.1 Transfection of HEK293T Cells

All steps carried out in sterile culture.

1. HEK293T cells are grown in DMEM containing FCS 10 % (v/v), glutamine, penicillin, and streptomycir. Maintain cells at 37 °C in a humidified environment with 5 % CO_2.

2. 24 h prior to transfection, cells are passaged by first washing in prewarmed (37 °C) serum-free medium and then incubating in prewarmed (37 °C) Trypsin–EDTA solution for several

minutes. Detachment of cells from tissue culture flask should be visible by the eye, but complete detachment should be confirmed microscopically.

3. Seed 2×10^6 cells in 6 cm tissue culture dishes in a total of 5 ml of DMEM supplemented with serum, glutamine, and antibiotics.

4. For transfection, dilute plasmids in serum-free medium at a ratio of 1 µg DNA:100 µl SFM in a 1.5 ml microfuge tube (*see* **Notes 12** and **13**). Briefly, vortex TurboFect Transfection Reagent and add 2 µl of transfection reagent per 1 µg of DNA to the diluted DNA. Mix immediately by pipetting or vortexing. Incubate transfection reagent/DNA mixture for 15–20 min at room temperature.

5. Add transfection reagent/DNA mixture dropwise to each tissue culture plate, and gently rock to ensure even distribution. Incubate at 37 °C in a 5 % CO_2.

3.2 Preparing Transfected Cell Lysates for Immunoprecipitation

1. 24 h following transfection, discard tissue culture medium and gently rinse tissue culture plate with 2 ml of ice-cold PBS. It is important to take care to not detach cells at this step. Discard PBS.

2. Place tissue culture plates on ice. Add 1 ml of ice-cold PBS and detach cells with pipetting or, if necessary for very adherent cells, use a cell scraper. Transfer cell suspension to a precooled 1.5 ml microfuge tube. Wash plate with 0.5 ml of PBS and transfer any remaining cells to the 1.5 ml microfuge tube and place on ice.

3. Pellet cells by centrifugation at $11,000 \times g$ for 1 min in a refrigerated benchtop centrifuge set at 4 °C.

4. Aspirate supernatant and using a pipette tip with the end removed, quickly resuspend pellet in 100 µl of 1 % SDS solution. The solution will be very viscous at this stage, and excessive pipetting is not recommended as the sample will stick to pipette tip.

5. Incubate sample in a heat block (95 °C) for 5 min to denature, and briefly centrifuge to collect sample and return to ice.

6. Disrupt cell lysate by sonication while keeping samples on ice (*see* **Note 14**). Immerse sonicator tip in sample without touching the sides of the microfuge tube. Sonicate for 5–10 s (30 % output and 50 % duty cycle) to disrupt lysate (*see* **Note 15**). Clean sonicator probe tip with 70 % ethanol between samples to prevent crossover, ensuring tip is completely dry before proceeding to the next sample.

7. Centrifuge for 10 min at maximum speed at 4 °C. Remove supernatant and transfer to a new 1.5 ml microfuge tube and place on ice.

8. Prepare input sample for SDS-PAGE (*see* Subheading 3.4) by adding 5 μl lysate to 5 μl 2× sample buffer (*see* **Note 16**). Incubate sample in a heat block (95 °C) for 5 min. Briefly centrifuge to collect sample and store at –20 °C until analysis.

3.3 Immunopre-cipitation

1. Dilute remaining lysate (prepared in Subheading 3.2) with 10 volumes of ice-cold RIPA buffer freshly supplemented with protease and phosphatase inhibitors (*see* **Note 17**).

2. Preclear lysate by adding 20 μl of Protein G agarose slurry to lysate and incubating for 30 min with end-to-end rotation at 4 °C (*see* **Note 18**). Preclearing helps to reduce nonspecific binding to the Protein G agarose beads.

3. Centrifuge for 2 min at $14,000 \times g$, 4 °C. Remove supernatant to new 1.5 ml microfuge tube containing 20 μl of Protein G agarose slurry.

4. Add 1 μg of anti-p65 antibody and incubate overnight with end-to-end rotation at 4 °C (*see* **Notes 19** and **20**).

5. Centrifuge for 10 s at $11,000 \times g$, 4 °C. Carefully remove and discard supernatant.

6. Wash beads by adding 1 ml of ice-cold RIPA buffer and inverting ten times. Pellet beads as in **step 5** and repeat wash step twice. Keep on ice during wash steps (*see* **Note 21**).

7. Elute immunoprecipitates by adding 20–40 μl of 2× sample buffer to beads and incubating sample in a heat block (95 °C) for 5 min. Vortex for 10 s and centrifuge for 2 min at $16,000 \times g$ at 4 °C.

8. Remove supernatant and analyze by SDS-PAGE and immunoblot.

3.4 SDS-PAGE

1. The following instructions are for casting and running 8 % SDS gels using the Mini-Protean® Tetra Cell electrophoresis system (Bio-Rad, Hercules, CA). Follow manufacturer's instructions for other systems.

2. Assemble clean glass plates in the casting frame using 1.0 mm spacer plates.

3. In a disposable tube, prepare an 8 % resolving gel by mixing 4.6 ml of water, 2.6 ml of 30 % bis-acrylamide solution (29:1), 2.6 ml of 1.5 M Tris pH 8.8, 100 μl of 10 % SDS, 100 μl of 10 % APS, and 6 μl of TEMED.

4. Pour the acrylamide solution between the glass plates, leaving enough space for the stacking gel. Gently overlay with isopropanol.

5. Do not discard the excess gel solution as this will act as a guide for polymerization.

6. Once polymerization is complete, pour off the isopropanol overlay from the gel and wash with water several times to remove unpolymerized acrylamide. Using the edge of a tissue or filter paper, ensure that the top of the resolving gel is dry before pouring the stacking gel.

7. Prepare the stacking gel by mixing 1.36 ml of water, 340 μl of 30 % bis-acrylamide solution (29:1), 260 μl of 1.0 M Tris pH 6.8, 20 μl of 10 % SDS, 20 μl of 10 % APS, and 2 μl of TEMED, in a disposable tube.

8. Pour the stacking gel solution between the glass plates on top of the polymerized resolving gel. Immediately insert a clean 15-well comb.

9. Prepare 1 L of 1× running buffer from the 5× stock.

10. Once polymerization is complete, carefully remove the comb, and using a wash bottle, rinse the wells with water to remove unpolymerized acrylamide.

11. Assemble the electrophoresis apparatus, and add 1× running buffer to the inner and outer chambers ensuring wells are submerged.

12. Thaw input samples prepared in Subheading 3.2 on ice.

13. Using a clean Hamilton syringe (gel-loading tips can also be used), add 5 μl of protein molecular weight standard to the first well. Rinse syringe at least three times with water between each sample to prevent crossover.

14. Load 10 μl of input (prepared in Subheading 3.2) and 20 μl of elution (prepared in Subheading 3.2) into separate wells in a predetermined order. Add loading buffer to any wells that are unused and adjacent to sample containing wells.

15. Connect the electrophoresis apparatus to an electric power supply, and run at 100–120 V at room temperature until the dye front extends into the running buffer.

3.5 Immunoblotting for Epitope-Tagged Polyubiquitinated NF-κB

1. The following instructions for protein transfer are for the Mini Trans-Blot electrophoretic transfer cell (Bio-Rad).

2. Prepare 1 L of 1× transfer buffer from a 5× stock.

3. Remove the gel from the electrophoresis apparatus, and carefully trim away the stacking gel and discard.

4. Equilibrate the gel in transfer buffer for 15 min prior to electrophoretic transfer.

5. During equilibration, cut four pieces of filter paper and a piece of nitrocellulose membrane large enough to cover the gel. Soak filter paper, nitrocellulose membrane, and two fiber pads in transfer buffer.

6. Open a transfer cassette and prepare the gel sandwich in the following order: fiber pad, two pieces of filter paper, gel,

nitrocellulose membrane, and two pieces of filter paper. Using a roller, remove any air bubbles that may have formed from the gel sandwich (*see* **Note 22**). Add final fiber pad to complete sandwich.

7. Place the sandwich in a transfer cassette ensuring the gel is closest to the negative electrode and the nitrocellulose membrane is closest to the positive electrode.

8. Place the cassette in a transfer module containing a magnetic stirrer and a cooling ice block, and fill with transfer buffer.

9. Connect the transfer module to an electric power supply, and run at 100 V for 1 h on a magnetic stirrer to ensure an even buffer temperature.

10. Disassemble transfer apparatus and examine nitrocellulose membrane. The protein molecular weight standard should be visible on the nitrocellulose membrane.

11. Stain membrane with Ponceau S solution to visualize protein transfer. Rinse with water to remove background staining and wash in PBS-T to remove Ponceau S.

12. Incubate the membrane in blocking buffer on a rocker for 1 h.

13. To detect HA-ubiquitin, incubate the membrane in anti-HA primary antibody, rocking, overnight at 4 °C (*see* **Note 2**).

14. Remove membrane from primary antibody and wash with PBS-T three times for 5 min in each change of buffer, rocking at room temperature.

15. Incubate the membrane in HRP-conjugated secondary antibody, rocking at room temperature for 1 h.

16. Remove membrane from secondary antibody, and wash with PBS-T three times for 5 min in each change of buffer, rocking at room temperature.

17. Prepare sufficient ECL substrate to cover membrane by mixing equal volumes of reagents 1 and 2.

18. Remove membrane from PBS-T and drain off excess buffer. On a level surface, incubate membrane protein side up with ECL HRP substrate for 2–5 min depending on reagent.

19. Remove excess reagent with capillary action by touching an absorbent material to the edge of the blot. Place membrane in between two sheets of acetate or transparent plastic wrap, and place in an X-ray film cassette.

20. In the dark, expose membrane to film and obtain a range of exposures.

21. Remove bound primary and secondary antibody with Western blot stripping buffer. Reprobe membrane with anti-p65 to assess immunoprecipitation and detect bound protein as per **steps 15–20**.

3.6 Detection of Endogenous Polyubiquitinated NF-κB in LPS-Stimulated RAW 264.7 Macrophage Cells

1. RAW 264.7 macrophage cells are grown in DMEM containing FCS 10 % (v/v), glutamine, penicillin, and streptomycin. Maintain cells at 37 °C in a humidified environment with 5 % CO_2.

2. 24 h before use, seed 6×10^6 RAW 264.7 cells in 10 cm tissue culture dishes in a total of 10 ml of DMEM supplemented with serum, glutamine, and antibiotics. One 10 cm dish will be sufficient to assay for one experimental condition.

3. Stimulate cells with 100 ng/ml LPS for 1 h; include 20 μM MG132 for the final 30 min. Incorporate an unstimulated, MG132-treated control.

4. Just prior to harvest, incubate cells with 10 mM NEM for 30 s (*see* **Note 23**).

5. Remove and discard medium; gently rinse tissue culture plate with 2 ml ice-cold PBS containing 10 mM NEM. It is important to take care to not detach cells at this step. Discard PBS.

6. Place tissue culture plates on ice. Add 2 ml of ice-cold PBS containing 10 mM NEM and detach cells with pipetting or if necessary for very adherent cells, use a cell scraper. Transfer cell suspension to a precooled 15 or 50 ml centrifuge tube. Wash plate with 1 ml of ice-cold PBS and transfer any remaining cells to the centrifuge tube and place on ice. If using multiple plates per condition, combine cell suspensions at this stage.

7. Pellet cells by centrifugation at $300 \times g$ for 5 min in a refrigerated centrifuge set to 4 °C.

8. Carefully remove and discard supernatant.

9. Resuspend pellet in 1 ml of ice-cold PBS containing 10 mM NEM, and transfer cell suspension to a precooled 1.5 ml microfuge tube.

10. Continue with lysis from **step 3** in Subheading 3.2.

11. Immunoprecipitation for endogenous polyubiquitinated NF-κB is identical to the method described in Subheading 3.2 with one addition. RIPA buffer is also supplemented with 20 mM NEM to block the action of deubiquitinases.

12. Analyze samples by SDS-PAGE (Subheading 3.4).

13. Continue with immunoblotting as for epitope-tagged polyubiquitinated NF-κB (Subheading 3.5) until **step 11**.

14. Prior to blocking, incubate membrane in 0.5 % glutaraldehyde/PBS pH 7.0 for 20 min (*see* **Notes 23** and **24**).

15. Incubate the membrane in blocking buffer on a rocker for 1 h.

16. Incubate the membrane in anti-ubiquitin primary antibody, rocking, overnight at 4 °C.

17. Continue with **step 14** Subheading 3.5 to detect endogenous polyubiquitinated p65.

3.7 In Vitro Ubiquitination Assay

1. Add 5 μg of GST-p65 to an in vitro ubiquitination assay reaction mix containing 1× ERS, 30 μg/ml UBE1, 160 μg /ml UbcH5a, 0.2 mg/ml Ubiquitin-FLAG, 3.3 mg/ml S-100 fraction, HeLa, 5 μM ubiquitin aldehyde, and 100 μM lactacystin, in a final volume of 20 μl.

2. Incubate reaction at 37 °C for 2 h.

3. To purify polyubiquitinated GST-p65, add reaction to 25 μl of glutathione agarose and 1 ml of RIPA buffer, and incubate at 4 °C for 2 h rotating.

4. Centrifuge for 10 s at $11,000 \times g$, 4 °C. Carefully remove and discard supernatant.

5. Wash agarose by adding 1 ml of ice-cold RIPA buffer and inverting ten times. Pellet beads as in **step 4** and repeat wash step twice. Keep on ice during wash steps (*see* **Note 21**).

6. Elute immunoprecipitates by adding 25 μl of 2× sample buffer to beads and incubating sample in a heat block (95 °C) for 5 min. Vortex for 10 s and centrifuge for 2 min at $16,000 \times g$, 4 °C.

7. Remove supernatant and analyze by SDS-PAGE and immunoblot (Subheadings 3.4 and 3.5).

4 Notes

1. DMEM supplemented with serum, glutamine, and antibiotics is stored at 4 °C and warmed to 37 °C prior to use.

2. Other epitope-tagged ubiquitin expression vectors can be used as an alternative to HA-ubiquitin. Use appropriate primary antibody directed against epitope of expression vector to detect tagged ubiquitin.

3. Linkage-specific ubiquitin expression vectors that can be used in place of wild-type ubiquitin are also available. These vectors will encode a version of ubiquitin with arginine substitutions at all lysine residues except the lysine of linkage of interest. Lysine to arginine mutation inhibits chain extensions, and therefore the role of specific ubiquitin linkages can be assessed. In addition, linkage-specific ubiquitin mutants are also available. These vectors encode a version of ubiquitin with a single mutation at the lysine of the linkage of interest. Many ubiquitin mutants are available from Addgene (www.addgene.org).

4. *N*-Ethylmaleimide, sodium fluoride (NaF), phenylmethanesulfonyl fluoride, sodium deoxycholate, sodium dodecyl sulfate, TEMED, and ammonium persulfate are extremely hazardous. Wear appropriate gloves and safety glasses and use solid form in a chemical fume hood.

5. 100 mM β-mercaptoethanol can be replaced with 100 mM dithiothreitol.

6. Add β-mercaptoethanol in a chemical fume hood just prior to use.

7. Other epitope-tagged versions of recombination ubiquitin can be also used. Use appropriate primary antibody directed against epitope to detect tagged ubiquitin.

8. Unpolymerized acrylamide is a potent neurotoxin. Extreme care should be taken when handling acrylamide. Wear appropriate gloves and safety glasses.

9. For convenience, a 5× running buffer stock, 125 mM Tris, 1.25 M glycine, and 0.5 % (w/v) SDS, can be made in advance and stored at room temperature. Use electrophoresis grade reagents.

10. For convenience, a 5× transfer buffer stock without methanol, 240 mM Tris, 195 mM glycine, and 0.1875 % SDS, can be made in advance and stored at room temperature. Prior to use, dilute to 1× with the addition of 20 % (v/v) methanol. Use electrophoresis grade reagents.

11. Advansta WesternBright ECL is sufficient for most applications; for low-abundance proteins or to detect endogenous ubiquitination of proteins, a more sensitive ECL substrate (WesternBright Sirius) may be needed.

12. We find a 1:1 ratio of epitope-tagged ubiquitin and NF-κB subunit expression vector to give consistent results.

13. As a negative control, set up transfection with epitope-tagged ubiquitin and empty expression vector.

14. Wear appropriate protective equipment including earmuffs when using a sonicator. Never touch sonicator probe tip when in active use. Sonication conditions should be optimized for each sonicator. Conditions provided here are for Bandelin SONOPULS ultrasonic homogenizer HD 2070 with the MS 73 microtip.

15. Inefficient sonication can lead to high background. Ensure that lysate is clear.

16. The concentration of SDS in the sample will interfere with many standard methods of protein quantification. We use equivalent volumes of sample and see little variability between samples.

17. Protease and phosphatase inhibitors are added to RIPA buffer fresh from stocks before use. Final concentrations are as follows: pepstatin (1 μg/ml), leupeptin (2 μg/ml), aprotinin (2 μg/ml) and PMSF (1 mM), Na_3VO_4 (1 mM), and NaF (1 mM).

18. Cut the ends off the pipette tips to allow the beads to pass through freely.

19. As an example, the method is described for p65. If another NF-κB subunit is being investigated, use an antibody directed against an epitope-tagged NF-κB subunit expression vector, or for an endogenous ubiquitination assay, use an antibody directed against the NF-κB subunit of interest.

20. The concentration will depend on the immunoprecipitating antibody.

21. To reduce potential variation, it is important that all of the liquid is removed from the beads following the last wash step. This can be achieved using a $26 \times 1/2$ gauge needle.

22. A 10 ml pipette or 15 ml centrifuge tube can be used in place of a roller.

23. Deubiquitinating enzymes or deubiquitinases can hydrolyze the amide bond between ubiquitin and the substrate protein making detection of ubiquitinated proteins difficult. NEM is an alkylating reagent that covalently modifies the active-site thiol residue of cysteine proteases that form the major class of deubiquitinases thereby preventing their action.

24. Do not use Tris–HCl containing buffer at this step, as glutaraldehyde is amine reactive.

References

1. Hershko A (1983) Ubiquitin: roles in protein modification and breakdown. Cell 34:11–12

2. Komander D, Rape M (2012) The ubiquitin code. Annu Rev Biochem 81:203–229

3. Pickart CM (2001) Mechanisms underlying ubiquitination. Annu Rev Biochem 70:503–533

4. Bernassola F, Karin M, Ciechanover A, Melino G (2008) The HECT family of E3 ubiquitin ligases: multiple players in cancer development. Cancer Cell 14:10–21

5. Petroski MD, Deshaies RJ (2005) Function and regulation of cullin-RING ubiquitin ligases. Nat Rev Mol Cell Biol 6:9–20

6. Chen ZJ, Sun LJ (2009) Nonproteolytic functions of ubiquitin in cell signaling. Mol Cell 33:275–286

7. Adhikari A, Chen ZJ (2009) Diversity of polyubiquitin chains. Dev Cell 16:485–486

8. Chen J, Chen ZJ (2013) Regulation of NF-kappaB by ubiquitination. Curr Opin Immunol 25:4–12

9. Spencer E, Jiang J, Chen ZJ (1999) Signal-induced ubiquitination of IkappaBalpha by the F-box protein Slimb/beta-TrCP. Genes Dev 13:284–294

10. Lamothe B, Besse A, Campos AD, Webster WK, Wu H, Darnay BG (2007) Site-specific Lys-63-linked tumor necrosis factor receptor-associated factor 6 auto-ubiquitination is a critical determinant of I kappa B kinase activation. J Biol Chem 282:4102–4112

11. Ea CK, Deng L, Xia ZP, Pineda G, Chen ZJ (2006) Activation of IKK by TNFalpha requires site-specific ubiquitination of RIP1 and polyubiquitin binding by NEMO. Mol Cell 22:245–257

12. Fong A, Sun SC (2002) Genetic evidence for the essential role of beta-transducin repeat-containing protein in the inducible processing of NF-kappa B2/p100. J Biol Chem 277:22111–22114

13. Carmody RJ, Ruan Q, Palmer S, Hilliard B, Chen YH (2007) Negative regulation of toll-like receptor signaling by NF-kappaB p50 ubiquitination blockade. Science 317:675–678

14. Saccani S, Marazzi I, Beg AA, Natoli G (2004) Degradation of promoter-bound p65/RelA is essential for the prompt termination of the nuclear factor kappaB response. J Exp Med 200:107–113

15. Geng H, Wittwer T, Dittrich-Breiholz O, Kracht M, Schmitz ML (2009) Phosphorylation of NF-kappaB p65 at Ser468 controls its

COMMD1-dependent ubiquitination and target gene-specific proteasomal elimination. EMBO Rep 10:381–386

16. Ryo A, Suizu F, Yoshida Y, Perrem K, Liou YC, Wulf G, Rottapel R, Yamaoka S, Lu KP (2003) Regulation of NF-kappaB signaling by Pin1-dependent prolyl isomerization and ubiquitin-mediated proteolysis of p65/RelA. Mol Cell 12:1413–1426

17. Maine GN, Mao X, Komarck CM, Burstein E (2007) COMMD1 promotes the ubiquitination of NF-kappaB subunits through a cullin-containing ubiquitin ligase. EMBO J 26:436–447

18. Mao X, Gluck N, Li D, Maine GN, Li H, Zaidi IW, Repaka A, Mayo MW, Burstein E (2009) GCN5 is a required cofactor for a ubiquitin ligase that targets NF-kappaB/RelA. Genes Dev 23:849–861

19. Li H, Wittwer T, Weber A, Schneider H, Moreno R, Maine GN, Kracht M, Schmitz ML, Burstein E (2011) Regulation of NF-kappaB activity by competition between RelA acetylation and ubiquitination. Oncogene 31: 611–623

20. Colleran A, Collins PE, O'Carroll C, Ahmed A, Mao X, McManus B, Kiely PA, Burstein E, Carmody RJ (2013) Deubiquitination of NF-kappaB by Ubiquitin-Specific Protease-7 promotes transcription. Proc Natl Acad Sci U S A 110:618–623

Chapter 22

Methodology to Study NF-κB/RelA Ubiquitination In Vivo

Haiying Li, Petro Starokadomskyy, and Ezra Burstein

Abstract

Nuclear factor-kappa B (NF-κB) is a family of transcription factors that regulate immune responses, cell proliferation, differentiation, and survival. Activity of the NF-κB pathway on a cellular level is tightly controlled through various mechanisms, one of which is the ubiquitin-dependent degradation of chromatin-bound NF-κB subunits. In general, the ubiquitination of NF-κB regulates the duration of gene transcription activated in response to inflammatory signals. In this article, we present protocols to examine the in vivo ubiquitination status of RelA, a critical protein of the NF-κB family.

Key words NF-κB, RelA, IκB, Immunoprecipitation, Ubiquitin, Ubiquitination

1 Introduction

1.1 Proteasome-Dependent RelA Ubiquitination and Degradation

The NF-κB pathway is stress-responsive signaling mechanism, which primarily regulates inflammation and cell survival in addition to variety of other cellular programs [1]. The NF-κB family in mammals includes five transcription factors: RelA (p65), RelB, c-Rel, NFKB1 (or p50/p105), and NFKB2 (or p52/100) [2], which form homo- and heterodimers. The so-called classical NF-κB signaling pathway depends to a great extent on the RelA-NFKB1/p50 heterodimer. Hence, in this article, we will concentrate on RelA regulation as a primary mediator of canonical NF-κB signaling.

In the basal state, the RelA-p50 complex is sequestered in the cytoplasm by interactions with IκB inhibitory proteins. Once cells get stimulated by various inflammatory signals as well as other stress conditions (oxidative stress, DNA damage, etc.), the inhibitory IκB protein undergoes degradation. The released RelA-p50 complex translocates to the nucleus, where it recognizes specific promoters, and activates expression of a spectrum of genes according to the character of the cell lineage and nature of the stimulus. In order to avoid an exaggerated response, which may lead to potentially harmful activity, the genes activated by NF-κB should be turned off after a certain period of time. The half-life of

Michael J. May (ed.), *NF-kappa B: Methods and Protocols*, Methods in Molecular Biology, vol. 1280,
DOI 10.1007/978-1-4939-2422-6_22, © Springer Science+Business Media New York 2015

RelA-p50 interaction with its correspondent promoters is a critical parameter of NF-κB signaling, and the duration of RelA-DNA interaction is tightly regulated [3].

Two main pathways contribute to the physical removal of the RelA-p50 complex from promoters. The first pathway involves the export of NF-κB from the nucleus to the cytosol, and this is mediated by newly synthetized IκB. After returning to the cytoplasm, IκB degradation can lead to reactivation of NF-κB which returns to the nucleus, creating an oscillatory pattern of transcription of NF-κB-responsive genes [4]. A second pathway consists of the ubiquitination and degradation of chromatin-bound active RelA subunits. RelA ubiquitination is controlled by the orchestrated action of several protein complexes, involving a Cul2-containing ubiquitin-ligase complex, regulated by COMMD1 and GCN5, as well as the deubiquitinase USP7. Following ubiquitination, RelA is degraded by the proteasome, which irreversibly terminates the expression of specific NF-κB-dependent genes.

1.2 Short Method Overview

In this article, we present detailed protocols for studying the ubiquitination of RelA. We used these approaches successfully in several previous studies, including the identification of COMMD1 and GCN5 as regulatory cofactors of Cul2-containing E3 ligase [5], the study of the role of mutually exclusive acetylation/ubiquitination processes in the regulation of RelA activity [6], and the characterization of the inflammatory status of *Commd1*-deficient mice [7].

We utilized two main approaches, which rely on protein denaturation to stabilize ubiquitinated RelA. In the first method, detection of RelA ubiquitination relies on RelA transfection and the precipitation of His_6-tagged ubiquitin under denatured conditions (Fig. 1). This method is good for proof-of-concept experiments. In the second approach, detection of endogenous RelA ubiquitination is performed through immunoprecipitation of nuclear RelA under denatured conditions (Fig. 2). This protocol allows detection of intranuclear changes in RelA regulation under physiological conditions, although it needs more material and is predictably more laborious.

2   Materials

2.1 Reagents

1. Anti-RelA antibody (Santa Cruz Biotechnology, Santa Cruz, CA, USA).

2. Anti-polyubiquitin antibody (Stressgen, San Diego, CA, USA).

3. Aprotinin.

4. Bradford protein assay.

5. $CaCl_2$.

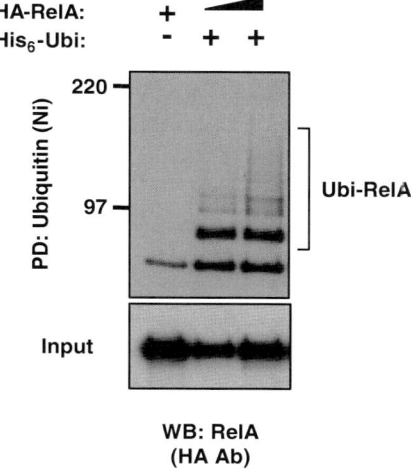

Fig. 1 Detection of RelA ubiquitination using overexpressed His$_6$-tagged ubiquitin. HA-tagged RelA was coexpressed with His$_6$-tagged ubiquitin in HEK293 cells. The ubiquitination status of RelA was assessed by precipitation of ubiquitin, followed by immunoblotting for RelA (using an HA antibody) according to the first protocol described

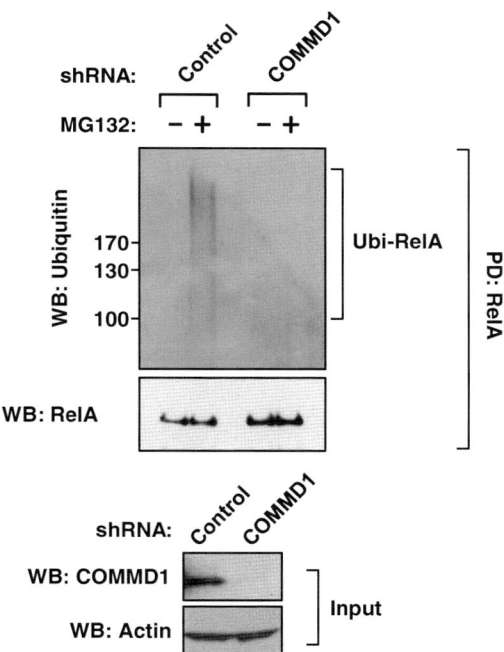

Fig. 2 COMMD1 is critical for endogenous RelA ubiquitination. Nuclear extracts were prepared from a U2OS cell line where COMMD1 had been silenced by stable expression of a specific short hairpin (shCOMMD1) or an irrelevant target control (shControl). In both cases, cells were treated with MG-132 for 3 h as indicated. Denatured immunoprecipitation of RelA was performed, and ubiquitinated RelA was detected by immunoblotting for ubiquitin according to the second protocol described

6. Complete Mini EDTA-free protease/phosphatase inhibitor tablets (Roche, Madison, WI, USA).

7. Dithiothreitol (DTT).

8. ECL Western blotting detection reagents (Perkin Elmer, Waltham, MA, USA).

9. Ethylenediaminetetraacetic acid (EDTA).

10. Glycerol.

11. HEPES.

12. KCl.

13. Leupeptin.

14. $MgCl_2$.

15. Proteasome inhibitor MG-132 (Boston Biochem, Boston, MA, USA).

16. NaCl.

17. Ni-NTA Agarose (Invitrogen, Carlsbad, CA, USA).

18. Nonfat dry milk.

19. Nonidet P-40

20. NuPAGE LDS sample buffer 4× (Invitrogen, Carlsbad, CA, USA).

21. PBS (phosphate buffered saline), 1× without calcium and magnesium.

22. Phenylmethylsulfonyl fluoride (PMSF).

23. Protein G Agarose (Invitrogen).

24. Pre-stained molecular weight marker mix for SDS-PAGE.

25. HA-RelA expressing eukaryotic plasmid.

26. His_6-tagged human ubiquitin expression vector.

27. Secondary IgG antibodies conjugated with horseradish peroxidase (anti-rabbit IgG and anti-mouse IgG).

28. Sodium dodecyl sulfate (SDS).

29. Sodium orthovanadate.

30. Triton X-100.

31. Tris base.

32. Tween-20.

33. Urea.

2.2 Equipment

1. Spectrophotometer.

2. Tabletop centrifuge with a fixed-angle rotor that can reach $15,000 \times g$.

3. Cell lifter, 3-cm blade.

4. Sonicator with ultrafine tip for small volumes (<1 mL) (we use a Branson Sonifier).

5. Filter paper.

6. Inverted microscope.

7. Film cassette.

8. Laminar flow hood and CO_2 incubators for cell culture.

9. Microcentrifuge tubes, 1.5 mL.

10. Rocking platform.

11. Small plastic container for Western membrane developing.

12. Serological pipettes and pipette aid.

13. Tissue culture plates, 10 cm.

14. Tube rotator.

15. X-ray film.

2.3 Cell Line(s)

The type of cell line will depend on the specific conditions of the experiment. Human embryonic kidney (HEK) 293 cells and human osteosarcoma cell line U2OS cells are cultured in complete DMEM medium with 10 % FBS.

2.4 Stock Solutions and Buffers

Prepare all solutions using ultrapure water (prepared by purifying deionized water to attain a sensitivity of 18 MΩ cm at 25 °C) and analytical grade reagents. Prepare and store all reagents at room temperature (RT, 22–25 °C), unless indicated otherwise. All waste materials should be disposed following all applicable regulations. Unless otherwise noted, the following solutions can be made ahead of time.

1. 5 M NaCl.

2. 1 M KCl.

3. 0.5 M EDTA.

4. 1 M HEPES, pH 7.2.

5. $CaCl_2$, 2 M (filtered through 0.45-μM membrane and cell culture tested).

6. 1 M Tris, pH 7.5.

7. 20 % SDS.

8. 1 M DTT (aliquot and store at –20 °C).

9. 100 mM PMSF in isopropanol (store at –20 °C).

10. 100 mM sodium orthovanadate (*see* **Note 1**).

11. Aprotinin, 1 mg/mL (store at –20 °C).

12. Leupeptin, 10 mg/mL (store at –20 °C).

13. 2× HBSS buffer: 50 mM HEPES acid, 1.5 mM Na_2HPO_4, and 280 mM NaCl. Adjust pH to 7.0. Sterilize and store at 4 °C.

14. Urea lysis buffer: 8 M urea, 300 mM NaCl, 0.5 % NP-40, 50 mM Tris (pH 8.0), 1 mM PMSF, 1 µg/mL aprotinin, and 10 µg/mL leupeptin. This solution must be made fresh just prior to use (*see* **Note 2**).

15. Ni-beads washing buffer: 8 M urea, 300 mM NaCl, 0.5 % NP-40, 50 mM Tris (pH 8.0).

16. Triton lysis buffer: 25 mM HEPES, 100 mM NaCl, 1 mM EDTA, 10 % (v/v) glycerol, 1 % (v/v) Triton X-100. This buffer can be made ahead of time and stored at RT. Prior to use, add the following to make "complete" lysis buffer: 1 mM PMSF, 10 mM DTT, and 1 mM sodium orthovanadate, protease/phosphatase inhibitor tablet (1 tablet per 10 mL of the lysis buffer). Complete lysis buffer can be stored at −20 °C for 4 weeks.

17. TSD buffer: 50 mM Tris pH 7.5, 2 % SDS, 5 mM DTT.

18. Buffer 1: 25 mM HEPES, 5 mM KCl, 0.5 mM $MgCl_2$, 1 mM PMSF, 1 mM DTT, 1 ng/mL aprotinin, 10 µg/mL leupeptin. Store at 4 °C.

19. Buffer 2: Buffer 1 + 2 % NP-40. Store at 4 °C.

20. Buffer 3: Buffer 1 + Buffer 2 (1:1). Store at 4 °C.

21. Buffer 4: 25 mM HEPES, 350 mM NaCl, 10 % sucrose, 1 mM PMSF, 1 mM DTT, 1 ng/mL aprotinin, 10 ng/mL leupeptin. Store at 4 °C.

3 Methods

3.1 Detection of Ubiquitinated RelA Through Precipitation of His₆-Tagged Ubiquitin Under Denatured Conditions

3.1.1 Cell Culture

Culture conditions and cell numbers will depend on the specific cell line being utilized. For HEK293, $2.5–3 \times 10^6$ cells are seeded in standard 10-cm tissue culture dishes 24 h prior to transfection.

3.1.2 Cell Transfection

For a 10-cm dish, we use 2–5 µg of pEBB-HA-RelA and 1 µg of pEBB-RGS-His₆-human ubiquitin. Traditional calcium phosphate transfection is sufficient for HEK293 cells [8].

1. Plasmid DNA is mixed with 61 µL of 2 M $CaCl_2$ and balanced to a total volume of 500 µL with sterilized H_2O.

2. Add equal volume of 2× HBSS to the mixture from **step 1**. To ensure calcium phosphate/DNA precipitation, pipette the reaction mixture up and down ten times. Incubate the solution for 5 min at room temperature.

3. Gently apply the transfection mixture to 10-cm dish evenly, and tilt the plate to mix culture medium with DNA precipitate.

4. After 8 h, replace with fresh complete DMEM to keep cells in optimal condition. Allow cells to grow for about 36–48 h before proceeding to the next step.

3.1.3 Preparation of Cell Lysates

1. Aspirate culture medium and rinse cells with 5 mL of 1× PBS. Cells are lysed directly in the culture dish by adding 500 μL of 8 M urea lysis buffer. The cell lysate is then scraped and transferred into a 1.5-mL Eppendorf centrifuge tube (*see* **Note 3**).

2. To complete the lysis process, the lysate is sonicated with 25 pulses of ultrasonic waves (output control 2.5, duty cycle 75 %) (*see* **Note 4**).

3. Centrifuge lysate at $16,000 \times g$ for 10 min at RT. Transfer clear lysate to a fresh tube and measure protein concentration by Bradford assay (*see* **Note 5**).

3.1.4 Nickel Pulldown

1. Aliquot equal amount of protein from the lysates into a fresh microfuge tube containing 30 μL of 50 % nickel beads slurry. Bring volume to 750–1,000 μL by adding 8 M urea lysis buffer. Rotate samples for 2 h at RT.

2. Wash beads three times with 1 mL of 8 M urea washing buffer, rotating for 5 min at each wash.

3. Wash beads 1 time with Triton X-100 lysis buffer to remove urea. Resuspend beads with 30 μL SDS-PAGE sample loading buffer containing 5 % β-mercaptoethanol. Boil samples at 100 °C for 10 min. Store at −20 °C if not ready to run SDS-PAGE (*see* **Note 6**).

3.2 Detecting Ubiquitination of Endogenous RelA

This procedure has been successful with several lines such as U2OS, HEK293, and mouse fibroblasts. Typically, one 10-cm culture dish is enough for this protocol. All the steps should be performed on ice. Prechill the microfuge rotor before starting.

3.2.1 Cell Culture

The type of cell line will depend on the specific conditions of the experiment. For U2OS cells, 3×10^6 cells are seeded in standard 10-cm tissue culture dishes 24 h prior to the experiment.

3.2.2 Preparation of Cell Lysate Samples for RelA Immunoprecipitation

1. The ubiquitinated form of endogenous RelA is generally quite unstable, so proteosomal blockade (20 μM of MG-132 for 3 h) is typically needed.

2. Discard medium completely and add 1 mL of ice-cold PBS to each 10-cm dish.

3. Scrape cells and transfer to microfuge tube. Pellet cells at $500 \times g$ for 5 min at 4 °C (*see* **Note 7**).

4. Resuspend cell pellet in 200 μL of Buffer 1 with gentle pipetting up and down.

5. Add 200 μL of Buffer 2 and rotate at 4 °C for 15 min.

6. Spin at $500 \times g$ for 1 min at 4 °C.

7. Transfer supernatant to a fresh microfuge tube (this is the cytoplasmic fraction) (*see* **Note 8**).

8. Add 100 μL of Buffer 3 to the nuclear pellet, and gently mix. Spin at $500 \times g$ for 1 min at 4 °C.

9. Resuspend the pellet in Buffer 4 and vortex. The volume should be twice that of the pellet itself (*see* **Note 9**).

10. Add TSD buffer and vortex. The volume should be equal to that of Buffer 4. Heat at 90 °C for 5 min.

11. Add Triton X-100 lysis buffer in this lysate in order to achieve a total volume that is suitable for sonication (>500 μL). Sonicate the samples (20–30 pulses with Branson Sonifier, output control 2, 90 % duty cycle) (*see* **Note 10**).

12. Remove cell debris by centrifugation of the lysate at $16,000 \times g$ for 10 min at 4 °C. Transfer supernatant to a fresh microfuge tube and determine protein concentration by Bradford assay. Set aside about 20 μL of the supernatant as nuclear input control.

13. Add additional Triton X-100 lysis buffer to sample in order to dilute TBS buffer 20-fold (i.e., SDS final concentration of 0.1 % so the samples be suitable for immunoprecipitation).

3.2.3 RelA
Immunoprecipitation

1. For immunoprecipitation, add lysate to a fresh tube (typically ~90 % of lysate is used for immunoprecipitation), and ensure that equal amounts of total protein are used in all samples in the experiment.

2. Preclear the cell lysate with 5 μL of Protein G Agarose slurry pre-equilibrated with Triton X-100 lysis buffer (*see* **Note 11**). Rotate for 30 min at 4 °C. Spin to pellet beads and transfer supernatant to fresh Eppendorf tube. Precleared lysate is rotated with RelA antibody (250 ng of antibody per mg cell lysate) at 4 °C for 1.5 h.

3. Add 15 μL of Protein G Agarose slurry (pre-equilibrated with Triton X-100 lysis buffer). Rotate at 4 °C for 1 h.

4. Microcentrifuge at $500 \times g$ for 2 min at 4 °C.

5. Wash agarose beads four times with 1 mL of Triton X-100 lysis buffer, and rotate at 4 °C for 5 min at each washing step.

6. Aspirate wash buffer thoroughly using a fine tip. Resuspend the beads with 30 μL of SDS-PAGE sample loading buffer. The samples can be stored at –20 °C at this point prior to Western blot analysis (*see* **Note 12**).

3.3 SDS-PAGE Separation

1. Samples that have been resuspended in SDS-PAGE sample loading buffer should be heated for 10 min at 80–100 °C (depending on the sample loading buffer being used). After heating, centrifuge at RT for 1 min at maximum speed.

2. Load 25 μg of input samples and maximal amount of immunoprecipitated sample onto an SDS-PAGE gel, and separate by electrophoresis (*see* **Note 13**).

3.4 Western Blotting

1. Transfer of proteins to nitrocellulose membrane using appropriate transfer apparatus.

2. After transfer, block membrane for 30 min at room temperature with 5 % solution of fat-free milk or BSA in TBST buffer with constant gentle agitation.

3. Incubate blocked membrane with primary antibody overnight at 4 °C on a horizontal rocker.

4. After washing the primary antibody three times with TBST, a secondary HRP-conjugated antibody is added to the membrane and incubated for 1 h with gentle agitation at RT. The membrane is then washed three times with TBST prior to proceeding to ECL-based detection (*see* **Note 14**).

4 Notes

1. Prepare a 100 mM sodium orthovanadate solution in double-distilled water. Set pH to 9.0 with HCl and boil until colorless. Cool down to room temperature, and again set pH to 9.0. Repeat this process until the solution remains at pH 9.0 after boiling and cooling. Bring up to the initial volume with water. Store in aliquots at –20 °C.

2. To prepare 8 M urea buffers, first make a 10 M urea stock solution. Make sure that urea is totally dissolved before proceeding to the next steps.

3. This entire step needs to be performed at RT to avoid precipitation of urea.

4. If lysate is still viscous upon pipetting, more sonication is needed. Incomplete sonication will lead to inaccurate protein concentration and stickiness with Ni beads.

5. Additional centrifugation at $16{,}000 \times g$ for 5 min after transfer to new tube will significantly reduce nonspecific background.

6. DTT is not a suitable reducing agent for this step and will cause migration problems during SDS-PAGE separation.

7. Perform this step quickly and gently to prevent cell from bursting.

8. Carefully remove the supernatant as the pellet is not very compact after the low-speed centrifugation step.

9. A small volume is preferred in order to minimize the amount of SDS in the lysate, as this will have to be diluted in subsequent steps.

10. Test if chromatin is completely shredded by pipetting. If the lysate is viscous upon pipetting, further sonication is required.

11. Pre-equilibrate Protein G Agarose in Triton X-100 lysis buffer by adding the required amount of beads (10 µL of 50 % bead slurry per sample) to 200–500 µL of Triton X-100 lysis buffer (a minimum of 10× volumes of beads). Wash the beads in this buffer a total of three times, by rotating the beads at 4 °C for 5 min followed by centrifugation (300 ×*g* for 2 min at 4 °C). After completing the washes, resuspend the beads as 50 % slurry in this buffer.

12. When aspirating the wash buffer, it is critical to avoid touching the beads while removing the buffer completely. Leaving variable amounts of wash buffer in the samples could cause uneven loading during SDS-PAGE separation.

13. To separate different forms of ubiquitinated RelA, long running time is strongly recommended. Depending on the electrophoresis system used, this can result in overheating of the gel. This can be avoided by submerging the running tank in iced water.

14. Ubiquitinated endogenous RelA is not very abundant and may be hard to detect. Long ECL exposures are typically needed, and therefore avoiding short exposures of the membrane is strongly recommended.

References

1. Karin M, Greten FR (2005) NF-κB: linking inflammation and immunity to cancer development and progression. Nat Rev Immunol 5(10):749–759

2. Hayden MS, Ghosh S (2008) Shared principles in NF-κB signaling. Cell 132(3):344–362

3. Karin M (2006) Nuclear factor-κB in cancer development and progression. Nature 441(7092):431–436

4. Nelson DE, Ihekwaba AE, Elliott M, Johnson JR, Gibney CA, Foreman BE, Nelson G, See V, Horton CA, Spiller DG, Edwards SW, McDowell HP, Unitt JF, Sullivan E, Grimley R, Benson N, Broomhead D, Kell DB, White MR (2004) Oscillations in NF-κB signaling control the dynamics of gene expression. Science 306(5696):704–708

5. Mao X, Gluck N, Li D, Maine GN, Li H, Zaidi IW, Repaka A, Mayo MW, Burstein E (2009) GCN5 is a required cofactor for a ubiquitin ligase that targets NF-κB/RelA. Genes Dev 23(7):849–861

6. Li H, Wittwer T, Weber A, Schneider H, Moreno R, Maine GN, Kracht M, Schmitz ML, Burstein E (2012) Regulation of NF-κB activity by competition between RelA acetylation and ubiquitination. Oncogene 31(5):611–623

7. Li H, Chan L, Bartuzi P, Melton SD, Weber A, Ben-Shlomo S, Varol C, Raetz M, Mao X, Starokadomskyy P, van Sommeren S, Mokadem M, Schneider H, Weisberg R, Westra HJ, Esko T, Metspalu A, Kumar V, Faubion WA, Yarovinsky F, Hofker M, Wijmenga C, Kracht M, Franke L, Aguirre V, Weersma RK, Gluck N, van de Sluis B, Burstein E (2014) Copper

metabolism domain-containing 1 represses genes that promote inflammation and protects mice from colitis and colitis-associated cancer. Gastroenterology 147(1):184–195.e3

8. Maine GN, Gluck N, Zaidi IW, Burstein E (2009) Bimolecular affinity purification (BAP): tandem affinity purification using two protein baits. Cold Spring Harb Protoc. doi: 10.1101/pdb.prot5318

Chapter 23

Using Sequential Immunoprecipitation and Mass Spectrometry to Identify Methylation of NF-κB

Tao Lu and George R. Stark

Abstract

Posttranslational modifications have long been known to play an essential role in the regulation of NF-κB activity. In the past few years, in addition to more traditional modifications such as phosphorylation, the p65 subunit of NF-κB has been found to be methylated at multiple sites. Here, we describe procedures for using immunoprecipitation and mass spectrometry to identify the methylation sites of p65.

Key words Immunoprecipitation, Mass spectrometry, Methylation, Posttranslational modification, p65, NF-κB

1 Introduction

The nuclear factor κB (NF-κB) family includes p65 (RelA), RelB, c-Rel, p50/p105 (NF-κB1), and p52/p100 (NF-κB2). Since p65 is the major functional subunit of the predominant NF-κB heterodimer (with p50), more attention has been devoted to how p65 is regulated by the posttranslational modifications. Several types of modification of p65 have long been known, including ubiquitination, phosphorylation, sumoylation, nitrosylation, and acetylation [1]. These regulatory modifications vary with different NF-κB pathways of activation, and even the same modifications may lead to different downstream effects in different situations [1].

In the past few years, six lysine (K) and one arginine (R) methylation sites have been identified on p65: K37, K218, K221, K310, K314, and K315 [2–5] and R30 (Fig. 1). We have used the combination of immunoprecipitation and mass spectrometry to identify monomethylated K218, dimethylated K221, and dimethylated R30 [2, 6]. For example, in an attempt to identify novel regulators of NF-κB by using a novel genetic approach, we [2, 7–9] discovered that the nuclear receptor binding Set domain protein 1 (NSD1) and the demethylase F-box leucine-rich protein 11 (FBXL11)

Michael J. May (ed.), *NF-kappa B: Methods and Protocols*, Methods in Molecular Biology, vol. 1280, DOI 10.1007/978-1-4939-2422-6_23, © Springer Science+Business Media New York 2015

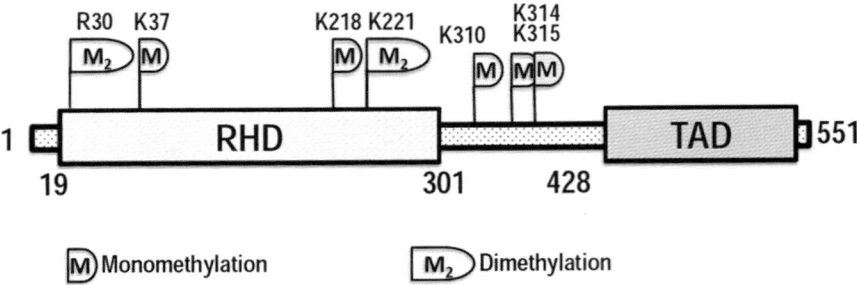

Fig. 1 Methylation of the p65 subunit of NF-κB. Figure shows a schematic diagram of the principal structural motifs of p65: Rel homology domain (RHD, amino acid 19–301), transactivation domain (TAD, amino acid 428–551), and the linker region (amino acid 302–427). The mapped sites are the known methylation modifications on either Y or R residues of the p65 subunit of NF-κB

regulate NF-κB through the reversible methylation of K218 and K221 of p65. Hur and coworkers [10] reported that the plant homeodomain finger protein 20 (PHF20, also called glioma-expressed antigen 2, GLEA2) promotes NF-κB transcriptional activity by interacting with methylated p65 at K218 and K221. The interaction between PHF20 and methylated p65 blocks the recruitment of phosphatase PP2A to maintain the p65 phosphorylation status. Another example is K37, which was identified by Ea et al. [3]. Upon activation of NF-κB by tumor necrosis factor-α (TNFα), Set domain-containing protein 9 (Set9) monomethylates p65 at K37, and this epigenetic modification regulates the promoter binding of p65 to a subgroup of genes, such as inhibitor of NF-κB (IκBα), interferon-induced protein 10 (IP-10), and TNFα. The induction of IP-10 and TNFα was greatly reduced in p65−/− MEF cells that express the K37Q mutant as compared to cells expressing wild-type p65 (wtp65) [3].

Why does p65 need methylation on multiple K sites? Recently, we compared the effects of methylation on the K37 and K218/221 sites of p65 [11], finding that mutation of K218/221 greatly reduced the expression of ~50 % of NF-κB inducible genes, whereas the K37Q mutation reduced the expression of only ~25 % of such genes. Chromatin immunoprecipitation sequencing (CHIP-seq) analysis showed that the mutation of K218/221Q greatly reduces the affinity of p65 for many promoters and that the K37Q mutation does not. Structural modeling shows that the newly introduced methyl groups of K218/221 interact directly with DNA to increase the affinity of p65 for specific κB sites. Therefore, the K218/221 and K37 mutants have dramatically different effects because methylations of these residues affect different genes by distinct mechanisms [11]. This phenomenon further highlights the importance and complexity of the posttranslational modifications of proteins.

To date, four approaches have been used to identify the K methylation of p65.

1.1 Combination of In Vitro Assay of p65 Fragments and Point Mutation

This method was used by Ea et al. [3] for the identification of methylation of K37. These authors first generated five GST-tagged p65 fragments, including GST-p65 (1–110), (101–210), (201–330), (325–350), and (411–441). They then used bacterially generated histone methyltransferase Set9 for an in vitro assay. This approach allowed them to show that only GST-p65 (1–110) was methylated. There are six lysine residues within this region, K28, K37, K56, K62, K79, and K93. To map the methylation site, Ea et al. substituted each of these lysines with non-methylatable arginines (R). A single point mutation at K37 completely abolished methylation by Set9. Therefore, Ea et al. concluded that K37 is the site of p65 that is methylated by Set9.

1.2 Combination of In Vitro Assay of p65 Fragments, Point Mutation, and Mass Spectrometry Analysis

Levy et al. [5] also did in vitro assay by mixing Set domain-containing protein D6 (SetD6) with two different p65 fragments: N-terminal p65 polypeptide encompassing amino acids 1–431 [p65 (1–431)] and C-terminal polypeptide (residues 430–531) [p65 (430–531)]. They proved that SetD6 methylated p65 (1–431) but not p65 (430–531). Replacement of individual K residues by R within p65 (1–431) identified K310 as the target site of SetD6. Further mass spectrometry analysis of SetD6-catalyzed methylation on p65 peptides spanning K310 [p65 (300–320)] demonstrated that SetD6 only adds a single methyl group to p65 at K310.

1.3 Combination of In Vitro Assay of p65 Fragments and Mass Spectrometry

This method was used by Yang et al. [4]. They first confirmed that p65 can be methylated by the histone-modifying enzyme Set9. They then bacterially expressed GST fusion proteins containing five different segments of p65, including GST-p65 (1–90), (91–180), (181–270), (271–364), and (365–551), and used them as substrates for the in vitro methylation assay. Of these five fragments, Yang et al. found that only GST-p65 (271–364) was methylated by Set9 in vitro. Later, by using mass spectrometry to analyze this particular fragment, they identified that K314 and K315 are the sites that were monomethylated.

1.4 Combination of Immunoprecipitation and Mass Spectrometry

The above three approaches all used in vitro reactions with ^{3}H-labeled S-adenosyl-L-methionine (SAM) as methyl group donor and p65 fragments as substrates. In contrast, we directly purified the entire p65 protein from cells in culture [2, 6] and then analyzed these samples by mass spectrometry. We believe that this approach more faithfully reflects the biological processes in live cells.

Below, we describe how to use this approach to identify methylation sites on the p65 subunit of NF-κB.

2   Materials

Unless stated, water used for preparation of buffers and solutions should have a resistivity of 18.2 MΩ and an organic content of less than five parts per billion (*see* **Note 1**).

2.1 Cell Culture and Treatment with Interleukin-1β (IL-1β)

1. Cells: 293C6 cells are 293 cells previously transfected with constructs encoding IL-1R1 and accessory protein and newly transfected with E-selectin-driven zeocin resistance and thymidine kinase genes [2]. 293C6-wtp65 cells are stable cells that were derived from 293C6 cells by expressing wtp65-Flag protein. These cells are maintained under the same conditions as 293C6 cells.

2. Dulbecco's Modified Eagle's Medium (DMEM) supplemented with 10 % (v/v) fetal bovine serum and 1 % of Thermo Scientific™ HyClone™ Penicillin-Streptomycin Solution (Fisher Scientific, Pittsburgh, PA, USA) (*see* **Note 2**).

3. Thermo Scientific™ HyClone™ Phosphate Buffered Saline (PBS) (10×) (Fisher Scientific, Pittsburgh, PA, USA). Dilute with water to 1× PBS, and aliquot to 500 ml per bottle. Autoclave and store at room temperature (*see* **Note 3**).

4. Thermo Scientific™ HyClone™ Trypsin (0.05 %) and ethylenediaminetetraacetic acid (EDTA) (1 mM) (Fisher Scientific, Pittsburgh, PA, USA). Trypsin derived from porcine pancreas; gamma irradiated; formulated without calcium and magnesium.

5. Falcon™ Gridded Tissue Culture Dish (15 cm) (Fisher Scientific, Pittsburgh, PA, USA).

6. Recombinant Human IL-1β (10 μg). Reconstitute IL-1β into serum-free DMEM medium to 10 μg/ml. Aliquot at 20 μl per tube, and store at −80 °C. Use at 1:1,000 dilution of 10 ng/ml final concentration for the treatment of cells.

2.2 Immunoprecipitation Sample Preparation for SDS-PAGE

1. Anti-p65 antibody.

2. EZview™ Red ANTI-FLAG® M2 Affinity Gel (Sigma, St. Louis, MO, USA).

3. Protein A/G resins (Thermo Fisher Scientific, Rockford, IL, USA).

4. Complete protease inhibitor cocktail tablets (Roche Applied Science, Indianapolis, IN, USA).

5. Immunoprecipitation buffer: 1 % Triton X-100, 50 mM Tris–HCl (pH 7.4), 150 mM NaCl, 1 mM EDTA, and add 1 tablet of complete protease inhibitor cocktail per 10 ml of buffer. Store as aliquots at −20 °C.

6. Immunoprecipitation washing buffer: 50 mM Tris–HCl (pH 7.4), 150 mM NaCl, 1 mM EDTA, 0.5 % Triton X-100, and store at 4 °C.

7. TBS buffer: 50 mM Tris–HCl, 150 mM NaCl (pH 7.4). Store at 4 °C.

8. FLAG® peptide (Sigma, St. Louis, MO, USA). Dissolve and make a final concentration of 5 mg/ml FLAG peptide in TBS described above, and store as aliquots at –20 °C.

9. Sample buffer (5×): 0.3 M Tris–HCl (pH 6.8), 5 % (w/v) sodium dodecyl sulfate (SDS), 50 % (v/v) glycerol, 100 mM dithiothreitol (DTT), 0.03 % (w/v) bromophenol blue. Store as aliquots at –20 °C. If further diluted at 1:5 in water, this can be used as 1× sample buffer.

10. Cell scrapers.

2.3 SDS Polyacrylamide Gel Electrophoresis (SDS-PAGE)

1. ProtoGel (30 %) (National Diagnostics, Atlanta, GA, USA). ProtoGel is stable for 24 months when stored tightly capped in a dark area at room temperature (20 °C) (*see* **Note 4**).

2. ProtoGel stacking buffer (4×) (National Diagnostics, Atlanta, GA, USA). ProtoGel stacking buffer forms a gel of 0.125 M Tris–HCl and 0.1 % SDS, pH 6.8, and is stable for 24 months when stored tightly capped in a dark area at room temperature.

3. ProtoGel resolving buffer (4×) (National Diagnostics, Atlanta, GA, USA). ProtoGel resolving buffer forms a gel of 0.375 M Tris–HCl and 0.1 % SDS, pH 8.8, and is stable for 24 months when stored tightly capped in a dark area at room temperature.

4. *N*,*N*,*N'*,*N'*-tetramethylethylenediamine (TEMED).

5. Ammonium persulfate (APS): prepare 10 % (w/v) solution, aliquot at 10 µl per tube, and store at –20 °C.

6. Running buffer (10×): dissolve 30.0 g of Tris base, 144.0 g of glycine, and 10.0 g of SDS in 1,000 ml of water. The pH of the buffer should be 8.3 and no pH adjustment is required. Store the running buffer at room temperature. Make 2 l of 1× running buffer for the mass spectrometry SDS-PAGE, and store at 4 °C in the cold room.

7. Prestained protein molecular weight marker.

2.4 Gel Fixation and Staining Reagents

1. Fixative: 50 % ethanol, 10 % glacial acetic acid, 40 % water; mix well and store at room temperature.

2. GelCode Blue Safe Protein Stain (Thermo Scientific, Rockford, IL USA).

2.5 Mass Spectrometry Reagents

1. 100 mM ammonium bicarbonate containing 50 % acetonitrile.

2. Acetonitrile.

3. 20 mM DTT.

4. 50 mM iodoacetamide in 100 mM ammonium bicarbonate.

5. 100 mM ammonium bicarbonate.

6. 50 mM ammonium bicarbonate containing sequencing grade modified trypsin.

7. 50 % acetonitrile in 5 % formic acid.

3   Methods

Carry out all procedures at room temperature unless otherwise specified.

3.1 Immunoprecipitation of p65-Flag Protein

1. Culture 2×15 cm plates of 293C6-p65-Flag cells till 90 % confluence.

2. Aspirate medium completely, and wash with 10 ml of 1× PBS (room temperature). Aspirate the 1× PBS, add another 5 ml of cold 1× PBS per plate, scrape and combine two plates of cells into a 50 ml Falcon tube, and add approximately 40 ml more 1× PBS to make a final 50 ml. Spin at $150 \times g$ for 10 min at 4 °C. Aspirate the PBS, and add 1 ml of immunoprecipitation buffer to resuspend the pellet. Keep the sample on ice for 20 min, and vortex every other 5 min for 3–5 s. Transfer the cell lysate into an Eppendorf tube. Spin at $10,000 \times g$, 10 min at 4 °C. Check protein concentration of supernatant by using Bradford protein assay reagent (Bio-Rad Laboratories, Inc., Hercules, CA, USA). Transfer the supernatant into the Eppendorf tube with prewashed beads (see below how to prepare beads).

3. Prewash 100 μl of EZview™ Red ANTI-FLAG® M2 Affinity Gel suspension in 15 ml of 1× PBS (cold), once. Spin at $1,200 \times g$ for 3 min at 4 °C. Aspirate 1× PBS, add 250 μl of cold 1× PBS to resuspend the beads, and spin at $10,000 \times g$ at 4 °C for 40 s. Aspirate the PBS, and add the supernatant to the Eppendorf with prewashed beads. Rotate at 4 °C overnight (*see* **Note 5**).

4. On the second day, wash beads with immunoprecipitation washing buffer for 5 min per time, four times in total by rotating the beads in a 50 ml Falcon tube filled up with immunoprecipitation washing buffer. Centrifuge the tube at $1,200 \times g$, for 3 min at 4 °C to pellet the beads. At the final step, resuspend the resin in 1 ml of immunoprecipitation washing buffer and transfer the resin suspension into an Eppendorf tube, spin at $10,000 \times g$ for 40 s, and aspirate the supernatant completely (*see* **Note 6**).

5. Dilute 10 μl of Flag peptide stock solution (5 mg/ml) into 90 μl of TBS, mix well, and then add this total 100 μl of Flag solution into the Eppendorf with the resin. Mix at 4 °C by rotating for 30 min. Spin at $10,000 \times g$ for 1 min. Then transfer all the supernatant to a new Eppendorf tube, add 5× sample

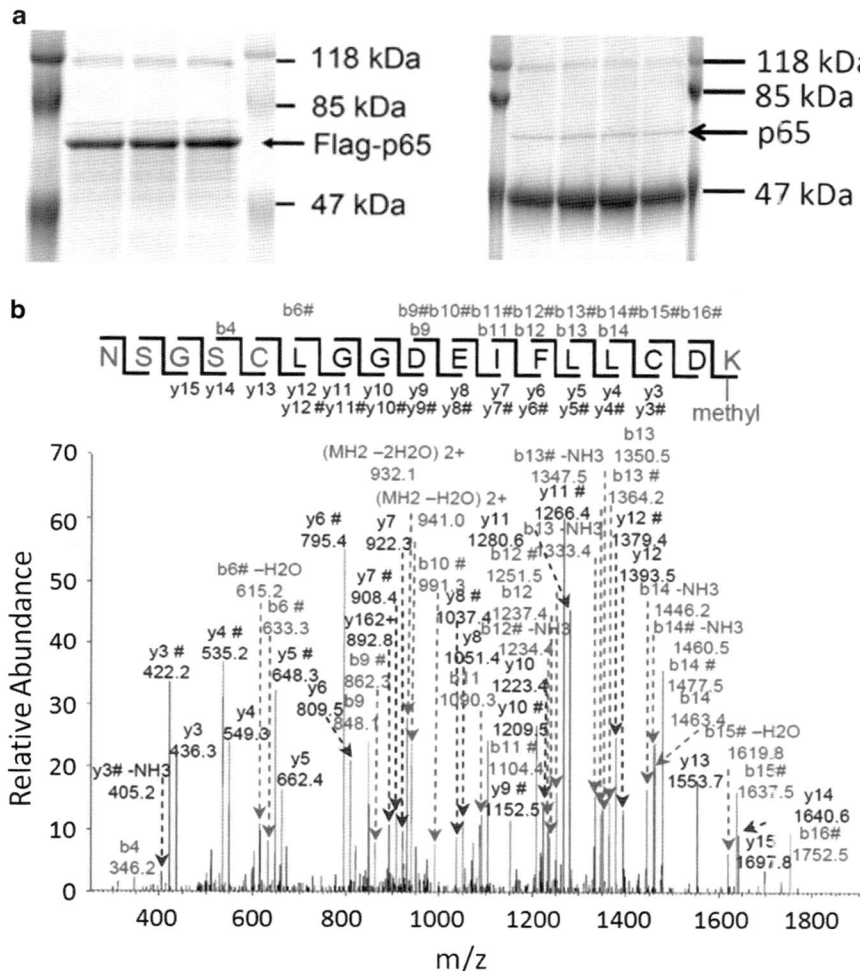

Fig. 2 Mass spectrometry (MS) shows that p65 is methylated on K218 on NF-κB activation. (**a**) GelCode Blue Safe Protein Stained gel. *Left panel* showing that p65-Flag is immunoprecipitated and isolated as a very strong band. *Right panel* showing endogenous p65 was purified and isolated as a very neat band. The same sample was loaded into multiple lanes. (**b**) Analysis of tryptic peptides derived from p65 suggests that K218 is mono-methylated. The pure single band was digested in the gel and samples were analyzed by LC-MS/MS. A mass shift of +14 was observed spanning peptide 202–218. Tandem MS analysis further suggested that K218 on the C-terminal side of the peptide is modified

buffer, and boil at 100 °C for 5 min. Samples are then loaded into multiple wells in a large 10 % SDS-PAGE gel (example image is shown in Fig. 2a, left panel) [2] (*see* **Note 7**).

3.2 Immunopre-cipitation of Endogenous p65

The immunoprecipitation method described above (Subheading 3.1) is used for purifying p65-Flag protein. We also used 293C6 cells and treat with IL-1β to isolate endogenously activated p65 directly.

1. Culture 293C6 cells in 2×15 cm plates to 95 % confluence in DMEM complete medium.

2. Treat cells with 10 ng/ml of IL-1β for 1 h. Immediately aspirate medium from the plates and wash with 10 ml pre-cold 1× PBS per plate. Aspirate the PBS completely, add 5 ml of cold 1× PBS again per plate, use cell scraper to scrape cells down, and transfer all the cells from 2×15 cm plates in PBS into a 15 ml Falcon tube (BD Bioscience, San Jose, CA, USA). Spin at $1,200 \times g$ for 10 min at 4 °C. Aspirate PBS and resuspend the pellet in 400 µl of immunoprecipitation buffer. Keep samples on ice for 20 min. Vortex every other 5 min for 4–5 s. Check protein concentration by using Bradford protein assay reagent (Bio-Rad Laboratories, Inc., Hercules, CA, USA) (*see* **Note 5**).

3. Prewash immobilized protein A/G resin with cold 1× PBS, and spin down at $10,000 \times g$ for 30 s at 4 °C. Aspirate all the PBS, then premix the resin with 100 µl of anti-p65 for 1 h, then further mix with cell lysates with equivalent amount of proteins at 4 °C, and rotate overnight.

4. Wash gel beads four times with 20 volumes of immunoprecipitation washing buffer with rotation at 4 °C for 5 min each time. At the last step, the gel beads are resuspended in an equal volume of 1× sample loading buffer and boiled at 100 °C for 5 min. Samples are then spun at $10,000 \times g$, for 1 min, and supernatants are then separated in a big 10 % SDS-PAGE gel (example is shown in Fig. 2a, right panel) (*see* **Note 7**).

3.3 SDS-PAGE

1. Assemble the glass plates [20 cm × 20 cm (outer plate) and 20 cm × 16 cm plate (inner plate)] in the gel-casting apparatus using spacer for a 1.5 mm gel.

2. Prepare a 10 % resolving gel mixture in a beaker by mixing 18.8 ml of water, 11.3 ml of ProtoGel resolving buffer, 15 ml of ProtoGel solution, 600 µl of APS, and 35 µl of TEMED. Pour the solution into the gel chamber leaving space for the stacking gel. Lay water onto the gel.

3. Check for polymerization by looking at the appearance of the clear thin line in between the water layer and the gel layer. This should occur around 15–20 min.

4. While waiting for the gel to polymerize, prepare the stacking gel mixture with 12.4 ml of water, 5 ml of ProtoGel stacking buffer, and 2.7 ml of ProtoGel, omitting APS and TEMED.

5. Pour off the water layer from the top of the polymerized gel. Add 120 µl of APS and 20 µl of TEMED to the premade stacking gel mixture. Then pour the solution onto the top of the separating gel, and insert a ten-well comb carefully (*see* **Note 8**).

6. Assemble the electrophoresis cassette in the tank when the gel has fully polymerized. Take out the comb gently, rinse all the

wells with water, and then pour 1× running buffer (pre-cold) into the inner and outer chambers.

7. Load the samples in several wells, and run the large SDS-PAGE at 200 voltage in the 4 °C cold room.

8. Stop running the gel while the samples have been fully separated over the entire length of gel. This is visualized by watching the line of the blue dye reaching about 1 cm from the bottom of the glass plate.

3.4   Gel Fixation

1. Gently take out the gel by separating the two glass plates. Soak the entire gel in sufficient volume of fixative (fully merged) for 20 min with gentle agitation on a platform shaker.

2. Pour off the fixative, add sufficient volume of water to fully merge the gel, and wash with gentle agitation on a platform shaker, for 15 min per time, four times in total.

3. Pour off the water, and add sufficient volume of GelCode Blue Safe Protein Stain, with gentle agitation on a platform shaker overnight.

4. The next day, the gel is destained with water for 20 min per time, four times in total with gentle agitation on a platform shaker at room temperature.

5. Target band can be carefully cut out on a clean glass plate. Keep the gel sample in Eppendorf tube and store at −20 °C before analysis with mass spectrometry method.

3.5   Protein In-Gel Digestion

1. Pieces cut from SDS-PAGE gels should be macerated and subjected to in-gel tryptic digestion.

2. The excised bands are washed twice with 100 mM ammonium bicarbonate containing 50 % acetonitrile for 1 h and twice with acetonitrile for 10 min, and then the proteins in the gels treated with 20 mM DTT at room temperature for 30 min followed by 50 mM iodoacetamide for 30 min in 100 mM ammonium bicarbonate.

3. After the treatment, the reagents are removed, and the gel pieces are washed with 100 mM ammonium bicarbonate and then dehydrated in acetonitrile.

4. The dried gel pieces are then re-swollen in 50 mM ammonium bicarbonate, containing sequencing grade modified trypsin for overnight digestion. Tryptic peptides are extracted from the gel with 50 % acetonitrile in 5 % formic acid.

3.6   Liquid Chromatography– Tandem MS (LC-MS/ MS) Analysis

1. Analysis of the proteolytic digest is performed using a LTQ Orbitrap XL linear ion trap mass spectrometer (Thermo Fisher Scientific) coupled with UltiMate 3000 HPLC system (Dionex).

2. The proteolytic digests are injected onto a reverse-phase C18 column (0.075 × 150 mm, Dionex) equilibrated with 0.1 % formic acid/4 % acetonitrile (v/v). A linear gradient of acetonitrile from 4 to 40 % in water in the presence of 0.1 % formic acid over a period of 45 min is used, at a flow rate of 300 nl/min. The spectra are acquired by data-dependent methods, consisting of a full scan (m/z 400–2,000) and then tandem MS on the five most abundant precursor ions. The previously selected precursor ions are scanned once during 30 s and then are excluded for 30 s.

3. The obtained data are submitted to Mascot software (Matrix Science) to search for methylated, dimethylated, and trimethylated lysine or arginine residues in the p65 protein. The tandem mass spectra of the possibly modified peptides are further interpreted manually (example shown in Fig. 2b) [2].

4   Notes

1. Water meeting these standards can be obtained using a water purification system such as the Milli-Q Synthesis system (Millipore, Billerica, MA, USA).

2. Tissue culture medium should be stored at 4 °C and then warmed to 37 °C in a water bath for at least 30 min prior to use. Trypsin/EDTA may be stored at 4 °C and warmed to 37 °C for use. Be sure to spray the bottles containing any medium or solutions with 70 % ethanol (v/v) before placing in the tissue culture hood.

3. This PBS is without calcium or magnesium to facilitate trypsinizing cells later on.

4. Unpolymerized acrylamide is a potentially harmful neurotoxin. Extreme care should be taken when handling acrylamide including the wearing of gloves and safety glasses.

5. 293C6 cells are used as control by following the same procedures.

6. It is important to wash the beads in large volume of immunoprecipitation buffer for multiple times to reduce the nonspecific binding significantly.

7. It is important to use a large SDS-PAGE gel to ensure the samples are separated over enough gel length, and the target band is nicely isolated.

8. Be sure to avoid any bubble in the gel, or under the wells of the comb, to ensure the samples to be run neatly.

Acknowledgments

Work in the authors' laboratories is supported by CA062220 (to G.R.S.) and 23-862-07TL (to T.L.).

References

1. Perkins D (2006) Posttranslational modifications regulating the activity and function of the nuclear factor κB pathway. Oncogene 25:6717–6730

2. Lu T, Jackson MW, Wang B, Yang M, Chance MR, Miyagi M, Gudkov AV, Stark GR (2010) Regulation of NF-κB by NSD1/FBXL11-dependent reversible lysine methylation of p65. Proc Natl Acad Sci U S A 107:46–51

3. Ea CK, Baltimore D (2009) Regulation of NF-κB activity through lysine monomethylation of p65. Proc Natl Acad Sci U S A 106:18972–18977

4. Yang XD, Huang B, Li M, Lamb A, Kelleher N, Chen LF (2009) Negative regulation of NF-κB action by Set9-mediated lysine methylation of the RelA subunit. EMBO J 28: 1055–1066

5. Levy D, Kuo AJ, Chang Y, Schaefer U, Kitson C, Cheung P, Espejo A, Zee BM, Liu CL, Tangsombatvisit S, Tennen RI, Kuo AY, Tanjing S, Cheung R, Chua KF, Utz PJ, Shi X, Prinjha RK, Lee K, Garcia BA, Bedford MT, Tarakhovsky A, Cheng X, Gozani O (2011) Lysine methylation of the NF-κB subunit RelA by SETD6 couples activity of the histone methyltransferase GLP at chromatin to tonic repression of NF-κB signaling. Nat Immunol 12: 29–36

6. Wei H, Wang B, Miyagi M, She Y, Gopalan B, Huang D, Ghosh G, Stark GR, Lu T (2013) PRMT5 dimethylates R30 of the p65 subunit to activate NF-κB. Proc Natl Acad Sci U S A 110:13516–13521

7. Lu T, Jackson MW, Singhi AD, Kandel ES, Yang M, Gudkov AV, Stark GR (2009) Validation-based insertional mutagenesis identifies lysine demethylase FBXL11 as a negative regulator of NF-κB. Proc Natl Acad Sci U S A 106:16339–16344

8. Lu T, Stark GR (2010) Use of forward genetics to discover novel regulators of NF-κB. Cold Spring Harb Perspect Biol 2(6):a001966

9. Stark GR, Wang Y, Lu T (2011) Lysine methylation of promoter-bound transcription factors and relevance to cancer. Cell Res 21: 375–380

10. Zhang T, Park KA, Li Y, Byun HS, Jeon J, Lee Y, Hong J, Kim J, Huang S, Choi S, Kim S, Sohn K, Ro H, Lu T, Stark GR, Shen H, Liu Z, Park J, Hur GM (2013) PHF20 regulates NF-κB signaling by disrupting recruitment of PP2A to p65. Nat Commun 4:2062

11. Lu T, Yang M, Huang DB, Wei H, Ozer GH, Ghosh G, Stark GR (2013) Role of lysine methylation of NF-κB in differential gene regulation. Proc Natl Acad Sci U S A 110: 13510–13515

Chapter 24

Methods to Detect NF-κB Acetylation and Methylation

JinJing Chen and Lin-Feng Chen

Abstract

Posttranslational modifications of NF-κB, including acetylation and methylation, have emerged as an important regulatory mechanism for determining the duration and strength of NF-κB nuclear activity as well as its transcriptional output. Within the seven NF-κB family proteins, the RelA subunit of NF-κB is the most studied for its regulation by lysine acetylation and methylation. Acetylation or methylation at different lysine residues modulates distinct functions of NF-κB, including DNA-binding and transcription activity, protein stability, and its interaction with NF-κB modulators. Here, we describe the experimental methods to monitor the in vitro and in vivo acetylated or methylated forms of NF-κB. These methods include radiolabeling the acetyl or methyl groups and immunoblotting with pan- or site-specific acetyl- or methyl-lysine antibodies. Radiolabeling is useful in the initial validation of the modifications. Immunoblotting with antibodies provides a rapid and powerful approach to detect and analyze the functions of these modifications in vitro and in vivo.

Key words Acetylation, Methylation, NF-κB, RelA, Posttranslational modification

1 Introduction

Posttranslational modifications (PTMs) play crucial roles in modulating the activation and termination of NF-κB signaling. These modifications, including ubiquitination, phosphorylation, acetylation, and methylation, target various cytoplasmic or nuclear proteins and modulate the strength and duration of the transcriptional outcomes in response to various extracellular signals. It is well known that PTMs (e.g., phosphorylation and ubiquitination) in the cytoplasm are essential for the activation of IκB kinases (IKKs) and the subsequent degradation of IκBα, allowing the rapid translocation of NF-κB into the nucleus, where NF-κB binds to the κB enhancer and stimulates the expression of hundreds of NF-κB target genes involved in different biological processes, including inflammation, proliferation, and cell survival [1].

Accumulating evidence indicates that PTMs within the nucleus are also important in determining the transcriptional outcome of NF-κB signaling [2, 3]. Within the nucleus, the RelA subunit of

Michael J. May (ed.), *NF-kappa B: Methods and Protocols*, Methods in Molecular Biology, vol. 1280,
DOI 10.1007/978-1-4939-2422-6_24, © Springer Science+Business Media New York 2015

NF-κB undergoes a variety of posttranslational modifications, and these modifications regulate the activity of nuclear NF-κB and fine-tune the expression of NF-κB target genes, adding another layer of complexity to the transcriptional regulation of NF-κB [3, 4].

Lysine acetylation and methylation have recently emerged as important modifications for the regulation of nuclear NF-κB function. Acetylation is a reversible event mediated by histone acetyltransferases (HAT) and histone deacetylases. These enzymes mediate the addition or removal of the acetyl group to or from lysine residues. Reversible acetylation of RelA regulates diverse functions of NF-κB, including DNA-binding activity, transcriptional activity, and its ability to associate with other proteins, and plays important roles in the NF-κB-mediated inflammatory response and cancer [3, 4]. Seven lysines, acetylated either by p300/CBP or PCAF, have been identified within RelA. These lysines include lysines (K) 122, 123, 218, 221, 310, 314, and 315 [5–7]. Acetylation of these different lysines modulates distinct functions of NF-κB. For example, acetylation of K221 enhances the DNA binding of NF-κB and, together with acetylation at K218, impairs its association with IκBα [6]. Acetylation of K122 and K123 by p300/CBP or PCAF seems to negatively regulate NF-κB-mediated transcription by reducing RelA binding to the κB enhancer [5]. Acetylation of K314 and K315 by p300 affects neither NF-κB shuttling, DNA binding, nor the induction of anti-apoptotic genes, but differentially regulates the expression of specific sets of NF-κB target genes [7, 8]. K310 is acetylated by p300/CBP and is deacetylated by SIRT1 [9]. Acetylation of K310 is required for the full transcriptional potential of NF-κB [6]. Acetylation of K310 also regulates the stability of RelA by preventing its methylation at lysine 314/315 [10]. Recent studies also indicate an important role of K310 acetylation in maintaining the sustained NF-κB activity in cancer cells [11–13].

Like lysine acetylation, lysine methylation has been extensively studied in recent years for its ability to control the nuclear NF-κB activity. The functional consequence of methylation depends on both position and state of the methylation site, since lysine can be mono-, di-, or tri-methylated. Several methyltransferases have been identified to methylate RelA at different lysines, including K37, K218, K221, K310, K314, and K315 [14–17]. Methylation of these lysines regulates different properties of NF-κB. For example, TNF-α- or LPS-induced monomethylation of RelA by Set9 negatively regulates the function of NF-κB by inducing the ubiquitination and degradation of promoter-bound RelA [18]. However, methylation of K37 by Set9 appears to be important for the activation of a subset of NF-κB target genes by stabilizing the binding of NF-κB to its enhancers [15]. RelA is also methylated by nuclear receptor-binding SET domain-containing protein 1 (NSD1) at K218 and K221 [16]. NSD1-mediated monomethylation of K218 and dimethylation of K221 enhance the transcriptional activity of NF-κB and the expression of NF-κB target genes. These methylated lysines are demethylated by F-box and leucine-rich repeat protein 11 (FBXL11) [16]. Demethylation of RelA by F3XL11 negatively regulates the transcriptional activity of NF-κB and decreases cell proliferation and colony formation of HT29 cancer cells [16].

Due to the essential role of lysine acetylation and methylation of RelA in fine-tuning the nuclear activity of NF-κB, it is important to develop sensitive and specific assays to detect these modifications. A variety of assays have been used to successfully detect the acetylation or methylation of RelA. These assays include radiolabeling the acetyl or methyl groups, immunoblotting with pan- or site-specific acetyl- or methyl-lysine antibodies, and mass spectrometry [6, 7, 16, 18, 19]. Radioactive labeled acetyl-CoA or S-adenosyl methionine (SAM) is widely used in in vitro assays utilizing the recombinant proteins. In these assays, the recombinant enzymes transfer radiolabeled acetyl or methyl groups from acetyl-CoA or SAM to the lysines in recombinant RelA [20]. With the availability of antibodies against acetylated or methylated lysines, immunoblotting has become an easy and powerful tool. These commercially available antibodies are able to detect modified RelA only when RelA is overexpressed with the enzymes [6, 18]. However, these antibodies cannot precisely determine the site and status of a modification and are not suitable for endogenous RelA. Several site-specific anti-acetylated or methylated RelA antibodies have been developed and used to detect the modifications of endogenous RelA in response to various stimuli [15–18, 21].

2   Materials

2.1   Cell Lines

1. HEK293T and A549 cells.

2. Cells are cultured in Dulbecco's Modified Eagle's Medium (DMEM) supplemented with 10 % fetal bovine serum (FBS), penicillin–streptomycin 100 U/ml–100 μg/ml, and 2 mM L-glutamine.

2.2   Antibodies

Anti-pan acetylated lysine antibodies are from Cell Signaling (Danvers, MA, USA). Site-specific anti-acetylated lysine-310 antibodies are from Cell Signaling and from Abcam (Cambridge, MA, USA). Anti-pan methylated lysine antibodies are from Abcam. Polyclonal antibodies against monomethylated lysines 314/315 RelA are from New England Peptide (Gardner, MA, USA) with a synthesized peptide corresponding to amino acids 308–320 of RelA (NH2-TFKSIMK[Me]K[Me]SPFSGC-COOH) as the antigen [18]. Anti-NF-κB/p65 antibodies are from Santa Cruz Biotech Labs (Santa Cruz, CA, USA).

2.3 Transient Transfection with Calcium Phosphate

1. 2.5 M calcium chloride ($CaCl_2$) solution.

2. 2× HBS buffer: 280 mM NaCl, 10 mM KCl, 1.5 mM $Na_2HPO_4 \cdot 2H_2O$, 12 mM dextrose, and 50 mM HEPES. Adjust pH to 7.05 using HCl and sterilize with 0.45 μm filters. Buffer can be aliquoted and stored at –80 °C.

2.4 In Vitro Acetylation Assay

1. Recombinant RelA: The recombinant RelA is commercially available and can be purchased from vendors such as Abnova (Taipei, Taiwan) (*see* **Note 1**).

2. p300 histone acetyltransferase (HAT): HA-tagged p300 immunoprecipitated from transiently transfected HEK293T cells has high HAT activity and can be used in in vitro acetylation assay.

3. [^{14}C]-Acetyl-CoA and unlabeled acetyl-CoA (PerkinElmer, Waltham, MA, USA) (*see* **Note 2**).

4. 5× HAT assay buffer: 250 mM Tris–HCl (pH 8.0), 50 % glycerol, 0.5 mM EDTA, and 5 mM dithiothreitol (DTT).

5. Amersham Amplify Fluorographic Reagent (GE Healthcare, Mickleton, NJ, USA).

2.5 In Vitro Methylation Assay

1. Recombinant RelA: As described in Subheading 2.4.

2. Methyltransferase Set9: Recombinant Set9 can be purchased from Millipore or purified from *E. coli*.

3. [^{3}H]-S-adenosyl-L-methionine ([^{3}H]-SAM) is from GE Healthcare; unlabeled S-adenosyl-L-methionine is from NEB (*see* **Note 3**).

4. 5× methylation buffer: 750 mM NaCl, 100 mM Tris–HCl (pH 7.5), 5 mM EDTA, 0.1 % Triton.

2.6 Immunoprecipitation Assay

1. Immunoprecipitation (IP) buffer: 50 mM HEPES (pH 7.4), 250 mM NaCl, 0.1 % NP-40, 1 mM EDTA (pH 8.0), 1 mM phenylmethylsulfonyl fluoride (PMSF), and complete protease inhibitor (Roche, Madison, WI, USA).

2. Preparation of protein G agarose (50 % slurry): centrifuge 500 μl protein G agarose beads at $1,000 \times g$ for 1 min at 4 °C, and carefully remove the supernatant without disturbing the beads. Wash beads three times with 1 ml of IP buffer. Drain the beads with a syringe and needle and measure the volume of beads. Add same volume of IP buffer to the beads to make 50 % slurry. Store the beads at 4 °C.

3. Preparation of anti-HA or anti-Flag antibody-conjugated agarose (Sigma, St. Louis, MO, USA): the anti-HA or anti-Flag antibody-conjugated agarose beads are prepared as described for protein G agarose beads. Store the beads at 4 °C.

4. Anti-T7 antibody-conjugated agarose are from Novagen (Madison, WI, USA) and are ready for use.

5. Syringe (1 ml) and needle (26G1/2).

6. 6× SDS loading buffer: 0.375 M Tris–HCl (pH 6.8), 12 % SDS, 60 % glycerol, 9 % 2-mercaptoethanol, 0.06 % bromophenol blue, and 1 mM DDT.

2.7 Chromatin Immunoprecipitation Assays (ChIPs)

1. 1 % formaldehyde: Dilute 37 % formaldehyde to 1 % with PBS.

2. Salmon Sperm DNA/Protein A agarose (Millipore, Billerica, MA, USA).

3. Low-salt wash buffer: 0.1 % SDS, 1 % Triton X-100, 2 mM EDTA, 20 mM Tris–HCl (pH 8.1), 150 mM NaCl.

4. High-salt wash buffer: 0.1 % SDS, 1 % Triton X-100, 2 mM EDTA, 20 mM Tris–HCl (pH 8.1), 500 mM NaCl.

5. LiCl wash buffer: 0.25 M LiCl, 1 % NP40, 1 % deoxycholate, 1 mM EDTA, 10 mM Tris–HCl (pH 8.0).

6. TE buffer: 10 mM Tris–HCl, 1 mM EDTA (pH 8.0).

7. Elution buffer: 25 mM Tris–HCl (pH 7.5), 10 mM EDTA, 0.5 % SDS.

8. Dilution buffer: 0.01 % SDS, 1.1 % Triton X-100, 1.2 mM EDTA, 16.7 mM Tris–HCl (pH 8.1), 167 mM NaCl.

9. PCR primers: human E-selectin – forward, 5′-AAGGCAT GGACAAAGGTGAAG-3′; reverse, 5′-TGTCCACATCCAG TAAAGAGGAAAT-3′; human TNF-α – forward, 5′-CGCTT CCTCCAGATGAGCTC-3′; reverse, 5′-TGCTGTCCTTGC TGAGGGA-3′.

3 Methods

3.1 Acetylation of RelA

Although time-consuming, radioactive labeling with [³H]-acetate provided the first direct evidence that RelA is acetylated in vivo; the weak signal from [³H] limits its further application to study the kinetics or stoichiometry of RelA acetylation. As many antibodies that react with acetylated histone or nonhistone proteins are now commercially available, immunoblotting has become a popular method to detect the acetylation of specific proteins. The anti-acetylated lysine antibodies recognize acetylated lysine residues in a sequence-independent manner and thus can be used for immunoblotting of acetylated nonhistone proteins. We have successfully used the antibody from Cell Signaling to detect the acetylation of overexpressed RelA in the presence of co-expressed p300 [6]. Both polyclonal and monoclonal anti-acetylated lysine antibodies from Cell Signaling recognize in vitro acetylated recombinant RelA or in vivo overexpressed RelA when co-expressed with p300.

3.1.1 Preparation of p300 for In Vitro Acetylation Assay

GST-p300 HAT domain fusion recombinant proteins have been shown to acetylate a variety of histone or nonhistone proteins and can be purchased from several vendors. However, we found that it barely acetylated recombinant RelA in vitro. We have been using p300 immunoprecipitated from transfected HEK293T cells as HAT for in vitro acetylation assay.

1. Seed HEK293T cells (2×10^5/ml) in 100 mm dishes.

2. When the cells reach 60–80 % confluence 16–24 h later, transfect each dish with 15 μg of HA-tagged p300 plasmid DNA with the calcium phosphate procedure (*see* **Note 4**).

3. 36 h after transfection, aspirate the medium and add 1 ml of IP buffer to cells, and lyse the cells at 4 °C for 10–15 min by leaving the dishes on a rocker.

4. Centrifuge the cell lysates at $14,000 \times g$ for 10 min at 4 °C and then collect the supernatants.

5. To preclear the supernatants, add 20 μl of protein G agarose bead slurry (50 % slurry) to the supernatants, and incubate for 30 min at 4 °C on a rotator (*see* **Note 5**).

6. Remove the agarose beads by centrifugation at $5,000 \times g$ for 5 min at 4 °C. Transfer the supernatant to a fresh centrifuge tube.

7. Immunoprecipitate HA-p300 from the precleared lysate with anti-HA antibody-conjugated agarose beads (60 μl of 50 % slurry for 1 ml of cell lysates) for 2 h at 4 °C.

8. After immunoprecipitation, collect the agarose beads by centrifugation at $14,000 \times g$ for 30–60 s, and wash the beads twice with ice-cold IP buffer.

9. Wash the beads once in 1× HAT assay buffer (diluted from 5× HAT assay buffer). Drain the wash buffer completely using a 1-ml syringe and a needle (26G1/2).

10. Add 200 μl of 1× HAT buffer to the beads and aliquot them into 1.5-ml Eppendorf tubes with 20 μl in each tube and store at −80 °C for later use (*see* **Note 6**).

11. Test the activity of p300 immunoprecipitates in an in vitro acetylation assay using histone H3 or H4 as substrates.

3.1.2 In Vitro Acetylation Assay

1. Thaw one tube of frozen p300 immunoprecipitates aliquot on ice and spin the tube at 4 °C. Empty 1× HAT buffer completely using a syringe and needle.

2. Add 4 μl of 5× HAT assay buffer, 1 μg of recombinant RelA protein, and 2 μl of [^{14}C]-acetyl-CoA or acetyl-CoA to the beads in the tube. Add distilled water to a total volume of 20 μl (*see* **Note 7**).

3. Spin the tube and mix the contents of the tube by tapping the bottom of the tube. Incubate the tube at 30 °C for 1 h with occasionally shaking (*see* **Note 8**).

4. Collect the supernatants by spinning the tube, and transfer the supernatant to a new 1.5-ml Eppendorf tube. Stop the reaction by adding 4 μl of 6× SDS loading buffer to the supernatant and boil for 5 min.

5. Run the samples by SDS-PAGE. When using [^{14}C]-acetyl-CoA as the acetyl group donor, fix the gel using a fixing solution (isopropanol–water–acetic acid (25:65:10, v/v) for approximately 30 min.

6. Soak the gel in Amersham Amplify Fluorographic Reagent (sufficient for the gel to be free floating) with agitation for 15–30 min.

7. Dry the gel and hold the gel in close contact with an appropriate X-ray film and at –70 °C to –80 °C (*see* **Note 9**).

8. Detect the radioactive signal by autoradiography.

9. When nonradioactive acetyl-CoA is used as the acetyl group donor, gel can be directly transferred to a nitrocellulose membrane, and the acetylation signal of RelA can be detected by immunoblotting with an anti-acetylated lysine antibody (*see* **Note 10**).

3.2 In Vivo Acetylation Assay

In vivo acetylated RelA can be detected by radiolabeling using radiolabeled sodium acetate or by immunoblotting using anti-acetylated lysine antibodies. Before the antibodies for acetylated RelA became available, radiolabeling using [^{3}H]-acetate was used to demonstrate the acetylated RelA in cultured cells [19]. However, due to the limited amount of proteins that can be labeled in the cells and the weak radioactive signals from [^{3}H], it would take several weeks before the acetylation signal can be seen from the X-ray film. In vivo labeling the acetylated RelA using radioisotope has been described previously [22]. Here, we focus on how to detect the acetylated RelA in vivo using anti-acetylated lysine antibodies.

3.2.1 Detection of Acetylation of Overexpressed RelA

1. Seed 2 ml of HEK293T cells (2×10^5/ml) in each well of a six-well plate and culture overnight at 37 °C.

2. Use two wells for each sample. Make a master mix for the transfection mixture. Dilute 1 μg of T7-tagged RelA and 4 μg of p300 expression vector DNA in 230 μl of distilled water in a 1.5-ml Eppendorf tube.

3. Add 20 μl of 2.5 M CaCl$_2$ to the mixture and mix evenly by pipetting up and down.

4. Slowly add 250 μl of 2× HBS buffer to the diluted DNA and mix gently by pipetting up and down. Leave the mixture sit at room temperature for 15 min.

5. Slowly add 250 µl of transfection mix to each well of the duplicates. Swirl the dish to ensure even distribution.

6. Thirty-six hours after transfection, aspirate the medium completely and add 350 µl of IP buffer to each well and leave the plate on a rocker at 4 °C for 10–15 min to promote cell lysis.

7. Collect supernatants from the cell lysates of two wells by centrifugation at $14,000 \times g$ for 10 min.

8. Preclear the cell lysates by adding 20 µl of protein G agarose beads (50 % slurry) to the supernatants, and incubate for 30 min at 4 °C on a rotator.

9. Remove the agarose beads by centrifugation at $5,000 \times g$ for 5 min at 4 °C. Transfer the supernatant to a fresh Eppendorf tube.

10. Add 20 µl of anti-T7 antibody-conjugated agarose beads (50 % slurry) to the supernatant. Gently mix the cell lysates and agarose beads for 2 h at 4 °C on a rotator.

11. Spin the Eppendorf tube for 30 s in the microcentrifuge at $14,000 \times g$, and then discard the supernatant.

12. Wash the beads three times with 500 µl of ice-cold IP buffer, and use a syringe and needle to drain the wash buffer completely.

13. Add 50 µl of 2× SDS sample buffer into the beads and boil the agarose beads for 5 min. Spin the Eppendorf tube and use the supernatant for SDS-PAGE.

14. Transfer the gel to a nitrocellulose membrane.

15. Incubate the membrane in a blocking solution for 30 min at room temperature with agitation.

16. Rinse the membrane briefly with wash buffer PBS-T twice.

17. Incubate the membrane with polyclonal anti-pan acetylated lysine antibodies (1:1,000 dilution) overnight with agitation.

18. Wash the membrane three times with wash buffer for 10 min each.

19. Incubate the membrane with diluted (1:5,000) secondary antibodies (anti-rabbit IgG antibody conjugated to horseradish peroxidase) for 30 min at room temperature with agitation.

20. Wash the membrane three times with wash buffer for 10 min each.

21. Add ECL detection reagents to the membrane following the manufacturer's instructions, and detect the signal by using a ChemiDoc imaging system or using an X-ray film developer.

3.2.2 Detection of Acetylation of Endogenous RelA by Immunoblotting and Immunoprecipitation

The anti-pan acetylated lysine antibodies from Cell Signaling are effective in detecting the acetylated RelA when RelA is co-expressed with p300 [6]. However, this antibody is not specific enough to detect acetylated endogenous RelA. Instead, site-specific anti-acetylated RelA antibodies could be used. For example, the anti-acetylated lysine-310 antibodies from Cell Signaling and Abcam are able to recognize acetylated endogenous RelA. For detection of acetylation of endogenous RelA, it is also important to treat the cells with a stimulus (e.g., TNF-α and LPS) to allow the translocation of RelA into nucleus, where most HATs localize. The acetylated RelA could be detected by directly immunoblotting the whole-cell extracts with anti-acetylated RelA antibodies [21]. But the levels of acetylated RelA might vary among different cells. Alternatively, immunoprecipitation of the acetylated RelA with anti-acetylated RelA antibodies would concentrate the acetylated RelA and enhance the detectable acetylated RelA signals.

1. Seed 2 ml of A549 cells (5×10^5) in each well of a six-well plate and culture overnight at 37 °C. Prepare two wells for each sample.

2. When the cells reach 90 % confluency, treat the cells with TNF-α (20 ng/ml) for the desired time and wash twice with cold PBS.

3. Add 350 μl of IP buffer to each well and lyse the cells for 30 min with agitation at 4 °C.

4. Use a cell scraper to collect the cell lysates from the two wells into 1.5-ml Eppendorf tubes.

5. Centrifuge the sample for 10 min at $14,000 \times g$. Transfer the supernatant to a new 1.5-ml Eppendorf tube.

6. Preclear the cell lysates by adding 20 μl of protein G agarose bead slurry (50 %) and incubate for 30 min at 4 °C on a rotator.

7. Remove the agarose beads by centrifugation at $5,000 \times g$ for 5 min at 4 °C. Transfer the supernatant to a fresh Eppendorf tube.

8. Add 5 μl of anti-NF-κB/p65 (acetyl K310) antibodies to the supernatant. Incubate at 4 °C for 3 h with gentle agitation.

9. Add 25 μl of protein G agarose beads (50 % slurry in IP buffer) to cell lysates. Gently mix the cell lysates and agarose beads overnight at 4 °C.

10. Collect the agarose beads by pulse centrifugation at $14,000 \times g$ for 30 s. Discard the supernatant fraction. Wash the beads three times with 500 μl of ice-cold IP buffer and use a syringe and needle to drain the wash buffer completely.

11. Resuspend the agarose beads in 30 µl of 2× SDS sample buffer and mix gently. Boil the agarose beads for 5 min. Collect the beads by centrifugation and use the supernatant for SDS-PAGE.

12. Resolve the samples on a 10 % SDS-PAGE gel and transfer the gel to a nitrocellulose membrane.

13. Immunoblot the membrane with diluted anti-RelA antibody (Santa Cruz, F-6, 1:500) overnight at 4 °C. Overnight incubation of the antibody is required for detection of a significant signal.

3.2.3 Detection of Chromatin-Associated Acetylated Endogenous RelA by ChIPs

1. Seed A549 cells (2×10^6) in 100-mm dishes and culture overnight at 37 °C.

2. When cells reach 90 % confluency, treat the cells with TNF-α (20 ng/ml) for indicated time points.

3. Wash the cells once with PBS to remove the TNF-α and cross-link proteins to DNA by adding 2 ml of 1 % formaldehyde to the cells and incubate for 10 min at room temperature.

4. Stop the fixation with 125 mM glycine for 5 min.

5. Use a cell scraper to collect the cells to a conical tube and pellet cells for 5 min, at $200 \times g$ at 4 °C.

6. Wash the cells once with cold PBS and resuspend the cell pellets in 500 µl of solution containing 1 % SDS, 50 mM Tris (pH 8.1), and 10 mM EDTA (5×10^5 cells/100 µl solution).

7. Lyse the cell with sonication under conditions optimized to generate DNA fragments with an average size of 200–1,000 base pairs.

8. Centrifuge samples for 10 min at $16,000 \times g$ at 4 °C, and transfer 200 µl of the sonicated cell supernatant to a new Eppendorf tube. The remaining supernatants can be stored at –80 °C for IP with a different antibody.

9. Dilute the supernatant by adding 1.8 ml of dilution buffer.

10. Add 20 µl of Salmon Sperm DNA/Protein A agarose beads to the cell lysates and rotate for 30 min at 4 °C to preclear the cell lysates.

11. Pellet the agaroses by spinning the tubes and transfer the supernatant into a new Eppendorf tube.

12. Add 5–10 µl of anti-acetylated lysine-310 RelA antibody, and incubate for 4 h at 4 °C with agitation, followed by the addition of 50 µl of Salmon Sperm DNA/Protein A agarose beads and incubate for overnight at 4 °C with agitation. An immunoprecipitation without adding any antibodies is used as a negative control.

13. Pellet the agarose beads by centrifugation at $900 \times g$ for 1 min at 4 °C. Discard the supernatant.

   (a) Wash the beads for 5 min on a rotating platform with 1 ml of each of the following buffers once in the order as given below: low-salt wash buffer, high-salt wash buffer, and LiCl wash buffer followed by 1× TE buffer twice.

   (b) Add 250 µl of the elution buffer to the agarose beads. Vortex briefly to mix and incubate at room temperature for 15 min with rotation. Spin down agarose, and carefully transfer the supernatant fraction to another tube and repeat elution. Combine supernatants (total volume $\approx$500 µl).

   (c) De-cross-link the protein and DNA cross-links by adding 20 µl of 5 M NaCl and incubate the samples at 65 °C O/N. Digest the samples with pronase for 1 h at 42 °C.

14. Purify the de-cross-linked DNA using Qiagen QuickSpin purification kit and elute in 50 µl DNA elution buffer.

15. Use 1–5 µl DNA eluates as template and 40 cycles of PCR amplification with Taq polymerase and proper primers for desired genes (e.g., E-selectin and TNF-α). The PCR products are analyzed on 2.5 % agarose gels (*see* **Note 11**).

3.3 Methylation of RelA

The approaches to assess RelA methylation are quite similar to those approaches used to assess RelA acetylation. [³H]-labeled SAM is often used in the in vitro methylation assay using recombinant RelA and methyltransferases. Immunoblotting with anti-pan methylated lysine antibodies or site-specific methylated lysine antibodies allows the rapid detection of in vitro or in vivo methylated RelA. We have used anti-pan methylated lysine antibodies from Abcam to detect Set9-mediated methylation of recombinant RelA or overexpressed RelA in cultured cells. It is not clear whether these antibodies will recognize other forms of methylated RelA mediated by methyltransferase rather than Set9. Several site-specific methylated RelA antibodies are used in the immunoblotting or ChIP assays to detect the stimulus-coupled methylation of endogenous RelA [10, 15–17]. Due to the complexity of lysine methylation, mass spectrometry has been successfully used to map the methylation sites and status [16–18].

3.3.1 In Vitro Methylation Assay

1. Prepare the following reaction in an Eppendorf tube: 1 µg of recombinant 6× His-tagged RelA or GST-RelA fusion proteins, 1 µg of recombinant Set9, and 6 µl of 5× methylation buffer. Depending on the reaction, use either 1 µM [³H]-labeled SAM or 40 µM nonradioactive SAM. Add H_2O to reach a final volume of 30 µl.

2. Spin the tube by centrifugation and then mix the tube by slightly tapping the bottom of the tube.

3. Incubate the samples at 30 °C for 60 min.

4. Stop the reaction by adding 6 μl of 6× SDS protein sample loading buffer and heat at 95 °C for 5 min.

5. Take 15 μl of the reaction products and resolve the samples on a 10 % SDS-PAGE gel.

6. When using [³H]-SAM as the methyl group donor, fix the gel using a fixing solution (isopropanol–water–acetic acid (25:65:10, v/v)) for 30 min.

7. Soak the gel in Amersham Amplify Fluorographic Reagent (sufficient for the gel to be free floating) with agitation for 15–30 min.

8. Dry the gel and hold the gel in close contact with an appropriate X-ray film and at –70 °C to –80 °C overnight.

9. Detect the radioactive signal by autoradiography.

10. When nonradioactive SAM is used as the methyl group donor, gel can be directly transferred to a nitrocellulose membrane, and the methylation signal of RelA can be detected by immunoblotting with an anti-methylated lysine antibodies (*see* **Note 12**).

3.4 Methylation of RelA In Vivo

The detection of methylation of overexpressed RelA is similar to that of acetylation (*see* Subheading 3.2.1).

3.4.1 Detection of Methylation of Overexpressed RelA

1. Seed 2 ml of HEK293T cells (2×10^5/ml) in each well of a six-well plate and culture overnight at 37 °C. Use two wells for each sample.

2. Transfect HEK293T cells with 0.5 μg of T7-tagged RelA and 2 μg of Set9 expression plasmids using calcium phosphate method into each well.

3. Thirty-six hours after transfection, lyse the cells with IP buffer.

4. Immunoprecipitate T7-RelA and methylation levels can be assessed by immunoblotting of immunoprecipitates with anti-pan-methyl-lysine antibodies (Abcam) or anti-methylated K314/315 antibodies.

3.4.2 Detection of Methylation of Endogenous RelA

The detection of methylation of endogenous RelA is similar to that of acetylation (*see* Subheading 3.2.2). Anti-methylated lysine 314/315 antibodies have been used to detect the TNF-α-induced methylation of endogenous RelA [18]. However, these antibodies are not suitable for straight immunoblotting with the cell lysates. This might be due to the low reactivity of the antibodies. Another possible reason is that in vivo methylated RelA is chromatin-bound and the amount of methylated RelA in the whole-cell extracts is relatively low. However, if these antibodies are used to immunoprecipitate methylated RelA, a significant amount of methylated RelA can be detected either from the TNF-α-treated whole-cell extracts or from chromatin.

4   Notes

1. Recombinant full-length RelA is difficult to obtain in the *Escherichia coli* (*E. coli*) cells due to the premature translation termination or partial proteolyses. However, under denaturing conditions, full-length RelA protein can be obtained from *E. coli* as described [23]. Recombinant RelA can also be purified using the baculovirus–insect cell expression system.

2. Since acetyl-CoA is unstable, it is usually prepared fresh from powder. Aqueous solutions stored in aliquots at –20 °C are stable for no longer than 2 weeks. Solutions can be stored at –80 °C for up to 6 months.

3. SAM is stored at –20 °C as a 32 mM solution dissolved in 0.005 M H_2SO_4 and 10 % ethanol (pH 7.5). Under this condition, SAM is stable for up to 6 months. SAM is unstable at 37 °C and should be replenished in reactions if the reactions run longer than 4 h.

4. We typically use calcium phosphate method for transient transfection of HEK293T cells. This method is cost-effective and displays excellent transfection efficiency in HEK293T cells. Other commercially available transfection reagents (e.g., Lipofectamine or FuGENE) can also be used.

5. Preclearing cell lysates reduces nonspecific binding of proteins to agarose beads and improves the quality of p300 immunoprecipitates.

6. The freshly made p300 is ready for use. One aliquot of 20 µl is good for one in vitro acetylation reaction. Since the 1× HAT buffer contains 10 % glycerol, HA-p300 immunoprecipitates can be stored in –80 °C for later use (up to several months) without losing much of the HAT activity.

7. Nonradioactive acetyl-CoA is used when acetylated RelA is detected using anti-acetylated antibodies.

8. Gently tap the bottom of the reaction tube to occasionally mix the beads with substrates for a more efficient acetylation reaction.

9. Exposure time varies with the acetylation signal. It usually takes 1–7 days. Phosphoimaging might also be used to help reduce the exposure time.

10. Anti-pan acetylated lysine antibodies from Cell Signaling or anti-acetylated lysine-310 antibodies from Cell Signaling or from Abcam can be used in the immunoblotting. Incubation with these antibodies overnight is required for the detection of acetylated RelA.

11. Quantitative real-time PCR can also be used to determine the amount of DNA in the eluates.

12. Anti-pan methylated lysine antibodies are from Abcam. Incubation with these antibodies overnight is required for the detection of methylation signals.

Acknowledgment

The work described in this article was supported in part by National Institutes of Health grants (RO1DK085158 and R21DK093865-01) to L.F.C. and by funds from University of Illinois at Urbana-Champaign.

References

1. Ghosh S, Karin M (2002) Missing pieces in the NF-kappaB puzzle. Cell 109(Suppl):S81–S96

2. Chen LF, Greene WC (2004) Shaping the nuclear action of NF-kB. Nat Rev Mol Cell Biol 5(5):392–401

3. Huang B, Yang XD, Lamb A, Chen LF (2010) Posttranslational modifications of NF-kappaB: another layer of regulation for NF-kappaB signaling pathway. Cell Signal 22(9):1282–1290

4. Perkins ND (2006) Post-translational modifications regulating the activity and function of the nuclear factor kappa B pathway. Oncogene 25(51):6717–6730

5. Kiernan R, Bres V, Ng RW, Coudart MP, El Messaoudi S, Sardet C et al (2003) Post-activation turn-off of NF-kB-dependent transcription is regulated by acetylation of p65. J Biol Chem 278(4):2758–2766

6. Chen LF, Mu Y, Greene WC (2002) Acetylation of RelA at discrete sites regulates distinct nuclear functions of NF-kB. EMBO J 21(23):6539–6548

7. Buerki C, Rothgiesser KM, Valovka T, Owen HR, Rehrauer H, Fey M et al (2008) Functional relevance of novel p300-mediated lysine 314 and 315 acetylation of RelA/p65. Nucleic Acids Res 36(5):1665–1680

8. Rothgiesser KM, Fey M, Hottiger MO (2010) Acetylation of p65 at lysine 314 is important for late NF-kappaB-dependent gene expression. BMC Genomics 11:22

9. Yeung F, Hoberg JE, Ramsey CS, Keller MD, Jones DR, Frye RA et al (2004) Modulation of NF-kappaB-dependent transcription and cell survival by the SIRT1 deacetylase. EMBO J 23(12):2369–2380

10. Yang XD, Tajkhorshid E, Chen LF (2010) Functional interplay between acetylation and methylation of the RelA subunit of NF-kappaB. Mol Cell Biol 30(9):2170–2180

11. Zou Z, Huang B, Wu X, Zhang H, Qi J, Bradner J et al (2013) Brd4 maintains constitutively active NF-kappaB in cancer cells by binding to acetylated RelA. Oncogene 33(18):2395–2404

12. Wu X, Qi J, Bradner JE, Xiao G, Chen LF (2013) Bromodomain and extraterminal (BET) protein inhibition suppresses human T cell leukemia virus 1 (HTLV-1) Tax protein-mediated tumorigenesis by inhibiting nuclear factor kappaB (NF-kappaB) signaling. J Biol Chem 288(50):36094–36105

13. Lee H, Herrmann A, Deng JH, Kujawski M, Niu G, Li Z et al (2009) Persistently activated Stat3 maintains constitutive NF-kappaB activity in tumors. Cancer Cell 15(4):283–293

14. Lu T, Yang M, Huang DB, Wei H, Ozer GH, Ghosh G et al (2013) Role of lysine methylation of NF-kappaB in differential gene regulation. Proc Natl Acad Sci U S A 110(33):13510–13515

15. Ea CK, Baltimore D (2009) Regulation of NF-kappaB activity through lysine monomethylation of p65. Proc Natl Acad Sci U S A 106(45):18972–18977

16. Lu T, Jackson MW, Wang B, Yang M, Chance MR, Miyagi M et al (2010) Regulation of NF-kappaB by NSD1/FBXL11-dependent reversible lysine methylation of p65. Proc Natl Acad Sci U S A 107(1):46–51

17. Levy D, Kuo AJ, Chang Y, Schaefer U, Kitson C, Cheung P et al (2010) Lysine methylation of the NF-kappaB subunit RelA by SETD6 couples activity of the histone methyltransferase GLP at chromatin to tonic repression of NF-kappaB signaling. Nat Immunol 12(1):29–36

18. Yang XD, Huang B, Li M, Lamb A, Kelleher NL, Chen LF (2009) Negative regulation of NF-kappaB action by Set9-mediated lysine

methylation of the RelA subunit. EMBO J 28(8):1055–1066

19. Chen LF, Fischle W, Verdin E, Greene WC (2001) Duration of nuclear NF-kB action regulated by reversible acetylation. Science 293(5535):1653–1657

20. Benson LJ, Annunziato AT (2004) In vitro analysis of histone acetyltransferase activity. Methods 33(1):45–52

21. Chen LF, Williams SA, Mu Y, Nakano H, Duerr JM, Buckbinder L et al (2005) NF-kappaB RelA phosphorylation regulates RelA acetylation. Mol Cell Biol 25(18):7966–7975

22. Chen LF, Greene WC (2005) Assessing acetylation of NF-kappaB. Methods 36(4):368–375

23. Thanos D, Maniatis T (1996) In vitro assembly of enhancer complexes. Methods Enzymol 274:162–173

Chapter 25

A Method for the Quantitative Analysis of Stimulation-Induced Nuclear Translocation of the p65 Subunit of NF-κB from Patient-Derived Dermal Fibroblasts

Alex W. Wessel and Eric P. Hanson

Abstract

Developmental and immune-mediated disease has been linked to genetic mutation of key signaling components involved in NF-κB activation that leads to impaired activation or regulation of the canonical IKK complex. We identify patients with suspected or known defects of the NF-κB signaling pathway through clinical phenotyping and genetic sequencing. To help understand how mutations cause disease, we quantitate the kinetics and dose-response of NF-κB activation signaling events in their cells. Following activation of the canonical IKK complex, phosphorylation of the inhibitor of NF-κB proteins (IκB) leads to their degradation and the subsequent translocation of NF-κB family members from the cell cytoplasm to the nucleus. Here, we provide a method to obtain patient-derived dermal fibroblasts and quantitatively assess the integrity of the signal transduction pathway from receptor activation to nuclear p65 translocation.

Key words NF-κB, p65, Nuclear translocation, Fibroblasts, IL-1R, TNFR, TLR, RLR, Microscopy

1 Introduction

Stimulation of pattern recognition receptors such as the Toll-like receptors, NOD-like receptors, and RIG-I-like receptors and of certain cytokine receptors such as the TNF receptor or IL-1 receptor activates NF-κB during the immune response and in the setting of inflammatory disease [1–4]. Recruitment of signaling components to these various receptors leads to activation of the canonical inhibitor of IkBa kinase (IKK) complex and NF-κB nuclear translocation (Fig. 1). The primary substrate of the activated IKK complex is the inhibitor of NF-κB proteins (IκB) whose phosphorylation and subsequent degradation permit the nuclear translocation of NF-κB family members which otherwise remain largely in the cytosolic fraction [5]. Activated NF-κB family members once in the nucleus are further posttranslationally modified to permit DNA binding that enhances

Michael J. May (ed.), *NF-kappa B: Methods and Protocols*, Methods in Molecular Biology, vol. 1280, DOI 10.1007/978-1-4939-2422-6_25, © Springer Science+Business Media New York 2015

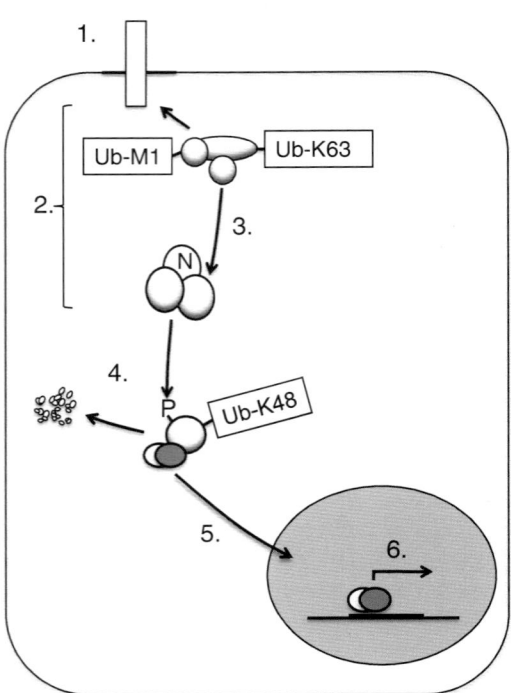

Fig. 1 Receptor-induced signal transduction leads to NF-κB activation. Endogenous Toll-like receptors, NOD-like receptors, TNF receptor superfamily members, and cytosolic nucleic acid sensors are among the many inducers of NF-κB activation that can be studied in primary dermal fibroblasts. Receptor stimulation (1) leads to recruitment of multiple signaling proteins, termed the signalosome, (2) through posttranslational modifications such as polyubiquitination. This leads to recruitment and activation (3) of the canonical IKK complex, which is comprised of NEMO (IKKγ), IKKβ, and IKKα. The IKK complex phosphorylates IκBα proteins on specific serine residues which then lead to K48-linked polyubiquitination of IkBa and its degradation by the 26S proteasome (4). IκBα degradation permits access to nuclear localization sequences on NF-κB members which translocate to the nucleus (5), bind DNA elements, and mediate gene expression (6). Quantitative measurement of p65 nuclear translocation following receptor activation permits evaluation of the integrity of steps 1–5, above

or, in some cases, represses gene expression due to specific activity at promoter and enhancer elements within the genome [5, 6].

Immunodeficiency, inflammatory disease, and/or developmental defects arise due to altered NF-κB activation [1–4, 7]. To understand how genetic mutation causes disease, we perform cellular phenotyping that includes analysis of signal transduction events in patient cells in vitro. Often, patients with Mendelian disorders in these pathways are young and severely affected, and so obtaining adequate materials to perform experiments is challenging. Dermal fibroblasts cultured from a single biopsy can be expanded greatly and cryopreserved, permitting abundant material for study.

The effects of hypomorphic or hypermorphic mutant proteins on signal transduction events may be modest, yet relatively small differences "upstream" may result in large "downstream" effects on gene expression or cell survival, due to multiple successive stages of signal amplification. To enhance the sensitivity of detecting differences between samples, it is important to develop robust assays that utilize quantitation and statistical analysis. For these, time-course and dose-response experiments are of great utility which require larger numbers of test conditions. The method we describe is performed in a 96-well format, permitting multiple technical replicates to be performed for each condition, enhancing the reproducibility of experimental results. We provide a method for measuring canonical IKK induction of nuclear translocation of p65 following stimulation of primary dermal fibroblasts. The method describes a procedure to obtain cells using a simple skin biopsy; however, it can be adapted to study many different types of adherent cells such as HeLa, monocyte-derived macrophages, or mesenchymal stem cells.

2 Materials

2.1 Skin Biopsy and Plating

1. 27-gauge needle.
2. 5-cc Luer-Lok disposable syringe.
3. 1 % lidocaine (10 mg/mL, Hospira, Inc, Lake Forest, IL, USA).
4. Betadine or chlorhexidine swab.
5. Disposable scalpel blade.
6. T75 tissue culture flask.
7. Dulbecco's Modified Eagle's Medium (DMEM).
8. Collagenase type IV solution (Gibco, Carlsbad, CA, USA). Make 1 mg/mL solution in DMEM.
9. Fibroblast culture medium: DMEM containing 15 % fetal calf serum (FCS) supplemented with penicillin (final 100 U/mL), streptomycin (final 100 μg/mL), MEM nonessential amino acids (Gibco) (stock concentration is 100×; use 10 mL per liter of medium), HEPES (final 10 mM), and sodium pyruvate 1 μM final. Combine components and then filter using a disposable tissue culture filter unit.

2.2 Cell Culture and Cell Stimulation

1. Lab benchtop centrifuge.
2. Fibroblast culture medium.
3. 96-well flat-bottom tissue culture-treated plate.
4. Phosphate buffered saline (PBS).
5. Primary dermal fibroblasts (*see* **Note 1**).
6. Recombinant human TNF-alpha (R&D Systems, Minneapolis, MN, USA) (*see* **Note 2**).

7. Trypsin-EDTA 0.25 %.

8. 37 °C, 5 % CO_2 incubator.

9. 50-mL conical tubes.

2.3 Immunofluo-
rescent Staining
and Image Acquisition

1. Permeabilization buffer: PBS containing 0.1 % Triton X-100 and 0.1 % saponin.

2. Alexa Fluor 488 goat anti-rabbit IgG antibody (Molecular Probes, Carlsbad, CA, USA).

3. Rabbit antihuman p65 antibody (Santa Cruz Biotechnology, Santa Cruz, CA).

4. 37 % formaldehyde.

5. Goat serum.

6. PBS.

7. ProLong Gold antifade reagent with DAPI (Invitrogen, Carlsbad, CA, USA) (*see* **Note 3**).

8. IncuCyte ZOOM (*see* **Note 4**) (Essen Biosciences, Ann Arbor, MI, USA).

2.4 Image Analysis

1. ImageJ or FIJI software.

2. Excel or Prism software.

---

3 Methods

To define the molecular mechanisms that lead to NF-κB activation, it is useful to compare the kinetics of receptor stimulation-induced signaling between cell lines that express mutant forms of known or suspected key signaling components of this pathway. Oftentimes to achieve this, cell lines are generated that transiently or stably over-express the mutant proteins found in individuals with immune-mediated disease. However, there is great utility in being able to study signaling characteristics from primary cells, as overexpression of certain molecules that function in signal transduction may para-doxically impair signaling, leading to spurious interpretations of their true function [8].

The shave skin biopsy is a straightforward procedure requiring minimal preparation that can be learned and performed by medically trained individuals with minimal experience (*see* **Note 5**). Prior to performing the procedure, informed consent from patients and control subjects must be obtained in accordance with institutional IRBs. In this procedure, a small piece of skin tissue is obtained which is then enzymatically treated to permit fibroblast growth on tissue culture dishes. Dermal fibroblasts undergo a finite number of divisions before reaching replicative senescence [9], and therefore it is important to plan cell preservation and storage to maximize utilization of the primary material and to ensure that experiments are performed using cells of similar passage number (*see* **Note 6**). Cells are cultured in medium and passaged by trypsinization and replating at a 1:3 ratio (the cells comprising one confluent flask are transferred onto three new flasks). For analysis of stimulation-induced p65 translocation, these cells are replated onto 96-well flat-bottom plates and stimulated with various different cytokines and recombinant pathogen-associated molecular patterns in the presence or absence of inhibitors or other controls (*see* **Note 7**). Following stimulation, cells are fixed and stained with specific antibodies to the NF-κB subunit p65. The kinetics and dose-response of NF-κB nuclear translocation from cells obtained from normal healthy controls are compared to that from individuals with primary immunodeficiency or Mendelian inflammatory disease in order to assess the integrity of signaling up to and including transcription factor nuclear translocation.

3.1 Skin Biopsy and Plating

1. Review procedure with the individual; obtain informed consent.

2. Swab the skin on the medial aspect of the forearm with antiseptic.

3. Fill a 5-cc syringe with 1 % lidocaine, and fit with a 27-gauge 1-in. needle.

4. Introduce the needle under the epidermis, injecting several microliters of lidocaine under the skin as the needle penetrates the skin. Eventually infuse approximately 50 μL of lidocaine under the skin, keeping the needle firmly in place with one hand (*see* **Note 8**).

5. Lift gently upward holding the syringe with the attached needle to create a tent of the skin (Fig. 2).

6. While the skin is under tension, slice horizontally immediately adjacent to the tip of the needle using a scalpel held in the opposite hand. A small chunk of the skin which contains cells from the epidermis and dermis will remain stuck to the needle as the skin is completely cut away.

7. Place the skin sample in a 15-mL conical tube containing 10 mL of DMEM (*see* **Note 9**).

8. Aspirate the DMEM, leaving the skin at the bottom of the conical tube. Add 1 mL of 1 % collagenase to the skin, and incubate at 37 °C 5 % CO_2 for 12 h.

9. Vortex the sample, add 7 mL of culture medium, and plate the entire contents onto a T75 flask (*see* **Note 10**).

3.2 Cell Culture and Cell Stimulation

1. Allow cells to proliferate. Cells will reach confluence in 1–3 weeks depending on the starting amount of cells in the initial sample.

2. Once cells have reached confluence, passage cells by trypsinization, splitting them 1:3 in early passages (1–5).

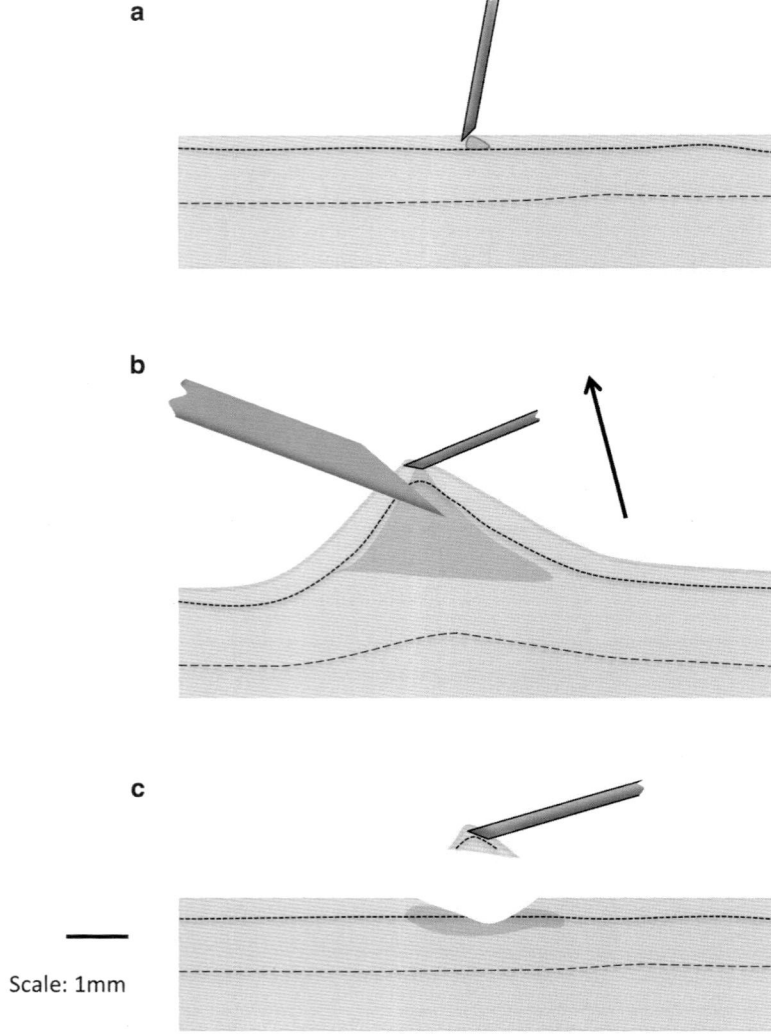

Fig. 2 The shave skin biopsy procedure. (**a**) The site to be biopsied is cleaned with topical antiseptic followed by injection of 1 % lidocaine subcutaneously as a sterile fine gauge needle is introduced into the dermal layer. (**b**) After injecting a small "blister" of lidocaine, the needle is angled parallel to the surface of the skin, while a gentle upward force is applied to create a "tented" area of the skin, the tip of which is sliced off. (**c**) The skin tissue sample containing dermal and epidermal layers is transferred from the tip of the needle to a tube containing the medium

3. Aspirate medium from the T75 flask of confluent fibroblasts and wash once with PBS.

4. Add trypsin to the T75 flask of confluent fibroblasts and incubate at 37 °C for 5 min until all the cells have lifted from the bottom of the flask.

5. Add 20 mL of the medium and transfer to a 50-mL conical tube.

6. Centrifuge for 5 min at $600 \times g$ to pellet the cells.

7. Aspirate the medium and resuspend in fresh medium. Count the cells.

8. Seed 3×10^3 fibroblasts per well of a 96-well tissue culture plate in 200 µL of medium and culture at 37 °C, 5 % CO_2 until the layer reaches 80–90 % confluence (*see* **Note 11**).

9. Aspirate the culture medium from wells and replace with 200 µL of warm medium.

10. Stimulate cells with TNF at a final concentration of 10 ng/mL in DMEM (*see* **Note 12**).

11. Incubate the fibroblasts at 37 °C, 5 % CO_2 for 15–60 min.

3.3 Immunofluo-rescent Staining and Cell Imaging

1. Add 200 µL of pre-warmed 2 % formaldehyde solution to each well of fibroblasts and incubate at 37 °C for 10 min.

2. Aspirate the medium and formaldehyde from each well and wash once with 500 µL of PBS.

3. Permeabilize the cells by adding 200 µL of permeabilization buffer to each well and incubate at room temperature for 40 min.

4. Aspirate each well and wash five times with 500 µL of PBS.

5. Add anti-p65 antibody diluted 1:200 in permeabilization buffer for 1 h at 4 °C (*see* **Note 13**).

6. Aspirate the primary antibody solution and wash five times with 500 µL of PBS.

7. Add 200 µL of 5 % goat serum in permeabilization buffer and incubate at room temperature for 20 min.

8. Aspirate each well and wash five times with 500 µL of PBS.

9. Add anti-rabbit antibody conjugated to Alexa Fluor 488, diluted 1:750 in P/W buffer. Mix by gentle rotation and incubate at room temperature for 1 h, protected from light.

10. Aspirate the secondary antibody and wash five times with 500 µL of PBS.

11. Using the IncuCyte ZOOM fluorescence microscope (*see* **Note 14**), load the 96-well plate onto the imaging platform.

12. Image each well using the phase contrast, green, and red channels to observe p65 signal (Figs. 3 and 4).

13. Save all images as .tif files.

3.4 Quantitation of Nuclear p65

1. Using ImageJ, open the .tif files of the blue (DAPI-nuclear stained, Fig. 6i) and green (p65 stained, Fig. 6iii) channels for the first pair of images.

2. Using the menu labeled "Image," select "Type" from the drop-down menu, and select 8 bit to convert the images from RGB color to 8-bit images, generating two new image files (Fig. 6ii, iv).

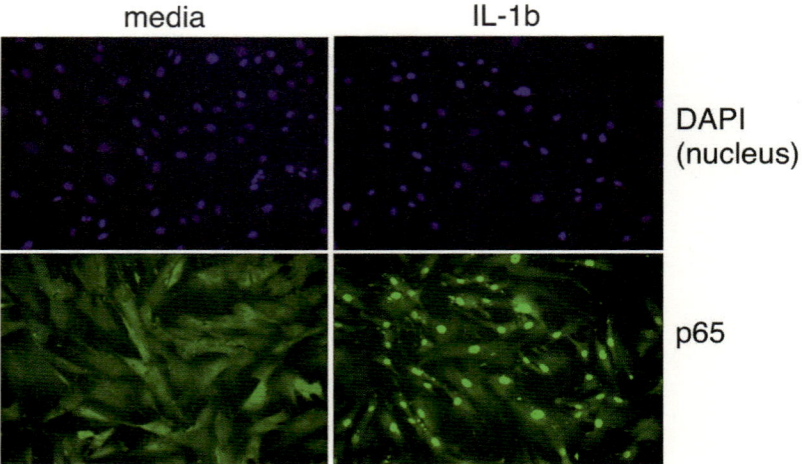

Fig. 3 Fluorescence microscopy of primary dermal fibroblasts cultured in the medium (*left panels*) and stimulated for 30 min with IL-1beta (*right panels*). Following stimulation, cells are fixed, permeabilized, and stained for NF-κB NF-κ p65 (*green*) and the cell nucleus (*blue*)

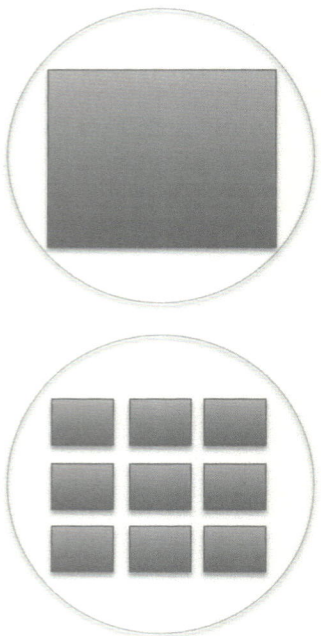

Fig. 4 Image acquisition. Images from a 96-well plate can be acquired using a 4× objective lens (*top*) or four or nine regularly spaced images per well are obtained using 10× or 20× (*bottom*)

3. Select the nuclear 8-bit image (Fig. 6ii), and from the ImageJ menu, select "Image" and then "Adjust Threshold" from the drop-down menu, selecting "Default" for the thresholding algorithm, "Red," and "dark background," to generate a nuclear mask (Fig. 6v).

4. Select "Analyze" from the top ImageJ menu and "Set Measurements" from the drop-down menu, and select the options that you would like to have calculated. Select "mean gray value" and "limit to threshold" (*see* **Note 15**). Essential step: select "Redirect to," and from the drop-down menu, select the file name corresponding to the 8-bit image of the p65 stain.

5. From the drop-down menu in ImageJ, select "Analyze," then "Analyze Particles." From the pop-up window, select "Show," then "Outlines," and select "Display Results." Select "OK" (Fig. 5).

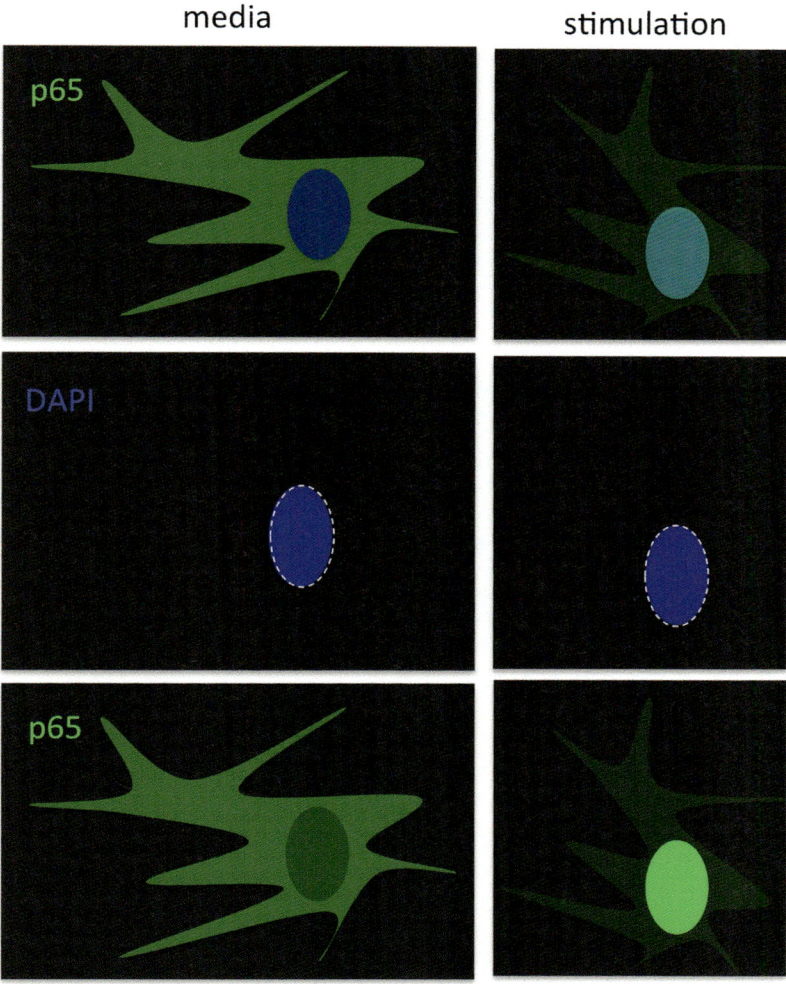

Fig. 5 Quantitation of nuclear p65. For quantitation, two single-color images are obtained corresponding to nuclear and p65 stains (*left*, above). Using ImageJ, a nuclear mask is created using the image of the stained nuclei (*dashed white oval*) which is then applied to the image corresponding to p65 staining so that the mean nuclear pixel intensity is calculated for each cell in the image. Hundreds or thousands of individual cell nuclear intensities can be determined per 10× or 4× image, respectively

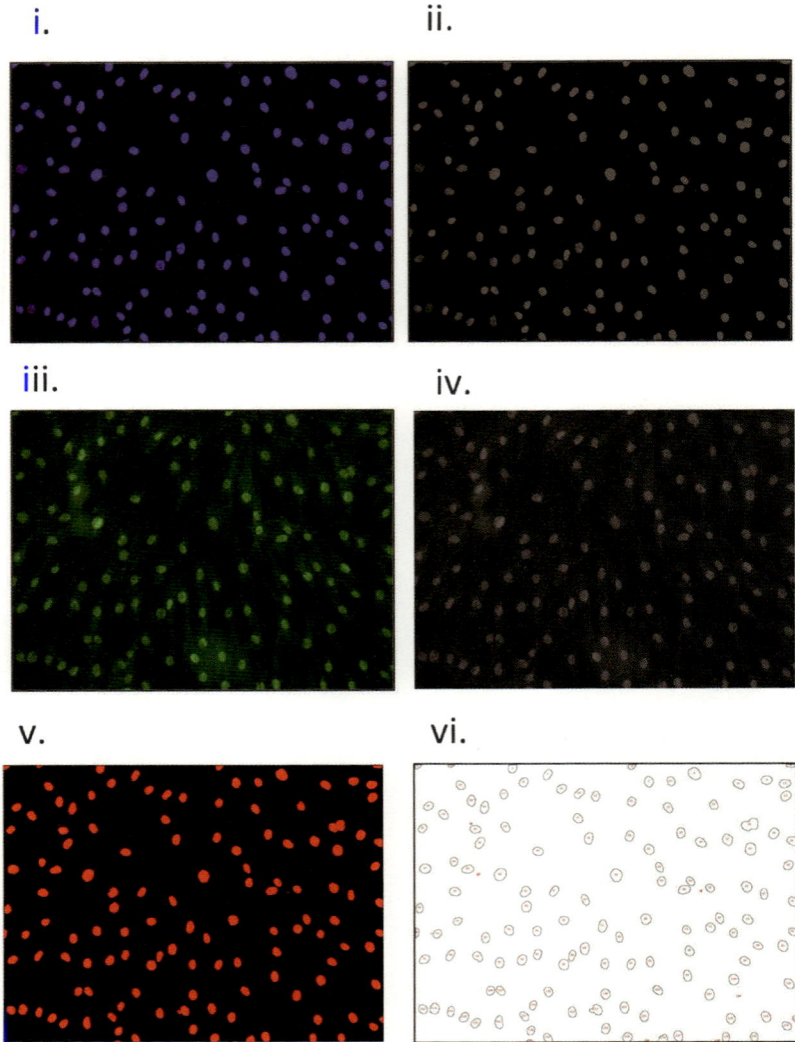

Fig. 6 Workflow of image analysis. An image of nuclear staining is opened (i) and converted to an 8-bit raw image (ii), as is done for the p65 stained image (iii–iv). Automatic thresholding of the 8-bit nuclear image will result in a thresholded image (v). Following particle analysis, a data file of numerical values and drawing with numbered nuclei is generated (vi), which permits visual correlation of numerical data with the corresponding individual cells

6. A result file is generated with multiple columns corresponding to the values chosen in Subheading 3.4, **step 4**. An image file named "Drawing of *name of image file.tif*" with the numbered outlines of the nuclear masks (Fig. 6vi) will appear simultaneously with the file of values, labeled "Results" (*see* **Note 16**).

7. Save the result file, or directly copy and paste these values into Excel or your graphing software of choice.

8. Repeat **steps 1–7** for all nuclear/p65 image pairs.

9. Graph the data or perform statistical analyses using Prism, Excel, or other software (Fig. 7).

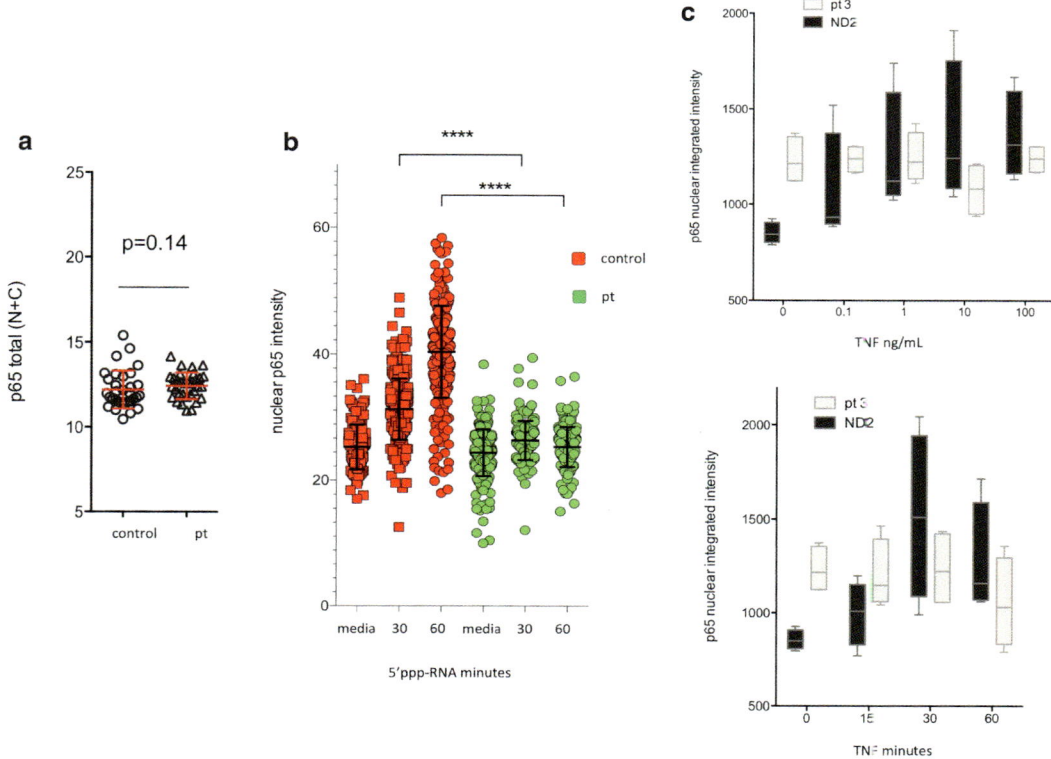

Fig. 7 Visual representation of data. (**a**) Total p65 intensity per well is calculated for all samples analyzed to determine that differential staining of samples did not occur, and each data point corresponds to a well of a 96-well plate. *N* nucleus, *C* cytoplasm, *pt* patient samples. (**b**) Data can be graphed from pooling multiple images, with each individual data point corresponding to the mean fluorescence intensity of each individual cell nucleus. (**c**) Alternatively, the total nuclear integrated intensity per well can be plotted using *n* replicates (corresponding to *n* individual wells of a 96-well plate) per condition

4   Notes

1. This protocol describes the use of primary dermal fibroblasts; however, other adherent cells such as monocyte-derived macrophages or dendritic cells can also be used, which require specific medium and culture conditions to permit pure culture.

2. Nuclear p65 accumulation can be readily observed in human dermal fibroblasts treated with IL-1β (10 μg/mL), LPS (100 ng/mL), poly(I:C) (10 μg/mL), and 5′ppp-RNA (Invivogen, San Diego, CA, USA).

3. Alternatively, if your imaging system only allows two-color (red/green) detection, you may want to stain the nuclei with either a red or green nucleic acid stain. Syto 21 Molecular Probes (Carlsbad, CA, USA) stock is 1:10,000 and is a suitable

green nuclear stain, and Syto 59 Molecular Probes (Carlsbad, CA, USA) is used at 1:1,000 and stains nuclei red. These are added simultaneously to the addition of the primary antibody.

4. Although developed primarily for live cell imaging, we use the IncuCyte ZOOM for convenient phase contrast and two-color fluorescent imaging of fixed cells. However, any fluorescence microscope can be used for this technique. If whole well or predetermined patterns of images are not automatically taken by an image acquisition device, we recommend having an investigator acquire images blinded to sample identity with a collaborator to perform image analysis in order to avoid inadvertent bias in image acquisition.

5. The biopsy site is chosen taking into consideration that scarring following wound healing is usually minimal but often leaves a residual area of hypopigmentation, which may be of cosmetic importance. Usually, we biopsy the medial (inner) aspect of the forearm, which is devoid of hair and has the skin that is under low tension.

6. We have found that one biopsy sample can be expanded to 20T75 tissue culture flasks within five passages to permit cryopreservation of 20 individual tubes. These can be stored in liquid nitrogen and thawed individually and used for experiments lasting several weeks before a new vial is thawed. This permits sequential experiments to be performed on cells that are of relatively equal passage number.

7. An important control is to stimulate some cells in the presence of inhibitors of canonical IKK activation or of p65 translocation. For that latter, MG-132 at 10 μM, added 1 h prior to stimulation, works well.

8. Just enough lidocaine to produce a small "blister" of fluid under the skin is sufficient. After about a minute, efficacy of anesthesia can be tested by jabbing the biopsy site with the scalpel tip. If done correctly, the only pain felt during the procedure should be from the first "jab" of the syringe into the skin.

9. The tissue sample can be kept in DMEM for up to 24 h at ambient temperature before the next step is performed. This potentially allows overnight shipment of sample from remote locations.

10. It is helpful to plate cells at sufficient density to allow rapid growth. If cells are plated very sparsely, they will take a very long time (months) to reach confluence, and culture may fail. If a tissue sample is particularly small or has decreased numbers of cells, centrifuge the sample ($400 \times g \times 10$ min) and resuspend in 1.5 mL of culture medium in one well of a six-well dish.

11. If cells are plated using a coverslip/microscopy slide method, empirically determine the number of cells to plate to obtain cell confluence 1–2 days after replating.

12. It is essential to stimulate all cells from one particular time point simultaneously. To achieve this, the use of a multichannel pipettor is essential. When adding the TNF containing the medium to the cells, it is preferable to add a larger volume of more dilute solution in order to minimize the effects of pipetting error.

13. Staining may be performed for longer, up to overnight, if convenient.

14. If using a conventional fluorescence microscope, add 20 μL antifade reagent with DAPI to each coverslip and place the coverslip onto a microscopy slide. Allow to dry overnight at room temperature, protected from light. Seal the edges of the glass coverslip with nail polish or other sealant and allow this to dry.

15. For some analyses, it may be useful to calculate the total nuclear intensity of p65 staining, rather than the mean value per nucleus, in which case "integrated intensity" should be selected.

16. It is useful to review each image alongside the corresponding data values. As a simple visual control, confirm that the numbered nuclei corresponding to the brightest p65 staining generated higher mean pixel intensity or integrated intensity values and that the nuclei with the dimmest p65 staining correspond to the lowest values. Depending on the quality of nuclear staining, small particles corresponding to debris may need to be excluded from the analysis by empirically modifying the default size range.

References

1. Hanson EP, Monaco-Shawver L, Solt LA, Madge LA, Banerjee PP, May MJ, Orange JS (2008) Hypomorphic nuclear factor-kappaB essential modulator mutation database and reconstitution system identifies phenotypic and immunologic diversity. J Allergy Clin Immunol 122(6):1169–1177.e1116. doi:10.1016/j.jaci.2008.08.018

2. Courtois G, Smahi A, Reichenbach J, Doffinger R, Cancrini C, Bonnet M, Puel A, Chable-Bessia C, Yamaoka S, Feinberg J, Dupuis-Girod S, Bodemer C, Livadiotti S, Novelli F, Rossi P, Fischer A, Israel A, Munnich A, Le Deist F, Casanova JL (2003) A hypermorphic IkappaBalpha mutation is associated with autosomal dominant anhidrotic ectodermal dysplasia and T cell immunodeficiency. J Clin Invest 112(7):1108–1115. doi:10.1172/JCI18714

3. Lahtela J, Nousiainen HO, Stefanovic V, Tallila J, Viskari H, Karikoski R, Gentile M, Saloranta C, Varilo T, Salonen R, Kestila M (2010) Mutant CHUK and severe fetal encasement malformation. N Engl J Med 363(17):1631–1637. doi:10.1056/NEJMoa0911698

4. Pannicke U, Baumann B, Fuchs S, Henneke P, Rensing-Ehl A, Rizzi M, Janda A, Hese K, Schlesier M, Holzmann K, Borte S, Laux C, Rump EM, Rosenberg A, Zelinski T, Schrezenmeier H, Wirth T, Ehl S, Schroeder ML, Schwarz K (2013) Deficiency of innate and acquired immunity caused by an IKBKB mutation. N Engl J Med 369(26):2504–2514. doi:10.1056/NEJMoa1309199

5. Ghosh S, May MJ, Kopp EB (1998) NF-kappa B and Rel proteins: evolutionarily conserved mediators of immune responses. Annu Rev Immunol 16:225–260. doi:10.1146/annurev.immunol.16.1.225

6. Karin M, Ben-Neriah Y (2000) Phosphorylation meets ubiquitination: the control of NF-[kappa]B activity. Annu Rev Immunol 18:621–663. doi:10.1146/annurev.immunol.18.1.621

7. Conte MI, Pescatore A, Paciolla M, Esposito E, Miano MG, Lioi MB, McAleer MA, Giardino G, Pignata C, Irvine AD, Scheuerle AE, Royer G, Hadj-Rabia S, Bodemer C, Bonnefont JP, Munnich A, Smahi A, Steffann J, Fusco F, Ursini MV (2014) Insight into

IKBKG/NEMO locus: report of new mutations and complex genomic rearrangements leading to incontinentia pigmenti disease. Hum Mutat 35(2):165–177. doi:10.1002/humu.22483

8. Li Y, Kang J, Friedman J, Tarassishin L, Ye J, Kovalenko A, Wallach D, Horwitz MS (1999) Identification of a cell protein (FIP-3) as a modulator of NF-kappaB activity and as a target of an adenovirus inhibitor of tumor necrosis factor alpha-induced apoptosis. Proc Natl Acad Sci U S A 96(3):1042–1047

9. Campisi J, d'Adda di Fagagna F (2007) Cellular senescence: when bad things happen to good cells. Nat Rev Mol Cell Biol 8(9):729–740. doi:10.1038/nrm2233

Methods for Assessing the In Vitro Transforming Activity of NF-κB Transcription Factor c-Rel and Related Proteins

Thomas D. Gilmore and Céline Gélinas

Abstract

Among NF-κB transcription factors, c-Rel and c-Rel-derived proteins, including v-Rel, are the only ones that have shown consistent and frank transforming activity in cell culture. In particular, viral, chicken, mouse, and human Rel proteins can rapidly transform primary chicken spleen and bone marrow cells. Overexpression of a human Rel protein missing a C-terminal transactivation domain can also enhance the transformed state of the human B-lymphoma cell line BJAB. As described in this chapter, these in vitro assays can be used to quantitatively assess the transforming activity of Rel proteins.

Key words c-Rel, v-Rel, NF-kappaB, Transformation assay, Oncogenesis, Chicken spleen cells, BJAB cells

1 Introduction

The avian and mammalian nuclear factor κB (NF-κB) family of transcription factors includes five members that can be divided into two subfamilies: the NF-κB proteins (p50/NF-κB1, p52/NF-κB2) and the Rel proteins (c-Rel, RelB, p65/RelA) (reviewed in [1]). These five proteins are related through an N-terminal DNA-binding/dimerization domain (the Rel homology domain [RHD]), and they can form diverse sets of homo- and heterodimers. However, the two subfamilies have distinct C-terminal halves. In particular, all Rel proteins have C-terminal transactivation domains, whereas NF-κB subfamily proteins do not have such domains.

Although increased nuclear NF-κB activity is associated with and required for many animal and human cancers (reviewed in [2]), NF-κB/Rel subunits have been shown to have direct oncogenic activity in only a limited number of circumstances (Table 1). Indeed, c-Rel and c-Rel-derived proteins appear to be the only true oncoproteins among NF-κB transcription factors. Well before NF-κB was discovered as a transcription factor involved in normal immune cell function, the v-*rel* oncogene of the avian reticuloendotheliosis virus strain T (Rev-T) retrovirus (which acquired a

Table 1
Oncogenic activity of NF-κB and Rel proteins

Protein	Assay
v-Rel	In vivo (chickens, mice); in vitro (chicken cells)
c-Rel (chicken, human, mouse)	In vitro (chicken, human cells)
p100/NF-κB2 truncations	In vivo (mice; ref. 39); in vitro (mouse fibroblasts; ref. 40)

portion of the turkey c-*rel* gene by a nonhomologous recombination event to create v-*rel*) was known to have potent transforming activity for chicken hematopoietic cells in vivo and in vitro (reviewed in [3]). Strikingly, young chickens injected with high titer stocks of avian retroviral vectors for expression of v-Rel can succumb within as few as 7 days with hundreds of tumor foci in their spleens and livers. It was later shown that transgenic mice with v-Rel expression controlled by a T-cell-specific promoter develop T-cell lymphomas [4]. Moreover, the *REL* (human c-*rel*) gene is amplified or mutated in a large number of human B-cell malignancies (reviewed in [5]).

v-Rel has a number of amino acid changes as compared to c-Rel, and several of these changes enhance its oncogenicity (reviewed in [3]). The primary change responsible for v-Rel's heightened oncogenicity as compared to c-Rel is a deletion of one of two C-terminal transactivation domains, which enables the protein to be a chronic, low-level activator of target gene transcription [3, 6]. That is, although full-length c-Rel proteins from chicken, mouse, and human can transform chicken spleen cells in culture, their transforming activities are enhanced by partial deletion of C-terminal transactivation domain sequences. In contrast, overexpression of RelA, p52, p50, or the upstream NF-κB-activating kinase IKKβ does not transform chicken spleen cells [7]; however, a hybrid protein with the RelA DNA-binding domain and v-Rel's C-terminal transactivation domain can transform primary chicken spleen cells [7].

Rel proteins appear to require cooperating changes to transform mammalian cells. v-Rel and c-Rel proteins have not been shown to transform primary mouse cells in vitro. Moreover, the long latency required for the development of v-Rel-directed T-cell lymphoma [4] or c-Rel-directed mammary carcinoma [8] in transgenic mice indicates that additional cellular changes are required for Rel proteins to effect mammalian cell transformation. Overexpression of a human c-Rel protein missing a C-terminal transactivation domain can, however, enhance the transformed state of the human B-lymphoma cell line BJAB [9].

Retroviral vectors for high-level expression of Rel proteins have been used in a number of assays to transform chicken hematopoietic cells in culture. v-Rel can transform a variety of cell types when primary chicken spleen or bone marrow cultures are infected in vitro (as first described in [10]), and such cells have been reported to have characteristics of immature and mature B and T cells, myeloid cells, erythroid cells, and pre-dendritic cells (reviewed in [3]). Such cells can be passaged for extended, and sometimes indefinite, times in vitro and can efficiently form tumors when injected into chickens. v-Rel can also weakly transform and extend the lifespan of primary chicken embryo fibroblast cultures [11–13]. In both hematopoietic and fibroblast cultures, the ability of v-Rel to extend their lifespan is likely to be due to v-Rel's ability to block apoptosis [14–18]. However, it is clear that v-Rel also provides proliferative and/or differentiation-blocking signals for avian hematopoietic cells [19], but whether the sets of genes that affect lifespan in culture, proliferation, and differentiation are distinct, or more likely, partially overlapping is not known. In this chapter, we describe cell-based assays for quantitatively assessing the transforming activity of Rel proteins in primary chicken spleen cells (Fig. 1) and the human BJAB B-lymphoma cell line (Fig. 2).

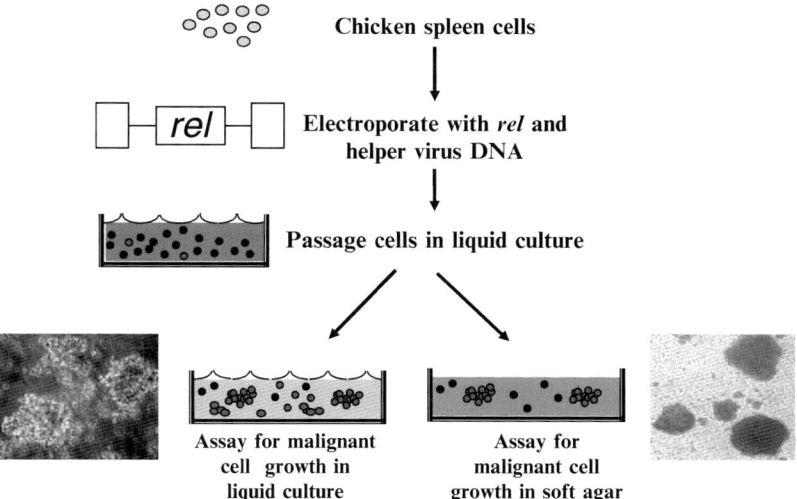

Fig. 1 Assays for transformation of chicken spleen cells by Rel proteins. As described in detail in the text, primary spleen cells are first isolated from young chickens. These cells are then immediately electroporated with an avian viral vector containing a *rel* gene along with helper virus plasmid DNA. Cells are passaged in culture and then assessed for transformation by outgrowth of transformed cells in liquid culture (indicated by *red cells* in the drawings) or by colony formation in soft agar (*red groups* of cells in image). Shown at the *left* is a microscopic image of human c-Rel-transformed cells, and on the *right* is a microscopic image of v-Rel-transformed colonies in soft agar (Color figure online)

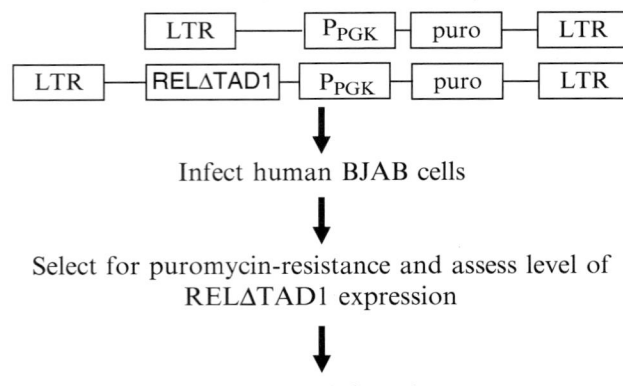

Infect human BJAB cells

Select for puromycin-resistance and assess level of
RELΔTAD1 expression

Analyze growth in soft agar

Fig. 2 Assay for transformation of human BJAB lymphoma cells by RELΔTAD1. Shown at the *top* are the structures of retroviral vectors: (1) control vector pMSCV containing the internal P_{PGK} promoter (phosphoglycerate kinase 1 promoter) driving the puromycin resistance gene (puro) and (2) the same vector with the upstream RELΔTAD1 gene (a highly oncogenic version of human c-Rel with a deletion of C-terminal transactivation domain 1 [9]; *see* also Fig. 3). As described in the text, virus stocks of these vectors are made by transfection of A293T cells. These virus stocks are then used to infect BJAB cells. Cells are selected for puromycin resistance, and high-level expression of RELΔTAD1 is confirmed by Western blotting [9]. Cells are then assessed for transformation by colony formation in soft agar

2   Materials

2.1   General Cell Culture

1. Human B-lymphoma BJAB cell line (*see* **Note 1**).
2. A293T cells (ATCC CRL-3216) (*see* **Note 2**).
3. Dulbecco's Modified Eagle's Medium (DMEM) (*see* **Note 3**).
4. Penicillin and streptomycin (*see* **Note 4**).
5. Fetal bovine serum (FBS) (*see* **Note 5**).
6. Puromycin at 500 μg/mL (*see* **Note 6**).
7. Tissue culture dishes (35 and 60 mm).
8. Sterile phosphate buffered saline (PBS).
9. Hemocytometer.
10. 2-, 5-, 10-, 25-mL sterile, individually wrapped disposable plastic pipettes.
11. 15- and 50-mL sterile plastic tubes.
12. 14-mL sterile snap-top tubes.
13. Desktop centrifuge (for 14-, 15-, and 50-mL plastic tubes).
14. Tissue culture incubators at 37 and 40.5 °C with 5–6 % CO_2.

**2.2   Preparation
of Chicken Spleen
Cells**

1. 18- to 22-day old chicks (SPAFAS-Charles River, CT, USA). With 10–12 chicks, one can do about 30 transformation assay samples (*see* **Note 7**).

2. Disposable laboratory gloves.

3. CO$_2$ source and chamber for euthanasia of chicks.

4. Two sheets of Labmat paper: one for chicks, one for dissection.

5. 70 % ethanol (in squirt bottle).

6. Sterile, autoclaved instruments for spleen removal and spleen mashing: scissors, scalpel, at least three forceps (straight and curved-tip), wire mesh screen (~4 by 4 cm).

7. Two sterile 125-mL Erlenmeyer flasks.

8. One sterile 250-mL Erlenmeyer flask.

9. Two sterile 100-mm Petri dishes (for intact spleens and spleen mashing).

10. Sterile 5¾ in. Pasteur pipettes.

11. Fresh, unopened box of autoclaved P20 and P200 plastic tips.

12. P20 and P200 Pipetman.

13. Two filled ice buckets.

2.3   Electroporation

1. Electroporation cuvettes (0.4 cm; Bio-Rad, Hercules, CA, USA).

2. Electroporator (e.g., Bio-Rad Gene Pulser with high capacitance extender).

3. Plasmids (pSW253 Rev-A helper virus DNA [20] and spleen necrosis virus (SNV)-derived Rel expression vectors) at approximately 1 µg/µL on ice (*see* **Note 8**).

4. DMEM-20: DMEM, 20 % FBS.

**2.4   BJAB Cell
Infection
and Transformation**

1. Polyethylenimine (PEI) at 1 µg/µL (Mol. Wt. 25,000 Da, Polysciences Inc., Warrington, PA, USA) (*see* **Note 9**).

2. Polybrene (Hexadimethrine bromide) at 20 µg/µL (*see* **Note 10**).

3. pCL10a1 helper virus plasmid [21] (*see* **Note 11**).

4. pMSCV [22] and pMSCV-RELΔTAD1 [9] (*see* **Note 12**).

**2.5   Soft Agar
Colony Assays**

1. Bacto Agar.

2. Petri dishes (60 mm).

3. Saran wrap.

3   Methods

**3.1   In Vitro
Transformation
of Chicken Spleen Cells**

1. Obtain ten to twelve 18- to 22-day old chicks. With 10–12 chicks, one can do about 30 transformation assay samples. Usually, one should perform at least 3–6 transformation assays

(i.e., 3–6 Petri dishes) for each gene being analyzed in order to obtain sufficient numbers for statistical analysis [23].

2. Sacrifice chickens using a CO_2 chamber. Depending on the size of the chamber, place 2–3 chicks in the chamber at a time. Turn on the CO_2 and wait until the animals are clearly moribund. To ensure their demise, decapitate the chicks (*see* **Note 7**).

3. In this step, all instruments (scalpels, forceps) should be autoclaved before use, and instruments should be liberally wetted with 70 % ethanol and flamed during the spleen removal procedure. To remove the spleen, first lay the sacrificed animal on its back on a piece of Labmat paper, with the head away from you. Liberally wet the ventral side of the animal with 70 % ethanol. With small sterile dissecting scissors, cut upwards along both sides of the breast bone, and pull back the skin flap to expose the body cavity. The spleen is located nearly dead center in the animal: that is, the spleen is under the top junction point of the two liver lobes, with the animal placed on its back. Using sterile forceps, pull the liver lobes apart to expose the spleen, which is a nearly perfectly round, pinkish organ, a bit smaller than a pea. Place curved-tip forceps under the spleen and gently remove the spleen by pulling up firmly (*see* **Note 13**).

4. Place all spleens in a single sterile 100-mm Petri dish.

5. Further sterilize a pre-autoclaved wire mesh screen (~4-by-4 cm) by soaking with 70 % ethanol and flaming over a Bunsen burner.

6. Place the wire mesh screen into a sterile 100-mm Petri dish, and add approximately 15 mL of DMEM containing 50 units/mL penicillin, 50 units/mL streptomycin, and 1–2 % FBS. Make sure that the top of mesh screen has been wetted with DMEM.

7. Using sterile forceps, place a spleen on the wire mesh screen. Using a sterile scalpel, mash the spleen through the mesh screen to obtain a cell suspension. Using forceps, remove integument of the spleen and discard. Repeat with each individual spleen. When finished with the last spleen, use the scalpel to wipe cells from the bottom of the screen into the culture medium in the Petri dish. Practice sterile technique throughout the procedure, in a tissue culture hood, with frequent flaming of instruments during the cell isolation procedure.

8. Using a sterile 5-mL pipette, transfer the cells and medium to a 50-mL plastic tube. To break up clumps of cells, first pipette the culture medium containing cells up and down several times with a 5-mL plastic disposable pipette. Repeat with a 2-mL pipette.

9. Pellet cells for 5 min at top speed in a desktop centrifuge. Remove and discard the supernatant by decanting.

10. To the cell pellet, add approximately 10–12 mL of DMEM/20 % FBS (DMEM-20). To resuspend the cells, pipette the cells up and down several times with a 5-mL pipette

and then with a 2-mL pipette. Allow large clumps to settle for about 1 min, and then transfer the supernatant (containing the cells for use) to a fresh 15-mL plastic centrifuge tube.

11. Count cells with a hemocytometer (dilute 0.1 mL of cells in 9.9 mL of PBS, and count cells in the 5-by-5 square). A count of 300 cells in that area corresponds to ~3×10^8 cells/mL. As necessary, adjust the cell mixture to 3×10^8 cells/mL by diluting with DMEM-20.

12. Add DMEM-20 to the cell mixture so that you have a total volume which is equal to 0.4 mL of cell mixture × number of samples you intend to analyze, e.g., if you have 30 samples, you will need 12 mL of cell mixture at a concentration of 3×10^8 cells/mL.

13. To the cell mixture, add pSW253 Rev-A helper virus plasmid [20] so that you will have 10 µg of plasmid per each 0.4-mL sample, e.g., add 300 µg of pSW253 for 30 samples in the 12 mL of cell suspension (*see* **Note 14**).

14. Resuspend the cells well by pipetting up and down using a 2-mL pipette, and then add 0.4 mL of the cell/pSW253 mixture to each individual 0.4 cm electroporation cuvette (aligned with lid labeled for each test sample). With a sterile P20 tip, add 20 µg of the appropriate SNV vector plasmid carrying the *rel* gene to each individual cuvette. For expediency, when adding the plasmid DNA, you can use the same P20 tip for all samples containing the same plasmid by just touching the DNA to the top of the medium in the cuvette. Cover each cuvette with the corresponding labeled lid. Overall, each 0.4-mL sample will contain approximately 1.2×10^7 cells, 10 µg pSW253, and 20 µg *rel* gene viral vector plasmid.

15. Incubate all electroporation cuvettes on ice for 10 min, with occasional swirling of the cuvettes.

16. Electroporate at 250 V and 975 µFarads with high capacitance extender turned ON. Be sure to swirl cuvettes with cells immediately before and after electroporation. After electroporation, there should be some pinkish denatured debris at the top of the liquid. The time constant reading is usually between 35 and 45 for each sample.

17. Incubate cuvettes for an additional 10 min on ice, with occasional swirling.

18. Remove cells from the cuvette with a sterile Pasteur pipette and place cells in 2 mL of pre-warmed DMEM-20 in a 35-mm tissue culture dish. When all samples have been seeded into plates, place the plates in a tissue culture incubator at 37–40.5 °C with 5–6 % CO_2.

19. Continue to Subheading 3.1.1 **step 1** below for soft agar assays or to Subheading 3.1.2 **step 1** for liquid outgrowth assays (*see* **Note 15**).

1. Four days later, transfer the cells from each 35-mm dish into a 14-mL snap-top plastic tube. Pellet the cells for about 5 min at medium speed in a desktop centrifuge.

2. After centrifugation, remove most of the culture medium by sterile aspiration (leaving about 0.3 mL of medium above the cells), and then add 5 mL of fresh DMEM-20 containing 0.3 % Bacto Agar (*see* **Note 16**). Mix well by pipetting up and down a few times with a sterile Pasteur pipette (avoiding bubbles). After briefly flaming the top of the tube, pour the cell/agar mix into a 60-mm Petri dish (*see* **Note 17**).

3. Allow the agar to solidify by incubating the Petri dishes for about 10–15 min at room temperature, and then place them in a 37–40.5 °C *humid* tissue culture incubator (i.e., lots of water in the bottom tray) (*see* **Note 18**). It is generally also good to loosely cover the Petri dishes with Saran wrap to avoid drying (*see* **Note 19**).

4. For the next 10–14 days, keep the incubator closed as much as possible so that the agar remains moist. Colonies grow best in agar that is rather sloppy and soft. Colonies can usually be counted with the naked eye or with a microscope at low magnification at 10–14 days after transfer to agar. For plasmid pGM282BS+ (wild-type v-*rel*) control [24], one can obtain 75–200 colonies per dish in a good assay. The efficiency of colony formation by wild-type v-Rel can be further increased by up to twofold by dividing individual transformation assays into multiple soft agar dishes (ref. 25; *see* **Note 17**).

5. These assays can be somewhat variable, and it is recommended to do 5–6 separate experiments with 3–6 dishes of each virus in order to obtain statistically significant results. The results can be presented as the number of colonies obtained (Fig. 3) or as a percentage of colonies obtained versus the positive control (e.g., representing the average number of v-Rel-generated colonies in a given experiment as 100).

1. Four to five days after electroporation, transfer cells to 14-mL snap-top tubes. Pellet cells for about 5 min at medium speed in a desktop centrifuge.

2. Using a sterile aspiration system in a tissue culture hood, remove most of the supernatant, leaving about 0.3–0.5 mL of the culture medium above the cell pellet. Add 2.5 mL of DMEM-20. Resuspend the cells by pipetting up and down a few times with a sterile Pasteur pipette. After briefly flaming the top of the tube, pour the cell suspension into a 35-mm dish. For the first cell passage, we recommend pouring the cells back into the exact same dish from which they were taken, as this seems to ensure optimal virus spread, perhaps from infected fibroblasts in the culture. For subsequent cell passages, you will have to use your judgment as to whether to

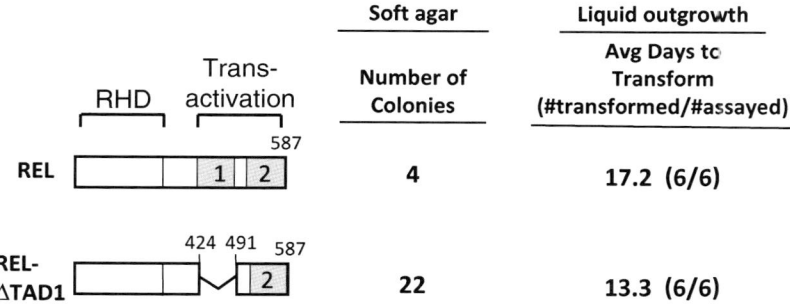

Fig. 3 Comparison of chicken spleen cell transformation by soft agar colony formation versus liquid outgrowth. The wild-type human c-Rel protein (REL) is a 587 aa protein with an N-terminal DNA-binding domain (RHD) and two C-terminal transactivation subdomains (*gray boxes*). RELΔTAD1 is a mutant in which transactivation subdomain 1 (amino acids 425–490) has been deleted [9, 28]. The abilities of these two proteins to transform chicken spleen cells were assessed. On the *left* are shown the average numbers of colonies obtained in a soft agar colony formation assay. On the *right* are the results of a liquid outgrowth assay in which six plates were assayed for each protein, and presented are the average number of days it took for cells to become transformed (and the number of plates transformed/number of plates assayed). In each case, RELΔTAD1 was more efficient at transforming than wild-type REL. Results are from Starczynowski et al. [28]

transfer the cells into new dishes based on whether there are too many fibroblasts attached and proliferating on the bottom of the dish that might acidify the medium.

3. Every 5 days, repeat the fluid changing with DMEM-20, as in Subheading 3.1.2 **steps 1** and **2**. Usually at the first passage (4–5 days after the assay is performed), cultures are acidified due to the growth of primary cells. By the tenth day (second passage), the cells are usually not acidifying the medium unless the cells have undergone transformation.

4. Score the cultures as positive (i.e., "transformed") when the cultures become acidified (turn yellow usually overnight) due to the rapid overgrowth of large clumps of transformed cells (for details, *see* refs. 26–28). v-Rel can transform these cultures in about 11–18 days. In the same assay, weakly oncogenic Rel proteins will take longer than v-Rel, and ones containing wild-type human c-Rel sometimes take as long as 30 days. Usually we do not continue assays beyond 30 days. The results are usually presented as the number of dishes that showed transformation with a given virus (e.g., 5/6 dishes) and the average days it took for those five dishes to become transformed [26–28] (*see* Fig. 3).

3.2 Creation of RELΔ TAD1-Expressing Human BJAB Cells

1. Preparation of virus for infection of BJAB cells (*see* **Note 20**).

    (a) Day One. Seed A293T cells in DMEM-10 in a 100-mm tissue culture dish, so that they will be ~50–60 % confluent on the next day.

    (b) Day Two. For each sample to a 14-mL snap-top tube, add 300 μL DMEM (no serum or pen-strep), 45 μL PEI

(1 μg/μL), 10 μg pCL10a1 helper virus plasmid DNA, and 5 μg pMSCV-RELΔTAD1 expression virus or pMSCV control (expressing only the *puro* gene) plasmid DNA. Incubate the mixture at room temperature for 15 min, with occasional swirling. Then add 3 mL of DMEM-10 to the DNA/PEI mixture. Aspirate the culture medium from the cells, and then wash twice with sterile PBS. Pour the 3.3 mL of DNA/PEI/DMEM-10 mixture onto the A293T cells (do so gently down the sides of the plates so as not to detach the cells). Incubate the A293T cells overnight (~24 h) in a 37 °C tissue culture incubator, with occasional swirling of the dish to make sure that the transfection mixture covers the cells.

(c) Day Three. Remove the culture medium/PEI mixture from the cells by aspiration. Then add 5 mL of fresh DMEM-10 to the dish. Incubate for 24 h at 37 °C in a tissue culture incubator.

(d) Day Four. Collect the 5 mL of virus-containing medium from the cell culture dish into a 14-mL snap-top tube, and then add 5 mL of fresh DMEM-10 to the dish. Next, spin down the virus-containing medium in a clinical centrifuge at top speed for 5 min in order to pellet any A293T cells that may have been taken up during virus harvest. After centrifugation, transfer the supernatant to a new 14-mL snap-top tube, being careful not to transfer any cells from the pellet at the bottom of the tube (i.e., do not go all the way to the bottom of the tube when transferring the supernatant). Place the tube on ice, and then either quick-freeze in a dry/ice ethanol bath or immediately place at –80 °C (do not leave the virus at room temperature on ice for long, as it will decrease the virus titer). Store virus stocks at –80 °C. Make sure that the tube is labeled with the name and date of the virus preparation.

(e) Day Five. Approximately 24 h after the first harvest, collect the 5 mL of virus-containing medium from the dish into a second 14-mL snap-top tube. Centrifuge, label, and store at –80 °C as in Subheading 3.2, **step 1d** above. The transfected A293T cells can now be discarded.

2. Spin infection and selection of BJAB cells.

(a) In a 14-mL snap-top tube, spin down $0.5–1 \times 10^6$ BJAB cells, which have been maintained in DMEM-20, in a clinical centrifuge at approximately medium speed for 5 min.

(b) After pelleting the cells, aspirate off the supernatant. To the cell pellet, add 2 mL of viral supernatant, resuspend the cells with a sterile Pasteur pipette, and then add polybrene to 8 μg/mL (*see* **Notes 21** and **22**).

3. Centrifuge the cells plus virus for 1–1.5 h at 2,500×*g* in a clinical centrifuge to infect by spin infection [29].

4. Remove the supernatant by sterile aspiration, and resuspend the cells in 2.5 mL of DMEM-20. Seed the cells in a 35-mm cell culture dish and incubate at 37 °C in a tissue culture incubator for 2–3 days.

5. To select infected cells based on puromycin resistance, centrifuge the cells and resuspend in 6 mL of DMEM-20 containing 2.5 µg/mL puromycin in 60-mm dishes. Puromycin-resistant cells should grow out after approximately 8–14 days.

6. Analyze puromycin-resistant cultures for expression of RELΔTAD1 protein by Western blotting with anti-c-Rel antiserum [9]. It is essential to have cells that are expressing high levels of RELΔTAD1 (i.e., at levels higher than the endogenous c-REL protein) [9], as we have found that lower levels of RELΔTAD1 do not enhance the growth of BJAB cells in soft agar (T.D. Gilmore, unpublished) (*see* **Note 23** and Fig. 4a).

3.3 In Vitro Soft Agar Assays of REL-Expressing Human BJAB Cells

1. Once you have established a BJAB cell culture that is expressing high levels of RELΔTAD1 and a negative control BJAB cell culture containing only pMSCV (Fig. 4a), you can compare their transformed states, as judged by growth in soft agar.

2. Use BJAB cells that are quite healthy, e.g., have been passaged 1–2 days prior to the experiment and may have acidified the medium to a slightly orange color (but not extremely yellow), and use cell cultures that do not contain many dead cells.

3. Before starting the experiment, prepare and label all the 14-mL snap-top tubes and 60-mm non-coated Petri dishes

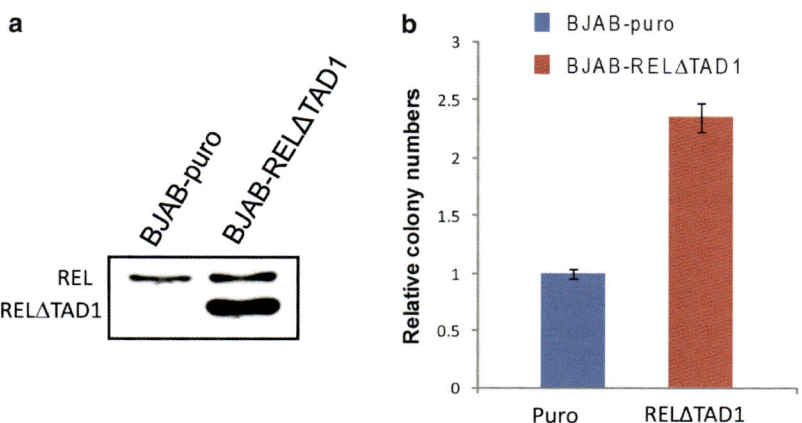

Fig. 4 RELΔTAD1 enhances the soft agar colony-forming ability of BJAB cells. (**a**) Western blot of extracts from puromycin-resistant BJAB cells showing overexpression of RELΔTAD1 (*right lane*), as compared to control cells infected with the MSCV vector alone (*left lane*). (**b**) Relative soft agar colony-forming ability of the two cell types in (**a**). For details on the calculation of relative colony formation, *see* **Note 26**. Results are from Chin et al. [9]

(i.e., not tissue culture dishes) for soft agar cell suspensions (indicating both the cell line and the number of cells to be seeded). Label the dishes on the top and the bottom, because the labels on the top can sometimes rub off during the experiment. For each cell line, cells will be seeded in a DMEM-20/ soft agar mixture at several different cell concentrations, and each cell concentration should be seeded in triplicate. For example, if assaying one cell line, and you are seeding cells at 250, 500, 1,000, and 2,000 cells per dish, then a total of 12 Petri dishes will be required (i.e., 4 cell concentrations × 3 dishes per concentration). Also, pre-warm DMEM-20 containing 0.3 % agar to ~37–40 °C in a water bath (*see* **step 6** below for total volume required).

4. Count each cell culture to be assayed using a hemocytometer to determine the cell concentration. For each cell dish, transfer the culture medium and cells to a sterile 50-mL disposable plastic tube, and mix cells repeatedly but avoid bubbles (e.g., pipette cell culture up and down in 50-mL tube at least five times until cells are homogeneous and in single cell suspension). Place a cover slip on top of the hemocytometer, and immediately load 20 µL of the cell mixture into the hemocytometer. Count the number of cells within several large squares (i.e., a large square is composed of 25 small squares), verifying the consistency of cell density among the large squares that you have counted. Calculate the average number of cells in a large square, and then multiply that average by 10^4 to obtain the number of cells per mL in the primary cell mixture in the 50-mL tube.

5. Using the cell concentrations determined in **step 4**, prepare a series of 14-mL snap-top tubes for each cell line and dilute each cell line to a concentration convenient for extracting relatively small volumes containing the desired number of cells (<500 µL to avoid significantly diluting the soft agar seeding mixture). Pipette appropriate volumes of dilutions (containing desired numbers of cells) into each of the labeled 14-mL snap-top tubes from **step 3**. For example, to obtain 250, 500, 1,000, and 2,000 cells, one could use the following scheme:

    Using the cell concentration of the original suspension (~10^6 cells/mL), pipette a calculated volume that would contain 10^5 cells (e.g., ~100 µL in this case) to a second tube containing 10 mL of DMEM-20 to get a new concentration of 10^4 cell/ml. Two hundred microliter of this dilution would contain ~2,000 cells, 100 µL would have ~1,000 cells, etc. Further dilute this mixture 1:10 in DMEM-20 in a third tube to obtain a concentration of 10^3 cells/mL, such that 500 µL of this dilution would contain ~500 cells and 250 µL ~250 cells. Be careful to avoid contamination when pipetting between tubes: first, always use an unopened packet of autoclaved pipette tips, and second, make sure that the pipette tip is the only part that touches the culture medium or the inner wall of the tubes. Also, before removing the volume of cells needed for a given dilution (e.g., 500 µL for 500 cells), make sure to mix each dilution tube well (i.e., pipette up and down), and also to mix each dilution by inverting the tightly capped tube several times before removing volumes to the labeled 14-mL snap-top tubes to be used for cell seeding. The more homogeneous the mixtures are, the more accurate the cell counts will be.

6. Five milliliters of DMEM-20 containing 0.3 % agar is necessary for each soft agar dish. Therefore, prepare a DMEM-20/0.3 % agar mixture as in **Note 16**.

7. Pipette 5 mL of the warm agar/DMEM-20 into each labeled snap-top tube, on top of the aliquots of cells. The agar/DMEM-20 mixture can be added to approximately 8–10 tubes at a time (return the agar mixture to the 37–40 °C bath between additions to the tubes). After adding the agar mixture to each snap-top tube, flame the mouth of the tube, and then pour the mixture into the corresponding 60-mm Petri dish, and swirl the dish briefly so that the agar mix covers the bottom of the dish. When all cells have been seeded, leave the Petri dishes at room temperature (without moving) for 10–20 min in order for the soft agar to solidify. The soft agar should assume a jellylike consistency in all dishes. When the soft agar has solidified, cover all dishes together loosely with a Saran wrap to prevent drying and place in a 37 °C tissue culture incubator, which is as humid as possible. As above, do not stretch the Saran wrap tightly across the Petri dishes or it may cause the lids to open partially, which will cause the dishes to dry out over the next couple of weeks in the incubator (*see* **Note 19**).

8. To allow colonies to form, incubate the soft agar dishes for 14 days in a humid tissue culture incubator. As much as possible, do not disturb the dishes during this period, i.e., do not take them out to look at them, as this can cause the agar to dry out and ruin the assay. After 14 days, remove the dishes and assess colony formation. Colonies should be visible and distinguishable to the naked eye. Either count the macroscopic colonies with the naked eye or under a low-power magnifying glass. It is easiest to count the macroscopic colonies by dotting them with a permanent marker on the back of the dish. Record the number of colonies formed for each sample.

9. To calculate soft agar colony formation, take the average number of colonies from the three samples at a given dilution and cell type. Occasionally, the numbers of colonies are not linear at the higher cell dilutions (e.g., 2,000 cells). If they are not

linear, calculate the relative colony formation at a dilution that appears to be in the linear range and normalize to the number of colonies formed by BJAB-MSCV cells (normalized as 1.0; *see* **Notes 24–26**) (Fig. 4b).

4   Notes

1. BJAB cells are not available from ATCC. However, they are a commonly used B-lymphoma cell line and have been distributed widely. BJAB cells are available from one of the authors (T.D.G.) upon request.

2. The A293T cell line, a simian virus 40 (SV40) large T antigen-containing clone of the human endothelial kidney HEK293 line, can be obtained from ATCC. Many academic labs also have these cells.

3. Tissue culture medium (e.g., DMEM) should be stored at 4 °C and warmed to 37 °C in a water bath for approximately 15–20 min prior to use.

4. Stock penicillin/streptomycin can be purchased from Life Technologies (Carlsbad, CA). We generally make 2.5-mL aliquots and freeze them at −20 °C for long-term storage. To obtain a final concentration of 50 units/mL penicillin and 50 units/ml streptomycin, one can add 2.5 mL of stock penicillin/streptomycin to 500 mL of DMEM-10 or DMEM-20.

5. High-quality FBS is essential for the optimal growth of primary chicken spleen cell colonies in agar. The lot of FBS can make the difference between 0 colonies and several hundreds of colonies or a liquid outgrowth assay that works well versus one that does not work at all. We have found that FBS from Biologos (Montgomery, IL) routinely works well in such assays, but others have also had success with certain lots of FBS from Gibco (Life Technologies, Carlsbad, CA). In any case, it is highly recommended to carefully screen and compare different lots of FBS for optimal efficiency in chicken spleen cell soft agar colony assays before purchase. In our experience, optimal FBS lots can increase the number of transformed colonies by up to fourfold compared to FBS lots that we previously considered to be good [25]. Once selected, FBS usually works well for approximately 2 years when stored at −80 °C. We find that new FBS lots must be evaluated using the soft agar assay each time before purchase.

6. We generally prepare stock solutions of puromycin at 500 μg/mL by dissolving the antibiotic in deionized water. Solutions are then filter sterilized (0.22 μm filter), and stock solutions may be stored at 4 or −20 °C for up to 1 year.

7. Ideally, chickens used for these assays should be virus-free and approximately 3 weeks old (18–22 days). This age appears to be optimal for obtaining sufficient numbers of target cells for transformation. We have tried 2- and 4-week-old birds, and they do not work as well in these assays. If the chicks are not going to be used until the afternoon, make sure that they have enough water once they arrive on site. Because these experiments require the use of live animals, it is essential that you obtain the approval of your Institutional Animal Care and Use Committee (IACUC) before starting.

8. pSW253 is a plasmid containing a replication-competent clone of reticuloendotheliosis virus strain A (Rev-A) [20]. Rev-A acts as a helper virus for the replication of SNV-derived expression vectors that are generally used to express the *rel* genes. The *rel* gene vectors can be single gene vectors [24, 30] or two gene vectors containing an internal ribosome entry sequence for expression of the second gene [27, 31].

9. Make up PEI at 1 mg/mL, pH 7.0, in distilled water. To do so, weigh out 10 mg of PEI and add 9.5 mL of sterile distilled water in a 14-ml snap-top tube. PEI can be difficult to get into solution. Therefore, you will need to first rock for about 1 h at room temperature. Then add ~1 drop of concentrated HCl, which will reduce the pH to ~2.0 (you can monitor with pH paper). Rock another 30–45 min at room temperature until the PEI crystals are in solution. Add 35–50 µL of 10 N NaOH, until the pH is about 7.0 (again, monitoring with pH paper). Once dissolved, filter sterilize through a 0.22 µm filter into a fresh 14-mL snap-top tube. Make 1-mL aliquots in sterile 1.5-mL microcentrifuge tubes, and store at 4 °C for short-term use or at –80 °C for long-term storage (good for approximately 1 year).

10. We generally make a polybrene stock at 20 µg/mL in sterile deionized water. From this stock, we will make small (25–50 µL) working aliquots at 2 µg/mL in sterile TD. Both solutions are stored at –20 °C. Working aliquots are used only once or twice before disposing.

11. pCL10a1 is a murine leukemia virus replication-defective packaging vector [21].

12. pMSCV is a widely used murine stem cell virus-derived vector [22]. *See* Fig. 2 for the general structures of the viral components.

13. To locate the chicken spleen, one can consult a book on chicken anatomy (e.g., ref. 32) or do a Google Images Search for a suitable image (e.g., refs. 33, 34). Avian hematopoietic cell transformation assays can also be performed using primary bone marrow-derived cells [14], but we have found it to be more convenient and less technically challenging to use spleen cell cultures.

14. We routinely use cesium chloride-prepared plasmids for chicken spleen cell assays, as they seem to work better than Qiagen column-prepared DNA. Alternatively, if you have a virus stock, you can infect cells for about 1 h (at Subheading 3.1, **step 12**) with virus in about 1 mL of medium containing ~10 µg/mL of polybrene. Incubate cells and virus in a 14-mL snap-top tube for 1 h at 37 °C in a tissue culture incubator (with occasional swirling, e.g., every 20 min). After infection, pellet the cells, resuspend, and seed directly in soft agar (or let cells incubate with virus overnight in about 3 mL of culture medium with serum before centrifugation and seeding into agar as described herein).

15. In general, we have found that the liquid outgrowth assay is more sensitive than the soft agar assay: that is, weakly trans-forming Rel proteins are more likely to score positive in the liquid outgrowth assay than in the soft agar assay [27]. The transforming ability of weakly transforming Rel proteins can also be enhanced by co-expression of the antiapoptotic protein Bcl-2 [27, 28]. To ensure that a given Rel protein is expressed in the electroporated spleen cell culture, we have performed Western blotting on one sample of electroporated cells at 5 days after electroporation (*see* ref. 28).

16. Preparation of culture medium containing 0.3 % agar. The following is for 100 mL or 20 assay dishes (scale up or down proportionately for different numbers of dishes/samples). However, no matter how much total culture medium/agar you are making, always prepare 30 mL of 3 % agar. To do so, weigh out 1 g of Bacto Agar into a 125-mL Erlenmeyer flask and add 30–32 mL of ddH$_2$O. Autoclave for 20–25 min, with slow exhaust. Immediately after autoclaving, place the flask in a 56 °C water bath, for at least 20 min. In a sterile 250-mL Erlenmeyer flask, add 70 mL of DMEM and 20 mL of FBS, and place the flask in a 37 °C water bath for about 15 min to equili-brate. Using sterile technique, add 10 mL of the 3 % agar solu-tion to the warmed 90 mL of DMEM-20, swirl well, and place the flask at 56 °C. (You must not add hot agar to cold DMEM-20 or the agar will congeal and the mixture will not be homog-enous.) About 7–10 min before adding to the cells, place the flask containing DMEM-20/0.3 % agar in a 37–40 °C water bath so that it will cool down a bit before adding to cells. This is a tricky part of the procedure: you do not want to kill the cells with hot agar, but you also do not want the agar to cool such that it begins to solidify. Often, it is a good idea to put the flask at 37–40 °C while the cells are undergoing centrifugation.

17. In our experience, we have been able to further increase the efficiency of colony formation in soft agar by up to twofold by dividing cells from individual transforming reactions at this stage into four 60-mm Petri dishes to avoid nutrient depletion

and excessive acidification due to high colony density. Under such conditions, wild-type v-Rel can yield on average 435 colonies per transforming reaction [25].

18. In our hands, the chicken spleen cell soft agar assays work best with the incubator temperature at 40.5 °C, probably because it accelerates the growth of the colonies in agar (not allowing the agar to dry out during the course of the assay).

19. Label the dishes on the bottoms, since the marking pen ink can sometimes smudge due to the Saran wrap. That is, lay the Saran wrap over the plates so that it is quite loose (almost tent-like), but do not stretch it tightly over the plates. (If you stretch the Saran wrap too tightly, it sometimes will open the plates a bit in the incubator, which can cause the agar to dry out; this will ruin the assay because cells will not expand into colonies if the agar is too dry.)

20. Because these procedures involve working with human lymphoma cells (BJAB) and retroviral vectors that can infect human cells (and these viruses may carry oncogenic genes), it is required by federal law and NIH guidelines that you take certain safety precautions and perform experiments at the BL-2 safety level.

    (a) Experiments should be approved by your Institutional Biosafety Committee.

    (b) Always wear a lab coat, safety glasses (or goggles), and gloves when working with virus-infected cultures and materials that have touched human cells or contain virus stocks.

    (c) All materials that have touched virus or virus-infected cells must be discarded in appropriate sterilization bags, which will later be autoclaved before disposal.

    (d) When finished aspirating culture medium or liquid from infected cell cultures or from other containers into the aspirator trap used for tissue culture, always disinfect the trap by adding bleach (10 %), and let it stand for at least 30 min. When emptying the trap into the sink, be sure to run many volumes of water (e.g., for 10–15 min) to flush trap contents down the drain.

    (e) Wipe up the tissue culture work area with 70 % ethanol when you are finished using that area.

21. The polybrene working stock is 2 µg/µL, therefore, add 8 µL to the 2 mL of virus inoculum.

22. When thawing a virus stock, it is best to thaw quickly at 37 °C in a water bath, and then use it as soon as it has thawed (do not thaw and leave at room temperature for a long time before use, or leave at 37 °C for an extended period of time before

use). Avoid thawing and refreezing retroviral stocks more than once (one freeze/thaw cycle can reduce the titer by two- to threefold, two freeze/thaw cycles can reduce the titer another five- to tenfold, i.e., render them useless).

23. We have found that high-level expression of RELΔTAD1 in BJAB cells is quite stable, once an appropriate cell culture/clone has been identified. The BJAB-RELΔTAD1 cells used in Chin et al. [9] have retained high-level expression of RELΔTAD1 for several years since they were isolated (e.g., *see* Fig. 4a).

24. Because the number of cells seeded is rather low (e.g., 250–2,000), it is often good to have a second measure of starting cell number. This can be done by measuring and comparing the protein content of your original cell mixture (i.e., before dilution) (as in ref. 9). To do so, remove a standard volume (e.g., 2 mL) of each of the pre-dilution cell line suspension. Centrifuge the cells, wash a few times with cold PBS, and lyse cells in a standard detergent lysis buffer (see ref. 9). Perform a standard protein assay (e.g., Bradford assay). The ratios of protein in different samples should be similar to the ratios of cells obtained by cell counting with the hemocytometer (Subheading 3.3, **step 4**, above).

25. The reason why BJAB cells are sensitive to the transforming activity of RELΔTAD1 is not known, and indeed, we have not seen a difference in soft agar colony formation in several other B-lymphoma cell lines when RELΔTAD1 is expressed. In soft agar assays similar to those we describe herein, BJAB cells have been shown to be susceptible to the oncogenic effects of other factors, including the Epstein–Barr virus LMP1 protein [35, 36] and small RNAs [37] and the AP12-MALT1 fusion protein encoded by chromosomal translocations in some human lymphomas [38].

26. To calculate the efficiency of soft agar colony formation by BJAB cells, see the following (idealized) example:

| Cell line | Avg. # colonies at the indicated number of cells seeded | | | |
	250	500	1,000	2,000
BJAB-MSCV	10	20	35	60
BJAB-RELΔTAD1	25	50	60	90
Relative colony #	2.5	2.5	1.7	1.5

Therefore, in this assay, we would use values from the dishes seeded with 250 and 500 cells and report a 2.5-fold increase in relative colony formation by BJAB-RELΔTAD cells (*see* Fig. 4b).

Acknowledgments

Research in the authors' laboratories on transformation of cells by Rel proteins was supported by NIH grants CA047763 (T.D.G.) and CA054999 (C.G.).

References

1. Gilmore TD (2006) Introduction to NF-κB: players, pathways, perspectives. Oncogene 25:6680–6684

2. Bassères DS, Baldwin AS (2006) Nuclear factor-κB and inhibitor of κB kinase pathways in oncogenic initiation and progression. Oncogene 25:6817–6830

3. Gilmore TD (1999) Multiple mutations contribute to the oncogenicity of the retroviral oncoprotein v-Rel. Oncogene 18:6925–6937

4. Carrasco D, Rizzo CA, Dorfman K, Bravo R (1996) The v-rel oncogene promotes malignant T-cell leukemia/lymphoma in transgenic mice. EMBO J 15:3640–3650

5. Courtois G, Gilmore TD (2006) Mutations in the NF-κB signaling pathway: implications for human disease. Oncogene 25:6831–6843

6. Fan Y, Gélinas C (2007) An optimal range of transcription potency is necessary for efficient cell transformation by c-Rel to ensure optimal nuclear localization and gene-specific activation. Oncogene 26:4038–4043

7. Fan Y, Rayet B, Gélinas C (2004) Divergent C-terminal transactivation domains of Rel/NF-κB proteins are critical determinants of their oncogenic potential. Oncogene 23:1030–1042

8. Romieu-Mourez R, Kim DW, Shin SM, Demicco EG, Landesman-Bollag E, Selden DC et al (2003) Mouse mammary tumor virus c-rel transgenic mice develop mammary tumors. Mol Cell Biol 23:5738–5754

9. Chin M, Herscovitch M, Zhang N, Waxman DJ, Gilmore TD (2009) Overexpression of an activated version of the REL oncoprotein enhances the transformed state of the human B-lymphoma BJAB cell line and alters its gene expression profile. Oncogene 28:2100–2111

10. Hoelzer JD, Lewis RB, Wasmuth CR, Bose HR Jr (1980) Hematopoietic cell transformation by reticuloendotheliosis virus: characterization of the genetic defect. Virology 100:462–472

11. Kralova J, Schatzle JD, Bargmann W, Bose HR Jr (1994) Transformation of avian fibroblasts overexpressing the c-rel proto-oncogene and a variant of c-rel lacking 40 C-terminal amino acids. J Virol 68:2073–2083

12. Moore BE, Bose HR Jr (1988) Expression of the v-rel oncogene product in reticuloendotheliosis virus-transformed fibroblasts. Virology 162:377–387

13. Morrison LE, Boehmelt G, Beug H, Enrietto P (1991) Expression of v-rel in a replication competent virus: transformation and biochemical characterization. Oncogene 6:1657–1666

14. Boehmelt G, Walker A, Kabrun N, Mellitzer G, Beug H, Zenke M et al (1992) Hormone-regulated v-rel estrogen receptor fusion protein: reversible induction of cell transformation and cellular gene expression. EMBO J 11:4641–4652

15. Capobianco AJ, Gilmore TD (1993) A conditional mutant of vRel containing sequences from the human estrogen receptor. Virology 193:160–170

16. White DW, Roy A, Gilmore TD (1995) The v-Rel oncoprotein blocks apoptosis and proteolysis of IκB-α in transformed chicken spleen cells. Oncogene 10:857–863

17. White DW, Gilmore TD (1996) Bcl-2 and CrmA have different effects on transformation, apoptosis, and the stability of IκB-α in chicken spleen cells transformed by temperature-sensitive v-Rel oncoproteins. Oncogene 13:891–899

18. Zong W-X, Farrell M, Bash J, Gélinas C (1997) v-Rel prevents apoptosis in transformed lymphoid cells and blocks TNFα-induced cell death. Oncogene 15:971–980

19. Madruga J, Briegel K, Diebold S, Boehmelt G, Vogl F, Zenke M (2000) Dendritic cells conditionally transformed by v-relER oncogene express lymphoid marker genes. Immunobiology 202:394–407

20. Watanabe S, Temin HM (1983) Construction of a helper cell line for avian reticuloendotheliosis virus cloning vectors. Mol Cell Biol 3:2241–2249

21. Naviaux RK, Costanzi E, Haas M, Verma IM (1996) The pCL vector system: rapid production of helper-free high-titer, recombinant retroviruses. J Virol 70:5701–5705

22. Hawley RG, Lieu FHL, Fong AZC, Hawley TS (1994) Versatile retroviral vectors for potential use in gene therapy. Gene Ther 1:136–138

23. Mosialos G, Hamer P, Capobianco AJ, Laursen R, Gilmore TD (1991) A protein kinase A recognition sequence is structurally linked to transformation by p59^{v-rel} and cytoplasmic retention of p68^{c-rel}. Mol Cell Biol 11:5867–5877

24. Sif S, Capobianco AJ, Gilmore TD (1993) The vRel oncoprotein increases expression from Sp1 site-containing promoters in chicken embryo fibroblasts. Oncogene 8:2501–2509

25. Rayet B, Fan Y, Gélinas C (2003) Mutations in the v-Rel transactivation domain indicate altered phosphorylation and identify a subset of NF-κB-regulated cell death inhibitors important for v-Rel transforming activity. Mol Cell Biol 23:1520–1533

26. White DW, Pitoc G, Gilmore TD (1996) Interaction of the v-Rel oncoprotein with NF-κB and IκB proteins: heterodimers of a transformation-defective v-Rel mutant and NF-κB p52 are functional in vitro and in vivo. Mol Cell Biol 16:1169–1178

27. Gilmore TD, Cormier C, Jean-Jacques J, Gapuzan M-E (2001) Malignant transformation of primary chicken spleen cells by human transcription factor c-Rel. Oncogene 20:7098–7103

28. Starczynowski DT, Reynolds JG, Gilmore TD (2003) Deletion of either C-terminal transactivation subdomain enhances the *in vitro* transforming activity of human transcription factor REL in chicken spleen cells. Oncogene 22:6928–6936

29. Gilmore TD, Jean-Jacques J, Richards R, Cormier C, Kim J, Kalaitzidis D (2003) Stable expression of the avian retroviral oncoprotein v-Rel in avian, mouse, and dog cell lines. Virology 316:9–16

30. Dougherty JP, Temin HM (1986) High mutation rate of a spleen necrosis virus-based retrovirus vector. Mol Cell Biol 6:4387–4395

31. Koo HM, Brown AM, Kaufman RJ, Prorock CM, Ron Y, Dougherty JP (1992) A spleen necrosis virus-based retroviral vector which expresses two genes from a dicistronic mRNA. Virology 186:669–675

32. Koch T, Rossa E, Skold BH, DeVries L (1973) Anatomy of the chicken and domestic birds. Press, Iowa State University

33. http://tbnranch.com/2012/01/15/chicken-anatomy/

34. http://www.poultryhub.org/physiology/body-systems/digestive-system/

35. Enberg I, Klein G, Biovanella BC, Stehlin J, McCormick KJ, Andersson-Anvret M et al (1983) Relationship between the amounts of EBV-DNA and EBNA per cell, clonability and tumorigenicity in two EBV-negative lymphoma lines and their EBV-converted cell lines. Int J Cancer 31:163–169

36. Wennborg A, Aman P, Saranath D, Pear W, Sümegi J, Klein G (1987) Conversion of the lymphoma cell line "BJAB" by Epstein-Barr virus into phenotypically altered sublines is accompanied by increased c-*myc* RNA levels. Int J Cancer 40:202–206

37. Yamamoto N, Takizawa T, Iwanaga Y, Shimizu N, Yamamoto N (2000) Malignant transformation of B lymphoma cell line BJAB by Epstein–Barr virus-encoded small RNAs. FEBS Lett 484:153–158

38. Ho L, Davis RE, Conne B, Chappuis R, Berczy M, Mhawech P et al (2005) MALT1 and the AP12-MALT1 fusion act between CD40 and IKK and confer NF-κB-dependent proliferative advantage and resistance against FAS-induced cell death in B cells. Blood 105:2891–2899

39. Ishikawa H, Carrasco D, Claudio E, Ryseck RP, Bravo R (1997) Gastric hyperplasia and increased proliferative responses of lymphocytes in mice lacking the COOH-terminal ankyrin domain of NF-κB2. J Exp Med 186:999–1014

40. Ciana P, Neri A, Cappellini C, Cavallo F, Pomati M, Chang C-C et al (1997) Constitutive expression of lymphoma-associated NFKB-2/Lyt-10 proteins is tumorigenic in murine fibroblasts. Oncogene 14:1805–1810

Chapter 27

Using RNA Interference in Lung Cancer Cells to Target the IKK-NF-κB Pathway

Daniela S. Bassères and Albert S. Baldwin

Abstract

RNA interference-based gene silencing has become a widely used technology to evaluate how inhibition of expression of individual proteins affects biological readout. Through the use of this technology, a lot has been learned about how different proteins function in a wide variety of biological contexts, including cancer. In this context, RNA interference-mediated gene silencing has contributed to further our understanding of how different proteins in the NF-κB signaling pathway (including the NF-κB members themselves) contribute to cancer. Here, we describe two RNA interference-based protocols in lung cancer cells targeting upstream activators of NF-κB transcription factor: the catalytic subunits of the IKK complex. The first protocol is designed to evaluate the impact of IKKα or IKKβ inhibition on NF-κB transcriptional activity, whereas the second protocol is designed to evaluate how siRNA-mediated IKK inhibition affects lung cancer cell proliferation.

Key words RNA interference, Transfection, Luciferase assay, BrdU incorporation assay

1 Introduction

Since its discovery in 1998 [1], RNA interference has become an important tool to evaluate gene function. Through the artificial introduction of small interfering RNAs (siRNAs) into mammalian cells, it is possible to achieve specific and efficient inhibition of expression of desired target genes [2]. siRNAs are recognized by the RNA-induced silencing complex (RISC), a protein complex that guides siRNAs to mRNAs with complementary sequences and, through the action of the protein Argonaute, mediates mRNA cleavage, thereby preventing translation [3]. The use of this technology to investigate the NF-κB pathway and function in cancer cells has proven to be very important (examples can be found in refs. 4–10). This allows not only for functional analysis of how individual proteins in the NF-κB pathway contribute to oncogenesis but also it validates that results obtained with pharmacological inhibitors of this pathway are not due to off-target effects.

Michael J. May (ed.), *NF-kappa B: Methods and Protocols*, Methods in Molecular Biology, vol. 1280, DOI 10.1007/978-1-4939-2422-6_27, © Springer Science+Business Media New York 2015

NF-κB is a dimeric transcription factor formed by members of a highly conserved family of proteins comprised of p65(RelA), RelB, c-Rel, p50 (NF-κB1), and p52(NF-κB2). NF-κB is expressed ubiquitously, and the most abundant NF-κB complexes in most cell types are heterodimers of p65 and p50 [11, 12].

Even though ubiquitously expressed, in most non-stimulated cells, NF-κB dimers are held by the inhibitor of κB (IκB) proteins in an inactive cytoplasmic form. Canonical activation of NF-κB involves phosphorylation of IκB by the IκB kinase (IKK) complex. The IKK complex is comprised of three subunits: IKKγ (Nemo), IKKα, and IKKβ. Nemo regulates assembly of the complex, whereas IKKα and IKKβ have catalytic activity, with IKKβ being the main component responsible for NF-κB activation. Upon phosphorylation by IKK, IκB proteins undergo rapid ubiquitination and proteasome-mediated degradation. The NF-κB subunits are then released and translocated to the nucleus where they regulate gene transcription [11, 12]. In the noncanonical pathway, IKKα phosphorylates the p100 precursor of p52, leading to its processing and the activation of a p52-RelB heterodimer [12].

NF-κB regulates genes involved in the oncogenic phenotype, including those involved in promoting cell proliferation, survival, angiogenesis, and stemness [11, 13]. Therefore, its deregulated activation can contribute to oncogenic transformation, and constitutive NF-κB activation has been detected in a large variety of human malignancies [11].

Here, we describe the transfection protocol used to evaluate in lung cancer cells the impact of inhibition of expression of the upstream catalytic kinases of the canonical NF-κB pathway IKKα and IKKβ on NF-κB activity. We also describe a protocol to evaluate how IKK knockdown affects cell proliferation.

2 Materials

All solutions should be prepared in ultrapure water prepared by purifying deionized water to attain a sensitivity of 18 MΩ cm at 25 °C.

2.1 Cell Culture Reagents

1. Lung cancer cell lines: lung cancer cell lines H358 (ATCC® CRL-5807™) and A549 (ATCC® CCL-185™) can be purchased from ATCC (Manassas, VA, USA) and maintained according to the provider's instructions.

2. Cell culture medium: sterile serum-free RPMI-1640 can be obtained from different vendors. To prepare, complete RPMI with 10 % fetal bovine serum (FBS), and add 50 mL of sterile heat-inactivated FBS (which can also be obtained from different vendors) to 500 mL of sterile RPMI-1640. This procedure should be performed under a laminar flow hood using aseptic technique [14].

3. 1× phosphate buffered saline (PBS): 137 mM NaCl, 2.7 mM KCl, 8 mM Na_2HPO_4, and 1.4 mM KH_2PO_4. Weigh 8 g NaCl, 0.2 g KCl, 1.14 g Na_2HPO_4 (anhydrous salt), and 0.19 g KH_2PO_4. Dissolve salts in 800 mL of ultrapure water, adjust the pH to 7.4 with HCl, and transfer to a cylinder. Adjust the volume to 1 L and transfer to a glass bottle. Autoclave on liquid cycle for 30 min. Alternatively, tissue culture grade sterile 1× PBS can be obtained from different vendors.

4. 0.2 M EDTA (disodium ethylenediaminetetraacetate·$2H_2O$) solution: weigh 37.2 g EDTA and dissolve in 400 mL of water in a 500 mL beaker. Adjust the pH to 8.0 with NaOH. (*NOTE*: EDTA will only enter into solution at pH~8.0). Adjust the volume to 500 mL and store at room temperature.

5. 0.05 % trypsin-EDTA: 0.05 % trypsin, 1 mM EDTA prepared in 1× PBS buffer. Weigh 0.5 g trypsin and dissolve in 800 mL of 1× PBS in a 1 L beaker. Add 5 mL of 0.2 M EDTA solution, and adjust the pH to 7.2–7.4 using either 1 M HCl or 1 M NaOH. Adjust the volume to 1 L and filter solution in a laminar flow hood (using a 500 mL 0.2 μm filter) into a sterile 1 L bottle. Incubate 1 mL of solution in a 35 mm dish at 37 °C overnight to test for sterility. Aliquot 45 and 12.5 mL into 50 and 15 mL sterile tubes, respectively, and store at –20 °C until use. Alternatively, sterile 0.05 % trypsin-EDTA solution can be purchased from different vendors.

6. Dishes and plates: 500 mL 0.2 μm filter units, standard 10 cm tissue culture dishes, 24-well plates, and 96-well plates, as well as white opaque 96-well plates.

7. 0.4 % Trypan Blue solution.

2.2 Transfection Reagents

1. IKKα and IKKβ siRNAs: we have successfully used siGENOME SMARTpool siRNAs for IKKα and IKKβ (Dharmacon/GE Healthcare, Mickleton, NJ, USA), each containing four different siRNAs targeting these kinases, as well as sets of the four separate siRNAs used in the SMARTpool for IKKα and IKKβ. This allows results obtained with the SMARTpools to be validated with individual siRNAs. For negative control, we recommend siGENOME Non-Targeting siRNA #3 (Dharmacon/GE Healthcare).

2. Sterile 5× siRNA buffer (Dharmacon/GE Healthcare). This buffer can also be prepared in the laboratory (300 mM KCl, 30 mM HEPES-pH 7.5, and 1 mM $MgCl_2$). This buffer should be diluted to 1× siRNA buffer with RNAse-free ultrapure water under aseptic conditions.

3. DharmaFECT 1 transfection reagent (Dharmacon/GE Healthcare).

4. Reporter plasmid vectors for NF-κB transcriptional activity: luciferase reporter vectors containing three copies of the

NF-κB binding site found in the major histocompatibility complex (MHC) class I promoter driving luciferase expression (3x-κB-WT vector) as well as a mutant version where the NF-κB binding sites were mutated (3X-κB-MUT vector) have been described [15] and are available from the Baldwin lab upon request.

5. Additional plasmids: pRL-TK (Promega, Madison, WI, USA) drives expression of *Renilla* luciferase. pcDNA3 vector or equivalent mammalian expression empty vector.

6. Lipofectamine LTX transfection reagent with plus reagent (Life Technologies, Carlsbad, CA, USA).

7. Opti-MEM 1 Reduced-Serum Medium (Life Technologies).

2.3 Dual-Luciferase Reporter Assay System

This system (Promega) allows for sequential measurement in a single sample of the activities of two different luciferases: the firefly (*Photinus pyralis*) luciferase (which in this experiment is expressed from the NF-κB reporter vectors) and *Renilla* (*Renilla reniformis*) luciferase (expressed from the pRL-TK vector used for transfection normalization). This is achieved by first measuring firefly luciferase activity using the Luciferase Assay Reagent II (LAR II) substrate, followed by quenching of firefly luminescence, and measurement of *Renilla* luciferase activity, both achieved using the Stop & Glo Reagent.

2.4 BrdU Cell Proliferation Assay

5-Bromo-2′-deoxyuridine (BrdU) is a thymidine analog that can be readily used during cellular DNA synthesis. Therefore, the amount of BrdU incorporated into DNA in a given amount of time is proportional to the level of DNA synthesis occurring in the cell and, therefore, of cellular proliferation. We describe a method using the EMD Millipore BrdU cell proliferation assay (Millipore, Billerica, MA, USA) that uses a solid-phase (cells grown in plates treated with BrdU for a fixed amount of time) colorimetric enzyme-linked immunoassay to quantitate the level of incorporated BrdU in cell samples.

1. The assay kit contains the following components: BrdU label, Fixative/Denaturing Solution, BrdU antibody, Antibody Diluent, Peroxidase Goat Anti-Mouse IgG, Conjugate Diluent, Substrate, Plate Wash Concentrate, and Stop Solution.

2. Prepare 1× Wash Buffer by diluting the 50× Wash Buffer provided in the kit with ultrapure water.

2.5 Equipments

1. Laminar flow hood.

2. CO_2 incubator.

3. Plate luminometer with injectors.

4. Microplate reader.

5. Hemocytometer.

6. Inverted microscope.

7. Swinging bucket tabletop centrifuge.

8. UV spectrophotometer.

3 Methods

3.1 Part 3.1: Targeting NF-κB Activity Through siRNA-Mediated Knockdown of IKKα or IKKβ Expression

3.1.1 Preparing Cells for the First Transfection

All procedures should be performed under a laminar flow hood unless otherwise noted, and the aseptic technique [14] should be followed. The following protocol works for both lung cancer cell lines H358 and A549.

1. Cells should be maintained in 10 cm dishes prior to transfection, at a cell density that allows for exponential growth.

2. The day before transfection, cells should be detached. In order to detach cells, the culture medium should be vacuum aspirated using a sterile Pasteur pipette or aspiration pipette. The 10 cm culture dish should be washed once with 2 mL sterile 1× PBS and after PBS removal, 2 mL of 0.05 % trypsin-EDTA solution should be added to the plate. The plate should be returned to the CO_2 incubator for 5 min to allow cells to detach.

3. After 5 min, cell detachment should be monitored by checking cells under an inverted microscope, and if necessary, cells should be incubated longer.

4. Afterwards, the plate should be returned to the laminar flow hood, and 8 mL of RPMI-1640 medium supplemented with 10 % FBS should be added to the plate and mixed by pipetting up and down.

5. The detached cells should then be transferred to 50 mL sterile conical tubes and centrifuged at $100 \times g$ for 5 min in a swinging bucket centrifuge.

6. The supernatant should be removed by aspiration and the cell pellet resuspended in 5 mL of RPMI-1640 medium supplemented with 10 % FBS.

7. 10 μL of the cell suspension should be mixed with 10 μL Trypan Blue solution, transferred to a hemocytometer, and the number of viable cells determined under an inverted microscope by counting cells that exclude the Trypan Blue dye.

8. A solution with 8×10^4 cells/mL for A549 cells or 15×10^4 cells/mL for H358 cells should be prepared in a 50 mL conical tube, and 0.5 mL should be plated in each well (4×10^4 cells/well for A549 cells or 7.5×10^4 cells/well for H358 cells) of a 24-well plate (*see* **Note 1**).

The resuspension protocol provided by the manufacturer should be followed as described below.

1. The siRNAs should be centrifuged before resuspension.

2. The siRNAs should be resuspended in 1× siRNA buffer. To make 1× siRNA buffer, add four parts of ultrapure sterile water to one part of sterile 5× buffer under a laminar flow hood. Follow the datasheet instructions to make a 20 μM stock solution. The volume of 1× siRNA buffer will depend on the amount (μmol) of siRNA purchased.

3. Pipette up and down five times to mix and incubate on a shaker at $100 \times g$ for 30 min.

4. Briefly spin the tube and confirm stock concentration by determining the optical density using a UV spectrophotometer.

5. Dilute the 20 μM siRNA stock to 2 μM with 1× siRNA buffer. Make enough for the experiment at hand and dilute fresh for each new experiment.

All steps should be performed under a laminar flow hood using aseptic technique [14] unless otherwise noted.

1. For each well to be transfected, mix 25 μL of 2 μM siRNA stock with 25 μL serum-free RPMI medium (multiplied by 7 for triplicate experiment—*see* **Note 2**). For untransfected controls, substitute 25 μL of 2 μM siRNA for 25 μL of 1× siRNA buffer.

2. In a separate tube, add 1 μL of DharmaFECT 1 to 49 μL of serum-free RPMI medium per transfected well. Make enough for all wells (*see* **Note 3**).

3. Incubate for 5 min at room temperature.

4. Add 50 μL of the DharmaFECT solution prepared in **step 2** to 50 μL of the siRNA solution prepared in **step 1** (multiplied by 3.5 or 7 for triplicate samples—*see* **Notes 2** and **3**) and mix by pipetting up and down gently.

5. Incubate for 20 min at room temperature. While incubating, aspirate the medium from the cells plated in 24-well plates the day before.

6. Add 400 μL of complete RPMI medium for each sample prepared in **step 4** (multiplied by 3.5 or 7 for triplicate samples—*see* **Notes 2** and **3**), and mix by pipetting up and down.

7. Add 500 μL of the transfection cocktail per well of the 24-well plate, gently swirl the plate, and place in CO_2 incubator overnight.

8. Replace the transfection medium with 500 μL of fresh medium 24 h after the first transfection.

*3.1.4 Preparing
the Plasmid Vectors
for Transfection*

Dilute the 3X-κB-WT and 3X-κB-MUT plasmid vectors to 0.1 μg/μL and the *Renilla* luciferase-expressing vector for normalization pRL-TK to 0.01 μg/μL. Dilute the negative control vector (pcDNA3 or equivalent) to 0.5 μg/μL. All dilutions should be made with ultrapure water.

*3.1.5 Performing
the Second Transfection*

1. The second transfection should be performed 48 h after the first transfection.

2. For the negative control sample, add 1 μL of 0.5 μg/μL pcDNA3 (500 ng) to 99 μL of Opti-MEM medium (multiplied by 3.5 for triplicate samples).

3. For all other samples, add 0.6 μL of 0.5 μg/μL pcDNA3 (300 ng), 2 μL of either the 3X-κB-WT or 3X-κB-MUT plasmid vectors (200 ng), and 1 μL of 0.01 μg/μL pRL-TK (10 ng) to 96.4 μL of Opti-MEM medium (multiplied by 3.5 for triplicate technical replicas and by the number of samples for each reporter vector—*see* **Note 4**).

4. Mix Plus Reagent and add 0.35 μL to each sample prepared in **steps 2** and **3** (multiplied by 3.5 for triplicate technical replicas and by the total number of samples in each case—*see* **Note 5**) and incubate for 5 min at room temperature.

5. Add 2.5 μL of lipofectamine LTX to each sample prepared in **step 4** (multiplied by 3.5 for triplicate technical replicas and by the total number of samples in each case—*see* **Note 5**) and incubate for 25 min at room temperature.

6. While incubating, change the medium of the 24-well plates, replacing with 500 μL of fresh complete RPMI-1640.

7. Add 100 μL of the transfection cocktails prepared in **step 4** to each appropriate well of the 24-well plate.

8. Gently swirl the plate and place in a CO_2 incubator for approximately 16 h (*see* **Note 6**).

*3.1.6 Preparing Lysates
for Luciferase Assay*

These steps can be performed on the benchtop. Aseptic technique is not required.

1. Remove plate from incubator and aspirate the culture medium.

2. Wash twice with 500 μL/well 1× PBS. Make sure to aspirate the last wash twice so there will be no leftover PBS in the well.

3. Add 200 μL/well of 1× Passive Lysis Buffer (PLB, diluted from 5× Passive Lysis Buffer provided in the Dual-Luciferase Assay kit). The dilution should be performed using ultrapure water.

4. Place the 24-well plate in a shaker and shake at high speed for approximately 1 h. At this point, the plates can be stored at −80 °C for posterior analysis if desired. Alternatively, the samples may be used immediately to measure luciferase activity.

3.1.7 Evaluating
Luciferase Activity

These steps can be performed on the benchtop. Aseptic technique is not required.

1. When you wish to proceed with the luciferase assay, warm the Luciferase Assay II (LAR II) Buffer and the Stop & Glo Buffer to room temperature and prepare the LAR II substrate and Stop & Glo Reagent as per manufacturer's instructions.

2. Transfer 20 µL of the cell lysate in each well to individual wells of a white opaque 96-well plate. Don't forget to make three blank wells by adding 20 µL of 1× PLB instead of cell lysates.

3. Bring the plate to the plate luminometer and place the first injector into the LAR II substrate and the second injector into the Stop & Glo substrate (both prepared in **step 1**).

4. Program the luminometer for dual read, prime the injectors, and program both injectors to inject 100 µL of susbtrate/well with a 10 s integration time and a 1.6 s delay after the second injection.

5. Assign the identity of each group of wells, including the blank wells.

6. If feasible, program the luminometer to display both luciferase readings and the average firefly/*Renilla* ratio.

7. Normalize the ratios obtained to the negative control sample (sample transfected with empty vector only).

3.2 Part 3.2:
Evaluating How NF-κB
Inhibition Affects Cell
Proliferation

This protocol is similar to the one described in Part 3.1 in that cells are transfected with siRNAs prior to analysis. However, here, cells are not transfected with luciferase vectors and the transfection format is different.

3.2.1 Preparing Cells
for siRNA Transfection

All procedures should be performed under a laminar flow hood unless otherwise noted and the aseptic technique [14] should be followed.

1. Follow **steps 1–7** of protocol in Subheading 3.1.1.

2. A solution with 2×10^5 cells/mL cells should be prepared in a 50 mL conical tube, and 0.1 mL should be plated in each well (2×10^4 cells/well) of a 96-well plate. Make replica plates for different timepoints (0, 48, 72, and 96 h post-transfection— *see* **Notes 7** and **8**).

3.2.2 Preparing
the siRNAs for Transfection

Follow the protocol described in Subheading 3.1.2.

3.2.3 Performing siRNA
Transfection for BrdU
Incorporation Analysis

All steps should be performed under a laminar flow hood using aseptic technique [14] unless otherwise noted. This procedure should be performed for all plates, except the 0 h timepoint plate, which will serve as a baseline for proliferation.

1. For each well to be transfected, mix 5 µL of 2 µM siRNA stock with 5 µL serum-free RPMI medium (multiplied by 3.5 for triplicate technical replicas and by 3—number of replica plates to be transfected). For untransfected controls, substitute 5 µL of 2 µM siRNA for 5 µL of 1× siRNA buffer.

2. In a separate tube, add 0.4 µL of DharmaFECT 1 to 9.6 µL of serum-free RPMI-1640 medium per transfected well. Make enough for all wells (*see* **Note 9**).

3. Incubate for 5 min at room temperature.

4. Add 10 µL of the DharmaFECT solution prepared in **step 2** to 10 µL of the siRNA solution prepared in **step 1** (multiplied by 3.5 for triplicate technical replicas and by 3 which is the number of replica plates to be transfected) and mix by pipetting up and down gently.

5. Incubate for 20 min at room temperature. While incubating, remove the medium from the cells plated in 96-well plates the day before.

6. Add 80 µL complete RPMI medium for each sample prepared in **step 4** (multiplied by 3.5 for triplicate technical replicas and by 3—number of replica plates to be transfected), and mix by pipetting up and down.

7. Add 100 µL of the transfection cocktail per well of the 96-well plate, gently swirl the plate, and place in a CO_2 incubator for the desired time frame (48, 72, or 96 h).

3.2.4 Preparing Transfected Cells for BrdU Incorporation Analysis

1. At the desired timepoint after transfection, change the plate medium adding fresh complete RPMI-1640 medium supplemented with 2 µL BrdU label/10 mL medium. The BrdU solution is provided with the BrdU incorporation kit. Incubate with BrdU for 2 h in a CO_2 incubator (*see* **Note 10**).

   All subsequent steps can be performed on a regular benchtop. Aseptic technique is not required.

2. Bring the Fixing Solution provided with the BrdU incorporation assay kit to room temperature.

3. Remove medium from plate, and tap the plate over a paper towel to remove excess medium.

4. Add 200 µl Fixing Solution in each well and incubate for 30 min at room temperature.

5. Invert the plate over a sink, and tap the plate onto a paper towel to remove the Fixing Solution.

6. The plate can be wrapped in paraffin and stored at 4 °C for up to a week.

3.2.5 BrdU Incorporation Analysis

When all the timepoints have been collected, the plates can be removed from 4 °C, and the non-isotopic enzyme immunoassay for the quantification of BrdU incorporation can be performed.

1. Prepare the required amount of 1× Wash Buffer (*see* **Note 11**).

2. Wash plate three times with 1× Wash Buffer. To wash plates, a plate washer can be used. Otherwise, washing can be done manually by completely filling the wells of the plate (~300 μL/ well) and then inverting the plate over a sink. After the last wash, tap the plate over a paper towel to remove excess 1× Wash Buffer.

3. Dilute the 100× anti-BrdU antibody in Antibody Diluent 1:100 (both provided with the assay), and pipette 100 μL/well of the diluted antibody.

4. Incubate plate for 1 h at room temperature.

7. Remove antibody solution by tapping the plate on a paper towel, and wash plate three times with 1× Wash Buffer as described in **step 2**.

5. Reconstitute the Peroxidase Goat Anti-Mouse IgG with the appropriate volume of 1× PBS (depends on the size of the assay obtained), and dilute in Conjugate Diluent according to the manufacturer's instructions (varies according to lot). Once diluted, this solution must be filtered using a 0.22 μm syringe filter.

6. Pipette 100 μL of this solution into each well and incubate for 30 min at room temperature.

7. Remove Conjugate Solution by tapping the plate on a paper towel, and wash plate three times with 1× Wash Buffer as described in **step 2**.

8. Flood the entire plate with deionized water. Invert plate and tap on paper towels to remove excess water.

9. Add 100 μL of Substrate Solution (provided in the assay) into each well, and incubate for 15 min at room temperature in the dark. Positive samples will turn blue.

10. Add 100 μL of Stop Solution to stop the reaction. Blue samples will turn yellow.

11. Measure absorbance in each well using a spectrophotometer plate reader at 450 nm. Read plate within 30 min after adding Stop Solution (*see* **Note 10**).

12. Assign the identity of each group of wells, including the blank wells (*see* **Note 12**).

13. Normalize the average absorbance of each sample to the average obtained in the untransfected control plate (0 h time-point). Plot the proliferation curve of untransfected samples, samples transfected with siRNA negative control #3, and samples transfected with each test siRNA as a function of time (0, 48, 72, and 96 h).

4   Notes

1. To determine the number of wells needed for an experiment and, therefore, the total volume of cell solution needed, multiply the number of siRNAs to be transfected by 3 (experiment should be performed in triplicate), and then multiply by 2 to account for the two different reporter vectors (3X-κB-WT and 3X-κB-MUT) that will be used in the second transfection. Finally, add the negative control samples (also multiplied by 3): one sample that should not be transfected with siRNAs or reporter vectors and two samples that should be transfected with reporter vectors only (3X-κB-WT or 3X-κB-MUT). The number of wells should then be: $(\#siRNAs \times 6 \times 2) + 9$.

2. Three technical replicates should be done for each condition after both transfections. In order to achieve this, each siRNA has to be transfected six times: three samples to be transfected in the second round with the 3X-κB-WT vector and three samples to be transfected with the 3X-κB-MUT vector. To account for pipetting errors, multiply required volumes by 7.

3. To make enough DharmaFECT solution, multiply the volumes required for a single well by the following calculation: Volume $\times \{[(\#$ siRNA x $7) + (\#$ controls $\times 3.5)] + 10$ %$\}$. The extra 10 % is included to account for pipetting errors.

4. The number of samples for each reporter vector can be calculated as follows: $(\#$ siRNAs $+ 1$ plasmid only control).

5. For the negative control sample, multiply the volume required by 3.5 for triplicate technical replicas. For the reporter vector samples, multiply the volume required by the number of samples calculated for each reporter vector (*see* **Note 4**) and by 3.5 for triplicate technical replicas.

6. Although not covered in this chapter, the efficiency of IKK inhibition by siRNA can be verified by standard Western blotting techniques.

7. To determine the number of wells needed for this experiment and, therefore, the total volume of cell solution needed, add the number of siRNAs to be transfected to two untransfected controls and multiply by 3 (experiment should be performed in triplicate) and by the number of replica plates used. One of the untransfected controls will be the negative control (without BrdU supplementation).

8. The easiest way to plate cells for this experiment is to use a multichannel pipette.

9. To make enough DharmaFECT solution, you should multiply the volumes required for a single well by the following calculation: $\{[(\#$ siRNA $\times 3.5) + (2$ negative controls (with and without BrdU supplementation) $\times 3.5)] \times \#$ of replica plates to be transfected$\}$.

10. The 0 h timepoint plate should be treated with BrdU and fixed 24 h after plating. These samples should not be transfected and will serve as a baseline for the level of proliferation. All plates should contain a negative control sample (in triplicate) not supplemented with BrdU.

11. To prepare enough 1× Wash Buffer, use the following calculation: 300 μL × # wells used/plate × 4 replica plates × 9 washes (three washes in three steps).

12. It is important to have blank wells, both for luciferase analysis, as well as for BrdU incorporation assay. To make blank wells for luciferase assay, just add 20 μL of 1× PLB buffer to each blank well. To make blank wells for BrdU assays, use negative control cells without BrdU label supplementation.

References

1. Fire A, Xu S, Montgomery MK, Kostas SA, Driver SE, Mello CC (1998) Potent and specific genetic interference by double-stranded RNA in *Caenorhabditis elegans*. Nature 391:806–811

2. Elbashir AM, Harborth J, Lendeckel W, Yalcin A, Weber K, Tuschl T (2001) Duplexes of 21-nucleotide RNAs mediate RNA interference in cultured mammalian cells. Nature 411:494–498

3. Van den Berg A, Mols J, Han J (2008) RISC-target interaction: cleavage and translational suppression. Biochim Biophys Acta 1779:668–677

4. Guo J, Fu YC, Becerra CR (2005) Dissecting role of regulatory factors in NF-kappaB pathway with siRNA. Acta Pharmacol Sin 26:780–788

5. Chua HL, Bhat-Nakshatri P, Clare SE, Morimiya A, Badve S, Nakshatri H (2007) NF-kappaB represses E-cadherin expression and enhances epithelial to mesenchymal transition of mammary epithelial cells: potential involvement of ZEB-1 and ZEB-2. Oncogene 26:711–724

6. Lu W, Zhang G, Zhang R, Flores LG 2nd, Huang Q, Gelovani JG, Li C (2010) Tumor site-specific silencing of NF-kappaB p65 by targeted hollow gold nanosphere-mediated photothermal transfection. Cancer Res 70:3177–3188

7. Gewurz BE, Towfic F, Mar JC, Shinners NP, Takasaki K, Zhao B, Cahir-McFarland ED, Quackenbush J, Xavier RJ, Kieff E (2012) Genome-wide siRNA screen for mediators of NF-κB activation. Proc Natl Acad Sci U S A 109:2467–2472

8. Warner N, Burberry A, Franchi L, Kim YG, McDonald C, Sartor MA, Núñez G (2013) A genome-wide siRNA screen reveals positive and negative regulators of the NOD2 and NF-κB signaling pathways. Sci Signal 6:rs3

9. Bassères DS, Ebbs A, Cogswell PC, Baldwin AS (2014) IKK is a therapeutic target in KRAS-Induced lung cancer with disrupted p53 activity. Genes Cancer 5:41–55

10. Kendellen MF, Bradford JW, Lawrence CL, Clark KS, Baldwin AS (2014) Canonical and non-canonical NF-κB signaling promotes breast cancer tumor-initiating cells. Oncogene 33:1297–1305

11. Basseres DS, Baldwin AS (2006) Nuclear factor-kappaB and inhibitor of kappaB kinase pathways in oncogenic initiation and progression. Oncogene 25:6817–6830

12. Hayden MS, Ghosh S (2012) NF-κB, the first quarter-century: remarkable progress and outstanding questions. Genes Dev 26:203–234

13. Bradford JW, Baldwin AS (2014) IKK/nuclear factor-kappaB and oncogenesis: roles in tumor-initiating cells and in the tumor microenvironment. Adv Cancer Res 121:125–145

14. Coté RJ (1998) Aseptic technique for cell culture. In: Bonifacino JS, Dasso M, Harford JB, Lippincott-Schwartz J, Yamada KM (eds) Current protocols in cell biology. Wiley, New York, pp 1.3.1–1.3.10

15. Mitchell T, Sugden B (1995) Stimulation of NF-kappa B-mediated transcription by mutant derivatives of the latent membrane protein of Epstein-Barr virus. J Virol 69:2968–2976

Chapter 28

Immunohistochemical Analysis of NF-κB in Human Tumor Tissue

Clint T. Allen and Carter Van Waes

Abstract

Immunohistochemistry is a valuable molecular technique based upon the principle of antibody specificity for target antigens in tissues with subsequent development of an amplified colorimetric signal. When staining specificity is ensured with the use of an isotype control, it allows for semiquantitative comparisons, tissue/cellular localization, and inference regarding activation status of proteins of interest. Here we describe a protocol for immunohistochemical analysis of NF-κB family members in fresh frozen human tumor samples.

Key words Immunohistochemistry, NF-κB, Frozen tissue, Tissue fixation, Isotype control, Peroxidase, Histoscore

1 Introduction

Signaling pathways that activate NF-κB/Rel family transcription factors induce IKK-mediated phosphorylation and subsequent proteasome-dependent degradation of IκBs, releasing NF-κB/Rel dimers to undergo further processing and translocate into the nucleus. Activation of NF-κB is normally transient and self-limiting in physiologic states, preventing the detrimental effects of prolonged inflammatory signaling [1]. However, NF-κB is constitutively activated in many cancers, including head and neck squamous cell carcinoma (HNSCC) [1, 2]. In HNSCC, NF-κB may be activated by a number of genetic alterations and tumor microenvironment stimuli, where it regulates a broad gene expression program that promotes cancer progression, metastatic phenotype, and resistance to therapy [2, 3]. Much work has focused on understanding canonical RelA-mediated mechanisms of activation and cellular transformation. However, recent evidence suggests other NF-κB/Rel family members, such as cRel and RelB, are localized to the nucleus, contribute to the malignant phenotype, and promote cell survival. While RelA promotes expression of prosurvival genes, cRel interacts with p53 family members ΔNp63 and p73 to inhibit

Michael J. May (ed.), *NF-kappa B: Methods and Protocols*, Methods in Molecular Biology, vol. 1280,
DOI 10.1007/978-1-4939-2422-6_28, © Springer Science+Business Media New York 2015

growth arrest and expression of apoptosis genes [4]. A lack of clinical response to proteasome inhibitors that block RelA activation is associated with nuclear localization of remaining NF-κB/Rel family members [5] and major upstream signals induce activation of both canonical and alternative NF-κB subunits [6]. These data suggest that study of all NF-κB/Rel family members is important.

Immunohistochemistry (IHC) represents a useful and validated mechanism of evaluating for the presence and activation of NF-κB subunits in tissue specimens. The assay is based in principle upon the ability to develop antibodies that bind a specific antigen or target of interest. Primary antigen-specific antibodies (direct method), or secondary antibodies against the primary antibody (indirect method), are then linked with an enzyme or fluorochrome that allows colorimetric detection of antibody/antigen complexes. By producing a visual stain in the section of interest and by using antibodies that target both total and phosphorylated levels of a protein of interest, IHC allows inferences to be drawn about the relative quantity, tissue/cellular localization, and possible activation status of NF-κB family members.

With protocol optimization, IHC can be performed on fresh frozen tissue specimens stored in a cryoembedding medium or on formalin-fixed paraffin-embedded (FFPE) tissue stored at room temperature. While FFPE tissue can be stored indefinitely at room temperature, formalin fixation cross-links proteins and potentially denatures epitopes, thus necessitating an antigen-retrieval step to prepare tissues for IHC that is not required with fresh frozen tissue. Due to concerns over antigen integrity with various antigen-retrieval techniques and difficulty detecting active phosphorylated subunits, our laboratory routinely performs IHC on fresh frozen tumor tissue from both murine preclinical studies and human clinical trials. Detailed here is our protocol for evaluation of intracellular NF-κB subunits in fresh frozen human tumor. Figure 1 demonstrates IHC

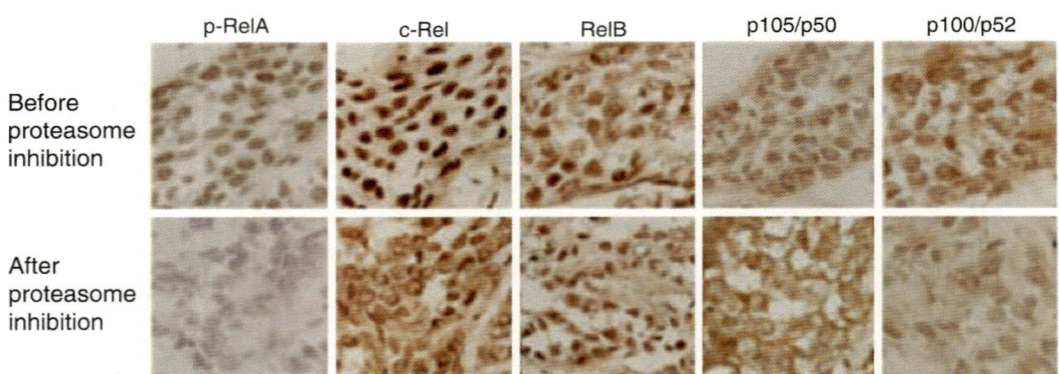

Fig. 1 Immunostaining of phospho-RelA (p-RelA; ser536), cRel, RelB, p50, and p52 in fresh frozen human tumor biopsies before and after treatment with proteasome inhibitor, demonstrating robust inhibition of nuclear phosphorylated RelA but inconsistent inhibition of other NF-κB subunits following therapy

of NF-κB/Rel subunits on fresh frozen human tumor biopsies from patients before and after treatment with proteasome inhibitor, illustrating effects on nuclear/cytoplasmic localization.

2   Materials

2.1   Cryofreezing and Sectioning Tumor Tissue Materials: An Appropriate Vessel and Embedding Medium Are Needed for Fresh Frozen Tumor Storage and Sectioning (See Note 1)

1. Tissue-Tek™ plastic base molds (Fisher, Pittsburgh, PA).

2. Tissue-Tek™ optimum cutting temperature (OCT) compound (Fisher).

3. Dry ice with or without ethanol for snap freezing tumor tissue in the OCT.

4. Facilities capable of cryosectioning—our laboratory has cryosectioning performed by a third-party pathology and tissue processing company (Histoserv, Germantown, MD) for all human tumor samples.

2.2   Staining Procedure Materials

1. Hydrophobic pen to create barrier around tumor sections on the glass slide. An example of such a pen is the ImmEdge Pen™ from Vector Labs, Burlingame, CA.

2. Tissue fixation in paraformaldehyde (*see* **Note 2**). Four percent paraformaldehyde in PBS: 2 mL of 10 % paraformaldehyde stock solution into 3 mL of non-sterile PBS (phosphate-buffered saline, Fisher) makes 5 mL of 4 % paraformaldehyde in PBS. Paraformaldehyde solutions should be prepared in a chemical hood to limit exposure. Adjust pH to 7.4 if necessary. Make fresh and do not store more than 1 week.

3. High-purity undiluted methanol for tissue permeation (*see* **Note 3**).

4. Wash buffer for washing steps throughout the protocol (*see* **Note 4**): 100 μL of Triton™ X-100 (Fisher) into 100 mL of TBS (Tris-buffered saline, Fisher) to make a 0.1 % Triton™ X-100 wash buffer.

5. Endogenous peroxidase quenching solution (*see* **Note 5**): 1 mL of 30 % H_2O_2 solution (Fisher) into 9.7 mL of methanol to make 10 mL of 3 % H_2O_2 in high-purity undiluted methanol solution.

6. Blocking solution to block nonspecific binding of secondary antibody: 500 μL of 5 % serum of the species in which the secondary antibody in 9.5 mL of 0.1 % Triton™ X-100 wash buffer makes 10 mL of blocking solution. As an example, make 5 % goat serum if the secondary antibody is a goat anti-rabbit IgG.

7. Diluted target-specific primary antibody in 3 % BSA/TBS solution (*see* **Note 6**): Dissolve 3 g of BSA into 100 mL of non-sterile TBS to make 3 % BSA in TBS solution. Add diluted antibody for volume desired. *Example: A 1:100 dilution of a*

rabbit anti-phospho-RelA (Ser536) IgG antibody into 100 μL total volume to use for staining would be 1 μL of stock antibody into 99 μL of 3 % BSA in TBS solution.

8. For each primary antibody selected, an isotype and species-specific isotype control antibody must be used in parallel on a separate section to ensure that the staining achieved is target specific. *Example: For the primary antibody selected above, the isotype control would be a nonspecific rabbit IgG antibody* (*see* **Note 7**).

9. Diluted secondary antibody in 3 % BSA/TBS solution (*see* **Note 8**). *Example: 100μL of a secondary antibody solution to detect the primary antibody listed above would be 1μL of biotinylated goat anti-rabbit IgG antibody into 99μL of 3 % BSA in TBS solution.*

10. Enzyme-conjugated streptavidin and biotin complex formation (*see* **Note 9**). Vectastain® ABC detection kits (Vector Labs) specific for the isotype and species of the primary antibody are available for purchase. Each kit is supplied with serum of the secondary antibody species, biotinylated secondary antibody, and solutions that contain biotin and streptavidin/peroxidase constructs. Follow kit instruction to prepare, which requires PBS and kit contents. *Example: For the primary antibody above, a Vectastain® ABC rabbit IgG peroxidase kit would be purchased.*

11. 3,3′-Diaminobenzidine (DAB) enzyme substrate (*see* **Note 10**). DAB peroxidase substrate kit (Vector Labs). Follow kit instructions to prepare, which requires dH_2O and kit contents.

12. Slide chamber full of dH_2O needed to quench DAB reaction.

13. Counterstain (*see* **Note 11**): Hematoxylin (Vector Labs) in slide chamber.

14. Graded alcohols (*see* **Note 12**): 95 % (190 proof) and 100 % (200 proof) ethanol (Fisher) in slide chamber.

15. Xylenes: ACS reagent grade xylenes (Fisher) in slide chamber.

16. Mounting agent (such as Permount™ [Fisher]) and glass coverslips.

3 Methods

3.1 Cryofreezing Tumor or Normal Control Tissue

1. Prepare bucket with dry ice for snap freezing specimens. An alternative is to put 250 mL of ethanol into a beaker in an ice bucket, cover the bottom of the bucket with dry ice, and allow to chill.

2. Pre-label plastic molds and external foil wrapper with permanent marker.

3. After harvesting tissue, place in plastic mold. Choose mold size that is appropriate for the size of the tumor being harvested.

4. Quickly cover the entire specimen with OCT compound ensuring the entire tumor specimen is covered. If bubbles are present, pop with needle before freezing. Place the mold with tumor covered in OCT on top of dry ice to freeze. Alternatively, float the tissue-containing mold in prechilled ethanol. Either method will freeze the OCT and tumor. Once the OCT is entirely white (takes 2–5 min usually), wrap the OCT-embedded tissue and mold in Parafilm, aluminum foil with an external label, and freeze at −80 °C.

3.2 Sectioning Tumor

1. Our laboratory has human tumor samples sectioned by a third-party company. Two sections typically will fit on one slide. Section thickness can be anywhere from 4 to 40 μm. In our experience, sections of 10–15 μm thickness provide optimum results for staining of NF-κB subunits.

2. Store sectioned tissue slides in −80 °C for storage until staining. In our experience, sectioned frozen tumor can remain at −80 °C for 3–6 months without significant degradation of tissue.

3.3 Staining Procedure

1. Remove slide(s) of interest for staining and briefly place the undersurface of the glass slide against the back of your hand to thaw the slide. This only takes seconds and you will see the frost on the slide disappear. Use a delicate wipe to clear excess moisture away from the top of the glass slide, taking care to not disrupt the tissue sections (*see* **Note 13**).

2. Use a hydrophobic pen to draw a complete circle around each section on the slide. This, along with surface tension, allows whatever liquid you place on section to be evenly distributed and prevents spill onto adjacent sections if you have more than one section and stain per slide.

3. Add fresh 4 % paraformaldehyde in PBS for 10 min at 4 °C. Use a volume sufficient to cover the section. For a 15–20 mm diameter circle created with the hydrophobic pen, 200–300 μL is a typically enough volume to cover the section. This will vary based on the size of your hydrophobic circle.

4. Remove the paraformaldehyde and permeabilize the cells with methanol for 5 min at −20 °C (*see* **Note 14**).

5. Remove methanol and perform wash step: Add wash buffer to cover the section for 5 min at room temperature (RT); repeat twice.

6. Remove wash buffer and quench endogenous peroxidase/further permeate cells by adding 3 % H_2O_2 in methanol for 10 min at RT.

7. Remove quench solution and wash sections twice.

8. Remove wash buffer and block nonspecific secondary antibody binding by adding blocking solution for 60 min at RT.

9. Remove blocking solution. DO NOT wash.

10. Add primary antibody at desired dilution in 3 % BSA/TBS overnight (8 h +) at 4 °C. Incubate section in a humidified chamber (*see* **Note 15**).

11. Remove primary antibody solution and wash sections three times.

12. Remove wash buffer and add secondary antibody solution in 3 % BSA/TBS for 30 min at RT. Incubate in a humidified chamber.

13. Prepare ABC complex reagents now (has to sit for 30 min before use). Follow kit instructions.

14. Remove secondary antibody solution and wash sections three times.

15. Remove wash buffer and add ABC complex solution for 30 min at RT.

16. Remove ABC complex solution and wash sections three times.

17. Add DAB to all sections and visualize sections closely. Compare sections where you expect positive staining to your isotype control stain. Quickly remove DAB solution from all sections and immerse slides in dH_2O in a slide chamber once the desired amount of staining is achieved (*see* **Note 16**).

18. Counterstain nucleus blue by immersing slides in hematoxylin in a slide chamber for 60 s.

19. Wash slides by immersion into dH_2O in a slide chamber for 5 min three times, and use fresh dH_2O for each wash.

20. Dehydrate sections by immersion into 95 % (190 proof) ethanol in a slide chamber for 10 s twice, then immersion into 100 % (200 proof) ethanol for 10 s twice.

21. Clear slides of debris and residual contaminants by immersion into xylenes in a slide chamber for 10 s twice.

22. Use a delicate wipe to carefully clean the slides around sections. Add a volume of mounting agent to cover the middle portion of each section on the slide and gently place a glass coverslip. Ideally enough mounting agent is added to spreads and cover the entire surface area of the slide that is covered by the glass coverslip. Use gentle pressure to squeeze out bubbles. Avoid smearing mounting agent as this will inhibit visualization of tissues under the microscope. If mounting goes poorly with smearing and bubbles, place section back in xylenes to clear the mounting agent and try again. Allow mounted slides to try overnight at RT. Once the mounting agent sets, the slides are stored at RT indefinitely for analysis.

3.4 Section Analysis

1. Interpretation of IHC staining requires comparison to the isotype control section. If the isotype control reveals little to no background staining, then staining on sections where antigen-specific primary antibodies were used is valid and can be considered target specific.

2. There are a number of ways to perform systematic analysis of stained tissue sections to attempt to account for heterogeneity of target expression and/or localization. Ultimately the method of analysis chosen depends upon the hypothesis being tested. Both cellular localization and phosphorylation status are important for evaluating the status of NF-κB subunits, so analyses such as those described below that account for cytoplasmic vs. nuclear localization and intensity of staining are commonly utilized.

3. Determining the histoscore [5] involves calculating the percentage of positive cells in a high-power field (HPF) on light microscopy and multiplying this by an assessment of staining intensity, typically on a scale of 0–3 with 0 being no staining and 3 being intense staining. Staining intensity can be calculated for cytoplasmic or nuclear staining. Histoscores are typically reported as average of 3–10 HPFs examined per section. The more HPFs counted and averaged per stained section, the less subjectivity introduced in stain scoring as more tumor tissue is included in the analysis.

4. As opposed to manually scoring sections as above, automated IHC scoring is available. Our laboratory uses ScanScope® hardware and Aperio Cell Quantification software (Aperio, Vista, CA) to count positive cells and score staining intensity of defined areas on a stained section.

4 Notes

1. Molds are designed to create a flat cutting surface for cryosectioning. Choose a size of plastic mold that is sufficient to allow the tumor specimen to be completely covered by OCT medium.

2. Tissue fixation is required to preserve tissue morphology and architecture, halt cellular processes to immobilize and preserve antigens of interest, and strengthen tissue to be able to withstand the staining procedure. A number of different fixation methods are available with formaldehyde and alcohol-based solutions most commonly used. In our experience, tissue architecture and antigen preservation of fresh frozen human tumor tissue are best with 4 % paraformaldehyde in PBS fixation. The best results are achieved when fresh stock is made weekly. Some antibody/antigen interactions may be denatured and require 50 % methanol/acetone fixation, but care must be taken during all steps as tissues or nuclei may not be well fixed to slide and may be lost in subsequent steps.

3. Methanol is used for tissue permeation for staining of intracellular NF-κB subunits. Alternatives are to rely on the cryosectioning alone to expose the intracellular compartment of cells and to use a mild detergent such as 0.2 % Triton™ X-100 in PBS (Sigma-Aldrich), but in our experience, staining of NF-κB subunits is better following methanol permeation of sections.

4. The wash buffer is a dilute Triton™ X-100 solution reducing surface tension to allow even distribution of the buffer when applied to tumor sections on slides, and it may also block some nonspecific staining by inhibiting hydrophobic antibody interactions.

5. A dilute H_2O_2 solution will quench the endogenous peroxidase present within the tissue that would otherwise generate a colorimetric reaction following incubation with DAB and give undesirable background staining.

6. Primary antibodies are selected based upon the target of interest. Several companies now make many antibodies directed at phosphorylated and nonphosphorylated epitopes present on NF-κB subunits. While choosing which primary antibody to use is often a matter of comfort with different vendors, practical approaches include investigating if antibodies offered have been validated for use in frozen IHC, which is listed on the vendor website, or choosing antibodies that others have used successfully in published experiments. Once an antibody is selected, experiments to determine optimum titration are often needed. Optimum titration is one that gives clear positive staining in tumor tissue at a concentration where the same concentration of nonspecific isotype control antibody gives little to no nonspecific background staining. Most antibodies specific for phosphorylated or nonphosphorylated NF-κB subunit epitopes provide optimum results at a 1:100 dilution, but optimum dilution can range from 1:50 to 1:500. Diluting the secondary antibody in a 3 % BSA in TBS solution blocks nonspecific background staining.

7. For each IHC experiment, one isotype control section is sufficient to ensure specific staining, even if multiple different target-specific primary antibodies are being used, as long as each primary is the same species and isoform.

8. An antibody specific for the isoform and species of the primary antibody is used as secondary antibody. This secondary antibody is also conjugated with biotin. Secondary antibody dilution of 1:100 produces consistent results. Again, diluting the secondary antibody in a 3 % BSA in TBS solution blocks nonspecific background staining.

9. Different systems that allow calorimetric detection of antibody complexes are available including peroxidase and alkaline phosphatase systems. To amplify calorimetric detection, solutions of biotin and enzyme-linked streptavidin are added to create

large biotin/streptavidin/enzyme complexes. A substrate is then added that, upon processing by the enzyme, produces visible stain. Peroxidase-based ABC detection systems (Vectastain® by Vector Labs) produce consistent results in our experience and are used in our laboratory.

10. 3,3′-Diaminobenzidine (DAB) is the substrate for peroxidase that produces a brown color on the section.

11. Counterstaining with hematoxylin stains the nucleus blue and allows for easier cellular visualization and determination of cytoplasmic vs. nuclear staining.

12. Graded alcohols are used to dehydrate the section, and xylenes are used to clear away debris and contaminants on the section before permanent mounting.

13. Throughout the protocol, do not let the tumor sections dry out. This leads to deterioration of tissue architecture and increases undesirable background staining.

14. Remove solution from sections carefully, taking care not to disrupt the tissue. Using a glass pipette attached to the suction tubing placed on the very perimeter of the hydrophobic circle works well to suck liquid off the top of tumor sections.

15. The humidification chamber prevents evaporation and concentration of incubating liquids. Humidification chambers can be purchased (e.g., *StainTray, Sigma-Aldrich*) or one can be made by cutting plastic 10 mL pipettes to wedge inside an empty pipette tip plastic container. Two cut plastic pipettes side by side create a flat surface to hold the slides. Line the bottom of the container with a dH_2O-soaked paper towel.

16. Timing how long to leave the DAB on the sections can be challenging. The longer you leave it on sections where you expect stain, the more intense stain that will develop. However, leaving DAB on the isotype control sections will eventually stain these sections as well. Ideally DAB is left in place long enough to see obvious brown staining on the slides where you expect stain but little to no brown staining on isotype control slides. If staining of the isotype control section occurs before or in parallel to staining on experiment sections, then procedures need to be modified to minimize this nonselective background staining. In our experience, steps in this protocol minimize background staining to allow several minutes of DAB exposure to isotype control sections before staining is seen.

Acknowledgments

Work developing and optimizing this protocol was supported in part by the NIH-Pfizer Clinical Research Training Program. We thank Dr. Zhong Chen for critical review of this protocol.

References

1. Karin M (2006) Nuclear factor-kappaB in cancer development and progression. Nature 441(7092): 431–436

2. Van Waes C (2007) Nuclear factor-kappaB in development, prevention, and therapy of cancer. Clin Cancer Res 13(4):1076–1082

3. Allen CT et al (2007) Role of activated nuclear factor-kappaB in the pathogenesis and therapy of squamous cell carcinoma of the head and neck. Head Neck 29(10):959–971

4. Lu H et al (2011) TNF-alpha promotes c-REL/DeltaNp63alpha interaction and TAp73 dissociation from key genes that mediate growth arrest and apoptosis in head and neck cancer. Cancer Res 71(21):6867–6877

5. Allen C et al (2008) Bortezomib-induced apoptosis with limited clinical response is accompanied by inhibition of canonical but not alternative nuclear factor-{kappa}B subunits in head and neck cancer. Clin Cancer Res 14(13): 4175–4185

6. Nottingham LK et al (2013) Aberrant IKKalpha and IKKbeta cooperatively activate NF-kappaB and induce EGFR/AP1 signaling to promote survival and migration of head and neck cancer. Oncogene 33:1135–1147

Chapter 29

Assessment of Canonical NF-κB Activity in Canine Diffuse Large B-Cell Lymphoma

Anita Gaurnier-Hausser and Nicola J. Mason

Abstract

Companion dogs with spontaneous malignancies are clinically relevant models in which to study the corresponding human diseases and potential therapies. In both dogs and people, non-Hodgkin's lymphoma (NHL) is the most common hematopoietic malignancy. Diffuse large B-cell lymphoma (DLBCL) is the most common NHL subtype in dogs and people, sharing similar biologic, behavioral, genetic, and molecular characteristics in both species. One such molecular characteristic is the constitutive activation of the canonical NF-κB pathway, which in health regulates the expression of target genes that control cellular proliferation, survival, and immune and inflammatory responses as well as multidrug resistance. We found that canine and human DLBCL patients share similar NF-κB activity profiles. Using the cell-permeable NBD peptide, which blocks NF-κB signaling, we inhibited constitutive NF-κB activity and induced apoptosis of primary canine malignant B cells in vitro. In addition, we found that NBD peptide administration to dogs with relapsed B-cell lymphoma inhibited the expression of NF-κB target genes and reduced tumor burden. In this chapter, we describe our methods for processing canine malignant lymphoid tissue. We also describe our methods for treating the lymphocytes isolated from this tissue with NBD peptide and evaluating constitutive canonical NF-κB activity in these cells via immunoblot and electrophoretic mobility shift assay (EMSA). We highlight the nuances of working with canine primary cells.

Key words Canine, NF-κB, Lymphoma, Immunoblot, EMSA

1 Introduction

The development of novel anticancer therapies requires feasible, clinically relevant large animal models to evaluate factors such as drug efficacy and toxicity. One such large animal model is the companion dog, which shares environmental, genetic, and physiologic similarities with people [1]. These similarities have been the basis for the recent use of companion dogs in biomedical research [1]. In 2005, the high-quality draft genome sequence of the dog revealed its close phylogenetic relationship with people [2]. This phylogenetic relationship further emphasizes the relevance of the canine model in identifying human cancer genes and evaluating response to novel therapies [3–5].

Michael J. May (ed.), *NF-kappa B: Methods and Protocols*, Methods in Molecular Biology, vol. 1280,
DOI 10.1007/978-1-4939-2422-6_29, © Springer Science+Business Media New York 2015

The advantages of considering the canine cancer model are numerous for human disease research and drug development. For one, this model allows the study of novel drug therapies in the context of a naturally occurring and progressing cancer [6]. Such information could be more difficult to acquire using traditional preclinical models or human trials alone [7, 8]. In addition, spontaneous tumors develop in humans and dogs in the presence of an intact immune system, enabling the canine model to more accurately model response to immune modulatory agents than xenogenic mouse models [8]. Further, inter- and intratumoral heterogeneity is present in both human and canine tumors. Therefore, both species are susceptible to acquiring drug resistance, disease recurrence, and metastasis [6, 9]. Molecular cytogenetic analyses also show that canine tumors are characterized by recurrent chromosome aberrations evolutionarily related to chromosome aberrations in the corresponding human tumors, suggesting a conserved disease pathogenesis between the two species [5].

Companion dogs are now recognized as a clinically relevant large animal model for human non-Hodgkin's lymphoma (NHL) [9–12]. The most common subtype of NHL in dogs is diffuse large B-cell lymphoma (DLBCL), which is similar on a biological, molecular, and genetic level with human activated B cell (ABC)-DLBCL [5, 10, 12, 13]. Accordingly, dogs with DLBCL are treated with the same cytotoxic agents as human ABC-DLBCL patients. Unfortunately, canine patients fail to maintain clinical remission, and 85–90 % relapse with lethal-drug-resistant disease, providing a good model for evaluating therapies to be used in the setting of gross, residual, chemorefractory DLBCL [11, 14].

In humans, ABC-DLBCL is the most clinically aggressive form of the NHL [15], characterized by constitutive canonical NF-κB activity, which promotes the transcription of NF-κB target genes that drive malignant lymphocyte proliferation and prevent apoptosis [16, 17]. NF-κB target genes thereby support lymphomagenesis and chemoresistance by encouraging malignant lymphocyte proliferation and survival [18, 19]. In health, NF-κB activity is typically under tight regulation, with family member dimers held inactive in the cytoplasm via binding to inhibitory IκB proteins [20]. Upon the appropriate infectious or inflammatory stimulus, the inhibitory IκB proteins become phosphorylated by an upstream phosphorylated and activated IKK complex. Phosphorylated IκB proteins are then targeted for ubiquitination and proteasomal degradation, allowing NF-κB dimers to translocate to the nucleus and initiate gene transcription [19].

Aberrant activity of regulatory proteins upstream of the IKK complex has been identified in patients with ABC-DLBCL [16, 21, 22]. The resultant constitutive IKK complex phosphorylation and activation promote ABC-DLBCL cell survival [18, 23]. Inhibiting constitutive NF-κB pathway activity induces cell cycle arrest, chemotherapeutic sensitivity, and apoptosis of ABC-DLBCL cell lines [18, 23, 24]. Targeting aberrant NF-κB activity may therefore be a promising therapeutic strategy for patients with diseases in which aberrant NF-κB activity drives disease initiation and progression, such as ABC-DLBCL. The NEMO-binding domain (NBD) peptide is a selective IKK complex inhibitor comprising 11 amino acids in the carboxy terminus of the catalytic IKKβ and IKKα subunits. This amino acid sequence binds to the scaffold protein NF-κB essential modulator (NEMO) [25–27]. NBD peptide inhibits the interaction of IKKα and IKKβ with NEMO, which prevents the assembly of the IKK complex [26]. Fusing NBD peptide to a protein transduction domain such as the *Drosophila* Antennapedia (pAnt) cell-penetrating peptide (CPP) enables the NBD peptide to cross the plasma membrane and enter cells. Once inside the cell, NBD peptide effectively inhibits constitutive NF-κB activity present in various tumor cell lines, which sensitizes these cells to TRAIL or TNFα-induced apoptosis [28–32]. A mutant peptide containing two tryptophan to alanine (W to A) substitutions within the NBD does not block the interaction of IKKα and IKKβ with NEMO and therefore does not inhibit dysregulated NF-κB activity [27].

In this chapter, we describe the methods to evaluate constitutive canonical NF-κB activity in dogs with diffuse large B-cell lymphoma (DLBCL). We have also successfully used these methods to evaluate NF-κB activity in dogs with peripheral T-cell lymphoma (PTCL) (unpublished data). Our methods to evaluate NF-κB activity may easily be extrapolated to other canine cancers. However, issues such as tissue processing methods and cell sensitivity may need to be adjusted accordingly. We describe our methods of processing canine malignant B-cell tumors. We also describe our methods to assay constitutive canonical NF-κB activity in the dog, namely, immunoblot and electrophoretic mobility shift assay (EMSA). Finally, we describe our methods to treat primary malignant canine B cells with NBD peptide.

2   Materials

All solutions are prepared using ultrapure water. Sodium azide is not added to any of the reagents. Media and buffers should be kept ice cold during use, unless otherwise noted.

2.1   Canine Lymph Node Processing and Cell Isolation

1. Roswell Park Memorial Institute medium (RPMI) supplemented with 10 % (v/v) fetal bovine serum (FBS), 1 % glutamine, and 1 % penicillin/streptomycin, stored at 4 °C. Unless otherwise noted, all RPMI used will contain these supplements (*see* **Note 1**).

2. Dulbecco's phosphate buffered saline (PBS) with calcium and magnesium (1×), stored at room temperature.

3. ACK lysis buffer (Invitrogen, Carlsbad, CA, USA), stored at room temperature.

4. 10-cm tissue culture dishes.

5. 50-mL conical centrifuge tubes.

6. Feather surgical blades (Fisher, Pittsburgh, PA, USA).

7. Bulb syringes (Becton Dickinson Labware, Franklin Lakes, NJ, USA).

8. Cell strainers (40 μm) designed to fit 50-mL conical centrifuge tubes (Becton Dickinson Labware, Franklin Lakes, NJ, USA).

9. Trypan blue, stored at room temperature.

2.2 Isolation of Canine PBMCs

1. Canine whole blood (*see* **Note 2**).

2. RPMI.

3. PBS.

4. 15-mL conical centrifuge tubes.

5. 1.5-mL microfuge tubes.

6. Ficoll-Paque (GE Healthcare, Piscataway, NJ, USA), stored at room temperature.

2.3 Treatment of Primary Canine Malignant B Cells with NBD Peptide

1. X-Vivo medium (Lonza, Walkersville, MD), stored at 4 °C.

2. PBS.

3. 15-mL conical centrifuge tubes.

4. 1.5-mL microfuge tubes.

5. Analytic scale.

6. Metal spatula.

7. NBD peptide, stored at –20 °C (*see* **Note 3**).

8. Dimethyl sulfoxide (DMSO), stored at room temperature.

2.4 Treatment of Canine PBMCs with TNF-α

1. X-Vivo medium.

2. PBS.

3. 15-mL conical centrifuge tubes (*see* **Note 4**).

4. 1.5-mL microfuge tubes.

5. Tumor necrosis factor-α, human recombinant, expressed in *E. coli* (Sigma, St. Louis, MO, USA) (*see* **Note 5**).

2.5 Sample Preparation for Immunoblot

1. PBS.

2. Immunoblot whole cell extract buffer working solution: 50 mM Tris–HCl, pH 7.6, 150 mM NaCl, 2.5 mM EDTA, 5 % (v/v) glycerol, and 1 % (v/v) NP-40, stored at 4 °C (*see* **Note 6**).

3. Protease inhibitor cocktail, stored at –20 °C (Sigma, St. Louis, MO, USA).

4. Phosphatase inhibitor cocktail 1, stored at 4 °C (Sigma) (*see* **Note 7**).

5. Phosphatase inhibitor cocktail 2, stored at 4 °C (Sigma).

6. Micro BCA assay kit (Thermo Scientific, Rockford, IL, USA).

7. 96-well microplate plates.

8. 96-well microplate sealing film.

2.6 SDS-Polyacrylamide Gel Electrophoresis (SDS-PAGE)

1. SDS-PAGE sample buffer stock solution (6×): 125 mM Tris–HCl, pH 6.7, 6 % (w/v) SDS, 10 % (v/v) 2-mercaptoethanol (BME), 2 % (v/v) glycerol, stored at −20 °C (*see* **Note 8**).

2. Resolving gel buffer (1×): 1.5 M Tris–HCl, pH 8.8, stored at 4 °C (*see* **Note 9**).

3. Stacking gel buffer (1×): 0.5 M Tris–HCl, pH 6.8, stored at 4 °C (*see* **Note 10**).

4. 40 % acrylamide/bis solution, 29:1 (3.3 % C), stored at 4 °C (Bio-Rad, Hercules, CA, USA).

5. N,N,N',N'-tetramethyl-ethylenediamine (TEMED), stored at room temperature.

6. Ammonium persulfate (APS) 10 % (w/v) solution in water (*see* **Note 11**).

7. Sodium dodecyl sulfate (SDS).

8. Butanol, stored at room temperature.

9. SDS-PAGE running buffer (10×): 250 mM Tris–HCl, 2 M glycine, 1 % (w/v) SDS, stored at room temperature (*see* **Note 12**).

10. Prestained protein ladder.

2.7 Immunoblotting for NF-κB

1. Transfer buffer (10×): 93 mM CAPS, 77.5 mM NaOH, 10 % (v/v) methanol, stored at 4 °C (*see* **Note 13**).

2. Immobilon-P polyvinylidene fluoride (PVDF) transfer membrane (Bio-Rad).

3. Extra thick filter paper.

4. Tris-buffered saline (TBS, low salt) (10×): 200 mM Tris–HCl, pH 7.6, 1.37 M NaCl, stored at room temperature (*see* **Note 14**).

5. TBST solution: TBS, 0.1 % Tween-20 (v/v) (*see* **Note 14**).

6. Coomassie Blue Stain (Bio-Rad).

7. Destain buffer: 20 % methanol, 7.5 % acetic acid (*see* **Note 15**).

8. Blocking buffer (milk): 5 % (w/v) nonfat dried milk (Sigma) in TBST, stored at 4 °C (*see* **Note 16**).

9. Blocking buffer (BSA): 5 % (w/v) BSA, stored at 4 °C in TBST (*see* **Note 17**).

10. Primary antibodies: (Table 1).

Table 1

Antibodies confirmed to work in the dog

Assay	Antibody	Clone	Human protein size[a]	Phosphorylation site	Dilution	Antibody incubation buffer	Supplier	Catalog #
Immunoblot	Rabbit antihuman p-IKKα/β	16A6	85, 87 kDa	Ser 176, Ser 180	1:1,000	5 % BSA in TBST (0.1 % Tween-20)	Cell Signaling	2697
Immunoblot	Rabbit antihuman IKKα	NA	85 kDa	NA	1:1,000	5 % BSA in TBST (0.1 % Tween-20)	Cell Signaling	2682
Immunoblot	Rabbit antihuman IKKβ	L570	87 kDa	NA	1:1,000	5 % BSA in TBST (0.1 % Tween-20)	Cell Signaling	2678
Immunoblot	Rabbit antihuman p-IκBα	14D4	40 kDa	Ser 32	1:1,000	5 % BSA in TBST (0.1 % Tween-20)	Cell Signaling	2859
Immunoblot	Mouse antihuman p-IκBα	5A5	40 kDa	Ser 32/Ser 36	1:1,000	5 % milk in TBST (0.1 % Tween-20)	Cell Signaling	9246
Immunoblot	Mouse antihuman IκBα	L35A5	39 kDa	NA	1:1,000	5 % milk in TBST (0.1 % Tween-20)	Cell Signaling	4814
Immunoblot	Rabbit antihuman p-p65	93H1	65 kDa	Ser 536	1:1,000	5 % BSA in TBST (0.1 % Tween-20)	Cell Signaling	3033
Immunoblot	Rabbit antihuman p65	C22B4	65 kDa	NA	1:1,000	5 % BSA in TBST (0.1 % Tween-20)	Cell Signaling	4764
Immunoblot	Rabbit antihuman β-actin	NA	45 kDa	NA	1:1,000	5 % BSA in TBST (0.1 % Tween-20)	Cell Signaling	4967
Immunoblot	Rabbit anti-histone H3	NA	17 kDa	NA	1:1,000	5 % BSA in milk (0.1 % Tween-20)	Cell Signaling	9715
Supershift	Rabbit antihuman p65	NA	NA	NA	1 µL	NA	Santa Cruz	sc-109X
Supershift	Rabbit antihuman p50	NA	NA	NA	1 µL	NA	Santa Cruz	sc-114X
Supershift	Rabbit antihuman c-Rel	NA	NA	NA	1 µL	NA	Santa Cruz	sc-70X
Supershift	Control Rabbit IgG	NA	NA	NA	1 µL	NA	Santa Cruz	sc-2027X

[a]A description of size differences between some of the human and canine proteins can be found in **Note 85**

11. Secondary antibodies: horseradish peroxidase-conjugated donkey anti-rabbit IgG and sheep anti-mouse IgG.

12. ECL Plus kit (Thermo Fisher Scientific, Rockford, IL, USA).

13. Autoradiography film.

14. Stripping buffer: 2 % SDS, 63 mM Tris–HCl, pH 6.8, 0.4 % BME (*see* **Note 18**).

2.8 Preparing Nuclear Extracts for Electrophoretic Mobility Shift Assay (EMSA)

1. PBS.

2. NAR-A buffer: 10 mM 4-(2-hydroxyethyl)-1-piperazineethanesulfonic acid (HEPES), pH 7.9, 10 mM KCl, 0.1 mM EDTA, pH 8.0, stored at 4 °C (*see* **Note 19**).

3. NAR-C buffer: 20 mM HEPES, pH 7.9, 0.4 M NaCl, 1 mM EDTA, pH 8.0, stored at 4 °C (*see* **Note 20**).

4. Dithiothreitol (DTT), stored at –20 °C.

5. Protease inhibitor cocktail.

6. Phosphatase inhibitor cocktail 1.

7. Phosphatase inhibitor cocktail 2.

8. 1 % (v/v) Nonidet P40 detergent solution, stored at room temperature.

2.9 Preparing and Labeling of EMSA Probes

1. Salt, Tris, EDTA buffer (STE): 150 mM NaCl, 10 mM Tris–HCl, pH 8.0, 1 mM EDTA, pH 8.0 (*see* **Note 21**).

2. Complementary consensus oligonucleotide probes for NF-κB (*see* **Note 22**).

3. Tris, EDTA buffer (TE): 10 mM Tris–HCl, pH 8.0, 1 mM EDTA (*see* **Note 23**).

4. ^{32}P-γ-labeled adenosine triphosphate (ATP; 3,000 Ci/mL) (GE Healthcare, Piscataway, NJ, USA), stored at 4 °C (*see* **Note 24**).

5. Polynucleotide kinase (PNK; 10,000 units/mL) (New England Biolabs, Ipswich, MA, USA), stored at –20 °C.

6. Mini Quick Spin Oligo Columns (Roche, Indianapolis, IN, USA).

2.10 Preparing Gel for EMSA

1. Tris, boric acid, EDTA buffer (TBE; 10×): 0.89 M Tris–HCl, pH 8.0, 0.89 M boric acid, 20 mM EDTA, pH 8.0, stored at room temperature (*see* **Note 25**).

2. 30 % percent acrylamide/bis solution (37.5:1 with 2.6 % C) stored at 4 °C (Bio-Rad).

3. TEMED.

4. APS.

2.11 Performing the EMSA and Supershift Assays

1. Binding buffer (2×): 40 mM Tris–HCl, pH 7.9, 100 mM NaCl, 2 mM EDTA, 20 % (v/v) glycerol, 2 mM DTT (added fresh), and 1 mg/mL BSA (added fresh), stored at 4 °C (*see* **Note 26**).

2. Poly dI:dC, stored at –20 °C (Roche) (*see* **Note 27**).

3. 1 M MgCl$_2$, stored at 4 °C (*see* **Note 28**).

4. EMSA sample buffer: TBE (2.5× diluted from 10× stock), 50 % (v/v) glycerol, 1.25 % (w/v) bromophenol blue, 1.25 % (w/v) xylene cyanol, stored at room temperature (*see* **Note 29**).

5. Extra thick filter paper.

6. HyBlot CL Autoradiography film (Denville Scientific, Metuchen, NJ, USA).

7. Antibodies for supershift: (Table 1).

3   Methods

All centrifugation steps should be performed in a 4 °C centrifuge at $200 \times g$ for 5 min. Tissue samples and buffers should be kept cold at all times, unless otherwise noted.

3.1 Canine Lymph Node Processing

1. Suspend lymph node (*see* **Note 30**) in a 50-mL conical centrifuge tube containing RPMI (*see* **Note 31**). Fill another tube with additional RPMI for washing steps only ("washing RPMI") (*see* **Note 32**).

2. Place tissue in a 10-cm sterile tissue culture dish containing RPMI (*see* **Note 33**).

3. Using sterile razor blades, dice the tissue until it becomes a paste-like consistency (*see* **Note 34**).

4. Remove the diced tissue from the Petri dish using a bulb syringe. Pass the tissue through a cell strainer fitted to the top of a sterile 50-mL conical centrifuge tube by rinsing with washing RPMI (*see* **Note 35**).

5. Continue to add washing RPMI to the tissue in the cell strainer until all tissue has passed into the tube. Repeat this straining step until all tissue from the dish has been passed through the cell strainer (*see* **Note 36**).

6. Centrifuge the processed tissue to pellet cells and remove any remaining adipose tissue.

7. Remove the supernatant and fill the tube with sterile PBS to wash the cells (*see* **Note 37**).

8. If the resulting pellet is pink/red after washing, red blood cells are contaminating the lymphocytes and will need to be lysed. If red blood cell lysis is required, remove the supernatant from

the cell pellet and add one volume (equivalent to the volume of cell pellet) of ACK lysis buffer. Gently swirl the tube for 60 s only; then fill the tube with RPMI *without* FBS. Centrifuge to pellet the lymphocytes (*see* **Note 38**).

9. Remove the supernatant, wash the cell pellet with at least 10–15 mL of RPMI, and resuspend as required. This tube will be referred to as the "stock lymphocyte suspension." This tube can be kept on ice for a short period of time (i.e., about 30 min), while preparing for NBD peptide treatment (*see* **Note 39**).

10. Use a hemocytometer to count the cells and determine the concentration of the stock lymphocyte suspension.

3.2 Isolation of Canine PBMCs

1. Obtain canine whole blood sample (*see* **Note 40**).

2. Pipette blood into a conical centrifuge tube and dilute 1:1 with cold PBS; mix by gently inverting the tube.

3. Add the appropriate volume of Ficoll solution to empty 15- or 50-mL conical centrifuge tubes (*see* **Note 41**).

4. Carefully layer diluted blood on top of the Ficoll (*see* **Note 42**).

5. Centrifuge the tubes at $1,000 \times g$ for 30 min at 4 °C.

6. After centrifugation, remove the tubes from the centrifuge immediately and carefully, to preserve the layers that have formed.

7. Remove the uppermost layer containing the plasma and platelets, and discard (*see* **Note 43**).

8. Remove the layer of PBMCs to a clean 15-mL conical tube (*see* **Note 44**). Fill this tube with cold PBS and centrifuge at $200 \times g$ for 10 min to wash the PBMCs.

9. Remove the supernatant, fill the tube with cold PBS again, and centrifuge at $200 \times g$ for 10 min.

10. Remove the supernatant and resuspend the PBMCs as required. This tube will be referred to as the "stock PBMC suspension." Keep this tube on ice until treatment with TNF-α (*see* **Note 45**).

11. Use a hemocytometer to determine the concentration of the stock PBMC suspension.

3.3 Treatment of Primary Canine Malignant B Cells with NBD Peptide

The methods described in this section apply to both the NBD and mutant peptides:

1. Determine the desired cell concentration during peptide treatment, as well as the total number of cells, and the total volume, per treatment condition (*see* **Note 46**).

2. Determine the desired NBD peptide concentration (*see* **Note 47**) and the treatment duration (*see* **Note 48**).

3. Using the concentration of the stock lymphocyte suspension calculated in Subheading 3.1, **step 10**, withdraw the required

volume of cells to obtain the desired cell number for each treatment condition. Pipette this cell volume into the appropriate number of conical tubes (*see* **Note 49**).

4. Leave these conical tubes on ice while preparing peptide dilutions (*see* **Note 50**).

5. Use an analytical balance (*see* **Note 51**) to weigh the appropriate amounts of each peptide into sterile 1.5-mL microcentrifuge tubes. These tubes will contain the peptide stock solutions. We recommend preparing multiple peptide stock solutions if multiple peptide concentrations will be tested (*see* **Note 52** and Fig. 1).

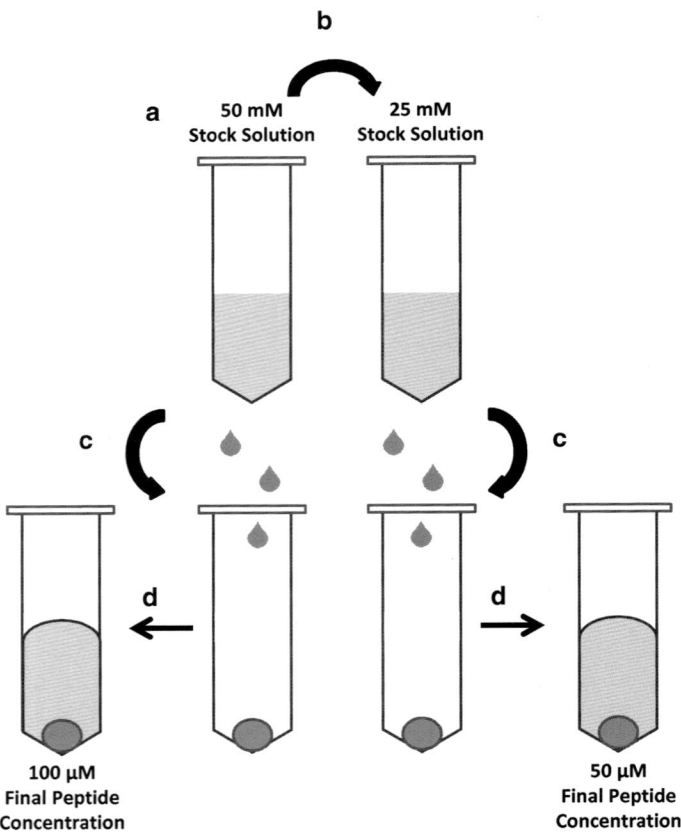

Fig. 1 Example dilution scheme for NBD and mutant peptides when multiple peptide stock solutions are required. (**a**) Determine the total volume of 50-mM peptide stock solution required and then calculate the amount of peptide that needs to be weighed out. A 50-mM peptide solution = 185-μg peptide/1 μL DMSO. (**b**) Dilute the 50-mM peptide stock solution 1:1 with DMSO to generate a 25-mM peptide stock solution. (**c**) Remove the appropriate volume of 50- and 25-mM peptide stock solutions and add to each cell pellet. These volumes will be the same. (**d**) Add the required volume of medium to each cell pellet to generate final peptide concentrations of 100 and 50 μM

6. Add the required volume of DMSO to the weighed peptide in each microcentrifuge tube to generate the peptide stock solutions (*see* **Note 53**). As the peptide stock solutions contain DMSO, keep them at room temperature (*see* **Note 54**).

7. To prepare the lymphocytes for treatment, centrifuge the conical tubes (**step 3**) and remove the supernatant. Add the required volume of peptide stock solution to the side of the tube directly above each cell pellet. To generate the "Vehicle–DMSO" tube, add a volume of DMSO, equivalent to the volume of peptide stock solution used, to the side of the tube directly above an additional cell pellet (*see* **Note 55**).

8. To each cell/peptide mix, add the required volume of X-Vivo medium to generate the desired final cell concentration. Resuspend cells gently but immediately.

9. Aliquot the peptide-treated cell suspensions into the wells of a 6-well tissue culture dish (*see* **Notes 56** and **57**).

3.4 Treatment of Canine PBMCs with TNF-α

1. Determine the desired PBMC concentration during TNF-α treatment, as well as the total number of cells, and the total volume, per treatment condition (*see* **Note 58**).

2. Using the concentration of the stock PBMC suspension calculated in Subheading 3.2, **step 11**, withdraw the required volume of cells to obtain the desired cell number for each treatment condition. Pipette this cell volume into the appropriate number of conical tubes (*see* **Note 59**).

3. Pellet the PBMCs by centrifugation and then remove the supernatant.

4. Remove the reconstituted stock (50 μg/mL) TNF-α from −80 °C storage immediately before use. Add the required volume of 2× final concentration TNF-α (*see* **Note 60**) directly to the cell pellet, and mix gently by pipetting.

5. To each cell pellet, add the same volume of X-Vivo medium to bring the TNF-α to the desired final concentration. Mix gently.

6. Treat the PBMCs at room temperature for 10 min and then immediately centrifuge the tubes at 4 °C to pellet the cells.

7. Remove the supernatant, fill the tube with cold PBS, and then centrifuge at 4 °C to wash cells.

8. Remove supernatant and proceed with protein extraction (Subheading 3.5, **step 4**).

3.5 Sample Preparation for Immunoblot

1. Prior to harvesting peptide-treated cells, add 1:50 protease inhibitor cocktail and 1:50 phosphatase inhibitor cocktails 2 and 3 to the required volume of immunoblot whole cell extract buffer working solution (*see* Subheading 2.5, **item 2**).

2. Remove peptide-treated cells from the 6-well dish into a conical centrifuge tube, and centrifuge to pellet cells (*see* **Note 61**).

3. Remove supernatant and fill tubes with ice-cold PBS. Centrifuge to wash and remove residual peptide, DMSO, and medium.

4. After centrifugation, remove the supernatant and add ice-cold immunoblot whole cell extract buffer to each cell pellet. Transfer to a microcentrifuge tube on ice (*see* **Note 62**).

5. Allow cells to lyse on ice for 10 min and then vortex briefly.

6. Centrifuge at maximum speed for 10 min at 4 °C.

7. Remove supernatant containing soluble proteins to a new microcentrifuge tube, and keep on ice during all subsequent steps to prevent protein degradation.

8. Measure protein concentration of each sample using the BCA assay kit. Generate a standard curve by serially diluting the 2 mg/mL BSA solution included in the kit (*see* **Note 63**).

9. Add 2 μL of each BSA dilution and 2 μL of each unknown to the microplate in triplicate (*see* **Note 64**).

10. Combine reagents A and B (included in the kit, 50:1) and add 200 μL of this working reagent to each well, including the blank (*see* **Note 65**).

11. Cover the microplate with microplate sealing film and incubate at 37 °C for 30 min.

12. Measure the absorbance at or near 562 nm using a microplate reader.

13. Calculate the experimental sample concentration using the standard curve generated by the BSA dilutions.

3.6 SDS-
Polyacrylamide Gel
Electrophoresis
(SDS-PAGE)

These methods are to cast and run a 10 % SDS gel using the min-iVE® Vertical Electrophoresis System (Hoefer, Holliston, MA, USA). For specific apparatus assembly instructions, follow the instructions and diagrams included with the apparatus. This procedure will allow preparation of two gels:

1. Clean glass plates in detergent followed by 70 % ethanol. Leave the plates at room temperature to dry.

2. Assemble the glass plates in the gel-casting apparatus using 0.75- or 1-mm spacers.

3. Check for leaks by pipetting water between the glass plates. Adjust the gel-casting apparatus to form a tighter seal if required.

4. Prepare a 10 % resolving gel by mixing 2.5 mL of resolving gel buffer, 2.5 mL of 40 % acrylamide/bis solution, 4.7 mL of water, and 200 μL of 10 % SDS; swirl gently to avoid creating air bubbles.

5. Add 100 µL of 10 % APS (*see* **Note 11**) and 10 µL of TEMED to the resolving gel solution; swirl gently then pipette immediately between glass plates; leave enough space at the top for the comb and stacking gel layer.

6. Cover the resolving gel layer with butanol (*see* **Note 66**), and monitor the remaining gel solution for polymerization.

7. While the resolving gel is polymerizing, prepare the stacking gel by mixing 2.5 mL of stacking gel buffer, 1.25 mL of 40 % acrylamide/bis solution, 6.15 mL of water, and 100 µL of 10 % SDS; swirl gently to avoid creating air bubbles. Do not add APS or TEMED at this point.

8. Pour off the butanol from the polymerized resolving gel, and wash the gel twice with water.

9. Add 40 µL of 10 % APS and 10 µL of TEMED to the prepared stacking gel solution; swirl gently and then pipette immediately between the glass plates.

10. Insert a 10-well comb into the stacking gel (*see* **Note 67**).

11. While the stacking gel is polymerizing, prepare sample tubes by adding the appropriate volumes of (1) sample, (2) SDS-PAGE sample buffer stock solution (6×), and (3) water to each tube (*see* **Note 68**). Prepare and keep samples on ice until ready to load the gel.

12. After the stacking gel has polymerized (about 30 min), assemble the electrophoresis cassette in the tank and fill the chambers with 1× SDS-PAGE running buffer.

13. Carefully remove the comb from the gel. Ensure that all wells are filled with running buffer and do not contain any air bubbles.

14. "Boil" the prepared samples by placing in a 95 °C heat block for 5 min and then centrifuge at top speed for 3 min (*see* **Note 69**).

15. Load the samples and protein ladder using gel-loading tips (*see* **Note 70**), connect to the power supply, and run at room temperature at 150 V until the dye front has reached the bottom of the gel.

3.7 Immunoblotting for NF-κB

These methods are to perform protein transfer using the Mighty Small Transfer Tank (GE Healthcare, Piscataway, NJ, USA). For specific apparatus assembly instructions, follow the instructions and diagrams included with the apparatus. All incubation steps are performed in plastic dishes on a shaker.

1. Prepare for protein transfer by chilling the transfer apparatus and the 1× transfer buffer (*see* **Note 71**).

2. Cut four pieces of filter paper and one piece of PVDF to the size of the gel (*see* **Note 72**).

3. Presoak the PVDF in 100 % methanol for 15 min and then immerse the PVDF in 1× transfer buffer in a different dish. Do not remove PVDF from this buffer until ready to assemble the transfer "sandwich" (*see* **Note 73**).

4. After completion of electrophoresis (*see* **Note 74**), remove the gel from the apparatus, and remove the stacking gel layer using a clean razor blade.

5. Immerse the gel in a dish containing 1× transfer buffer (*see* **Note 75**).

6. Immerse the four pieces of filter paper and two sponges in transfer buffer in a separate dish.

7. Open the transfer cassette and place on a flat surface. Layer the following components on the black side of the cassette, in this order: one sponge, two-piece filter paper, gel, PVDF, two-piece filter paper, one sponge (*see* **Note 76**).

8. Wrap rubber bands vertically around the left- and right-hand sides of the transfer cassette to keep it closed, making sure that the transfer sandwich components do not move or slide.

9. Insert the transfer cassette in the transfer chamber containing cold transfer buffer so the gel is facing the negative (black) electrode and the PVDF is facing the red (positive) electrode.

10. Transfer at 90 V for 3 h (*see* **Note 77**).

11. After protein transfer, remove the transfer cassette from the transfer chamber, and place black side down on the bench top (*see* **Note 78**).

12. Gently remove the sponge and filter papers, and transfer the PVDF (*see* **Note 79**) immediately into a dish containing blocking buffer (*see* **Note 80**). Transfer efficiency can be determined prior to this blocking step if desired (*see* **Note 80**).

13. Block PVDF for 1 h at room temperature with gentle shaking (*see* **Note 81**).

14. After blocking, remove the PVDF to a clean dish and wash vigorously with TBST three times for 10 min in each change of buffer (*see* **Note 82**).

15. Move the PVDF to a new dish and add primary antibody (*see* **Note 83**). Cover dish with parafilm and incubate overnight at 4 °C degrees with gentle shaking.

16. Remove the PVDF from the primary antibody solution, place in a new dish, and rinse with an initial copious volume of TBST to remove as much unbound antibody as possible. Wash vigorously with TBST three times for 10 min in each change of buffer.

17. Move PVDF to a new dish and incubate with the appropriate HRP-conjugated secondary antibody for 1 h at room temperature.

18. Remove the PVDF from the secondary antibody solution, place in a new dish, and rinse with an initial copious volume of TBST to remove as much unbound antibody as possible. Wash vigorously with TBST three times for 10 min in each change of buffer.

19. Mix the ECL Plus Substrates A and B in a 40:1 ratio. Using tweezers, to hold the membrane, blot the bottom edge of the PVDF on a clean paper towel to remove residual TBST. Place the PVDF protein-surface faceup on a piece of laboratory wrap on a flat surface to ensure even distribution of ECL Plus reagent over the PVDF surface.

20. Evenly cover the PVDF with the ECL Plus and then incubate at room temperature for 5 min.

21. Using tweezers, dab the bottom edge of PVDF on a clean paper towel to remove excess ECL Plus. Place the PVDF at the top of a long piece of laboratory wrap. Fold the bottom half of the laboratory wrap upward, so that it smoothly covers the PVDF. Pull taut to remove air bubbles between the PVDF and the laboratory wrap, which may obscure band visualization. Tape the PVDF/laboratory wrap inside an autoradiograph cassette (*see* **Note 84**).

22. Expose PVDF to film in a darkroom over a range of exposure times to obtain optimal band visualization and intensity (*see* **Note 85** and Fig. 2).

23. If the PVDF will be probed for another antigen, a stripping step is required. Remove the PVDF to a new dish, and wash with an initial copious volume of TBST to remove as much developing reagent as possible. Wash vigorously with TBST three times for 10 min in each change of buffer.

24. Move PVDF to a new dish and cover with 25 mL of stripping buffer. Incubate for 30 min at room temperature (*see* **Note 86**).

25. Pour off stripping buffer (*see* **Note 87**) and place PVDF in a new dish. Wash with an initial copious volume of TBST to remove as much stripping buffer as possible. Wash vigorously with TBST three times for 10 min in each change of buffer in a covered container until the BME can no longer be detected.

26. Repeat the immunoblotting procedure, starting with the blocking step (*see* **Note 88**).

3.8 Preparing Nuclear Extracts for Electrophoretic Mobility Shift Assay (EMSA)

1. Prior to harvesting peptide-treated cells, add 1:50 protease inhibitor cocktail, 1:50 phosphatase inhibitor cocktails 2 and 3, and 1 mM DTT to the required volume of NAR-A and NAR-C buffers (*see* **Notes 89** and **90**).

2. Remove peptide-treated cells from the 6-well dish into a conical centrifuge tube, and centrifuge to pellet cells (*see* **Note 60**).

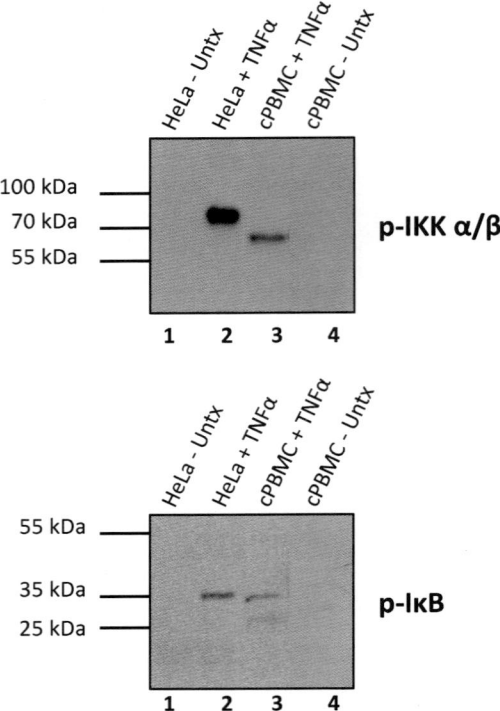

Fig. 2 Canine and human p-IKK α/β and p-IκBα size distinctions on immunoblot. Human HeLa cells and canine peripheral blood mononuclear cells (cPBMCs) were treated with or without 20 ng/mL TNF-α to activate the NF-κB pathway. Whole cell extracts were prepared, electrophoresed, and probed with rabbit anti-human p-IKK α/β (*top*) and rabbit antihuman p-IκBα (*bottom*). p-IKK α/β and p-IκBα are not detected in untreated HeLa cells or untreated cPBMCs, although they are detected in both cell types after TNF-α treatment. Note that the human p-IKK α/β and human p-IκBα run at slightly higher molecular weights than the corresponding canine molecules (compare *Lanes 2* and *3*)

3. Remove supernatant and fill tubes with ice-cold PBS. Centrifuge to wash and remove residual peptide, DMSO, and medium.

4. After centrifugation, remove the supernatant, and add 100 μL of ice-cold NAR-A buffer to cell pellet. Mix gently, and leave cells to swell on ice for 20–30 min.

5. Add 15–20 μL of 1 % NP-40 to each sample. Leave samples at room temperature for 5 min and then vortex strongly for 30 s (*see* **Note 91**).

6. Pellet nuclei by centrifuging at 3,000 × *g* for 90 s.

7. Remove the supernatant (*see* **Note 92**) and transfer to a new microfuge tube. Centrifuge the supernatant at full speed for 90 min at 4 °C. After centrifugation, recover supernatant (*see* **Note 93**) and discard the pellet (*see* **Note 94**).

8. While the cytosolic fraction is centrifuging (**step 7**), wash the nuclear pellet (**step 6**) twice in 1 mL of ice-cold NAR-A (*see* **Note 95**) by centrifuging at $4,000 \times g$ for 90 s.

9. Remove the supernatant and resuspend the nuclear pellet in 30 μL of ice-cold NAR-C (*see* **Note 96**).

10. Vortex nuclear pellets at full speed at 4 °C for 60 min (*see* **Note 97**).

11. Centrifuge nuclear lysates at full speed at room temperature for 20 min.

12. Harvest the supernatant, which contains the extracted nuclear proteins (*see* **Note 98**).

3.9 Preparing and Labeling of EMSA Probes

1. Store radioactive probe at 4 °C in a radiation-safe box, and use within 2 weeks of the reference date of the ^{32}P-γ ATP used for labeling.

2. Mix equimolar amounts of each complementary oligonucleotide in STE buffer in a 1.5-mL microcentrifuge tube. The final concentration of each oligonucleotide in the annealed probe should be 50 ng/μL.

3. Heat tube in a heat block at 90 °C for 10 min. Remove the heat block and tube from heat supply, and allow to cool slowly to room temperature.

4. Set up a 10-μL labeling reaction containing: 1 μL of annealed probe, 1 μL of 10× PNK buffer, 2 μL of water, 5 μL of ^{32}P-γ ATP, and 1 μL of PNK.

5. Incubate the labeling reaction in a heat block at 37 °C for 1 h. After incubation, add 40 μL of room temperature TE.

6. Resuspend the Sephadex matrix of a Quick Spin Oligo Column by inverting and tapping the column multiple times.

7. Keeping the column vertical, remove the column's cap and snap off the bottom tip.

8. Remove excess column buffer, and pack the column by placing in a 1.5-mL microcentrifuge tube and spinning for 1 min at $1,000 \times g$ at room temperature; start timing after the centrifuge has reached speed.

9. After spinning, discard the collection tube and its contents (*see* **Note 99**).

10. Place the prepared column in a new 1.5-mL microcentrifuge tube, and carefully apply the radiolabeled annealed probe to the center of the column bed.

11. Centrifuge at room temperature for 4 min at $100 \times g$.

12. Store the eluate (purified radiolabeled probe) appropriately, and discard the column as ^{32}P-γ ATP-contaminated solid waste.

3.10 Preparing Gel for EMSA

These methods use the Model V16 Vertical Gel Electrophoresis apparatus from Gibco/BRL, which makes a gel with 20 wells:

1. Wash glass plates by scrubbing both sides with detergent, followed by rinsing in water and then 70 % ethanol. Leave washed plates at room temperature for 10 min to dry before use.

2. Assemble plates in a sealed casting boot using 1-mm-thick spacers.

3. Pipette water into the gel-casting chamber to check for leaks. Adjust to form a tighter seal if needed.

4. Prepare gel in a beaker on a stir plate by mixing 14.7 mL of water, 2 mL of TBE, 3.3 mL of 30 % acrylamide/bis solution, 200 µL APS solution, and 15 µL TEMED. Stir gently for 30 s.

5. Pour the solution into the gel chamber and insert the comb. Check for polymerization in the remaining gel solution in the beaker. Polymerization should be complete after approximately 30–45 min.

6. After polymerization is complete, disassemble the casting apparatus and clamp the gel plates (containing the gel) into the electrophoresis apparatus.

7. Fill the upper and lower chambers with 0.5× TBE running buffer.

8. Add 2 µL of 5× sample buffer to the first well and pre-run the gel at 200 V for 1 h (*see* **Note 100**).

3.11 Performing the EMSA and Supershift Assays

1. Set up a 40-µL binding reaction in a 1.5-mL microcentrifuge tube containing 2–10 µg of nuclear extract, 1 µL poly-dI:dC, 5 mM $MgCl_2$, and the appropriate volumes of water and 2× binding buffer (*see* **Note 101**).

2. Incubate the binding reaction on ice for 10 min.

3. If performing an EMSA supershift assay, add appropriate antibody to sample tubes before proceeding to Subheading 3.11.4 below (*see* **Note 102**).

4. Add 1 µL of labeled probe to each sample and incubate at room temperature for 20 min.

5. Add 2–3 µL of EMSA sample buffer and load samples onto the pre-run gel using gel-loading tips.

6. Run the gel at 200 V at room temperature until the dye front has reached the bottom of the gel (*see* **Note 103**).

7. When electrophoresis is complete, cut a piece of filter paper to the size of the gel. Detach one of the glass plates (*see* **Note 104**) leaving the gel attached to the other plate. Place the filter paper on the gel (while the gel is still attached to the glass plate). Carefully remove the gel from the glass plate by peeling back the filter paper (to which the gel should remain attached) starting at one corner of the filter paper-gel.

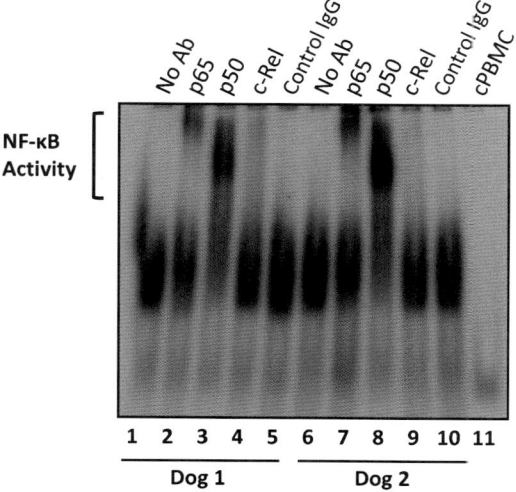

Fig. 3 Electrophoretic mobility shift assay (EMSA) supershif:. Nuclear extracts were prepared from a malignant lymph node of a dog with diffuse large B-cell lymphoma (DLBCL; dog 1) and peripheral T-cell lymphoma (PTCL; dog 2). An EMSA supershift was performed using 2.5 μg of each nuclear extract and an NF-κB consensus binding site oligonucleotide. Samples were incubated with 1 μL of rabbit antihuman p-65, p-50, or c-Rel antibody prior to electrophoresis. Controls included nuclear extract incubated with a nonspecific rabbit IgG (*Lanes 5* and *10*) or no IgG (*Lanes 1* and *6*). NF-KB transcription factor activity in malignant lymphocytes was compared to that in healthy canine peripheral blood mononuclear cells (PBMCs; *Lane 11*). Note the lack of a detectable supershifted NF-κB band in the No Ab lane, the faint c-Rel supershift in *Lane 4*, and lack of detectable c-Rel supershift in *Lane 9*

8. Place the filter paper-gel onto a gel dryer so that it is gel side-up. Cover with laboratory wrap (*see* **Note 105**).

9. Dry the gel (the gel will dry onto the filter paper) and then place into an autoradiography cassette.

10. Expose the gel to film in a darkroom over a range of exposure times to obtain optimal band visualizatior. and intensity (*see* **Note 106** and Fig. 3).

4   Notes

1. Spray all bottles containing medium or solutions with 70 % ethanol (v/v), and wipe down prior to placing in tissue culture cabinets.

2. Collect whole blood samples in sodium heparin ("green top") tubes to prevent coagulation.

3. We have successfully used NBD peptide stored at −20 °C for up to 1 year. We have not attempted to use peptide stored at

this temperature for longer than 1 year. NBD peptide can be obtained from commercial sources including EMD Millipore (Billerica, MA, USA) and Enzo Life Sciences (Farmingdale, NY, USA). NBD peptide can also be custom ordered from peptide synthesis facilities. All of our experiments were conducted using peptide from this source.

4. We treated cells for a short length of time (i.e., 15 min). Therefore, we used a 15-mL conical tube rather than a tissue culture dish, which minimized cell loss due to adhesion to plastic.

5. Reconstitute the contents of the vial of TNF-α (10 μg) to a concentration of 50 μg/mL by adding 500 μL water, following the manufacturer's protocol. Store at −80 °C in one-time-use (i.e., 5 μL) aliquots. Thaw on ice immediately before use. Discard any used portion.

6. First prepare a 1 M Tris–HCL stock solution by adding 6.06 g of Tris-base to 35 mL of water, adjusting the pH to 7.6 and then adding water to a final volume of 50 mL. Prepare a 5-M NaCl stock solution by adding 2.92 g of NaCl to 10 mL of water. To prepare the working solution, combine 500 μL of 1 M Tris, 300 μL of 5 M NaCl, 50 μL of 0.5 M EDTA, pH 8.0, 500 μL of 100 % glycerol, and 1 mL of 10 % NP-40. Add water to 10 mL.

7. This solution contains DMSO, so is solid after storing at 4 °C. Allow time to warm to room temperature before use.

8. First prepare a-500 mM Tris–HCL stock solution by adding 1.2 g of Tris-base to 15 mL of water, adjusting the pH to 6.7 and then adding water to a final volume of 20 mL. To prepare the working solution, combine 2.5 mL of 500 mM Tris–HCL, 0.6 g of SDS, 1 mL BME, and 2 mL of 10 % glycerol. Add water to 10 mL and then add a pinch of bromophenol blue. Wear a mask when weighing out SDS to prevent inhalation. Add BME in a chemical fume hood. This working solution may be aliquoted in microcentrifuge tubes and stored at −20 °C.

9. Add 9.1 g of Tris-base to 35 mL of water, adjust pH to 8.8, and then add water to a final volume of 50 mL.

10. Add 3.0 g of Tris-base to 35 mL of water, adjust pH to 6.8, and then add water to a final volume of 50 mL.

11. Prepare 10 % APS solution immediately before use, and keep on ice protected from light while using. Do not store the prepared solution.

12. To prepare the 10× stock solution, add 30.2 g of Tris-base and 144 g of glycine to a 1-L glass beaker with a stir bar. Add water to 900 mL and stir. Carefully weigh and add 10 g of SDS to beaker, and stir slowly to avoid creating excessive bubbles. Wear a face mask when working with SDS to prevent inhalation.

To prepare the 1× working solution, dilute the stock solution 1:9 with water. Store both stock and working solutions at room temperature.

13. To prepare the 10× stock solution, add 20.6 g of CAPS and 3.1 g of NaOH to a 1-L glass beaker with a stir bar. Add water to 1 L and stir. To prepare the 1× working solution, add 100 mL of stock solution and 100 mL of methanol to 800 mL of water and stir. Ensure working solution is prechilled to 4 °C prior to transfer to prevent protein degradation. Do not reuse the working solution. CAPS is an alkaline buffer whose pH will drop markedly after completion of the transfer procedure.

14. To prepare the 10× stock solution, add 24.2 g of Tris-base and 80 g of NaCl to a 1-L glass beaker with a stir bar. Add water to 900 mL and stir. Adjust pH to 7.6 and then add water to 1 L. To prepare the 1× high-salt TBST working solution, add 100 mL of stock solution and 21.2 g of NaCl to 1-L water. Stir and then add 1 mL of 100 % Tween-20. We have found that the high-salt TBST reduces nonspecific binding when used to wash membranes during immunoblotting. Note that the high-salt buffer may also reduce the band intensity of phosphoproteins. If band intensity of phosphoproteins is low, try washing with 1× TBST without additional NaCl added.

15. Add 200 mL of methanol to 600 mL of water and then add 75 mL of acetic acid. Add water to 1 L.

16. Add 2.5 g of nonfat dry milk to a 50-mL conical centrifuge tube containing 30 mL of 1× high-salt TBST. Invert to mix, and then fill to 50 mL with additional 1× high-salt TBST. Prepare fresh for each experiment.

17. Add 2.5 g of BSA to a 50-mL conical centrifuge tube containing 30 mL of 1× high-salt TBST. Allow BSA to dissolve in buffer at room temperature. Do not agitate tube. After BSA has fully dissolved (about 15 min), fill to 50 mL with additional 1× high-salt TBST. Prepare fresh for each experiment.

18. First prepare a 1-M Tris–HCL stock solution by adding 6.06 g of Tris-base to 35 mL of water, adjusting the pH to 6.8 and then adding water to a final volume of 50 mL. To prepare the working solution, combine 3.13 mL of the stock solution with 10 mL of 10 % SDS. Add water to 50 mL and then add 200 μL of BME.

19. NAR-A and NAR-C buffers were originally described by Dignam et al. in the journal *Nucleic Acids Research* [33]. "NAR" is an abbreviation of the journal name. First prepare a 100-mM HEPES stock solution by adding 1.19 g of HEPES to 40 mL of water, adjusting the pH to 7.9 and then adding water to 50 mL. Prepare a 2-M stock solution of KCl by adding 3 g of KCl to 20 mL of water. To prepare the NAR-A

working solution, combine 5 mL of 100 mM HEPES with 250 µL 2 M KCl. Add water to 50 mL, and then add 10 µL of 0.5 M EDTA.

20. To prepare the NAR-C working solution, combine 10 mL of 100 mM HEPES, pH 7.9 with 1.17 g of NaCl. Add water to 50 mL and then add 100 µL of 0.5 M EDTA.

21. First prepare a 1-M Tris–HCL stock solution by adding 6.06 g of Tris-base to 35 mL of water, adjusting the pH to 8.0 and then adding water to a final volume of 50 mL. To prepare the STE working solution, combine 100 µL of the stock solution and 300 µL of 5-M NaCl stock solution to 9 mL of water. Add 20 µL of 0.5 M EDTA, pH 8.0. Add water to 10 mL.

22. The consensus NF-κB oligo pair is 5′-AGTTGAGGGGACT TTCCCAGG-3′ and 5′-GCCTGGGAAAGTCCCCTAACT-3′. We purchased oligonucleotides from Integrated DNA Technologies (Coralville, IA, USA).

23. Add 100 µL of the 1-M Tris stock solution prepared in **Note 21** to 9 mL of water. Add 20 µL of 0.5 M EDTA, pH 8.0, and then add water to 10 mL.

24. Perform all work with radioactive isotopes in accordance with all prevailing safety and disposal regulations.

25. First prepare the 10× stock solution by adding 108 g of Tris-base and 55 g of boric acid to a 1-L glass beaker with a stir bar. Add 40 mL of 0.5 M EDTA, pH 8.0, to beaker and stir and then add water to 1 L. To prepare the 0.5× working solution running buffer, dilute the stock solution with water prior to use. Store stock and working solution running buffers at room temperature.

26. First prepare a 1-M Tris–HCL stock solution by adding 1.2 g of Tris-base to 9 mL of water, adjusting the pH to 7.9 and then adding water to a final volume of 10 mL. To prepare the 2× binding buffer, add 800 µL of the 1-M Tris solution, 0.12 g of NaCl, and 80 µL of 0.5 M EDTA, pH 8.0, to 10 mL of water. Pipette to mix; then add 4 mL of 100 % glycerol. Pipette well to mix; then add water to 20 mL. Do not store the 2× binding buffer with DTT or BSA. Add DTT and BSA fresh to an aliquot of 2× binding buffer when ready to use.

27. Prepare poly dI:dC by diluting 10 absorbance units (A_{260}) into 0.5 mL of water and then storing in 10-µL aliquots at –20 °C. Thaw on ice when ready to use, keep cold, and do not refreeze or reuse.

28. Add 0.95 g of $MgCl_2$ to 10 mL of water and stir.

29. Add 5 mL of 10× TBE, prepared in **Note 25**, to 5 mL of water and 10 mL of 100 % glycerol. Pipette well to mix; then add a pinch of bromophenol blue and xylene cyanol.

30. Collect all tissue samples following approval of the relevant Institutional Animal Care and Use Committee (IACUC). If working with a small tissue sample (i.e., that obtained from a core biopsy), suspend tissue in a microfuge tube and dice in a smaller dish or tissue culture plate (*see* Subheading 3.1.2 and 3.1.3). *DO NOT* pass small tissue samples through a cell strainer (*see* Subheading 3.1.4 and 3.1.5). Instead, proceed to **step 6** after dicing. We have also not attempted to lyse red blood cells of core biopsy samples using ACK lysis buffer or any other method (*see* Subheading 3.1.8).

31. If lymph node is too large to easily fit in a 50-mL conical centrifuge tube, or if the lymph node displaces more than about 10 mL of medium in the tube, divide the lymph node into multiple sections and process each section separately. Keep tissue in RPMI on ice at all times, removing only briefly for processing.

32. Reserving a separate volume of RPMI for washing prevents possible contamination of the stock RPMI bottle.

33. Add enough RPMI to the dish so that the tissue will be completely immersed in RPMI during and after processing.

34. At this point, adipose tissue may need to be trimmed from the lymph node using the razor blades. If lymph node sample is small, or if tissue is necrotic and not firm, adipose tissue may be separated from lymph node tissue by centrifugation (adipose tissue will layer at the top of the tube).

35. If the tissue has not been diced into a fine-enough consistency, it may clog the cell strainer such that RPMI will not pass through the strainer into the tube. If this occurs, use the bulb syringe to gently stir the tissue in the strainer. When stirring, avoid puncturing the fine mesh of the strainer. If the tissue still does not flow easily through the strainer into the tube, the tissue may need to be diced to a finer consistency. If further dicing is required, remove the tissue from the cell strainer using the bulb syringe (adding additional RPMI if needed), and place the tissue back in the dish. Continue to dice the tissue in RPMI until the tissue becomes the required consistency to easily pass through the cell strainer.

36. Depending upon the volume of the tissue sample, more than one 50-mL tube and/or cell strainer may be required. Use a separate cell strainer for each 50-mL tube. To achieve maximum cell yield, add a final rinse of washing RPMI to the tissue culture dish, and pass through the cell strainer.

37. Be careful that tissue is not lost while removing medium. After centrifugation, compaction of cell pellets will vary depending on the quality/health of the lymph node tissue (i.e., tissue will

vary in the level of necrosis). Large pellets may require a second wash step.

38. Red blood cell contamination typically occurs when the lymph node tissue sample is hemorrhagic and necrotic. This contamination may interfere with subsequent cell counting if red blood cells cannot easily be distinguished microscopically from lymphocytes. Canine red blood cells appear as biconcave disks and are about 7 μM in diameter, while canine lymphocytes are roughly twice this size. In addition, a discolored (pink/red) lymphocyte pellet will result in a discolored cell lysate during protein extraction. A discolored lysate may interfere with determination of protein concentration via the BCA assay. We therefore recommend lysing the contaminating red blood cells prior to cell counting and protein extraction. However, the malignant lymphocytes are fragile and subject to lysis themselves if kept in ACK lysis buffer for longer than the recommended 60 s. In addition, over-manipulation of these cells may induce NF-κB activation. Therefore, use a minimum number of lysis steps, for a minimum length of time. Treat cells gently while in ACK lysis buffer. Mix by gentle swirling; *do not* pipette to mix. If the sample size is large enough, you may use a Ficoll gradient (*see* Subheading 3.2) to remove contaminating red blood cells; however, the same issues regarding cell fragility and NF-κB activation also apply. We do not recommend using a Ficoll gradient if sample size is small, as cell yield will notably decrease.

39. We recommend using cells immediately after completion of lymph node processing. However, you may keep cells in a 37 °C tissue culture incubator for several hours if additional preparation time is required. We do not recommend freezing cells with the intention of assaying for NF-κB activity or treating with NBD peptide in the future. The freezing and/or thawing process may induce cell stress and activate the NF-κB pathway; therefore, NF-κB pathway activity observed in previously frozen cells may not represent the constitutive activity present in vivo in the malignant cells. If you must freeze cells prior to use, we recommend an efficient snap-freezing protocol to minimize cell damage and/or stress (i.e., liquid nitrogen and ethanol). Otherwise, we recommend using freshly isolated, nonfrozen lymphocytes to obtain a true measure of NF-κB activity.

40. Collect all tissue samples following the approval of the relevant Institutional Animal Care and Use Committee (IACUC). The Ficoll-Paque protocol described here is suitable for isolating PBMCs from canine defibrinated or anticoagulated whole blood.

41. We used a 2:3 ratio of Ficoll-to-PBS-diluted blood. It is important that the conical centrifuge tubes NOT be filled to capacity. For example, a 15-mL tube should contain a maximum of 4 mL of Ficoll and 6 mL of PBS-diluted blood. A 50-mL tube should contain a maximum of 16 mL of Ficoll and 24 mL of PBS-diluted blood. However, we recommend using 15-mL tubes if possible. The smaller diameter of the 15-mL tube allows easier identification and removal of the thin, translucent PBMC layer that forms after centrifugation.

42. Perform this step very slowly and carefully so the diluted blood stays as a discrete layer *on top* of the Ficoll. Layering techniques vary, but we have found that adding blood dropwise to the side of the Ficoll-containing tube with an automatic pipettor set to low speed minimizes the disruption of the Ficoll. We also recommend tilting the Ficoll-containing tube at a 45° angle while layering the blood. With time, red blood cells *will* begin to settle to the bottom of the Ficoll-containing tube. However, it is ideal to centrifuge the samples *before* too much red blood cell settling occurs.

43. The thin, translucent PBMC layer is directly below the plasma/platelet layer and is easy to disrupt. When removing the plasma/platelet layer, avoid accidentally removing PBMCs at the same time. It is important to remove as much of the uppermost layer as possible so the PBMCs are not contaminated with platelets.

44. When removing the PBMC layer, be careful to avoid removing the Ficoll layer directly below, as this may result in granulocyte contamination. We recommend using a 1,000-μL micropipette tip to remove the PBMC layer. It may be easier to hold the tube at eye level when removing the PBMCs, as the PBMC layer may be thin.

45. PBMCs may have become contaminated with other cell types during the Ficoll separation, so perform cell counting carefully (*see* **Note 38**). The platelets are relatively easy to recognize microscopically, as they are much smaller than the lymphocytes.

46. In our experiments, the total number of cells and the total volume for each treatment condition varied depending upon the yield of viable cells after lymph node processing. However, we found that using a cell concentration of 4×10^6 cells/mL was optimal.

47. We have treated primary malignant canine lymphocytes with peptide concentrations ranging from 25 to 200 μM. We did not observe changes in NF-κB pathway activity via immunoblot or via EMSA using 50-μM or less NBD peptide. We did observe inhibition of NF-κB pathway activity using 100-μM NBD peptide, while 100-μM mutant peptide had no detectable

effect on NF-κB pathway activity. In addition, we observed a precipitate forming almost immediately around cells treated with 200-µM NBD peptide. We do not know the effect of this precipitate, if any, on NBD peptide uptake or activity, or on cell viability. For each concentration of NBD peptide tested, we tested an equivalent concentration of mutant peptide. We also included untreated and vehicle controls (DMSO only) for each time point.

48. We have treated primary malignant canine lymphocytes with peptide for times ranging from 20 min to 24 h. For most tissue samples, we found that a longer treatment length (6–12 h) resulted in the greatest inhibition of NF-κB pathway activity along with the most measureable difference in apoptosis between NBD peptide- and control-treated cells. Note that NF-κB pathway inhibition may precede apoptosis induction by several hours or more. We noted pathway inhibition in as little as 40 min. Longer treatment lengths (i.e., >12 h) tended to result in excessive death of cells treated with either peptide. This excessive cell death may be attributable to the culturing of primary cells for long periods of time in the absence of exogenously added growth factors. We avoided growth factor addition, as it may have interfered with NBD peptide activity. Note that because the NF-κB pathway promotes malignant cell survival, some cell death should occur after treatment with NBD peptide. However, the excessive cell death observed at longer time points made it impossible to evaluate dose–response effects, or to compare apoptosis induced by NBD vs. mutant peptide. It is also difficult to evaluate peptide-induced NF-κB pathway inhibition at longer time points because cell stress induced by lack of growth factor stimulation may *induce additional* NF-κB pathway activity. This additional activity may counteract the effects of the NBD peptide.

49. We recommend assigning one conical tube for each peptide concentration to be tested. If treating cells with the same peptide concentration over multiple time points, we recommend that each conical tube contains the appropriate number of cells to account for the number of time points. For example, to treat 4×10^6 cells with 100-µM peptide for 6, 12, and 24 h, pipette 12×10^6 cells into the conical tube. If possible, we also recommend including "extra" cells in each conical tube to account for pipetting errors. For example, if testing 3 time points, pipette enough cells to account for 3.5 time points. In the above example, 14×10^6 cells.

50. Do not prepare a storage stock of either peptide in DMSO. The peptides become oxidized after several hours in DMSO at any temperature, including –80 °C. Peptide oxidation causes inactivity. Dissolve the peptides immediately before use only.

51. The peptides used in our experiments had an electrostatic charge, which caused them to adhere to surfaces such as metal spatulas and microcentrifuge tubes. Be careful when weighing out the peptides to prevent peptide loss and ensure accurate measurement. We recommend using a closed analytical balance to minimize air current. Do not weigh out the peptide in a tissue culture hood. Use an ethanol-sterilized metal spatula to transfer the NBD peptide from the stock tube to a tarred 1.5-mL microfuge tube. Leave the microfuge tube resting on the balance during this transfer to ensure accurate measurement. Note that the peptides may "cling" electrostatically to the inside walls of the microfuge tube. We have only used peptides from the source listed in **Note 3**. We therefore cannot comment on the nuances of weighing peptides obtained from other sources. Purification methods, as well as the specific protein transduction domain linked to the peptides, may be factors. We used peptides linked to the *Drosophila* Antennapedia (AntP) protein transduction domain.

52. Dissolve both peptides in DMSO. The molecular weight of the AntP-linked NBD and mutant peptides is 3,700. Therefore, a 1-M solution of either peptide contains 3.7 mg of peptide per 1 μL DMSO. We have used peptide stock solutions as concentrated as 50 mM. When an experiment required testing multiple final concentrations of peptide (i.e., 50, 100, and 200 μM), we generated multiple stock solutions. This ensured that an equal volume of peptide stock solution would be added to cells, regardless of treatment condition (Fig. 1).

53. It may be necessary to centrifuge the weighed peptide prior to adding DMSO, which will "pellet" any peptide adhering to the side of the tube and facilitate dissolving the peptide. After adding the DMSO, centrifuge the tube again. Do not pipette to mix the peptide with the DMSO, as the peptide does not immediately go into solution at this concentration. Pipetting may cause clumping of the peptide in the pipette tip, loss of peptide, and an inaccurate peptide stock solution concentration. Centrifuging the peptide in DMSO helps the peptide dissolve without the need for pipetting.

54. DMSO will become solid if placed on ice, which will negatively affect peptide activity. Prepare the peptide stock solutions immediately before use.

55. The "required volume of peptide stock solution" is that necessary to generate the desired final concentration of peptide in each tube, after addition of medium. If using multiple stock solutions to test multiple peptide concentrations, the same volume of peptide stock solution will be used (*see* **Note 52** and Fig. 1). Therefore, a single volume of DMSO is required to generate the vehicle control, regardless of the number of peptide

concentrations being tested. As DMSO may be cytotoxic in certain conditions, a uniform volume of DMSO across all treatment conditions will allow its potential effects to be recognized and/or accounted for as baseline/background effects.

56. Dish size may vary depending on the total cell number and total volume for each treatment condition. If treatment time will be relatively short (i.e., under 60 min), treat cells in a 15- or 50-mL conical centrifuge tube, placed in a 37 °C incubator with the cap loosened. Cell viability will decrease if cells are treated in a conical tube for longer than about 60 min, and cell clumping at the bottom of the tube may affect experimental results (i.e., peptide accessibility to cells and peptide uptake).

57. In some experiments, we wanted to determine whether NBD peptide also induced apoptosis of healthy canine PBMCs. For those experiments, we treated the PBMCs and the primary malignant lymphocytes side by side, using identical conditions (i.e., peptide and cell concentrations, treatment length, medium, etc.).

58. We treated PBMCs at a concentration of 4×10^6 cells/mL.

59. You will need at least two conical tubes, one for cells treated with TNF-α and one for untreated cells.

60. We recommend a final concentration of TNF-α of 20 ng/mL.

61. When the peptide-treated cells are examined microscopically, a precipitate may be visible, especially at higher peptide concentrations (i.e., greater than 100 μM). We also observed NBD peptide-treated cells adhering loosely to the bottom of the dish. Cell adhesion was minimal in mutant peptide-treated or control wells. Dislodge adherent cells from the dish by gentle pipetting. If necessary, rinse the dish with additional medium to further dislodge adherent cells. Note that rinsing the dish with additional medium will dilute the culture supernatant, which must be taken into consideration if the supernatant will be further analyzed. If cells have been treated for a short period of time in a conical centrifuge tube (i.e., 60 min or less), centrifuge the tubes to pellet the cells.

62. It can be difficult to determine the appropriate volume of immunoblot whole cell extract buffer to add to each cell pellet. When measuring protein concentrations (i.e., via BCA assay, *see* Subheading 3.5.8), if sample ODs fall outside the OD range generated by the BSA standard curve, you will need to further dilute samples with additional extract buffer and repeat the BCA assay. When in doubt, we recommend using a smaller volume of extract buffer to generate a higher protein concentration. The volume of a dilute protein sample required to generate a detectable band via immunoblot may not be feasible.

In addition, protein degradation occurs more rapidly in more dilute samples. When lysing 4–8×10^6 cells, we generally used 30–50-µL extract buffer to start, with further dilution/additional BCA assay sometimes required.

63. Aliquot and store the included 2 mg/mL BSA stock in single-use tubes at –80 °C. We have found that serial dilution of the BSA stock in PCR tubes followed by transfer to microplate wells produces more accurate, reproducible data than performing the serial dilutions directly in the microplate wells. We used 1:1 serial dilutions of the 2 mg/mL BSA stock to generate the following standard curve points (mg/mL): 2.0, 1.0, 0.5, 0.25, and 0.125.

64. When adding protein samples to the wells of the microplate, pipette samples directly into the corner of each well. This prevents sample loss due to retention in the pipette tip.

65. Prepare the working reagent immediately before use, and gently invert the tube to mix. Include a well with 2 µL of extract buffer as a blank.

66. The butanol layer will "compress" the resolving gel layer so that the top of the gel and the boundary between the resolving and stacking gel layers will be smooth. If butanol is not available, water may also be used to cover the resolving gel layer.

67. We did not achieve satisfactory band resolution using a 15-well comb. Bands were narrower horizontally, but thicker vertically. These vertically thicker bands are an issue when trying to distinguish between two proteins of close molecular weight, for example, IKKα (85 kDa) and IKKβ (87 kDa).

68. We have found that addition of 25–40 µg of whole cell extract per well produces a signal of appropriate strength when using the antibodies noted in Table 1 to evaluate protein expression in canine primary malignant lymphocytes. However, each experimenter should determine the optimal amount of whole cell extract to be used. After determining this amount, snap-freeze lysates (i.e., dry ice and ethanol) and store as single-use aliquots at –80 °C to prevent protein degradation.

69. Note that some commercially available protein ladders should not be boiled prior to electrophoresis; however, others may require boiling.

70. It will be easier to determine the band size if the "smile effect" is minimized. We therefore recommend adding protein ladder to both the first and last lanes of the gel, if space allows. Note that empty lanes anywhere in the gel will result in lateral diffusion of sample. If you add protein ladder to the first and last lanes of the gel, note that the gel will appear vertically symmetrical after electrophoresis. Therefore, it is important to distinguish

between the left- and right-hand sides of the gel by, for example, cutting a piece of gel from one top corner of the stacking layer.

71. We used a transfer apparatus fitted with a cooling unit. About 30 min before transfer, we poured the 1× transfer buffer working solution into the transfer apparatus chamber and turned on the cooling unit to chill. Using a stir bar, we stirred the transfer buffer continuously during protein transfer. Alternatively, you can perform protein transfer in a cold room.

72. Do not use a piece of PVDF larger than the gel because the larger PVDF will more easily allow air bubble formation between the gel and the PVDF during protein transfer.

73. Presoaking the hydrophobic PVDF in methanol will hydrate the PVDF. When the PVDF is initially immersed in the 1× transfer buffer after the methanol soak, the transfer buffer may bead up on the PVDF surface. Use clean tweezers to fully immerse the PVDF in the transfer buffer until the PVDF becomes fully hydrated.

74. After the power supply has been turned off, the lack of current running through the gel may cause resolved bands to become diffuse. If more time is needed after electrophoresis to prepare for protein transfer, continue to run the gel at a very low voltage to maintain some current flow. We recommend about 15 V. However, watch the protein ladder carefully to ensure that bands of interest do not run off the gel.

75. Do not add the gel to the dish containing the PVDF, as the gel may stick to the PVDF and tear. Keep the gel immersed in transfer buffer until ready to assemble the transfer sandwich so the gel does not dry out. However, soaking the gel for longer than 5–10 min may result in the diffusion of resolved bands.

76. Do not allow any components of the sandwich to become dry during this layering. Bubbles (air pockets) may form between the gel and the PVDF, which may prevent protein transfer. Remove air bubbles by gently rolling them out using a glass stir rod that has been wet with transfer buffer to avoid creating friction and tearing the gel or PVDF.

77. Alternatively, you can transfer proteins overnight. We recommend transferring overnight at 45 V and then transferring the following morning for 1 h at 90 V.

78. Disassemble the transfer cassette in the same orientation in which it was assembled, so the orientation of the gel and PVDF will be clear.

79. Use clean tweezers to remove the PVDF from the gel, only touching the PVDF in a corner where there will be no protein bands. When removing the PVDF from the gel, the "protein surface" of the PVDF (the surface that was touching the gel)

must be facing upward in the dish of blocking buffer. From this point forward, keep the PVDF in this orientation (i.e., protein-surface faceup). We recommend consistently cutting the same corner of each PVDF membrane, to ensure correct orientation.

80. Assess transfer efficiency by noting the transfer of the colored protein ladder onto the PVDF. If desired, stain the PVDF with Ponceau S for several minutes at room temperature prior to blocking. Ponceau S will stain protein bands on the PVDF. Note that Ponceau S will not stain less abundant proteins, but it will give you a rough idea of transfer efficiency. Remove Ponceau S from the PVDF by washing several times with the appropriate blocking buffer, until all residual stain is removed. In addition, you can stain gels with Coomassie blue after protein transfer to detect residual bands on the gel. Cover and soak gel in Coomassie blue for 30–60 min and then wash with destain buffer until protein bands are visible on the gel. Add several Kimwipes to the corner of the dish when destaining to facilitate removal of stain from the gel.

81. You can also block membranes overnight at 4 °C with gentle shaking. Cover the dish with parafilm to prevent buffer contamination and evaporation.

82. Move PVDF to a new dish for each washing and antibody incubation step to ensure complete removal of residual wash buffer and antibody. If the protocol for the primary antibody requires the same buffer for blocking and primary antibody incubation, omit this washing step.

83. See Table 1 for correct primary antibody dilution buffers. We routinely probed for p-IKBα and p-IKKα/β simultaneously by cutting the membrane horizontally at a point midway between the two antigen bands. Each section of PVDF can be incubated separately with the appropriate primary antibody. If the antigen of interest requires detection of both phosphorylated and total protein, we recommend probing for the phosphorylated protein, first, stripping the membrane (*see* Subheading 3.7.23–3.7.26) and, then, re-probing for total protein. In our experience, the amount of phosphorylated protein is much less than the amount of total protein, causing the phosphorylated protein to be more difficult to detect.

84. We recommend taping the wrapped PVDF to an inside corner of the autoradiography cassette. When developing, place film over the PVDF so that the film touches two inside corners of the cassette. After developing, the film may then be placed back over the PVDF in the same orientation as it was during the initial exposure step. This will help accurately align the film with the PVDF to determine sizes of detected protein bands.

85. The molecular weights of human IKKα and IKKβ are 85 and 87 kDa, respectively. Antibodies directed against phosphorylated IKK molecules recognize both p-IKKα and p-IKKβ, so the phosphorylated forms of these molecules are difficult to resolve via immunoblot. The differences in size that we observed between human and canine p-IKKα/β are likely not attributable to differences in molecular weight between the human and canine molecules. The predicted molecular weights of the two transcript variants of canine IKKα range from 82 to 85 kDa. Similarly, the predicted molecular weights of the four transcript variants of canine IKKβ range from 80 to 87 kDa. These ranges would not produce the size differences we observed between human and canine p-IKKα and p-IKKβ in our immunoblots. It is not as clear why the human p-IκB-α band runs at a slight higher molecular weight than the corresponding canine band, though we observed this result repeatedly. Human IκB-α should run at 39 kDa. The predicted molecular weights of the four transcript variants of canine IκB-α range from 40 to 53 kDa so would be expected to (possibly) run at a slightly *higher* molecular weight than the corresponding human molecule.

86. Prepare stripping buffer in a ventilated fume hood to avoid directly breathing BME. When incubating the membrane with the stripping buffer, use a covered container.

87. Ensure proper disposal of BME-containing buffers in accordance with federal, state, and local environmental control regulations.

88. Stripping efficiency may vary depending on factors such as the abundance of the antigen(s) of interest. You can check stripping efficiency by developing the stripped membrane with ECL Plus prior to the blocking step. Bands visible on the developed film indicate primary–secondary antibody complexes that remain bound to the PVDF. Additional stripping may be required. Note that this extra developing step will *not* detect primary antibody that remains bound to the PVDF in the absence of bound secondary antibody. Therefore, "old" bands may still appear after re-probing the membrane if the same secondary antibody is used.

89. Add protease and phosphatase inhibitor cocktails and DTT to a working aliquot of NAR-A and NAR-C immediately before use. Do not store either buffer with these cocktails or with DTT.

90. The amount of protein extracted from cell nuclei is notably less than that extracted from whole cells. The number of viable lymphocytes extracted from some of our lymph node tissue samples was too small to perform nuclear protein extraction after peptide treatment. For these smaller tissue samples, we isolated whole cell protein as in Subheading 3.5.

91. These steps lyse plasma membranes but leave nuclear membranes intact. You can evaluate lysis efficiency by microscopically visualizing a small aliquot of the lysed sample. Intact nuclei should be visible. If lysis of plasma membranes does not appear complete, additional NAR-A, 1 % NP-40, and vortexing may be required. However, because of the dying/necrotic cells often present in the canine primary malignant tissue samples, we have found erroneous nuclear lysis at this stage to be more common than lack of plasma membrane lysis.

92. This supernatant represents the cytosolic extract. While preparing this supernatant fraction for centrifugation, keep the corresponding nuclear pellet (still in the microfuge tube) on ice. Proceed with washing the nuclear pellet as soon as possible.

93. We recommend immediately determining the protein concentration of this supernatant fraction and snap freezing (dry ice and ethanol) in single-use aliquots at −80 °C to prevent protein degradation or changes in protein phosphorylation (i.e., protein phosphorylation may be lost over time, even in the presence of phosphatase inhibitors, if samples are not stored appropriately in a timely manner).

94. This pellet contains non-soluble cytosolic material.

95. This nuclear pellet is "sticky" and difficult to dissociate by pipetting during this washing step. Avoid drawing the entire nuclear pellet (which will clump together) into the pipette tip, as sample may be lost or damaged/lysed.

96. As with NAR-A, add protease and phosphatase inhibitors and DTT to a working aliquot of NAR-C immediately before use. Do not store either buffer with these cocktails or with DTT.

97. Set the tube shaker to the highest setting, and place in a refrigerator or cold room. If a tube shaker is not available, you can tape tubes securely to a vortex and shake at the highest speed. The high-salt concentration of NAR-C will extract the content of the nuclei during this shaking step.

98. We recommend immediately determining the protein concentration of this supernatant fraction (see **Note 91**). Alternatively, you can perform a second nuclear extraction step, if desired, to ensure complete nuclear protein extraction. If performing a second nuclear extraction, after removing the supernatant (nuclear proteins) in Subheading 3.8.12, add 30 μL of NAR-C to the remaining pellet. Proceed with vortexing and centrifugation (Subheading 3.8.10 and 3.8.11). Depending on the tissue sample, we were sometimes able to extract additional nuclear protein with this second lysis step. If you wish to assess the purity of the nuclear and/or cytoplasmic fractions, they may be probed for actin and histone H3 by immunoblot (Table 1).

99. To avoid the resin drying out, use the prepared column immediately.

100. This pre-running step removes any unpolymerized acrylamide, which improves band resolution.

101. Add DTT and BSA to the 2× binding buffer to final concentrations of 2 mM and 0.1 mg/mL, respectively. Each experimenter should determine the optimum amount of nuclear extract to use. You may need to adjust the volume of the binding reaction according to the concentration of the nuclear extracts (i.e., more concentrated extracts will require a smaller total binding reaction volume). The electrophoresis system we used can comfortably fit a total volume of about 55 μL in each well.

102. The protocol for the EMSA supershift assay is identical to that for the regular EMSA, except that you will need to add an antibody against the transcription factor of interest to the binding reaction. Antibody binding to the transcription factor/DNA complex results in an increase in the molecular weight of the binding complex. This increase will cause a mobility shift of the complex during electrophoresis. An EMSA supershift will allow distinction between p65, p50, and c-Rel activity. If a supershift will be performed, add 1 μL of the appropriate antibody or control IgG (Table 1) to the binding reaction after completion of Subheading 3.11.2. Leave the binding reaction on ice for 1 h before proceeding to probe addition in Subheading 3.11.4.

103. If performing an EMSA supershift assay, we run the gel at 100 V at room temperature until the dye front has reached the bottom of the gel (approximately 5 h).

104. After the electrophoresis is complete, it is not uncommon for small air bubbles to create spaces between the gel and the glass plates, resulting in a treelike branching pattern throughout the gel. This does not indicate problems with the electrophoresis and does not impact band resolution. However, be careful when detaching one of the glass plates from the gel, as the gel may be easily torn during this step.

105. Smooth the laboratory wrap carefully over the top of the gel after placing on the bed of the gel drier so that a seal will form during the drying process. If the laboratory wrap folds under the filter paper, a seal will not form and the gel will not be properly dried.

106. Use a dark room with safe lights to place film on the gel inside the cassette. You may need to test a number of exposures. For the majority of our experiments, a minimum 12-h exposure time was required. The longest exposure time we required was 72 h. For especially weak bands, we inserted an

enhancer screen into the cassette and stored the cassette, with film and enhancer screen, at −80 °C. If cassettes are stored at −80 °C, warm them to room temperature for approximately 30 min before developing film.

References

1. Rowell JL, McCarthy DO, Alvarez CE (2011) Dog models of naturally occurring cancer. Trends Mol Med 17:380–388

2. Lindblad-Toh K, Wade CM, Mikkelsen TS, Karlsson EK, Jaffe DB, Kamal M et al (2005) Genome sequence, comparative analysis and haplotype structure of the domestic dog. Nature 438:803–819

3. Dickerson EB, Fosmire S, Padilla ML, Modiano JF, Helfand SC (2002) Potential to target dysregulated interleukin-2 receptor expression in canine lymphoid and hematopoietic malignancies as a model for human cancer. J Immunother 25:36–45

4. Honigberg LA, Smith AM, Sirisawada M, Verner E, Loury D, Chang B et al (2010) The Bruton tyrosine kinase inhibitor PCI-32765 blocks B-cell activation and is efficacious in models of autoimmune disease and B-cell malignancy. Proc Natl Acad Sci U S A 107: 13075–13080

5. Thomas R, Smith KC, Ostrander EA, Galibert F, Breen M (2003) Chromosome aberrations in canine multicentric lymphomas detected with comparative genomic hybridisation and a panel of single locus probes. Br J Cancer 89:1530–1537

6. Hansen K, Khanna C (2004) Spontaneous and genetically engineered animal models: use in preclinical cancer drug development. Eur J Cancer 40:858–880

7. Gordon IK, Khanna C (2010) Modeling opportunities in comparative oncology for drug development. ILAR J 51:214–220

8. Gordon I, Paoloni M, Mazcko C, Khanna C (2009) The comparative oncology trials consortium: using spontaneously occurring cancers in dogs to inform the cancer drug development pathway. PLoS Med 6:1–5

9. Paoloni M, Khanna C (2008) Translation of new cancer treatments from pet dogs to humans. Nat Rev Cancer 8:147–156

10. Breen M, Modiano JF (2008) Evolutionarily conserved cytogenetic changes in hematological malignancies of dogs and humans—man and his best friend share more than companionship. Chromosome Res 16:145–154

11. Paoloni MC, Khanna C (2007) Comparative oncology today. Vet Clin North Am Small Anim Pract 37:1023–1032

12. MacEwen EG (1990) Spontaneous tumors in dogs and cats: models for the study of cancer biology and treatment. Cancer Metastasis Rev 9:125–136

13. Fournel-Fleury C, Magnol JP, Bricaire P, Marchal T, Chabanne L, Delverdier A et al (1997) Cytohistological and immunological classification of canine malignant lymphomas: comparison with human non-Hodgkin's lymphomas. J Comp Pathol 117:35–59

14. Vail DMME, Young KM (2001) Canine lymphoma and lymphoid leukemia. In: Withrow SJ, MacEwen EG (eds) Small animal clinical oncology. WB Saunders Co, Philadelphia, PA, pp 558–590

15. Nyman H, Adde M, Karjalainen-Lindsberg ML, Taskinen M, Berglund M, Amini RM et al (2007) Prognostic impact of immunohistochemically defined germinal center phenotype in diffuse large B cell lymphoma patients treated with immunochemotherapy. Blood 109:4930–4935

16. Davis RE, Ngo VN, Lenz G, Tolar P, Young RM, Romesser PB et al (2010) Chronic active B-cell receptor signalling in diffuse large B-cell lymphoma. Nature 463:88–92

17. Feuerhake F, Kutok JL, Monti S, Chen W, LaCasce AS, Cattoretti G et al (2005) NFkappaB activity, function, and target-gene signatures in primary mediastinal large B-cell lymphoma and diffuse large B-cell lymphoma subtypes. Blood 106:1392–1399

18. Davis RE, Brown KD, Siebenlist U, Staudt LM (2001) Constitutive nuclear factor kappaB activity is required for survival of activated B cell-like diffuse large B cell lymphoma cells. J Exp Med 194:1861–1874

19. Karin M, Lin A (2002) NF-kappaB at the crossroads of life and death. Nat Immunol 3:221–227

20. Gilmore TD (2006) Introduction to NF-kappaB: players, pathways, perspectives. Oncogene 25:6680–6684

21. Compagno M, Lim WK, Grunn A, Nandula SV, Brahmachary M, Shen Q et al (2009) Mutations of multiple genes cause deregulation of NF-kappaB in diffuse large B-cell lymphoma. Nature 459:717–721

22. Lenz G, Davis RE, Ngo VN, Lam L, George TC, Wright GW et al (2008) Oncogenic

CARD11 mutations in human diffuse large B cell lymphoma. Science 319:1676–1679

23. Lam LT, Davis RE, Pierce J, Hepperle M, Xu Y, Hottelet M et al (2005) Small molecule inhibitors of IkappaB kinase are selectively toxic for subgroups of diffuse large B-cell lymphoma defined by gene expression profiling. Clin Cancer Res 11:28–40

24. Ferch U, Kloo B, Gewies A, Pfander V, Duwel M, Peschel C et al (2009) Inhibition of MALT1 protease activity is selectively toxic for activated B cell-like diffuse large B cell lymphoma cells. J Exp Med 206:2313–2320

25. May MJ, D'Acquisto F, Madge LA, Glockner J, Pober JS, Ghosh S (2000) Selective inhibition of NfkappaB activation by a peptide that blocks the interaction of NEMO with the IkappaB kinase complex. Science 289:1550–1554

26. May MJ, Marienfeld RB, Ghosh S (2002) Characterization of the Ikappa B-kinase NEMO binding domain. J Biol Chem 277: 45992–46000

27. Solt LA, Madge LA, May MJ (2009) NEMO-binding domains of both IKKalpha and IKKbeta regulate IkappaB kinase complex assembly and classical NF-kappaB activation. J Biol Chem 284:27596–27608

28. Aggarwal S, Takada Y, Singh S, Myers JN, Aggarwal BB (2004) Inhibition of growth and survival of human head and neck squamous cell carcinoma cells by curcumin via modulation of nuclear factor-κB signaling. Int J Cancer 111:679–692

29. Kiessling MK, Klemke CD, Kaminski MM, Galani IE, Krammer PH, Gulow K (2009) Inhibition of constitutively activated nuclear factor-kappaB induces reactive oxygen species- and iron-dependent cell death in cutaneous T-cell lymphoma. Cancer Res 69:2365–2374

30. Shishodia S, Amin HM, Lai R, Aggarwal BB (2005) Curcumin (diferuloylmethane) inhibits constitutive NF-kappaB activation, induces G1/S arrest, suppresses proliferation, and induces apoptosis in mantle cell lymphoma. Biochem Pharmacol 70:700–713

31. Thomas RP, Farrow BJ, Kim S, May MJ, Hellmich MR, Evers BM (2002) Selective targeting of the nuclear factor-kappaB pathway enhances tumor necrosis factor-related apoptosis-inducing ligand-mediated pancreatic cancer cell death. Surgery 132:127–134

32. Ianaro A, Tersigni M, Belardo G, Di Martino S, Napolitano M, Palmieri G et al (2009) NEMO-binding domain peptide inhibits proliferation of human melanoma cells. Cancer Lett 274:331–336

33. Dignam JD, Lebovitz RM, Roeder RG (1983) Accurate transcription initiation by RNA polymerase II in a soluble extract from isolated mammalian nuclei. Nucleic Acids Res 11:1475–1489

Chapter 30

NEMO-Binding Domain Peptide Inhibition of Inflammatory Signal-Induced NF-κB Activation In Vivo

Kelly A. McCorkell and Michael J. May

Abstract

NF-κB comprises a family of transcription factors that regulate the expression of diverse gene families essential for inflammatory and immune responses as well as cell survival and cell death pathways. Aberrant NF-κB transcriptional activity plays pivotal roles in a large number of human pathologies, including a variety of cancers and chronic inflammatory diseases. Therefore, there has been a large increase in studies aimed at identifying and testing drugs or small molecule inhibitors that would specifically block NF-κB activation in inflammatory diseases and cancer. In this chapter, we describe an in vivo system to test the inhibitory effects of the NEMO-binding domain (NBD) peptide on NF-κB activation specifically in the vascular endothelium and lymphocytes in mice. We demonstrate that pretreatment of mice with the NBD peptide reduces the NF-κB induced gene expression of cell adhesion molecules and DNA-binding activity following systemic LPS stimulation. These methods can be further used to test alternate inhibitors for effects on NF-κB signaling in murine endothelium and immune cells.

Key words NF-κB, NEMO-binding domain (NBD), Lipopolysaccharide (LPS), Vascular cell adhesion molecule 1 (VCAM-1), Endothelial (E)-selectin, Vascular endothelium, Lymphocytes, SDS-PAGE immunoblotting, Electrophoretic mobility shift assay

1   Introduction

NF-κB transcription factors play an essential function in a myriad of cellular processes, including inflammation, immune responses, cellular proliferation, differentiation, and survival. In response to pro-inflammatory cytokines, stress, or pathogenic insults, the classical NF-κB pathway is rapidly and transiently activated to induce gene expression of growth factors, regulators of apoptosis, cytokines and chemokines, and cell adhesion molecules [1, 2]. Since NF-κB regulates such a diverse range of gene targets, it is not surprising that dysregulated activation of NF-κB is highly associated with a multitude of pathologies especially cancer and chronic inflammatory diseases such as arthritis, atherosclerosis, inflammatory bowel disease, and inflammation-associated metabolic diseases [3–6].

Michael J. May (ed.), *NF-kappa B: Methods and Protocols*, Methods in Molecular Biology, vol. 1280,
DOI 10.1007/978-1-4939-2422-6_30, © Springer Science+Business Media New York 2015

Given the central role NF-κB plays in human disease, there has been increased research efforts devoted to developing methods for the targeted inhibition of NF-κB signaling for therapeutic treatment of cancers and inflammatory diseases. One well-established approach for the successful and selective inhibition of NF-κB is the development of the cell permeable NEMO-binding domain (NBD) peptide. The NBD peptide contains the six-amino acid NEMO-binding motif found in the COOH terminus of the catalytic IKKα and IKKβ subunits. The peptide competes with IKKα and IKKβ for binding to NEMO, blocking their association and inhibiting classical NF-κB activation. Further biochemical studies showed that a mutated version of the NBD peptide, which contains two alanine residues substituted for tryptophan within the NBD has no effect on the association of the IKK subunits with NEMO and allow for NF-κB activation [7, 8]. The cell permeable NBD peptide has been shown to block inflammatory induced NF-κB activation and gene expression but not basal NF-κB signaling in a variety of cellular and rodent models of acute and chronic inflammation, including cerulein-induced acute pancreatitis [9], arthritis [10, 11], colitis [12], and Duchene muscular dystrophy [13]. Most recently, the NBD peptide has entered phase I clinical trials for the treatment of canine diffuse large B-cell lymphoma (DLBCL), and the cell permeable NBD peptide has been shown to be safe and effective in blocking constitutive NF-κB signaling following systemic administration [14].

In this chapter we describe an in vivo model used to test the effectiveness of the wild-type NBD peptide to block lipopolysaccharide (LPS)-induced NF-κB activation in the vascular endothelium and immune cells. Classical NF-κB plays a well-described regulatory role during the activation of the endothelium that aids in the recruitment, rolling, adhesion, and finally extravasation of leukocytes to sites of inflammation. To assess the NBD inhibition of LPS-activated NF-κB within the vascular endothelium in vivo, we describe methods by which mice receive either systemic administration of LPS alone or in conjunctions with the NBD peptide, followed by the excision of the murine aorta to examine the protein expression of NF-κB regulated cell adhesion molecules, VCAM-1 and E-selectin [15]. LPS treatment activates NF-κB resulting in the induction and increased expression of VCAM-1 and E-selectin compared to untreated mice. When pretreated with wild-type NBD peptide, the LPS-stimulated upregulation of both cell adhesion molecules is diminished demonstrating a block in NF-κB activation (Fig. 1).

To demonstrate the efficacy of the NBD peptide to block LPS-activated NF-κB in immune cells in vivo, we describe methods by

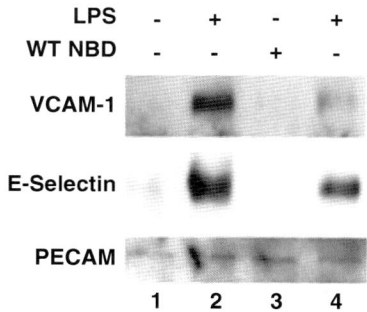

Fig. 1 NBD peptide reduction of LPS-induced expression of cell adhesion molecules in the mouse aorta. Immunoblots of VCAM-1 (*upper panel*) and E-selectin (*middle panel*) expression in protein lysates obtained from the excised aortas of an untreated mouse (*lane 1*), and mice treated with LPS alone (*lane 2*), NBD peptide alone (*lane 3*), or a mouse pretreated with NBD peptide for 15 min prior to LPS treatment (*lane 4*). PECAM expression (*lower panel*) used as a control for equal tissue lysate loading of the endothelial portion of aortic tissue

which NF-κB proteins are isolated from murine splenic lymphocytes, incubated with radiolabeled oligonucleotides corresponding to the κB consensus sequences found in promoter of their target genes, and lastly assessed for their DNA-binding activity via an electrophoretic mobility shift assay (EMSA). In untreated lymphocytes, NF-κB is sequestered in the cytoplasm and do not bind to the κB sequence containing probe. Following LPS stimulation, NF-κB is activated and migrates to the nucleus where it binds to the κB sites in the promoters of target genes. Activated NF-κB proteins isolated from lymphocytes of LPS-treated mice similarly bind to radiolabeled oligonucleotides that correspond to these κB consensus sequences and are visualized as shifted bands in an EMSA due to their slower mobility of the NF-κB:probe complex compared to protein lysates of untreated cells. Pretreatment of mice with wild-type NBD peptide reduces LPS-induced DNA-binding activity of NF-κB protein. In contrast, pretreatment with the mutant version of the NBD peptide fails to block any LPS activation of NF-κB (Fig. 2).

Although this chapter specifically describes the methods used to determine the effect of the NBD peptide on LPS-driven NF-κB signaling, the methods provide two distinct ways to assess NF-κB activity in vivo. These methods could be extended or modified to test the ability of alternate stimuli to activate NF-κB in murine endothelium and in immune cells. These methods could also be followed to test the efficacy of other inhibitors to block activation of the classical NF-κB in mice.

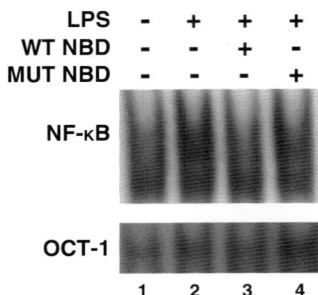

Fig. 2 NBD peptide reduction of LPS-induced DNA-binding activity of NF-κB in murine lymphocytes. EMSA gel of NF-κB consensus binding site oligonucleotides (*top panel*) incubated with whole cell lysates from splenocytes of an untreated mouse (*lane 1*) or mice treated with LPS alone (*lane 2*) or LPS following a 15 min pretreatment with WT NBD peptide (*lane 3*) or MUT NBD peptide (*lane 4*). The binding of Oct-1 consensus binding site oligonucleotides (*bottom panel*) used as a loading control

2   Materials

2.1   In Vivo Treatment of Mice

1. Mice: One C57BL/6NJ mouse (8 weeks or older) per treatment maintained in accordance with guidelines of the Institutional Animal Care and Use Committee (IACUC) and recommendations in the Guide and Use of Laboratory Animals of the National Institutes of Health (NIH).

2. Lipopolysaccharides from Escherichia coli 055:B5 (Sigma, St. Louis, MO, USA): Dissolve 10 mg in 2 ml of sterile saline (*see* **Note 1**).

3. Wild-type and mutant NBD peptides (*see* **Note 2**): Dissolve 100 µg in 10 µl sterile dimethylsulfoxide (DMSO) and bring volume up to final concentration of 100 µg in 100 µl for each mouse using sterile saline (*see* **Note 3**).

4. Insulin Syringe: Sterile-wrapped 1 cc insulin syringe with 26-G and 0.5-in. needle for each injection.

2.2   Dissection and Tissue Extraction

1. 70 % ethanol.

2. Phosphate-buffered saline (PBS): Pre-chill 10 ml per mouse on ice in 50 ml tubes.

3. Syringes: Sterile-wrapped 10 ml syringe with a 27-G and 1-in. needle (*see* **Note 4**).

4. Scissors.

5. Forceps.

6. Styrofoam board for pinning mouse.

7. Push pins.

8. Kim-wipes.

9. 1.5 ml tube prefilled with 1 ml PBS per mouse.

2.3 Aortic Tissue Dissociation

1. Dissection Scope (*see* **Note 5**).

2. Vannas Micro-dissection Spring Scissors: Angled on edge with a 3 mm cutting edge and 0.15 mm tip width (*see* **Note 6**).

3. Forceps: Dumont #5 forceps (*see* **Note 7**).

4. 25 ml beaker containing 15 ml of PBS.

2.4 Aortic Protein Preparation

1. TNT lysis buffer: 50 mM Tris–HCl (pH 6.8), 150 mM NaCl, 1 % (v/v) Triton X-100 pre-chilled on ice (*see* **Note 8**).

2. Complete protease inhibitor cocktail tablets (Roche Applied Science, Indianapolis, IN, USA). Stored at 4 °C (*see* **Note 9**).

3. Phosphatase Inhibitors: 1 mM Sodium Fluoride (NaF), 1 mM beta-glycerophosphate (*see* **Note 10**). Stored at –20 °C.

4. Coomassie Plus protein assay reagent kit containing BSA (2 mg/ml) protein standard (Pierce, Rockford, IL, USA). Reagent stored at 4 °C and BSA stored at –20 °C once aliquoted (*see* **Note 11**).

5. Sample Buffer (5×): 0.3 M Tris–HCl (pH 6.8), 5 % (w/v) sodium dodecyl sulfate (SDS), 50 % (v/v) glycerol, 100 mM DTT, 0.03 % (w/v) bromophenol blue. Stored at –20 °C in 0.5 ml aliquots (*see* **Note 12**).

2.5 SDS-Polyacrylamide Gel Electrophoresis (SDS-PAGE)

1. Separating buffer (4×): 1.5 M Tris–HCl, pH 8.8, 0.4 % (w/v) SDS. Stored at room temperature.

2. Stacking buffer (4×): 0.5 M Tris–HCl, pH 6 8, 0.4 % (w/v) SDS. Stored at room temperature.

3. Thirty percent acrylamide/bis solution (37.5:1 with 2.6 % C) stored at 4 °C (*see* **Note 13**).

4. *N,N,N,N′*-tetramethyl-ethylenediamine (TEMED): Stored at room temperature.

5. Ammonium persulfate (APS): Prepare in 10 % (w/v) and store at 4 °C (*see* **Note 14**).

6. Methanol.

7. Running buffer: 25 mM Tris, 250 mM, glycine, 0.1 % (w/v) SDS. Prepare a 5× stock stored at room temperature (*see* **Note 15**).

8. Spacer plates, short plates, and combs.

9. Prestained protein molecular weight standard markers.

2.6 Immunoblotting for Cell Adhesion Molecules

1. Transfer buffer: 25 mM Tris, 192 mM glycine, 20 % (v/v) methanol (*see* **Note 16**).

2. Ice pack.

3. Methanol at room temperature.

4. Polyvinylidene fluoride (PVDF) transfer membrane.

5. Thick cellulose chromatography filter paper.

6. Tris-buffered saline with Tween (TBS-T): 25 mM Tris–HCl, pH 8.0, 140 mM NaCl, 3 mM KCl, 0.05 % Tween-30 (*see* **Note 17**).

7. Blocking buffer: 5 % (w/v) nonfat dried milk in TBS-T.

8. Primary Antibodies: Dilute rabbit anti-VCAM-1 at 1:500 in blocking buffer, rabbit anti-E-selectin at 1:500 in blocking buffer, and goat anti-PECAM (all from Santa Cruz Biotechnology, Santa Cruz, CA, USA) at 1:500 in blocking buffer (*see* **Note 18**).

9. Secondary antibodies: Dilute Horse radish peroxidase-conjugated donkey anti-rabbit IgG, donkey anti-mouse IgG, and donkey anti-goat IgG in blocking buffer 1:5,000 immediately before use (*see* **Note 19**).

10. Western blotting luminal reagent kit. Stored at 4 °C.

11. Autoradiography Film.

2.7 Single Cell Suspension of Splenocytes

1. PBS.

2. 40 μm cell strainers and 50 ml conical tubes.

3. Syringe Plunger: Plungers (one per mouse) removed from 3 cc syringes.

4. ACK Lysis: 0.15 mM NH_4Cl, 10 mM $KHCO_3$, 0.1 mM EDTA. Store at 4 °C.

2.8 Preparation of Whole Cell Lysates of Splenocytes

1. Whole Cell Lysis Buffer: 20 mM Hepes pH 7.6/7.9, 420 mM NaCl, 25 % glycerol, 1.5 mm $MgCl_2$, 0.2 mM EDTA, complete protease inhibitor cocktail tablets (*see* **Note 9**), and 5 mM Dithiothreitol (DTT) (*see* **Note 20**).

2.9 Preparation of EMSA Probes

1. Complementary single-stranded consensus oligonucleotides for NF-κB (*see* **Note 21**).

2. Double-stranded Oct-1 consensus oligonucleotides (Santa Cruz Biotechnology, Santa Cruz, CA, USA).

3. Sodium Chloride-Tris-EDTA (STE) buffer: 10 mM Tris–HCl pH 8.0, 150 mM NaCl, 1 mM EDTA. Stored at room temperature.

4. T4 Polynucleotide kinase (PNK): 10,000 units/ml. Stored at −20 °C.

5. Tris, EDTA buffer (TE): 1 mM Tris–HCl, pH 8.0, 1 mM EDTA. Stored at room temperature.

6. ^{32}P-γ-labeled adenosine triphosphate (ATP): 3,000 Ci/ml (*see* **Note 22**). Stored at 4 °C.

7. Mini Quick spin oligo columns (Roche Life Sciences, Madison, WI, USA).

2.10 Preparation of the EMSA Gel

1. Tris-Boric Acid-EDTA (TBE; 10×) buffer: 89 mM Tris–HCl pH 8.0, 89 mM boric acid, 0.5 mM EDTA. Stored at room temperature.

2. Thirty percent acrylamide/Bis solution (37.5:1 acrylamide:bis with 2.6 %C) stored at 4 °C.

3. Ammonium persulfate: 10 % (w/v) prepared immediately before use.

4. *N*,*N*,*N*,*N*′-tetramethyl-ethylenediamine (TEMED).

5. TBE Running Buffer: Dilute 10× TBE buffer to a final concentration of 0.5× in water.

2.11 Performing the EMSA

1. Binding Buffer (2×): 40 mm Tris–HCl, pH 7.9, 100 mm NaCl, 2 mm EDTA, 20 % glycerol, 0.2 % Nondiet P-40, 2 mm DTT, 100 μg/ml bovine serum albumin (BSA).

2. MgCl$_2$ (100 mM). Stored at room temperature.

3. Poly-dI:dC stored in aliquots at –20 °C (*see* **Note 23**).

4. Radiolabeled EMSA probes: Prepared in advance (*see* Subheading 2.9).

5. Electrophoresis sample buffer (5×): TBE (2.5× diluted from 10× stock), 50 % (v/v) glycerol, 1.25 % (w/v) bromophenol blue, 1.25 % (w/v) xylene cyanol.

6. Thick chromatography filter paper.

7. BioMax X-ray Film (Kodak, Rochester, NY, USA).

3 Methods

3.1 Preparation of Peptides

1. Using an analytical balance, weigh an appropriate amount of wild-type and mutant NBD peptide into separate 1.5 ml tubes and dissolve in sterile DMSO (*see* **Note 3**) in a Biosafety cabinet.

2. Add PBS to the two tubes of peptides for a final concentration of 100 μg/100 μl.

3.2 Pretreatment of Mice with NBD Peptides

1. Fill a sterile 1 ml insulin syringe fitted with a 26-G and 0.5-in. needle with 0.1 ml of the designated peptide or vehicle control (*see* **Note 24**).

2. Restrain the appropriate C57BL/6NJ mouse using the scruff method and tilt the animal with its head slightly down to expose the ventral side of the abdomen.

3. Wipe the injection area with 70 % ethanol.

4. Insert the needle of the filled syringe, bevel side up, into the lower right side of the abdomen at 30 % angle.

5. Before injecting, pull back on the plunger and aspirate the needle/syringe to ensure proper position (*see* **Note 25**).

6. Administer the appropriate treatment of peptide.

7. Repeat the methods to administer the next peptide control using a new sterile syringe and needle for the injection of each mouse.

8. Keep two C57BL/6NJ mice untreated for each experimental group, one as an untreated mock-injected control and one for an LPS alone treatment.

9. Leave mice to rest for 15 min prior to LPS and Control treatments.

3.3 LPS Treatments

1. Thaw an aliquot of previously dissolved LPS (5 mg/ml) per mouse.

2. Add 180 μl of sterile PBS to each aliquot of LPS in a Biosafety cabinet.

3. Following the **steps 3–6** of Subheading 3.2, inject an untreated C57BL/6NJ mouse, a wild-type NBD, and a mutant NBD pretreated C57BL/6NJ mouse with 200 μl of LPS via intraperitoneal injections.

4. For an untreated mock-injected control, inject one C57BL/6NJ mouse with 200 μl PBS.

5. Continuously observe mice for 4 h until euthanization (*see* **Note 26**).

3.4 Euthanization and Dissection of Aorta and Spleen

1. Fill a 10 ml syringe with 10 ml of pre-chilled PBS.

2. Attach a 27-G needle.

3. Euthanize one mouse via CO_2 inhalation according to institutional regulations. Work with one mouse at a time.

4. Immerse the mouse in a 200 ml beaker filled with 100 ml of 70 % Ethanol.

5. Pin the mouse down to a piece of Styrofoam with its ventral side up using push pins to hold down each paw.

6. Using scissors and forceps, make a nick into the skin about 1 cm up from the groin.

7. Angle the scissors into the nick and cut a straight line towards head through both the skin and abdominal wall until you get to the diaphragm.

8. Open the thoracic cavity by cutting a v-shape through the diaphragm wall and rib cage cutting all the way through to the neck.

9. Starting from the original nick, make a second v-shaped cut through the abdominal wall towards the tail.

10. Pull the sides of skin and excess fat to the side to uncover the internal organs of the abdomen and thoracic cavity.

11. Make a nick in the hepatic portal vein and cover with several kim-wipes to soak up the blood.

12. Pierce the left ventricle with the PBS-filled syringe and perfuse with ice-cold PBS using the prefilled 10 ml syringe and 27-G needle.

13. Remove the spleen and place it in a 1.5 ml tube filled with 1 ml of PBS and keep on ice.

14. Remove the ribs, lungs, gastrointestinal, and reproductive organs, leaving the heart and kidneys intact.

15. Move the pinned mouse under dissection microscope.

16. Using the mini-Vannas scissors and Dumont forceps, dissect free the aorta by removing the fat overlying the aorta starting at the iliac bifurcation and moving up towards the kidneys (*see* **Note 27**).

17. Being careful to not cut the aorta itself, remove all fat surrounding the aorta attachments to each kidney but leaving the aorta still attached to the kidneys.

18. Dissect the aorta free from the diaphragm to the heart by removing the surrounding fat (*see* **Note 28**).

19. Puncture a hole in one wall of the aorta above the iliac bifurcation with the scissors by holding the aorta taut with the forceps. Slide the scissors up the aortic wall opening the aorta lengthwise.

20. Cut the aortic arch free from all tissue in the neck by sliding the mini-Vannas scissors along the curvature of the arch to the heart.

21. Cut the aorta free from the mouse starting with a cut across the arch as it enters the heart, then the attachments to the kidney, and a final cut at the iliac bifurcation.

22. Swirl the detached and opened aorta in a beaker of PBS to remove excess blood, dab on a kim-wipe quickly, and place in a 1.5 ml tube on ice (*see* **Note 29**).

23. To the 1.5 ml tube containing each aorta, add 200 μl of TNT lysis buffer containing complete protease inhibitor cocktail and phosphatase inhibitors.

24. Repeat the euthanization and dissection for each mouse in the experiment.

3.5 Preparing Aortic Samples for Immunoblotting

1. Incubate the 1.5 ml tube containing the aorta in TNT lysis buffer (*see* Subheading 3.4, **step 23**) on ice for 1 h, vortexing every 15 min.

2. Spin the 1.5 ml tubes for 10 min at $16,000 \times g$ in a refrigerated centrifuge at 4 °C.

3. Remove supernatant to a new 1.5 ml tube and place on ice while determining protein concentration (*see* **Note 11**).

4. Set up protein standard curve by dilution from the stock of BSA (2 mg/ml) to achieve concentrations of 0.5, 1, 2.5, 5, 10, and 20 μg/ml in water. Pipette duplicate aliquots of 150 μl of water (blank) and the protein standards into a 96-well plate.

5. Dilute 2 μl of each aortic lysate in 500 μl of water and vortex.

6. Pipette 150 μl of each diluted sample in duplicate into new wells of the 96-well plate containing the blanks and standards.

7. Add 150 μl of the Coomassie Plus reagent to each well and read the plate using a microplate reader. Calculate the protein concentration in each sample manually by constructing a graph using the standard curve or by using software specific for the plate reader used (*see* **Note 11**).

8. Prepare samples for SDS-PAGE (*see* Subheading 3.6) in sample buffer with each containing 20 μg of protein (*see* **Note 30**).

9. The remaining lysates should be frozen for short time frames at −20 °C for a future immunoblot. For longer term storage, remaining lysates should be stored at −80 °C.

3.6 SDS-PAGE

1. These directions are for casting and running 10 % SDS gels using the Bio-Rad Mini Protean Electrophoresis apparatus. Directions for other systems are similar and a potential alternative to making gels would be purchasing precast 10 % gels (*see* **Note 31**).

2. Clean the back spacer plates and front short plates before use. Plates should be wiped down with 70 % ethanol to avoid gel sticking unevenly to the plates.

3. Assemble the glass plates in the gel-casting apparatus placing a short front plate with a back spacer plate lined up evenly.

4. In a 50 ml conical tube, prepare a 10 % resolving gel by mixing 6.25 ml water, 3.75 ml 4× separating buffer, 5 ml 30 % acrylamide/bis solution, 50 μl APS solution, and 20 μl TEMED (*see* **Note 32**). Mix by inversion.

5. Pipette the solution into the gel chamber leaving space for the stacking gel. Layer 1 ml of methanol onto the resolving gel while the gel solidifies.

6. Check the leftover resolving gel mixture in the 50 ml tube for polymerization. Polymerization can take between 5 and 10 min.

7. Begin preparation of the stacking gel in a 50 ml conical with exception to the addition of TEMED, by mixing 3.05 ml of water, 1.25 ml of 4× stacking buffer, 0.65 ml of 30 % acrylamide/bis solution, and 25 μl of APS solution.

8. Once the gel is polymerized, place a piece of paper towel along the top of each gel chamber and invert the casting mold to absorb the methanol out of the gel chamber in preparation of the stacking gel.

9. Add 10 μl of TEMED to the stacking gel solution and mix by inversion.

10. Pour stacking gel on top of the resolving gel and insert a 10-well comb (see **Note 33**). The stacking will polymerize in 5–10 min.

11. While waiting for the stacking gel to polymerize, prepare 1 l of 1× running buffer from the 5× stock.

12. Assemble the electrophoresis cassette in the tank when gel has fully polymerized. Two gels will be required to create an inner chamber, using a dam plate if only one gel is required. Fill inner chamber entirely and 1/3 of the outer chamber with 1× running buffer.

13. Remove the comb by pulling straight up on both sides of the comb to keep wells intact.

14. Place the samples from Subheading 3.5, **step 9**, in a heat block (95 °C) for 5 min and then spin at top speed in a benchtop microcentrifuge for 2 min.

15. Using a gel-loading tip, add 10 μl of protein molecular weight standard to the first well. Continue the addition of samples to subsequent wells using gel-loading tips.

16. Connect the power supply and run at 150–180 V until the samples have separated over the entire length of the gel (visualized by watching the separation of the pre-stained markers, see **Note 34**). The gel is run at room temperature.

3.7 Immunoblotting for VCAM-1 and E-Selectin

1. These directions for protein transfer are for the Bio-Rad Mini-Trans-Blot® Electrophoretic Transfer Cell wet transfer apparatus.

2. Fill the transfer chamber with prefilled complete transfer buffer (containing 20 % methanol) and place an ice pack in the outer chamber.

3. Cut two pieces of chromatography filter paper and a single piece of PVDF membrane large enough to cover the gel (see **Note 35**). Saturate the PVDF membrane with 100 % methanol and immediately immerse membrane in transfer buffer. Do not allow membrane to dry.

4. Fill a dish large enough to hold the transfer cassette with complete transfer buffer. Immerse the membrane, filter paper, and foam pads in transfer buffer.

5. Remove the gel from the electrophoresis apparatus and trim away the stacking gel using a clean razor blade.

6. Assemble the gel transfer "sandwich" (keeping all materials immersed in transfer buffer) in the following order: sponge, filter paper, gel, PVDF membrane, filter paper, and sponge.

7. Remove any air bubbles from the assembly using a roller (*see* **Note 36**).

8. Place the sandwich in a cassette and insert the cassette into the inner chamber of the transfer apparatus with the gel side of the sandwich closest to the negative (black) electrode and the membrane side of the sandwich closest to the positive (red) electrode (*see* **Note 37**).

9. Run the transfer at 125 V for 1 h 20 min at 4 °C (*see* **Note 38**).

10. When transfer is complete the markers should be visible on the membrane. Disassemble the cassette and sandwich starting at one corner and place the PVDF membrane in blocking buffer on a rocker for 30 min at room temperature.

11. Place the membrane in a bag containing VCAM-1 primary antibody and seal the bag making sure to remove all air bubbles. Rock the membranes overnight at 4 °C (*see* **Note 39**).

12. The next day, remove membranes from primary antibodies and wash with TBS-T four times for 5 min each. Place dishes containing membranes on a rocker for each wash changing buffer between the washes (*see* **Note 40**).

13. Dilute the appropriate HRP-conjugated secondary antibody in blocking buffer and incubate with the membrane for 1 h on a rocker at room temperature.

14. Remove the membranes from secondary antibody and wash with TBS-T four times for 5 min, changing buffer between washes.

15. During the final wash, mix 1 ml of each of the luminal reagents for chemiluminescence detection. Immediately after discarding the TBS-T from the final wash, transfer the blots using forceps to a piece of saran or laboratory wrap on a flattened surface placing blots protein side up. Carefully pipette the mixed luminal reagent onto the blot. Incubate at room temperature for 3 min (*see* **Note 41**).

16. Pick up the membrane using forceps and gently dab off the luminal using Kim-Wipes. Place the membrane between two sheets of acetate (*see* **Note 42**) and tape into place inside an X-ray film cassette.

17. Place a piece of film over the membrane and test a range of exposure times to obtain optimal band intensity. A typical result is shown in Fig. 1.

18. Following exposure for optimal band intensity, the membrane can be removed from the film cassette and incubated in TBS-T for a 10 min wash.

19. Following the TBS-T wash, place the membrane in a bag containing anti-E-selectin primary antibody and seal the bag making sure to remove all air bubbles. Rock the membranes overnight at 4 °C (*see* **Note 43**).

20. Repeat **steps 12–18** to wash membranes, incubate with the appropriate HRP-conjugated secondary antibody, and prepare for chemiluminescence detection for detection of E-selectin.

21. Probe the blots for the last time, incubating the membrane with the anti-PECAM primary antibody as a loading control for protein expression of the endothelium (*see* **Note 44**).

3.8 Preparing Single Cell Suspensions from Spleen

1. Uncap a 50 ml conical tube for each spleen harvested during dissection (*see* Subheading 3.4, **step 13**).

2. Unwrap and place a 40 μm cell strainer onto the top of the 50 ml tube.

3. Pour out each spleen into its own individual cell strainer.

4. Pipette 5 ml of PBS over each spleen.

5. Using the handle portion of the plunger removed from a 3 cc syringe, scrape the spleen against the bottom of the cell strainer to dissociate tissue into single cells. Repeat the dissociation of each spleen using a new plunger.

6. Pipette 10 ml of PBS through the cell strainer and recap tubes.

7. Centrifuge samples at $300 \times g$ for 5 min at room temperature.

8. Decant supernatant and resuspend in 1 ml ACK Lysis buffer. Incubate for 10 min at room temperature to lyse the red blood cells (*see* **Note 45**).

9. Following the incubation, fill the 50 ml tube with PBS and centrifuge samples at $300 \times g$ for 5 min.

10. Aspirate the supernatant and snap-freeze pellets on dry ice/ethanol. At this point the cell pellets can be processed immediately for whole cell extracts or can be stored at –80 °C for preparation of whole cell extracts at a later time (*see* Subheading 3.9).

3.9 Preparing Whole Cell Splenic Lysates for EMSA

1. Resuspend the splenic cell pellets (*see* Subheading 3.8, **step 10**) in 200 μl of Whole Cell Lysis Buffer and transfer to a 1.5 ml tube. Incubate the samples on ice for 30 min.

2. Pellet the cell debris by spinning in a refrigerated microcentrifuge at $16,000 \times g$ for 30 min.

3. Transfer the supernatant containing the protein to a new 1.5 ml tube and either use for EMSA (*see* Subheading 3.10) immediately or snap-freeze on dry ice/ethanol. Samples should be stored at –80 °C.

3.10 Preparing and ³²P-Labeling the NF-κB Probe for EMSA

1. The probe should be prepared in advance and stored at 4 °C in a radiation-safe box. Probes should be used within 2 weeks of the reference date of the ^{32}P-γATP used for labeling (approximately one half-life of ^{32}P).

2. Mix equimolar amounts of each complementary oligonucleotide (*see* **Note 21**) in a 1.5 ml tube. For example, mix 25 μl of one oligonucleotide (100 ng/μl) with 25 μl of the complimentary strand (100 ng/μl) and 50 μl of STE buffer. The final concentration of each oligonucleotide in the annealed probe will be 50 ng/μl.

3. Heat tube in a heat block at 90 °C for 10 min. Remove the entire heat block containing the tube to the bench and cool slowly to room temperature (*see* **Note 46**).

4. Set up a 10 μl reaction containing 1 μl of annealed probe, 1 μl of 10× PNK Buffer, 2 μl of H_2O, 5 μl of ^{32}P-γATP, and 1 μl of PNK.

5. Incubate in a heat block at 37 °C for 1 h. After incubation, add 40 μl of TE at room temperature.

6. Resuspend the Sephadex matrix of a Quick Spin Oligo column by inverting the column and flicking several times. Keeping the column upright, remove the cap and then snap off the bottom tip.

7. To pack the column, place it in a 1.5 ml tube and spin for 1 min at $1,000 \times g$ at room temperature. Discard the collection tube and eluted buffer and place column in a new 1.5 ml tube.

8. Apply the sample carefully to the center of the packed column and centrifuge at $1,000 \times g$ for 4 min at room temperature.

9. Transfer the elute, now containing the purified probe, to a new 1.5 ml tube and discard the column as ^{32}P contaminated solid waste.

3.11 Preparing a TBE Gel for EMSA

1. These directions are for casting a TBE gel using the Model V16 Vertical Gel Electrophoresis apparatus from Gibco/BRL. This makes a gel with 20 wells. Directions for other systems are similar.

2. Wash the glass plates and spray with 70 % ethanol to clean plates and prevent gel from sticking to plates.

3. Assemble the plate in a sealed casting boot using spacers to make a 1 mm thick gel.

4. Prepare the gel by mixing 14.7 ml of water, 2 ml of 10× TBE, 3.3 ml of 30 % acrylamide/bis solution, 200 μl of APS solution, and 15 μl of TEMED. Mix by inversion.

5. Pipette 10 ml of the solution in the chamber and check for leaks. Finish pipetting the solution into the chamber over filling

it to the top. Quickly place the comb in the chamber, making sure it is completely down.

6. Wait for polymerization, watching the unpoured gel solution leftover in the beaker to polymerize. This should occur within 30–45 min.

7. Disassemble the casting apparatus by sliding off the casting boot starting at one corner and clamp the gel into the gel electrophoresis apparatus (*see* **Note 47**). Fill the upper and lower chambers with 0.5× TBE (running buffer).

8. Add 2 μl of 5× sample buffer to one well and then pre-run the gel at 200 V until the dye front has run 9 cm (*see* **Note 48**).

3.12 Performing the EMSA

1. Determine the protein concentration in the samples as described in Subheading 3.5, **steps 5–8**.

2. Set up a 10–20 μl binding reaction in a 1.5 ml microcentrifuge tube containing 5 μg of whole cell lysates from spleen, 1 μl of poly-dI:dC, 5 mM MgCl$_2$ diluted from 100 mM stock, and the appropriate volume of 2× binding buffer. Incubate on ice for 10 min.

3. Add 1 μl of labeled probe to each sample and incubate at room temperature for 20 min.

4. Add 2–4 μl of 5× electrophoresis sample buffer and then load the samples onto the gel using gel-loading tips (*see* **Note 49**).

5. Run gel at 200 V at room temperature until the dye front has reached the bottom of the gel. This occurs approximately after 1 h 45 min.

6. Cut two pieces of chromatography filter to the exact size of the gel and keep together to double the thickness.

7. Disassemble the gel apparatus in a sink by first unclamping gel from the apparatus and pouring out buffer. Pull out the spacers and detach the short glass plate while leaving the gel attached to the large glass plate.

8. Place the filter paper on the gel and carefully peel the gel off the glass plate. The gel will attach to the filter paper.

9. Place the filter paper with the gels onto the bed of a gel dryer and cover with saran or laboratory wrap (*see* **Note 50**). Dry the gel and then place it into an X-ray cassette.

10. In a dark room under safe lights, place a piece of BioMax X-ray film on the gel and close the cassette. A number of exposures should be tested. If the exposure is weak, then insert enhancer screens into the cassette and store the cassette containing the film in the dark at −80 °C until a suitable exposure has been obtained. A typical result from an experiment assaying the effects of WT and MUT NBD peptide on LPS-induced NF-κB is shown in Fig. 2.

4   Notes

1. LPS is prepared as a stock of 5 mg/ml and aliquoted into separate tubes of 20 µl and stored at –20 °C to avoid repeated freeze thaws. Dilute as working stock, immediately prior to injections, by adding 180 µl of sterile PBS in a laminar flow hood. One aliquot is used for each mouse to be injected. Each mouse is then injected with the entire 200 µl of solution for a total treatment of 100 µg.

2. The WT NBD peptide and MUT control peptide can be obtained from commercial sources including Calbiochem (San Diego, CA) and Biomol (Plymouth Meeting, PA). We have had good experience obtaining peptides from Biomol. Peptides should be stored at –20 °C in a sealed box containing desiccant. We have successfully used peptides for experiments that have been stored in this manner for over 3 years.

3. Do not make a storage stock of peptides in DMSO. The NBD peptide becomes oxidized after several hours in DMSO at any temperature including –80 °C. Oxidation of the peptide renders it inactive. The NBD peptide and MUT control should only be dissolved in DMSO immediately before use.

4. Using 10 ml syringes with separate detached needles is easier to fill prior to attaching a separate needle. Additionally, using syringes with detached needles creates less syringe waste since one syringe can be reused by refilling it with 10 ml of PBS followed by the addition of a new sterile needle for each mouse.

5. Adventitial fat surrounding the aorta can be removed in long pieces by gliding the mini-Vannas scissors along the exterior wall of the aorta lengthwise rather than cutting off small pieces. The dissection scope is best set at the lowest magnification to give you the longest view of the aorta. We use a magnification of 0.8× which allows us to work with 1/4 of the aorta length within each field of view.

6. Alternative mini-Vannas scissors with small cutting edges (typically 5 mm or lower) can be used for the dissection and removal of the aorta, including models that are straight; however we find the angled edge allows easier manipulation and more comfortable hand position during the dissection.

7. Alternative fine-tipped forceps can be used; however serrated edges should be avoided because they more easily rip into the walls of the aorta. We find the Dumont #5 size provides the easiest manipulation for handling the aorta without breaking the aorta or tearing holes in the walls of the aorta.

8. TNT lysis buffer can be stored at room temperature or at 4 °C; however after several weeks buffer becomes cloudy and should

not be used for cell lysis. We often prepare fresh TNT after one month of storage regardless of its appearance.

9. Dissolve one tablet of a mini-complete protease inhibitor cocktail in 400 μl of water to make a 25× stock. This can be stored at −20 °C in aliquots if not used immediately. Frozen aliquots should not be used with repeated freeze thaws. Protease inhibitors should not be stored in TNT lysis buffer and should be added fresh the same day as performing cell lysis. Use 40 μl of 25× stock per 1 ml of TNT.

10. Phosphatase inhibitors must be added fresh to TNT lysis buffer the same day as performing cell lysis and should not be stored in TNT lysis buffer. Both NaF and β-glycerophosphate can be prepared as 1 M stocks and stored in small aliquots at −20 °C. Aliquots should only be thawed once and discarded; therefore we make 10 μl aliquots.

11. Using the Coomassie Plus Protein Reagent, protein concentrations can be determined using a microplate reader equipped with a 595-nm filter. We use a Bio-Rad Model 680 microplate reader utilizing Microplate Manager-III software (Bio-Rad, Hercules, CA, USA).

12. A 1 M Tris–HCl pH 6.8 stock cannot be used during the preparation of Sample Buffer (5×) because the required volume of 1 M Tris will be higher than volume availability. We use a 2 M Tris–HCl pH 6.8 stock for preparation of Sample Buffer.

13. Unpolymerized acrylamide is a potentially harmful neurotoxin. Extreme care should be taken when handling acrylamide including the wearing of gloves and safety glasses.

14. 10 % stocks of APS solution can be stored at 4 °C for up 2 weeks without affecting polymerization.

15. We make a large stock of 20 l of 5× running buffer which can be stored at room temperature for up to 3 months.

16. We make a 20 l stock of transfer buffer containing Tris and Glycine but without the Methanol. On the day of the immunoblotting, we make 1 l of complete transfer buffer using 800 ml of the transfer buffer stock and 200 ml methanol and pre-chill at 4 °C at least 1 h prior to setting up the transfer. This 1 l will be sufficient to fill the transfer apparatus and have enough for setting up the transfer "sandwich."

17. For convenience during the repeated use of TBS-T, we make a 20 l stock of 1× TBS-T which can be stored for up to 3 months at room temperature.

18. Primary antibodies diluted in blocking buffer containing 0.1 % (w/v) sodium azide can be reused up to three times.

19. Sodium azide inhibits the enzymatic activity of HRP and should not be used for the preparation of secondary antibodies

in blocking buffer. Secondary antibodies should be prepared freshly and not stored for later or multiple uses since there is no sodium azide.

20. DTT is prepared as 1 M stock and stored at –20 °C in small aliquots for single use.

21. The consensus NF-κB oligo pair is 5′-AGTTGAGGGGACTT TCCCAGG-3′ and 5′-GCCTGGGAAAGTCCCCTAACT-3′. We routinely purchase oligonucleotides from Integrated DNA Technologies Inc. (Coralville, IA, USA).

22. All work with radioactive isotopes should be performed in accordance with all prevailing safety and disposal regulations.

23. Poly-dI:dC is prepared by diluting ten Absorbance Units (A_{260}) into 0.5 ml of H_2O and then storing in 10 μl aliquots at –20 °C.

24. Some insulin syringes have "dead" space that must be accounted for within your measurement. Each brand has different amounts of dead space so the amount of dead space for the particular insulin syringe must be looked up.

25. If you are in the proper position for injections, there should be nothing being aspirated into the syringe and the injection can proceed. If any blood is aspirated into the syringe, then your needle has punctured a blood vessel. Any green or brown aspirate means the needle is placed in the intestine. Do not continue with the injection until you have a fresh syringe and needle.

26. LPS is an endotoxin and when administered at high doses, such as those used in this protocol, results in sepsis and will show various symptoms of septic shock, including shaking, lower activity levels, and hunching. Mice should not be left for more than 4 h to avoid the risk of animal suffering and death.

27. As the aorta is dissected free, there is some flexibility to pull up on the aorta to slide the scissors along the walls, but be gentle in moving the aorta. If the aorta breaks, the pieces recoil. The aorta is much more difficult to dissect if it breaks, because you need to pull taut one end to keep exposing the aortic wall from the fat.

28. The aortic arch is covered by the left brachiocephalic vein coming out of the heart, so make sure to pull down on the aorta and cut through the vein first to expose the underlying ventral side of the aorta coming into the curvature of the aortic arch.

29. It is important to try to rinse off as much blood as possible from the aorta prior to putting the opened aorta into lysis buffer, since trace amounts of blood can cause misreading of protein concentrations during the Coomassie protein assay.

30. Protein concentrations can be increased if the signal is too weak as long as it fits within in the volume allowed by the well thickness. Protein concentrations can also be decreased if the

signal is too strong. We have found that 20 μg of protein from aortic tissue provides a sufficient signal for VCAM-1, E-selectin, and PECAM.

31. Precast gels can be obtained from several commercial sources. These are an expensive alternative to casting gels but provide reproducible high-quality results.

32. These volumes are sufficient for making two 0.75 mm thick gels.

33. Once TEMED is added to the stacking gel, the polymerization reaction begins, so the combs should be placed into the gel apparatus immediately following the pouring of the stacking gel.

34. Run the gel until the dye front emerges into the running buffer. This will provide enough separation to distinguish VCAM-1, E-selectin, and PECAM proteins following chemiluminescence detection.

35. Chromatography filter paper cut to pieces of 3″×4″ provide sufficient coverage of the gel and fit in the transfer cassette without overhanging. A membrane size of 2.5″×3.5″ will cover the separating portion of the gel.

36. A cut down 5 or 10 ml plastic pipette or a 15 ml tube can be used as a roller.

37. If using the Bio-Rad apparatus, the cassette has a black side and an opaque side. The black side can be used as a convenient guide to assemble the sandwich and align the cassette with the negative electrode in the transfer tank.

38. For convenience, we set up the apparatus in a chromatography refrigerator to transfer at 4 °C. Alternatively a 4 °C cold room can be used.

39. We have found the Impulse Sealer from Hualian Packaging Machinery Co. Ltd. (Wenzhou City, China) to be ideal for immunoblotting purposes.

40. Lids from pipette tip boxes work well as dishes for incubating membranes with TBS-T and secondary antibody solutions.

41. It is not necessary to perform this step in a dark room. However, the subsequent exposure to film should be performed under a safety light in a dark room.

42. Plastic sheet protectors available from most stationary stores can be cut to size and used for holding membranes for exposure instead of acetate sheets.

43. We re-probe membranes for detection of multiple proteins without performing stripping procedures when proteins of interest are different sizes. For example, VCAM-1 (64kD), E-selectin (120 kD), and PECAM (140 kD) can each be distinguished by size without removal of the signal leftover from previous chemiluminescence detection. If detecting proteins of

similar size, membranes must be stripped using commercially available stripping buffers prior to placing membranes into a different primary antibody.

44. Alternative primary antibodies such as Tubulin can be used as a loading control for expression of total protein.

45. Do not leave spleen cells in ACK lysis buffer for longer than 10 min; this increases cell death and makes cells hard to resuspend in lysis buffer.

46. Alternatively, this reaction could be performed in a PCR machine, beginning with a 90 °C incubation stage of 10 min and then bringing the temperature down using a gradient program that steps the incubation temperature down over time.

47. Binder clips can be used to clamp the gel plates to the gel apparatus. In order to avoid any possible leakage between the plates and the black gasket, two binder clips should be used on each side.

48. Pre-running the gel removes unpolymerized acrylamide. This greatly improves the resolution of the EMSA compared with non-pre-run gels.

49. These samples are used directly following the addition of 5× electrophoresis sample buffer and should not be heated as heating could affect the binding of proteins with the radiolabeled DNA probes.

50. Do not let the saran or laboratory wrap fold under the filter paper when placed on the bed of the gel drier. If this happens, the gel drier will not be able to extract liquid from the gel and it will not be dried properly.

Acknowledgments

Work in the authors' laboratory was supported by NIH RO1 HL080612 and RO1 HL096642.

References

1. Delhalle S, Blasius R, Dicato M et al (2004) A beginner's guide to NF-κB signaling pathways. Ann N Y Acad Sci 1030:1–13
2. Gilmore T.D. (2008) www.NF-kB.org
3. Courtois G, Gilmore TD (2006) Mutations in the NF-κB signaling pathway: implications for human disease. Oncogene 25:6831–6843
4. Karin M (2006) Nuclear factor-κB in cancer development and progression. Nature 441: 431–436
5. Tak PP, Firestein GS (2001) NF-κB: a key role in inflammatory disease. J Clin Invest 107:7–11
6. Baker RG, Hayden MS, Ghosh S (2011) NF-κB, inflammation and metabolic disease. Cell Metab 13:11–22
7. May MJ, D'Acquisto F, Madge LA et al (2000) Selective inhibition of NF-kappaB activation by a peptide that blocks the interaction of NEMO with the IkappaB kinase complex. Science 289: 1550–1554
8. May MJ, Marienfeld RB, Ghosh S (2002) Characterization of the IkappabB-kinase NEMO binding domain. J Biol Chem 277: 45992–46000

9. Ethridge RT, Hashimoto K, Chung DT et al (2002) Selective inhibition of NF-kappaB attenuates the severity of cerulean-induced acute pancreatitis. J Am Coll Surg 195:497–505

10. Dai S, Hirayama T, Abbas S et al (2004) The IkappaB kinase (IKK) inhibitor, NEMO-binding domain peptide, blocks osteoclastogenesis and bone erosion in inflammatory arthritis. J Biol Chem 279:37219–37222

11. Min SY, Yan M, Du Y et al (2013) Intra-articular nuclear factor-kB blockade ameliorates collagen-induced arthritis in mice by eliciting regulatory T cells and macrophages. Clin Exp Immunol 172(2):217–227

12. Shibata W, Maeda S, Hikiba Y et al (2007) Cutting edge: the IkappaB kinase (IKK) inhibitor, NEMO-binding domain peptide, blocks inflammatory injury in murine colitis. J Immunol 179(5):2681–2685

13. Peterson JM, Kline W, Canan BD et al (2011) Peptide-based inhibition of NF-kappaB rescues diaphragm muscle contractile dysfunction in a murine model of Duchenne muscular dystrophy. Mol Med 17:508–515

14. Habineza NG, Gaurnier-Hausser A, Patel R et al (2014) A phase I clinical trial of systemically delivered NEMO binding domain peptide in dogs with spontaneous activated B-cell like diffuse large B-cell lymphoma. PLoS One 9(5):e95404

15. de Martin R, Hoeth M, Hofer-Warbinek R et al (2000) The transcription factor NF-κB and the regulation of vascular cell function. Arterioscler Thromb Vasc Biol 20(11):e83–e88

Chapter 31

Regulation of NF-κB Signaling in Osteoclasts and Myeloid Progenitors

Gaurav Swarnkar and Yousef Abu-Amer

Abstract

The transcription factor nuclear factor kappa-light-chain-enhancer of activated B cells (NF-κB) is crucial for immune responses and skeletal development. Work in recent years has shown that various members of the NF-κB family are viable targets to regulate activity and survival of bone cells and hence bone metabolism. In this regard, deletion of upstream kinases or distal NF-κB subunits resulted with bone deformities. Thus, it has become increasingly apparent that detailed investigation of NF-κB in bone cells may provide opportunities to design new therapeutic modalities. In this chapter we present modified methodology describing efficient approaches to regulate the NF-κB pathway in vitro and in vivo to assess its function in bone cells and tissues.

Key words NF-κB, NEMO, IKK, Osteoclast, RANKL

1 Introduction

Various studies have established that the transcription factor NF-κB mediates receptor activator of nuclear factor kappa-B ligand (RANKL)-induced osteoclastogenesis and thus plays a major role in normal skeletal functions and development [1–5]. Deletion or modulation of various components of this pathway results in abnormal skeletal development [5–11]. According to current findings binding of RANKL to its receptor, Receptor Activator of Nuclear Factor κ-B (RANK) triggers formation of an IKK complex (IκB kinase) containing various adaptors and kinases, including TNF receptor associated factor (TRAF6), TGF-β–activated kinase 1 (TAK1), and NF-kappa-B essential modulator (IKKγ/NEMO) which leads to phosphorylation and activation of IKK2 [12–16]. Active IKK2 phosphorylates the inhibitory protein IκB, which in turn undergoes rapid proteasomal degradation resulting in accumulation of p65/RelA and p50. These NF-κB subunits form dimers and translocate to the nucleus to activate target genes [1].

Michael J. May (ed.), *NF-kappa B: Methods and Protocols*, Methods in Molecular Biology, vol. 1280,
DOI 10.1007/978-1-4939-2422-6_31, © Springer Science+Business Media New York 2015

However, work with primary precursors of bone cells such as monocytes/macrophages is challenging, as these cells are not readily transfectable. In this chapter, we present modified methodology describing efficient approaches to regulate the NF-κB pathway in vitro and in vivo to assess its function in bone cells and tissues.

2 Materials

2.1 Isolation of Bone Marrow Monocytic Cells (BMMs)

1. 4- to 6-week-old mouse.
2. Sterile Dulbecco's PBS (DPBS).
3. Minimal Essential Medium, alpha (α-MEM).
4. α-MEM containing 10 % fetal bovine serum (FBS).
5. 70 μM cell strainers (BD, Franklin Lakes, NJ, USA).
6. M-CSF (R&D Systems, Minneapolis, MN, USA).
7. RANKL (R&D Systems).
8. Leukocyte acid phosphatase kit (Sigma-Aldrich, St. Louis, MO, USA).

2.2 NEMO Binding Domain Binding Peptide (NBD-BP) Synthesis and FPLC-Purification

1. BL21 *E. coli* bacteria transformed with a pTAT-cDNA-NBD plasmid.
2. Isopropyl-beta-D-thiogalactopyranoside (IPTG).
3. Buffer Z: 8 M urea, 100 mM NaCl, 20 mM HEPES pH 8.0.
4. Imidazole.
5. Buffer Z containing imidazole: 8 M urea, 100 mM NaCl, 20 mM HEPES (pH 8.0), 10–20 mM imidazole.
6. Ni-NTA column for purification of His-tagged proteins.
7. SDS-PAGE gels and running system.
8. Coomassie blue stain.
9. Slide-A-Lyzer Dialysis Cassettes (Thermo Scientific, Rockford, IL, USA).
10. Dialysis buffer: 137 mM NaCl, 20 mM HEPES pH 8.0.
11. PBS.
12. BSA.
13. Mono Q 10/10 column (GE Healthcare Life Sciences, Mickleton, NJ, USA).
14. FPLC system such as the AKTA from GE Healthcare Life Sciences.
15. Buffer A: 50 mM NaCl, 20 mM HEPES pH 8.0.
16. Buffer B: 1 M NaCl, 20 mM HEPES pH 8.0.
17. PD-10 disposable G-25 Sephadex gravity columns (GE Healthcare Life Sciences).

2.3   Testing the NBD-BPs

1. Lysis buffer: 40 mM Tris–HCl, pH 8.0, 500 mM NaCl, 0.1 % Nonidet P-40, 6 mM EDTA, 6 mM EGTA, 5 mM β-glycerophosphate, 5 mM NaF, 1 mM NaVO$_4$.

2. Protease Inhibitor Cocktail (Roche Applied Science, Madison, WI, USA).

3. Cell scrapers.

4. BCA protein assay (Thermo Scientific).

5. Isotype matched IgG antibody for clearing the immunoprecipitation.

6. Protein A/G beads.

7. Anti-IKK1 and anti-IKK2 antibodies (Cell Signaling Technology, Danvers, MA, USA).

8. Kinase buffer: 50 mM Tris–HCl, pH 8.0, 100 mM NaCl, 10 mM MgCl$_2$, 1 mM dithiothreitol, 10 μM ATP, 5 mM β-glycerophosphate, 5 mM NaF, 1 mM NaVO$_4$.

9. GST-IκBα protein substrate.

10. 2× SDS protein sample buffer.

11. Anti-IκBα and anti-p-IκBα antibodies (Cell Signaling Technology).

12. Hypotonic lysis buffer A: 10 mM HEPES, pH 7.8, 10 mM KCl, 1.5 mM MgCl$_2$, 0.5 mM dithiothreitol, 0.5 mM 4-(2-aminoethyl)benzene-sulfonyl fluoride (AEBSF), 5 μg/ml leupeptin.

13. Nonidet P-40.

14. Nuclear extraction buffer B: 20 mM HEPES, pH 7.8, 420 mM NaCl, 1.2 mM MgCl$_2$, 0.2 mM EDTA, 25 % glycerol, 0.5 mM dithiothreitol, 0.5 mM AEBSF, 5 μg/ml pepstatin A, 5 μg/ml leupeptin.

15. p32 isotope end-labeled double-stranded oligonucleotide probe containing the sequence 5′-AAA CAG GGG GCT TTC CCT CCT C-3′ derived from the κB3 site of the TNF promoter.

16. Binding buffer: 20 mM HEPES, pH 7.8, 100 mM NaCl, 0.5 mM dithiothreitol, 1 μg poly (dI-dC), and 10 % glycerol.

17. 4 % native polyacrylamide gel.

2.4   Constitutive Expression of IKK2-SSEE in Osteoclasts

1. QuickChange II Site Directed Mutagenesis Kit (Stratagene, La Jolla, CA, USA).

2. Mutation primers for PCR: (IKKβ_S177_181E_f, GAGCTGGATCAGGGCGAACTGTGCACGGA ATTTGTGGGGACTCTGC, and IKKβ_S177_181E_r, GCAGAGTCCCCACAAATTCCGTGCACA GTTCGCCCTGATCCAGCTC).

3. Retroviral pMX-flag-IKK2-WT, pMX-GFP plasmids.

4. Plat-E retroviral packaging cells.

5. DMEM containing 10 % FBS.

6. Serum-free DMEM.

7. XtremeGENE 9 transfection reagent (GE Healthcare Life Sciences).

8. 0.45 μM syringe filters.

9. Polybrene (8 μg/ml).

10. Accutase cell detachment medium (GE Healthcare Life Sciences).

2.5 In Vivo Effects of IKK2-SSEE Expression on Osteoclastogenesis

1. R26StopIKK2-SSEE-floxed mouse.

2. 10 % neutral formalin.

3. 70 % ethanol.

4. MicroCT instrument such as the μCT 40 (Scanco Medical, Brüttisellen, Switzerland).

5. 10 % EDTA, pH 7.0 to de-calcify bones.

6. Graded alcohol solutions and xylene to dehydrate decalcified bones.

7. Paraffin for embedding tissue.

8. Hematoxylin and eosin stain.

9. Tartrate-resistant acid phosphatase (TRAP).

2.6 SUMOylation of NEMO

1. pMx retroviral vector.

2. Protein lysis buffer: 20 mM Tris–HCl, pH 8.0, 150 mM NaCl, 1 mM EDTA, 1 mM EGTA, 1 mM β-glycerophosphate, 1 % Triton, 1 mM NaF, 1 mM NaVO$_4$.

3 Methods

3.1 Modulating NFκB Signaling During Osteoclastogenesis

3.1.1 Isolation of Bone Marrow Monocytic Cells (BMMs)

1. Day 0: Dissect the long bones (femur and tibia) from a 4- to 6-week-old mouse and place them in petri plate with sterile DPBS on ice. Clean the muscles and other tissues from the bone.

2. Poke 3–4 holes in a 0.6 ml tube using a 20-G needle and place it inside a 1.5 ml tube and add 400 μl of α-MEM without FBS.

3. Cut the ends of the femur and tibia carefully and place them vertically inside the 0.5 ml tube from **step 2**.

4. Centrifuge at 12,000–16,000×*g* at 4 °C for 3 min. During the centrifugation place a 70 μM cell strainer on a 50 ml tube.

5. The bone marrow will pass from the 0.6 ml tubes into the 1.5 tubes during centrifugation. Discard the 0.6 ml tubes containing the bones (devoid of marrow).

6. Resuspend the bone marrow cell (BMMs) pellet in α-MEM and filter through the 70 µM cell strainer. The cells will be collected in the 50 ml tube leaving the debris in the strainer.

7. Rinse the 1.5 ml tube and strainer once again with fresh α-MEM and add to the 50 ml tube through the strainer to increase the yield.

3.1.2 Differentiation of BMMs to Osteoclast

1. Plate the cells in 150 mm petri plate in α-MEM containing FBS and 50 ng/ml M-CSF (R&D # 416-ML-010) (*see* **Note 1**).

2. Day 1: After 24 h, collect the cells in a sterile 50 ml tube. Rinse the plate two times with sterile DPBS to collect all the non-adherent cells. These cells will be differentiated to osteoclasts (OC).

3. Centrifuge at $300 \times g$ for 5 min. Resuspend the cells in 5 ml of α-MEM with FBS.

4. Count and plate the cells in a 96-well plate at different densities (e.g., 10,000; 20,000; and 40,000 cells/well). Add 50 ng/ml M-CSF and 50 ng/ml RANKL to wells containing the cells. Replenish MCSF and RANKL again on day 3.

5. On day 4 or 5, mature multinucleated OC will be readily visible in the plate. Fix and stain the cells for tartrate-resistant acid phosphatase (TRAP) using the leukocytes acid phosphatase kit (Fig. 1).

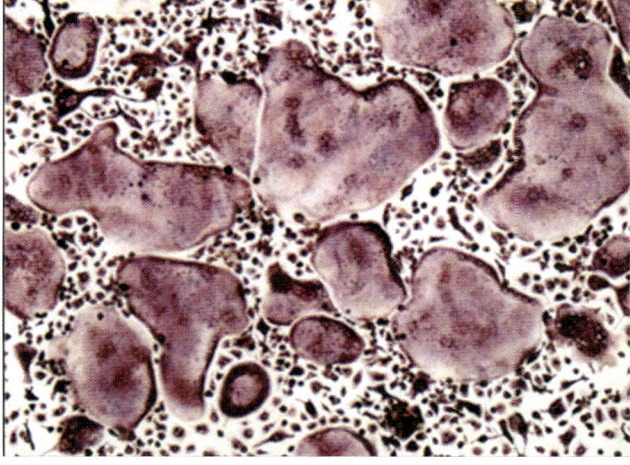

Fig. 1 Mature multinucleated TRAP positive Osteoclast from bone marrow cells

3.2 NEMO Binding Domain Binding Peptide (NBD-BP): Synthesis and FPLC-Purification of Three TAT-NBD-BPs

The NBD was discovered by May et al. [17] as a specific region in IKK1 and IKK2 that facilitates binding of NEMO. Subsequently, a corresponding peptide was generated and widely used as an inhibitor of NEMO binding to the IKKs and hence inhibited NF-κB activation in vitro and in a wide range of inflammatory animal models [18–24]. Given that BMMs and osteoclasts are difficult to transfect, we have developed and adopted a TAT-based system to enable delivery of the NBD-BP into these cells and study the role of NF-κB signaling in OCs and regulation of bone resorption [21]. We describe here the protocol to synthesize, purify, and test three NBD-BPs: (1) Functional wild-type TAT-NBD (YGRKKRRQRRR-G-TTLDWSWLQME), (2) Negative control mutant TAT-NBD (YGRKKRRQRRR-G-TTLDASALQME), (3) GFP-conjugated TAT-NBD to trace distribution (GFP-YGRKKRRQRRR-G-TTLDWSWLQME) (*see* **Note 2**).

3.2.1 Purification

1. Start a 200 ml overnight culture of BL21 *E. coli* expressing His-tagged pTAT-cDNA-NBD plasmids of interest in LB broth.

2. Next day, inoculate 1 l of LB broth with the 200 ml overnight culture and incubate at 37 °C with shaking until OD_{600} reaches 0.4–0.8.

3. Add 400 µM IPTG and rotate for 4–6 h at 37 °C.

4. Spin the cells at $2,300 \times g$ for 5 min. Wash the pellet with 50 ml of PBS and spin again.

5. Resuspend the pellet in 10 ml of buffer Z then sonicate on ice for 3×15 s pulses or until turbid. Clarify by centrifugation at $13,000 \times g$ for 10 min at 4 °C. Save the supernatant.

6. Bring supernatant up to 10–20 mM imidazole and add at 4 °C or RT to a pre-equilibrated 3–10 ml Ni-NTA column in Buffer Z containing 10–20 mM imidazole. Allow flow to proceed by gravity or apply slight air-pressure via a syringe as required. Save the flow-through (FT).

7. Wash the column with 50 ml of Buffer Z containing imidazole.

8. Elute the His-TAT-NBD protein by stepwise addition of 5–10 ml each of 100 mM, 250 mM, 500 mM, and 1 M imidazole in Buffer Z.

9. Run samples from the supernatant, flow-through, wash, and fractions on an SDS-PAGE gel and stain with Coomassie Blue to visualize the purified peptides.

3.2.2 Dialysis

1. Dialyze the peptides overnight in a "Slide-A-Lyzer" from Pierce against dialysis buffer or PBS in a large volume (~4 l). Use an oversized Slide-A-Lyzer to increase the surface area:volume ratio, and decrease the time to reach equilibrium.

2. Change the buffer twice.

3. After dialysis, spin out the insoluble particles and determine the protein concentration and integrity by running a sample on an SDS-PAGE gel along with protein standards (i.e., a range of concentrations of BSA) and then staining the gel with Coomassie Blue. It is important to know how much peptide was lost to dialysis (i.e., remained insoluble), so check the samples from before and after dialysis.

3.2.3 Ion Exchange Chromatography/ FPLC

1. Dilute the pooled fractions 1:1 in Buffer Z so that the buffer contains 4 M urea, 50 mM NaCl, 20 mM HEPES (pH 8.0).

2. Inject the sample into a 10/10 or larger Mono-Q column attached to an FPLC system pre-equilibrated in Buffer A (*see* **Note 3**).

3. Wash the column with 40 ml of Buffer A and elute the HIS-TAT- protein by separate steps of 150, 250, 350, and 500 mM NaCl by gradient addition of Buffer B.

4. Desalt the fractions on a PD-10 disposable G-25 Sephadex gravity column (*see* **Note 4**). This will result in a 1:1.4 dilution; however, a sharper peak is obtained with a 1 M NaCl step, and therefore you can afford the resulting dilution caused here.

5. Check the Supernatant, FT, and fractions by Coomassie blue-stained SDS-PAGE and compare to a standard, such as BSA, for protein concentration on same gel.

6. Flash-freeze in aliquots of 250–500 μl in 10–15 % glycerol on dry ice/liquid N_2 and store at –80 °C.

7. Thaw a test vial, spin at $13,000 \times g$ at 4 °C for 10 min. Run the supernatant on an SDS-PAGE gel and compare to pre-freeze sample for the drop out of solution rate of specific HIS-TAT-NBD proteins [25–29] (*see* **Note 5**).

3.3 Determining the Effects of the NBD-BPs

3.3.1 Testing the Effects of the NBD-BPs on Osteoclastogenesis

1. For osteoclastogenesis, add 50–200 μM of NBD binding peptides to the BMM cultures (Subheading 3.1).

2. Add NBD-BPs again after 2 days (day 3 of OC culture) with RANKL and M-CSF.

3. Fix and stain the cells for tartrate-resistant acid phosphatase (TRAP) expression using leukocytes acid phosphatase kit (Fig. 2).

3.3.2 Testing the Effects of the NBD-BPs on IKK Activity (Kinase Assay)

1. Isolate the BMMs as described in Subheading 3.1. Plate BMMs in a 95 mm tissue culture plate (p100) at a density of $2–3 \times 10^6$ cells per plate. Add 50 ng/ml M-CSF to the plate.

2. When the cells adhere to the plate and reach 80–90 % confluence, remove the medium and wash with 10 ml of sterile DPBS. Add 10 ml of serum-free α-MEM to the plates with cells and starve the cells for 2–3 h.

Control   **WT-NBD**   **Mutant-NBD**

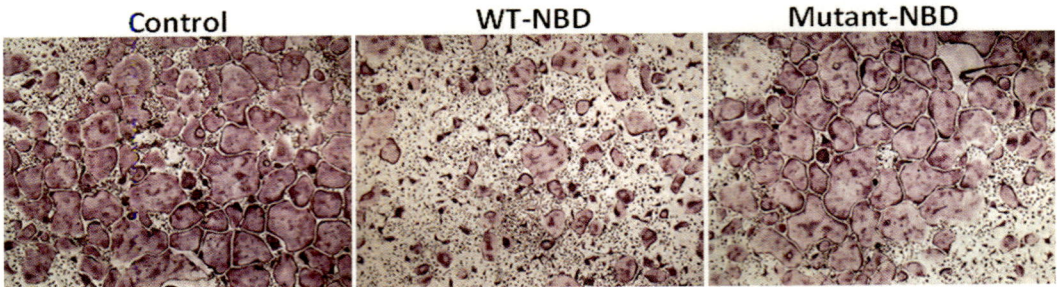

Fig. 2 Osteoclastogenesis in the presence of WT-NBD and NBD-negative control (NBD, NEMO binding domain binding peptide)

3. After starvation, treat the cells with 0, 50, and 200 μM of NBD-BP and negative control (Mutated) NBD-BP for 1 h followed by RANKL stimulation (40 ng/ml) for 30 min.

4. Remove the medium from the plates and wash twice with ice-cold PBS. Remove all the DPBS from the plates (aspirate with a vacuum to minimize leftover DPBS). Handle the plates on ice or in a cold room.

5. Collect the cells in lysis buffer containing the protease inhibitor cocktail. For each plate, use 500 μl of lysis buffer. To collect the lysate, scrape the cells using a cell scraper while keeping the plates on ice.

6. Transfer the cell lysate to a 1.5 ml tube and keep it on ice. Disrupt using 20-G needles for four times, place on a nutator at 4 °C for 15–20 min, and then centrifuge the 1.5 ml tube at $16,000 \times g$ for 10 min.

7. Keep the supernatant and discard the pellet. Measure the protein concentration of the supernatant using a BCA protein assay.

8. Pre-clear the lysate using respective IgG (50 ng/tube) with 30 μl of Protein A/G beads for 15–30 min. Spin down at $16,000 \times g$ for 10 min and keep the supernatant.

9. Add 2 μl of anti-IKKα, anti-IKKβ, or isotype-matched IgG antibody to it and rotate the tubes for 1–2 h at 4 °C. Add 60–75 μl of protein A/G beads to the lysate and rotate at 4 °C overnight.

10. Next day, spin down and wash the beads twice with lysis buffer, then wash beads with 1 ml of kinase buffer twice.

11. Perform kinase assays at 30 °C for 30 min in kinase buffer containing immunoprecipitates and 5 μg of a GST-IκBα substrate.

12. Terminate the reaction by adding 2× SDS protein sample buffer. Boil for 5 min, spin, and load on 10 % SDS-PAGE and dry. Phosphorylation of GST-IκB is then detected by immunoblot using anti-IκBα and anti-p-IκBα (Fig. 3).

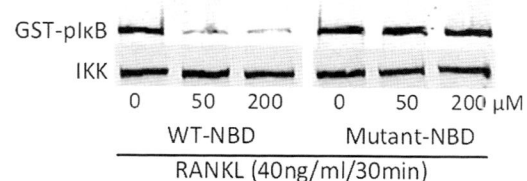

GST-pIκB

IKK

| 0 | 50 | 200 | 0 | 50 | 200 μM |

WT-NBD   Mutant-NBD

RANKL (40ng/ml/30min)

Fig. 3 NBD peptide inhibits IKK kinase activity (RANKL, Receptor activator of nuclear factor kappa-B ligand; NBD, NEMO binding domain binding peptide)

3.3.3 Testing the Effects of the NBD-BPs on NFκB Activation by EMSA

1. Plate the BMMs in p100 plates at a density of $2–3 \times 10^6$ cells/plate. Add 50 ng/ml M-CSF to the plate.

2. Once the cells adhere to the plate at approximately 80–90 % confluence, remove the medium and wash the cells with 10 ml of sterile DPBS. Add 10 ml of serum-free medium to the p100 plates with cells. Starve the cells for 2–3 h.

3. After starvation, treat the cells with 0, 50, and 200 μM of NBD-BP or the negative control (Mutated) NBD-BP for 1 h followed by RANKL stimulation (40 ng/ml) for 30 min.

4. To prepare the nuclear lysate, wash the cells twice in ice-cold PBS. Resuspend the cells in hypotonic lysis buffer A and incubate on ice for 15 min. Add Nonidet P-40 to a final concentration of 0.64 %.

5. Pellet down the nuclei by centrifuging at $9,000 \times g$ for 10 min and carefully remove the cytosolic fraction.

6. Resuspend the nuclei in nuclear extraction buffer B. Vortex for 30 s and rotate for 30 min at 4 °C.

7. Centrifuge the samples at $9,000 \times g$ for 10 min and transfer the nuclear proteins in the supernatant to fresh tubes and measure protein content.

8. Incubate the nuclear extracts (10 μg) with the p32 isotope end-labeled double-stranded oligonucleotide. Perform the reaction in a total of 20 μl of binding buffer for 30 min at room temperature.

9. Fractionate samples on a 4 % native polyacrylamide gel and visualize by exposing the dried gel to film (Fig. 4, *see* **Note 6**).

3.4 Constitutive Activation of IKK2 (IKK2-SSEE) in Osteoclasts In Vitro

The IκB kinase (IKK) complex activates NF-κB downstream of RANK. Upstream signals like RANKL lead to association of two catalytically active kinases, IKKα and IKKβ, with IKKγ/NEMO. This association is required for activation of IKK through phosphorylation of two serines in the activation loop of IKK2. IKK then phosphorylates IκB, leading it for proteasomal degradation and allowing NF-κB to enter the nucleus and regulate gene transcription. IKK2 is necessary for RANKL-mediated osteoclastogenesis,

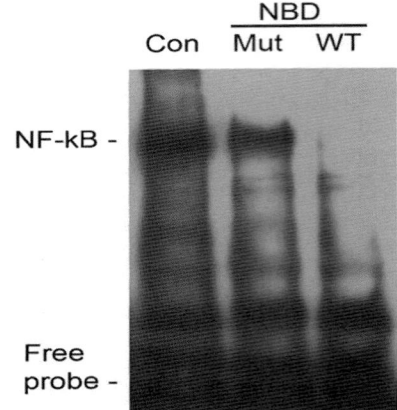

Fig. 4 NBD peptide inhibits NFκB activation

but its activation also is sufficient for osteoclast formation. Using an active mutated IKK2 (IKK2-SSEE) where the serine resides has been mutated to glutamic acid can induce osteoclast formation even in the absence of RANKL [30].

1. Generate IKK2-SSEE mutations using the QuickChange II Site Directed Mutagenesis Kit (Stratagene, La Jolla, CA, USA) using retroviral pMX-flag-IKK2 as a template.

2. To generate retroviral particles, use Plat-E cells which are stably expressing retroviral structural proteins gag-pol and env for transient production of high-titer retrovirus. Briefly, plate 5×10^6 Plat-E cells/p100 in DMEM with 10 % FBS. Make three plates each for transduction with pMX-GFP, pMX-IKK2-WT, and pMX-IKK2-SSEE.

3. Next day, transfect the Plat-E cells with pMX-GFP, pMX-IKK2-WT, and pMX-IKK2-SSEE. Specifically, in a sterile tube add 485 µl serum-free DMEM, 15 µl of XtremeGENE 9, and 5 µg of DNA (pMX-GFP, pMX-IKK2-WT, and pMX-IKK2-SSEE). Wait for 15–20 min and then add it to the plate in a dropwise manner. Swirl the plate slowly to spread the DNA evenly. Incubate the plate in a CO_2 incubator.

4. Next day, check for transfection efficiency of pMX-GFP under a UV-microscope. Normally this should be >90 % transfection efficiency. Change the medium to α-MEM with 10 % FBS.

5. After 24 h collect the α-MEM containing retroviral particles and filter it using a 0.45 µM syringe filter (day 1 collection). Add fresh α-MEM with 10 % FBS and collect again after 24 h (day 2 collection).

6. The virus containing medium can be stored on ice or at 4 °C. Freeze-thawing the virus reduces its efficiency.

7. On day 2 (the same day the first virus collection is performed) euthanize a mouse, isolate the BMMs, and plate them in petri plates as described in Subheading 3.1.

8. Next day collect all the non-adherent cells from the petri plate. Wash the plate twice to collect all the cells. Centrifuge the cells for 5 min at $500 \times g$ and resuspend them in fresh α-MEM with 10 % FBS.

9. Count the cells and plate them at a density of 5×10^6 per p100 plate in the virus containing α-MEM collected earlier for 2 days. Add polybrene (8 μg/ml) and 100 ng/ml M-CSF to the plate.

10. Next day collect the floating and adherent cells using Accutase cell detachment medium, centrifuge and resuspend in fresh medium.

11. At this stage the cells can be plated for osteoclastogenesis in the presence and absence of RANKL (50 ng/μl) and M-CSF (50 ng/ml) as described earlier. Also, plate the remaining cells in p100 plates in the presence of M-CSF (50 ng/μl) to detect expression of transduced proteins.

12. After 4–5 days fix and stain the OCs for TRAP (Subheading 3.1). Mature OCs will be observed in the IKK2-SSEE expressing cells even in the absence of RANKL, but not in GFP or WT-IKK2 expressing cells (Fig. 5a).

13. Collect the protein by lysing the cells from the p100 plates in lysis buffer (Subheading 3.3). Measure the protein concentration and immunoblot for FLAG-IKK2 and beta-actin (Fig. 5b).

3.5 Determining the In Vivo Effects of IKK2-SSEE Expression on Osteoclastogenesis

For in vivo experiments the R26StopIKK2-SSEE-floxed mouse [30] is used. In these mice a cDNA, encoding IKK2 containing two serine to glutamate substitution in the activation loop of the kinase domain, preceded by a loxP-flanked STOP cassette, was cloned into the ubiquitously expressed ROSA26 locus.

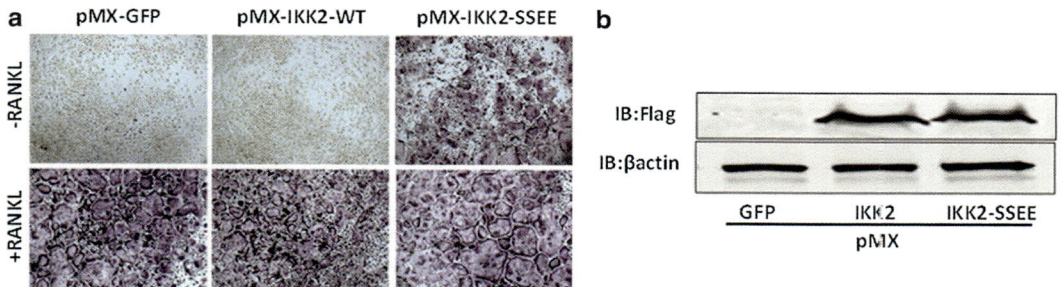

Fig. 5 Constitutively active IKK2 (IKK2SSEE) induces RANKL-independent osteoclastogenesis. RANKL-treated cells differentiate into osteoclasts and are included as positive controls. *Right panel* represents immunoblots (IB) for Flag-IKK2 and beta-actin in cell lysates expressing pMX-GFP, pMX-FLAG-IKK2WT, and pMX-FLAG-IKK2SSEE (pMX, retroviral expression vector; pMX-GFP, pMX-FLAG-IKK2WT, pMX-FLAG-IKK2SSEE, retroviral vector expressing green fluorescent protein (GFP), wild type IKK2, and constitutively active IKK2, respectively)

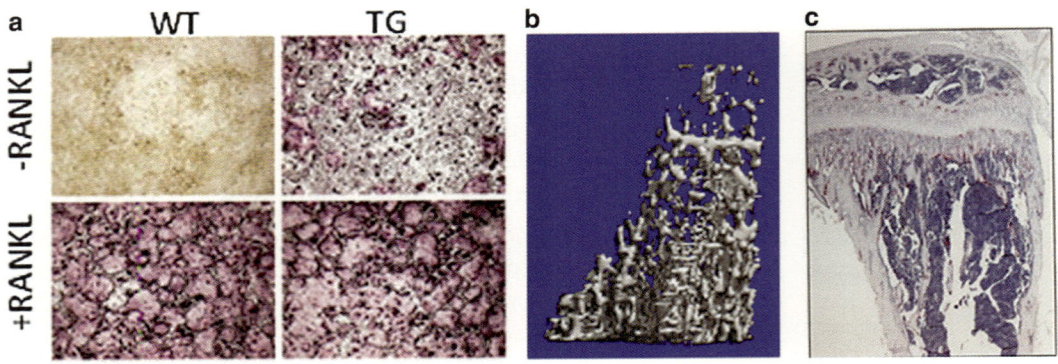

Fig. 6 IKK2SSEE transgene induces osteoclastogenesis and stimulates bone loss in mice. (**a**) Bone marrow macrophages (BMMs) were isolated from wild type (WT) and IKK2SSEE transgenic (TG) mice and cultured in osteoclastogenic conditions. (**b**) Representative images for microCT analysis. (**c**) Representative images for tartrate-resistant acid phosphatase (TRAP) immunostaining of bone section

When bred to mice that express Cre recombinase, the resulting offspring will have the STOP cassette deleted in the cre-expressing tissue(s), resulting in expression of IKK2-SSEE. Expression of IKK2-SSEE leads to constitutively active NF-κB transcription factor activity.

3.5.1 Generating CD11b-cre-IKK2-SSEE-Floxed Mice (TG)

1. Cross CD11b cre mice with IKK2SSEE floxed mice to generate CD11b-cre-IKK2SSEE-transgenic mice (TG) where the NFκB signaling is constitutively active in the myeloid lineage.

2. To characterize the TG mice, isolate BMMs and differentiate osteoclast as described earlier (Subheading 3.1) from both 6- to 8-week-old WT and transgenic mice (TG). After 4–5 days of culture, fix and stain the osteoclasts for TRAP (Fig. 6a).

3. To measure the bone density parameters (using microCT), isolate the bones (femur and tibia) from 6- to 8-week-old WT and TG animals. Fix the bones in 10 % neutral formalin for 16–24 h. After fixation wash the bone with DPBS twice and put them in 70 % ethanol.

4. Scan the mouse bones using microCT instrument. The proximal metaphysis regions of tibias can be scanned to assess trabecular bone morphology (with the following parameters: 55 kVp, 145 mA, standard resolution, 16.4-mm diameter, 16-mm voxel size, 300-ms integration time). Identify the area of interest just distal to the growth plate and scan a height of 480 mm (100 slices). Using 3D analysis tools, measure bone mineral density (BMD), fraction of bone volume known as bone volume over total volume (BV/TV), trabecular thickness (Tb.Th), trabecular separation (Tb.Sp), and trabecular number (Tb.N). These analyses will show the change in the bone parameter for a TG mice compared with the WT mice (Fig. 6b).

5. For Histology of bone sections, decalcify the fixed bones using 10 % EDTA, pH 7.0, for 14 days with gentle rocking and daily replacement of solution. Dehydrate the decalcified bones in graded alcohol, clear through xylene, and embed in paraffin. Cut 5 μM thick longitudinal sections of paraffin blocks and then stain them with hematoxylin and eosin or histochemically with tartrate-resistant acid phosphatase (TRAP) to determine osteoclasts (Fig. 6c).

3.6 SUMOylation of NEMO to Modulate NFκB Activity

Protein modification by SUMO (small ubiquitin-like modifier) is an important regulatory mechanism for multiple cellular processes. It has been shown earlier that SUMOylation of NEMO at different positions regulates NFκB signaling by regulating NEMO stability [31–33]. Based upon this we modulated osteoclastogenesis by generating wild type and mutated NEMO (at positions K270 and K302), without or with SUMO fusion.

3.6.1 Generation of SUMOylated NEMO

1. To generate SUMOylated-NEMO, clone RFP/GFP tagged NEMO-SUMO1/2 fusion in pMx-retroviral vector. SUMO-1 binds at K302 and SUMO2 at K270.

2. Generate pMX-NEMO and pMX-NEMO-SUMO1/2 mutants (NEMO-K270A and NEMO-K302A with or without SUMO1/2) using QuickChange II Site Directed Mutagenesis Kit.

3. Transfect Plat-E cells with pMX-GFP, pMX-NEMO-WT, pMX-NEMO-K270A, pMX-NEMO-K302A, pMX-NEMO-WT-SUMO1, pMX-NEMO-WT-SUMO2, pMX-NEMO-K270A-SUMO1, pMX-NEMO-K270A-SUMO2, pMX-NEMO-K302A-SUMO1, pMX-NEMO-K302A-SUMO2 to generate retro viral particles as described in Subheading 3.4.

4. Transduce the wild type BMMs with the resulting viruses.

5. Next day, plate the cells at a density of 20,000 cells /well of 96 well plate for osteoclastogenesis in α-MEM with 10 % FBS, 50 ng/ml M-CSF and 50 ng/ml RANKL, as described earlier (in Subheading 3.1).

6. Fix and Stain the cells for TRAP.

3.6.2 Western Blot for p-IκB and IκB to Analyze the Activation of NFκB Signaling

1. 1 day post infection of virus in BMMs, plate the cells at a density of $2–3 \times 10^6$/p100 plate in α-MEM containing 10 % FBS and MCSF.

2. When the cells are 80–90 % confluent, wash the cells with sterile PBS and starve the cells in serum-free α-MEM for 2–3 h.

3. After starvation stimulate the cells with RANKL (50 ng/ml) for 15 min and wash with ice-cold DPBS and lyse the cells in Protein Lysis Buffer containing the protease inhibitor cocktail.

4. Measure the protein concentration and immunoblot for phosphorylated (p)-IκB and IκB using anti-p-IκB and anti-IκB antibodies.

4   Notes

1. Do not use tissue culture plates but use sterile bacterial culture petri plates to plate the cells on day 0.

2. As an alternative to preparing the NBD-BPs they are commercially available from a number of vendors including Millipore (Billerica, MA, USA). Dissolve the NBD binding peptides in cell culture grade DMSO (50 mg/ml).

3. To equilibrate and elute a Mono Q (10/10) column do the following: (1) Wash with low-salt buffer, (2), wash with high-salt buffer, (3) equilibrate with low-salt buffer 10 min, (4) inject sample, (5) wash with 40 ml of buffer A, and (6) elude with 1 M NaCl (buffer B), vol. 10 ml.

4. To desalt using a PD-10 (cold) column: (1) pour off excess liquid, and remove bottom cap, (2) equilibrate in 25 ml of PBS, (3) add 2.5 ml of sample, (4) discard the eluate, (5) elute with 3.5 ml of elution buffer solution (Buffer A plus protease inhibitor cocktail), and (6) collect fractions.

5. The ability of the NBD-BPs to enter cells can be tested as follows: Label ~5–25 µg of NBD-BPs with FITC (Thermo Scientific) in 300 µl for 2 h at RT in the dark. Then inject into a gel filtration column (S-12, S-6, and S-200) in PBS and collect appropriate fractions. Add 100–400 µl of purified TAT-FITC fusion protein to ~1×10^6 cells in medium/FBS (non-adherent cells are best experimentally). Check for transduction at t = 0′, 15′, 30′, 45′, 60′ on FACS (FL-1). Analyze cells directly in medium or fix. Add TAT fusion proteins directly to tissue culture medium plus ~5–10 % FBS.

6. Alternatively, a non-radioactive method can be used by utilizing biotin-labeled DNA. At the end of **step 7**, transfer the gel to membrane and proceed with detection using standard chemiluminescence protocols.

Acknowledgments

Work in the authors' laboratory was supported by R01 AR049192, AR054329 (NIH/NIAMS), and 85100 from the Shriners Hospital for Children.

References

1. Abu-Amer Y (2013) NF-kappaB signaling and bone resorption. Osteoporos Int 24:2377–2386

2. Shiotani A, Takami M, Itoh K, Shibasaki Y, Sasaki T (2002) Regulation of osteoclast differentiation and function by receptor activator

of NFkB ligand and osteoprotegerin. Anat Rec 268:137–146

3. Teitelbaum SL (2000) Bone resorption by osteoclasts. Science 289:1504–1508

4. Boyce BF, Yao Z, Xing L (2010) Functions of nuclear factor kappaB in bone. Ann N Y Acad Sci 1192:367–375

5. Franzoso G, Carlson L, Xing L, Poljak L, Shores E, Brown K, Leonardi A, Tran T, Boyce B, Siebenlist U (1997) Requirement for NF-kappaB in osteoclast and B-cell development. Genes Dev 11:3482–3496

6. Iotsova V, Caamano J, Loy J, Yang Y, Lewin A, Bravo R (1997) Osteopetrosis in mice lacking NF-kappaB1 and NF-kappaB2. Nat Med 3:1285–1289

7. Kong YY, Yoshida H, Sarosi I, Tan HL, Timms E, Capparelli C, Morony S, Oliveira-dos-Santos AJ, Van G, Itie A, Khoo W, Wakeham A, Dunstan CR, Lacey DL, Mak TW, Boyle WJ, Penninger JM (1999) OPGL is a key regulator of osteoclastogenesis, lymphocyte development and lymph-node organogenesis. Nature 397:315–323

8. Yasuda H, Shima N, Nakagawa N, Yamaguchi K, Kinosaki M, Mochizuki S, Tomoyasu A, Yano K, Goto M, Murakami A, Tsuda E, Morinaga T, Higashio K, Udagawa N, Takahashi N, Suda T (1998) Osteoclast differentiation factor is a ligand for osteoprotegerin/osteoclastogenesis-inhibitory factor and is identical to TRANCE/RANKL. Proc Natl Acad Sci U S A 95:3597–3602

9. Boyce BF, Xing L, Franzoso G, Siebenlist U (1999) Required and nonessential functions of nuclear factor-kappa B in bone cells. Bone 25:137–139

10. Abu-Amer Y (2005) Advances in osteoclast differentiation and function. Curr Drug Targets Immune Endocr Metabol Disord 5:347–355

11. Lamothe B, Lai Y, Xie M, Schneider MD, Darnay BG (2013) TAK1 is essential for osteoclast differentiation and is an important modulator of cell death by apoptosis and necroptosis. Mol Cell Biol 33:582–595

12. Hayden MS, Ghosh S (2004) Signaling to NF-κB. Genes Dev 18:2195–2224

13. Karin M, Yamamoto Y, Wang M (2004) The IKK NF-κB system: a treasure trove for drug development. Nat Rev Drug Discov 3:17–26

14. Siebenlist U, Franzoso G (2001) Structure, regulation and function of NF-κB. Proc Natl Acad Sci U S A 89:4333–4337

15. Ting AY, Endy D (2002) Signal transduction: decoding NF-κB signaling. Science 298:1189–1190

16. Wong B, Lee S, Vologodskaia M, Steinman R, Choi Y (1998) The TRAF family of signal transducers mediates NF-κB activation by the TRANCE receptor. J Biol Chem 273:28335–28359

17. May MJ, D'Acquisto F, Macge LA, Glockner J, Pober JS, Ghosh S (2000) Selective inhibition of NF-kappaB activation by a peptide that blocks the interaction of NEMO with the IkappaB kinase complex. Science 289:1550–1554

18. Strickland I, Ghosh S (2006) Use of cell permeable NBD peptides for suppression of inflammation. Ann Rheum Dis 65(Suppl 3):75–82

19. von Bismarck P, Winoto-Morbach S, Herzberg M, Uhlig U, Schutze S, Lucius R, Krause MF (2012) IKK NBD peptide inhibits LPS induced pulmonary inflammation and alters sphingolipid metabolism in a murine model. Pulm Pharmacol Ther 25:228–235

20. Cheng MX, Gong JP, Chen Y, Liu ZJ, Tu B, Liu CA (2012) NBD peptides protect against ischemia reperfusion after orthotopic liver transplantation in rats. J Surg Res 176:666–671

21. Dai S, Hirayama T, Abbas S, Abu-Amer Y (2004) The IkappaB kinase (IKK) inhibitor, NEMO-binding domain peptide, blocks osteoclastogenesis and bone erosion in inflammatory arthritis. J Biol Chem 279:37219–37222

22. Choi M, Rolle S, Wellner M, Cardoso MC, Scheidereit C, Luft FC, Kettritz R (2003) Inhibition of NF-κB by a TAT-NEMO-binding domain peptide accelerates constitutive apoptosis and abrogates LPS-delayed neutrophil apoptosis. Blood 102:2259–2267

23. Clohisy JC, Yamanaka Y, Faccio R, Abu-Amer Y (2006) Inhibition of IKK activation, through sequestering NEMO, blocks PMMA-induced osteoclastogenesis and calvarial inflammatory osteolysis. J Orthop Res 24:1358–1365

24. Shibata W, Maeda S, Hikiba Y, Yanai A, Ohmae T, Sakamoto K, Nakagawa H, Ogura K, Omata M (2007) Cutting edge: the I{kappa}B kinase (IKK) inhibitor, NEMO-binding domain peptide, blocks inflammatory injury in murine colitis. J Immunol 179:2681–2685

25. Abu-Amer Y, Dowdy SF, Ross FP, Clohisy JC, Teitelbaum SL (2001) TAT fusion proteins containing tyrosine 42-deleted IkappaBalpha arrest osteoclastogenesis. J Biol Chem 276:30499–30503

26. Nagahara H, Vocero-Akbani AM, Snyder EL, Ho A, Latham DG, Lissy NA, Becker-Hapak M, Ezhevsky SA, Dowdy SF (1998) Transduction of full-length TAT fusion

proteins into mammalian cells: TAT-p27Kip1 induces cell migration. Nat Med 4:1449–1452

27. Schwarze S, Ho A, Vocero-Akbani A, Dowdy S (1999) In vivo protein transduction: delivery of a biologically active protein into the mouse. Science 285:1569–1572

28. Schwarze S, Hruska K, Dowdy S (2000) Protein transduction: unrestricted delivery into all cells? Trends Cell Biol 10:290–295

29. Wadia JS, Stan RV, Dowdy SF (2004) Transducible TAT-HA fusogenic peptide enhances escape of TAT-fusion proteins after lipid raft macropinocytosis. Nat Med 10:310–315

30. Otero JE, Chen T, Zhang K, Abu-Amer Y (2012) Constitutively active canonical NF-kappaB pathway induces severe bone loss in mice. PLoS One 7:e38694

31. Bloor S, Ryzhakov G, Wagner S, Butler PJ, Smith DL, Krumbach R, Dikic I, Randow F (2008) Signal processing by its coil zipper domain activates IKK gamma. Proc Natl Acad Sci U S A 105:1279–1284

32. Huang TT, Wuerzberger-Davis SM, Wu ZH, Miyamoto S (2003) Sequential modification of NEMO/IKKgamma by SUMO-1 and ubiquitin mediates NF-kappaB activation by genotoxic stress. Cell 115:565–576

33. Mabb AM, Wuerzberger-Davis SM, Miyamoto S (2006) PIASy mediates NEMO sumoylation and NF-kappaB activation in response to genotoxic stress. Nat Cell Biol 8:986–993

Chapter 32

NF-κB Activation with Aging: Characterization and Therapeutic Inhibition

Jing Zhao, Xuesen Li, Sara McGowan, Laura J. Niedernhofer, and Paul D. Robbins

Abstract

Aging is a condition characterized by progressive decline in tissue homeostasis due, at least in part, to the accumulation of replicative, oxidative, and genotoxic stress over time. The activity of the transcription factor NF-κB is upregulated in both naturally aged mice and multiple progeroid mouse models of accelerated aging. Suppressing NF-κB activity genetically or pharmacologically has been shown to delay the onset and progression of aging pathology and therefore prolong the healthspan in progeroid mouse models. Here, we describe the methods for measuring aging endpoints along with NF-κB activation in mice, as well as after pharmacologic intervention to prevent NF-κB activation using a NEMO-binding domain (NBD)–protein transduction domain (PTD) fusion peptide.

Key words NF-κB, Aging, Mouse model of aging, Progeria, Stress response, Aging endpoints

1 Introduction

Aging is defined as the loss of tissue homeostasis, possibly mediated by the accumulation of senescent somatic cells and dysfunctional stem cells [1–3]. Aging is the greatest risk factor for many chronic diseases including neurodegenerative disease, cardiovascular disease, type 2 diabetes, and osteoporosis [4]. Telomere shortening, DNA damage, and reactive oxygen species (ROS) all play important roles in driving aging, either by increasing genome instability or generating damaged molecules including carbonylated proteins and peroxidative lipids [2, 3, 5, 6]. p16^{INK4a}, a cyclin-dependent kinase inhibitor (CDKI) and tumor suppressor, is upregulated with aging in numerous murine and human tissues, such as kidney, liver, skeletal muscle, and adipose tissue [7, 8]. Importantly, clearance of senescent cells expressing p16 successfully delays the onset of age-related phenotypes and prolongs the healthspan in a *BubR1*$^{H/H}$ progeroid mouse model, strongly suggesting the fundamental role of p16-positive cells in driving age-related pathology [8].

Michael J. May (ed.), *NF-kappa B: Methods and Protocols*, Methods in Molecular Biology, vol. 1280, DOI 10.1007/978-1-4939-2422-6_32, © Springer Science+Business Media New York 2015

Studying aging in normal mice is time-consuming and expensive, as it takes more than 2 years to observe signs of aging. Therefore, mouse models of human progeroid syndromes with accelerated aging provide the opportunity to study aging more rapidly and economically. In our studies, we have used a novel progeroid mouse model expressing approximately 10 % of the normal level of the ERCC1–XPF protein complex, termed $Ercc1^{-/\Delta}$ mice, which show a dramatically shortened lifespan of approximately 6 months [9]. ERCC1–XPF, functioning as a structure-specific endonuclease, is required for nucleotide excision repair (NER), interstrand crosslink (ICL) repair, and the repair of some double-strand DNA breaks (DSB) [10–13]. Patients with mutations in *XPF* exhibit a progeroid appearance with accelerated liver and kidney dysfunction, osteoporosis, loss of vision and hearing, hypertension, sarcopenia, neurodegeneration, and skin atrophy [14]. $Ercc1^{-/\Delta}$ mice phenocopy these age-related symptoms observed in XP-F patients, providing an ideal model to examine the underlying mechanisms of aging and aging-related diseases [9, 14].

The transcription factor nuclear factor-κB (NF-κB) is a pivotal factor in regulating cell proliferation, apoptosis, and inflammation upon exposure to proinflammatory, genotoxic, and oxidative stress [15]. Defects in the regulation of NF-κB signaling lead to inflammatory diseases, cancers, and aging-related degenerative diseases [16]. In mammals, the NF-κB family includes five structurally related proteins, RelA (p65), RelB, c-Rel, p105 (p50), and p100 (p52), forming homodimers or heterodimers through Rel homology domains [17]. In the absence of stimulation, the conserved nuclear localization signal (NLS) of the NF-κB dimer is masked by the inhibitory IκB protein, which sequesters NF-κB dimer in the cytoplasm [16]. Upon exposure to inflammatory or genotoxic stress, the upstream IκB kinase (IKK) is activated and phosphorylates IκB, which then triggers the polyubiquitination and degradation of IκB by the 26S proteasome [17]. NF-κB dimers with accessible NLS then translocate into the nucleus, bind target DNA sequences, and regulate gene expression [17]. The IKK complex is composed of three subunits, the catalytic subunits IKKα and IKKβ and a regulatory subunit termed NF-κB essential modulator (NEMO or IKKγ) [18]. The NEMO-binding domain in the C-terminus of IKKβ is required for the kinase activity of IKK complex and associates with the N-terminus of NEMO [19]. The PTD–NBD peptide, which contains a protein transduction domain (PTD) and an 11-amino acid NEMO-binding domain (NBD) derived from IKKβ C-terminal sequence, was developed as a highly selective inhibitor of the IKK complex and disrupts the association between IKKβ and NEMO [19]. An NBD peptide fused to a cationic PTD containing eight lysines (8K-NBD) transduces numerous cell types and inhibits NF-κB activation in a variety of mouse models of disease including delayed-type hypersensitivity (DTH), chronic murine colitis, and Duchenne muscular dystrophy (DMD) [20–24].

There is accumulating evidence that NF-κB activity is upregulated with aging. NF-κB DNA binding activity and nuclear localization have been reported to increase in multiple organs with aging, including liver, kidney, brain, and skin [25–28]. NF-κB also appears to play a direct role in driving mammalian aging since inhibition of NF-κB in the skin of a transgenic mouse reversed expression of aging-related genes and hallmarks of skin aging [29]. Our own studies demonstrate that decreasing NF-κB activity by genetic depletion or pharmacologic inhibition delays the age at onset and progression of age-associated symptoms and pathology in the $Ercc1^{-/\Delta}$ progeroid mouse model [30]. Importantly, it appears as if reduction in NF-κB activity reduces the level of mitochondrial ROS, the levels of oxidative DNA damage, and the extent of cellular senescence in tissues such as the liver. A reduction of NF-κB activity also prolongs longevity and delays aging symptoms in a Hutchinson–Gilford progeria syndrome (HGPS) murine model with nuclear lamina defects [31]. Herein, we provide general protocols to measure NF-κB activation in vivo, as well as characterization of aging endpoints in mice with altered NF-κB activity.

2   Materials

2.1   Treatment with a PTD–NBD Peptide in a Mouse Model of Accelerated Aging

1. Progeroid mouse model: The $Ercc1^{-/\Delta}$ mouse model was developed in Dr. Jan H.J. Hoeijmakers Laboratory, Molecular Genetics Department, Erasmus Medical Center Rotterdam (the Netherlands).

2. 8K-NBD   peptide: KKKKKKKKGGTALDWSWLQTE (University of Pittsburgh Peptide Synthesis Core Facility).

3. 8K-mNBD   (mutated   NBD)   peptide: KKKKKKKKGGTALDASALQTE (University of Pittsburgh Peptide Synthesis Core Facility).

4. BD 1/2 cc tuberculin syringe (27 G 1/2).

2.2   Analysis of NF-κB Activation with Aging: Immunoblotting of Phospho-p65 and Phospho-IκBα

1. Lysis buffer: RIPA buffer (Cell Signaling, Danvers, MA, USA); Protease inhibitor cocktail (Sigma-Aldrich); Halt phosphatase inhibitor cocktail (Thermo Scientific, Waltham, MA, USA). Keep lysis buffer ice-cold.

2. SDS-sample loading buffer 4× (Bioworld, Irving, TX, USA).

3. 10× transfer buffer (2 L): 288 g of glycine; 60.4 g of Tris base. Add ddH$_2$O to the final volume of 2 L. To prepare fresh 1× transfer buffer (1 L), add 100 ml of 10× transfer buffer, 200 ml of methanol, and 700 ml of ddH$_2$O.

4. 10× running buffer (2 L): 60 g of Tris base; 288 g of glycine; 20 g of SDS. Add ddH$_2$O to the final volume of 2 L and pH should be 8.3, no adjustment needed.

5. PBST: 1× PBS (137 mM NaCl, 2.7 mM KCl, 10 mM Na$_2$HPO$_4$, 2 mM KH$_2$PO$_4$, pH 7.4) with 0.1 % Tween 20.

6. Blocking buffer: 5 % bovine serum albumin (BSA) in PBST.

7. FastPrep Lysing Matrix tubes (MP Biomedicals, Santa Ana, CA, USA).

8. DC protein assay kit (Bio-Rad, Hercules, CA, USA).

9. Mini-PROTEAN TGX stain-free gels (Bio-Rad).

10. Antibody: phospho-p65 (Ser536) rabbit mAb (Cell Signaling), 1:1,000 dilution; NF-κB p65 XP rabbit mAb (Cell Signaling), 1:1,000 dilution; phospho-IκBα (Ser32/36) mouse mAb (Cell Signaling), 1:1,000 dilution; IκBα rabbit Ab (Santa Cruz Biotechnology, Santa Cruz, CA, USA), 1:1,000 dilution; GAPDH XP rabbit mAb (Cell Signaling), 1:10,000 dilution.

11. SuperSignal West Pico Chemiluminescent Substrate (Thermo Scientific).

12. Restore Western blot stripping buffer (Thermo Scientific).

13. Eppendorf Model 5415R microcentrifuges.

14. FastPrep-24 instrument (Homogenizer; MP Biomedicals).

2.3 Analysis of NF-κB Activation with Aging in NF-κB^{EGFP} Reporter Mice: Fluorescent Microscopy

1. Mice: *Ercc1$^{-/Δ}$*NF-κB^{EGFP} and NF-κB^{EGFP} mice.

2. 4 % paraformaldehyde (PFA) solution (1 L): Mix 40 g of paraformaldehyde powder into 800 ml of 1× PBS at 60 °C. Stir and increase pH by NaOH for complete dissolution. Cool down and filter out precipitates. Add 1× PBS to the total volume of 1 L. Adjust pH to 7.4. Aliquot and store at −20 °C for future use.

3. 30 % sucrose: 30 g of sucrose; 70 ml of ddH$_2$O.

4. O.C.T. compound.

5. Tissue-Tek Cryomold (standard or intermediate).

6. Fisherbrand Superfrost Plus Microscope slides.

7. Cover glass (22 × 50 mm).

8. VECTASHIELD mounting medium with DAPI (Hard Set; Vector Laboratories, Burlingame, CA, USA).

9. Olympus Fluoview FV 1000 confocal microscope.

10. Leica CM1950 cryostat.

2.4 Measurements of Biomarkers of Aging: Quantitative Real-Time Polymerase Chain Reaction (qRT-PCR) of p16 and p21

1. TRI Reagent solution (Applied Biosystems, Grand Island, NY, USA).

2. 2-Propanol.

3. Ethanol (200 proof; molecular biology grade ethanol).

4. DEPC-treated and nuclease-free water.

5. Power SYBR green RNA-to-Ct 1-step kit (Applied Biosystems).

6. MicroAmp Fast 96-well reaction plate (0.1 ml; Applied Biosystems).

7. MicroAmp Optical Adhesive Film Kit (Applied Biosystems).

8. qRT-PCR primers: *Cdkn1a* (p21) forward: GTCAGGCTGGTCTGCCTCCG.

   Cdkn1a reverse: CGGTCCCGTGGACAGTGAGCAG.
   Cdkn2a (p16) forward: ACTCCAAGAGAGGGTTTTC.
   Cdkn2a reverse: ATCATCATCACCTGGTCC.
   Actb (β-actin) forward: GATGTATGAAGGCTTTGGTC.
   Actb reverse: TGTGCACTTTTATTGGTCTC.

9. NanoDrop 2000 spectrophotometer (Thermo Scientific).

10. StepOnePlus real-time PCR system (Applied Biosystems).

2.5 Senescence-Associated β-Galactosidase (SA-βgal) Assay

1. 10 % buffered formalin phosphate.

2. 0.2 M citric acid/Na phosphate buffer (100 ml): 36.85 ml of 0.1 M citric acid; 63.15 ml of 0.2 M sodium phosphate. Adjust pH to 5.8.

3. SA-βgal staining solution (20 ml): 1 ml of 20 mg/ml X-gal in dimethylformamide; 4 ml of 0.2 M citric acid/Na phosphate buffer with pH 5.8; 1 ml of 100 mM potassium ferrocyanide; 1 ml of 100 mM potassium ferricyanide; 0.6 ml of 5 M sodium chloride; 0.04 ml of 1 M magnesium chloride; 12.4 ml of water. Confirm again pH = 5.8.

4. pH meter.

5. Lab oven.

6. Bright-field microscopy.

2.6 Analysis of Immunohisto-chemistry (IHC) and Aging Pathology

1. Permeabilization buffer: 0.25 % Triton X-100 in PBS (PBST).

2. Blocking buffer and antibody dilution buffer: 1 % BSA in PBST.

3. Antibodies: Anti-phospho-Histone $H_2A.X$ (Ser139) (Millipore, Billerica, MA, USA), 1:500 dilution for IHC; Alexa Fluor 633 goat anti-mouse IgG (H+L) (Life Technologies, Grand Island, NY, USA), 1:500 dilution.

3 Methods

3.1 Treatment with NBD Peptides in Mouse Model of Aging

1. Prepare 8K-NBD and mutant control 8K-mNBD stock solution at 40 mM in DMSO. Aliquot and store at −80 °C for future use (*see* **Note 1**).

2. Dilute NBD peptides in PBS to bring DMSO concentration to no more than 10 % for intraperitoneal (i.p.) injection (*see* **Note 2**).

3. Treat age- and gender-matched $Ercc1^{-/\Delta}$ littermate pairs with 10 mg/kg of 8K-NBD or 8K-mNBD peptide as a control three times per week i.p. starting from 5 weeks of age or a time point prior to the onset of aging symptoms.

4. Terminate NBD treatment and sacrifice mice by CO_2 euthanasia when aging symptoms occur and harvest tissues for further analysis (liver, kidney, lung, etc.).

3.2 Analysis of NF-κB Activation with Aging: Immunoblotting of p-p65 and p-kBα

1. Harvest fresh tissues (liver) and place into a 1.7 ml Eppendorf tube.

2. Place tubes in liquid nitrogen for snap freezing. Keep tissues in liquid nitrogen till the end of the harvest.

3. Store tissues in −80 °C freezer for further use (*see* **Note 3**).

3.2.1 Prepare Tissue Lysate for Immunoblotting

4. Prepare tissue lysis buffer and keep ice-cold (*see* **Note 4**).

5. Thaw out tissue samples on ice. Transfer 50–100 mg of tissues into a Lysing Matrix D microtube and add 1 ml of ice-cold lysis buffer for homogenization.

6. Homogenize tissues three times in a FastPrep-24 homogenizer at speed 6.0 for 20 s each. Cool samples down on ice for 5 min between each homogenization to avoid heat generated by mechanical friction.

7. Centrifuge for 20 min at maximum speed at 4 °C using a benchtop microcentrifuge. Collect supernatant and transfer to a new Eppendorf tube for immunoblotting. Discard lipid layer on the top and cell debris on the bottom (*see* **Note 5**).

8. Store tissue lysate at −80 °C for future use or on ice for immediate use.

3.2.2 Immunoblotting for p-p65 and p-kBα

1. Perform Lowry protein assay to quantify protein concentration (*see* **Note 6**).

2. Determine protein concentration of each sample and prepare 50 μg (p-p65) or 30 μg (p-IκBα) of protein for electrophoresis (*see* **Note 7**). Dilute samples 3:1 in a 4× sample buffer supplemented with 20 % 2-mercaptoethanol. Mix well and heat for 5 min at 95 °C in a heat block and then fast spin down the steam on the lids (*see* **Note 8**).

3. Assemble a precast Mini-PROTEAN TGX stain-free gel in the tank and fill with 1× running buffer. Make sure there is no leakage from the inner chamber. Rinse each well with running buffer using a syringe.

4. Load 5 μl of protein standard and samples prepared above to the gel using gel-loading tips.

5. Run the gel at 120–150 V for 1 h at room temperature or until markers separate well.

6. Wet transfer: Prepare 1× transfer buffer and precool at 4 °C for 10 min. Prepare wet transfer sandwich (sponge, filter paper, gel, membrane, filter paper, and sponge) in the transfer buffer and make sure to roll out air bubbles trapped between the gel and membrane by using a roller or a pipette. Clamp the sandwich into the transfer cassette with gel facing the black side and membrane facing the transparent side. Insert the cassette into a transfer tank with an ice box and perform transfer at 100 V for 1 h in cold room.

7. Block the membrane in blocking buffer for 1 h on a shaker at room temperature.

8. Incubate the membrane in primary antibody diluted in blocking buffer on a rocker overnight at 4 °C or 2 h at room temperature.

9. Wash the membrane three times with PBST, 10 min each.

10. Incubate the membrane in HRP-conjugated secondary antibody diluted in blocking buffer on a shaker for 1 h at room temperature.

11. Wash the membrane three times with PBST, 10 min each.

12. Incubate the membrane in the mix of detection reagents A and B, 1 ml of each, for 2 min at room temperature.

13. Get rid of excess reagents and wrap the membrane in a plastic wrap.

14. Expose to X-ray film in a dark room.

15. If needed, strip the membrane in a stripping buffer on an agitator for 20 min at room temperature.

16. Wash with PBST three times, 5 min each, and repeat **steps 7–14**.

3.3 Analysis of NF-κB Activation with Aging in NF-κB-EGFP Reporter Mice: Fluorescent Microscopy

1. Harvest target tissues (kidney, muscle, pancreas, liver, fat, and spleen) from $Ercc1^{-/\Delta}$NF-κB^{EGFP} and naturally aged NF-κB^{EGFP} mice for analysis.

2. Fix tissues in 4 % PFA overnight at 4 °C and then transfer to 30 % sucrose. Incubate at 4 °C for 24 h to completely substitute water with sucrose (*see* **Note 9**).

3. The following day, discard 30 % sucrose and place tissues on Whatman filter paper to get rid of excess sucrose.

4. Freeze tissues in a 2-methylbutane/dry ice bath for 30–60 s and then store the tissues at −80 °C for future use (*see* **Note 10**).

5. O.C.T. imbedding: Imbed tissues in O.C.T. compound in a cryomold and adjust the orientation of tissues properly based on the experimental needs. Place cryomolds on dry ice to freeze down till O.C.T. compound completely turns to white and tissues are ready for cryosection.

6. Perform tissue sectioning at 4–8 µm at –20 °C by using a cryostat (*see* **Note 11**) and mount tissues onto glass slides (*see* **Note 12**). Slides can be stored at –80 °C for future use.

7. Wash tissue slides with PBS at room temperature twice, 5 min each.

8. Mount the slides with glass coverslips by VECTASHIELD mounting medium with DAPI and store at 4 °C for future imaging.

9. Bring slides to room temperature 30 min before fluorescent microscopy imaging to allow the aqueous mountant to harden.

10. Obtain five images per sample and quantify the percentage of EGFP-positive cells using MetaMorph software for the analysis of NF-κB activity.

3.4 Measurements of Biomarkers of Aging: Quantitative RT-PCR of p16 and p21

1. Harvest snap-frozen tissues (liver, lung, and kidney) for quantitative RT-PCR of senescence biomarkers, such as p16 and p21. Store at –80 °C for future use.

2. Homogenize tissue samples: Weigh 100 mg of tissue and transfer to a Lysing Matrix D homogenization tube. Suspend and homogenize the tissues in 1 ml of TRI Reagent by using a FastPrep-24 homogenizer at speed 6.0 for 20 s. Incubate tissue homogenate in TRI Reagent 15 min at room temperature for complete disruption of nucleoprotein complex (*see* **Note 13**).

3. Transfer tissue homogenate to a 1.7 ml Eppendorf tube and centrifuge at $12,000 \times g$ 10 min at 4 °C to spin down insoluble fraction. Discard the pellets and transfer the supernatant to a new Eppendorf tube.

4. Add 200 µl of chloroform (adjust the volume proportionally) to 1 ml of TRI Reagent and mix vigorously for 15 s (*see* **Note 13**). Incubate for 15 min at room temperature.

5. Centrifuge protein (organic phase) and DNA (interphase) fraction down at $12,000 \times g$ for 15 min at 4 °C and carefully transfer the aqueous phase (RNA fraction) to a new Eppendorf tube (*see* **Note 14**).

6. RNA precipitation: Add 500 µl of 2-propanol and mix moderately for 5–10 s. Incubate the mixture for 5–10 min at room temperature and pellet RNA down at $12,000 \times g$ for 8 min at 4 °C (*see* **Note 15**).

7. Carefully remove supernatant and add 75 % ethanol (molecular biology grade ethanol diluted in DEPC-treated water). Invert tubes 5–10 times for washing.

8. Spin down the pellet at $12,000 \times g$ for 5 min at 4 °C and carefully remove ethanol (*see* **Note 16**).

9. Air-dry RNA pellet and dissolve RNA in appropriate amount of DEPC-treated water (*see* **Note 17**).

10. Determine RNA concentration by using NanoDrop and then calculate the volume of each samples needed for one-step quantitative RT-PCR.

11. Prepare RT-PCR reaction mix (20 μl) in a MicroAmp Fast Optical 96-well reaction plate following the guidelines provided by the company (*see* **Note 18**):

Power SYBR green RT-PCR mix (2×)	10 μl
Forward and reverse primer (200 nM)	X μl
RT enzyme mix (125×)	0.16 μl
RNA template (100 ng)	Y μl
Water	to 20 μl

12. Seal the plate tightly with an optical adhesive film and vortex the plate to mix well.

13. Centrifuge at $1000 \times g$ for 3 min.

14. Run RT-PCR following the condition recommended by the product guideline.

15. Analyze RT-PCR results by normalizing data to an internal control, such as β-actin.

3.5 Senescence-Associated β-Galactosidase (SA-βgal) Assay

SA-βgal has been identified as a biomarker of cellular senescence since 1995 and is now the most well recognized senescence biomarker [32]. Lysosomal β-galactosidase achieves its optimal enzymatic activity at pH 4.0 in physical condition, and its activity becomes detectable at a suboptimal condition of pH 6.0 in senescent cells due to increased lysosomal mass [33, 34].

1. Harvest tissues (liver, kidney, fat, pancreas, and lung) for SA-βgal and fix in 10 % formalin for 4 h at 4 °C (*see* **Note 19**). Then transfer to 30 % sucrose overnight at 4 °C.

2. The next day, remove 30 % sucrose and place tissues on Whatman filter papers to get rid of excess sucrose. Freeze tissues in a 2-methylbutane/dry ice bath and store at −80 °C in a Ziploc bag.

3. O.C.T. imbedding: Imbed tissues in O.C.T. compound as described in Subheading 3.3.

4. Precool tissues imbedded in O.C.T. at −20 °C for 30 min. Perform cryosectioning at 5–7 μm by using a cryostat (*see* **Note 10**). Keep slides at −80 °C before staining (*see* **Note 20**).

5. SA-βgal staining: Wash tissue slides and bring the slides back to room temperature in a Coplin jar with PBS × 3 times. Add 50 ml of SA-βgal staining solution with pH 5.8 (*see* **Notes 21–24**) into the Coplin jar and incubate for 16–24 h in an oven without CO_2 injection at 37 °C (*see* **Note 25**).

6. Wash slides with PBS × 3 times and mount with a coverslip using VECTASHIELD mounting medium with DAPI.

7. Quantification: View blue-stained SA-βgal positive cells using a bright-field microscope. Count the number of SA-βgal positive cells in each field and quantify at least five fields for the calculation of mean ± SD.

3.6 Measurements of Aging Phenotype in Ercc1^{-/Δ} Mice

Ercc1^{-/Δ} mice are featured by aggressive neurodegenerative changes with aging which represents a group of symptoms including dystonia, trembling, and ataxia [35].

3.6.1 Measurement of Neurodegenerative Symptoms

1. Dystonia is a manifestation of neurological disorder, which can be tested by lifting the mouse up by the tail, and observe the dystonic postures characterized by sustained pathological muscle contraction of extremities.

2. Trembling should be assessed at the state of both rest and climbing.

3. Ataxia is an indicator of impaired cerebellar function featured by imbalanced gait. Release mice for free movements and observe wobbling gait and widened pace [36].

3.6.2 Kyphosis

Examine mice for hunchback at a stretching position and slide fingers down along the spine to detect abnormal spinal curvature.

3.6.3 Sarcopenia

Place the mice on the mouse cage lid and let it grasp the grid. Then lift the mouse up by holding the tail to assess the extent of muscle wasting.

3.6.4 Urinary Incontinence

Check tail base to see if there is any sign of urine.

3.6.5 Body Condition

Place mouse on a flat surface to evaluate the fat/flesh that covers the lower spine and pubic bone. Inspect for prominent bones in the sacroiliac region.

3.7 Analysis of Age-Related Pathology

Persistent DNA damage foci are one of the features of cellular senescence and aging. H_2AX is a histone variant of histone H_2A, recruited and phosphorylated at serine 139 by ATM/ATR in response to DNA double-strand break [37]. Thus, the phosphorylation form of H_2AX, termed as γH_2AX, is considered as a biomarker of senescence combined with the use of SA-βgal to detect cellular senescence and aging.

3.7.1 Immunohistochemistry of γH₂AX Foci

1. Process tissues and perform sectioning as described in Subheading 3.3, **steps 1–6**.

2. Wash slides with PBS at room temperature twice, 5 min each.

3. Permeabilization: Permeabilize cell and nuclear membrane with 0.25 % Triton X-100 in PBS at room temperature for 10 min.

4. Block tissue slides in 1 % BSA in PBST for 1 h at room temperature (*see* **Note 26**).

5. Wash slides with PBS × 3 times, 5 min each.

6. Incubate slides in primary antibody (anti-phospho-H$_2$AX diluted at 1:500) diluted in blocking buffer in a humidified chamber or a Coplin jar overnight at 4 °C or 2 h at room temperature.

7. Wash slides with PBS × 3 times for 5 min.

8. Incubate slides in fluorophore-conjugated secondary antibody (Alexa Fluor 633 goat anti-mouse IgG diluted at 1:500) diluted in blocking buffer for 1 h at room temperature.

9. Wash slides with PBS × 3 times, 5 min each.

10. Mount tissues with glass coverslips using a drop of VECTASHIELD mounting medium with counterstain DAPI.

11. Slides can be stored at 4 °C for up to a month. For long-term storage, preserve slides at –20 °C.

12. Bring slides to room temperature for 20 min before imaging to allow the aqueous mounting medium to harden. Obtain five fields or 200 cells per sample by using a confocal microscope at 60× magnification and quantify the average number of γ-H$_2$AX foci per cell.

3.7.2 Lipofuscin Staining

Lipofuscin, another biomarker of aging, is also termed "aging pigment" [38, 39]. The autofluorescent lipofuscin pigment tends to accumulate with senescence and aging in postmitotic senescent cells and aged tissues and is mainly consist of oxidized proteins and lipids [40, 41] presenting in cytosol and lysosome [42].

1. Process tissues and sectioning as described in Subheading 3.3, **steps 1–6**.

2. Wash slides with PBS at room temperature twice, 5 min each.

3. Mount tissues with coverslips using VECTASHIELD mounting medium with counterstain DAPI.

4. Bring slides to room temperature for 20 min before imaging. The lipofuscin autofluorescence can be imaged by using a fluorescence microscope. Obtain five fields per sample at 20× magnification and quantify the average fluorescent area by using MetaMorph software.

4 Notes

1. Dissolving NBD peptides in DMSO instead of water improves its activity. Make sure to aliquot and store the stock solution at –80 °C for long-term preservation.

2. DMSO concentration higher than 10 % could be toxic to mice by i.p. injection. Therefore, prepare NBD peptides injection solution in PBS with no higher than 10 % DMSO.

3. Use relatively fresh tissues to detect phosphorylated protein in tissues. Storage of less than 2 months is recommended.

4. Keep lysis buffer and tissue lysate ice-cold all the time to avoid possible protein degradation and minimize phosphatase activity.

5. Incomplete removal of tissue lipid could cause smeared bands or lanes. Therefore, carefully transfer the supernatant after centrifuge without mixing lipid residual in.

6. Dilute protein-rich tissues, such as liver and kidney, 10- to 20-fold for Lowry protein assay so that the sample concentration falls into the range of standard curve.

7. If p-p65 cannot be detected by using 50 μg of protein, increase protein amount to 80–100 μg.

8. Always vortex and spin down samples before and after boiling in the heat block.

9. Complete substitution of water with sucrose in tissues gives the best cellular morphology. Tissues should sink down to the bottom after 24 h incubation in 30 % sucrose.

10. For long-term storage, tissues for IHC should be stored in a Ziploc bag to prevent freezer artifact at −80 °C.

11. While cryosectioning, 5 μm is recommended to get the best structure and morphology especially in liver.

12. Use Superfrost plus microscope slides to facilitate the electrostatic adherence of tissue sections to the slides, to prevent sections from peeling off.

13. TRI Reagent and chloroform are corrosive reagents. Extra care should be taken while handling them. Carefully working in a chemical fume hood is recommended, and always wear a lab coat, nitrile gloves, and safety goggles.

14. Disturbance of interphase could cause DNA contamination. Be very cautious when transferring the RNA aqueous phase to a new tube to avoid any possible DNA or protein contamination. If sample is suspected to be contaminated, include a step of DNase treatment to remove DNA residues.

15. Always place the tube hinge to face outside of the centrifuge in case small amount of RNA pellet is invisible after spin down.

16. To better remove ethanol residual, another spin down could be performed at $12,000 \times g$ for 5 min. Complete removal of ethanol residual is required not to hinder the following PCR reaction.

17. Do not overdry RNA pellet, which decreases RNA solubility significantly.

18. Power SYBR green RT-PCR mix needs to be inverted a few times before use. Protect the mix from light during thawing.

19. For SA-βgal staining, over-fixation impairs the enzymatic activity of β-galactosidase dramatically. If 2 % PFA is used, less than 2 h fixation is recommended. Cryosection of 5 μm is recommended to obtain the best image of SA-βgal staining.

20. Preserving tissues at −80 °C for more than 4 weeks is not recommended for SA-βgal staining due to the loss of enzymatic activity.

21. The stock solution of citric acid/Na phosphate buffer, sodium chloride, and magnesium chloride can be prepared in a large scale and stored at room temperature for months.

22. Potassium ferrocyanide and ferricyanide need to be protected from light and stored at 4 °C for long-term use.

23. X-gal solution is very unstable. Therefore, it needs to be prepared freshly right before use every time.

24. For SA-βgal staining, it is very important to make sure the pH of the solution is 5.8 for liver and lung staining. Staining works best for kidney at pH 6 with incubation time of 6–8 h. Confirm the pH value of SA-βgal solution every time before staining.

25. Use an oven or an incubator without CO_2, which dramatically decreases the pH value of staining solution and therefore interferes the staining.

26. For immunofluorescence, it is recommended to use a blocking buffer from the same species as secondary antibody to get the best result.

Acknowledgments

This work was supported by National Institutes of Health (NIH) grants AG024827, AR051456, and AG043376 to P.D.R. and ES016114, CA103730, and AG43376 to L.J.N. and by the Ellison Medical Foundation (AG-NS-0303-05) to L.J.N. The authors would like to thank Rafael R. Flores for critical review of this manuscript.

References

1. Kirkwood TB (2005) Understanding the odd science of aging. Cell 120(4):437–447

2. Collado M, Blasco MA, Serrano M (2007) Cellular senescence in cancer and aging. Cell 130(2):223–233

3. Lombard DB et al (2005) DNA repair, genome stability, and aging. Cell 120(4):497–512

4. Chung HY et al (2009) Molecular inflammation: underpinnings of aging and age-related diseases. Ageing Res Rev 8(1):18–30

5. Green DR, Galluzzi L, Kroemer G (2011) Mitochondria and the autophagy-inflammation-cell death axis in organismal aging. Science 333(6046):1109–1112

6. Liu L, Trimarchi JR, Smith PJ, Keefe DL (2002) Mitochondrial dysfunction leads to telomere attrition and genomic instability. Aging Cell 1(1):40–46

7. Krishnamurthy J et al (2004) Ink4a/Arf expression is a biomarker of aging. J Clin Invest 114(9):1299–1307

8. Baker DJ et al (2011) Clearance of p16Ink4a-positive senescent cells delays ageing-associated disorders. Nature 479(7372):232–236

9. Gregg SQ et al (2012) A mouse model of accelerated liver aging caused by a defect in DNA repair. Hepatology 55(2):609–621

10. Sijbers AM et al (1996) Xeroderma pigmentosum group F caused by a defect in a structure-specific DNA repair endonuclease. Cell 86(5):811–822

11. Sijbers AM et al (1996) Mutational analysis of the human nucleotide excision repair gene ERCC1. Nucleic Acids Res 24(17):3370–3380

12. Weeda G et al (1997) Disruption of mouse ERCC1 results in a novel repair syndrome with growth failure, nuclear abnormalities and senescence. Curr Biol 7(6):427–439

13. Ahmad A et al (2008) ERCC1-XPF endonuclease facilitates DNA double-strand break repair. Mol Cell Biol 28(16):5082–5092

14. Niedernhofer LJ et al (2006) A new progeroid syndrome reveals that genotoxic stress suppresses the somatotroph axis. Nature 444(7122):1038–1043

15. Gloire G, Legrand-Poels S, Piette J (2006) NF-kappaB activation by reactive oxygen species: fifteen years later. Biochem Pharmacol 72(11):1493–1505

16. Perkins ND (2007) Integrating cell-signalling pathways with NF-kappaB and IKK function. Nat Rev Mol Cell Biol 8(1):49–62

17. Hayden MS, West AP, Ghosh S (2006) NF-kappaB and the immune response. Oncogene 25(51):6758–6780

18. Hayden MS, Ghosh S (2004) Signaling to NF-kappaB. Genes Dev 18(18):2195–2224

19. May MJ (2000) Selective inhibition of NF-kappa B activation by a peptide that blocks the interaction of NEMO with the Ikappa B kinase complex. Science 289(5484):1550–1554

20. Khaja K, Robbins P (2010) Comparison of functional protein transduction domains using the NEMO binding domain peptide. Pharmaceuticals 3(1):110–124

21. Mi Z, Mai J, Lu X, Robbins PD (2000) Characterization of a class of cationic peptides able to facilitate efficient protein transduction in vitro and in vivo. Mol Ther 2(4):339–347

22. Mai JC, Shen H, Watkins SC, Cheng T, Robbins PD (2002) Efficiency of protein transduction is cell type-dependent and is enhanced by dextran sulfate. J Biol Chem 277(33):30208–30218

23. Dave SH et al (2007) Amelioration of chronic murine colitis by peptide-mediated transduction of the IkappaB kinase inhibitor NEMO binding domain peptide. J Immunol 179(11):7852–7859

24. Reay DP et al (2011) Systemic delivery of NEMO binding domain/IKKgamma inhibitory peptide to young mdx mice improves dystrophic skeletal muscle histopathology. Neurobiol Dis 43(3):598–608

25. Helenius M, Kyrylenko S, Vehvilainen P, Salminen A (2001) Characterization of aging-associated up-regulation of constitutive nuclear factor-kappa B binding activity. Antioxid Redox Signal 3(1):147–156

26. Kim HJ, Kim KW, Yu BP, Chung HY (2000) The effect of age on cyclooxygenase-2 gene expression: NF-kappaB activation and IkappaBalpha degradation. Free Radic Biol Med 28(5):683–692

27. Korhonen P, Helenius M, Salminen A (1997) Age-related changes in the regulation of transcription factor NF-kappa B in rat brain. Neurosci Lett 225(1):61–64

28. Bregegere F, Milner Y, Friguet B (2006) The ubiquitin-proteasome system at the crossroads of stress-response and ageing pathways: a handle for skin care? Ageing Res Rev 5(1):60–90

29. Adler AS et al (2007) Motif module map reveals enforcement of aging by continual NF-kappaB activity. Genes Dev 21(24):3244–3257

30. Tilstra JS et al (2012) NF-kappaB inhibition delays DNA damage-induced senescence and aging in mice. J Clin Invest 122(7):2601–2612

31. Osorio FG et al (2012) Nuclear lamina defects cause ATM-dependent NF-kappaB activation and link accelerated aging to a systemic inflammatory response. Genes Dev 26(20):2311–2324

32. Dimri GP et al (1995) A biomarker that identifies senescent human cells in culture and in aging skin in vivo. Proc Natl Acad Sci U S A 92(20):9363–9367

33. Lee BY et al (2006) Senescence-associated β-galactosidase is lysosomal β-galactosidase. Aging Cell 5(2):187–195

34. Kurz DJ, Decary S, Hong Y, Erusalimsky JD (2000) Senescence-associated (beta)-galactosidase reflects an increase in lysosomal mass during replicative ageing of human endothelial cells. J Cell Sci 113(20):3613–3622

35. Gregg SQ, Robinson AR, Niedernhofer LJ (2011) Physiological consequences of defects in ERCC1–XPF DNA repair endonuclease. DNA Repair 10(7):781–791

36. Schmahmann JD (2004) Disorders of the cerebellum: ataxia, dysmetria of thought, and the cerebellar cognitive affective syndrome. J Neuropsychiatry Clin Neurosci 16(3):12

37. Rogakou EP, Pilch DR, Orr AH, Ivanova VS, Bonner WM (1998) DNA double-stranded breaks induce histone H2AX phosphorylation on serine 139. J Biol Chem 273(10):5858–5868

38. Gutteridge JM (1984) Age pigments: role of iron and copper salts in the formation of fluorescent lipid complexes. Mech Ageing Dev 25(1–2):205–214

39. Koistinaho J, Hartikainen K, Hatanpaa K, Hervonen A (1989) Age pigments in different populations of peripheral neurons in vivo and in vitro. Adv Exp Med Biol 266:49–59

40. Bourre JM, Haltia M, Daudu O, Monge M, Baumann N (1979) Infantile form of so-called neuronal ceroid lipofuscinosis: lipid biochemical studies, fatty acid analysis of cerebroside sulfatides and sphingomyelin, myelin density profile and lipid composition. Eur Neurol 18(5):312–321

41. Granier LA, Langley K, Leray C, Sarlieve LL (2000) Phospholipid composition in late infantile neuronal ceroid lipofuscinosis. Eur J Clin Invest 30(11):1011–1017

42. Brunk UT, Terman A (2002) The mitochondrial-lysosomal axis theory of aging: accumulation of damaged mitochondria as a result of imperfect autophagocytosis. Eur J Biochem 269(8):1996–2002

The "Sneaking-Ligand" Approach: Cell-Type Specific Inhibition of the Classical NF-κB Pathway

Bettina Sehnert, Harald Burkhardt, Stefan Dübel, and Reinhard E. Voll

Abstract

The intracellular delivery of molecules across the plasma membrane represents a major obstacle. The conjugation of cell-permeable peptides (CPPs) to proteins promotes the uptake and internalization. However, uptake of CPPs is receptor independent and not cell-type specific. Recently, we established the "sneaking-ligand" approach which is based on multimodular recombinant fusion proteins that consist of three modules connected with serine-glycine linkers. Module one is responsible for receptor-mediated endocytosis; module two supports translocation into the cytoplasm so that the effector module three can interact with its binding partner in the cytoplasm. For NF-κB inhibition, we described an NF-κB inhibitor that targets selectively the activated endothelium via an oligopeptide motif. Upon E-selectin-mediated endocytosis, the *Pseudomonas exotoxin A* domain II (ETAII) translocates the NEMO-binding peptide to the cytoplasm interfering with IκB kinase complex assembly. Inflammatory autoimmune diseases are triggered, but also resolved by a variety of cell types. Therefore, the inhibition of NF-κB should be restricted to those cells that are crucially involved in the pathogenesis of inflammatory diseases. A general blockade of NF-κB may result in severe immunosuppression and possibly in organ dysfunction or damage. The "sneaking-ligand" approach could minimize the risks of therapeutic interventions and identify disease-relevant cell types. Here we describe the recombinant expression and purification of the E-selectin-specific "sneaking-ligand construct" (SLC1) and its ability to inhibit cytokine-induced NF-κB activation in vitro.

Key words NF-κB, NEMO-binding peptide, Endothelium, Inhibitor, E-selectin, Sneaking-ligand approach

1 Introduction

The nuclear factor (NF)-κB is crucial for the coordinated transcriptional control of numerous proinflammatory mediators including cytokines, chemokines, enzymes, and cell adhesion molecules [1, 2]. Furthermore, NF-κB has an evolving role in resolution of inflammation depending on the cell type and the disease phase [3]. However, ubiquitous pharmacologic suppression of NF-κB activity is likely to cause severe side effects including profound immunosuppression, liver cell apoptosis, and other organ dysfunctions [4–6]. Moreover, the NEMO-binding peptide

Michael J. May (ed.), *NF-kappa B: Methods and Protocols*, Methods in Molecular Biology, vol. 1280,
DOI 10.1007/978-1-4939-2422-6_33, © Springer Science+Business Media New York 2015

(NBP) coupled to cell-penetrating peptides (CCPs) like transactivator of transcription protein of the Human Immunodeficiency Virus Type-1 (TAT) or Antennapedia (Antp) transduction domain enters all cells via a receptor- and endocytosis-independent mechanism to inhibit NF-κB activation [7]. Hence, inhibition of NF-κB should be restricted to certain cell types contributing to disease pathogenesis in order to maintain NF-κB function in other cells or organs.

To develop an efficient therapeutic NF-κB inhibitor, a class of recombinant fusion proteins was created that interferes with the NF-κB-pathway. The NF-κB inhibitor is designated as "sneaking-ligand construct" (SLC1) and inhibits NF-κB activation selectively only in activated endothelial cells [8]. This cell type was chosen because the endothelium plays a pivotal role in the inflammatory response, acting as a gate keeper that either prevents or allows transmigration of neutrophils, monocytes, as well as T and B lymphocytes into sites of inflammation [2, 9].

In this method, recombinant SLC1 is expressed in *E. coli* and constructed of three functional modules designed for (1) specific ligand binding [10–12], (2) subsequent sneaking into the cytosolic compartment [13, 14], and (3) NF-κB blockade [7] (Fig. 1a, b). E-selectin-mediated endocytosis and uptake of SLC1 into late endosomes initiate the transport to the cytoplasm in the retrograde manner [13]. Subsequently, the C-terminal located NBP interacts with NEMO and blocks IKK assembly (Fig. 2). The complete amino acid sequence of SLC1 is presented in Fig. 3. The Strep-tag II (WSHPQFEK) at the N-terminus and the His_6Tag at the C-terminus are integrated for purification and detection of these sneaking-ligand fusion proteins (SLFPs). In this chapter, we describe the prokaryotic recombinant expression of SLC1 in *E. coli*, followed by denatured affinity purification and refolding. SDS-PAGE and Western blot analysis illustrate full-length expression of the SLFPs. In the last part, the E-selectin-specific inhibition of cytokine-induced NF-κB activation after SLC1 treatment is described.

2   Materials

2.1 Recombinant Expression in E. coli

1. Erlenmeyer flasks—5 l, 1 l.
2. Incubator shaker.
3. Photometer.
4. Centrifuge beakers (500 ml).
5. Centrifuges (for 500 and 50 ml beakers).
6. 2YT medium (*see* **Note 1**).
7. Chloramphenicol: 25 mg/ml stock solution prepared in 100 % ethanol (*see* **Note 2**).
8. 1 M IPTG stock solution (*see* **Note 3**).

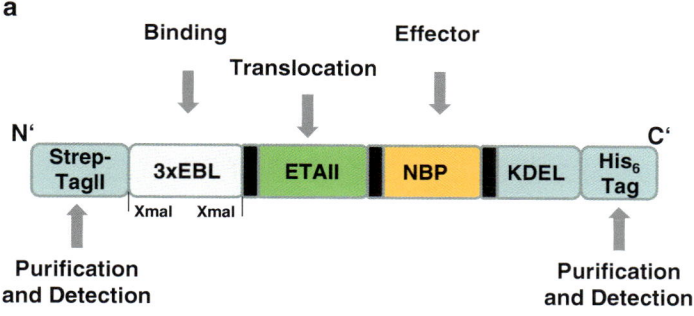

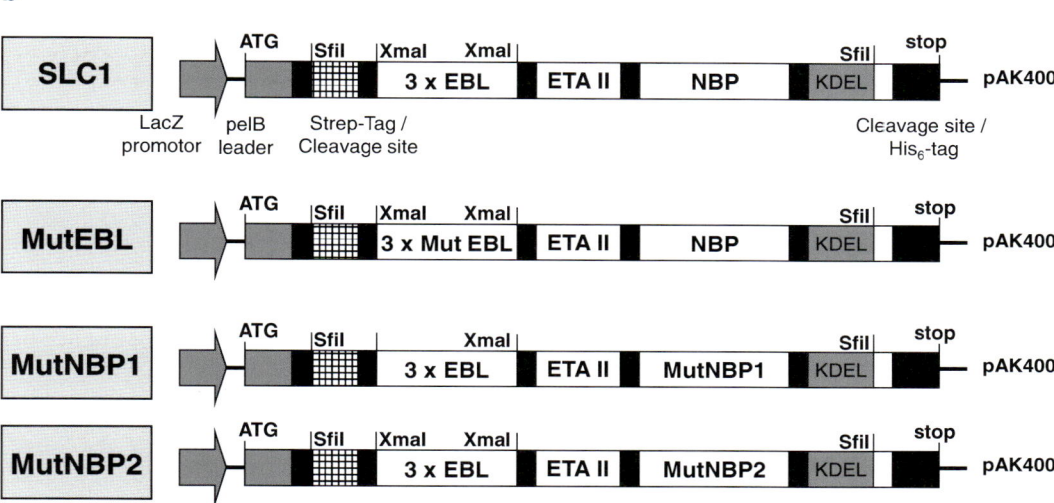

Fig. 1 (**a**) Schematic structure of the multimodular sneaking-ligand construct SLC1 encompasses (1) the receptor-specific binding domain (3xEBL), (2) the translocation domain of *Pseudomonas exotoxin A* (ETAII), and (3) the effector domain (NBP). KDEL sequence is responsible for a retrograde-directed transport. Strep-tag II and His₆Tag are integrated for detection and affinity purification. (**b**) Illustration of functional and nonfunctional SLFPs. EBL, three repeats of an E-selectin-specific peptide (*DITWDQLWDLMK*) connected with a S₄G linker: MutEBL (*WKLDTLDMIQD*) [10], ETAII (translocation domain of *Pseudomonas exotoxin A domain II*) [14], and NBP (NEMO-binding peptide encompassing amino acids 644–756 from IKK2) [7]. Amino acids that were identified for NEMO interaction were mutated: MutNBP1 (*FTALDASALQTE*) and MutNBP2 (scrambled peptide: *DLAWQTFLTES*). Sfil restriction sites are used to clone the multimodular synthetic gene into the pAK400 plasmid for expression in *E. coli* HB2151

    9. 30 % glucose solution (*see* **Note 4**).

   10. DMSO stock of SLFP plasmid (*see* **Note 5**).

2.2 Preparation of Inclusion Bodies

1. Ultrasonic homogenizer. We use a SONOPULS ultrasonic homogenizer from Bandelin Electronic (Berlin, Germany), but any similar device can be used (*see* **Note 6**).

2. Centrifuge tubes (50 ml).

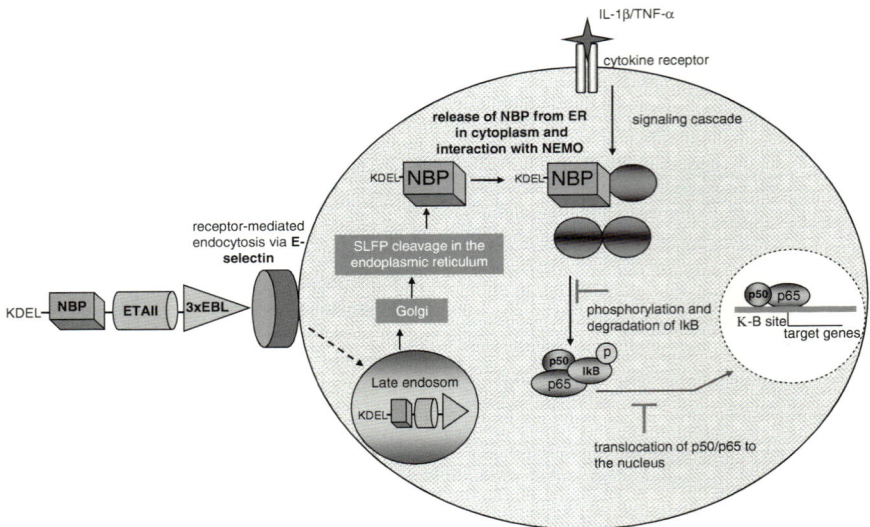

Fig. 2 The "sneaking-ligand" approach: E-selectin-binding peptide (EBL) confers high avidity to E-selectin enabling receptor-mediated endocytosis. KDEL-bearing constructs are further transported by retrograde trafficking from the Golgi to the ER. The ETAII domain possesses a furin cleavage site after Arg279 [13]. A furin-mediated process generates two fragments, whereas the C-terminal NBP is transported across the ER membrane into the cytoplasm. NBP interacts with NEMO, blocks IKK assembly with NEMO, and inhibits NF-κB activation

Amino acid sequence of mature SLC 1

MAS WSHPQFEK GAL EVLFQGP GSDITWDQLWDLMKS
SSSGDITWDQLWDLMKSSSSGDITWDQLWDLMKPGS
SSSGSSSSGAAA EGGSLAALTAHQACHLPLETFTRHR
QPRGWEQLEQCGYPVQRLVALYLAARLSWNQVDQVI
RNALASPGSGGDLGEAIREQPEQARLALTLAAAESER
FVRQGTGNDEAGAASAD SSSSGSSSSGAAA VRLQEK
RQKELWNLLKIACSKVRGPVSGSPDSMNASRLSQPG
QLMSQPSTASNSLPEPAKKSEELVAEAHNLCTLLENAI
QDTVREQDQSFTALDWSWLQTEEEEHSCLEQA SSSS
SGSSSSG KDEL SG LVPRGS AAAGAD HHHHHH

☐	E-selectin binding motif
🟩	Pseudomonas exotoxin domain II
☐	NBP
🟦	tags for detection and purification
☐	Motif for ER retention
☐	cleavage site

Fig. 3 Complete amino acid sequence of the mature E-selectin-specific SLC1: The Strep-tag II (WSHPQFEK) is followed by a 3C protease cleavage site. Three repeats of the E-selectin-binding peptide *DITWDQLWDLMK* are connected with S_4G linkers. Moreover, this binding domain is flanked with Xmal sites resulting in a PGS amino acid sequence. The ETAII translocation domain is flanked with a $(S_4G)_2$AAA linker. The NBP domain is in front of a $(S_4G)_2$ linker and the KDEL retention signal. Finally, a thrombin cleavage site and the His_6Tag are located at the C-terminus

3. *E. coli* resuspension buffer: 50 mM sodium phosphate, 500 mM NaCl, 5 mM β-Me (β-mercaptoethanol), protease inhibitor tablet, pH 7.5 (*see* **Note 7**).

4. Lysis buffer: 50 mM sodium phosphate, 150 mM NaCl, 1 % (v/v) Triton X-100, 0.2 % (w/v) sodium deoxycholate, 5 mM β-Me, protease inhibitor tablet, pH 8.0.

5. Inclusion bodies washing buffer: 50 mM sodium phosphate, 150 mM NaCl, 0.2 % (w/v) sodium deoxycholate, 5 mM β-Me, protease inhibitor tablet, pH 8.0.

6. Solubilization buffer: 6 M guanidinium hydrochloride, 50 mM sodium phosphate, 500 mM NaCl, 3 mM β-Me, pH 8.0.

2.3 Affinity Chromatography

1. Laboratory roller.

2. Empty polypropylene columns (0.8 × 4 cm).

3. TALON *Metal Affinity* Resins (Takara Bio Europe/Clontech, Saint-Germain-en-Lay, France), supplied as 50 % (w/v) slurry.

4. 15 and 50 ml tubes.

5. 1.5 ml tubes.

6. Cobalt washing buffer: 6 M guanidinium hydrochloride, 50 mM sodium phosphate, 500 mM NaCl, 3 mM β-Me, pH 7.8.

7. Elution buffer: 6 M guanidinium hydrochloride, 50 mM sodium phosphate, 500 mM NaCl, 3 mM β-Me, 150 mM imidazole, pH 7.8.

2.4 Preparation of Refolding Buffers

1. Cobalt washing buffer: 6 M guanidinium hydrochloride, 50 mM sodium phosphate, 500 mM NaCl, 3 mM β-Me, pH 7.8.

2. For refolding: 100 mM reduced glutathione (GSSH), 10 mM oxidized glutathione (*see* **Note 8**).

3. Stock solution of refolding buffer I (10×): 1 M Tris, 20 mM EDTA, pH 8.0. Sterilized by autoclaving.

4. Stock solution of refolding buffer II (10×): 500 mM Tris, 1.5 M NaCl, 20 mM EDTA, pH 7.5. Sterilized by autoclaving.

8. Refolding buffer I (1×): 500 mM arginine, 2 mM β-Me, pH 7.5 in 1× refolding buffer I.

5. Dialysis membrane tubing with a 12–16 kDa cutoff.

2.5 Preparing an SDS Gel

1. Electrophoresis chamber (*see* **Note 9**).

2. 10× electrophoresis running buffer: 250 mM Tris, 2.5 M glycine, 1 % (w/v) SDS (*see* **Note 10**).

3. Reagents for separation and stacking gels: autoclaved Milli-Q water, 1 M Tris–HCl, pH 8.8, 10 % (w/v) SDS (sodium dodecyl sulfate), 30 % acrylamide/bis solution (37.5:1); 4× stacking gel buffer (0.5 M Tris–HCl, pH 6.8, 0.4 % SDS).

4. 10 % ammonium persulfate (APS) (*see* **Note 11**).

5. *N,N,N,N'*-tetramethylethylenediamine (TEMED).

6. Saturated butanol solution stored at room temperature (*see* **Note 12**).

7. 5× Laemmli gel loading buffer: 60 mM Tris–HCl, pH 6.8, 2 % (w/v) SDS, 10 % glycerol, 0.01 % bromophenol blue, 5 % (v/v) β-Me. Aliquots stored at –20 °C.

8. Prestained protein molecular weight marker (e.g., *see* ColorPlus Broad Range 10–230 kDa, NEB, Frankfurt, Germany).

9. Heating block.

2.6 Western Blot Analysis

1. Semidry blotting system (*see* **Note 13**).

2. Power supply.

3. Rocking platform.

4. BioTrace™ PVDF Transfer Membrane 0.45 µm (Pall Life Sciences, Dreieich, Germany) (*see* **Note 14**).

5. Whatman paper 17 Chr, 460 × 570 mm, 0.92 mm thick (*see* **Note 15**).

6. 100 % methanol.

7. Blotting buffer: 25 mM Tris, 192 mM glycine, 0.3 % SDS, 20 % methanol (*see* **Note 16**).

8. Phosphate buffered saline (PBS).

9. Blocking solution: 5 % milk powder (low-fat, blotting grade) in PBS.

10. TBST washing buffer (1×): 10 mM Tris, 150 mM sodium chloride, 0.05 % Tween-20, pH 8.0 (*see* **Note 17**).

11. Staining trays.

12. Antibodies: mouse anti-Strep-tag II HRP (Biozol, Eching, Germany), rabbit anti-IKK1/2 (H-470) (Santa Cruz Biotechnology, Santa Cruz, CA, USA), goat anti-rabbit IgG (H + L)-HRP (Bio-Rad, Munich, Germany).

13. ECL Western blotting substrate (Thermo Fisher Scientific, Schwerte, Germany) (*see* **Note 18**).

14. Chemiluminescence detection with chemiluminescence imaging fusion FXF7 instrument (Peqlab, Erlangen, Germany) (*see* **Note 19**).

2.7 Cell Culture

1. CHO medium: RPMI 1640, 10 % fetal calf serum (FCS), 1 % glutamine, 1 % sodium pyruvate, 1 % penicillin/streptomycin (*see* **Note 20**).

2. CHO-E-selectin medium: MEM, 10 % FCS, 1 % penicillin/streptomycin, 0.6 mg/ml G418.

3. PBS.

4. Solution of trypsin (0.25 %) and ethylenediaminetetraacetic acid (EDTA) (1 mM).

5. Tissue culture flask and dishes (75 cm², 6- and 24-well).

6. Recombinant mouse IL-1β (R&D Systems GmbH, Wiesbaden, Germany) (*see* **Note 21**).

7. SLC1 and control constructs (*see* **Note 22**).

8. Antennapedia NEMO-binding peptide (Antp-NBP) and Antp-MutNBP (*see* **Note 23**).

2.8 NF-κB Luciferase Reporter Assay

1. RPMI 1640 medium and MEM without serum and antibiotics (*see* **Note 24**).

2. Pre-warmed PBS.

3. FuGENE HD transfection reagent (Roche, Mannheim, Germany). Store at 4 °C (*see* **Note 25**).

4. NF-κB firefly luciferase reporter construct (*see* **Note 26**).

5. Luciferase assay system kit (Promega, Mannheim, Germany) (*see* **Note 27**).

6. Ice-cold PBS.

7. Orbital shaker.

8. 1.5 ml microfuge tubes.

9. Plate luminometer such as the Sirius instruments available from Berthold Detection Systems GmbH (Pforzheim, Germany).

3 Methods

The transport of molecules across the plasma membrane for interaction with intracellular located proteins represents many obstacles. The development of cell-permeable peptides which import biologically active peptides into cells independently of any receptor carries mostly short sequences of antennapedia or TAT [15]. However, these methods for delivery of molecules into the cells are unspecific and enter every cell type.

Here we describe the engineering of recombinant fusion proteins that are structural analogs of toxins such as the *Pseudomonas exotoxin A*. In general, the individual domains of *sneaking-ligand* fusion proteins (SLFPs) could be amplified from cDNA with the respective gene primers. The domains are connected with serine-glycine linkers by assembly PCR [16]. Another way to generate artificial genes for recombinant expression is to use gene synthesis services supplied by companies (*see* **Note 28**). An advantage of this method is to adapt the codon usage for prokaryotic or eukaryotic expression.

The described NF-κB inhibitor binds specifically to E-selectin which is highly expressed on only activated vascular endothelial cells [9]. Therefore, nucleotide sequences corresponding to the E-selectin-specific peptide AF 10166 [10] or the scrambled peptide

AF 11678 [10], followed by the ETA II domain (amino acids 253–364) [14] and the NBP (amino acids 644–756 from IKK2) [7] and mutated NBP [7] (Fig. 1b), were artificially designed including the respective restriction sites and linkers (Fig. 3). The functional SLC1 gene fragment and the nonfunctional SLFPs (MutEBL, MutNBP1, MutNBP2) are cloned via the SfiI restriction site in the expression vector pAK400 [8, 16]. SLC1 consists of 1,213 base pairs that results in a molecular weight of 39 kDa after translation.

The protein expression in the Gram-negative bacterium *Escherichia coli* (*E. coli*) represents an attractive host for handling and yield [17]. We use the *E. coli* strain HB2151 for SLFP expression. There is no guarantee that the recombinant product is expressed in a soluble and biological active form, and this is strongly dependent on the amino acid sequence and molecular weight of the fusion proteins. Also, the recombinant gene product of the endothelium-specific SLFP (SLC1) as well as the nonfunctional mutants was recovered after accumulation in inclusion bodies of *E. coli*. In the methods below, we describe the conditions of SLFP expression, followed by pre-purification of inclusion bodies and metal ion affinity chromatography. For renaturing, the purified SLFP fractions undergo a refolding process by removing the denaturing agents using stepwise dialysis.

To analyze the purity and yield of the isolated and refolded SLFPs, a SDS-PAGE is performed. SLC1 and the nonfunctional SLFPs migrate at approximately 39 kDa under reducing conditions. The SDS gel shows minor contamination with foreign proteins (Fig. 4a). Full-length expression is checked by Western blot analysis. The electroblotted SLFPs are detectable with an

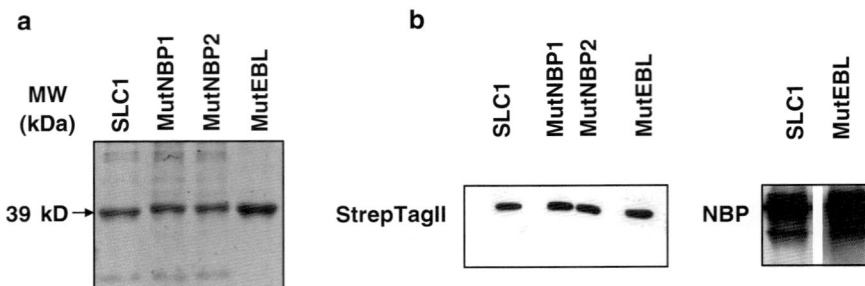

Fig. 4 SDS-PAGE and Western blot analysis of purified and refolded recombinant sneaking-ligand fusion proteins: (**a**) the purity of the SLFPs is analyzed by SDS-PAGE. Twenty microliters of the respective SLFPs are mixed with 5× Laemmli loading gel buffer and loaded on a 14 % SDS gel. SLC1, MutEBL, MutNBP1, and MutNBP2 migrate at the expected molecular weight of 39 kDa. (**b**) After transfer to the PVDF membrane, the bands were visualized using a HRP-labeled anti-Strep-tag II antibody or an anti-IκB kinase antibody. With both antibodies, a band of the expected molecular weight is detectable confirming full-length expression of the SLFPs

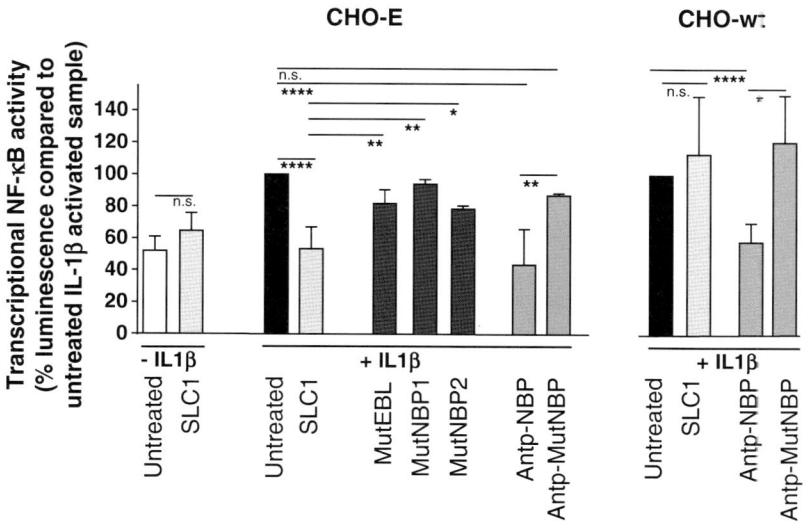

Fig. 5 NF-κB-dependent luciferase reporter gene assay: CHO cells expressing mouse E-selectin (CHO-E) or wild-type CHO cells were transient transfected with the NF-κB-dependent luciferase reporter vector DNA (pB2xLuc). Cells were treated with 500 nM of SLC1, MutNBP1, MutNBP2, or MutEBL for 90 min at 37 °C followed by cytokine treatment to activate NF-κB. The CCPs Ant-NBP (10 μM) and Antp-MutNBP2 (10 μM) were used as controls. Lysates are prepared and firefly luciferase activity is measured in a luminometer. The reporter activity (RLU) is normalized to the protein content and displayed as % luminescence compared to untreated IL-1β-activated sample. SLC1 reduces luciferase expression only in CHO-E cells. Antp-NBP reduces the transcriptional NF-κB activity in CHO-E and in CHO-wt cells. The nonfunctional SLFPs have no impact on NF-κB inhibition

antibody against the N-terminal Strep-tag. The NBP domain located at the C-terminus is visible by an anti-IκB kinase (IKK) antibody (Fig. 4b).

Cytokine activation induces the assembly of the IKK complex that results in the phosphorylation of IκBα and subsequent degradation of IκB. NF-κB activation tightly regulates the expression of proinflammatory mediators (1). An NF-κB luciferase reporter gene assay can be used to analyze the transcriptional activity of NF-κB. E-selectin-mediated uptake of SLC1 reduces NF-κB transcriptional activity significantly in CHO-E cells, but not in wild-type CHO cells (CHO-wt) (Fig. 5). The internalization of the cell-permeable Antp-NBP inhibits NF-κB activation in CHO-E cells and also in CHO-wt cells. Thus, this reporter gene assay emphasizes the specific uptake of active SLC1 exclusively in E-selectin-expressing cells leading to an inhibition of the classical NF-κB pathway.

3.1 Bacterial Culture for SLC1 Expression

1. Inoculate a culture of 200 ml of 2YT medium supplemented with 25 μg/ml chloramphenicol and 1 % (v/v) glucose with the DMSO stock of SLC1, and incubate at 37 °C overnight with shaking (180 rpm) (*see* **Note 29**).

2. On the next day, measure the OD of the pre-culture at 600 nm.

3. Dilute the culture with chloramphenicol-supplemented 2YT medium to an OD_{600nm} of 0.3, and further incubate at 37 °C under shaking (180 rpm) until the OD reaches 0.8 (*see* **Note 30**).

4. Induce SLC1 expression by adding IPTG to a final concentration of 1 mM for 4 h.

5. After 4 h at 37 °C with shaking, harvest the bacterial cells by centrifugation at $3,000 \times g$ for 10 min at 4 °C (*see* **Note 31**).

6. Pool the bacteria cell pellets and store at −20 °C until purification (*see* **Note 32**).

3.2 Purification of Inclusion Bodies

1. Resuspend 10 g of a cell pellet in 100 ml of *E. coli* resuspension buffer (*see* **Note 33**); split into 3× 33 ml portions for sonication.

2. Disrupt the *E. coli* cells in solution using an ultrasonic homogenizer. Use an amplitude of 70 % for a 20 s pulse. After an interval of 2 s, repeat the pulsation (**step 2**) two more times (*see* **Note 34**).

3. Centrifuge the suspension at $3,000 \times g$ for 20 min at 4 °C (*see* **Note 35**).

4. Further resuspend each pellet vigorously in 40 ml of Triton-X100-containing lysis buffer.

5. Sonicate the suspension again as described above, followed by centrifugation.

6. Repeat the sonication step with Triton-X100 buffer twice.

7. Add 40 ml of inclusion bodies washing buffer to the unsoluble pellet, and homogenize again by sonication (*see* **Note 36**).

8. Centrifuge the suspension at $3,000 \times g$ for 20 min; then add 40 ml of inclusion bodies washing buffer to the pellet, and centrifuge again. This is done for twice to remove Triton-X100.

9. Solubilize the washed inclusion bodies (IBs) in a total volume of 150 ml of solubilization buffer by stirring for 12 h at room temperature (*see* **Note 37**).

10. Centrifuge the denatured inclusion bodies at $3,000 \times g$ for 30 min at 4 °C, and carefully transfer the supernatant to a new tube and store at −20 °C.

3.3 Affinity Column Purification of Denatured SLC1 Fusion Protein

1. Transfer 2 ml of the cobalt resin slurry to a 15 ml Falcon tube filled with cobalt washing buffer.

2. Centrifuge the resin at $700 \times g$ at 4 °C for 5 min, and then carefully remove the supernatant.

3. Add 50 ml of solubilized inclusion bodies to the resin, and incubate rolling at 4 °C for 30 min (*see* **Note 38**).

4. Centrifuge the slurry-extract mix at $700 \times g$ at 4 °C for 5 min, and decant the supernatant carefully.

5. Wash nonspecific proteins away by adding 30 ml of cobalt washing buffer. Gently invert the tubes, and then incubate rolling at room temperature for 10 min.

6. Centrifuge the mix again and repeat **step 5** two more times (*see* **Note 39**).

7. Resuspend the resin in 5 ml of cobalt washing buffer, and transfer into an empty plastic column.

8. After the buffer is passed through the column, add 5×1.2 ml of elution buffer containing 150 mM imidazole, and collect each fraction in a 1.5 ml tube.

9. Measure the absorbance of each fraction at A_{280nm} (*see* **Note 40**).

3.4 Refolding the Native Structure of SLC1

1. Refolding the SLC1 is performed in the presence of oxidized glutathione and reduced glutathione by dialysis. The final concentration of the SLC1 protein in the refolding solution should be 300 μg/ml (*see* **Note 41**).

2. Calculate the final volume, and for the outstanding volume, a mixture of 0.5 mM oxidized glutathione (GSSG) and 5 mM reduced glutathione (GSSH) in cobalt washing buffer should be prepared (*see* **Note 42**).

3. Add the glutathione-containing buffer to the SLC1 solution dropwise while stirring at 4 °C.

4. After stirring for 1 h, transfer this solution into a dialysis membrane tubing and dialyze in precooled refolding buffer I for at least 12 h at 4 °C (*see* **Note 43**).

5. Move the tubing into refolding buffer II and dialyze three times for 12 h at 4 °C.

6. After refolding dialysis, transfer the SLC1 protein into a tube and centrifuge at $3,000 \times g$ for 5 min at 4 °C to remove precipitates.

7. Transfer the supernatant containing the renatured SLC1 fusion protein to a new tube, and store at 4 °C (*see* **Note 44**).

3.5 Separating SLFPs by SDS Gel

1. Electrophoretic separation of SLFP is performed using a vertical electrophoresis system such as the *Eco-Mini* by Biometra (Biometra, Jena, Germany). Other systems are also suitable and could be used.

2. Assemble a spacer plate with a thickness of 1.5 mm and a notched plate, and put into the electrophoresis module (*see* **Note 45**).

3. Tighten the clamps of the electrophoresis module and insert the module into the casting stand (*see* **Notes 46** and **47**).

4. The SLFPs are separated in a 14 % SDS gel. For a 1.5 mm gel, make 10.5 ml of separation gel buffer as follows: 4.0 ml of 1 M Tris–HCl, pH 8.8, 1.4 ml of autoclaved Milli-Q water, 4.9 ml of 30 % acrylamide/bis solution, 100 μl of 10 % (w/v) SDS, 10 μl of TEMED, and 35 μl of 10 % APS.

5. Pour the separation gel mix between the glass plates using a plastic pipette (*see* **Note 48**). Overlay the solution with a thin layer of water-saturated *n*-butanol.

6. Allow the gel to polymerize for 60 min (*see* **Note 49**).

7. Remove the saturated *n*-butanol under running double-distilled water (*see* **Note 50**).

8. Prepare a 4.5 % stacking gel as follows: 640 μl stacking gel buffer, 1.35 ml of autoclaved Milli-Q water, 380 μl of 30 % acrylamide/bis solution, 6 μl TEMED, and 26 μl of 10 % APS.

9. Layer the stacking gel mix over the separation gel until the notched glass plate is reached.

10. Insert a 1.5 mm comb between the glass plates, and allow the gel to polymerize for 60 min (*see* **Note 51**).

11. Remove the electrophoresis module from the casting stand, and insert it into the buffer tank.

12. Fill the buffer tank with running buffer to the indicated maximum level line.

13. Fill the chamber between the glass plates and electrophoresis module with running buffer.

14. Remove the comb carefully and rinse the sample pockets with running buffer to remove unpolymerized gel.

15. Add 5 μl of gel loading buffer to 20 μl of SLC1 sample, and heat the sample at 95 °C for 5 min.

16. Spin down the samples and load carefully into the gel pockets.

17. Run at 100 V until the samples enter the separation gel, and then turn to 150 V. Electrophoresis is complete when the dye front migrates to 5 mm from the bottom of the gel.

3.6 Analyzing Refolded SLC1 by Western Blot

1. Separate the SLFP in a 14 % SDS-PAGE (see above).

2. When the run is finished, soak six sheets of Whatman paper in blotting buffer for 5 min, and incubate the PVDF membrane for 1 min in methanol for activation. Soak the gel in blotting buffer.

3. Place three paper sheets onto the electroblotter, followed by the membrane; then put the gel carefully onto the membrane followed by three paper sheets (*see* **Note 52**).

4. Put the lid on the blotting system and run for 80 min at 70 mA.

5. Open the blotting system and carefully remove the upper layers of paper and gel.

6. Place the membrane into a staining tray filled with 20 ml of blocking solution, and incubate on a rocking shaker overnight at 4 °C.

7. On the next day, decant the blocking solution and incubate the membrane with 15 ml of the diluted horseradish-labeled mouse anti-Strep-tag II antibody (1:3,000 in 1 % milk powder in PBS) while shaking at room temperature for 60 min.

8. For detecting the C-terminus of the SLC1, incubate the membrane with the anti-IKK1/2 antibody (1:5,000 in 1 % milk powder in PBS).

9. Wash the membrane three times for 10 min with 20 ml of TBST.

10. To detect anti-IKK binding, add 20 ml of goat anti-rabbit IgG–HRP antibody (1:3,000 in 1 % milk powder in PBS), and shake for 1 h at room temperature.

11. Wash the membrane again three times for 10 min with 20 ml of TBST and 10 min with PBS.

12. Place the membrane into ECL solution and incubate while shaking for 1 min (*see* **Note 53**).

13. Drain the excess liquid on some paper sheets, and place the membrane directly on the detection plate of a digital imaging system. Visualize the chemiluminescence signal using appropriate software (*see* **Note 19**).

3.7 Transfection with pB2xLuc Vector DNA and Treatment of E-Selectin-Expressing CHO Cells with SLC1

The experiments described are performed using adherent Chinese hamster ovary fibroblasts (wild-type CHO cells) and CHO cells stably transfected with full-length mouse E-selectin (CHO-E) (*see* **Note 54**).

1. Grow CHO wild-type and CHO-E cells in their appropriate medium in 75 cm^2 culture flasks.

2. Passage the cells when they reach 80 % confluence by first washing with pre-warmed PBS and incubating for 5 min at 37 °C in 1 ml of trypsin/EDTA, which is also pre-warmed.

3. Verify complete detachment/microscopically.

4. Stop the trypsinization by adding 9 ml of growth medium. For subculturing, split CHO cells 1:30 in growth medium.

5. For transfection experiments, plate CHO and CHO-E cells in 24-well culture plates at a density of 1×10^4 cells in 1 ml of growth medium (*see* **Note 55**). Set up cells to test each SLPF in triplicate wells.

6. Visually inspect the CHO cells and transfect when they have reached 60 % confluence.

7. To analyze NF-κB activation, we use the plasmid pB2xLuc that contains an NF-κB-driven *luciferase* gene. Numerous NF-κB-driven luciferase reporter plasmids are commercially available.

8. Wash the cells once with PBS and transfect using a reagent such as FuGENE® HD following the specific manufacturer's protocols. For FuGENE® HD, use a reagent–vector ratio of 3:1. For cells in 24-well plates, this would require that 1.5 μl FuGENE® HD reagent is mixed with 0.5 μg pB2xLuc plasmid DNA.

9. To each well, add 500 μl of medium without supplements; then add 100 μl of reagent/DNA mix dropwise. Control wells should receive the mix without DNA.

10. After 6 h, add 500 μl of growth medium, and place the plates in a 37 °C CO_2 incubator for 24 h.

11. Aspirate the medium and wash the cells once with growth medium. Incubate the cells for 90 min at 37 °C with 500 nM of either SLC1 or the control SLFPs.

12. For treatment with the cell-permeable Antp-NBP or Antp-MutNBP peptides, these should be diluted to a final concentration of 10 μM (*see* **Note 56**).

13. Activate NF-κB by adding 8 ng rmIL-1β in 100 μl growth medium per well, and further incubate for 4 h in the 37 °C CO_2 incubator (*see* **Note 57**).

3.8 Luciferase Reporter Assay

1. Remove the medium and wash the cells once in PBS. Add 150 μl of 1× lysis buffer per well (*see* **Note 27**). Suspend the cells vigorously before incubating the plate for 20 min at room temperature on an orbital shaker.

2. Transfer the lysed cells into 1.5 ml microfuge tubes, and spin down for 5 min at $13,000 \times g$ at 4 °C.

3. Transfer the supernatant to a new tube and place on ice. Add 5 μl of each lysate into a 96-well microtiter luminometer plate, and insert into a plate luminometer.

4. Set up the luminometer program so that 50 μl of firefly luciferase substrate is added to each well (*see* **Note 58**). Start the measurement and record the luciferase activity in relative light units (RLU).

5. Normalize the luciferase activity stated in relative light units (RLU) to the protein concentration in each cell lysate. Protein concentration can be determined using a number of standard approaches, but we use the bicinchoninic acid (BCA) method (Thermo Fisher Scientific, Schwerte, Germany) (*see* **Note 59**).

4 Notes

1. 2YT medium (16 g of Bacto tryptone; 10 g of Bacto yeast extract; 5 g of NaCl per liter) is prepared, autoclaved at 121 °C for 20 min, and stored at 4 °C.

2. Chloramphenicol stock solution is stored at –20 °C.

3. IPTG (isopropylthio-β-D-galactosidase; MW = 283.3) is dissolved in autoclaved Milli-Q water and filter sterilized (0.22 μm filter). Two milliliter aliquots are stored at –20 °C.

4. Dissolve 30 g glucose in 100 ml of Milli-Q water by stirring, and sterilize by autoclaving. Solution is stored at room temperature.

5. Heat shock transformation is used to integrate 1 ng pAK400-SLC1 plasmid DNA (Fig. 1b) or the respective control constructs in 100 μl chemically competent *E. coli* HB2151. The bacterial cells are plated on an LB–chloramphenicol (25 μg/ml) agar plate and incubated for 16 h at 37 °C. A bacterial colony is transferred into 5 ml of 2YT medium supplemented with 25 μg/ml chloramphenicol and incubated for 16 h at 37 °C under shaking at 180 rpm. On the next day, add 75 μl of DMSO to 1 ml of culture, invert the tube several times and incubate for 1 h at 4 °C, transfer it to –20 °C for 1 h, and then place at –80 °C for long-term storage.

6. Bacterial cells are lysed with a Bandelin SONOPULS ultrasonic homogenizer. For volumes up to 40 ml, a microtip sonotrode *MS73* is used.

7. It is important to use EDTA-free protease inhibitors because EDTA chelates metals (cobalt resin) and inhibits the binding of the SLFPs to the resin. We use complete EDTA-free protease inhibitor tablets (Roche, Mannheim, Germany). One tablet is dissolved in 250 ml of buffer immediately before use.

8. 100 mM reduced glutathione (GSSH) and 10 mM oxidized glutathione are prepared in autoclaved water and frozen in aliquots at –20 °C.

9. Here we use electrophoresis chambers of Biometra (Jena, Germany) and describe the method for this system. Any other system could also be used.

10. This electrophoresis chamber has to be filled with 2.2 l of running buffer. A 10× stock solution of running buffer is prepared in advance and is stored at room temperature for up to 3 months.

11. APS should be dissolved in Milli-Q water and prepared freshly for each run.

12. Mix equal volumes of butanol and Milli-Q water. The saturated butanol is the upper phase. Overlay immediately the

unpolymerized separating gel with a thin layer of saturated butanol. Saturated butanol can be stored at room temperature for 1 year.

13. In the described method, we use the *Fast Blot* semidry blotting system from Biometra. Any other system could also be used.

14. The PVDF membrane should be cut into 10 cm × 10 cm pieces. Wear cloves during handling to reduce fingerprints, and use forceps to move the membrane.

15. Several sheets of Whatman paper should be cut into 10 cm × 10 cm pieces and stored.

16. Methanol to final volume of 20 % should be added immediately before use. Prepare only the volume you need to reduce chemical waste.

17. TBST washing buffer can be prepared as 10× stock and stored at room temperature for at least 2 months.

18. For chemiluminescence detection, we used ECL Western blotting substrate from Thermo Fisher Scientific. Substrates from other companies can also be used.

19. Chemiluminescence signals were detected using the imaging fusion FXF7 instrument from Peqlab (Germany). Western blot could also detect by exposing to film and developing using a film processor.

20. All tissue culture media and reagents (trypsin/EDTA, PBS) should be stored at 4 °C and pre-warmed to 37 °C before use. The hood and all bottles should be disinfected before use.

21. Recombinant mouse IL-1β should be reconstituted at 100 μg/ml in sterile PBS containing 0.1 % human or bovine serum albumin and stored in 20 μl aliquots at –80 °C. Thawed aliquots can be stored at 4 °C for 4 weeks.

22. For cell culture experiments, the SLFPs should be dialyzed against PBS. If using SLFPs in Tris-containing refolding buffer II, use the same buffer as a control.

23. The wild-type Antp-NBP and the Antp-MutNBP peptide [7] are used as controls and can be commercially obtained from several sources including Calbiochem (San Diego, CA, USA) and Biomol (Hamburg, Germany). Alternatively, peptides can be ordered from peptide synthesis companies or facilities. Lyophilized peptides should be stored at –20 °C in a sealed box containing desiccant.

24. Cell culture medium without any supplements and antibiotics should be used to prepare the transfection-DNA mix according to the manufacturer's instructions.

25. Several transfection reagents were tested. However, in the described experiment, we found that FuGENE HD worked best and gave comparable and reproducible results.

26. In the described experiment, the firefly pB2xLuc construct was used [18]. Plasmid DNA should be prepared using an endotoxin-free plasmid maxi kit such as that from QIAGEN (Valencia, CA, USA).

27. The luciferase assay system kit contains a luciferase assay substrate, an assay buffer, and a 5× reporter lysis buffer. The lyophilized luciferase assay substrate is reconstituted with luciferase assay buffer. This luciferase assay reagent should be stored in aliquots at −80 °C.

28. Artificial genes can be cloned by gene synthesis services. The codon usage can be adapted for prokaryotic or eukaryotic expression. In our studies, the SLC1 gene was adapted for prokaryotic expression (see Fig. 3) and the company supplied SLC1 in a standard vector. For SLC1 cloning, the supplied vector was cleaved by SfiI restriction enzyme, and the digested SLC1 insert was cloned into a SfiI-digested pAK00 vector [16].

29. 2YT medium should be pre-warmed at room temperature and 500 μl of DMSO stock added to each pre-culture. Do not refreeze DMSO stocks. Expression and purification of the respective nonfunctional SLFP (MutEBL, MutNBP1, MutNBP2) should be induced in the same manner.

30. The final volume of the culture in a 5 l shaking flask should be 2 l.

31. The centrifuge should be precooled at 4 °C. The centrifuge beakers should be filled and centrifuged and then the supernatant decanted. The beakers should be refilled with the bacterial culture and centrifuged again.

32. The supernatant should be removed and the cell pellet scraped out of the beaker with a Stripette and transferred into a 50 ml tube. Thirty milliliter of PBS should be poured into one beaker, and the remaining pellet should be resuspended. At the end, the cell suspension should be transferred into a 50 ml tube and centrifuged for at least 20 min at $3,000 \times g$ at 4 °C. When the supernatant is clear, it should be removed and the bacterial cell pellet stored at −20 °C. The pellets could be stored for at least 2 years.

33. β-Me and protease inhibitor tablets should be added immediately before use to the buffers in this method. E. coli resuspension buffer and lysis buffer should be precooled on ice. Pellets have to be vigorously resuspended to enhance cell lysis.

34. Sonication is performed with all samples on ice.

35. Centrifugation time should be increased until supernatant is clear.

36. Inclusion washing buffer can be stored at room temperature. When using this buffer, the centrifugation steps are performed at 16 °C; otherwise, the suspensions get jellylike.

37. If the solubilized inclusion bodies become jellylike, add more solubilization buffer.

38. The solubilized inclusion bodies should be pre-warmed to room temperature before the resin is added. This increases binding affinity.

39. Check the absorbance at 280 nm of supernatant after the third washing step. The supernatant should be protein free; otherwise, washing steps should be repeated.

40. We calculate the extinction coefficient of each SLFP using the online ProtParam tool from the ExPASy portal. For example, the extinction coefficient of SLC1 is 1.7 meaning that an absorbance$_{280nm}$ of 1.7 corresponds to 1 mg/ml.

41. The optimal protein concentration for refolding is amino acid dependent. The range could be 100–1,000 µg/ml. The optimal refolding buffers and the protein concentration for SLC1 should be determined in pilot experiments.

42. The elution fractions with positive absorbance are pooled and measured again. Then the final volume is calculated to obtain 300 µg/ml SLC1 ($= A_{280nm}$ 0.5). An example would be 4.2 ml of pooled SLC1 elution fractions have an A_{280nm} of 0.9. To reach an A_{280nm} of 0.5, the final volume is 7.56 ml. 2.6 ml of cobalt washing buffer supplemented with 380 µl GSSG (1/20 volume of final volume) and 380 µl of GSSH (1/20 volume of final volume) is prepared freshly and added in the cool room under stirring to the pooled fractions.

43. If precipitates in the SLC solution are visible between the dialysis steps, take it out of the membrane tubing, spin down at $3,000 \times g$ for 10 min at 4 °C, and transfer the SLC-containing supernatant into a new dialysis membrane tube.

44. The SLCs are stored at 4 °C not longer than 4 weeks.

45. The Eco-Mini system from Biometra produces gels with a size of 9.4×8.0 cm. The plates have to be cleaned with water and 70 % ethanol and dried.

46. Instead of preparing two gels, you can insert one dummy plate into the electrophoresis module.

47. Before preparing the gel, insert the comb and mark the height of the separating gel. This should be 5–10 mm below the comb teeth. Also check for leakage. Pipette water into the gel-casting chamber. If the water leaks out, restart the plate assembly.

48. It is important to pipette the separating gel slowly between the plates without air bubbles.

49. The separating gel is polymerized when a sharp dividing line between the gel and overlay is visible.

50. The water has to be removed using filter paper before adding the stacking gel solution.

51. Pre-wet the comb in 10 % APS to improve the formation of the sample pockets.

52. Carefully smooth the gel out to remove any air bubbles between the gel and the membrane.

53. Here we freshly mixed 5 ml of detection reagent 1 with 5 ml of detection reagent 2 and incubate the membrane under shaking for 1 min.

54. A stabile CHO cell line expressing mouse E-selectin was kindly provided by D. Vestweber, Münster, Germany [19, 20]. CHO cells are cultured in 5 % CO_2 and CHO-E cell in 10 % CO_2.

55. It is an advantage to plate different cell densities, e.g., 1×10^4 or 5×10^4 cells in 1 ml of growth medium, inspect the cells visually the next day, and use the cells that are approximately 60 % confluent.

56. Antp-NBP and Antp-MutNBP have to be dissolved freshly for every experiment in DMSO to a stock concentration of 20–50 mM and then further diluted with culture medium.

57. Recombinant mouse IL-1β diluted in growth medium can be stored at 4 °C for 4 weeks.

58. The luciferase assay reagent is thawed to room temperature freshly and should not be refrozen. It should be also protected from light.

59. The RLU values are normalized against the total protein content. This normalization method was chosen because of the difficulties to efficiently transfect these cells.

Acknowledgments

The work in the author's laboratory was supported by the German Research Foundation (SFB 643 project B3 and A8; FOR 832, project 7; Sachbeihilfe DU337/3-2; and BU 584/4-1), BMBF 01EO0803 Grant to the Center of Chronic Immunodeficiency, and the Federal State of Hessen (LOEWE-Project: Fraunhofer IME-Project-Group Translational Medicine and Pharmacology, Goethe University Frankfurt).

References

1. Denk A et al (2001) Activation of NF-kappa B via the Ikappa B kinase complex is both essential and sufficient for proinflammatory gene expression in primary endothelial cells. J Biol Chem 276:28451–28458
2. Oppenheimer-Marks N, Lipsky PE (1996) Adhesion molecules as targets for the treatment of autoimmune diseases. Clin Immunol Immunopathol 79:203–210
3. Lawrence T, Gilroy DW, Colville-Nash PR et al (2001) Possible new role for NF-kappaB in the resolution of inflammation. Nat Med 7:1291–1297
4. Auphan N, DiDonato JA, Rosette C et al (1995) Immunosuppression by glucocorticoids: inhibition of NF-kappa B activity through induction of I kappa B synthesis. Science 270:286–290
5. Senftleben U (2008) Anti-inflammatory interventions of NF-kappaB signaling: potential applications and risks. Biochem Pharmacol 75:1567–1579
6. Yamamoto Y, Gaynor RB (2001) Therapeutic potential of inhibition of the NF-kappaB pathway in the treatment of inflammation and cancer. J Clin Invest 107:135–142
7. May MJ, D'Acquisto F, Madge LA et al (2000) Selective inhibition of NF-kappaB activation by a peptide that blocks the interaction of NEMO with the IkappaB kinase complex. Science 289:1550–1554
8. Sehnert B, Burkhardt H, Wessels JT et al (2013) NF-kappaB inhibitor targeted to activated endothelium demonstrates a critical role of endothelial NF-kappaB in immune-mediated diseases. Proc Natl Acad Sci U S A 110:16556–16561
9. Pober JS, Sessa WC (2007) Evolving functions of endothelial cells in inflammation. Nat Rev Immunol 7:803–815
10. Martens CL, Cwirla SE, Lee RY et al (1995) Peptides which bind to E-selectin and block neutrophil adhesion. J Biol Chem 270:21129–21136
11. Shamay Y, Paulin D, Ashkenasy G et al (2009) E-selectin binding peptide-polymer-drug conjugates and their selective cytotoxicity against vascular endothelial cells. Biomaterials 30:6460–6468
12. Zinn KR, Chaudhuri TR, Smyth CA et al (1999) Specific targeting of activated endothelium in rat adjuvant arthritis with a 99mTc-radiolabeled E-selectin-binding peptide. Arthritis Rheum 42:641–649
13. Weldon JE, Pastan I (2011) A guide to taming a toxin–recombinant immunotoxins constructed from Pseudomonas exotoxin A for the treatment of cancer. FEBS J 278:4683–4700
14. Wick MJ, Hamood AN, Iglewski BH (1990) Analysis of the structure-function relationship of Pseudomonas aeruginosa exotoxin A. Mol Microbiol 4:527–535
15. Dunican DJ, Doherty P (2001) Designing cell-permeant phosphopeptides to modulate intracellular signaling pathways. Biopolymers 60(1):45–60
16. Krebber A, Bornhauser S, Burmester J et al (1997) Reliable cloning of functional antibody variable domains from hybridomas and spleen cell repertoires employing a reengineered phage display system. J Immunol Methods 201(1):35–55
17. Sahdev S, Khattar SK, Saini KS (2008) Production of active eukaryotic proteins through bacterial expression systems: a review of the existing biotechnology strategies. Mol Cell Biochem 307:249–264
18. Saksela K, Baltimore D (1993) Negative regulation of immunoglobulin kappa light-chain gene transcription by a short sequence homologous to the murine B1 repetitive element. Mol Cell Biol 13:3698–3705
19. Blanks JE, Moll T, Eytner R et al (1998) Stimulation of P-selectin glycoprotein ligand-1 on mouse neutrophils activates beta 2-integrin mediated cell attachment to ICAM-1. Eur J Immunol 28:433–443
20. Zöllner O, Lenter MC, Blanks JE et al (1997) L-selectin from human, but not from mouse neutrophils binds directly to E-selectin. J Cell Biol 136:707–716

Chapter 34

Sneaking-Ligand Fusion Proteins Attenuate Serum Transfer Arthritis by Endothelium-Targeted NF-κB Inhibition

Bettina Sehnert, Harald Burkhardt, Michael J. May, Jochen Zwerina, and Reinhard E. Voll

Abstract

The nuclear transcription factor κB (NF-κB) is a crucial mediator of the inflammatory and immune response. The contribution of dysregulated NF-κB is established in the pathogenesis of arthritis. Accordingly, NF-κB represents an attractive molecular target for the development of therapeutic interventions in inflammatory diseases. However, ubiquitous pharmacologic suppression of NF-κB activity is limited by the hazards of toxic side effects and profound immunosuppression. Cell type-specific NF-κB inhibition with the "sneaking-ligand" approach could identify disease-relevant cell types and improve risk-benefit ratios of therapeutic interventions. Vascular endothelial cells act as a gatekeeper and are crucial for leukocyte recruitment into sites of inflammation. The endothelium-specific NF-κB inhibitor SLC1 ameliorates serum transfer arthritis in mice and protects against inflammation and cartilage destruction. In this chapter, we describe the SLC1 treatment schedule in the *K/BxN* serum transfer arthritis and present the evaluation system to analyze arthritis severity and histopathological alterations.

Key words NF-κB, Endothelium, Sneaking-ligand approach, Mouse model, Arthritis

1 Introduction

Approximately 1 % of the population is affected by rheumatoid arthritis (RA), an autoimmune disease associated with autoreactive lymphocytes, activated macrophages, and synoviocytes leading to massive cartilage and bone destruction [1, 2]. A key event in inflammatory disorders is the recruitment of leukocytes into the target tissue, and in this regard, vascular endothelial cell activation is crucial. The activation is hallmarked by the expression of various pro-inflammatory mediators including cytokines, chemokines, enzymes, and cell adhesion molecules [3]. Moreover, this multi-cascade process is regulated by the nuclear factor (NF)-κB signaling pathway (Fig. 1). Dysregulated NF-κB activation has been implicated in the pathogenesis of immune-mediated inflammatory diseases like RA [4, 5].

Michael J. May (ed.), *NF-kappa B: Methods and Protocols*, Methods in Molecular Biology, vol. 1280,
DOI 10.1007/978-1-4939-2422-6_34, © Springer Science+Business Media New York 2015

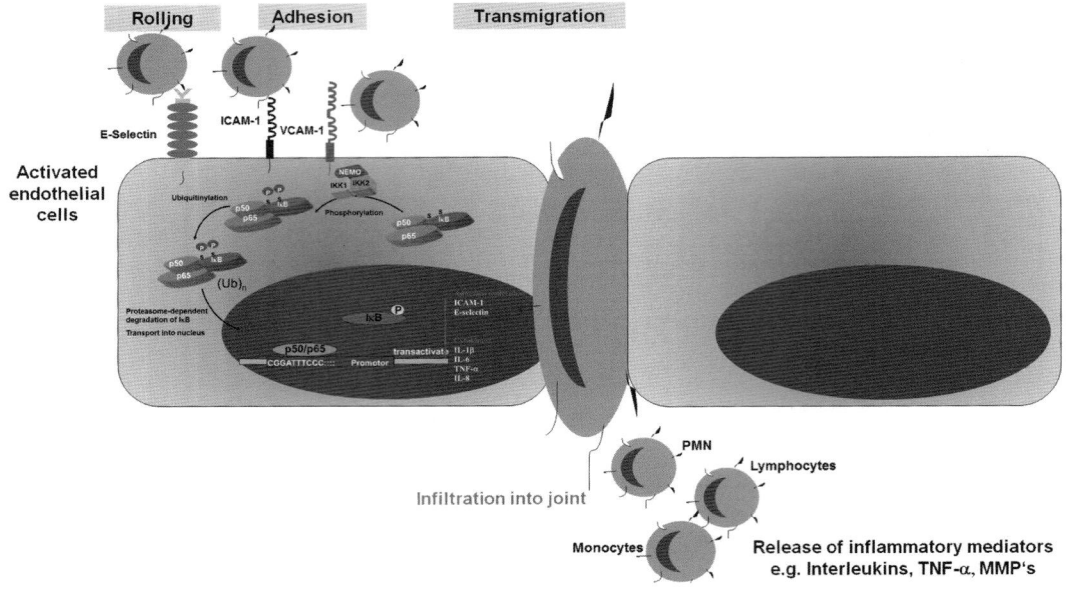

Fig. 1 The contribution of the classical NF-κB pathway in endothelial cells: p50/p65-mediated activation of endothelial cells leads to the expression of adhesion molecules, e.g., E-selectin, ICAM-1, and VCAM-1. Multicascade events including rolling, adhesion, and diapedesis of leukocytes are subsequently initiated and are responsible for the infiltration of leukocytes into the inflamed joint tissue. A recombinant E-selectin-specific NF-κB inhibitor SLC1 [8] selectively blocks the p50-/p65-initiated transcription of target genes and inhibits the extravasation of inflammatory cells into the joint

The main component of the classical NF-κB pathway is the IKK complex that includes the kinases IKK1 and IKK2 and the regulatory subunit NEMO. The association of NEMO with IKK2 initiates the phosphorylation of IkBα and prepares IkBα for proteasomal degradation [6]. Thus far cell-permeable NEMO-binding peptides (NBPs) have been used to study the role of cytokine-induced NF-κB signaling [7]. An approach to investigate the significance of certain cell types in the pathogenesis of inflammatory autoimmune diseases is the "*sneaking-ligand approach*" [8]. Moreover, this approach represents an attractive tool for therapeutic interventions in immune-mediated disease.

In this chapter, we describe the application of a recombinant endothelium-specific sneaking-ligand protein (SLC1) in the KxB/N serum transfer model (Fig. 2). Mouse models have been widely used to study the pathogenesis of RA [9]. Serum from K/BxN mice containing antibodies against glucose-6-phosphoisomerase (anti-G_6PI) causes an acute polyarthritis [10, 11]. This aggressive serum transfer-induced arthritis (STIA) shares features with human RA, including polyarticular manifestation, leukocyte invasion, pannus formation, synovitis, as well as cartilage and bone erosions. Here we illustrate (1) the induction of arthritis

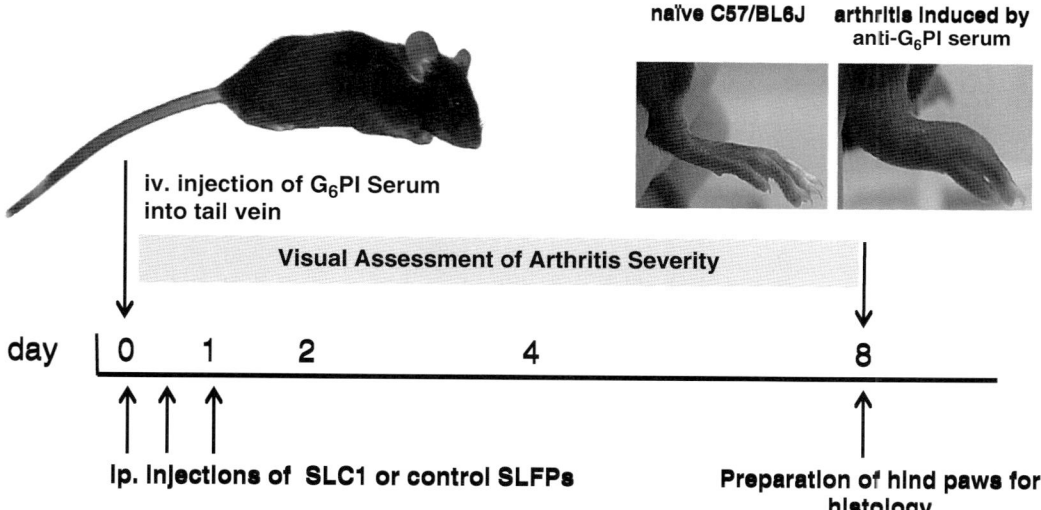

Fig. 2 Treatment schedule to modulate aggressive arthritis induced by K/BxN sera: anti-G_6PI serum is injected intravenously in C57/BL6J mice resulting in a massive joint swelling of the entire paw and ankle at day 8. The treatment protocol includes three intraperitoneal injections ($t=0$ h, $t=7$ h, $t=24$ h) of SLC1 or respective controls (PBS, MutNBP2; *see* **Note 3**). Compared to a naïve C57/BL6J mouse (*left* paw), the image of the paw shows the massive swelling after arthritis induction (*right* paw) 8 days after serum transfer. The grade of arthritis severity is evaluated by visual assessment with two independent observers. At day 8, the hind paws were removed and prepared for histological analysis

in naïve mice and the applied SLC1 treatment protocol, (2) the visual scoring system to assess clinical arthritis severity, and (3) the grading system to analyze the histopathology.

2 Materials

2.1 Anti-G_6PI Serum Transfer Arthritis

1. 4- to 6-week-old female C57BL/6J mice.

2. K/BxN serum (*see* **Note 1**).

3. Mouse restrainer.

4. Heat lamp.

5. 1 ml syringes BD Ultra-Fine™ (Beckton Dickinson, Franklin Lakes, NJ, USA) with 12.7 mm ($1/2''$) × 30G needle (*see* **Note 2**).

6. 1 ml syringe.

7. Needles 0.45×12 mm, 26G × 1/2.

8. Sneaking-ligand fusion proteins (SLFP) (*see* **Note 3**).

9. Cell-permeable Antennapedia NEMO-binding peptide (Ant-NBP) (*see* **Note 4**).

2.2 Visual Assessment of Arthritis Severity

1. Two independent observers (*see* **Note 5**).

2.3 Caliper Measurement of Paw Diameter

1. Digital caliper.

2.4 Preparation of Hind Paws

1. Dissecting instruments: surgical tweezers, anatomically tweezers with hooks, dissecting scissors, and shears.
2. 70 % ethanol (*see* **Note 6**).
3. Tissue cassettes with integral lid (L 40 × B 28 × H 6.8 mm) (*see* **Note 7**).

2.5 Fixation Solution and Decalcification Solution

1. PFA fixation solution: 4 % paraformaldehyde in PBS, pH 7.5 (*see* **Note 8**).
2. Decalcification solution: 10 % EDTA, 100 mM Tris–HCl pH 7.5, autoclaved (*see* **Note 9**).
3. Beakers (500 ml and 1 l).
4. Filters (0.45 μm).

2.6 Hematoxylin/ Eosin staining

1. Paraffin-embedded tissue sections (*see* **Note 10**).
2. Slide rack.
3. Xylene (*see* **Note 11**).
4. 100 % ethanol.
5. 95 % ethanol.
6. 70 % ethanol.
7. ddH$_2$O.
8. Mayer's hematoxylin solution and eosin solution (0.5 %) (*see* **Note 12**).
9. Microscope slides: size 76 × 26 × 1 mm.
10. Coverslips: 24 mm × 55 mm, thickness 0.08–0.12 mm.
11. Water-based mounting medium (e.g., Faramount, Dako, Carpinteria, CA, USA).

2.7 Toluidine Blue Staining

1. Paraffin-embedded tissue sections (*see* **Note 10**).
2. 0.1 % toluidine blue dye solution.
3. Xylene.
4. 100 % ethanol.
5. 95 % ethanol.
6. ddH$_2$O.
7. Microscope slide coverslips.
8. Water-based mounting medium such as Faramount (Dako).

2.8 Assessment of Histological Tissue Sections

1. Two independent observers.

2. Light microscope.

---

3 Methods

An attractive approach for the development of new therapeutic interventions is to define the role of certain cell types involved in the pathogenesis of autoimmune diseases. In inflammation, the extravasation of cells from the bloodstream across the endothelium into the target tissue is a key event that is transcriptionally regulated through activation in endothelial cells of the classical NF-κB pathway (Fig. 1) [8]. Therefore, targeting classical NF-κB signaling specifically in endothelial cells improves the understanding of the underlying pathogenic mechanisms in immune-mediated diseases.

The use of mouse models to study pathological processes in arthritis is common. One model that resembles the human RA is the serum transfer-induced arthritis. Naïve C57/BL6J mice injected with K/BxN serum develop arthritis [12]. K/BxN mice develop a B and T cell response against the ubiquitous glycolytic enzyme G_6PI, and the presence of high anti-G_6PI titers is detectable. Immune complexes accumulate in the joints causing an enrichment of destructive immune system components.

In the method described here, mice injected with K/BxN serum develop swollen hind and front paws from day 4. Systemic treatment with the E-selectin-specific NF-κB inhibitor SLC1 is performed simultaneously with arthritis induction, as well as 7 and 24 h later (Fig. 2). SLC1 treatment reduces the swelling compared to the control fusion protein MutNBP2 and to the PBS mice (Fig. 3).

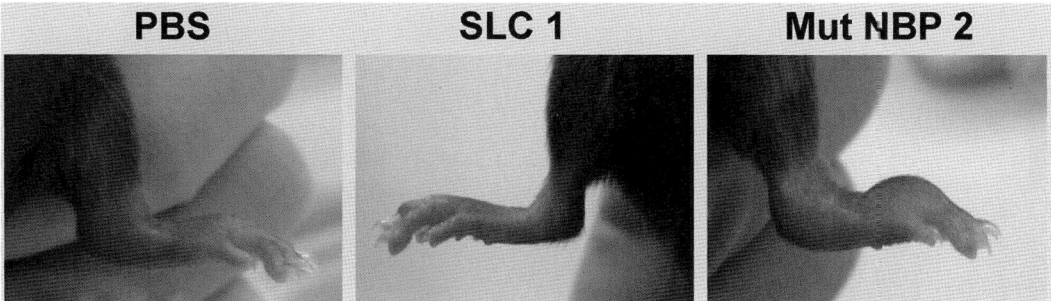

Fig. 3 Hind paws of arthritic mice: representative photographs of mice that received K/BxN serum and the indicated treatment modalities (PBS, SLC1, MutNBP2). The joint swelling here is recorded 3 days after serum transfer. Compared to PBS and MutNBP2 mice, the swelling is less pronounced in SLC1-treated mice

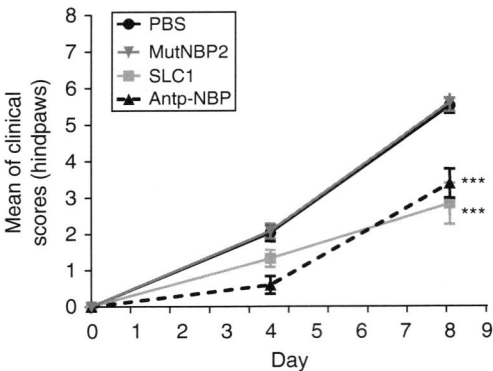

Fig. 4 Clinical manifestation is defined by a graded scoring system. After K/BxN serum transfer, two independent observers inspected the mice at day 0, day 4, and day 8 and evaluated the severity of arthritis to a graded scale (0–4). The score values of the hind paws per mouse are summed, and then the values are averaged per group. The clinical score is diminished in SLC1-treated mice compared to PBS and MutNBP2 treatment. SLC1 attenuates arthritis severity similar to Antp-NBP treatment. PBS, $n=19$; SLC1, $n=13$; MutNBP2, $n=9$; and Antp-NBP, $n=9$. Data are presented as mean ± SEM. P values were calculated by one-way ANOVA followed by Bonferroni's multiple comparison posttest: ***$P < 0.001$

To evaluate the severity of arthritis, each paw is inspected according to a graded scale (0–4) at day 0, day 4, and day 8 by two independent valuers. The determined severity score values are lower in SLC1-treated mice (Fig. 4). The effect of SLC1 in reducing joint swelling is comparable to Antp-NBP treatment (Fig. 4). However, Ant-NBP has to be used in a 20-fold molar excess. Further, the extent of swelling can be measured with a caliper. This procedure also confirms the visually obtained arthritis score. The paw diameter is smaller after three injections of the endothelium-specific NF-κB inhibitor (SLC1) compared to MutNBP2 treatment (Fig. 5). The paw diameter is also lowered in Antp-NBP-treated mice.

Furthermore, the outlined method gives insights into the analysis of the joint histology. Hence, mice were sacrificed at the end of the experiment at day 8, and the hind paws are prepared (Fig. 6) for histology. Hematoxylin and eosin staining (H&E) is a standard staining histology to investigate the inflammatory activity. Hematoxylin stains cell nuclei blue, while eosin stains cytoplasm and depicts inflammatory infiltrates. In SLC1-treated mice, less infiltration of inflammatory cells is observed compared to the PBS and MutNBP2 group (Fig. 7). To investigate the extent of cartilage destruction, sections are incubated with toluidine blue (Tb) solution that stains cartilage proteoglycan subunits. Endothelium-specific NF-κB inhibition leads to a suppressed cartilage breakdown group (Fig. 7). The next defined method shows that a three-grade classification system allows us to define inflammation and cartilage

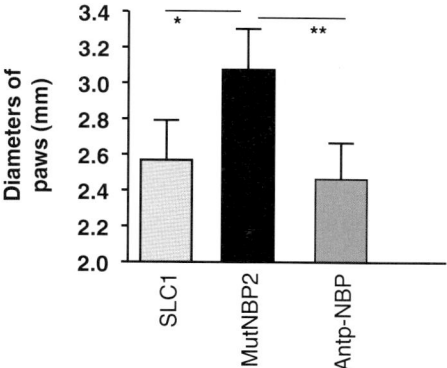

Fig. 5 Caliper measurement of hind paw swelling. The diameters of the mid-foot joints and ankles of the paws are measured with a caliper, and the mean values were calculated. Endothelium-specific NF-κB inhibition (SLC1) results in less swollen joints compared to MutNBP2 treatment. The control treatment with Antp-NBP also reduces inflammatory swelling. Data are presented as mean±SD. P values were calculated by one-way ANOVA followed by Bonferroni's multiple comparison posttest: **P<0.01, *P<0.05

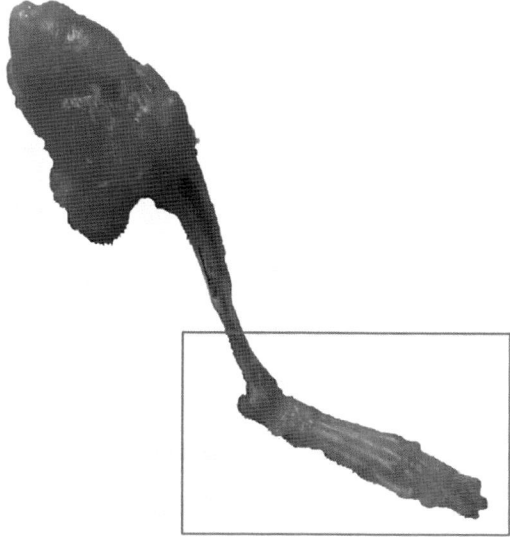

Fig. 6 Dissected mouse hind paw for paraffin embedding. Skin, muscles, and tendons have to be precisely dissected before fixation and decalcification. The part indicated by the *frame* is used for analysis and placed into the tissue cassette for fixation and decalcification. The joint is placed planar side down into the embedding mold

erosion. To minimize the subjectivity of the evaluation, a second observer assesses the sections. Overall, these combined methods illustrate that endothelium-specific NF-κB inhibition treatment protects against inflammation and cartilage erosions (Fig. 8).

PBS   **SLC1**   **MutNBP 2**

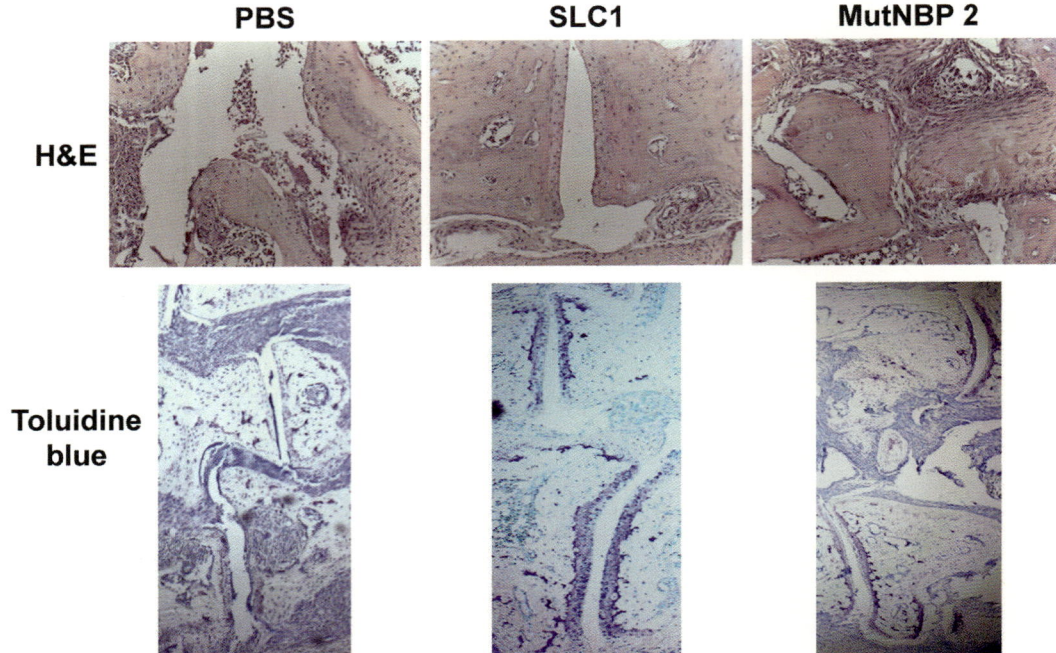

H&E

**Toluidine
blue**

Fig. 7 Endothelium-specific NF-κB inhibition improves histopathological parameters. Representative images of H&E-stained and Tb-stained sections of arthritic paws are shown. H&E sections of PBS and MutNBP2-treated mice present an accumulation of inflammatory cells, whereas in SLC1 mice less leukocyte infiltration is visible. Tb-stained sections show a loss of cartilage with pannus infiltration in PBS and MutNBP2-treated mice. SLC1 mice showed a well-preserved cartilage structure

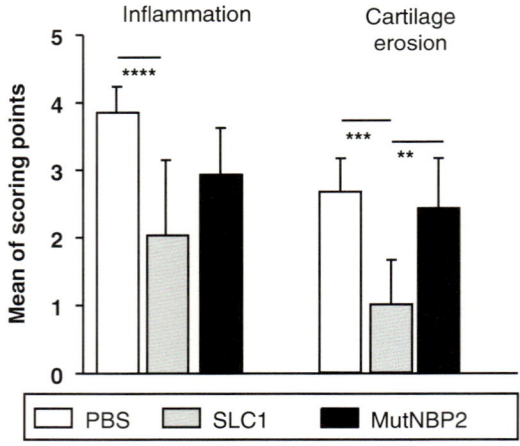

Fig. 8 Histological assessment of H&E and Tb-stained section. The two observers evaluated the stained sections according to a graded scale (0–3) in respect to inflammatory areas and cartilage breakdown. The mean of scoring points reveals that SLC1 protects against inflammatory response and induced structural damage of cartilage. Data are presented as mean ± SD. P values were calculated by one-way ANOVA followed by Bonferroni's multiple comparison posttest: ****$P < 0.0001$, ***$P < 0.001$, **$P < 0.01$

3.1 Induction of G₆PI Serum Transfer Arthritis and SLC Treatment

1. Fix a mouse in the restrainer and slowly inject 200 μl of G_6PI serum intravenously into the vein tail (*see* **Notes 13** and **14**).

2. Inject 100 μg of SLC1 or control fusion protein in a volume of 500 μl intraperitoneally (*see* **Notes 3** and **15**).

3. Repeat the SLC1 treatment in the same manner 7 and 24 h after the first injection.

3.2 Clinical Assessment of Arthritis in Mice

1. Visual scoring of the clinical symptoms of arthritis should be performed by two independent observers (*see* **Note 5**).

2. Restrain the mouse by hand to allow examination of the paws.

3. Evaluate the hind and front paws of each mouse visually according the following scoring system ranging from 0 to 4:

   0 = no swelling and redness.

   1: mild swelling and redness of one joint (mid-foot, tarsals, metatarsals, ankle joint, or digit).

   2: moderate swelling and redness in more than one joint (mid-foot, tarsals, metatarsals, ankle joint, or digit).

   3: swelling and redness in the entire paw extending from the ankle to the metatarsal joints.

   4: severe swelling and redness in the entire paw extending from the ankle to the metatarsal joints and deformity and/or ankylosis.

4. Repeat the scoring of paws two to three times a week (*see* **Note 16**).

5. Sum the scores for each mouse hind paws yielding a maximum score of 8 per mouse. The arthritic score is then expressed as the average of all mice per group.

3.3 Measurement of Paw-Swelling Diameter

1. At the end of the experiment at day 8, sacrifice the mice under the specific protocols approved by the Institutional Animal Ethics Committee.

2. Using a caliper, measure the thickness of the mid-foot and ankle of each mouse (*see* **Note 17**).

3. Calculate the average of the values per group.

3.4 Preparing Hind Paws for Histological Analysis

1. Euthanize the mice 8 days after serum transfer.

2. Separate the hind legs from the mouse torso. Remove the skin, muscles, and tendons (*see* **Note 18**).

3. Dissect the whole paw (thighbone, ankle, tarsals, metatarsal, digits) from muscles, skin, and tendons.

4. Place the joint in an embedding cassette for fixation.

5. Add the 4 % PFA fixation solution and incubate overnight at 4 °C (*see* **Note 19**).

6. Change the solution to EDTA to decalcify the bones at room temperature under gentle shaking on an orbital shaker (*see* **Note 20**).

3.5 Hematoxylin–Eosin Staining to Detect Inflammatory Infiltrates

1. After decalcification, embed the joints in paraffin, cut 5 μm sections using a microtome, and mount the sections onto microscope slides (*see* **Note 10**).

2. Dry the slides at 65 °C for 1 h in a drying oven, and then put into a rack to cool down to room temperature for de-paraffinization and rehydration.

3. To remove paraffin, insert the rack into four consecutive staining dishes filled with xylene for 10 min each.

4. Change the rack to a dish with 100 % ethanol for 5 min.

5. Sequentially transfer the rack to dishes containing 95 % ethanol, 80 % ethanol, and 70 % ethanol for 5 min each.

6. Fill the last staining dish with tap water and incubate for 5 min to remove ethanol.

7. Fill one container with Mayer's hematoxylin solution and stain the sections for 10 min.

8. To remove hematoxylin, rinse the sections with tap water for 10 min.

9. Dip in ddH$_2$O.

10. Incubate rack in eosin for 5 min.

11. Wash the sections under running tap water for 30 s.

12. Dehydrate the sections using ethanol with increasing concentrations as follows: dip the rack five times into 70 and 95 % ethanol. Afterward, incubate the section for 3 min in 100 % ethanol and then in xylene.

13. Distribute a small amount (approximately three to five drops) of the mounting medium on a coverslip and then carefully cover the section (*see* **Note 21**).

14. Let the mounting medium harden overnight at room temperature in the dark.

15. Next day, seal the edges of the coverslip with nail polish and allow to air dry.

16. The stained slides can be stored at room temperature.

3.6 Toluidine Blue Staining Procedure to Detect Cartilage Degradation

1. The paraffin-embedded joint sections are de-paraffinized and rehydrated as described above (Subheading 3.5, **steps 1–5**).

2. Incubate the slides in 0.1 % toluidine blue solution for 10 min, and then dip them five times in ddH$_2$O.

3. To obtain durable preparations, dip the slides into 95 % ethanol, 100 % ethanol, and xylene. Each time, dip the rack five times into the solutions for 1 min.

4. Mount the slides and apply the coverslip as described above (Subheading 3.5, **steps 12–15**).

3.7 Assessment of Inflammation and Loss of Cartilage in Histological Sections

1. A visual scoring system performed microscopically by two independent observers should be used for joint histological assessment.

2. Use the following scoring system from 0 to 3 for *inflammatory infiltrates*: 0 = no inflammation, 1 = mild change, 2 = moderate change, and 3 = severe change.

3. Use the following scoring system from 0 to 3 for *cartilage destruction* (i.e., proteoglycan loss and matrix dissolution): 0 = fully stained cartilage, 1 = destained cartilage, 2 = destained cartilage with synovial cell invasion, and 3 = complete loss of cartilage.

4. The scoring points should then be expressed as the mean average of all mice per group.

4   Notes

1. Serum from K/BxN 6 to 12-week-old mice is collected over a period of several weeks, pooled, and stored at –20 °C. The effectiveness of the serum to induce arthritis has to be tested in preliminary experiments. For STIA experiments, usually 100–200 μl serum is injected per mouse.

2. BD Ultra-Fine™ with a 12.7 mm needle is used for intravenous injections of the serum.

3. The recombinant expression in *E. coli* and affinity purification of the "sneaking-ligand fusion proteins" (SLFPs) are described in [8]. Do not store the SLFPs longer than 4 weeks at 4 °C. The nonfunctional SLFP is designated as MutNBP2 [8].

4. The cell-permeable Antennapedia-linked NEMO-binding peptide (Ant-NBP) is commercially available from Calbiochem (San Diego, CA, USA) or can be synthetized by service companies.

5. It is important to evaluate the paw swelling according to the graded scale objectively. Therefore, two independent blinded observers familiar with the arthritis scoring system are necessary. The scorers should work independently from one another.

6. The ethanol solution and all buffers used in this chapter are prepared with purified water using a water purification system such as the Milli-Q synthesis system (Millipore, Billerica, MA, USA).

7. After removing the skin, tendons, and muscles, the hind paws should be immediately put into labeled tissue cassettes and placed into the PFA-containing beaker for fixation.

8. 4 % (w/v) paraformaldehyde should be dissolved in 1× PBS. The solution should be stirred at 60 °C under a ventilated fume hood and 1 M NaOH added until PFA is dissolved. Cool the PFA solution down to room temperature, and adjust the pH with diluted HCl to pH 7.5. Sterile filter the PFA solution using vacuum 0.45 μm filter units.

9. After fixation, the cassettes should be transferred to a beaker filled with decalcification solution.

10. Standard paraffin-embedding procedure and section preparation using a microtome to cut 5 μm sections should be followed. Alternatively, if available, joint samples can be submitted to a pathology core or slide-making service for paraffin embedding, cutting sections, and generating slides.

11. To minimize the use of xylene, alternative organic solvent products such as Roti®-Histol (Carl Roth, Karlsruhe, Germany) can be used.

12. Staining solutions (Mayer's hematoxylin solution and eosin solution, 0.5 %) are available as ready-to-use formulations from suppliers including Sigma, BIOMOL, or Millipore.

13. The cage is placed under a heating lamp for 10 min to warm the mouse and dilate the veins prior to intravenous injections. It should be taken care that the mice are not overheated.

14. While fixing the mouse in the restrainer, grasp the tail at midlength between the thumb and forefinger. Place the needle parallel to the lateral vein and inject the serum slowly. Remove the needle carefully out of the vein to inhibit a leakage of the injected serum.

15. After injecting the serum, 100 μg of SLC1 or control fusion protein is injected intraperitoneally. Tilt the mouse so that the head is hanging downward and the abdomen is exposed. Insert the needle to a depth of about half a centimeter into the abdomen at about a 30° angle. Inject the SLFP intraperitoneally in a volume of 500 μl with a 0.45×12 mm, 26G×1/2 needle.

16. Avoid excessive handling of the animals as this can reduce the severity or incidence of arthritis.

17. Do not compress the paw swelling.

18. Skinning is easier to do when you have first cut the mouse nails. This also prevents dulling the microtome blades.

19. Embedding cassettes should be placed in a beaker and completely covered with PFA solution.

20. PFA solution should be removed and refilled with PBS for washing. Then the cassettes should be transferred to a beaker

filled with EDTA solution. Change the EDTA twice a week for 2 weeks. After a washing step with ddH$_2$O, the cassettes can be stored in 70 % ethanol at room temperature. We have found that the quality of the embedded joint section is not reduced after a storage of 6 months.

21. Numerous mounting media are commercially available. We use either Cytoseal 60 (Thermo Scientific, Waltham, MA, USA) or Entellan (Merck, Whitehouse Station, NJ, USA). The mounting medium must be at room temperature, and it is important to remove excess buffer by absorbing the buffer carefully at the edge of the slide using a paper towel.

Acknowledgments

The work in the author's laboratory was supported by the German Research Foundation (SFB 643 project B3 and A8; FOR 832, project 7; Sachbeihilfe DU337/3-2 and BU 584/4-1), BMBF 01EO0803 Grant to the Centre of Chronic Immunodeficiency, the Federal State of Hessen (LOEWE-Project: Fraunhofer IME-Project-Group Translational Medicine and Pharmacology, Goethe University Frankfurt), the German Federal Ministry of Education and Research ArthroMark (project 4, 01 EC 1009C), and National Institutes of Health/National Heart, Lung, and Blood Institute Grant RO1HL096642.

References

1. Firestein GS (2003) Evolving concepts of rheumatoid arthritis. Nature 423:356–361

2. McInnes IB, Schett G (2011) The pathogenesis of rheumatoid arthritis. N Engl J Med 365:2205–2219

3. Williams MR, Azcutia V, Newton G et al (2011) Emerging mechanisms of neutrophil recruitment across endothelium. Trends Immunol 32:461–469

4. Makarov SS (2001) NF-kappa B in rheumatoid arthritis: a pivotal regulator of inflammation, hyperplasia, and tissue destruction. Arthritis Res 3:200–206

5. Simmonds RE, Foxwell BM (2008) Signalling, inflammation and arthritis: NF-kappaB and its relevance to arthritis and inflammation. Rheumatology (Oxford) 47:584–590

6. Hayden MS, Ghosh S (2008) Shared principles in NF-kappaB signaling. Cell 132: 344–362

7. May MJ, Marienfeld RB, Ghosh S (2002) Characterization of the Ikappa B-kinase NEMO binding domain. J Biol Chem 277:45992–46000

8. Sehnert B, Burkhardt H, Wessels JT et al (2013) NF-kappaB inhibitor targeted to activated endothelium demonstrates a critical role of endothelial NF-kappaB in immune-mediated diseases. Proc Natl Acad Sci U S A 110:16556–16561

9. Terato K, Hasty KA, Reife RA et al (1992) Induction of arthritis with monoclonal antibodies to collagen. J Immunol 148:2103–2108

10. Nandakumar KS, Holmdahl R (2006) Antibody-induced arthritis: disease mechanisms and genes involved at the effector phase of arthritis. Arthritis Res Ther 8:223

11. Nimmerjahn F, Ravetch JV (2007) Fc-receptors as regulators of immunity. Adv Immunol 96:179–204

12. Maccioni M, Zeder-Lutz G, Huang H et al (2002) Arthritogenic monoclonal antibodies from K/BxN mice. J Exp Med 195: 1071–1077

Chapter 35

Analysis of NF-κB Activation in Mouse Intestinal Epithelial Cells

Helena Shaked, Monica Guma, and Michael Karin

Abstract

Nuclear factor kappa B (NF-κB) is a key transcription factor controlling inflammation, innate immunity, and tissue integrity. NF-κB is activated by IκB kinase (IKK) in response to pro-inflammatory stimuli but is also found to be chronically activated in many inflammatory diseases accompanied by tissue destruction. To study the effects of chronic NF-κB activation in intestinal epithelium, we generated IKKβ(EE)^{IEC} transgenic mice which express constitutively active form of IKKβ in their intestinal epithelial cells (IEC). In this chapter, we describe three different methods that we applied for analysis of NF-κB activation in IEC of IKKβ(EE)^{IEC} transgenic mice: immunohistochemistry (IHC), nuclear fractionation, and chromatin immunoprecipitation (ChIP). These methods can be also applied to analyze NF-κB activation in mouse intestinal tissue in general.

Key words NF-κB, IKKβ, p65, Intestinal epithelial cells, Immunohistochemistry, Nuclear fractionation, Chromatin immunoprecipitation

1 Introduction

NF-κB activity is of particular importance for maintenance of epithelial barriers [1, 2], but it was also proposed that NF-κB activation in epithelial cells can lead to production of inflammatory chemokines that recruit immune cells to the tissue, thereby initiating an inflammatory amplification cascade [3]. Indeed, persistent activation of NF-κB signaling pathways is often associated with chronic inflammatory diseases, including inflammatory bowel disease [4, 5]. It is unclear, however, whether the persistent activation of NF-κB in epithelial cells is sufficient for driving chronic inflammation and tissue damage. NF-κB is regulated by the activated IκB kinase (IKK) complex consisting of the IKKα and IKKβ catalytic subunits and IKKγ/NEMO regulatory subunit [6–8]. To determine the effects of chronic epithelial activation of NF-κB, we generated IKKβ(EE)^{IEC} mice that express constitutively activated IKKβ in their intestinal epithelial cells (IECs) [9].

Michael J. May (ed.), *NF-kappa B: Methods and Protocols*, Methods in Molecular Biology, vol. 1280,
DOI 10.1007/978-1-4939-2422-6_35, © Springer Science+Business Media New York 2015

Analysis of NF-κB activation in intestinal epithelium can pose certain technical challenges, comparing to cell lines, due to high rate of intestinal tissue deterioration and high protease content, which can result in reduced signal. We find it critical to harvest the tissue immediately after sacrificing the mouse, keep it on ice during all the procedures, and proceed to fixation (when applicable) as fast as possible. Here we present the techniques that we successfully applied to analyze NF-κB activity in intestinal epithelium of IKKβ(EE)[IEC] mice [9, 10]: immunohistochemistry (IHC), nuclear fractionation, and chromatin immunoprecipitation (ChIP).

2   Materials

Prepare all solutions using ultrapure water and analytical grade reagents. Prepare and store all reagents at room temperature unless indicated otherwise.

2.1 Immunohisto-chemistry

1. Phosphate buffered saline (PBS) – may be prepared from 10× stock with sterile water.
2. Formalin 10 % (w/v) in PBS.
3. Ethanol 100 %.
4. Ethanol 95 % (v/v) in water.
5. Ethanol 70 % (v/v) in water.
6. Xylene.
7. Target retrieval buffer (S1700, Dako, Carpinteria, CA, USA).
8. 1 % bovine serum albumin (BSA) in PBS – prepare fresh and keep at 4 °C.
9. 3 % hydrogen peroxide (H_2O_2).
10. Blocking solution: 5 % goat serum in 1 % BSA in PBS – prepare fresh.
11. Antibodies for p65 and phospho-p65 (Cell Signaling, Danvers, MA, USA).
12. Biotinylated anti-rabbit secondary antibody.
13. Streptavidin-HRP conjugate.
14. DAB developing substrate (Vector Labs, Burlingame, CA, USA).
15. Hematoxylin.
16. Bluing solution: 0.2 % ammonium hydrochloride in distilled water.

2.2 Intestinal Epithelial Cell Isolation

1. RPMI 1640 medium.
2. 0.5 M EDTA.

2.3 Nuclear Fractionation and Western Blotting

1. 15 mM EDTA in RPMI 1640 medium supplemented with 10 % fetal bovine serum.

2. Buffer A: 10 mM HEPES, 1.5 mM $MgCl_2$, 10 mM KCl, 0.5 mM DTT, 0.05 % NP40 (or 0.05 % Igepal or Tergitol) pH 7.9.

3. Buffer B: 5 mM HEPES, 1.5 mM $MgCl_2$, 0.2 mM EDTA, 0.5 mM DTT, 26 % glycerol (v/v), pH 7.9.

4. Bradford protein assay (BioRad, Hercules, CA, USA).

5. Protease inhibitor cocktail in tablets (Roche, Madison, WI, USA).

6. Laemmli loading buffer, 2×: 120 mM Tris–HCl pH 6.8, 4 % SDS, 20 % glycerol, 10 % β-mercaptoethanol, 0.02 % bromophenol blue.

7. Gradient acrylamide mini-gel, precast (BioRad).

8. Protein standards.

9. Running buffer: 25 mM Tris–HCL, 192 mM glycine, 0.1 % SDS.

10. Transfer buffer: 25 mM Tris–HCl, 192 mM glycine, 20 % methanol, 0.04 % SDS.

11. PBS-T: Phosphate buffered saline (PBS) (may be prepared from 10× stock with sterile water), 0.05 % Tween-20.

12. Blocking buffer: 5 % dry milk in PBS-T.

13. Rabbit anti-p65 antibody, rabbit anti-HDAC1 antibody, and mouse anti-tubulin antibody (all from Santa Cruz Biotechnology, Santa Cruz, CA, USA); secondary HRP-conjugated anti-mouse and anti-rabbit antibodies (Cell Signaling, Danvers, MA, USA).

14. ECL substrate (BioRad).

15. "Restore Western Blot" stripping buffer (Thermo Scientific, Waltham, MA, USA).

2.4 ChIP Analysis

1. PBS-may be prepared from 10× stock with sterile water.

2. 1 % formaldehyde in PBS, prepare fresh using 37 % formaldehyde (*see* **Note 1**).

3. 2.5 M glycine, filter to avoid contamination during prolonged storage.

4. Anti-rabbit Dynabeads (Life Technologies).

5. Blocking solution: 0.5 % BSA in PBS, prepare fresh.

6. Anti-p65 antibody (rabbit polyclonal, Santa Cruz Biotechnology).

7. Lysis buffer 1 (LB1): 50 mM HEPES-KOH, pH 7.5, 140 mM NaCl, 1 mM EDTA, 10 % glycerol, 0.5 % NP-40, 0.25 % Triton X-100; filter and keep at 4 °C; supplement with protease inhibitor cocktail before use.

8. Lysis buffer 2 (LB2): 10 mM Tris–HCl, pH 8.0, 200 mM NaCl, 1 mM EDTA, 0.5 mM EGTA; filter and keep at 4 °C; supplement with protease inhibitors.

9. Lysis buffer 3 (LB3): 10 mM Tris–HCl, pH 8.0, 100 mM NaCl, 1 mM EDTA, 0.5 mM EGTA, 0.5 % N-lauroylsarcosine; filter and keep at 4 °C; supplement with protease inhibitors.

10. 0.5 % DOC in LB3: 10 mM Tris–HCl, pH 8.0, 100 mM NaCl, 1 mM EDTA, 0.5 mM EGTA, 0.5 % N-lauroylsarcosine, 0.5 % Na-deoxycholate; filter and keep at 4 °C; supplement with protease inhibitors.

11. RIPA buffer: 50 mM HEPES-KOH, pH 7.5, 500 mM LiCl, 1 mM EDTA, 1.0 % NP-40, 0.7 % Na-deoxycholate; filter and keep at 4 °C.

12. TE: 10 mM Tris–HCl, pH 8.0, 1 mM EDTA.

13. TE-NaCl: 10 mM Tris–HCl, pH 8.0, 1 mM EDTA, 50 mM NaCl.

14. Elution buffer: 50 mM Tris–HCl, pH 8.0, 10 mM EDTA, 1.0 % SDS.

15. Proteinase K solution, 20 mg/ml (Roche).

16. QIAquick PCR purification kit (Qiagen, Valencia, CA, USA).

17. 2 % agarose gel with ethidium bromide: boil 1 g agarose with 50 ml of water, add 5 μl of 10 mg/ml ethidium bromide in the chemical hood, and cast into mini-gel apparatus with a 10-well comb.

18. 6× loading dye.

19. 100 bp DNA ladder.

20. Real-time PCR reagent mix (BioRad).

21. Primers specific to promoters of NF-κB target genes and control DNA regions (*see* Table 1 for sequences of primers used in [9, 10]).

2.5 Equipment

1. Magnetic particle collector for 1.5-ml tubes (DynaMag, Life Technologies, Carlsbad, CA, USA).

2. Sonicator equipped with microtip.

3 Methods

3.1 Immuno-histochemistry on Formalin-Fixed Paraffin-Embedded Tissue

Nuclear translocation of p65 is a hallmark of canonical NF-κB activation. For analysis of protein localization in vivo, immunohistochemistry (IHC) remains the most widely used and powerful technique. Here we used antibodies for p65 and phospho-p65 to visualize nuclear p65 (Fig. 1 and *see* Note 2).

Table 1
Sequences of the primers used for ChIP with p65 antibody in IKKβ(EE)^{IEC} mice [9, 10]

Gene	Forward primer	Reverse primer
Prl2B1	TCTGCATCCCAAAGTCCTTC	GTAACAGCCCGGAAACAAGA
TNF	CCCCAACTTTCCAAACCCTCTGC	CCTCCTGGCTAGTCCCTTGCTGTC
MIP2	CAGGGCAGTAGAATGAGGCAGG	AGGCTGAAGTGTGGCTGGAGTC
MCSF	GGGCCTCTGGGGTGTAGTAT	CCGAGGCAAACTTTCACTTT
MCP1	TTTCCACGCTCTTATCCTACTCTGC	TTGTCTGTTTCCCTCTCACTTCAC
CCL20	AGGCAGGAAGTTTTCCCTGT	CACAAGAAGGCGTGTTCTCA
NOS2	GCATGAGGATACACCACA	TGCATAACTGTTCCCAAAGG

The corresponding DNA region for each pair of primers can be located in the mouse genome using UCSC genome browser (http://genome.ucsc.edu/cgi-bin/hgPcr?command=start); it will also show where this region is located relatively to the transcription start site of the corresponding gene. Note that Prl2B1 is a control, nonspecific gene

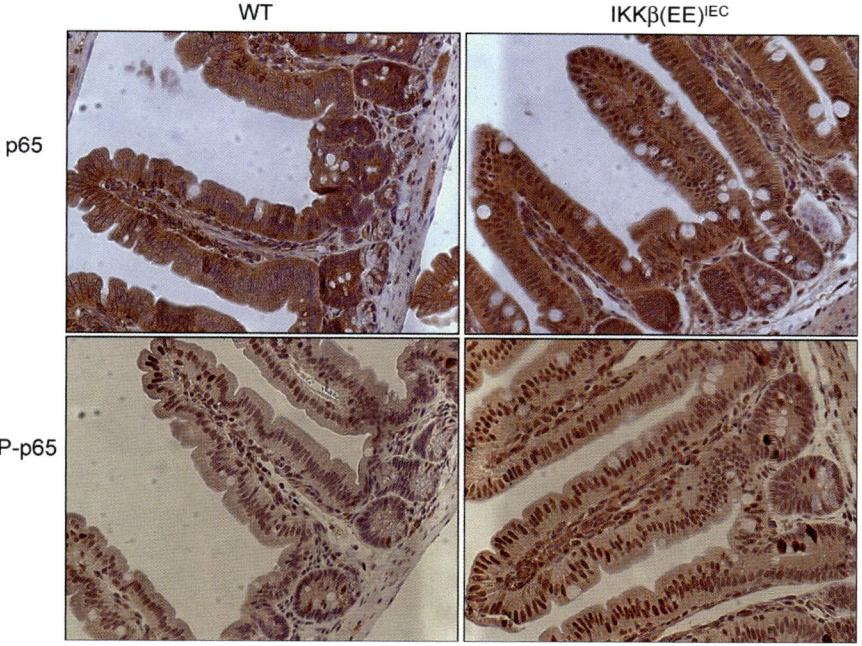

Fig. 1 IHC using p65 and phospho-p65 antibodies. Note that nuclei in intestinal epithelial cells stained *brown* in IKKβ(EE)^{IEC} sections, versus *purple* color in WT sections

1. Euthanize the mouse using the approved procedure, perform midline incision, and retract skin (*see* **Note 3**).

2. Cut small intestine ~0.5 cm below stomach. Hold intestine at the top with forceps and slowly pull it out of peritoneal cavity. Cut intestine ~0.5 cm above cecum, then cut in the middle, and place both halves in a 6-well dish with cold PBS, on ice.

3. Flush the intestine immediately with ice-cold 10 % formalin in PBS, using a 20- or 50-ml syringe with 18-G needle (*see* **Note 4**). If holes in the intestinal wall occur, push the syringe in up to the problem area or reinsert the syringe after the hole and continue to flush with formalin solution until the entire intestine is flushed and all its content is washed out. Repeat with the second half.

4. Place the intestine on a glass plate over a polystyrene box with ice, to keep it cold. Cut it longitudinally with fine scissors and open flat. Remove any residual fecal content with Kimwipes™, make a "Swiss roll" out of the intestine or a part of it, and immediately place it in 10 % formalin in PBS (*see* **Note 5**). Repeat with the second half. Fix for 24 h and then replace the formalin with 70 % ethanol in H_2O.

5. Embed in paraffin and prepare 5 μm sections.

6. Preheat a coupling jar with target retrieval buffer at 96 °C. Perform standard *deparaffinization and rehydration* procedures as follows: place the slides in a rack and perform the following washes:

   Xylene—3×5 min.

   Ethanol 100 %—2×5 min.

   Ethanol 95 %—2×5 min.

   Ethanol 70 %—5 min.

   Rinse briefly in water and then keep in water until ready for antigen retrieval. From this point on till dehydration, never let the slides to dry.

7. Antigen retrieval: place the slides in the preheated retrieval buffer at 96 °C for 45 min, and then cool down for 20 min.

8. Wash with PBS, drain the slides for a second, wipe around the tissue with tissue paper, and circle the tissue on the slide with a Dako pen and rinse with PBS again. From this point on, all the incubations will be performed using a humidified chamber, and the washes, using a Coplin jar.

9. Incubate the slides in 3 % H_2O_2 for 10 min.

10. Block the slides with blocking solution for 1 h.

11. For each intestinal roll, apply 50–100 μl of the anti-p65 or anti-phospho-p65 antibody, diluted 1:50 (dilution should be optimized for each antibody and tissue) in 1 % BSA/PBS and place overnight at 4 °C.

12. Wash with PBS, 3×5 min, and apply the secondary biotin-conjugated antibody, specific for the source of the primary antibody, diluted 1:250 in 1 % BSA/PBS (*see* **Note 6**). Incubate for 1 hr at room temperature.

13. Wash 3×5 min with PBS and apply the streptavidin-HRP conjugate, diluted 1:500 in 1 % BSA/PBS. Incubate for 30 min at room temperature.

14. Wash 3×5 min with PBS and develop with DAB as a substrate, for 10–30 min, carefully monitoring the color developing under the microscope every 5 min to prevent unspecific staining caused by prolonged incubation.

15. Wash briefly with water and counterstain with hematoxylin. Wash the slides extensively under running water until clean water comes out. Then incubate with bluing solution for 1 min and briefly wash with water.

16. Dehydrate the slides: place the slides in a rack and perform the following washes:

    Ethanol 70 %—5 min.

    Ethanol 95 %—2×5 min.

    Ethanol 100 %—2×5 min.

    Xylene—3×5 min.

17. Apply a drop of a mounting solution on each intestinal roll and carefully cover the slides with coverslips, avoiding air bubbles. Let slides dry for several hours and observe under the microscope.

3.2 IEC Isolation and Nuclear Fractionation

Nuclear fractionation and analysis by Western blot (WB) are used to examine nuclear translocation of p65 but can also be employed to analyze nuclear translocation of other NF-κB proteins such as p50 or c-Rel.

1. Cut small intestine ~0.5 cm below stomach. Hold intestine at the top with forceps and slowly pull it out of peritoneal cavity. Cut intestine ~0.5 cm above cecum, then cut out a fragment 12–15 cm long in the area of interest (proximal, middle, or distal), and place it in a 6-well dish with cold PBS, on ice.

2. Flush the intestine immediately with ice-cold PBS, using a 20- or 50-ml syringe with 18-G needle. If holes in the intestinal wall occur, push the syringe in up to the problem area or reinsert the syringe after the hole and continue to flush with PBS solution until the entire intestine is flushed and all its content is washed out.

3. Place the intestine on a glass plate over a polystyrene box with ice, to keep it cold.

4. Remove Peyer's patches by careful dissection. They are on the antimesenteric side of the gut wall. You should remove 6–8 PP per small intestine.

5. Cut it longitudinally with fine scissors and open flat. Remove any residual fecal content with tissue paper.

6. Place in Petri dish and shake to remove feces and mucus. Repeat this process 3–4 times in new Petri dishes with fresh medium.

7. Cut into 0.5 cm pieces and place them into 50-ml tube with 25 ml of 2 mM EDTA in RPMI.

8. Place tubes horizontally in 37 °C orbital shaking incubator for 20 min.

9. Filter intestine pieces using steel mesh filter. Place intestine pieces back into 50-ml conical tube and add 25 ml of 2 mM EDTA/RPMI. Place on shaker for 20 min.

10. Filter the cells through 70 μm cell strainer and spin at $300 \times g$ for 5 min at 4 °C (*see* **Note** 7).

11. For cytoplasm isolation, prepare 1 ml of buffer A with added inhibitor cocktail and store on ice.

12. Add 200 μl of buffer A per small pellet of IEC on ice and resuspend thoroughly. Leave on ice for 10 min.

13. Centrifuge at 4 °C at $960 \times g$ for 10 min.

14. Remove and reserve supernatant (this will contain everything except large plasma membrane pieces, DNA, nucleoli), remove out 10 μl for Bradford assay.

15. Resuspend the pellet in 200 μl of buffer A for washing.

16. On ice, resuspend pellet in 75.2 μl of buffer B and add 4.8 μl of 5 M NaCl to give 300 mM NaCl (high salt helps lyse membranes and forces DNA into solution).

17. Leave on ice for 30 min.

18. Centrifuge at $24,000 \times g$ for 20 min at 4 °C.

19. Aliquot supernatant (this will contain nuclear fraction), remove 10 μl for Bradford assay, and store at −70 °C.

20. Analyze protein concentration using Bradford assay [11].

21. Run in gradient polyacrylamide gel for 30 min at 200 V, along with size markers.

22. Transfer to 3×7 in. nitrocellulose membrane using wet blotting system, for 80 min at 100 V.

23. Incubate the blot with blocking buffer for 30 min at room temperature on rotating platform.

24. Incubate the blot with the primary (rabbit anti-p65) antibody diluted 1:500 in blocking buffer, overnight at 4 °C on rotating platform.

25. Wash three times for 10 min with PBS-T on rotating platform.

26. Incubate with the secondary antibody (HRP-conjugated anti-rabbit) diluted 1:2,000 in blocking buffer, for 1 hr at room temperature on rotating platform.

27. Wash three times as in **step 25**.

28. Develop using ECL substrate for 5 min and expose to film.

29. To verify the efficiency of fractionation, strip the blot using stripping buffer for 15 min, repeat the Western blot procedure (**steps 23–28**) using anti-HDAC1 as a primary antibody for a nuclear-specific control, and then strip again and repeat the **steps 23–28** again using anti-tubulin as a primary antibody, for a cytoplasm-specific control, and anti-mouse-HRP as a secondary antibody.

3.3 Chromatin Immunoprecipitation

Chromatin immunoprecipitation (ChIP) is an invaluable method to study protein-DNA interactions in living cells, and in particular it allows one to detect interactions between a transcription factor of interest and its target gene promoters. First, proteins are cross-linked to DNA covalently but reversibly with formaldehyde, then the cross-linked chromatin is sheared by sonication, and the protein of interest is precipitated with a specific antibody along with the bound DNA fragments; the DNA is then recovered by reversing the cross-links and protein digesting and analyzed by quantitative PCR using primers specific for the binding sites. For NF-κB, many target genes and their promoters are well characterized; however, it should be noted that, as with any transcription factor, target genes vary between different cell types and upon different conditions. Several typical NF-κB target genes are listed below, along with primer sequences for their promoter region. In addition, before performing ChIP, it is important to optimize conditions for NF-κB activation, preferably by confirming NF-κB nuclear translocation and expression of its target genes.

3.3.1 Formaldehyde Cross-Linking

1. Prepare a 15-ml conical tube with 10 ml of fixative (*see* **Note 1**).

2. Euthanize the mouse using the approved procedure, perform midline incision, and retract skin (*see* **Note 8**).

3. Cut small intestine ~0.5 cm below stomach. Hold intestine at the top with forceps and slowly pull it out of peritoneal cavity. Cut intestine ~0.5 cm above cecum, then cut out a fragment 12–15 cm long in the area of interest (proximal, middle, or distal), and place it in a 6-well dish with cold PBS, on ice.

4. Flush the intestine immediately with ice-cold PBS, using a 20- or 50-ml syringe with 18-G needle. If holes in the intestinal wall occur, push the syringe in up to the problem area or reinsert the syringe after the hole and continue to flush with formalin solution until the entire intestine is flushed and all its content is washed out.

5. Place the intestine on a glass plate over a polystyrene box with ice, to keep it cold. Cut it longitudinally with fine scissors and open flat. Remove any residual fecal content with lab tissue paper.

6. Using a thin spatula, gently scrap the villi into the tube with the fixative. Using 10-ml serologic pipette, pipette up and down several times, until the tissue is resuspended homogeneously.

7. Incubate for 10 min in room temperature on tube rotator.

8. Add 0.5 ml of 2.5 M glycine to quench the fixation.

9. Spin down at $700 \times g$ for 5 min at 4 °C.

10. Wash with PBS and remove supernatant. At this step, tissue can be flash-frozen in liquid nitrogen and stored at –80 °C.

3.3.2 Preparing Magnetic Beads

The following steps should be performed at 4 °C. The amount of beads and antibody listed here is for two samples. For one sample, this should be reduced by half. For more than two samples, more tubes with beads and antibody should be set up accordingly.

1. Resuspend Dynabeads by pipetting and add 100 μl of Dynabeads to a 1.5-ml microfuge tube. Add 1 ml of blocking buffer.

2. Collect beads using Dynal magnet: place tubes in magnetic rack. Allow beads to attach to side of the tube (it will take approximately 30 s) and carefully remove the supernatant.

3. Wash beads by adding 1 ml of blocking buffer, removing the tube from the magnet and gently inverting until the beads are evenly resuspended. Collect beads as in the previous step.

4. Wash beads again as in the previous step.

5. Add 250 μl of blocking buffer and 10 μg of antibody. Incubate for 6 hrs or overnight on tube rotator.

6. Collect beads as in **step 2** and wash them three times with blocking buffer as in **step 3**.

7. Add 40 μl of blocking buffer.

3.3.3 Cell Sonication

1. Remove frozen tissue pellets from –80 °C, if using frozen tissue. Resuspend in 10 ml of LB1. Rock at 4 °C on platform rocker for 10 min.

2. Spin at $1,350 \times g$ for 5 min at 4 °C in a tabletop centrifuge. Discard supernatant.

3. Resuspend each pellet in 10 ml of LB2. Rock gently at room temperature for 10 min.

4. Pellet nuclei in tabletop centrifuge by spinning at $1,350 \times g$ for 5 min at 4 °C. Discard supernatant.

5. Resuspend each pellet in 250–280 μl (*see* **Note 9**) of 0.5 % DOC/LB3 (*see* **Note 10**) and transfer to 1.5-ml tube. Place tube in an ice-water bath using plastic microfuge floating rack.

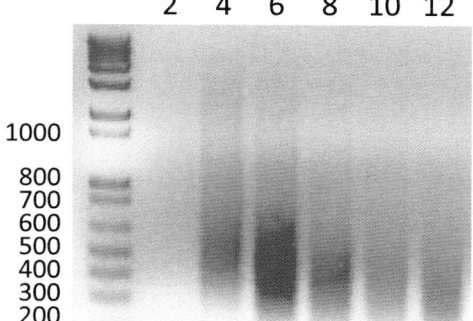

Fig. 2 Sonication optimization: the lysate was sonicated, with 20 μl saved aside after every two cycles as indicated; cross-links were reversed in each sample; and DNA was purified. Note that too short sonication results in inability to recover DNA due to too long DNA molecules; sonication more than six cycles, although further decreases the fragment length, is unnecessary and may even lead to some loss of material. Therefore six cycles, 2 min total, were chosen as the optimal sonication length

6. Sonicate the sample six times, 20 s each (*see* **Note 11** and Fig. 2), with 20 s interval, to obtain 200–800 bp DNA fragments (*see* **Note 12**). Keep the tubes on ice all the time.

7. Add 4 volumes of LB3 to bring the DOC concentration to 0.1 %. Add 1/20 volume of 20 % Triton X-100. Mix and spin at $20,000 \times g$ for 10 min at 4 °C to pellet debris and transfer the supernatant to another 1.5-ml tube (*see* **Note 13**).

8. Save 20 μl of cell lysate from each sample as "input DNA." Store at −20 °C. At least one input DNA aliquot should be kept per batch of sonicated lysate. Sonication efficacy can be examined after cross-link reversal and DNA purification, by running a sample in 2 % agarose gel.

3.3.4 Immuno-precipitation

1. Add 20 μl of antibody-coated beads from **step 3.3.2.7** to the lysate and incubate for 16–24 h on tube rotator (*see* **Note 14**).

2. Wash beads five times with 1 ml of RIPA buffer as in **step 3.3.2.3**. It should be performed at 4 °C.

3. Add 1 ml of TE-NaCl, resuspend by pipetting, and transfer the beads to a new 1.5-ml tube (*see* **Note 15**).

4. Collect the beads as previously. Spin briefly and remove any residual buffer.

3.3.5 Elution

1. Add 100 μl elution buffer. Incubate for 15–30 min at 65 °C with shaking or vortex every 2 min during the incubation.

2. Spin down beads at $16,000 \times g$ for 1 min at room temperature.

3. Transfer the supernatant to a new tube. Material can be frozen at −20 °C and stored overnight.

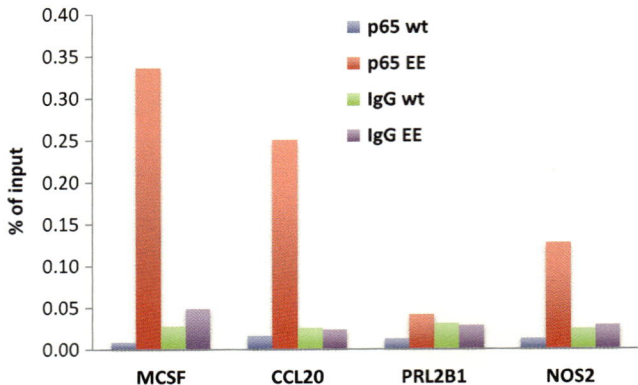

Fig. 3 ChIP was performed on villi from WT and IKKβ(EE)^{IEC} mice with p65 antibody and control IgG and analyzed via qPCR using primers for indicated promoter regions

3.3.6 Cross-Link Reversal and DNA Purification

1. Thaw 20 μl of input DNA reserved after sonication (**step 3.3.3.8**), add 80 μl of elution buffer, and mix.

2. Incubate both input and IP (from **step 3.3.5.3**) samples for 6–16 h at 65 °C.

3. Add 30 μl of TE and 1.3 μl of 20 mg/ml proteinase K, incubate for 1–2 h at 55 °C.

4. Purify DNA with Qiagen PCR purification kit; in the final step, preheat DNA/DNase-free water at 65 °C and elute DNA from the columns with 100 μl of this water.

3.3.7 PCR Analysis

1. Perform quantitative PCR (qPCR) analysis using IP sample and several dilutions of input sample (1:20, 1:50, 1:100) with primers for p65 target regions and unspecific regions. Calculate amount of bound DNA regions as % of input (*see* **Note 16** and Fig. 3).

4 Notes

1. Formaldehyde is flammable and highly toxic by inhalation and contact. It should be used with appropriate safety measures, such as protective gloves, glasses, and clothing and adequate ventilation. Waste should be disposed of according to regulations for hazardous waste. 1 % solution should be prepared in chemical hood.

2. Because cytosolic p65 is present in high amounts in intestinal epithelial cells in unchallenged wild-type mice, we found helpful to also perform staining for the phosphorylated form of p65, which exclusively localizes to the nucleus (Fig. 1).

3. To preserve nuclear localization of p65, tissue should be fixed immediately after sacrificing the mouse and the procedure should be performed as quickly as possible.

4. We found that flushing the intestines with ice-cold fixative immediately after their removal is preferable to transcardial perfusion of the anesthetized animal. However, this procedure should be done with special caution, using gloves and safety glasses to avoid splashing of formalin to eyes.

5. For the "Swiss roll," each segment is rolled up longitudinally, with the mucosa outward, using a wooden stick [12].

6. **Steps 12–13** (incubation with secondary, biotin-conjugated antibody, and then streptavidin-HRP-conjugated antibody) can be substituted by EnVision system (Dako).

7. Centrifugation at lower speed ($300 \times g$) will help to remove lymphocytes. If you like to split the pellet in small aliquots for nuclear fractionation, resuspend the pellet in 5–6 ml, aliquot them in 1.5-ml Eppendorf, and centrifuge at $300 \times g$ for 5 min.

8. As with visualization of nuclear NF-κB using IHC, fixation for ChIP should be performed immediately after sacrificing the mouse and as quickly as possible, to avoid NF-κB dissociation from the DNA and exiting the nucleus.

9. Smaller volume can result in foaming, which can lead to protein loss; however, too big volume will necessitate longer sonication.

10. We found that relatively high concentration of DOC increases the efficacy of sonication, thereby shortening the time to obtain desired fragment length; however, it increases the risk of foaming; therefore, sonication should be performed with caution. The lysate will be diluted after the sonication to decrease the DOC concentration to the one optimal for the immunoprecipitation reaction.

11. Sonication length should be optimized to obtain 200–600 bp fragments (Fig. 2).

12. To decrease foaming, initially set output power to minimal and increase manually to final power during first burst (if there is significant foaming, all bubbles can be removed by brief centrifugation followed by gentle resuspension of all material, leaving no foam bubbles). Avoid overheating the sample, which may reverse the cross-links and decrease the signal.

13. Untypically large pellet indicates poor sonication.

14. Lysate can be split for precipitation with two different antibodies, i.e., p65 and any other antibody of interest or IgG control.

15. This step is aimed to avoid unspecific DNA that may be adsorbed to the tube walls and later released at the elution step.

16. ChIP results are usually presented as "% of input," and it is expected that there will be enrichment for the specific versus unspecific DNA region (Fig. 3).

Acknowledgments

This work was supported by NIH grant AR064834 (M.G.) and AI043477 (M.K.).

References

1. Karin M, Lin A (2002) NF-kappaB at the crossroads of life and death. Nat Immunol 3:221–227

2. Pasparakis M (2009) Regulation of tissue homeostasis by NF-kappaB signalling: implications for inflammatory diseases. Nat Rev Immunol 9:778–788

3. Barnes PJ, Karin M (1997) Nuclear factor-kappaB: a pivotal transcription factor in chronic inflammatory diseases. N Engl J Med 336:1066–1071

4. Neurath MF, Pettersson S, Meyer zum Buschenfelde KH, Strober W (1996) Local administration of antisense phosphorothioate oligonucleotides to the p65 subunit of NF-kappa B abrogates established experimental colitis in mice. Nat Med 2:998–1004

5. Kaser A, Zeissig S, Blumberg RS (2010) Inflammatory bowel disease. Annu Rev Immunol 28:573–621

6. Mercurio F, Zhu H, Murray BW, Shevchenko A, Bennett BL, Li J, Young DB, Barbosa M, Mann M, Manning A, Rao A (1997) IKK-1 and IKK-2: cytokine-activated IkappaB kinases essential for NF-kappaB activation. Science 278:860–866

7. DiDonato JA, Hayakawa M, Rothwarf DM, Zandi E, Karin M (1997) A cytokine-responsive IkappaB kinase that activates the transcription factor NF-kappaB. Nature 388:548–554

8. Rothwarf DM, Karin M (1999) The NF-kappa B activation pathway: a paradigm in information transfer from membrane to nucleus. Sci STKE 1999:RE1

9. Guma M, Stepniak D, Shaked H, Spehlmann ME, Shenouda S, Cheroutre H, Vicente-Suarez I, Eckmann L, Kagnoff MF, Karin M (2011) Constitutive intestinal NF-kappaB does not trigger destructive inflammation unless accompanied by MAPK activation. J Exp Med 208:1889–1900

10. Shaked H, Hofseth LJ, Chumanevich A, Chumanevich AA, Wang J, Wang Y, Taniguchi K, Guma M, Shenouda S, Clevers H, Harris CC, Karin M (2012) Chronic epithelial NF-kappaB activation accelerates APC loss and intestinal tumor initiation through iNOS up-regulation. Proc Natl Acad Sci U S A 109: 14007–14012

11. Noble JE, Bailey MJ (2009) Quantitation of protein. Methods Enzymol 463:73–95

12. Moolenbeek C, Ruitenberg EJ (1981) The "Swiss roll": a simple technique for histological studies of the rodent intestine. Lab Anim 15:57–59

Part V

Bioinformatics and Modeling Approaches to Study NF-κB

Chapter 36

Characterizing the DNA Binding Site Specificity of NF-κB with Protein-Binding Microarrays (PBMs)

Trevor Siggers, Thomas D. Gilmore, Brian Barron, and Ashley Penvose

Abstract

NF-κB transcription factors control a wide array of important cellular and organismal processes in eukaryotes. All NF-κB transcription factors bind to DNA target sites as dimers. In vertebrates, there are five NF-κB sub-units, p50, p52, RelA (p65), c-Rel, and RelB, that can form almost all combinations of homodimers and heterodimers, which recognize distinct, but overlapping, target sequences. In this chapter, we describe the use of protein-binding microarrays (PBMs), a high-throughput method to measure the binding of proteins to different DNA sequences. PBM datasets allow for sensitive comparisons of NF-κB dimer DNA-binding differences and can aid in the computational and experimental prediction of NF-κB target genes

Key words NF-kappaB, DNA binding, Protein-binding microarrays, Site specificity

1 Introduction

Nuclear factor κB (NF-κB) proteins comprise an evolutionarily conserved family of transcription factors (TFs). NF-κB TFs bind as dimers to short target DNA sites called κB sites. In vertebrates, there are five NF-κB proteins, c-Rel, RelA/p65, RelB, p50, and p52, that can form various homo- and heterodimeric species [1]. All NF-κB proteins have an approximately 300-amino acid DNA-binding/dimerization domain in their N-terminal half called the Rel homology domain (RHD). The RHD is necessary and sufficient for DNA target site binding by NF-κB proteins. Although the RHD is reasonably well conserved among all NF-κB proteins both within and across species, sequence similarity can nevertheless be used to subdivide NF-κB proteins into two subfamilies: the "NF-κB" proteins (p50, p52) and the "Rel" proteins (c-Rel, RelA/p65, RelB). Furthermore, as discussed herein, the NF-κB and Rel subfamily proteins show distinct preferences for binding to target DNA sites.

NF-κB dimers bind to DNA to regulate genes involved in a wide range of cellular responses [2]. The DNA binding characteristics of

Michael J. May (ed.), *NF-kappa B: Methods and Protocols*, Methods in Molecular Biology, vol. 1280,
DOI 10.1007/978-1-4939-2422-6_36, © Springer Science+Business Media New York 2015

NF-κB dimers play an important role in gene regulation. Differences in NF-κB dimer DNA binding site affinity and specificity contribute to dimer-specific functions (reviewed in [3]). Cooperative binding of NF-κB dimers with other TFs at composite DNA elements increases specificity in gene targeting and provides a mechanism to integrate information relayed by each protein [4, 5]. Recent studies have also drawn attention to the role of DNA as an allosteric regulator of DNA-bound NF-κB [6–8]. Differences in the DNA sequence bound by NF-κB dimers or NF-κB dimer-containing complexes (e.g., p52:Bcl3 [7]) can directly influence the ability of NF-κB dimers to recruit regulatory cofactors and regulate target gene expression. The varied ways in which DNA interactions impact NF-κB function highlight the need for methods to assay the target site-specific DNA binding of NF-κB dimers and dimer-containing complexes to advance our understanding of NF-κB function.

The most commonly used approaches for detecting and examining protein-DNA interactions are electrophoretic mobility shift assays (EMSA) and DNase I footprinting [9–11]. These approaches are powerful but are generally "low throughput," enabling, at best, tens of distinct DNA sequences to be examined in a single experiment. In the early 1990s, the SELEX (systematic evolution of ligands by exponential enrichment) technique provided the first high-throughput (HT) approach to assess protein binding to thousands of DNA sequences simultaneously and enabled an unbiased characterization of protein-DNA-binding specificity [12, 13]. In the last 10 years, many new techniques have emerged that utilize microarray or sequencing-based approaches for the HT characterization of protein-DNA binding (reviewed in [14]). The detailed DNA-binding datasets enabled by these methods are leading to new insights into DNA binding site diversity (e.g., nontraditional binding sequences, DNA binding via multiple structural modes, etc.) and are providing a wealth of data for genomic analyses of gene regulatory elements [14].

PBMs are in in vitro, microarray-based technique to measure the binding of proteins or proteins complexes to thousands of DNA sequences in a single experiment [15–19]. In PBMs, a microarray slide containing double-strand DNA (dsDNA) sequences is probed with a protein of interest (e.g., an NF-κB dimer), and the protein-DNA complex is then detected with a fluorescently labeled antibody (Fig. 1). The amount of protein bound to each probe sequence (i.e., microarray spot, *see* Fig. 2a) can then be quantified using a microarray scanner and image analysis pipeline.

PBM technology has been used to examine the binding of hundreds of proteins from both prokaryotes and eukaryotes [18, 20–30]. To date, PBMs have been used to characterize the DNA-binding specificity of a number of mammalian NF-κB dimers, three NF-κB homodimers from *Drosophila melanogaster*, as well as an NF-κB ortholog from the sea anemone *Nematostella vectensis* (Table 1).

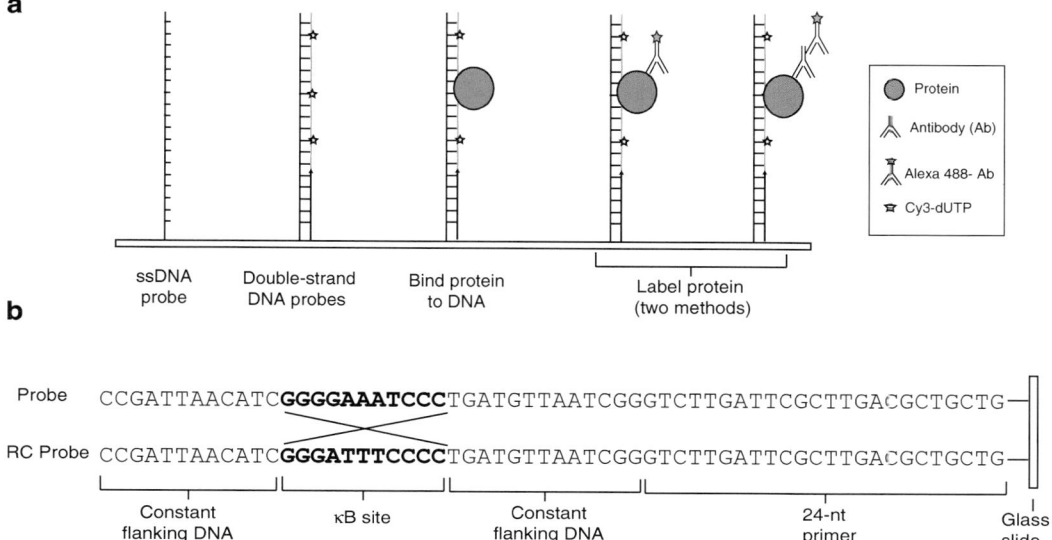

Fig. 1 Protein-binding microarray (PBM) overview. (**a**) Shown is a schematic of steps involved in the PBM experiment: (1) double-strand the single-stranded DNA (ssDNA), (2) bind protein to DNA probes, and (3) label the bound protein with Alexa Fluor 488 (A488)-conjugated antibodies (Ab). Two methods of labeling the bound protein are shown: a one-step procedure with a labeled primary Ab and a two-step procedure with a primary Ab and a labeled secondary Ab. (**b**) Schematic of probe sequences in the forward (Probe) and reverse complement (RC Probe) directions. The NF-κB site (i.e., κB site) in each probe is set at a fixed position from the glass slide. The 24-nucleotide primer sequence common to all probes is shown

PBM-based analyses of multiple mammalian NF-κB dimers reaffirmed the basic binding specificity divisions between NF-κB dimer species (Fig. 2b) while also identifying many nontraditional binding sequences recognized by NF-κB dimers (Fig. 2c, d) [18, 20, 30]. In this chapter, we describe the application of PBMs to the high-throughput examination of DNA binding by NF-κB dimers.

2 Materials

2.1 Double-Stranding Microarray Probes

1. Agilent CGH microarray (*see* **Note 1**).

2. GenePix 4400A scanner (with blue laser kit) (Molecular Devices, Sunnyvale, CA, USA).

3. GenePix Pro 7 Acquisition and Analysis Software (Molecular Devices).

4. Microarray hybridization chamber kit (stainless steel hybridization chamber and tweezers) (Agilent, Santa Clara, CA, USA).

5. Gasket slide, one chamber per slide (Agilent).

6. Two glass staining dishes (0.7 L), with cover.

7. Glass staining slide rack.

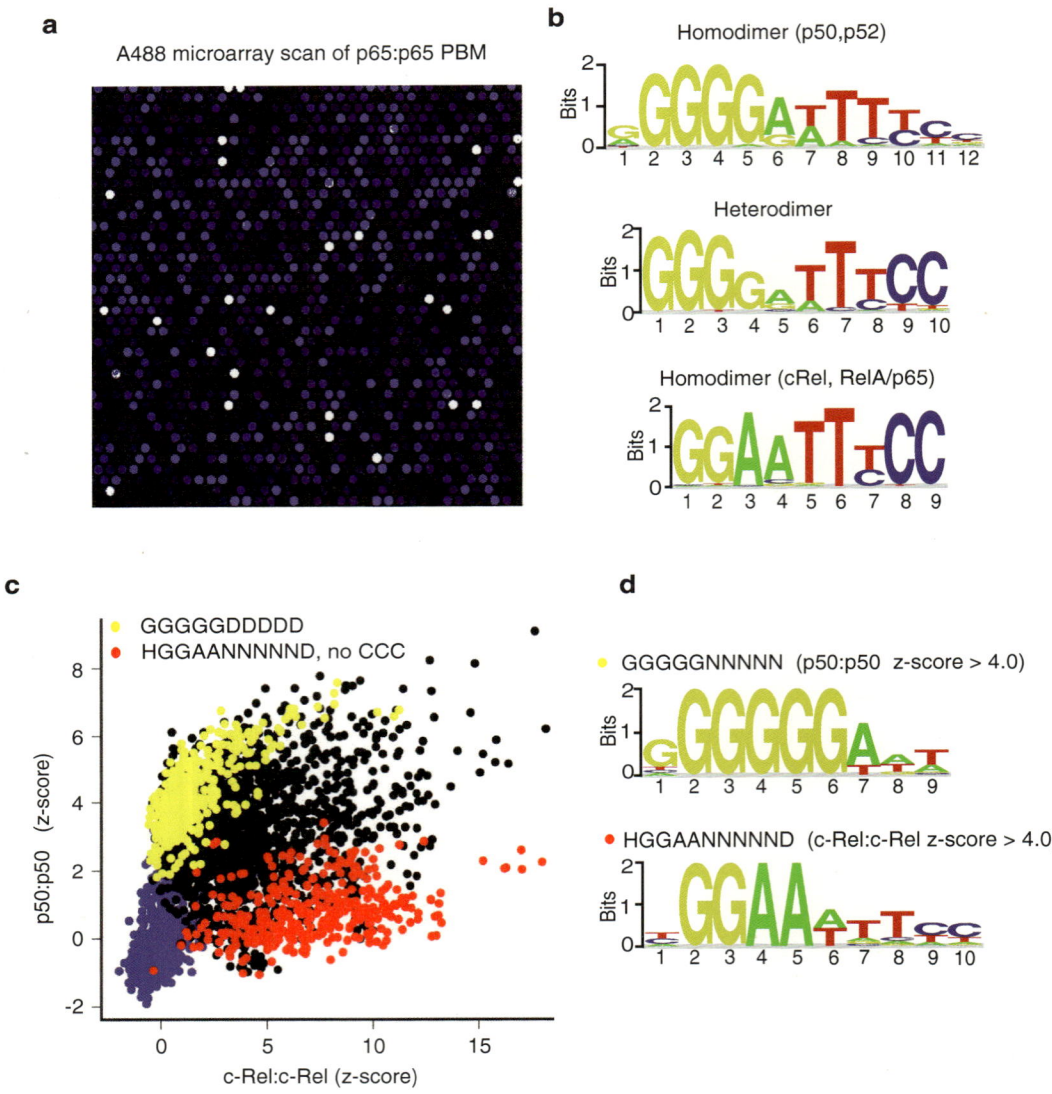

Fig. 2 NF-κB dimer binding specificity assessed by PBM. (**a**) Portion of microarray scan for a PBM bound by human p65 homodimer and labeled with Alexa Fluor 488 (A488). Spot intensity correlates with the amount of p65 protein bound. (**b**) Representative DNA-binding site motifs determined for the three basic binding-specificity classes of NF-κB dimer: p50 and p52 homodimers, all heterodimers, and c-Rel and RelA/p65 homodimers. (**c**) Pairwise comparison of mouse p50:p50 and c-Rel:c-Rel homodimers binding to 3,285 10-bp κB sites (*black*) and a background set of 1,200 random 10-bp sites (*blue*). κB sites bound preferentially by p50:p50 (*yellow*) or c-Rel:c-Rel (*red*) are shown. Highlighted κB sites conform to consensus pattern 5′-GGGGGNNNNN-3′ (*yellow*) or 5′-HGGAANNNNND-3′ (H = not G, D = not C, NNNNN = all 5-bp sequences except those containing CCC triplet) (red). (**d**) Binding site motifs for κB sites subsets highlighted in (**c**). Results in (**b, c,** and **d**) are from [20]

Table 1
Summary of NF-κB dimers assayed by PBM experiment

Dimer	Species	Protein (aa, tag)	Prep	1° Ab	2° Ab	Probe seqs	Ref
p50	Hu	p50 (2–400)	a	p50	Cy5	256, 10-bp sites	[18]
p50	Hu	p50 (2–400, CT-His)	b	His	Cy5	803, 11-bp sites	[30]
p50	Hu	p50 (1–453)	c	no Ab	--	Ig-κB & mutants	[29]
p52	Hu	p52 (4–332, CT-His)	a	His	Cy5	256, 10-bp sites	[18]
p52	Hu	p52 (4–332, CT-His)	b	His	Cy5	803, 11-bp sites	[30]
p52	Hu	p52 (4–332, CT-His)	d	A488-His	--	3,300, 10-bp sites	[20]
p65	Hu	p65 (1–307, CT-His)	b	His	Cy5	803, 11-bp sites	[30]
p65	Hu	p65 (1–307, CT-His)	d	p65	Cy5	803, 11-bp sites	[30]
p65:p50	Hu	p65 (1–307, CT-His); p50 (2–400, CT-His)	b	His	Cy5	803, 11-bp sites	[30]
p65:p50	Hu	p65 (1–307); p50 (7–356, CT-His)	b	p65	A488	803, 11-bp sites	[20]
p65:p52	Hu	p65 (1–307, CT-His); p52 (4–332, CT-His)	b	His	Cy5	803, 11-bp sites	[20]
RelB:p50	Hu	RelB (120–401, CT-His); p50 (2–400, CT-His)	b	His	Cy5	803, 11-bp sites	[20]
RelB:p52	Hu	RelB (12–401, CT-His); p50 (2–400, CT-His)	b	His	Cy5	803, 11-bp sites	[20]
RelB:p52	Hu	RelB (120–401); p52 (4–332, CT-His)	b	A488-His	A488	3,300, 10-bp sites	[20]
c-Rel:p50	Hu	c-Rel (1–285, CT-His); p50 (2–400, CT-His)	b	His	Cy5	803, 11-bp sites	[20]
c-Rel:p50	Hu	c-Rel (1–285); p50 (7–356, CT-His)	b	p50	A488	3,300, 10-bp sites	[20]
c-Rel:p52	Hu	c-Rel (1–285, CT-His); p52 (2–332, CT-His)	b	His	Cy5	803, 11-bp sites	[20]
p50	Mo	p50 (1–429, CT-FLAG)	d	p50	A488	3,300, 10-bp sites	[20]
p65	Mo	p65 (1–314)	d	p65	A488	3,300, 10-bp sites	[20]
c-Rel	Mo	c-Rel (1–282)	d	p65	A488	3,300, 10-bp sites	[20]
p65:p50	Mo	p65 (1–314); p50 (1–429, CT-His)	d	p65	A488	3,300, 10-bp sites	[20]
RelB:p50	Mo	RelB (120–401); 50 (1–429, CT-His)	d	p50	A488	3,300, 10-bp sites	[20]

(continued)

Table 1
(continued)

Dimer	Species	Protein (aa, tag)	Prep	1° Ab	2° Ab	Probe seqs	Ref
Dif	Fly	Dif (17–378, CT-His)	b	His	Cy5	182, 10-bp sites	[27]
Relish	Fly	Relish (117–434, CT-His)	b	His	Cy5	182, 10-bp sites	[27]
Dorsal	Fly	Dorsal (16–384, CT-His)	b	His	Cy5	182, 10-bp sites	[27]
NvNF-κB	Sea An.	NvNF-κB (1–440, CT-His)	b	His	Cy5	803, 11-mer sites	[26]

(*Dimer, Species*) NF-κB dimer names and species. *Hu* Human, *Mo* Mouse, *Fly Drosophila melanogaster, NvNF-κB Sea An* NF-κB ortholog from Sea Anemone (*Nematostella vectensis*). (*Protein*) Clone used for NF-κB members. *CT-His* C-terminal Histidine-tag. (*Prep*) Protein preparation: (**a**): Expressed in *E. coli* (strain BL21(DE3)LysS) using pET32a (Novagen, Madison, WI, USA) expression vectors. Purified by DNA-oligo affinity purification using dsDNA with NF-κB binding site [18]. (**b**): Expressed in *E. coli* (strain BL21(DE3)LysS) using pET21d (Novagen) expression vectors. Purified using a two-step process: (1) Ni-NTA His-bind column (Merck, Whitehouse Station, NJ, USA) and (2) DNA-oligo affinity purification using dsDNA with NF-κB binding site. Heterodimers were made by mixing subunits 1:1, denaturing with 8 M urea, and refolding by dialyzing out the urea [19]. (**c**) Purchased from Promega; (**d**) Homodimers were made by expressing protein in *E. coli* (strain BL21(DE3)LysS) using pET11a vectors; heterodimers were made by co-expressing each subunit from a bi-cistronic bacterial expression plasmid [31]. All dimers were purified by ion exchange chromatography [20]. (*1° Ab*) Primary Ab. *p50* anti-p50 (Santa Cruz Biotechnology, Santa Cruz, CA, USA), *His* anti-His (Santa Cruz), *A488-His* A488 Penta-His (Qiagen, Valencia, CA, USA), *p65* anti-p65 (Santa Cruz); c-Rel: anti-c-Rel (Santa Cruz), and *No Ab*: protein directly labeled with fluorophore. (*2° Ab*) Secondary Ab. *Cy5*: Cy5 anti-rabbit IgG (Jackson Labs, Bar Harbor, ME, USA), *A488*: A488 anti-rabbit IgG (Invitrogen, Carlsbad, CA, USA). (*Probe seqs*) Probes sequences analyzed: 256 variants of the κB consensus sequence GGRRNNYYCC; 803 variants of the consensus sequence RGGRNNHHYYB; 3,300 top-scoring 10-bp κB sites determined by universal PBM experiment; 182 variants of the consensus sequence GGRDNNHHBS

8. Microscope slide box.

9. Hybridization incubator.

10. Vacuum desiccator.

11. 80-place microcentrifuge tube rack (SIMPORT UniRack, Beloeil, QC, Canada).

12. Dust cover (*see* **Note 2**).

13. Water bath (set at 37 °C).

14. Stir plate and stir bars.

15. Dust Off XL canned air (VWR, Arlington Heights, IL, USA).

16. 70 % ethanol solution (in spray bottle).

17. Water in spray bottle.

18. 1.5-mL microcentrifuge tubes.

19. Phosphate-buffered saline (PBS) (pH 7.4) (*see* **Note 3**).

20. 10 % (v/v) Triton X-100 (0.2 μm filter sterilized).

21. Thermo Sequenase cycle sequencing kit (contains 10× Thermo Sequenase buffer and DNA polymerase with pyrophosphatase, 4 U/μL) (Affymetrix, Santa Clara, CA, USA).

22. dNTP mix (25 mM each) (*see* **Note 4**).

23. Cy3-conjugated dUTP (GE Healthcare, Mickleton, NJ, USA).

24. Primer sequence (24-nt, desalted) (EuroFins, Lancaster, PA, USA) (*see* **Note 1** and Fig. 1).

25. Wash buffer: PBS, pH 7.4; 0.01 % TX-100.

2.2 Protein-Binding Microarray (PBM) Experiment

Equipment and reagents listed above in Subheading 2.1 are not listed again here.

1. Gasket slide, four chambers per slide (Agilent) (*see* **Note 5**).

2. Coplin staining jars.

3. Nylon syringe filters (0.45 μm), with Leur-Lok.

4. Syringe with Leur-Lok (30 mL).

5. Nalgene disposable filter units (0.2 μm).

6. Two 1-L Nalgene bottles.

7. Bovine serum albumin (BSA).

8. ssDNA from salmon testes (Sigma, St. Louis, MO, USA).

9. Nonfat dry milk.

10. 20 % (v/v) Tween 20 (0.2 μm filter sterilized).

11. 1 M Tris–HCl, pH 7.4 (0.2 μm filter sterilized).

12. 1 M dithiothreitol (DTT) (0.2 μm filter sterilized).

13. 1 M NaCl (0.2 μm filter sterilized).

14. Protein samples (*see* **Note 16** and Table 1).

15. Antibodies (*see* **Note 18** and Table 1).

16. Benchtop orbital shaker.

17. 2 % milk in PBS: 0.2 g dried milk, 10 mL of PBS, pH 7.4.

18. 2 % milk in sterile water: 0.2 g dried milk; 10 mL of sterile water.

19. Five separate PBM wash solutions: (1) 70 mL of PBS; (2) 280 mL of 0.01 % Triton X-100/PBS: 280 mL of PBS, 280 μL of 10 % Triton X-100; (3) 70 mL of 0.1 % Tween /PBS: 70 mL of PBS, 350 μL of 20 % Tween-20; (4) 70 mL of 0.5 % Tween/PBS: 70 mL of PBS, 1,750 μL of 20 % Tween-20; (5) 210 mL of 0.05 % Tween/PBS: 210 mL of PBS, 525 μL of 20 % Tween-20.

20. 0.5 % Tween-20/PBS: 500 mL of PBS, 1.25 mL of 20 % Tween-20.

21. Binding reaction mixture: 55 μg/ml salmon testes DNA, 0.2 mg/ml BSA, 90 μL of 2 % milk in water, 10 mM Tris–HCl, pH 7.4, 2 mM DTT, 0.02 % Triton X-100, 60 mM NaCl, 100–200 nM protein sample, sterile water to final volume of 175 μL.

22. Primary Ab mixture (per chamber): 20 μg/ml Ab, 2 % milk/PBS to final volume of 175 μL.

23. Secondary Ab mixture (per chamber): 20 μg/ml Alexa Fluor 488-conjugated Ab, 2 % milk/PBS to final volume of 175 μL.

3   Methods

3.1   Double-Stranding Microarray Probes

In this step, the ssDNA Agilent microarrays are made double stranded by primer extension from a 24-nucleotide primer sequence common to each probe sequence (Fig. 1). Briefly, the microarray slide and primer extension mixture are heated to 85 °C in a hybridization oven, and the temperature is then slowly reduced in a stepwise fashion to the final primer extension temperature of 60 °C. When the primer extension has been completed, the microarray slide is washed and imaged in a microarray scanner. The double-stranding process takes ~4 h; the microarray scan takes ~20 min (*see* **Note 6**).

1. Place two 80-well microcentrifuge tube racks into the hybridization oven, and set oven to 85 °C (temperature should stabilize after ~1 h) (*see* **Note 7**). The exterior of the hybridization oven should be made light impermeable to avoid photobleaching of the Cy3 fluorophore (this can be done by covering the hybridization oven window with thick cardboard, held in place with heavy-duty tape).

2. On ice, thaw the primers, dNTPs, and Cy3-conjugated dUTP.

3. After the hybridization oven temperature has stabilized at 85 °C, prepare the 900 μL primer extension mixture in a 1.5-mL microcentrifuge tube: sterile water (775.3 μL), 1× Thermo Sequenase reaction buffer (90 μL of 10× stock solution, *see* **Note 8**), 1.17 μM primer (10.5 μL of 100 μM stock), 163 μM dNTPs (14.7 μL of 10 mM stock—stock solution is 2.5 mM of *each* dNTP), 1.63 μM Cy3-conjugated dUTP (1.47 μL of 1 mM stock), and 0.036 U/μL Thermo Sequenase polymerase (8 μL of 4 U/μL solution). Pipette gently ten times to mix thoroughly.

4. Warm primer extension mixture, steel hybridization chamber, and single-chamber gasket slide in the 85 °C hybridization oven for 20 min (*see* **Note 9**).

5. Warm Agilent microarray slide in the 85 °C hybridization oven for 5 min. Make sure that DNA side of array is facing up (*see* **Notes 10** and **11**).

6. Assemble the microarray, gasket slide, and primer extension mixture in the steel hybridization chamber and return to 85 °C oven. Assembly should follow the Agilent microarray instruction manual, except that bubbles should be prevented (*see* **Note 12**). The hybridization chamber should be laid flat during the incubation—it is *not* rotated. It is *critical* that the assembly step be done quickly to avoid temperature drop of equipment and mixture (*see* **Note 13**).

7. Manually decrease the oven temperature in a stepwise fashion to the final primer extension temperature, as follows: 10 min at 85 °C, 10 min at 75 °C, 10 min at 65 °C, and 90 min at 60 °C.

8. Warm 1 L of wash buffer in a water bath at 37 °C (~1 h, can be done concurrently with 90 min primer extension step).

9. Just prior to the completion of the primer extension step, fill two glass staining dishes with 500 mL of pre-warmed, 37 °C wash buffer. Place dish #1 on the benchtop. Place dish #2 on a magnetic stir plate and insert glass slide rack and small stir bar.

10. When the primer extension step is finished, remove the hybridization chamber from the oven. Disassemble the steel chamber on the benchtop, carefully remove the microarray/gasket slide "sandwich," and separate the microarray slide from the gasket slide while immersed in wash buffer (dish #1) according to Agilent instructions (see **Note 17** for more details on disassembly cautions) (Fig. 3c). To wash the slide, shake it briefly up and down in wash buffer.

11. Place the microarray into dish #2 in the slide rack with the DNA side facing toward the center. Cover the staining dish with an empty ice bucket to limit exposure to light, which can cause photobleaching of Cy3-dUTP. Stir for 10 min at a speed just high enough to cause a small vortex in the wash buffer.

12. Fill dish #1 with 500 mL of room temperature PBS (*with no detergent*). Lift glass slide rack and microarrays from dish #2 and place into dish #1 containing PBS only. Place on stir plate, cover with an ice bucket, and stir for 3 min as described in the previous step.

13. Remove the microarray from the slide rack in staining dish #1 by grabbing the microarray slide at top two corners, using index finger and thumb, and slowly remove the microarray slide from staining dish so that slide "de-wets" and is free of buffer (Fig. 3d). As you remove the microarray, keep the DNA side facing down at a slight incline to the buffer surface.

14. Scan the microarray in GenePix scanner at 2.5 μm resolution. The laser and filter setting should be set for Cy3 label (Ex: 535 nm, Em: green filter). Save scanned image as ".TIF" file for subsequent analysis.

15. The microarray can be stored in a microscope slide box in the dark (i.e., in a drawer) for weeks until needed for the protein-binding microarray experiment.

3.2 Protein-Binding Microarray (PBM) Experiment

In this step, the protein will be applied to the microarray slide, bound protein will be detected by fluorescent antibody probing, and then the slide will be imaged using a microarray scanner. Fluorescent detection of bound protein can be achieved either with

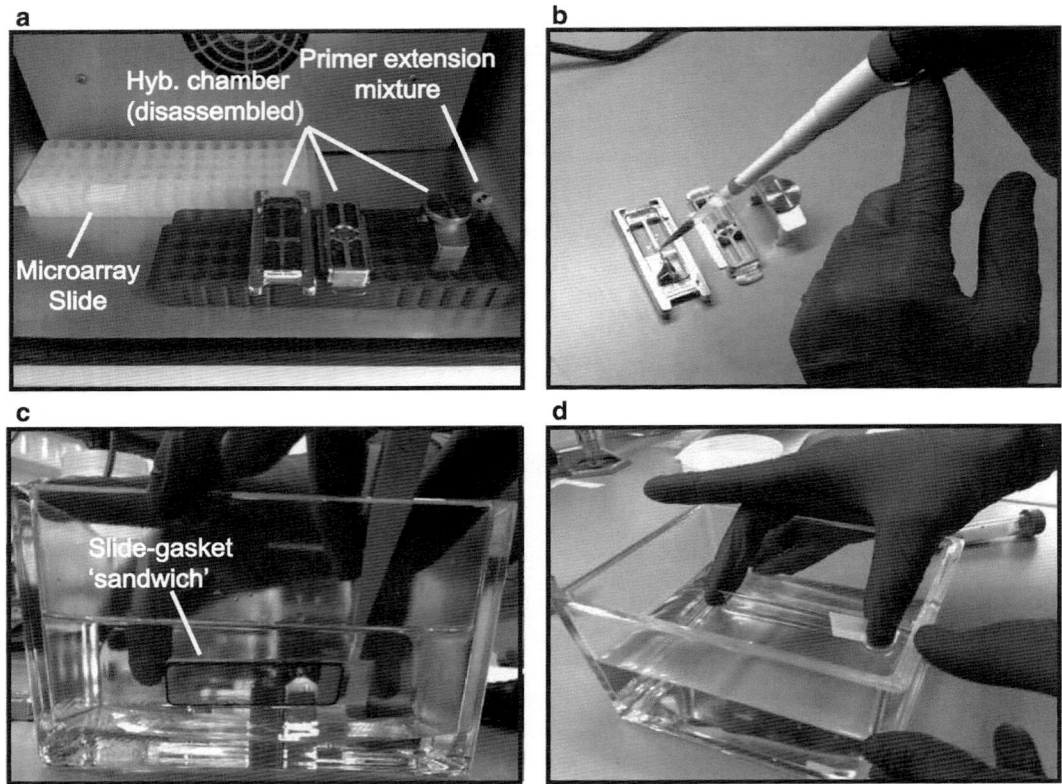

Fig. 3 Steps in the PBM experiment. (**a**) Suggested layout for microarray slide, hybridization chamber, gasket slide (sitting in chamber base), and primer extension mixture as described in **Note 11**. (**b**) Dispensing mixture into gasket slide prior to the assembly of hybridization chamber. (**c**) Disassembling the microarray slide-gasket "sandwich:" sandwich is fully immersed in the buffer in staining dish, tweezer edge is inserted between gasket and slide, tweezers are rotated/twisted to pry open sandwich causing gasket slide (on bottom) to fall to bottom of staining dish, and microarray slide remains firmly held by thumb and forefinger. Buffer is stained blue for illustration. (**d**) "De-wetting" the microarray slide: slide is held by thumb and forefinger fully immersed in PBS in staining dish, slide is slowly removed from buffer (DNA side down) at a slight incline, and slide should emerge effectively dry, ready for the next step (i.e., chamber assembly or microarray scanning) (color figure online)

a fluorophore-conjugated primary antibody (one-step labeling) or via a two-step approach using a primary antibody and a fluorophore-conjugated secondary antibody [20] (two-step labeling) (Fig. 1a). The selection of the labeling method will depend on your protein sample (i.e., tagged or untagged) and available antibodies (*see* **Notes 16** and **18**). Published examples of NF-κB dimer clones, protein preparation procedures, and antibodies in PBM experiments are listed in Table 1. Here, we describe the two-step approach for a four-chambered PBM that has been used successfully to analyze untagged NF-κB dimers from mouse and human [20]. Alterations to the protocol needed for the one-step labeling are described. The PBM experiment takes ~4.5 h; the microarray scans (assuming scans at 3–4 separate intensity levels) take an additional ~1–1.5 h.

1. Prepare 10 mL of 2 % milk in PBS and 10 mL of 2 % milk in sterile water; these will be used as blocking reagents in antibody and protein-binding steps, respectively. Filter 2 % milk solution using 0.45-μm syringe filter. Milk takes a while to dissolve (~1 h); while waiting, carry out **steps 2** and **3**.

2. Prepare the five PBM wash solutions: (1) 70 mL of PBS, (2) 280 mL of 0.01 % Triton X-100/PBS, (3) 70 mL of 0.1 % Tween /PBS, (4) 70 mL of 0.5 % Tween/PBS, and (5) 210 mL of 0.05 % Tween/PBS. Cover buffers throughout the experiment (i.e., parafilm over the beaker tops) to keep dust and debris out. Buffers should be prepared fresh on the day of the experiment.

3. Pre-wet *double-stranded* Agilent microarray in a Coplin jar filled with 0.01 % Triton X-100/PBS for >5 min on rotating shaker (125 rpm).

4. Fill one glass staining dish with 500 mL of PBS (staining dish #1) and a second dish with 500 mL of 0.5 % Tween-20/PBS (staining dish #2). These two wash buffers will be used throughout the experiment, and so they should be kept covered to reduce accumulation of dust and debris.

5. Prepare the hybridization chamber and gasket slide with 2 % milk blocking solution (*see* **Note 14**).

6. Block microarray with milk. Rinse the microarray slide and assemble with gasket slide and blocking solution in hybridization chamber (*see* **Note 15**). Incubate the assembled chamber in the dark (i.e., in a drawer) for 1 h.

7. Prepare binding mixtures. While the microarray slide is blocking in milk/PBS, prepare the protein-binding mixtures for each separate gasket. The total reaction volume for 4×180K-format microarrays is 175 μL (for 8×60K format adjust components for total volume of 75 μL) (*see* **Note 16**). Binding mixtures should sit at room temperature ~30 min before adding, applying to microarray slide.

8. After the microarray blocking step, disassemble steel hybridization chamber (*see* **Note 17**) and wash slide: 0.1 % Tween-20/PBS for 5 min then 0.01 % Triton X-100/PBS for 2 min. All microarray wash steps are performed in Coplin jars with 70 mL of wash buffer on orbital shaker set at 125 rpm.

9. Prepare hybridization chambers with binding mixtures. During wash steps, clean the gasket slide with sterile water, rinse with 70 % ethanol, remove any liquid with Dust Off XL canned air, and prepare the hybridization chamber and gasket slide with binding mixture: disassemble the steel hybridization chamber on benchtop, place gasket slide in chamber base, pipette 175 μL of binding mixture into each gasket chamber, and cover with microarray dust cover.

10. Apply binding mixture to microarrays. After the final wash, rinse the microarray slide, and assemble with gasket slide and binding mixture in the hybridization chamber (*see* **Notes 12** and **15** for notes on chamber assembly). Incubate the assembled chamber in the dark (i.e., in a drawer) for 1 h.

11. Prepare primary and secondary antibody mixtures. During the protein-binding incubation, prepare the primary and secondary antibody (Ab) mixtures (*see* **Note 18**). Allow antibody mixtures to sit at room temperature for ~30 min in the dark (to avoid fluorophore photobleaching) prior to use.

12. After the protein-binding step, disassemble the steel hybridization chamber (*see* **Note 17** for notes on chamber disassembly), and wash slide: 0.5 % Tween-20/PBS for 3 min; 0.01 % Triton X-100/PBS for 2 min.

13. Prepare hybridization chambers with primary Ab mixtures. During wash steps, clean the gasket slide with sterile water, rinse with 70 % ethanol, remove any liquid with Dust Off XL canned air, and prepare the hybridization chamber and gasket slide with the primary Ab mixture: disassemble the steel hybridization chamber on benchtop, place gasket slide in chamber base, pipette 175 µL of antibody mixture into each gasket chamber, and cover with the microarray dust cover.

14. Apply primary antibody to microarrays. After the final wash, rinse the microarray slide, and assemble with gasket slide and primary Ab mixture in the hybridization chamber (*see* **Notes 12** and **15** for notes on chamber assembly). Incubate the assembled chamber in the dark (i.e., in a drawer) for 20 min.

15. After primary Ab step, disassemble the steel hybridization chamber (*see* **Note 17** for notes on chamber disassembly), and wash slide using 0.05 % Tween-20/PBS for 3 min then 0.01 % Triton X-100/PBS for 2 min.

16. Prepare hybridization chambers with secondary Ab mixtures. During wash steps, clean the gasket slide with sterile water, rinse with 70 % ethanol, remove any liquid with Dust Off XL canned air, and prepare the hybridization chamber and gasket slide with secondary Ab mixture: disassemble the steel hybridization chamber on benchtop, place gasket slide in chamber base, pipette 175 µL of antibody mixture into each gasket chamber, and cover with microarray dust cover.

17. Apply secondary Ab to microarrays. After the final wash, rinse microarray slide, and assemble with gasket slide and secondary Ab mixture in hybridization chamber (*see* **Notes 12** and **15** for notes on chamber assembly). Incubate the assembled chamber in the dark (i.e., in a drawer) for 20 min.

18. After secondary Ab step, disassemble steel hybridization chamber (*see* **Note 17** for notes on chamber disassembly), and wash

slide twice in 0.05 % Tween-20/PBS for 3 min each time then PBS only for 2 min.

19. Rinse microarray slide in staining dish #1 (PBS only) to de-wet (*see* **Note 13**).

20. Scan microarray in GenePix scanner at 2.5 μm resolution. Laser and filter setting should be set for Alexa Fluor 488 label (Ex: 488 nm, Em: Blue Filter). Save scanned image as ".TIF" file for subsequent analysis. Scans should be taken at multiple sensitivities to ensure that all spots (which may vary greatly in intensity) are properly visualized. Start at a low sensitivity where the brightest spots are below the signal saturation window, and increase until the dimmest spots are visible. Using the GenePix 4100A scanner, set the laser power to 100, and adjust the photomultiplier gain (generally starting and 400 and proceeding to 700–800). Save scanned images at each sensitivity level as a ".TIF" file for subsequent analysis.

21. Store microscope slide box in the dark (i.e., in a drawer).

3.3   Data Analysis   This step involves analyzing the microarray images and quantifying the protein binding to each DNA probe sequence. Probe fluorescence values are proportional to the relative binding affinity of the protein for the different probe sequences (*see* **Note 19**) [15, 20]. In practice, many different software packages and microarray normalization procedures could be used to determine the microarray probe fluorescence values. In this chapter, we describe the approach developed in the Bulyk lab specifically for analyzing PBM data that require GenePix microarray analysis software and custom Perl scripts. The Perl scripts designed to analyze the widely used universal PBM design are available from the Bulyk lab website ("Universal PBM analysis suite": http://the_brain.bwh.harvard.edu/software.html) and can be readily modified to analyze custom microarray designs.

1. Use GenePix 7.0 software to determine the median, background-subtracted fluorescence intensity values for all microarray spots. Open one of the ".TIF" image files, open a ".gal" file (i.e., file extension ".gal") using "Load Array List," and align the resulting grid with the microarray spots. Gal files are generated when you design a microarray and are available through the Agilent website. Manually flag any "bad" spots that you wish to disregard; by visual inspection, you can readily identify problem spots due to scratches, bubbles, or dust. "Analyze" the data, and save it such that the data for each microarray block/chamber is saved as an individual ".gpr" file (i.e., file extension ".gpr"). The gpr files contain the background-subtracted median intensity values for all microarray probe features (i.e., spots). The "median intensity" determined

using the GenePix software is the median of the pixel intensity for each microarray probe spot (*see* GenePix documentation for additional detail); the median value across *replicate probes* will be determined later. Perform the image analysis procedure for the scans done at all scanner sensitivities/intensities. Also perform this procedure for the Cy3-dUTP scan (*see* **Note 20**).

2. Combine the microarray scans performed at different sensitivities using "masliner" software [32]. The masliner software performs a linear regression procedure to determine the intensity of saturated spots in the higher-intensity scan using the values from a lower-intensity scan. Masliner can be downloaded for free (for academic users) from the Church lab website (http://arep.med.harvard.edu/masliner/supplement.htm). For the rest of the protocol, probe values (or intensities) will refer to masliner-adjusted, background-subtracted, median fluorescence intensities.

3. Remove probes that have been flagged as "bad" in either the A488 or Cy3 scans from further analysis.

4. Perform a spatial normalization across all probes. Determine the median probe intensity across the whole microarray (i.e., the global median) and a local neighborhood median for each probe based on a 15×15 grid centered on each microarray probe. For probes near the edge, use an offset 15×15 grid. Normalize each probe intensity by the ratio of global median/neighborhood median.

5. Combine probe intensities for replicate PBM experiments using a quantile normalization procedure.

6. Determine the median intensity over all replicate probes (and from replicate experiments), in both the forward and reverse complement direction. As a quality check, we require that the median intensity values determined separately for all probes in the forward and reverse complement directions (Fig. 1) are within 40 % of each other (*see* **Note 21**). These final median intensity values, averaged over replicate probes, provide a robust quantification of the relative binding preference for the protein to each DNA probe sequence.

7. Binding site motifs can be determined for top scoring (i.e., high affinity) κB sites to visualize sequence specificity (e.g., Fig. 2b). Motifs from top scoring sites can be determined by running the Priority 2.1.0 motif finding algorithm sequences [33] and turned into graphical sequence logos using enoLOGOS [34]. To identify specificity differences between related dimers, pairwise comparison of binding to all κB sites can be examined using a simple scatterplot (e.g., Fig. 2c). Specific subsets of factor-preferred (i.e., off-diagonal) sites can then be identified and examined further (e.g., turned into a binding motif to highlight unique preferences, Fig. 2c, d).

4 Notes

1. Design of the microarray. Agilent CGH microarrays are manufactured as single-stranded DNA (ssDNA) microarrays. For use in a PBM experiment, the DNA probes are made double stranded in the lab using a primer extension reaction from a 24-nucleotide (nt) primer sequence common to each probe (Fig. 1). Agilent microarrays are available in different formats that vary in the number of probe features (i.e., spots) and the number of separate chambers (e.g., the 4×180K-format has four chambers with the same set of 177,400 user-defined probe features in each chamber, the 8×60K-format has eight chambers with 61,979 user-defined features). Multi-chambered microarrays allow multiple binding experiments to be conducted simultaneously on a single microarray slide.

   The probe sequences on Agilent microarrays are 60-nt long. To make the ssDNA probes double stranded (discussed below), probe sequences must be designed with a common 24-nt primer sequence placed at their 3′ end; this design allows for 36 nt of "usable" sequence per probe (Fig. 1b). PBM experiments have been performed using synthetic and genomic sequences (e.g., [20, 35]). A widely used PBM design is the universal PBM [15, 36] that allows an unbiased characterization of protein binding to all possible 8-bp DNA sequences. This design has the advantage of being completely comprehensive and unbiased and has been used successfully to determine NF-κB binding motifs [20]. However, due to the length of the NF-κB binding site (canonically 9–12 bp), we suggest using targeted selection of synthetic or genome-derived probe sequences that contain a single NF-κB binding site sequence in each probe (Fig. 1b, and [20]). Protein binding can be affected by proximity to the glass slide to which the DNA probe is attached; therefore, probe sequences should be designed such that the NF-κB binding sites are placed at a constant position along the probe sequence (Fig. 1b). To ensure that the orientation of the probe sequence does not affect the results, probes should be included in both forward and reverse complement (RC) directions (Fig. 1b). To ensure the measurement robustness, we suggest four replicate probes be included for each orientation of the probe (i.e., eight replicates in total, four in each orientation).

2. The dust cover is required to keep dust and debris off the gasket slide and buffers immediately prior to assembly of hybridization chamber. Anything that fully covers the hybridization chamber device is sufficient, such as a disposable pipette box lid or cover fashioned from tinfoil.

3. To make 3.5 L of PBS: 3 L sterile water, 28 g of NaCl, 0.7 g of KCl, 5.04 g of Na_2HPO_4, and 0.84 g of KH_2PO_4. Mix ~30 min on stir plate to dissolve salts. Add sterile water to a final volume of 3.5 L. Adjust pH to 7.4 with HCl. Autoclave to sterilize.

4. Make 10 mM stock solution (2.5 mM of each dNTP). Store at -20 °C.

5. The gasket slide will depend on the microarray format chosen for the given experiment (i.e., the number of chambers required). In this chapter, we describe a PBM experiment that uses a microarray with a four-chambered gasket slide. Agilent suggests only using gasket slides only one time; however, we have found that they can be used successfully a number of times (~10 times). After use, wash the gasket slide with sterile distilled water and 70 % ethanol, dry with canned air, and store at room temperature.

6. Some key details to keep in mind throughout the experiment. First, critical to the success of the primer extension reaction is that the components (i.e., chamber, gasket slide, microarray slide, mixture) remain as close to 85 °C as possible to prevent mis-priming. The most critical part of this procedure, there-fore, is the assembly of the hybridization chamber and primer extension mixture on the benchtop, at which time the compo-nents will cool—the quicker this assembly is carried out, the better. Second, to ensure a clean PBM experiment, compo-nents should be covered and kept free of dust/debris as much as possible. This includes all buffers (in beakers and staining dishes) and the hybridization chamber and gasket slide prior to assembly. Finally, in all steps, when assembling the hybridiza-tion chamber (i.e., putting together the gasket slide, mixtures, and microarray slide), care should be taken to prevent bubbles forming in the chambers. This PBM hybridization chamber is *not* rotated during incubation steps, as in some microarray experiments; therefore, bubbles will prevent access to the underlying DNA probes. The key step in preventing bubbles is to *slowly* lower the microarray slide down onto the gasket slide and mixtures when assembling the hybridization chamber.

7. Microcentrifuge tube racks are used as heat insulators when assembling the hybridization chamber on the benchtop, thereby preventing equipment from cooling down. Having the microarray slide elevated on a microcentrifuge tube rack also makes it easier to pick up when assembling the hybridization chamber, which is a time-sensitive step. Arrange the two racks so that they are parallel to the front of the oven, one behind the other (Fig. 3a).

8. Thermo Sequenase 10× reaction buffer is provided with the Thermo Sequenase cycle sequencing kit, but can also be prepared

in lab. 10× Thermo Sequenase reaction buffer: 26 mL of 1 M Tris–HCl, pH 9.5; 60 mL of sterile water. Dissolve 6.18 g of $MgCl_2$, and add sterile water to the final volume of 100 mL. Filter sterilize with 0.2 μm Nalgene filter. Store at room temperature.

9. The hybridization chamber should be partially assembled with the gasket slide sitting gasket side up in the chamber base, as per Agilent microarray instruction manual (http://www.cfg-biotech.com/microarray/manuals/Agilent_hyb_chamber_user_guide_00061330.pdf). Place the hybridization chamber components (along with gasket slide) and primer extension mixture on the microcentrifuge tube rack closest to the oven door (Fig. 3a).

10. When multiple microarrays are purchased, they are often shipped together in a vacuum-sealed slide container. After opening this container to use the first array, the other microarray slides should be stored in a vacuum desiccator, in the dark (this can be done by covering the slide box with tinfoil before placing it in the desiccator).

11. Place the microarray slide with the DNA side up (the side with "Agilent" printed on it) on the microcentrifuge tube rack at the back of the oven. Place the microarray slide perpendicular to the rack so that the edges of the slide extend beyond the rack; this makes it easier to pick up the microarray slide when you are assembling the hybridization chamber in the next step (Fig. 3a).

12. Hybridization chamber assembly should follow the Agilent microarray instruction manual, with the exception that bubbles should be prevented. The hybridization chamber is not rotated during the PBM experiment—in contrast to the standard Agilent microarray procedure; therefore, there is no mixing, and bubbles occurring in the chambers will prevent access to the underlying DNA probes. To prevent bubbles, (1) carefully lower the microarray slide down onto the gasket slide and dispensed mixture, and (2) if bubbles occur, then it can be coaxed to the chamber periphery by gently tapping the chamber on the benchtop while holding the chamber edgewise. However, during the primer extension step, when the liquid is hot, we suggest *not* performing this "tapping" step as it wastes critical time (i.e., the chamber cools), and the reduced surface tension of the hot liquid causes more bubbles to form.

13. Care must be taken to keep reagents as close to 85 °C as possible during the assembly process. A stepwise strategy that works well is as follows: (1) Remove the tube rack at the front of the oven with the hybridization chamber, gasket slide, and primer extension mixture, and place it on the benchtop and shut the oven door; (2) Pipette the 900 μL primer extension

mixture onto the gasket slide; (3) Remove the microarray slide from the oven, shut the oven door, and assemble the microarray slide and hybridization chamber according to Agilent instructions; and (4) return assembled hybridization chamber to the 85 °C oven.

14. Disassemble steel hybridization chamber on the benchtop, place gasket slide in the chamber base, and pipette 175 μL of 2 % milk/PBS blocking solution into separate gasket chambers according to Agilent instructions. The volume of blocking solution will depend on the choice of gasket and microarray format: 4 × 180K format microarrays have four, 175 μL chambers; 8 × 60K format microarrays have eight 75 μL chambers. In this chapter, volumes are adjusted to the 4 × 180K format.

15. Hybridization chamber assembly. Remove Coplin jar from the orbital shaker. Lift the microarray slide from the Coplin jar (with gloved fingers or Agilent tweezers) and immediately immerse the slide, DNA side down, in staining dish #1 with PBS. Note that during this step you should hold the microarray slide so that it does not touch the bottom of the staining dish! Agitate the microarray slide up and down to wash away any debris. Holding the microarray slide by the top and bottom edges (i.e., the short edges) with your thumb and index finger, slowly lift the microarray edgewise from the staining dish so that the microarray slide "de-wets" and is free of buffer (Fig. 3d). As you lift the microarray slide, keep the DNA side facing down at a slight incline. Immediately assemble the microarray, gasket slide, and applied solution in the steel hybridization chamber. The assembly should follow the Agilent microarray instruction manual. If bubbles occur in a chamber and are covering any part of the arrayed DNA probes, they can often be moved to the edge of the gasket, where no DNA is present, by gently tapping the chamber on the desktop.

16. The buffer conditions (i.e., salt concentration, pH, etc.) can vary depending on the details of your particular experiments as is done with more standard binding assays, such as an EMSA. In this chapter, we describe a buffer that was previously used for PBM experiments on mouse and human NF-κB dimers [20]. Most published PBM experiments have been performed using bacterially expressed and purified NF-κB dimers. Protein clones, protein preparation procedures, and antibodies used in these published PBM studies are listed in Table 1. Protein sample concentrations can be determined using the standard Bradford assay; when achievable, a final protein concentration of 100–200 nM in the PBM binding step is desirable and has worked well with many different proteins and sample preparations. NF-κB dimers can be made with protein tags (GST-tag, unpublished data, T. Siggers and A.

Penvose; Histidine-tag [20]) or untagged [20] (Table 1). In short, protein samples and buffer conditions that work in more traditional DNA-binding assays (e.g., EMSAs) should generally work in PBM experiments. Finally, we note that other protein preparation procedures have also been used successfully in PBM experiments for non-NF-κB proteins, such as (1) in vitro transcription and translation (IVT) without subsequent purification (e.g., [21, 23]) and (2) crude nuclear lysate without subsequent purification [22].

17. Hybridization chamber disassembly. On the benchtop, disassemble the hybridization chamber, lift microarray/gasket slide "sandwich" from chamber base, immerse it in a staining dish #2 (0.5 % Tween-20/PBS), and separate the two slides according to Agilent instructions. Shake the microarray slide briefly up and down in buffer to wash the array. To prevent the microarray slide from lifting when removing the top of the steel chamber, use tweezers to hold the microarray slide down—there are two long openings in the top of the chamber where you press the microarray with the tweezers to hold it down. Similarly, when removing the microarray/gasket "sandwich" from the chamber base, you can prevent the "sandwich" from sticking and prematurely separating by lifting the chamber and poking the gasket slide from underneath using tweezers. It is critical that the microarray slide and gasket chamber do not separate until they are immersed in buffer in staining dish #2, as separation can cause solution to spill over between chambers. Note that the microarray slide will not "de-wet" when removed from this buffer due to the presence of detergents. Clean the gasket slide with sterile water, and rinse with 70 % ethanol; remove any liquid with Dust Off XL canned air.

18. Proteins are labeled with the Alexa Fluor 488 (A488) or Cy5 for visualization (Table 1). In this chapter, we describe labeling with A488 as we have found this fluorophore to be the most sensitive. There are two methods for labeling DNA-bound protein: (1) A488-conjugated primary antibody (one-step labeling) or (2) primary antibody followed by an A488-conjugated secondary antibody (two-step labeling) (Fig. 1 and Table 1). In this chapter, we describe the two-step labeling approach. To adjust protocol for one-step labeling, after incubation with primary Ab, do not apply secondary Ab, and proceed directly to **step 19** (final slide washes). A list of antibodies that have been used successfully to label NF-κB dimers in PBM experiments is provided (Table 1).

19. Over a concentration range in which high-affinity spots are not saturated with protein, a linear relationship will hold between protein-DNA binding free energies and the log fluorescence values [15, 35]. However, as with any equilibrium binding

experiment performed at a single protein concentration, an actual binding constant (i.e., Kd value) cannot be determined. A more involved approach aimed at determining Kd values has been described where multiple PBM experiments are performed at a series of protein concentrations [35]; this work has a detailed discussion of the relation between probe fluorescence and binding affinity.

20. The Cy3 scans are used for manual identification of microarray probes that did not double-strand properly (i.e., due to a bubble or buffer leak); problem areas will look black and can be manually flagged as bad. In original PBM papers using the universal PBM design, Cy3 scans were used in a linear regression procedure to estimate the double-stranding efficiency and normalize binding data accordingly [15, 36]. However, for custom microarrays, there is often not enough sequence variability across the different probe sequences to accurately estimate the expected Cy3 intensities; and we, therefore, use the Cy3 intensity values (i.e., the image) to manually screen for problem areas (i.e., damaged probes or scratches).

21. Disagreement between probes in the forward or reverse complement direction can occur due to (1) primer extension problems through G-rich sequences (i.e., when G is the template base, as opposed to A,T, or C) or (2) additional binding sites that occur in the flanking DNA sequences, which will be at different distances from the glass slide in the forward and reverse complement probes.

Acknowledgments

Research in the authors' laboratories on DNA binding by NF-kB proteins was supported by NIH grant K22AI09379 (T.S.), NSF grant MCB-090461 (T.D.G), and NSF grant IOS-1354935 (T.S. and T.D.G).

References

1. Ghosh S, May MJ, Kopp EB (1998) NF-kappa B and Rel proteins: evolutionarily conserved mediators of immune responses. Annu Rev Immunol 16:225–260

2. Pahl HL (1999) Activators and target genes of Rel/NF-kappaB transcription factors. Oncogene 18(49):6853–6866

3. Smale ST (2012) Dimer-specific regulatory mechanisms within the NF-kappaB family of transcription factors. Immunol Rev 246(1): 193–204

4. Natoli G (2006) Tuning up inflammation: how DNA sequence and chromatin organization control the induction of inflammatory genes by NF-kappaB. FEBS Lett 580(12):2843–2849

5. Akira S, Kishimoto T (1997) NF-IL6 and NF-kappa B in cytokine gene regulation. Adv Immunol 65:1–46

6. Leung TH, Hoffmann A, Baltimore D (2004) One nucleotide in a kappaB site can determine cofactor specificity for NF-kappaB dimers. Cell 118(4):453–464

7. Wang VY, Huang W, Asagiri M, Spann N et al (2012) The transcriptional specificity of NF-kappaB dimers is coded within the kappaB DNA response elements. Cell Rep 2(4):824–839

8. Mrinal N, Tomar A, Nagaraju J (2011) Role of sequence encoded kappaB DNA geometry in gene regulation by Dorsal. Nucleic Acids Res 39(22):9574–9591

9. Carey MF, Peterson CL, Smale ST (2013) Electrophoretic mobility-shift assays. Cold Spring Harb Protoc 2013(7):636–639

10. Carey MF, Peterson CL, Smale ST (2013) DNase I footprinting. Cold Spring Harb Protoc 2013(5):469–478

11. Carey MF, Peterson CL, Smale ST (2012) Experimental strategies for the identification of DNA-binding proteins. Cold Spring Harb Protoc 2012(1):18–33

12. Kinzler KW, Vogelstein B (1990) The GLI gene encodes a nuclear protein which binds specific sequences in the human genome. Mol Cell Biol 10(2):634–642

13. Tuerk C, Gold L (1990) Systematic evolution of ligands by exponential enrichment: RNA ligands to bacteriophage T4 DNA polymerase. Science 249(4968):505–510

14. Siggers T, Gordan R (2013) Protein-DNA binding: complexities and multi-protein codes. Nucleic Acids Res. doi:10.1093/nar/gkt1112

15. Berger MF, Philippakis AA, Qureshi AM, He FS et al (2006) Compact, universal DNA microarrays to comprehensively determine transcription-factor binding site specificities. Nat Biotechnol 24(11):1429–1435

16. Bulyk ML, Gentalen E, Lockhart DJ, Church GM (1999) Quantifying DNA-protein interactions by double-stranded DNA arrays. Nat Biotechnol 17(6):573–577

17. Mukherjee S, Berger MF, Jona G, Wang XS et al (2004) Rapid analysis of the DNA-binding specificities of transcription factors with DNA microarrays. Nat Genet 36(12):1331–1339

18. Linnell J, Mott R, Field S, Kwiatkowski DP et al (2004) Quantitative high-throughput analysis of transcription factor binding specificities. Nucleic Acids Res 32(4):e44

19. Field S, Udalova I, Ragoussis J (2007) Accuracy and reproducibility of protein-DNA microarray technology. Adv Biochem Eng Biotechnol 104:87–110

20. Siggers T, Chang AB, Teixeira A, Wong D et al (2012) Principles of dimer-specific gene regulation revealed by a comprehensive characterization of NF-kappaB family DNA binding. Nat Immunol 13(1):95–102

21. Badis G, Berger MF, Philippakis AA, Talukder S et al (2009) Diversity and complexity in DNA recognition by transcription factors. Science 324(5935):1720–1723

22. Bolotin E, Chellappa K, Hwang-Verslues W, Schnabl JM et al (2011) Nuclear receptor HNF4alpha binding sequences are widespread in Alu repeats. BMC Genomics 12:560

23. Zhu C, Byers KJ, McCord RP, Shi Z et al (2009) High-resolution DNA-binding specificity analysis of yeast transcription factors. Genome Res 19(4):556–566

24. Berger MF, Badis G, Gehrke AR, Talukder S et al (2008) Variation in homeodomain DNA binding revealed by high-resolution analysis of sequence preferences. Cell 133(7):1266–1276

25. Gordan R, Murphy KF, McCord RP, Zhu C et al (2011) Curated collection of yeast transcription factor DNA binding specificity data reveals novel structural and gene regulatory insights. Genome Biol 12(12):R125

26. Ryzhakov G, Teixeira A, Saliba D, Blazek K et al (2013) Cross-species analysis reveals evolving and conserved features of the nuclear factor kappaB (NF-kappaB) proteins. J Biol Chem 288(16):11546–11554

27. Copley RR, Totrov M, Linnell J, Field S et al (2007) Functional conservation of Rel binding sites in drosophilid genomes. Genome Res 17(9):1327–1335

28. Grove CA, De Masi F, Barrasa MI, Newburger DE et al (2009) A multiparameter network reveals extensive divergence between C. elegans bHLH transcription factors. Cell 138(2):314–327

29. Wang JK, Li TX, Bai YF, Lu ZH (2003) Evaluating the binding affinities of NF-kappaB p50 homodimer to the wild-type and single-nucleotide mutant Ig-kappa3 sites by the unimolecular dsDNA microarray. Anal Biochem 316(2):192–201

30. Wong D, Teixeira A, Oikonomopoulos S, Humburg P et al (2011) Extensive characterization of NF-kappaB binding uncovers non-canonical motifs and advances the interpretation of genetic functional traits. Genome Biol 12(7):R70

31. Rucker P, Torti FM, Torti SV (1997) Recombinant ferritin: modulation of subunit stoichiometry in bacterial expression systems. Protein Eng 10(8):967–973

32. Dudley AM, Aach J, Steffen MA, Church GM (2002) Measuring absolute expression with microarrays with a calibrated reference sample and an extended signal intensity range. Proc Natl Acad Sci U S A 99(11):7554–7559

33. Gordan R, Narlikar L, Hartemink AJ (2010) Finding regulatory DNA motifs using alignment-free evolutionary conservation information. Nucleic Acids Res 38(6):90

34. Workman CT, Yin Y, Corcoran DL, Ideker T et al (2005) enoLOGOS: a versatile web tool for energy normalized sequence logos. Nucleic Acids Res 33(Web Server issue):W389–W392

35. Siggers T, Duyzend MH, Reddy J, Khan S, Bulyk ML (2011) Non-DNA-binding cofactors enhance DNA-binding specificity of a transcriptional regulatory complex. Mol Syst Biol 7:555

36. Berger MF, Bulyk ML (2009) Universal protein-binding microarrays for the comprehensive characterization of the DNA-binding specificities of transcription factors. Nat Protoc 4(3):393–411

Chapter 37

Methods for Analyzing the Evolutionary Relationship of NF-κB Proteins Using Free, Web-Driven Bioinformatics and Phylogenetic Tools

John R. Finnerty and Thomas D. Gilmore

Abstract

Phylogenetic analysis enables one to reconstruct the functional evolution of proteins. Current understanding of NF-κB signaling derives primarily from studies of a relatively small number of laboratory models—mainly vertebrates and insects—that represent a tiny fraction of animal evolution. As such, NF-κB has been the subject of limited phylogenetic analysis. The recent discovery of NF-κB proteins in "basal" marine animals (e.g., sponges, sea anemones, corals) and NF-κB-like proteins in non-metazoan lineages extends the origin of NF-κB signaling by several hundred million years and provides the opportunity to investigate the early evolution of this pathway using phylogenetic approaches. Here, we describe a combination of bioinformatic and phylogenetic analyses based on menu-driven, open-source computer programs that are readily accessible to molecular biologists without formal training in phylogenetic methods. These phylogenetically based comparisons of NF-κB proteins are powerful in that they reveal deep conservation and repeated instances of parallel evolution in the sequence and structure of NF-κB in distant animal groups, which suggest that important functional constraints limit the evolution of this protein.

Key words NF-kappaB, Evolution, Motifs, Phylogenetic analysis

1 Introduction

Nuclear factor κB (NF-κB) names an evolutionarily conserved family of transcription factors that control a variety of key developmental and immunity-related processes across a phylogenetically broad range of animals [1]. Studies from "basal" animal phyla suggest that the founding member of this protein family evolved at or near the base of animal evolution and subsequently underwent multiple gene duplications in triploblastic animals [2–5]. Consistent with this hypothesis, triploblastic animals have generally been found to possess multiple NF-κB proteins (Fig. 1). For example, among deuterostomes, there are five NF-κB proteins in humans and two in the sea squirt, *Ciona intestinalis*. Among spiralian protostomes [6], two NF-κB proteins have been identified in the snail

Michael J. May (ed.), *NF-kappa B: Methods and Protocols*, Methods in Molecular Biology, vol. 1280,
DOI 10.1007/978-1-4939-2422-6_37, © Springer Science+Business Media New York 2015

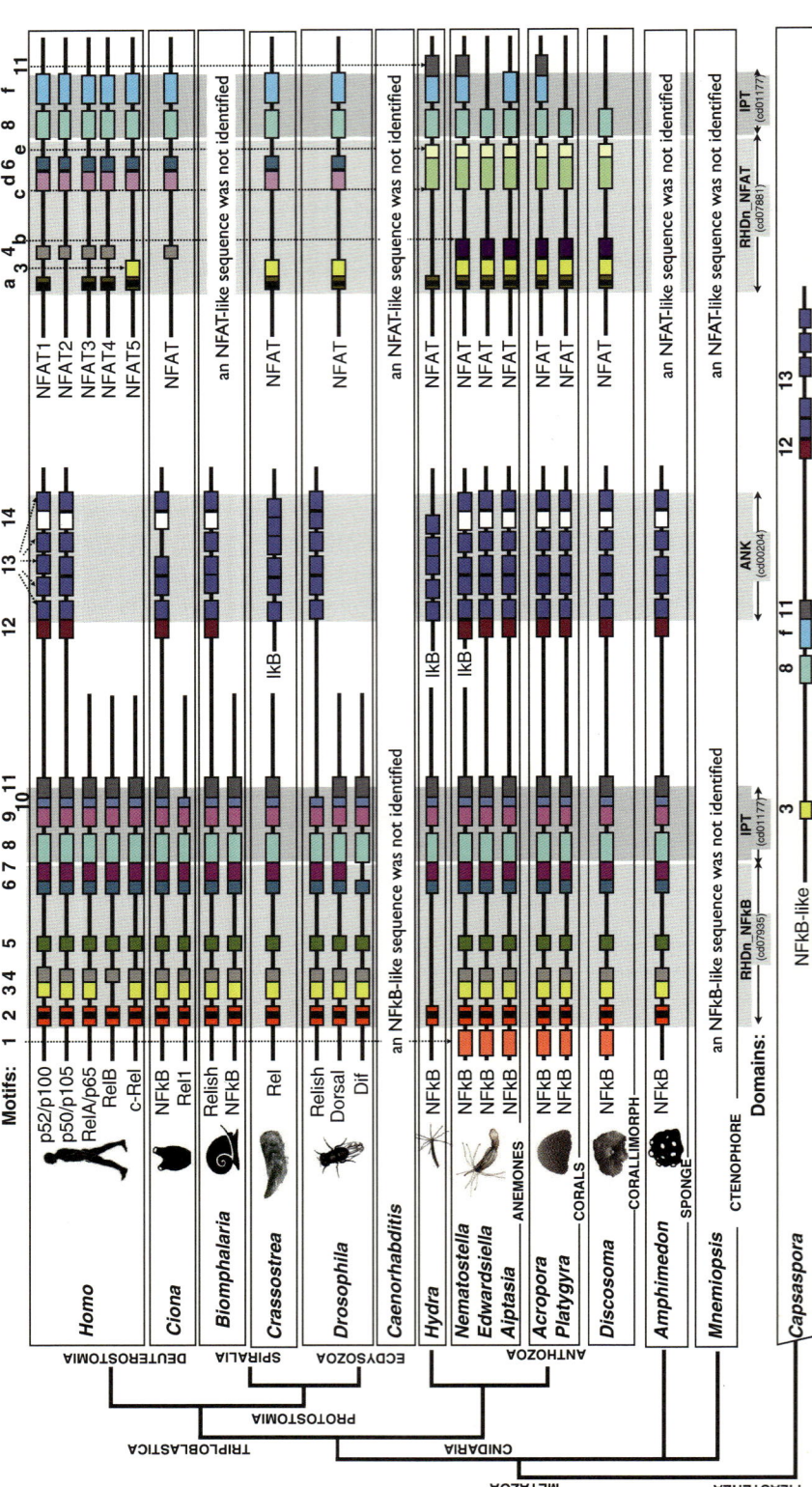

Fig. 1 NF-κB and NFAT proteins from major animal lineages. The phylogenetic relationship among these major lineages, which is consistent with molecular phylogenetic and phylogenomic studies [7, 12, 13, 15], is depicted at the far left. The genomes and/or transcriptomes of species selected to represent these higher taxa (Table 1) were searched for NF-κB (*left*) and NFAT (*right*) proteins. Distinct IκB proteins are also shown for the oyster *Crassostrea gigas*, the sea anemone, *Nematostella vectensis*, and the freshwater cnidarian *Hydra magnipapillata*, as these taxa appear to lack an NF-κB protein with a terminal ankyrin repeat region [2]. Conserved motifs identified using MEME [22] are represented by *colored boxes* whose width is proportional to their length in amino acids. By definition, every occurrence of a given motif, both between proteins and within proteins, is equal in length. The inter-motif regions do vary in length, but, in order that the motifs might be aligned in register to facilitate visual comparison of motif architecture across proteins, this variation is not shown. All numbered motifs (1–14) are found among the NF-κB proteins, while motifs 3, 4, 6, 8, and 11 are also found among the NFAT proteins. The *lettered motifs* (a–f) are found only in the NFAT proteins. The NF-κB-like protein from the filasterean *Capsaspora* is placed in the center of the figure because it possesses both NF-κB-specific (12, 13) and NFAT-specific (f) motifs, as well as motifs shared by both protein families (3, 8). Gray shading delineates the location of conserved protein domains [34] relative to the motifs identified by MEME, which was determined for human p52/p100 and NFAT5. Black regions in motif 2 and motif 4 correspond to a homologous cluster of DNA-binding sites in the NF-κB and NFAT proteins, respectively (*see* Fig. 8 for an alignment of this region)

Biomphalaria glabrata. However, the oyster *Crassostrea gigas*, a fellow mollusk, appears to possess only a single NF-κB protein. Among the ecdysozoan protostomes [7], *Drosophila melanogaster* possesses three NF-κB-like proteins, while the nematode *Caenorhabditis elegans* [8] has lost NF-κB proteins entirely.

Among non-triploblastic animals—the so-called "basal" animal phyla—single NF-κB proteins with extensive sequence similarities to the NF-κB proteins of triploblastic animals have been identified in multiple cnidarian species [2, 5] and one sponge [4]. Molecular and biochemical studies of NF-κB from the starlet sea anemone, *Nematostella vectensis*, have demonstrated extensive functional similarities to the NF-κB proteins of triploblastic animals [9–11]. Despite the availability of a sequenced genome and transcriptome, NF-κB proteins have not been found in Ctenophora, a phylum of gelatinous marine animals that may be the sister group to all other animals [12].

Among these diverse animal phyla, NF-κB proteins can be recognized by an N-terminal DNA-binding/dimerization domain (the Rel homology domain [RHD]). NF-κB proteins have been further divided into two subfamilies—the NF-κB and the Rel proteins—based on distinct C-terminal regions as well as differences within the RHD. All Rel proteins (e.g., human RelA, RelB, and c-Rel; *Drosophila* Dorsal, Dif) have C-terminal transactivation domains, but the sequences in these transactivation domains are not generally conserved either within or between species. In contrast, NF-κB subfamily proteins (e.g., human p52/p100 and p50/p105; *Drosophila* Relish) usually have C-terminal domains that contain multiple ankyrin repeats, which must be removed by proteolysis for the protein to become an active DNA-binding protein. Generally, within a species, all NF-κB and Rel proteins can form homo- and heterodimers, and these different complexes can have different target site specificities and gene targets. A related protein family, the NFAT proteins, also has a DNA-binding RHD, but NFAT proteins do not form dimers with Rel/NF-κB proteins.

Outside of the Metazoa, evidence for the existence of bona fide NF-κB proteins is less clear. Although no RHD-containing proteins were found in the sequenced genome of the choanoflagellate *Monosiga brevicollis* [13], other choanoflagellates do appear to have RHD-containing NF-κB-like genes/proteins [14]. There is an NF-κB-like protein encoded in the sequenced genome of the filasterean *Capsaspora owczarzaki* [15]. No NF-κB proteins have been discovered in fungi.

If we adhere to a phylogenetically based definition of gene families, then in animals, it is clear that there are separate NFAT and NF-κB families and that they share a common ancestry through the RHD. If they are sister gene families, the origin of the NF-κB and NFAT families can be traced to the moment when the ancestral NF-κB/NFAT sequence duplicated. This hypothetical ancestral condition is found in modern day cnidarians: one NF-κB gene and one NFAT gene that presumably derived from a single ancestral gene. If that is how these gene families are related, then taxa that possess a bona fide NF-κB gene should also possess an NFAT gene (unless they subsequently lost that gene). If certain basal animal taxa (e.g., sponges) and out groups to the Metazoa (e.g., *Capsaspora*) possess only one RHD-containing protein, that protein might be a direct descendant of the original NF-κB/NFAT ancestor. That ancestral protein may have been much more NF-κB-like than NFAT like (*see Capsaspora* in Fig. 1), but, strictly speaking, it would not be a member of the NF-κB family. While more similar to NF-κB than to NFAT, it would be equally closely related to NF-κB and NFAT proteins. For that reason, we should be cautious about calling the *Capsaspora* sequence "NF-κB."

Given the recent profusion of genome and transcriptome sequencing projects from basal metazoan lineages and non-metazoan groups that are closely related to animals (e.g., choano-flagellates, filastereans), we now have sufficient sequence data to reconstruct the broad functional evolution of this key family of eukaryotic transcription factors. To do so, we must reconstruct the phylogeny of this protein family and track changes in the sequence and structure of NF-κB proteins over the course of evolution, with the goal of relating specific changes in NF-κB (and its molecular partners) to particular selection pressures and phenotypic outcomes in the organisms that possess these proteins. Until recently, the bioinformatic and phylogenetic tools and expertise required for this endeavor would have been beyond the computational capabilities of many molecular biologists and biochemists. That is, complicated and finicky software packages would have to be installed and compiled on a powerful personal computer or a computer cluster to which a researcher had access, and the analyses would often have to be run through a command-line interface. However, a number of powerful bioinformatic and phylogenetic analyses can now be accessed via web-driven interfaces and run remotely on supercomputing clusters, greatly reducing the computational time. Here, we describe a computational approach to identifying conserved sequence motifs and reconstructing the phylogenetic relationships of NF-κB proteins from a wide range of taxa. The approach uses free web-driven software, and all of the relevant input and output files are available upon request from the corresponding author, so that interested researchers can repeat our analyses and build upon them.

Our approach consists of seven steps (Fig. 2). First, we compiled sequences from NF-κB proteins and NFAT proteins from species with published sequenced genomes and/or transcriptomes. We selected species to represent the oldest evolutionary lineages

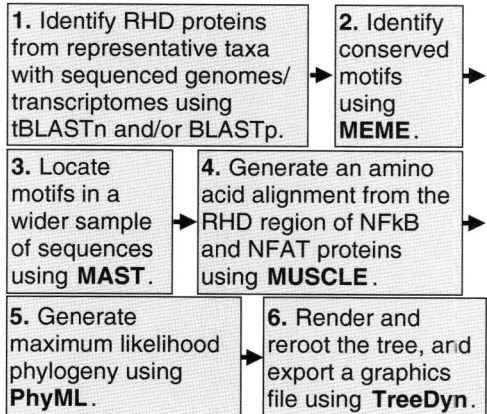

Fig. 2 Overview of the bioinformatic and phylogenetic workflow described in this paper. Five free software applications (MEME, MAST, MUSCLE, PhyML, and TreeDyn) located at two web portals are used

of animals (deuterostomes, ecdysozoans, spiralians, cnidarians, and poriferans) as well as closely related out groups. Second, we interrogated these sequences for the presence of conserved motifs. Third, we located these motifs in a wider sample of sequences. Fourth, we assembled an amino acid alignment. Fifth, we constructed a phylogeny from this alignment using a maximum likelihood approach. Sixth, we evaluated the support for particular nodes on this tree. Finally, we rooted and re-rendered the tree in a format suitable for publication.

The approach described here reveals that bona fide NF-κB proteins (and NFAT proteins) can be traced to the common ancestor of humans and cnidarians, and perhaps the more ancient common ancestor of humans and sponges, but not much earlier. The RHD is comprised of numerous smaller motifs that are highly conserved across taxa, but are clearly distinct between NF-κB and NFAT proteins. The arrangement of these motifs in NF-κB and NFAT proteins has been highly conserved in all major animal lineages since the first appearance of these proteins (some 600 million years ago). However, lineage-specific differences clearly distinguish the cnidarian NF-κB and NFAT proteins from those of triploblastic animals. Additionally, there are numerous instances of parallel evolutionary changes occurring in both the motif architecture and the identity of amino acids at functionally important locations within these conserved motifs. These parallel changes likely represent phylogenetically widespread evolutionary constraints in protein sequence/structure that operate to limit the evolution of NF-κB proteins.

2   Materials

2.1   Computer

1. A personal computer with internet access that can run a modern browser.

2.2   Sequence Databases

1. National Center for Biotechnology Information.
2. EdwardsiellaBase—*Edwardsiella lineata* Genomics Database [16, 17] (*see* **Note 1**).
3. PcarnBase—A Transcriptomic Database for *Platygyra carnosus* [18, 19].
4. *Acropora digitifera* genome browser [20, 21].

2.3   Motif Analysis Software

1. MEME. Multiple Em for Motif Elicitation [22, 23].
2. MAST. Motif Alignment and Search Tool [24, 25].

2.4   Alignment Software

1. MUSCLE [26] at Phylogeny.fr [27] (*see* **Note 2**).

2.5   Maximum Likelihood Phylogeny Estimation Software

1. PhyML 3 [28] at Phylogeny.fr [27] (*see* **Note 3**).

2.6   Tree Viewers and Dynamic Editors

1. TreeDyn [27, 29] at Phylogeny.fr [27].

3   Methods

3.1   Compilation of RHD Proteins from Major Metazoan Lineages and Out-Groups

1. Based on your research focus, identify the organismal lineages you wish to sample. In the example presented here, our goal was to reconstruct the broad pattern of diversification of NF-κB proteins over the course of animal evolution. For that reason, we sampled the major metazoan lineages (Deuterostomia, Ecdysozoa, Lophotrochozoa, Cnidaria, Ctenophora, Porifera) in addition to closely related non-metazoan taxa (Choanoflagellata, Filasterea; *see* **Note 4**).

2. Based on your research focus, identify the gene families and subfamilies you wish to sample. To place the origin of NF-κB in its appropriate phylogenetic context, it is necessary to include related RHD proteins, namely, the NFAT proteins, which can serve as an out group. It is also important to include representatives from both NF-κB subfamilies: the NF-κB-type and the Rel-type proteins.

3. At NCBI [30], obtain the amino acid sequences of RHD proteins to serve as query sequences for identifying putative homologues in less well-studied taxa. For example, the human p105, RelA/p65, and NFAT5 proteins (NCBI Reference Sequences

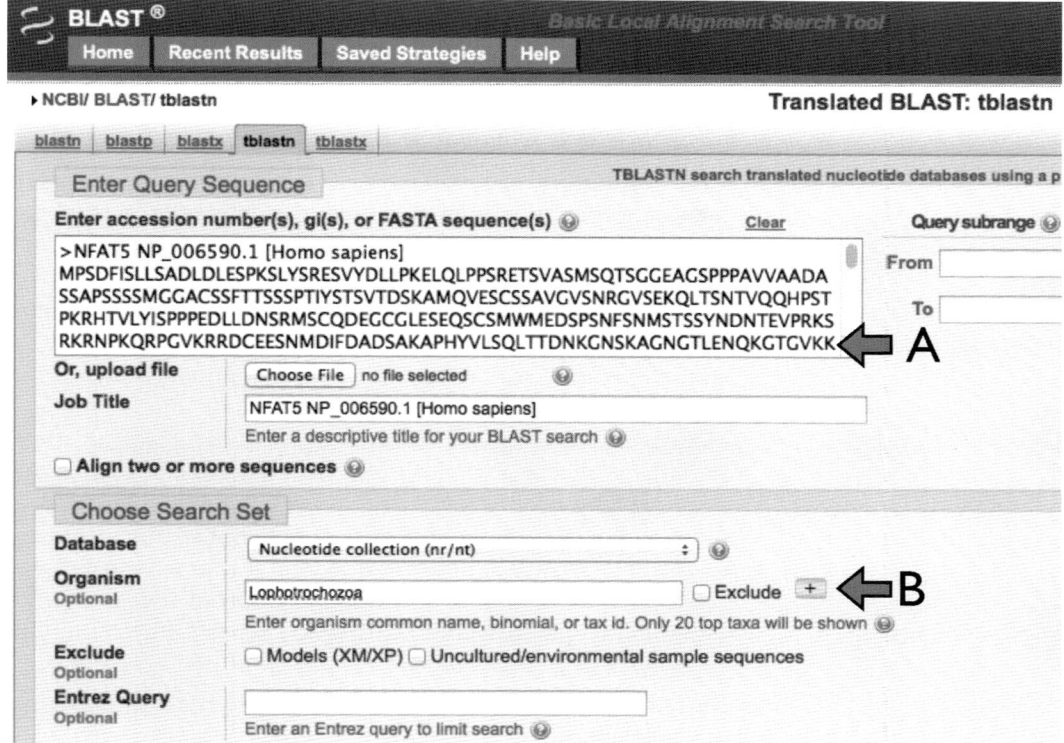

Fig. 3 BLAST settings used at NCBI. (*A*) The text box into which the query sequence is pasted. (*B*) Taxonomic restrictions placed upon the search. In this example, only those matching sequences from lophotrochozoan animals will be returned by the BLAST search

NP_003989.2, NP_001230914, and NP_006590.1) were used to search for NF-κB, Rel, and NFAT proteins, respectively.

4. Through NCBI's BLAST suite, conduct taxonomically restricted BLASTp or tBLASTn searches in order to identify homologues of NF-κB, Rel, and NFAT proteins from all organismal lineages under study. BLASTp can be used for organisms that boast a complete repertoire of protein sequences, but tBLASTn will enable you to identify homologues in those taxa for which only DNA or RNA sequences are in the database. On the BLAST interface page, under "Choose Search Set," enter the name of a taxon to which you wish to restrict a given search (e.g., "Lophotrochozoa"; Fig. 3).

5. Choose a species to represent each higher taxon (e.g., the fruit fly *Drosophila melanogaster* can represent insects or arthropods). It is best to select species with fully sequenced genomes and/or well-characterized transcriptomes so that you can obtain all of the RHD proteins encoded in the genome.

6. As preliminary confirmation of homology, perform a reciprocal BLAST search. Utilize the putative RHD sequence you have

identified in the uncharacterized taxon as a query sequence to search the genome of the well-characterized species from which the original query sequence derived (e.g., *Homo sapiens*). Verify that your top hits in this reciprocal BLAST search are RHD proteins, ideally including the original query sequence.

7. Keep careful records of your BLAST searches and results including the query sequence used, the database searched, and the search protocol utilized (e.g., tBLASTn).

3.2 Identification of Conserved Motifs

1. Compile a "training set" of protein sequences that will be used to identify possible conserved motifs. Do not include closely related protein sequences (e.g., orthologous proteins from recently diverged sister species). The goal of this analysis is to identify regions that are highly conserved as a result of strong stabilizing selection. Inclusion of closely related sequences will result in false positives (i.e., regions of apparent conservation that are due to recent common ancestry, not strong stabilizing selection). In the example provided here, the two most closely related sequences in our training set are from species that are estimated to have diverged from each other 240–360 million years ago [17]. All sequences should be in FASTA format.

2. Run MEME [23] on the training set. Paste the sequences into the appropriate text box on the sequence submission form (Fig. 4). By default, MEME limits its search to the three most statistically significant motifs it identifies (below a certain maximum allowable *E*-value), and it identifies only one occurrence of each motif in each sequence. In our experience, it is usually best to set the "maximum number of motifs" to ~20, and in the case of proteins such as NF-κB, which are known to contain repeated sequence elements (i.e., ankyrin repeats), it is essential to allow the program to identify "any number of repetitions" of a given motif in a single protein sequence. Select "start search." Save the MEME output.

3. If you wish to evaluate closely related protein sequences for the presence of conserved motifs, compile a more inclusive set of sequences in a text file, and search these sequences for the motifs previously identified by MEME. Unlike the training set, this "search set" can contain closely related sequences because the motifs have already been defined using distantly related sequences. All sequences should be in FASTA format.

4. Run MAST [25]. On the data submission form (Fig. 5), select the MEME output text file as "your motif file." Select the sequences you wish to analyze as "your FASTA sequence file." Hit the "start search" button. Save the MAST output. At this time, the visual representation of conserved motifs generated by MAST cannot be exported in a standard graphics file format. The webpage can be exported as a .pdf document; the

Multiple Em for Motif Elicitation

Version 4.9.1

Use this form to submit DNA or protein sequences to MEME. MEME will analyze your sequences for similarities among them and produce a description (motif) for each pattern it discovers.

Data Submission Form

Required

Your e-mail address:

irf3@bu.edu

Re-enter **e-mail address:**

irf3@bu.edu

Please enter the sequences which you believe share one or more
motifs. The sequences may contain no more than **60000** characters
total total in any of a large number of formats.

Enter the name of a file containing the sequences here:

Choose File | no file selected | Clear

or
the actual sequences here (Sample Protein Input Sequences):

>NFkB[Ciona intestinalis] NP_001071772.1
MSDQSLVLHQSTRENMFPEENGEPYLEIIENPKSRGF
RFRYTCEGPSHGGIPGGSSDKNKKTFPAVKICNYQGY
ARIVVQLVTNEENPRLHPHSLVGKQCQNGICTVQCGP
KDMTATFPNLGIQHVTKKNVATILEERYIAAEMQLSSIN ←C

How do you think the occurrences of a single motif are distributed among the sequences?

○ **One per sequence**

○ **Zero or one** per sequence

⦿ **Any number** of repetitions ←A

MEME will find the optimum width of each motif within the limits you specify here:

6 | **Minimum** width (>= 2)

50 | **Maximum** width (<= 300)

20 | Maximum number of motifs to find ←B

Fig. 4 MEME interface showing the search conditions specified in this study. (*A*) The default number of occurrences of a single motif was increased to "any number." (*B*) The maximum number of motifs was increased to 20. (*C*) The amino acid sequences in FASTA format have been pasted into the sequence input box

image can be saved as a graphics file using a screen-capture utility, or the motif depictions can be redrawn using a graphics or presentation program, as we have done here (Fig. 1).

3.3 Prepare Sequences for Alignment and Phylogenetic Analysis

1. Using the conserved motifs as a guide, remove nonhomologous regions of the proteins that are to be aligned (*see* **Note 5**). In the example provided here, we focused our phylogenetic analysis on the Rel homology domain (which is shared by NF-κB and NFAT proteins). Therefore, we truncated the protein sequences immediately after motif 11 (in the case of the NF-κB proteins) or 31 amino acids downstream of motif f (in the case of NFAT proteins; *see* Fig. 1). For demonstration purposes, we culled the dataset to 28 sequences. We intentionally eliminated the NF-κB sequence from *Hydra* because it is highly divergent, and therefore, it could generate a phylogenetic artifact known as "long branch attraction" (*see* **Note 6**). Save the file in FASTA format.

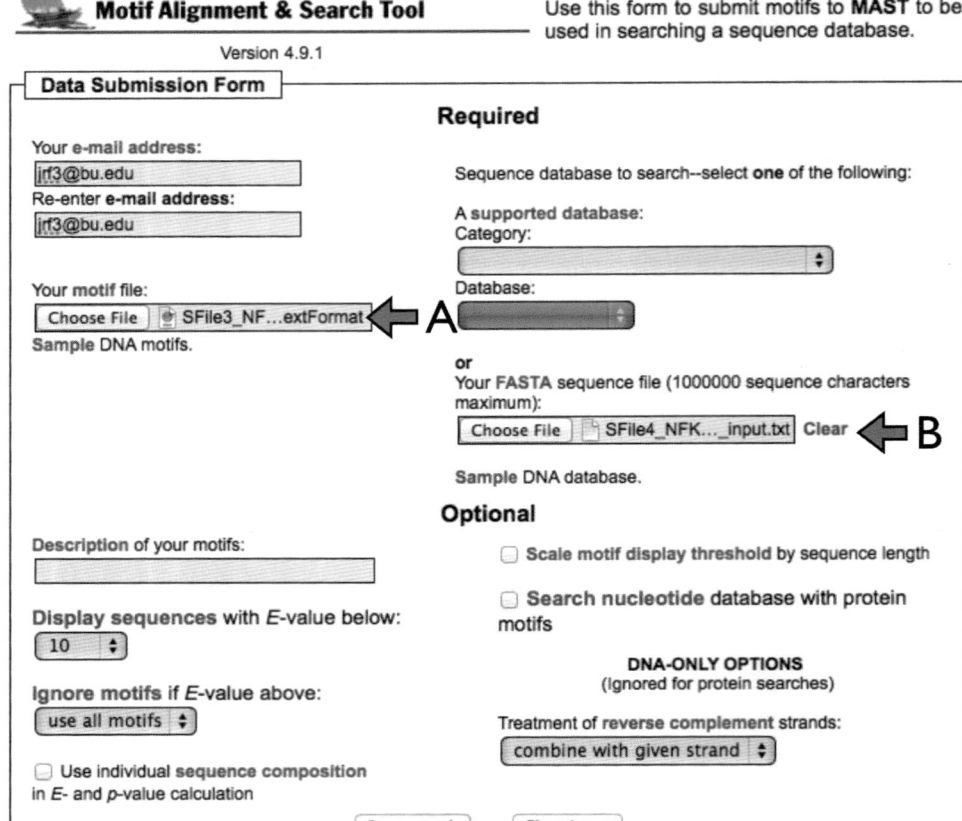

Fig. 5 MAST interface showing the search conditions specified in this study. (*A*) For "motif file," indicate the file containing the motifs identified by MEME. (*B*) For "Your FASTA sequence file," indicate the file containing the amino acid sequences that you intend to search for the presence of the already identified motifs

3.4 Generate a Rooted Phylogenetic Tree with Support Indices at the Nodes

1. At Phylogeny.fr (http://www.phylogeny.fr), select "One Click" under Phylogeny Analysis in the toolbar.

2. On "Data and Settings" page (Fig. 6), deselect "Use the Gblocks program to eliminate poorly aligned positions and divergent regions."

3. Paste the input sequences into the text box (Fig. 6).

4. Hit the Submit button.

5. If desired, perform dynamic editing on the Tree Rendering using the tools provided. For example, you can rename the sequences, re-root the tree at a particular internal branch, or color or rotate the branches. None of these changes alter the

Fig. 6 User interface for the "One Click" Mode at Phylogeny.fr. (*A*) The text box into which the amino acid sequences are pasted. (*B*) Deselect the option to use Gblocks

topology of the tree, but re-rooting the tree changes the interpretation by indicating which internal branch is the most ancient branch on the tree.

6. Export the tree in a graphics format (e.g., .png or .svg) to prepare a figure for publication (as in Fig. 7).

4 Notes

1. For certain "non-model" species, extensive datasets produced via next-generation sequencing have been deposited at databases other than NCBI. In this report, we conducted BLAST searches at these other databases (Table 1) so that our analysis would include RHD-containing proteins from the corals *Acropora millepora* and *Platygyra carnosus* and the sea anemone *Edwardsiella lineata*.

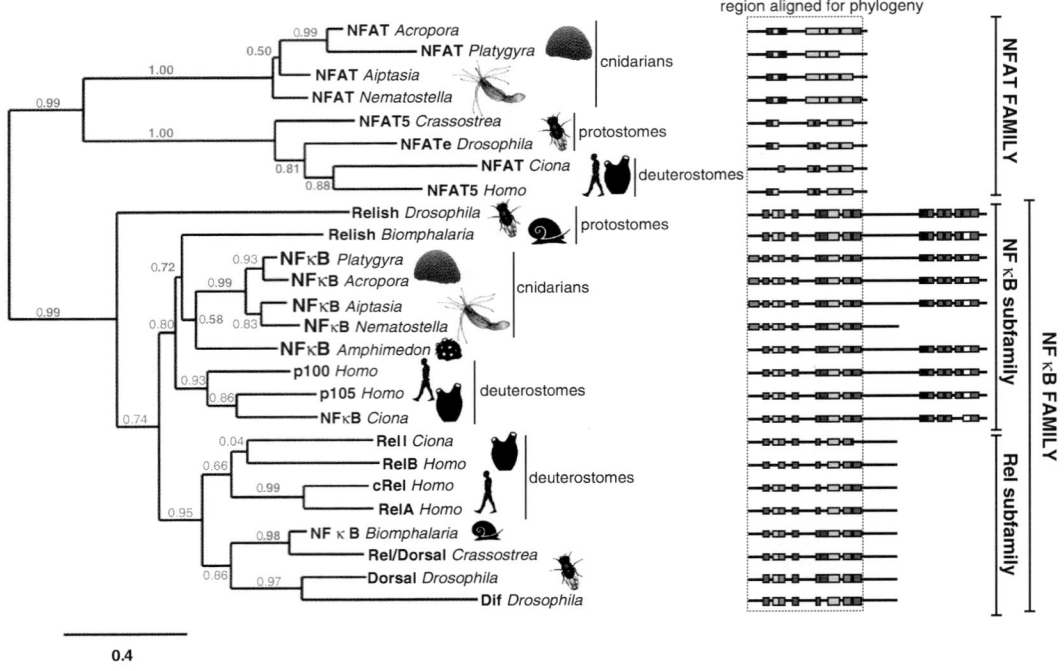

Fig. 7 Phylogeny of selected RHD-containing proteins based on maximum likelihood analysis of amino acid sequences from the RHD itself. To the *right* of the sequence names are the motif diagrams produced using MEME. The *dashed box* indicates the region of the protein used in the phylogenetic analysis. *Red numbers* at the nodes indicate the support for the given node using approximate log-likelihood ratio tests [35]. The highest possible support for a given grouping is 1.00. The lengths of horizontal branches are proportionate to the number of amino acid substitutions that have occurred along that branch. The scale at the *lower left* indicates branch length in units of expected amino acid substitutions per residue changes per residue. The NFAT family and the NF-κB and Rel subfamilies are well supported according to the aLRT values at their respective nodes (0.99, 0.80, and 0.95, respectively), although the *Drosophila* Relish protein did not group with other members of the NF-κB family with which it is traditionally placed. Within the NFAT family, there is a strong support for the monophyletic origin of cnidarian sequences (the corals *Acropora* and *Platygyra* and the anemones *Nematostella* and *Edwardsiella*) and triploblast sequences (human, sea squirt, fruit fly, and oyster). Within the NF-κB family, there is a strong support for the grouping of cnidarian (0.99) as well as deuterostome sequences (0.93), although the protostome Relish sequences (*Drosophila* and *Biomphalaria*) do not group together. Within the Rel subfamily, the deuterostome sequences and protostome sequences form mutually exclusive groups, as expected. Of note, there are no basal metazoan sequences (sponge or cnidarian) in the portion of the tree comprising the Rel subfamily

2. A variety of free computer programs for multiple sequence alignment (MSA), phylogeny construction, and tree manipulation are available via the web. Here, we take advantage of the "One Click" mode at Phylogeny.fr, which caters to the non-specialist and involves no shuttling of input and output between programs. Given an input sequence file in FASTA format, the sequences are aligned using MUSCLE; a maximum likelihood phylogeny is constructed using PhyML, and the resultant tree can be viewed and edited using TreeDyn.

Table 1
Taxa investigated for the presence of NF-κB and NFAT proteins

Higher taxa		Species	Database searched	NF-κB proteins	NFAT proteins
Metazoa					
Triploblastica	Deuterostomia	*Ciona intestinalis*	NCBI	2	1
		Homo sapiens	NCBI	5	5
	Ecdysozoa	*Drosophila melanogaster*	NCBI	3	1
		Caenorhabditis elegans	NCBI	0	0
	Spiralia	*Biomphalaria glabrata*	NCBI	2	0
		Crassostrea gigas	NCBI	1	1
Cnidaria	Actinaria	*Aiptasia pallida*	NCBI	1	1
	(Anemones)	*Edwardsiella lineata*	[17]	1	1
		Nematostella vectensis	NCBI	1	1
	Scleractinia	*Acropora digitifera*	[21]	1	1
	(Hard corals)	*Platygyra carnosus*	[19]	1	1
	Corallimorpharia	*Discosoma* sp.	Unpubl	1	1
	(Mushroom corals)				
Porifera	(Sponges)	*Amphimedon queenslandica*	NCBI	1	0
Ctenophora	(Comb jellies)	*Mnemiopsis leidyi*	NCBI	0	0
Choanoflagellata		*Monosiga brevicollis*	NCBI	0	0
Filasterea		*Capsaspora owczarzaki*	NCBI	1*	

* may not represent a bona fide NF-kappaB protein

TreeDyn also allows the tree to be exported in a variety of graphics file formats.

3. There are a number of alternative approaches for reconstructing phylogenetic networks from molecular sequences, and there is an enormous literature comparing their relative merits [31]. Here we employ the maximum likelihood approach, because simulation studies have shown it to be a robust method for the recovery of the true tree. PhyML is one among several online computer programs for reconstructing maximum likelihood trees from molecular sequences [28].

4. Over 99 % of extant animals belong to the Triploblastica (=Bilateria), and this clade comprises three major lineages [7, 32]: Deuterostomia, Ecdysozoa, and Lophotrochozoa (=Spiralia). Among diploblastic animals, the Cnidaria (sea anemones, corals, jellyfishes, and hydras) are supported as the sister group of triploblasts in a number of recent phylogenomic studies [7, 12, 13, 15]. Animal phyla that diverged prior to the cnidarian-triploblast common ancestor include the Porifera (sponges) and the Ctenophora (comb jellies). Two closely related out groups to the Metazoa were sampled in this study: Choanoflagellata and the Filasterea.

	Gene	Species	*DNA-binding sites ★ ★★ ★★ ★★★★	E-value
Motif 1 (NF-κB specific)	NFkB1	*Homo*	YLQILEQPKQRGFRFRYVCEGPSHGGLPG	4.22E-34
	NFkB2	*Homo*	YLVIVEQPKQRGFRFRYGCEGPSHGGLPG	2.87E-34
	RelA	*Homo*	YVEIIEQPKQRGMRFRYKCEGRSAGSIPG	1.65E-33
	RelB	*Homo*	HLVITEQPKQRGMRFRYECEGRSAGSILG	1.16E-30
	cRel	*Homo*	YIEIIEQPRQRGMRFRYKCEGRSAGSIPG	3.94E-32
	NFkB	*Ciona*	YLEIIENPKSRGFRFRYTCEGPSHGGIPG	7.5E-33
	rel1	*Ciona*	VLEIVEQPKQRGMRFRYECEGRSAGSIPG	3.53E-32
	Dorsal	*Drosophila*	YVKITEQPAGKALRFRYECEGRSAGSIPG	2.18E-26
	Dif	*Drosophila*	HLRIVEEPTSNIIRFRYKCEGRTAGSIPG	2.37E-24
	NFkB	*Biomophalaria*	YVEILEQPKSRGLRFRYECEGRSAGSVPG	3.18E-31
	Relish	*Biomophalaria*	YVVITEQPQQRGFRFRYECEGPSHGGLQG	5.26E-31
	NFkB	*Hydra*	YLKIERQPRKYGYRFRYKTEGVCHGGILA	5.94E-22
	NFkB-C	*Nematostella*	YLEILEQPKPRGFRFRYPCEGPSHGGLPG	6.68E-36
	NFkB-S	*Nematostella*	YLEILEQPKPRGFRFRYPSEGPSHGGLPG	6.72E-32
	NFkB	*Edwardsiella*	YLEILEQPRPRGFRFRYPCEGPSHGGLPG	1.71E-34
	NFkB	*Aiptasia*	YLEILEQPKSRGFRFRYPCEGPSHGGLPG	3.51E-36
	NFkB	*Acropora*	YLEILEQPKQRGFRFRYPCEGPSHGGLPG	3.25E-37
	NFkB	*Platygyra*	YMEILEQPKQRGFRFRYPCEGPSHGGLPG	6.68E-36
	NFkB	*Discosoma*	YLEILEQPKQRGFRFRYPCEGPSHGGLPG	3.25E-37
	NFkB	*Amphimedon*	RLEIVEQPKSRGFRFRYDCEGQSHGGLPG	1.4E-31
Motif A (NFAT-specific)	NFAT1	*Homo*	HRAHYETEG-SRGAVKAPTGGH	6.12E-19
	NFAT3	*Homo*	HRAHYETEG-SRGAVKAAPGGH	2.14E-18
	NFAT4	*Homo*	HRAHYETEG-SRGAVKASTGGH	4.45E-20
	NFAT5	*Homo*	HRARYLTEG-SRGSVKDRTQQG	8.52E-16
	NFAT	*Nematostella*	PEENYTSEG-CRGPIHGSSDNT	2.91E-18
	NFAT	*Edwardsiella*	YRARYESEG-CRGPIHGSSDNT	4.62E-27
	NFAT	*Aiptasia*	YRARYESEG-CRGPIHGSSEST	6.3E-25
	NFAT	*Acropora*	YRARYESEG-CRGPIHGSQENT	4.62E-27
	NFAT	*Platygyra*	YRARYESEG-CRGPIHGSQDNT	1.6E-27
	NFAT	*Discosoma*	YRARYESEG-CRGPIHGSKDNT	6.63E-27
	NFAT	*Hydra*	YRARYESEG-CKGPIHGSTENC	2.35E-24

★ ★ ★★   ★★★

Fig. 8 An alignment of amino acid sequences in Motif 1, which is NF-κB-specific, and motif A, which is NFAT specific. These overlapping motifs encompass a conserved cluster of known DNA-binding sites (gray shading*), including nine residues in the case of Motif 1 (see cd07935 in NCBI Conserved Domains Database) and seven residues in the case of Motif A (cd07881). The *E*-value represents the probability that a sequence with as much similarity to the consensus motif would have occurred by chance

5. Standard methods for phylogenetic analyses of proteins depend upon the accurate alignment of homologous residues in the proteins under consideration. At best, the nonhomologous regions are uninformative, but at worst, these regions can confound multiple alignment programs. For that reason, we recommend the removal of nonhomologous protein regions prior to alignment.

6. Long branch attraction is a phenomenon where two or more highly distinctive but unrelated sequences cluster together in the tree to produce "anomalous phylogenetic groupings" [33]. The effect can be minimized by removing highly divergent sequences from a phylogenetic analysis, and where possible,

substituting a more slowly evolving sequence that represents the same gene from the same taxon. For example, in this analysis, we could substitute a less divergent NF-κB sequence from another hydrozoan (other than *Hydra*) if it was available.

Acknowledgments

Research in the authors' laboratories on the evolution of NF-κB is supported by NSF grants MCB-0920461 and IOS-1354935.

References

1. Gilmore TD (2006) Introduction to NF-κB: players, pathways, perspectives. Oncogene 25:6680–6684

2. Sullivan JC, Kalaitzidis D, Gilmore TD, Finnerty JR (2007) Rel homology domain-containing transcription factors in the cnidarian *Nematostella vectensis*. Dev Genes Evol 217:63–72

3. Gilmore TD, Wolenski FS (2012) NF-κB: where did it come from and why? Immunol Rev 246:14–35

4. Gauthier M, Degnan BM (2008) The transcription factor NF-κB in the demosponge *Amphimedon queenslandica*: insights on the evolutionary origin of the Rel homology domain. Dev Genes Evol 218:23–32

5. Miller DJ, Hemmrich G, Ball EE, Hayward DC, Khalturin K, Funayama N, Agata K, Bosch TC (2007) The innate immune repertoire in cnidaria—ancestral complexity and stochastic gene loss. Genome Biol 8:R59

6. Giribet G (2008) Assembling the lophotrochozoan (=spiralian) tree of life. Philos Trans R Soc Lond B Biol Sci 363:1513–1522

7. Aguinaldo AM, Turbeville JM, Linford LS, Rivera MC, Garey JR, Raff RA, Lake JA (1997) Evidence for a clade of nematodes, arthropods and other moulting animals. Nature 387:489–493

8. Irazoqui JE, Urbach JM, Ausubel FM (2010) Evolution of host innate defence: insights from *Caenorhabditis elegans* and primitive invertebrates. Nat Rev Immunol 10:47–58

9. Ryzhakov G, Teixeira A, Saliba D, Blazek K, Muta T, Ragoussis J, Udalova IA (2013) Cross-species analysis reveals evolving and conserved features of the nuclear factor κB (NF-κB) proteins. J Biol Chem 288:11546–11554

10. Sullivan JC, Wolenski FS, Reitzel AM, French CE, Traylor-Knowles N, Gilmore TD, Finnerty JR (2009) Two alleles of NF-κB in the sea anemone *Nematostella vectensis* are widely dispersed in nature and encode proteins with distinct activities. PLoS One 4:e7311

11. Wolenski FS, Garbati MR, Lubinski TJ, Traylor-Knowles N, Dresselhaus E, Stefanik DJ, Goucher H, Finnerty JR, Gilmore TD (2011) Characterization of the core elements of the NF-κB signaling pathway of the sea anemone *Nematostella vectensis*. Mol Cell Biol 31:1076–1087

12. Ryan JF, Pang K, Schnitzler CE, Nguyen AD, Moreland RT, Simmons DK, Koch BJ, Francis WR, Havlak P, Program NCS et al (2013) The genome of the ctenophore *Mnemiopsis leidyi* and its implications for cell type evolution. Science 342:1242592

13. King N, Westbrook MJ, Young SL, Kuo A, Abedin M, Chapman J, Fairclough S, Hellsten U, Isogai Y, Letunic I et al (2008) The genome of the choanoflagellate *Monosiga brevicollis* and the origin of metazoans. Nature 451:783–788

14. Richter DJ (2013) The gene content of diverse choanoflagellate illuminates animal origins. University of California, Berkeley, CA

15. Suga H, Chen Z, de Mendoza A, Sebe-Pedros A, Brown MW, Kramer E, Carr M, Kerner P, Vervoort M, Sanchez-Pons N et al (2013) The *Capsaspora* genome reveals a complex unicellular prehistory of animals. Nat Commun 4:2325

16. EdwardsiellaBase—*Edwardsiella lineata* Genomics Database [http://www.edwardsiellabase.org]

17. Stefanik DJ, Lubinski TJ, Granger BR, Byrd AL, Reitzel AM, DeFilippo L, Lorenc A, Finnerty JR (2014) Production of a reference transcriptome and transcriptomic database (EdwardsiellaBase) for the lined sea anemone, *Edwardsiella lineata*, a parasitic cnidarian. BMC Genomics 15:71

18. Sun J, Chen Q, Lun JC, Xu J, Qiu JW (2013) PcarnBase: development of a transcriptomic database for the brain coral *Platygyra carnosus*. Mar Biotechnol (NY) 15:244–251

19. PCarnBase—Transcriptomic Database for *Platygyra carnosus* [http://www.comp.hkbu.edu.hk/~db/PcarnBase/]

20. Shinzato C, Shoguchi E, Kawashima T, Hamada M, Hisata K, Tanaka M, Fujie M, Fujiwara M, Koyanagi R, Ikuta T et al (2011) Using the Acropora digitifera genome to understand coral responses to environmental change. Nature 476:320–323

21. *Acropora digitifera* genome browser [http://marinegenomics.oist.jp/acropora_digitifera]

22. Bailey TL, Elkan C (1994) Fitting a mixture model by expectation maximization to discover motifs in biopolymers. Proceedings of the second International conference on intelligent systems for molecular biology. pp 28–36

23. MEME. Multiple Em for Motif Elicitation [http://meme.nbcr.net/meme/cgi-bin/meme.cgi]

24. Bailey TL, Gribskov M (1998) Combining evidence using p-values: application to sequence homology searches. Bioinformatics 14:48–54

25. MAST. Motif Alignment and Search Tool [http://meme.nbcr.net/meme/cgi-bin/mast.cgi]

26. Edgar RC (2004) MUSCLE: multiple sequence alignment with high accuracy and high throughput. Nucleic Acids Res 32:1792–1797

27. Phylogeny.fr Robust Phylogenetic Analysis for the Non-Specialist/TreeDyn 198.3 [http://phylogeny.lirmm.fr/phylo_cgi/one_task.cgi?task_type=treedyn]

28. Criscuolo A (2011) morePhyML: improving the phylogenetic tree space exploration with PhyML 3. Mol Phylogenet Evol 61:944–948

29. Chevenet F, Brun C, Banuls AL, Jacq B, Christen R (2006) TreeDyn: towards dynamic graphics and annotations for analyses of trees. BMC Bioinformatics 7:439

30. National Center for Biotechnology Information [http://www.ncbi.nlm.nih.gov/]

31. Hall BG (2005) Comparison of the accuracies of several phylogenetic methods using protein and DNA sequences. Mol Biol Evol 22:792–802

32. Halanych KM, Bacheller JD, Aguinaldo AM, Liva SM, Hillis DM, Lake JA (1995) Evidence from 18S ribosomal DNA that the lophophorates are protostome animals. Science 267:1641–1643

33. Anderson FE, Swofford DL (2004) Should we be worried about long-branch attraction in real data sets? Investigations using metazoan 18S rDNA. Mol Phylogenet Evol 33:440–451

34. Marchler-Bauer A, Lu S, Anderson JB, Chitsaz F, Derbyshire MK, DeWeese-Scott C, Fong JH, Geer LY, Geer RC, Gonzales NR et al (2011) CDD: a Conserved Domain Database for the functional annotation of proteins. Nucleic Acids Res 39:D225–D229

35. Anisimova M, Gascuel O (2006) Approximate likelihood ratio test for branches: a fast, accurate and powerful alternative. Syst Biol 55:539–552

Chapter 38

Studying NF-κB Signaling with Mathematical Models

Simon Mitchell, Rachel Tsui, and Alexander Hoffmann

Abstract

Mathematical modeling of NF-κB signaling can be employed to understand how the network of molecular interactions leads to signaling phenomena observed experimentally. Model construction is a challenging process; however, existing models can be utilized and can provide a great deal of insight quickly and inexpensively. The simulation of various inputs and the identification of potential therapeutic targets using the mathematical model are detailed here.

Key words Computational, Modeling, Simulation, Mathematical, Systems, Dynamical, Theoretical

1 Introduction

Mathematical models can be used as a virtual laboratory, to perform in silico experimentation and to replace, complement, and improve experimental approaches in the wet lab. The NF-κB signaling network is complex, and understanding the individual components and interactions in isolation only provides limited progress towards the goal of understanding its functions and enabling effective clinical intervention. Combining knowledge of network components with knowledge of their interactions through simulations allows emergent behavior to be predicted and explained in a rigorous manner.

There are many different modeling approaches and software packages available (including MATLAB, Mathematica, and COPASI); however, the principles used in constructing a model to ensure utility are consistent, regardless of the framework used. In this chapter we detail how a model of the NF-κB signaling networks can be constructed, validated, and utilized to gain the best insight from experimental studies and inform future experimentation. There is much knowledge to be gained from existing models, and constructing a new model should only be considered after attempting to utilize existing models. Therefore, in Section 3.2, a published model [1] is used to gain insight into the NF-κB signaling network.

Michael J. May (ed.), *NF-kappa B: Methods and Protocols*, Methods in Molecular Biology, vol. 1280,
DOI 10.1007/978-1-4939-2422-6_38, © Springer Science+Business Media New York 2015

2   Materials

1. A computer with access to appropriate software (detailed below).

2. Experimentally derived parameters, often based on extensive searches or expert understanding of the literature.

3. Experimental or clinical findings, which the model is required to reproduce or account for.

3   Methods

Here we describe a computational approach to improve understanding of NF-κB signaling. The general approach to model construction, validation, and application can be applied to any biological system; however, we illustrate this general approach with the specific steps to simulate NF-κB signaling (*see* **Note 1**).

***3.1 Model
Construction
(see Notes 2 and 3)***

1. *Define the question to be addressed by the model.* Modeling can provide insight into a wide variety of biological questions. The type of question that the model is required to answer must be identified first as this informs the model-building process. For example, a model designed to investigate physiological scale processes such as the control of fever may not provide insight into the importance of dimerization of Rel proteins.

   NF-κB signaling models [2] can provide insight into many of the important open questions in immune signaling. Some phenomena that can be investigated using such models are dose–response relationships [3, 4], dynamic control as revealed by time-course studies [1, 5, 6], the impact of cross talk [7, 8], analysis of sensitivities, and identification of points of control [9, 10], which may provide viable therapeutic targets. Questions beyond the scope of these models include the control of gene expression profiles, the physiological functions of cytokines, or the cellular decisions to divide or die.

   When the required scope, detail, and type of model (*see* below **steps 2–4**) have been identified, and an appropriate model has been found or constructed, a fundamental question that can be addressed is whether the known molecular mechanisms represented in the model are sufficient to account for cell biological or physiological level phenomena. When the answer is yes, the model can be used to explore the emergent system properties. When the answer is no, the model can direct experimentation to identify additional molecular mechanisms that render it sufficient.

2. *Define the required scope of the model.* The scope of an appropriate model is largely determined by the question of interest.

The scope determines what the input(s) and output(s) of the model are. Inputs and outputs are generally measureable quantities of metabolites ("molecular species"). All molecular species upstream of the input and downstream of the output are outside the scope of the model. The scope of a model could extend to a whole cell or organism or be restricted to small regulatory circuits networks. While it is tempting to choose an ambitious scope, this can lead to a poor model that is insufficiently determined by insufficient data. It is best practice to ensure that the initial scope of modeling work is the minimal required to provide insight and only expand the model's scope once initial models have been shown to be predictive. The practice of generating preliminary data before embarking on extensive work has been common in wet lab studies and remains important with computational approaches.

NF-κB signaling is a highly complex biological system with a wide variety of potentially important cross-talking pathways. To simulate every system that could potentially control NF-κB, via the IKK hub, is unfeasible as this would extend to a large proportion of cellular pathways. As a result, IKK activity profiles were chosen as inputs to the NF-κB simulation. Upstream signaling from tumor necrosis factor (TNF) receptor-mediated IKK activation was not included in initial models was later added as an additional regulatory module [4], thereby expanding the scope of the resulting model. NF-κB levels were identified as appropriate output from the model as this was highly informative without introducing the challenges associated with simulating the extensive NF-κB-induced gene expression profiles.

3. *Define the model detail required.* Within the identified scope of the model, the level of detail included must be decided; this can be considered the "graininess" of the model. This is a measure of how closely the underlying biological mechanisms identify to the mathematical representation constructed. It is often necessary to group a multiple-step process into a single-compound reaction. This is straightforward when one reaction has the highest control over the rate (the rate-limiting step) in all relevant conditions.

Components required to recreate the behavior being investigated, and answer the question of interest, should be included without adding poorly understood components that add unnecessary complexity or reliance on under-determined parameters. The key components of NF-κB signaling were identified as:

(a) NF-κB and its localization

(b) Multiple IκB dimers

(c) IKK activity curves

(d) mRNA levels for each protein

The scope chosen permits detailed investigation into the effect of temporal features of IKK activity on NF-κB activation. By including mRNAs we were able to investigate transcriptional regulation and utilize the large amount of mRNA data for model parameterization and validation. The multiple monomers, which combine to form the family of NF-κB dimers, were not included in initial models, as understanding general temporal NF-κB profiles was the priority.

IKK was modeled as an enzyme that degrades IκB directly, whereas the underlying process actually involves ubiquitination and proteasomal degradation [11]. By assuming that IKK-mediated phosphorylation of IκB is the rate-limiting step of IκB degradation, the intricate ubiquitin-dependent proteasomal degradation pathway could be represented by a single reaction.

4. *Decide on the type of model required.* We distinguish here between three types of mechanistic models. (We do not consider statistical models here.) (1) Logical modeling does not rely on kinetic parameters and may be most appropriate if the system is poorly characterized, but can provide insights about steady-state control. (2) A differential equation-based approach will make best use of kinetic parameter information and can provide insights on the dynamic time evolution of quantitative concentrations and fluxes. (3) A stochastic model accounts for the stochastic nature of individual molecular reactions and is the most detailed. While the primary consideration in choosing the appropriate model strategy is the biological question of interest, the trade-off between the need for high-quality, highly detailed data and accuracy of output contributes to choosing the right modeling strategy [12].

The outputs from differential equation-based models are deterministic representations of a system's average behavior. This deterministic result may differ from any single small-scale (e.g., single cell) experiment or simulation due to the effect of the noise present in all biological systems. This limitation of deterministic modeling is usually unimportant as most commonly used experimental techniques also produce data of this type. For simulations of systems with small numbers of molecules, stochastic modeling techniques are most appropriate as they capture the importance of noise in these systems. Recent studies of newly divided cells (sibling analysis) have shown that cell-to-cell variability is mainly attributed to extrinsic variability such as initial conditions and rate constants rather than intrinsic noise [13]. This type of cell-to-cell variability as a result of noise can be incorporated into an ordinary differential equation (ODE) modeling framework through sampling initial conditions.

For modeling of NF-κB signaling, an ODE-based approach is possible due to many of the individual components of the networks being well characterized and a number of relevant knockout experimental systems being available for model validation.

5. *Investigate if an appropriate model already exists.* This can be done through a standard literature search or through querying a repository of models such as the BioModels Database (www. Biomodels.net) [14]. If an appropriate model exists (i.e., one that answers the question of interest), then this should be utilized and Section 3.2 details this process. For investigation of the temporal control of NF-κB activation provided by the isoforms of IκBs, models from our laboratory are appropriate and employed in Section 3.2.

6. *Identify data available for model construction.* For the construction of a new model, its quality, and therefore the quality of the insight it provides, improves with the amount and quality of data available. Data useful for modeling can fall into two categories: (a) physicochemical results, which are used in model parameterization, and (b) emergent properties/physiological results, which are used in parameter fitting and model validation.

   There are multiple sources of parameters for modeling including existing literature, databases of reaction kinetics (such as SABIO-RK [15]), and quantitative experimental techniques performed in the wet lab.

   If more high-quality data than expected are available, then the scope and graininess of the proposed model (**steps 3** and **4**) can be increased to make best use of these data. Similarly if the data required to parameterize a model of the scope/graininess desired are not found, then these should be decreased. Constructing a smaller, good quality model that can later be expanded is preferable to attempting an over ambitious simulation that may not be feasible.

   For NF-κB signaling models, a wide variety of experimentally derived parameters were identified. Protein and mRNA half-lives, steady-state concentrations, and binding affinities were all measured for many of the interactions to be modeled. NF-κB activity time courses in IκB knockout systems were generated for the purpose of parameter fitting and model validation.

7. *Construct a network diagram of the system.* A network diagram of the system being modeled is a useful first step for model construction. An accurate network diagram represents a consensus of the current understanding and ensures that the mathematical formulation being constructed closely represents the underlying biochemical interactions. Care should be taken that each metabolite and reaction to be modeled is represented by a single shape or line in the diagram. Identifying diagrammatic entities with mathematical entities (equations or terms within equations) allows the model to be used as a protocol for model construction. While any representation that accurately represents both the biological and mathematical systems is appropriate, Systems Biology Graphical Notation (SBGN) [16] provides a standardized visual language that may aid communication.

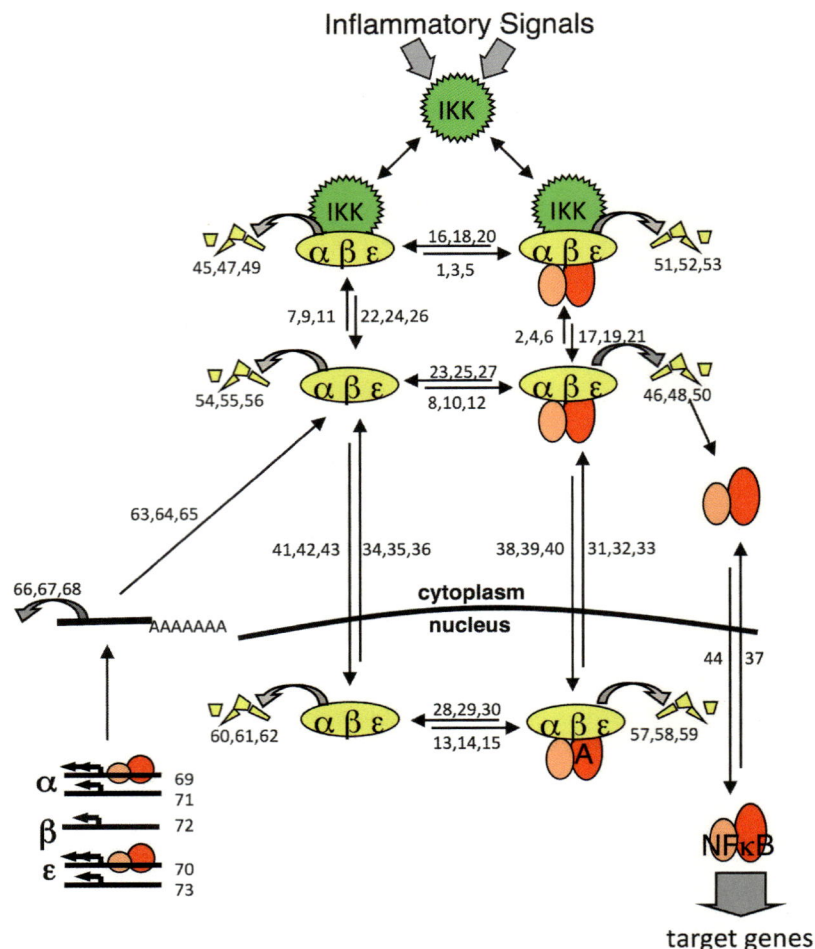

Fig. 1 Diagram representing NF-κB signaling used in the creation of computational simulations. Each colored shape is a metabolite (also known as "molecular species") in the system and an ODE in the mathematical representation. Each *arrow* is a reaction in the system, mathematically represented by terms in the ODEs of metabolites involved in the reaction. The inputs to the system are IKK activation curves, and the outputs are concentrations of free NF-κB, which is capable of binding DNA

A diagrammatic representation of the NF-κB signaling network is given in Fig. 1. This representation of the biochemical network was used to construct the computational model we described in 2002 [3].

8. *Identify an appropriate software framework.* A number of different computational environments and software packages exist for model construction and analysis including MATLAB, COPASI [17], Berkeley Madonna [18], and the SimBiology MATLAB toolbox. Choosing appropriate tools is key to efficient model construction and maximizing the models utility and reusability. Broadly speaking, constructing a model

directly in a programming language such as MATLAB, C++, or Python provides the most versatility for advanced analysis and construction of models that do not conform to common biochemical behavior. Constructing models in such an environment requires more specialized technical expertise and can restrict the models' utility for those without specific programming knowledge. COPASI and the SimBiology MATLAB toolbox provide a more accessible model-building environment, while also providing some checks to ensure that the model constructed is mathematically sensible and biochemically valid. These environments allow those without programming experience to construct and utilize models, but may restrict the model and its analysis to commonly used methods. Berkeley Madonna provides a powerful ODE solving graphical user interface without specifically limiting the system to biochemical simulations.

MATLAB was chosen as the environment for modeling of NF-κB signaling due to the complexity of some of the behavior and inputs required to recreate accurately experimentally observed NF-κB responses. There are well-characterized delays in transcriptional activation of IκBs in response to NF-κB activation; some modeling environments do not support explicit delays, but these can be implemented in MATLAB. To make best use of the experimentally derived IKK activity time courses as inputs to the model, interpolated input curves were required, and this was most easily implemented in MATLAB. To ensure that the complex model constructed for NF-κB signaling was accessible to those without programming experience, a web-based version of the model, which provided an intuitive interface for performing simulations, was published (http://signalingsystems.org/webmodel/).

9. *Construct the mathematical representation.* Using the previously constructed network diagram as a guide, the mathematical representation of the biochemical network should now be input into the software. For ODE-based modeling, as used for NF-κB signaling, this requires construction of an ODE for each metabolite in the system. The terms of each metabolite's ODE represent the reactions that alter the concentration of that metabolite.

For example, the terms of the ODE representing the change in concentration of free IκB as a result of its binding to NF-κB are given in Fig. 2. Mass action kinetics were used for most reactions in the NF-κB signaling network and should be used unless there are experimentally derived data suggesting a more intricate mechanism. Terms should be added to the appropriate ODE for each reaction it is involved in using the parameters identified.

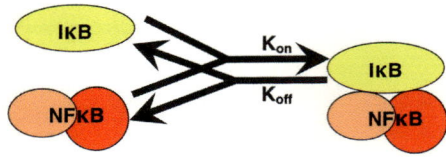

ODE: $\frac{d[I\kappa B]}{dt} = -K_{on}[I\kappa B][NF\kappa B] + K_{off}[I\kappa B\text{-}NF\kappa B]$

MATLAB: delta(IkB) = - kon * IkBa * NFkB + koff * IkBaNFkB

Fig. 2 A diagrammatic representation of a complex binding reaction within the NF-κB signaling system. The reaction is reversible, permitting binding and release of reaction's. An ODE for free IκB was written using the diagram as a guide; binding depletes free IκB, while decomplexation increases free IκB. The MATLAB code representation of the ODE is given

10. *Estimate any parameters that were not experimentally determined.*
    The lack of accurate, experimentally derived, kinetic parameters is a common challenge for the construction of computational models. Any parameter that could not be identified from a search of the literature, or from a database, or through experimentation, must be estimated or derived.

    The most basic form of parameter deduction is used when there are parameters available for all but one of the reactions in which a metabolite partakes. For example, if a metabolite's half-life and steady-state concentration are known, then the expression rate constant can be deduced. This is done by setting the rate of change to zero to represent the steady state, substituting the known parameters, and solving the equation below for the remaining unknown parameter:

$$\frac{d[X]}{dt} = -k_{deg}[X] + k_{exp} = 0$$

Some parameters can be estimated using knowledge of similar systems; for example, parameters from homologous proteins can often provide a guide for parameter estimation.

If some parameters still remain unidentified, then they must be fitted to their most likely value. Many parameters can be constrained within ranges using knowledge from similar systems and common biochemical limits.

If some parameters still remain unknown or broadly constrained, then a parameter-fitting methodology should be employed to find their most appropriate values. There exist a number of algorithms such as Hooke and Jeeves, particle swarm, gradient descent, etc. Many of these algorithms are incorporated into software packages such as COPASI [19].

Despite being complex, the NF-κB signaling model was already highly constrained by previously published parameters. Multiple IκB isoform knockout cell lines were used to provide data for multiple conditions that the model could be fitted to; this ensured that the parameters chosen were valid in a variety of conditions [3]. A number of fitting techniques were trialed to assess which gave the best fit when measured as standard deviation from the experimental results. A random search method was found to perform well, and the parameters identified by this search were subsequently adjusted to improve the qualitative fit regarding frequency and amplitude of oscillations. Standard fitting methods, based around a distance metric (such as root mean square), do not perform well with the NF-κB signaling simulation. The NF-κB response to stimuli is often highly dynamic with some oscillatory behavior; common fitting algorithms can fail to find good fits that are slightly offset in initial time or frequency of oscillation. Feature-based fitting techniques have been found to perform well for the NF-κB signaling network [4, 8]. These methods prioritize features such as a maximal peak at a specified time point that matches experimental observations.

3.2 Utilizing Models

3.2.1 Predicting Response to Stimuli

Once a model has been constructed that closely matches experimental data, it can be used as a tool to provide insight, to make testable predictions, and to target better experimental studies.

Here we demonstrate how the model of Werner et al. [1] can be used to make predictions on how temporally different IKK activity profiles in various genotypes affect the stimulus-specific gene expression program:

1. Select the TNF input curve, by choosing curve 1, from the array of defined input curves. These input values have been quantitated and normalized from IKK immunoprecipitation kinase assays.

```
ikk_curves      = {'TNFp15' 'TNFc' 'LPSp45'};
ikk_curve_num   = 1;
```

2. Run the simulation.

3. Select the lipopolysaccharide (LPS) input curve (curve 3) from the array.

```
ikk_curve_num   = 3;
```

4. Run the simulation and compare the results.

The input curves chosen using the above method can be seen in Fig. 3a; the resulting simulated NF-κB profile can be seen in Fig. 3b. This predicted response could inform experimental protocol to ensure that the correct time points are chosen to observe the dynamics. The simulated response prediction was closely matched in mouse embryonic fibroblasts (MEFs) exposed to TNF and LPS by Werner et al. [1].

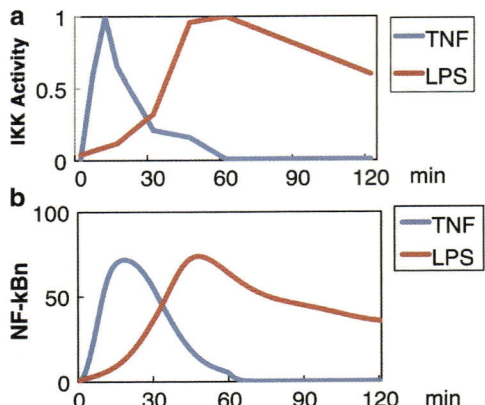

Fig. 3 (**a**) IKK activity input curves for TNF and LPS, quantitated and normalized from IKK immunoprecipitation kinase assays. (**b**) The simulated free NF-κB resulting from each input curve

3.2.2 Investigating Sensitivity to Identify Therapeutic Targets

The simulation can identify potential therapeutic targets for most effectively controlling the systems response to stimuli. A good therapeutic target is able to make large predictable changes to the system with only a small perturbation applied. A target with these properties leads to lower drug dosages and fewer side effects. The effect of perturbations that simulate therapeutic intervention can be easily tested in the simulation by adjusting parameters and measuring the output:

1. Specify the amount you wish to vary the parameters by in a variable. For an order of magnitude variation in a parameter,

   ```
   delta = 10;
   ```

2. Multiply each parameter you wish to investigate by the previously defined variable. Often multiple parameters are changed simultaneously as the same process affects various isoforms. To change the rate of IKK-mediated degradation of all IκB isoforms,

   ```
   params(78)=params(78)*(delta);
   params(79)=params(79)*(delta);
   params(80)=params(80)*(delta);
   ```

3. Run the simulation with the increased parameter; this will also plot the NF-κB response.

4. Divide each parameter by the perturbation variable.

   ```
   params(78)=params(78)/(delta);
   params(79)=params(79)/(delta);
   params(80)=params(80)/(delta);
   ```

5. Run the simulation again to plot the response with the decreased parameter.

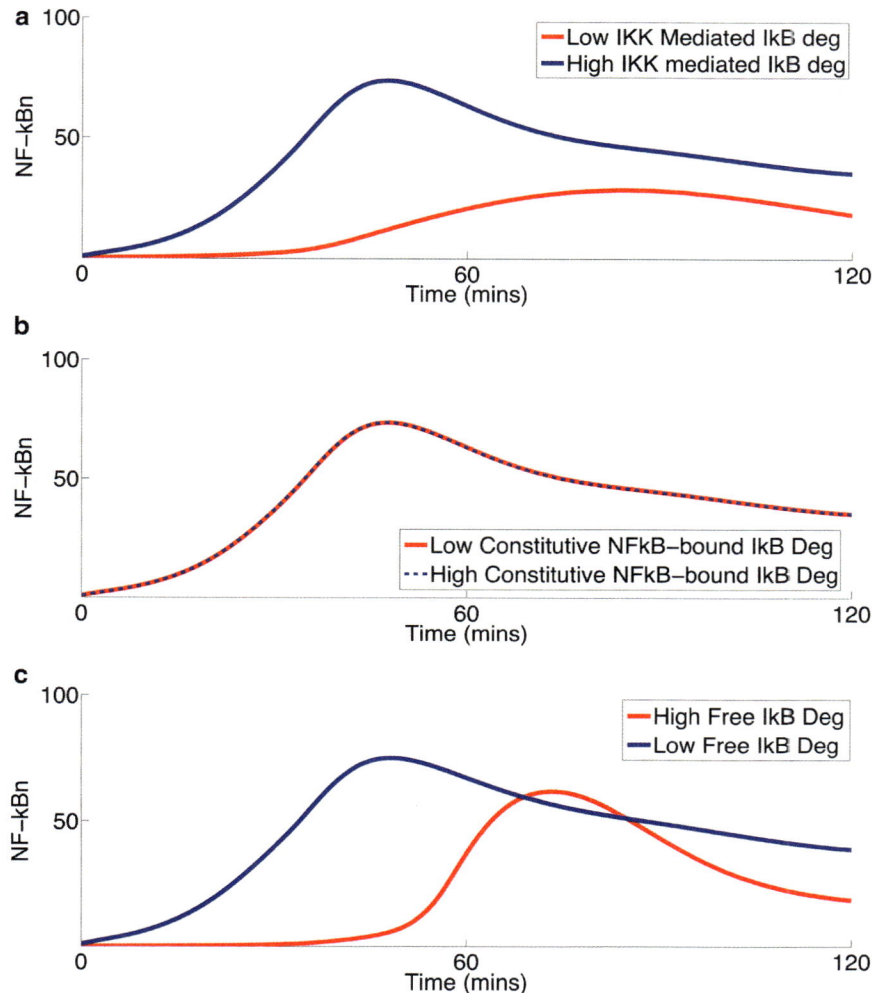

Fig. 4 Simulated NF-κB in response to LPS input curves with reaction rates increased an order of magnitude above and below basal for (**a**) IKK-mediated IκB degradation, (**b**) IKK-independent NF-κB-bound IκB degradation, and (**c**) free IκB degradation

Repeating this process for three sets of parameters representing the basal and IKK-mediated rate of IκB-NF-κB degradation and the rate of free IκB degradation gives the plots in Fig. 4.

As expected, increasing the rate of IKK-mediated degradation has a strong effect on the response of NF-κB (Fig. 4a). Also as expected, increased IKK-mediated IκB degradation results in an increase in free NF-κB. Interestingly, the amplitude of the response is not only reduced, but the time at which the response is maximal is shifted later with a lower degradation rate. As a result of the strong control provided by the rate of IKK activity, this process has been highly studied.

Figure 4b shows the basal (not IKK mediated) rate of NF-κB-IκB degradation. These parameters had no effect on the simulated NF-κB response. Therefore, attempting to use this process as a therapeutic intervention is unlikely to be successful.

Figure 4c shows perturbation of free IκB degradation; this was found to have a strong effect on the response. Decreasing free IκB degradation rate resulted in a slightly decreased peak of NF-κB response, but also a delayed response. The peak NF-κB response was found to be narrower and returned to lower levels more quickly when this parameter was decreased. Free IκB degradation rate has been much less actively studied, as it is a less intuitive point of intervention; however, through simulations we are able to identify this reaction as a point of strong control that should be investigated further.

3.2.3 Comparing Single-Cell Results with Population Level Results

The simulation of NF-κB signaling is a detailed representation of the intracellular environment; however, to draw conclusions relevant at a physiological level, then cell-to-cell variability must be considered. Here we create a simple physiologically relevant model by repeated simulations with variable delays applied to the output:

1. Define a mean and variance for the delay.

```
meanValue=65;
variance=1200;
```

2. Calculate μ and σ for a log-normal distribution.

```
mu=log((meanValue^2)/sqrt(variance+meanValue^2));
sigma = sqrt(log(variance/(meanValue^2)+1));
```

3. Define the range of repeated simulations you wish to run. Here we use a logarithmic range

```
logspaceVals=logspace(0,2,6);
```

4. For each value chosen, run the simulation that number of times. Also create a vector consisting of a concatenation of the basal NF-κB level until the delayed time point, the simulation output, and the basal NF-κB level for the remainder of the vector. Store this vector in a variable and plot the average of these vectors to get the population level dynamics. Plot each of these averages on the same figure.

```
for i=logspaceVals
    for j=1:i
        delay=floor(lognrnd(mu,sigma));
        plot(time+(delay),nfkb_timecourse);
        concatVector=[ones(delay,1) …
        *nfkb_timecourse(1,1);…
         nfkb_timecourse;…
```

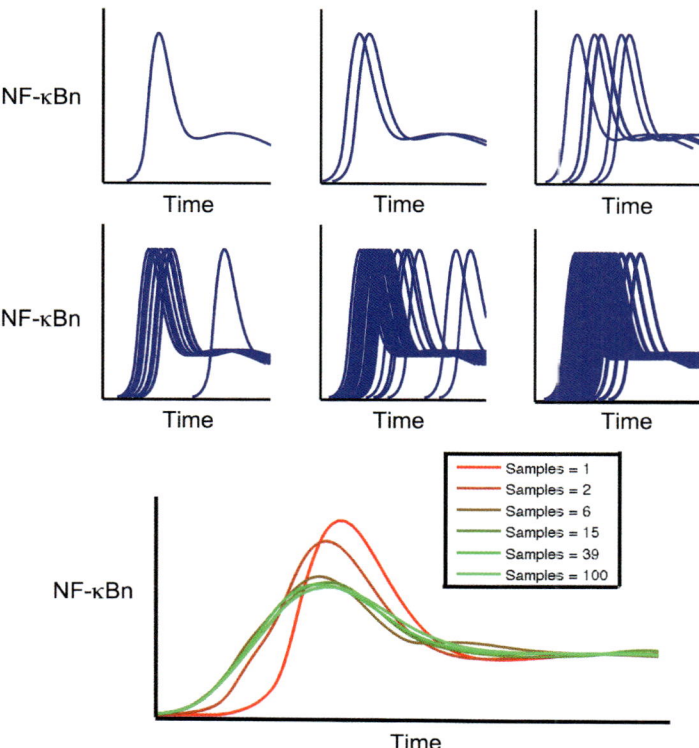

Fig. 5 NF-κB in response to LPS in repeated simulations sampled with a log-normally distributed delay. *Top*: Each curve represents a single simulation. *Bottom*: Average NF-κB concentration of multiple individual simulations

```
            ones(500-(SIM_TIME...
            +delay),1)*nfkb_timecourse(1,1)];
            averageMatrix=[averageMatrix,...
            concatVector];
    end
    figure(meanCurve);
    hold on;
    meanPlot=mean(averageMatrix,2);
    plot(meanPlot);
end
```

Figure 5 demonstrates that the average behavior of multiple single-cell simulations tends toward a less dynamic time course. This is in agreement with experimental results and represents a simple way of making the mechanistic cellular-scale model applicable at larger scales.

4   Notes

1. The web-based NF-κB signaling model available at http://
signalingsystems.org/webmodel/ can provide a convenient
way to produce preliminary results. Figures can be generated
without any modeling expertise and presented to modeling
experts in order to motivate more in-depth investigations or
model expansion.

2. While the goal is generally to produce models that reproduce
experimental findings and generate predictions that extend the
experimental work, it is important to note that an important
function of the model is to provide a sufficiency test. Akin to
the in vitro reconstituted biochemical systems that led to the
discovery and characterization of a multitude of replication
and transcription or translation factors and mechanisms, mathematical models of signaling allow one to ask whether the
known factors and molecular mechanisms are sufficient to
account for cell biological or physiological phenomena. In other
words, the model determines whether mechanistic knowledge
"adds up." If it does not, the model can direct the experimental discovery and characterization of the missing factor or
mechanism. This utility of modeling is often overlooked in the
field of purely theoretical computational biology, but is often
a major contribution in work that combines both experimental and modeling approaches.

3. To integrate computational studies into an interdisciplinary
team effort requires careful consideration of phasing the different aspects of a project. The ideal situation, with model
building and experimentation being conducted simultaneously, and iteratively informing each other, is often difficult to
realize successfully within the time frame available for projects
and relies on very close collaboration between researchers
from different backgrounds.

   Modeling work undertaken prior to wet lab work can be a
powerful tool for making predictions and generating hypoth-
eses. This type of exploratory modeling produces the most
exciting findings, but it can subsequently take many years for
the experimental work validating the findings to be completed.
Satisfying the needs of computational biologists may require
the publication of exclusively computational studies, though
they often lack the impact of studies combining theoretical
and experimental work.

   The alternative approach is to perform modeling work
after key experimental results have been obtained. The model
can then be used to provide mechanistic insights about the
experimental findings and direct additional, highly quantitative

experimentation that improves the quality of the work. This approach results in combined experimental and theoretical publications, which can have high impact; however, the full predictive potential of modeling is not explored, unless model predictions also lead to extensions of the scope of experimentation.

References

1. Werner SL, Barken D, Hoffmann A (2005) Stimulus specificity of gene expression programs determined by temporal control of IKK activity. Science 309:1857–1861

2. Basak S, Behar M, Hoffmann A (2012) Lessons from mathematically modeling the NF-κB pathway. Immunol Rev 246:221–238

3. Hoffmann A, Levchenko A, Scott ML, Baltimore D (2002) The IkappaB-NF-kappaB signaling module: temporal control and selective gene activation. Science 298:1241–1245

4. Werner SL, Kearns JD, Zadorozhnaya V, Lynch C, O'Dea E, Boldin MP, Ma A, Baltimore D, Hoffmann A (2008) Encoding NF-kappaB temporal control in response to TNF: distinct roles for the negative regulators IkappaBalpha and A20. Genes Dev 22:2093–2101

5. Kearns JD, Basak S, Werner SL, Huang CS, Hoffmann A (2006) IkappaBepsilon provides negative feedback to control NF-kappaB oscillations, signaling dynamics, and inflammatory gene expression. J Cell Biol 173:659–664

6. Shih VF-SF, Kearns JD, Basak S, Savinova OV, Ghosh G, Hoffmann A (2009) Kinetic control of negative feedback regulators of NF-kappaB/RelA determines their pathogen- and cytokine-receptor signaling specificity. Proc Natl Acad Sci U S A 106:9619–9624

7. Basak S, Kim H, Kearns JD, Tergaonkar V, O'Dea E, Werner SL, Benedict CA, Ware CF, Ghosh G, Verma IM, Hoffmann A (2007) A fourth IkappaB protein within the NF-kappaB signaling module. Cell 128:369–381

8. Shih VF-SF, Davis-Turak J, Macal M, Huang JQ, Ponomarenko J, Kearns JD, Yu T, Fagerlund R, Asagiri M, Zuniga EI, Hoffmann A (2012) Control of RelB during dendritic cell activation integrates canonical and noncanonical NF-κB pathways. Nat Immunol 13:1162–1170

9. O'Dea EL, Kearns JD, Hoffmann A (2008) UV as an amplifier rather than inducer of NF-kappaB activity. Mol Cell 30:632–641

10. Behar M, Hoffmann A (2013) Tunable signal processing through a kinase control cycle: the IKK signaling node. Biophys J 105:231–241

11. DiDonato JA, Hayakawa M, Rothwarf DM, Zandi E, Karin M (1997) A cytokine-responsive IkappaB kinase that activates the transcription factor NF-kappaB. Nature 388:548–554

12. Kirschner DE, Hunt CA, Marino S, Fallahi-Sichani M, Linderman JJ (2014) Tuneable resolution as a systems biology approach for multi-scale, multicompartment computational models. Wiley Interdiscip Rev Syst Biol Med 6:289–309

13. Spencer SL, Gaudet S, Albeck JG, Burke JM, Sorger PK (2009) Non-genetic origins of cell-to-cell variability in TRAIL-induced apoptosis. Nature 459:428–432

14. Li C, Donizelli M, Rodriguez N, Dharuri H, Endler L, Chelliah V, Li L, He E, Henry A, Stefan MI, Snoep JL, Hucka M, Le Novère N, Laibe C (2010) BioModels database: an enhanced, curated and annotated resource for published quantitative kinetic models. BMC Syst Biol 4:92

15. Wittig U, Kania R, Golebiewski M, Rey M, Shi L, Jong L, Algaa E, Weidemann A, Sauer-Danzwith H, Mir S, Krebs O, Bittkowski M, Wetsch E, Rojas I, Müller W (2012) SABIO-RK–database for biochemical reaction kinetics. Nucleic Acids Res 40:D790

16. Le Novère N, Hucka M, Mi H, Moodie S, Schreiber F, Sorokin A, Demir E, Wegner K, Aladjem MI, Wimalaratne SM, Bergman FT, Gauges R, Ghazal P, Kawaji H, Li L, Matsuoka Y, Villéger A, Boyd SE, Calzone L, Courtot M, Dogrusoz U, Freeman TC, Funahashi A, Ghosh S, Jouraku A, Kim S, Kolpakov F, Luna A, Sahle S, Schmidt E, Watterson S, Wu G, Goryanin I, Kell DB, Sander C, Sauro H, Snoep JL, Kohn K, Kitano H (2009) The systems biology graphical notation. Nat Biotechnol 27:735–741

17. Hoops S, Sahle S, Gauges R, Lee C, Pahle J, Simus N, Singhal M, Xu L, Mendes P, Kummer U (2006) COPASI—a COmplex PAthway SImulator. Bioinformatics 22:3067–3074

18. Macey R, Oster G (2001) Berkeley Madonna: modeling and analysis of dynamic systems

19. Mendes P, Kell D (1998) Non-linear optimization of biochemical pathways: applications to metabolic engineering and parameter estimation. Bioinformatics 14:869–883

INDEX

Printed by Printforce, the Netherlands